Inductively Coupled Plasmas in Analytical Atomic Spectrometry

Inductively Coupled Plasmas in Analytical Atomic Spectrometry

Edited by

Akbar Montaser
and
D. W. Golightly

Akbar Montaser
Department of Chemistry
George Washington University
Washington, D.C. 20052

D. W. Golightly
Branch of Analytical Chemistry
U.S. Geological Survey
Reston, Virginia 22092

Library of Congress Cataloging-in-Publication Data

Inductively coupled plasmas in analytical atomic
   spectrometry.

   Includes bibliographies and index.
   1. Plasma spectroscopy. I. Montaser, Akbar,
1946-   . II. Golightly, D. W.
QD96.P62I423 1987   543'.0858   86-32557
ISBN 0-89573-334-X

© 1987 VCH Publishers, Inc.

Printed in the United States of America.

ISBN 0-89573-334-X VCH Publishers
ISBN 3-527-26529-5 VCH Verlagsgesellschaft

Distributed in North America by:

VCH Publishers, Inc.
220 East 23rd Street, Suite 909
New York, New York 10010

Distributed Worldwide by:

VCH Verlagsgesellschaft mbH
P.O. Box 1260/1280
D-6940 Weinheim
Federal Republic of Germany

DEDICATION

The editors' contributions to this book are dedicated to Shirin, Azadeh, Rassa, Marilyn, and Yvonne. Without their patience and understanding, this work would not have been possible.

ACKNOWLEDGMENTS

The valuable contributions by Marilyn Brodine Golightly, both in reading and in entering most of the manuscripts into a wordprocessing system are gratefully acknowledged.

A. Montaser's contribution was sponsored in part by the U.S. Department of Energy, under contract number DE-AS05-84-ER-13172, and by the Donors of the Petroleum Research Fund, administered by the American Chemical Society.

Foreword

The most significant series of events that occurred during the past two decades in the field of analytical atomic spectroscopy was the emergence of various atmospheric-pressure, flame-like plasmas as vaporization, atomization, excitation, and ionization sources for analytical atomic spectroscopy. These flame-like plasmas included various DC transferred plasmas, plasma jets and "plasmatrons," capacitively-coupled, high-frequency plasmas, microwave-induced capillary plasmas, and inductively-coupled plasmas (ICP). Judged on the basis of their acceptance by the analytical spectroscopy community, by their growth in usage, and by the number of commercial instruments sold, the ICPs undoubtedly have had the greatest impact to date. Also, among all of the analytically useful, flame-like plasmas, only the ICP versions have so far demonstrated their flexibility and analytical merit as excitation cells for atomic emission, as atomization cells for atomic fluorescence, and as ionization cells for mass spectroscopy. Individually or collectively, these ICP spectroscopies are now having a major impact on the structure of the field of analytical atomic spectroscopy and on the way elemental analyses are being performed. Chemical analysts accustomed to performing multielemental determinations in other ways are finding the ICP spectroscopies straightforward and capable of producing analytical data at an unusually high rate.

The rapid growth and acceptance of the ICP spectroscopies by practicing analytical chemists and spectroscopists has in recent years underscored the need of a comprehensive reference book on the theory and practice of this rapidly expanding field. That consideration alone makes the appearance of such a book very timely, but there is still another equally compelling reason for the publication of a reference book. This reason is directly related to the proverbial ages of new analytical methodologies. Laitinen[1] was the first to relate Shakespeare's seven ages of man[2] to a common set of seven phases through which every new analytical method passes on its way to old age and senescence.

Subsequently, Barnes[3] and Fassel[4] adapted these phases to ICP–AES in a slightly different form. My version of these ages is presented below:

The Seven Ages of a New Physical Analytical Method

1. Conception of the idea
2. Design and construction of the first operating apparatus to verify scientific principles
3. Successful demonstration of the idea; publication of initial papers
4. Design and construction of finely tuned facility to identify experimental parameters that lead to useful analytical figures of merit, eg, sensitivity, selectivity, accuracy and precision
5. Attainment of maturity as represented by general acceptance by the analytical community and by the development of standardized operating conditions and procedures; automation of operations and computer management of data
6. Improved understanding of scientific principles
7. Old age and senescence

For this division of a life cycle, the three ICP spectroscopies have now passed through the fourth age and the atomic emission version has reached the maturity represented by age five. The first four ages may be logically identified as the revolutionary phase. L'vov[5] has compared this phase to the process of sinking a new screw that results in the appearance of a new analytical measurement concept and opens up new avenues of research for others to follow. The further turning of this screw in its continuing spiral, representing the ages 5 through 7, are then expected to lead to evolutionary improvements. For the ICP spectroscopies, these improvements are expected to impact their modes of operation, their analytical figures of merit, and their scope of application. It is during the virtually imperceptible transition from the revolutionary to the evolutionary phases that the appearance of a comprehensive reference book is most timely for several reasons. First, the rapid growth in applications during the fifth age is continuing to create a shortage of personnel familiar with the theory and practice of the ICP spectroscopies. For these individuals a comprehensive guide to the state of the art is indispensible. Second, all three ICP spectroscopies involve the interfacing of several complex hardware systems that are based on quite different scientific disciplines. The current and past published literature, which tends to be fragmented into scholarly specialized discourses, falls short of providing analytical spectroscopists with an integrated guide on how to devise evolutionary improvements and how to apply the new techniques in the most effective manner. In this respect as well, this book fills a well defined need.

Velmer A. Fassel
Iowa State University

REFERENCES

1. H. A. Laitinen, The Seven Ages of an Analytical Method, *Anal. Chem.* 45, 2305 (1973).
2. W. Shakespeare, As You Like It, Act II, Scene VII.
3. R. M. Barnes, Inductively Coupled Plasma Atomic Emission Spectroscopy: A Review, *Trends in Anal. Chem.* 1, 51–55 (1981).
4. V. A. Fassel, Analytical Inductively Coupled Plasma Spectroscopies—Past, Present, and Future, *Fresenius Z. Anal. Chem.* 324, 511–518 (1986).
5. B. V. L'vov, Twenty-five Years of Furnace Atomic Absorption Spectroscopy, *Spectrochim. Acta* 39B, 149–157 (1984).

Preface

Since its introduction in 1964, the inductively coupled plasma (ICP) has steadily risen to the forefront as a subject of both fundamental and applied research in analytical atomic spectroscopy. This discharge, considered by many to be the most significant development in analytical atomic spectrometry for the past two decades, is routinely applied to the quantitative determination of elemental compositions of extremely diverse materials in many laboratories throughout the world. Today, ICP spectrometry is a highly developed measurement technique, many facets of which are presented in this book.

Inductively Coupled Plasmas in Analytical Atomic Spectrometry is the work of 25 scientists who have contributed concepts, details, and insights that are based on personal experiences. Every chapter offers tutorial and background materials, including suggestions for particularly relevant reviews or papers useful to the student or analyst just beginning work in a particular area. References cited give the title of each publication to further assist the reader in deciding the potential relevancy of a specific article to a topic of special interest.

The book encompasses four basic units. The first unit addresses atomic emission spectrometry (AES) and includes seven chapters on basic concepts, optical instrumentation, ICP equipment, the analytical performance of ICP–AES, spectral interferences, high-resolution spectrometry, and fundamental properties of ICPs. The second unit details the use of ICP–atomic fluorescence spectrometry and ICP–mass spectrometry. In the third unit, sample introduction is discussed in three chapters for liquids, solids, and gases. Low-gas-flow torches and ICP discharges in gases other than argon are treated in separate chapters in the same unit. The final unit focuses on recent applications of ICP–AES. Because of the excellent review article on computer modeling by Boulos [M. I. Boulos, *Pure Apl. Chem.* 57, 1321–1352 (1985)], this subject is not treated extensively in this book.

The rigorous editing necessary to influence the content, the style, and the clarity of this type of book regrettably antagonized most of our authors and, furthermore, gained the editors reputations as "nitpickers." For obvious reasons, the chapters represent the authors' ideas and convictions. We are sincerely grateful to all the chapter authors who generously contributed their knowledge, skills, and time to this book.

Akbar Montaser
D. W. Golightly

Contents

4   Common Radio Frequency Generators, Torches, and Sample Introduction Systems   123

Stanley Greenfield

PART II. COMPLEMENTARY ICP TECHNIQUES

9   Atomic Fluorescence Spectrometry with the Inductively Coupled Plasma   323

Nicolò Omenetto and James D. Winefordner

Inductively Coupled Plasmas in Analytical Atomic Spectrometry

The Significance of Inductively Coupled Plasmas Relative to Other Laboratory Plasmas in Analytical Spectrometry

AKBAR MONTASER

Department of Chemistry
George Washington University
Washington, D.C.

DANOLD W. GOLIGHTLY

Branch of Analytical Chemistry
U.S. Geological Survey
Reston, Virginia

1.1 CHEMICAL ANALYSIS OF THINGS AS THEY ARE

In 1933, G.E.F. Lundell wrote[1]: "There is no dearth of methods that are entirely satisfactory for the determination of elements when they occur alone." The current popularity of inductively coupled plasmas (ICPs) in analytical spectrometry stems from the remarkable selectivities and sensitivities that make

Table 1.1. Ideal Requirements[15] for Methods of Elemental Analysis

1. Applicable to all elements.
2. Simultaneous or rapid sequential multielement determination capability at the major, minor, trace, and ultratrace concentration levels without change of operating conditions.
3. No interelement interference effects.
4. Applicable to the analysis of microliter- or microgram-sized samples.
5. Applicable to the analysis of solids, liquids, and gases with minimal preliminary sample preparation or manipulation.
6. Capable of providing rapid analyses; amenable to process control.
7. Acceptable precision and accuracy.

them so applicable to "the chemical analysis of things as they are." By any reasonable measure, the three major ICP-based methods, that is, atomic emission spectrometry (ICP–AES),[2–4] atomic fluorescence spectrometry (ICP–AFS),[5,6] and mass spectrometry (ICP–MS),[7–10] have approached a set of ideal requirements[11] for elemental analysis (Table 1.1). Classical techniques, such as, flame, arc, and spark spectrometries, although currently in use in various laboratories, have been less successful in achieving the capabilities listed in this table.

Flamelike electrical discharges, such as the direct-current plasma (DCP) and the microwave-induced plasma (MIP), in contrast to the ICP, do not possess annular channels for injecting the sample through the core of the discharge, thereby reducing efficiencies of vaporization, atomization, ionization, and excitation due to diminished plasma–sample interaction. This unique property, that is, the axial channel of the ICP, has been mainly responsible for certain desirable characteristics, such as relative freedom from matrix interferences, detection limits in the picogram-to-nanogram range, a linear dynamic range of 4 to 5 orders of magnitude, and precision of measurement of 1 to 3% relative standard deviation. Moreover, the common ICP-based techniques are fundamentally capable of isotopic analysis, although this potential has been demonstrated for ICP–AES and ICP–MS only.

1.2 GROWTH OF APPLICATIONS AND RESEARCH ON INDUCTIVELY COUPLED PLASMAS

The success of the ICP-based techniques may be gauged by two primary indices: (a) the number of ICP installations, and (b) the growth in the number of ICP-related publications. Since 1974, when the first ICP–AES system was introduced into the marketplace, approximately 6000 ICP facilities have been installed by various manufacturers (Table 1.2). The majority of these companies are involved with ICP–AES, while the ICP–MS is marketed by two manufacturers and the ICP–AFS system is available from only one company. Most ICPs are sustained in atmospheric-pressure argon at a flow rate of approximately

Table 1.2. Some Manufacturers of ICP Instrumentation

1. ICP–AES

   Thermo Jarrell Ash Corporation

   Applied Research Laboratories

   Baird Corporation

   Hilger Analytical, Ltd.

   Hitachi, Ltd.

   Instruments S.A./Jobin Yvon

   Kontron GmbH

   Labtest Equipment, Ltd.

   Leeman Labs, Inc.

   Perkin–Elmer Corporation

   Philips Industrie S.A.

   PRA International, Inc.

   RF Plasma Products, Inc.

   Spectro, Inc.

   Spex Industries, Inc.

2. ICP–AFS

   Baird Corporation

3. ICP–MS

   Sciex, Incorporated

   VG Isotopes Ltd.

Source: Adopted from the 1984 issue of *Annual Reports on Analytical Atomic Spectroscopy*, The Royal Society of Chemistry.

20 L/min, although low-gas-flow torches requiring 6 to 8 L/min of argon are now available from companies such as Applied Research Laboratories, Labtest Equipment Company, RF Plasma Products, and Sheritt–Gordon Mines, Ltd. Alternatively, the cost of generating an ICP can be reduced by forming mixed-gas and molecular gas ICPs on certain existing facilities. In fact, the Baird Corporation has recently introduced an air-ICP for process control applications.

The unique analytical characteristics of the ICP, combined with the commercial availability of ICP equipment, have led to the publication of a large number of papers (Figure 1.1). The approximate number of publications for the period from 1979 to 1985 have been classified into four categories by R. F. Browner.[12] Despite the large number of publications, significant problems exist in two areas. First, the introduction of samples into plasmas continues to be the limiting factor of analytical plasma spectrometry.[13,14] For example, with most pneumatic nebulizers, only 1 to 5% of the sample solution is injected into the discharge, and if solid sampling techniques are used, reproducibility of measurements is deteriorated. Relatedly, the commercial availability of solid sample introduction equipment is currently limited compared to solution or gaseous injection. Second, diagnostic and mechanistic studies of processes prevailing in

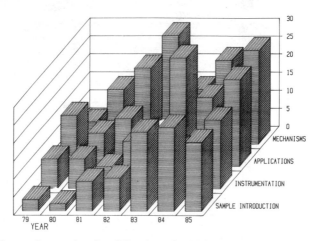

Figure 1.1. Rate of growth of publications in ICP-related research for four major categories. (From Reference 12, with permission.)

the ICP have improved our knowledge of the discharge, yet as shown in Chapter 8, they are not adequately conclusive to guide analytical spectroscopists in devising a more efficient source.

1.3   APPROACH TO STUDIES OF MECHANISMS AND STRUCTURES OF PLASMAS

The complexities of the processes in the ICP require the use of a particular strategy for studying mechanisms and structures of laboratory ICPs, a strategy not entirely followed in past work. The steps in a suggested strategy, proposed by V. A. Fassel,[15] are listed in Table 1.3. The progression of study involves

Table 1.3. Suggested Strategy[15] for Studying Excitation and Ionization Mechanisms and the Physical Structure of Laboratory ICPs

Step 1. Understanding of a "uniform," pure ICP.

Step 2. Understanding of a "structured," pure Ar ICP.

Step 3. Understanding of excitation and ionization behavior of free atoms of analyte introduced selectively into axial channel, or outer flow, or both.

Step 4. Understanding of "structured" Ar ICP upon the introduction of aqueous aerosol.

Step 5. Understanding of "structured" Ar ICP upon the introduction of aqueous aerosols containing analyte species.

Step 6. Understanding of the transport or diffusion of energetic species from the induction zone into the axial channel.

successive understandings of mechanisms and structures of the ICP from the simplest case to the most complex situation. The simplest case, step 1, is represented by a uniform, pure Ar ICP that exhibits no axial channel. By introducing pure argon gas into the axial channel, a "structured" pure argon ICP is formed in step 2. The picture is then gradually made more complicated by injecting gaseous analyte (step 3), followed by pure aqueous aerosol (step 4), and by analyte-containing aqueous aerosols (step 5). The most complicated step (step 6) is the study of transport, or diffusion, of energetic species from the induction zone into the axial channel. Obviously, the study of mixed-gas and molecular gas ICPs, discussed in Chapter 15, is even more complicated, but the use of a similar strategy should facilitate the understanding of excitation-ionization mechanisms and transport processes in these discharges, or in other plasmas.

1.4   OTHER COMMON OR NEW ATMOSPHERIC-PRESSURE PLASMAS

Although the unique characteristics of ICPs and the widespread commercial availability of ICP equipment set this discharge apart from other atmospheric-pressure plasmas, the analyst should not neglect the potentials of other plasma sources, which are experiencing continued development. However, the discussion in this book is limited to the ICP because of two major factors. First, the inclusion of other topics would have significantly increased the size of the book, and because of the involvement of additional authors, publication would have been unreasonably delayed. Even for the present work, two years have passed since the conception of the volume by the editors to the completion of the exhaustive editing process. Second, most aspects of the work on DCP[16,17] and MIP[18–20] have been comprehensively reviewed. However, it is appropriate to briefly cite some of the features of the common and new discharges.

1.4.1   Direct-Current Plasmas

Direct-current plasmas are commercially available from Applied Research Laboratories. The three-electrode version of this discharge (Figure 1.2) is being used in many laboratories. Unfortunately, compared to the MIP and ICP discharges, fewer publications have appeared for DCP-AES during the past 10 years. The last comprehensive review[17] on advances in DCP instrumentation and applications dates back to 1977. The recent development of the conical DCP[21] should enhance interest in this discharge because of increased sample penetration into the core of the discharge.

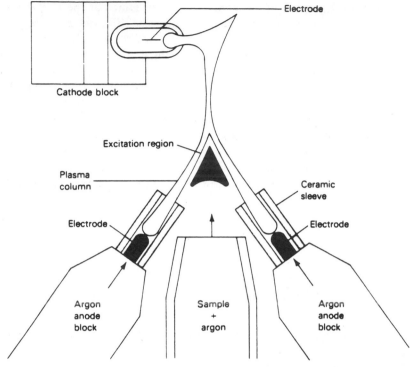

Figure 1.2. The commonly used direct-current plasma. The plasma, formed between three electrodes arranged in an inverted-Y configuration, operates at an argon flow rate of about 7 L/min using a single-phase, 20-A dc power supply. The electrodes, made from graphite (anodes) and tungsten (cathode), are water-cooled and are enclosed in ceramic sleeves, but they must be adjusted every few hours. Analytical observation is made under the arc column. In contrast to the Ar ICP, fewer strong ion lines are observed from the DCP. (From Applied Research Laboratories, with permission.)

1.4.2 Microwave-Induced Plasmas

Various versions of MIPs have proved to be excellent sources for chromatographic applications,[20] and the work on MIPs has led to several interesting developments. Thus, for example, the introduction of the Beenakker cavity (Figure 1.3) has allowed stable operation of the discharge at atmospheric pressure.[18–20] After the initial publication[22] on the tangential-flow torch, several torches have been described[23–26] to form toroidal MIPs generated at low (100 W) to moderate (500 W) powers. Caruso and associates[27,28] have fabricated a laminar, low-gas-flow torch that can sustain a MIP with flow rates as low as 5 mL/min, a very suitable detector for gas-chromatographic (GC) applications. Microwave-induced plasmas generated by a surfatron[29–34] (Figure 1.4) seem to be easier to operate and to tune than those formed in the

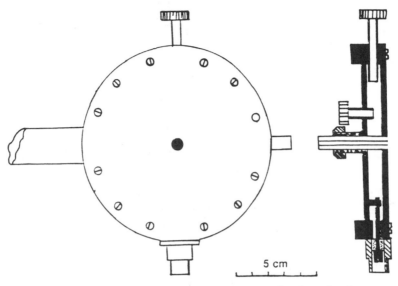

Figure 1.3. The Beenakker cavity, surrounding a quartz or alumina tube shown as a dark spot at the center of the cavity, may be used to form both Ar and He MIPs at atmospheric pressure for power levels of 100 to 500 W. The two adjustable screws shown on the right are used to minimize the reflected power. Energy is coupled into the cavity by a coupling loop that enters through the side of the cavity. The cavity is made from copper. (From Reference 20, with permission.)

Beenakker cavity. Regardless of the type of cavity used, the MIPs, although annular in certain cases, are usually in contact with the plasma containment tube, thus causing gradual erosion of the inside wall of the torch. Such a disadvantage is not exhibited by the ICP (ie, the ICP is both an annular and a suspended discharge). The capacitively coupled microwave plasmas (CMP) require minimum impedance matching,[35,36] but because the microwave energy is conducted through a coaxial waveguide to the tip of an electrode, the plasma is not as free from contamination as the ICP.

1.4.3 The Three-Electrode Argon Plasma Alternating-Current Arc

The electrode-plasma interaction is also a problem with the newly developed plasma that is referred to by Piepmeier and associates[37,38] as the three-electrode argon plasma arc (TEAPA). The discharge is generated in a torch (Figure 1.5), which resembles the ICP torch. The sample introduction tube is surrounded by three equally spaced, thoriated-tungsten electrodes that are used to generate an annular plasma. The most recent version of the TEAPA operates with a total argon flow[38] of approximately 4 L/min and an ac power of 0.5 to 2 kW. One should note that, during operation of this plasma, the lengths of the electrodes decrease at a rate of less than 1 mm/hr[37] and that tungsten atomic emission

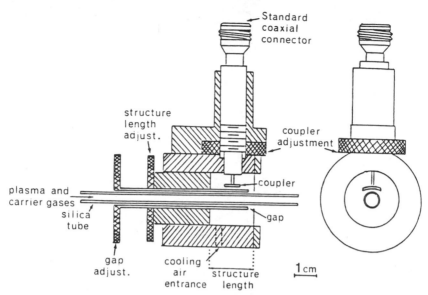

Figure 1.4. Schematic drawing of the surfatron cavity using a quartz tube with an outer diameter of up to 7 mm. The outer surface of the tube may be coated with aluminum oxide to prevent arcing between the discharge tube and the coupler. Alternatively, an alumina tube may be used instead of the quartz tube. To minimize the reflected power to less than 1 W, at a forward power of 130 W, both the "structure length" and the "coupler" must be adjusted. The cavity is made from copper, except for the cross-hatched sections which are made from brass. (From Reference 31, with permission.)

interferes with certain determinations.[38] Because the ac power supply is relatively simple, the capital cost of equipment for operating a TEAPA is low compared to that for the ICP. Further development is required before the analytical capabilities of this discharge may be compared with that of the ICP.

1.4.4   The Shielded Plasma Source

The shielded plasma source,[39,40] (Figure 1.6) is a new spectrochemical source that has been investigated to a limited extent. The plasma is generated in an inductive plasma tube[40] that consists of an induction coil surrounding a quartz enclosure containing an ionizable gas, such as argon. The induction coil is connected to a high-current power supply operated at frequencies of 400 kHz to 5 MHz. To protect the quartz tube from the high gas temperature of the plasma, 12 water-cooled copper shields are arranged symmetrically inside the length of the induction tube. Because the field of the induction coil must penetrate into the central cavity to allow plasma formation, the shields are spaced apart. To prevent electrical arcing between the induction coil windings, as a result of ultraviolet radiation from the plasma, chevron-shaped shields are used. Such an

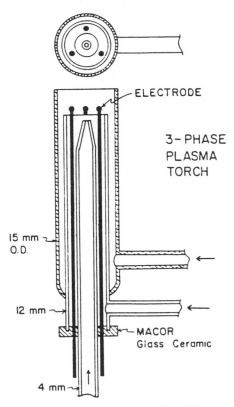

ELECTRODE

3-PHASE
PLASMA
TORCH

15 mm
O.D.

12 mm

MACOR
Glass Ceramic

4 mm

Figure 1.5. Top and side views of the torch used to form a three-electrode argon plasma source. The three quartz tubes are similar in size to those of a Fassel-type ICP torch. The three thoriated-tungsten electrodes are held by Macor glass-ceramic plates. The electrodes are spaced 6 mm from each other and are cooled by a flowing stream of argon. The electrode tips are located approximately 2 mm above the intermediate quartz tube. A three-phase, 2-kVA power supply that can provide 23 A per phase is sufficient to generate the ac arc plasma. (From Reference 38, with permission.)

interlocking arrangement between adjacent shields blocks transmission of ultraviolet radiation to the quartz tube and to the induction coil, while preserving the gap between the shields. With this configuration, the diameter of the discharge decreases at higher power levels, thus isolating the plasma from the induction tube. End-on viewing must be used to monitor the discharge.

The shielded plasma source has been formed in a stationary gas at atmospheric pressure at a power level of about 20 kW.[39,40] For such conditions, temperatures of 10,000 K and electron number densities of 10^{16} cm^{-3} have been reported for the discharge.[40] Although solid refractory samples such as yttrium oxide have been directly vaporized in the plasma, definitive statements on the capabilities of this source await the conclusion of investigations by Bieniewski and Hull.[40]

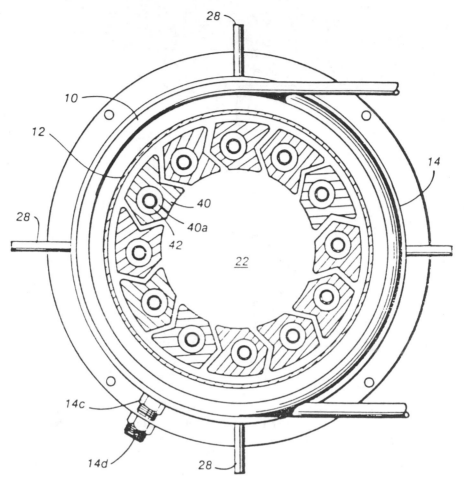

Figure 1.6. Top view of the plasma induction tube. The tubular quartz enclosure (12) is mounted on a water-cooled base (14). The quartz tube is surrounded by a water-cooled copper induction coil (10). Twelve chevron-shaped shields (40) are located inside the quartz tube. Each chevron-shaped shield is cooled by the counterflow of water through the water supply tube (42) and the central bore (40a) for each shield. The plasma tube includes a set of four gas intake tubes (28) which open into the lower end of the cavity (22). The plasma arc is formed in the cavity. (From Reference 39, with permission.)

1.4.5 *Atmospheric-Pressure Afterglow Discharges*

In contrast to the atmospheric-pressure plasmas discussed above, the atmospheric-pressure afterglow discharges[41-48] possess low gas temperatures, but they are excellent detectors for gas chromatography. Fassel and associates[41-48] have been the chief proponents of these discharges, which are sustained in a flowing stream of nitrogen, argon, or helium. The most advanced version of

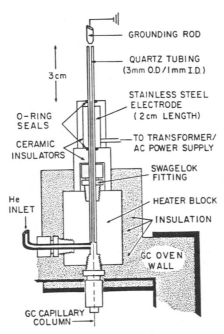

Figure 1.7. Schematic diagram of the discharge tube assembly for the formation of an atmospheric-pressure helium afterglow. The primary discharge is generated in the quartz tube, which is surrounded by a 2-cm-long, cylindrical electrode made from stainless steel. The afterglow is observed above the cylindrical electrode over a distance of approximately 3 cm. (From Reference 48, with permission.)

these discharges is the helium afterglow,[48] recently introduced into the marketplace by Cetac Technologies (Ames, Iowa). A significant advantage of this helium afterglow detector for GC application is that effluents are introduced beyond the primary discharge region, thus preserving the stability of the afterglow without the need for solvent venting.

A schematic diagram of the apparatus used for forming the helium afterglow is shown in Figure 1.7. The primary, electrodeless discharge is formed in a quartz tube of 1 mm inside diameter, which is surrounded by a cylindrical, stainless steel electrode, approximately 2 cm long. The outlet of the GC capillary column traverses through a heated block and resides flush with the top of the cylindrical electrode. The helium afterglow is observed above this region. To form the primary discharge and the afterglow, the cylindrical electrode must be connected to one pole of an alternating-current power supply, while a grounded, stainless steel rod, located at the top of the discharge tube, serves to complete the circuit. The helium used to sustain the discharge is introduced into the discharge tube at a rate of 80 mL/min through an inlet located in the heated block. The discharge is operated at approximately 45 W and a frequency of 26 to 27 kHz, and no external means of discharge initiation, such as the use of a Tesla coil, is required. Because contaminations of the discharge tube and gas lines with air

Table 1.4. Detection Limits (ng) for Three
Element-Selective Spectrometric
Gas-Chromatographic Detectors[48]

Element	Detector		
	Helium afterglow	MIP	Ar ICP
As	0.020	0.155	—
Br	0.015	0.106	50
C	0.010[a]	0.012	12
Cl	0.008	0.155	50
F	0.02	0.064	—
Hg	0.0005	0.060	—
I	0.002	0.056	4
P	0.030	0.056	0.6
S	0.005	0.140	6

[a]Limited by residual carbon in helium.

leads to warmup times greater than 1 hr, the entire discharge tube assembly must be sealed and pressurized when not in use.

The helium afterglow discharge provides absolute detection limits in the picogram range for certain hard-to-excite elements (Table 1.4). These detection limits are definitely superior to the results achieved so far for the He–MIP and the Ar–ICP. Further development of the ICP discharges in helium[49–52] may lead to similar improvements for the ICP-based methods.

The foregoing overview provides a brief summary of some of the potentials and drawbacks of ICPs and other discharges for elemental analysis. The next 15 chapters will address various aspects of instrumentation, fundamentals, and analytical applications of ICP-based methods. These discussions also highlight the existing deficiencies and the future directions in this field.

ACKNOWLEDGMENT

This work (at G.W.U.) was sponsored in part by the U.S. Department of Energy, under contract number DE-AS05-84-ER-13172. Acknowledgment is made to the Donors of the Petroleum Research Fund, administered by the American Chemical Society, for the partial support of this research.

BIBLIOGRAPHY

Thompson, Michael, and J. Nicholas Walsh. "A Handbook of Inductively Coupled Plasma Spectrometry." Blackie, Glasgow, 1983.
Trassy, Christian, and Jean-Michel Mermet. "Les Applications Analytiques des Plasmas Haute-Frequence, Technique et Documentation." Lavoisier, Paris, 1984.

REFERENCES

1. G.E.F. Lundell, The Chemical Analysis of Things as They Are, *Ind. Eng. Chem., Anal. Ed.* 5, 221–225 (1933).
2. S. Greenfield, I.L.I. Jones, and C. T. Berry, High-Pressure Plasmas as Spectroscopic Emission Sources, *Analyst* 89, 713–720 (1964).
3. R. H. Wendt and V. A. Fassel, Induction-Coupled Plasma Spectrometric Excitation Source, *Anal. Chem.* 37, 920–922 (1965).
4. G. W. Dickenson and V. A. Fassel, Emission Spectrometric Detection of Elements at Nanogram per Milliliter Levels Using Induction Coupled Plasma Excitation, *Anal. Chem.* 41, 1021–1024 (1969).
5. A. Montaser and V. A. Fassel, Inductively Coupled Plasmas as Atomization Cells for Atomic Fluorescence Spectrometry, *Anal. Chem.* 48, 1490–1499 (1976).
6. D. R. Demers and C. D. Allemand, Atomic Fluorescence Spectrometry with an Inductively Coupled Plasma as Atomization Cell and Pulsed Hollow Cathode Lamps for Excitation, *Anal. Chem.* 53, 1915–1921 (1981).
7. R. S. Houk, V. A. Fassel, G. D. Flesch, H. J. Svec, A. L. Gray, and C. E. Taylor, Inductively Coupled Argon Plasmas as Ion Sources for Mass Spectrometric Determination of Trace Elements, *Anal. Chem.* 52, 2283–2289 (1980).
8. R. S. Houk, H. J. Svec, and V. A. Fassel, Mass Spectrometric Evidence for Suprathermal Ionization in an Inductively Coupled Argon Plasma, *Appl. Spectrosc.* 35, 380–384 (1981).
9. A. R. Date and A. L. Gray, Plasma Source Mass Spectrometry Using an Inductively Coupled Plasma and a High-Resolution Quadrupole Mass Filter, *Analyst* 106, 1255–1267 (1981).
10. R. S. Houk, A. Montaser, and V. A. Fassel, Mass Spectra and Ionization Temperatures in an Argon–Nitrogen Inductively Coupled Plasma, *Appl. Spectrosc.* 37, 425–428 (1983).
11. V. A. Fassel, There Must be an Easier Way: Some Reminiscences, *Spectrochim. Acta* 40B, 1281–1292 (1985).
12. R. F. Browner, private communication, 1986.
13. R. F. Browner and A. W. Boorn, Sample Introduction: The Achilles Heel of Atomic Spectroscopy, *Anal. Chem.* 56, 786A–798A (1984).
14. L. de Galan, New Directions in Optical Atomic Spectrometry, *Anal. Chem.* 58, 697A–707A (1986).
15. V. A. Fassel, Where Do We Go from Here? Plenary lecture at the 1986 Winter Conference on Plasma Spectrochemistry, January 2–8, 1986, Kailua-Kona, Hawaii.
16. (a) M. H. Miller, D. Eastwood, and M. S. Hendrick, Excitation of Analytes and Enhancement of Emission Intensities in a DC Plasma Jet: A Critical Review Leading to Proposed Mechanistic Models, *Spectrochim. Acta* 39B, 13–56 (1984). (b) M. H. Miller, E. Keating, D. Eastwood, and M. S. Hendrick, Measured and Modeled Enhancement of Transition Metal Emission in the DC Plasma Jet, *Spectrochim. Acta* 40B, 593–616 (1985).
17. C. D. Keirs and T. J. Vickers, DC Plasma Arcs for Elemental Analysis, *Appl. Spectrosc.* 31, 273–283 (1977).
18. See, for example, M. P. Matousek, B. J. Orr, and M. Selby, Microwave-Induced Plasmas: Implementation and Applications, *Prog. Anal. At. Spectrosc.* 7, 275–314 (1984), and references therein.
19. See, for example, S. R. Goode and K. N. Baughman, A Review of Instrumentation Used to Generate Microwave-Induced Plasmas, *Appl. Spectrosc.* 38, 755–763 (1984), and references therein.
20. See, for example, T. H. Risby and Y. Talmi, Microwave-Induced Electrical Discharge Detectors for Gas Chromatography, *CRC Crit. Rev. Anal. Chem.* 14, 231–265 (1983), and references therein.
21. G. A. Meyer, Conical Three-Electrode DC Plasma for Spectrochemical Analysis. Presented at the 1986 Winter Conference on Plasma Spectrochemistry, January 2–8, 1986, Kailua-Kona, Hawaii.
22. A. Bollo-Kamara and E. G. Codding, Considerations in the Design of a Microwave Induced Plasma Utilizing the TM010 Cavity for Optical Emission Spectroscopy, *Spectrochim. Acta* 36B, 973–982 (1981).
23. D. Kollotzek, P. Tschopel, and G. Tolg, Three-Filament and Toroidal Microwave-Induced Plasmas as Radiation Sources for Emission Spectrometric Analysis of Solutions and Gaseous Samples—II. Analytical Performance, *Spectrochim. Acta* 39B, 625–636 (1984) and references therein.

24. S. R. Goode, B. Chambers, and N. P. Buddin, Use of a Tangential-Flow Torch with a Microwave-Induced Plasma Emission Detector for Gas Chromatography, *Spectrochim. Acta* 40B, 329–333 (1985) and references therein.
25. D. L. Hass and J. A. Caruso, Characterization of a Moderate-Power Microwave-Induced Plasma for Direct Solution Nebulization of Metal Ions, *Anal. Chem.* 56, 2014–2019 (1984) and references therein.
26. K. G. Michelwicz, J. J. Urh, and J. W. Carnahan, A Microwave-Induced Plasma for the Maintenance of Moderate Power Plasmas of Helium, Argon, Nitrogen, and Air, *Spectrochim. Acta* 40B, 493–499 (1985), and references therein.
27. M. L. Bruce, J. M. Workman, J. A. Caruso, and D. J. Lahti, A Low-Flow Laminar-Flow Torch for Microwave-Induced Plasma Emission Spectrometry, *Appl. Spectrosc.* 39, 935–942 (1985), and references therein.
28. M. L. Bruce and J. A. Caruso, The Laminar-Flow Torch for Gas Chromatography: He Microwave Plasma Detection of Pyrethroids and Dioxins, *Appl. Spectrosc.* 39, 942–949 (1985), and references therein.
29. M. Moisan, P. Leprince, C. Beaudry, and E. Bloyet, Devices and Methods of Using HF Waves to Energize a Column of Gas Enclosed in an Insulating Casting, U.S. Patent 4,049,940 (1977).
30. J. Hubert. M. Moisan, and A. Richard, A New Microwave Plasma at Atmospheric Pressure, *Spectrochim. Acta* 33, 1–10 (1979).
31. M. H. Abdallah, S. Coulombe, J. M. Mermet, and J. Hubert, An Assessment of an Atmospheric Pressure Helium Microwave Plasma Produced by a Surfatron as an Excitation Source in Atomic Emission Spectroscopy, *Spectrochim. Acta* 37B, 583–592 (1982).
32. M. Moisan, C. M. Ferreira, Y. Haglaoui, D. Henry, J. Hubert, R. Pantel, A. Ricard, and Z. Zakryewski, Properties and Applications of Surface-Wave-Produced Plasmas, *Rev. Phys. Appl.* 17, 707–727 (1982).
33. P. S. Moussounda, P. Ranson, and J. M. Mermet, Spatially Resolved Spectroscopic Diagnostics of an Argon MIP Produced by Surface Wave Propagation (Surfatron), *Spectrochim. Acta* 40B, 641–651 (1985).
34. M. Selby and G. M. Hieftje, Everybody Goes Surfatron. Presented at the 1986 Winter Conference on Plasma Spectrochemistry, January 2–8, 1986, Kailua-Kona, Hawaii.
35. H. Feuerbacher, A New CMP Excitation Source for Optical Emission Spectroscopy, *ICP Inf. Newsl.* 6, 571–575 (1981).
36. J. Dahman, Capacitively Coupled Microwave Plasma: A Status Report, *ICP Inf. Newsl.* 6, 576–595 (1981).
37. T. R. Mattoon and E. H. Piepmeier, Three-Phase Argon Plasma Arc for Atomic Emission Spectrometry, *Anal. Chem.* 55, 1045–1050 (1983).
38. R. A. Masters and E. H. Piepmeier, Sample Entraining Three-Electrode Argon Plasma Source for Atomic Emission Spectroscopy, *Spectrochim. Acta* 40B, 85–91 (1985).
39. D.E. Hull, Induction Plasma Tube. U.S. Patent 4,431,901 (1984).
40. T. M. Bieniewski and D. E. Hull, Shielded Plasma Sources as a New Spectrochemical Tool. Presented at the 1986 Winter Conference on Plasma Spectrochemistry, January 2–8, 1986, Kailua-Kona, Hawaii.
41. A. P. D'Silva, G. W. Rice, and V. A. Fassel, Atmospheric Pressure Active Nitrogen (APAN)—A New Source for Analytical Spectroscopy, *Appl. Spectrosc.* 34, 578–584 (1980).
42. G. W. Rice, J. J. Richard, A. P. D'Silva, and V. A. Fassel, Atmospheric Pressure Active Nitrogen Afterglow as a Detector for Gas Chromatography, *Anal. Chem.* 53, 1519–1522 (1981).
43. G. W. Rice, J. J. Richard, A. P. D'Silva, and V. A. Fassel, Gas Chromatography–Atmospheric Pressure Active Nitrogen (GC-APAN)—A New Method for Organomercury Speciation in Environmental Samples, *J. Assoc. Off. Anal. Chem.* 65, 14–19 (1982).
44. G. W. Rice, J. J. Richard, A. P. D'Silva, and V. A. Fassel, Comparison of Analytical Figures of Merit of an Active Nitrogen Afterglow and a Flame Ionization Detector for Gas Chromatography, *Anal. Chim. Acta* 142, 47–54 (1982).
45. G. W. Rice, A. P. D'Silva, and V. A. Fassel, Analytically Useful Spectra Excited in an Atmospheric Pressure Active Nitrogen Afterglow, *Appl. Spectrosc.* 38, 149–154 (1984).
46. G. W. Rice, A. P. D'Silva, and V. A. Fassel, Molecular Chemiluminescence from Mercury Halides Excited in an Atmospheric-Pressure Active-Nitrogen Afterglow, *Appl. Spectrosc.* 39, 554–556 (1985).
47. G. W. Rice, A. P. D'Silva, and V. A. Fassel, An Atmospheric-Pressure, Argon-Afterglow Detector for Gas Chromatography, *Anal. Chim. Acta* 166, 27–38 (1984).

48. G. W. Rice, A. P. D'Silva, and V. A. Fassel, A New He Discharge-Afterglow and Its Application as a Gas Chromatographic Detector, *Spectrochim. Acta* 40B, 1573–1584 (1985).
49. S. Chan and A. Montaser, A Helium Inductively Coupled Plasma for Atomic Emission Spectrometry, *Spectrochim. Acta* 40B, 1467–1472 (1985).
50. S. Chan, R. L. Van Hoven, and A. Montaser, Generation of a Helium Inductively Coupled Plasma in a Low-Gas-Flow Torch, *Anal. Chem.* 58, 2342–2343 (1986).
51. S. Chan and A. Montaser, A Helium Inductively Coupled Plasma: Background Spectra Emitted in the Red and Near-Infrared Spectral Regions, *Appl. Spectrosc.* (in press).
52. S. Chan and A. Montaser, Characterization of an Annual Helium Inductively Coupled Plasma Generated in a Low-Gas-Flow Torch, Spectrochim Acta (in press).

2

Basic Concepts in Atomic Emission Spectroscopy

MYRON MILLER

Whiting School of Engineering
Johns Hopkins University
Baltimore, Maryland

Department of Mechanical Engineering
U.S. Naval Academy
Annapolis, Maryland

2.1 INTRODUCTION

2.1.1 Scope and Orientation

Emission from spectroscopic plasmas is treated at an introductory level, without undue emphasis on any particular light source. It should be a useful starting point for a spectroscopist interested in investigating inductively coupled plasmas (ICP). Equations are presented without derivation, but generally with discussion of spectrochemical implications or validity ranges. Some typical numerical examples are provided.

Fundamentals of light emission can be reviewed using compact general notation.[1-8] Selectivity in plasma regime is exercised from the outset by not introducing the coronal model for low-density or astrophysical plasmas[1-8] or by not considering the continuum-dominated emission from high-density plasmas, or the emission from multiply ionized plasmas aimed at controlled nuclear fusion.[1-8]

A narrowly ICP-oriented approach is deemed too device specific, and also requires discussions of non-LTE excitation rates that are presently subjects of research and debate. To conform to the family of spectrochemical light sources, this chapter confines itself to plasmas at atmospheric pressure and possessing mean kinetic energies in the range of 0.4 to 2 eV corresponding to 4600 to 23,000 K. The major plasma constituent is presumed to be argon, and explicit discussion is devoted to argon atomic and first ionic populations or their corresponding emission lines. It should be recognized, however, that the most comprehensive theoretical treatments of plasma emission and excited state population rates have addressed hydrogen or hydrogenic plasmas,[1–8] both as a matter of tractability and also in response to astrophysical and fusion interests.

2.1.2   Background and Referral

Several texts and classical papers rigorously treat emission from low-energy plasmas. The reader is referred to the chapter appendix for details.[1–56]

2.1.3   Organization

Emission theory needed by spectrochemists is likely to be more rudimentary than is sought by spectrochemical plasma modelers. The analytical chemist relies on a few linearity assumptions (whose validity can be tested operationally) to relate the signals from samples, blanks, and standards. Optimization of analytical signal-to-noise (S/N) ratios is achieved by empirically tuning parameters such as slit widths, flow rates, and observation height. Fiduciary spectra are provided by published atlases of ICP line strengths.[43,57] This segment of the ICP community rarely requires insights about energy distribution functions or competing excitation or relaxation rates.

Accordingly, we start with the simplest, most idealized, plasma model, and work toward more general and realistic, but theoretically less tractible, models. The approach is outlined in Table 2.1.

2.1.4   Types of Emissions Treated

Isolated atomic lines and lines from singly ionized atomic species are the spectroscopic features usually of interest for chemical analyses by, and diagnostic studies of, low-power plasmas at atmospheric pressure. Prominent emission lines are, therefore, the main focus of this chapter. Continuum emissions, from free–free and free–bound interactions of plasma electrons, influence line-to-background ratios and are also considered. Transitions of these types are illustrated in Figure 2.1. Molecular spectra[58] are not treated because of space limitations and because (a) molecular emissions generally characterize outer boundaries of the plasma and (b) molecular transition probabilities usually are not reliably known.[1]

Table 2.1. Organization of Major Subjects in Chapter 2.

Section	Emission and source model	Topics reviewed and comments
2.2	Spectroscopic signal proportional to analyte abundance	Linearity between emission and analyte concentration and robustness of excitation conditions, rather than LTE, are prerequisites for spectrochemical usefulness
2.3	Optically thin emission, integrated over profile wavelength, from a laminar, homogeneous plasma in LTE	Emission intensity in optically thin limit; emitting population density via Saha–Boltzmann and dissociation equilibrium; role of slit width; continuum intensities; gA values
2.4	Optically thin, well-resolved line profiles, emitted from a homogeneous, laminar plasma in LTE	The shape factor L; line broadening and line shifts due to plasma microfields; convolution integrals; slit strategies; optimizing signal-to-noise ratios
2.5	Emission from inhomogeneous or nonlaminar geometry plasmas	Reabsorption in laminar boundary layers, perhaps undetected due to swamping by instrumental profile; nonlinear emission response to hot spots; Abel inversion; demixing along gradients
2.6	Optically thick emission from LTE plasmas	Emissivity, optical depth, and absorptivity; Planck function; radiative trapping; radiative transfer
2.7	Thermodynamic equilibrium: theoretical criteria and experimental tests	Thermalizing role of electron impacts, difficulty in equilibrating ground and resonance states; CTE, LTE, PLTE, non-LTE; measured temperatures and the distributions they represent; comments on non-LTE rates

Line emissions from plasmas are affected by the collective influence of perturbing long-range and short-range fields of ions, atoms and electrons (Figure 2.1). Compared to the dilute case (low pressure and low electron number density, which in the limit behave as isolated atoms), the energy levels in the dense case are spread by time- and space-varying plasma microfields. Superimposed on this spreading, which spectroscopically is observed as pressure broadening, are systematic shifts in the mean energies of the levels. Neighboring states that can interact via dipole-allowed transitions tend to repel each other, raising the energy of some levels and depressing others, thereby shifting line wavelengths. The highest energy states, the Rydberg states, spread the most because their exposed orbitals are the least screened, partaking most of the character of free electrons in the continuum. When perturbations blend the energies of these uppermost levels, as predicted by the Inglis–Teller formula,[1,2,59] the lines merge with the continuum. For this reason, and because of lowering of potentials due to the plasma environment,[1–4,60] the ionization

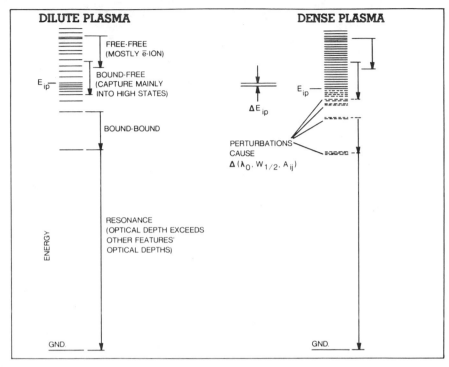

Figure 2.1. Schematic representation of electron transitions corresponding to the main types of plasma radiation. Bound–free radiation, resulting in the recombination of an atom–ion pair, is far stronger in the UV-visible than the free-free radiation from electron–ion interaction without capture into a bound state. Resonance lines are almost always the strongest and narrowest atomic or ionic lines. For VUV resonance lines, thermal equilibration between resonance and ground states may be incomplete at $n_e < 10^{15}$ cm^{-3}. In a dense plasma, perturbation of the emitter, mainly by electron collisions, causes emission lines to broaden and to shift and may slightly change the line transition probability. Lowering of the ionization potential ΔE_{ip} is attributable to impact-induced merging of high-lying states and to the reduction in potentials due to the plasma environment.

limit is lowered. The detailed population distributions of emiting states for all plasma constituents are coupled by the electron number density.

For simplicity, we first consider wavelength-integrated line emission (Section 2.3). Line shapes and shifts are presumed to be averaged over by use of spectral bandpasses much broader than the line widths. Line widths and shifts, caused by the plasma environment are dealt with in Section 2.4.

2.2   ESSENTIAL AND DESIRABLE PROPERTIES OF SPECTROCHEMICAL LIGHT SOURCES

Suitability of a plasma as a light source for spectrochemical analysis does not depend on its attaining local thermodynamic equilibrium (LTE). The properties that a luminous, controlled plasma must possess to be used for practical analysis are: (a) repeatability of signal and (b) analyte concentration that is linearly related to the observed spectroscopic signal. Linearity is expected if the analyte line is not self-absorbed (saturated or radiatively trapped) and if atomic excitation conditions (which in LTE plasmas are specified by temperature) do not detectably depend on sample concentration. Decoupling of sample concentration and plasma temperature permits linear interpolation using a pair of standards whose concentrations bracket that of the samples.

Non-saturation and plasma robustness can be ensured operationally without recourse to emission theory. If doubling analyte concentration does not cause doubling of detector response, for a non-saturating detector, the spectrochemist can either serially dilute the sample or choose a weaker line in the analyte spectrum. Invariance of excitation conditions becomes tenuous if the sample contains easily ionized elements in greater than trace concentrations. These elements, whose emission spectra are easily excited, if present in matrices can perceptibly perturb plasma shape, size, excitation temperature, and electron number density. Matching the matrices of the standards, blanks, and sample, and reducing the interpolation range, can operationally circumvent "matrix interferences," which can increase or reduce analyte signal.[42,44]

If LTE is not essential for spectrochemical analyses, is it at least desirable? The answer depends entirely on one's goals. The Saha-Boltzmann statistics governing emitting state populations, in combination with continuum intensity dependence on the square of electron number density, act together to bound signal-to-background (S/B) ratios in LTE plasmas. The present generation of sensitive spectrochemical light sources, including the ICP and DCP, encourage non-LTE populations in pursuit of improved S/B ratios. However, if one's objective is not primarily to maximize S/B ratios but rather to quantitatively describe conditions within the plasma, the task is manifestly easier if LTE prevails. For an example of well-understood LTE emission plasmas, consider the wall-stabilized, atmospheric argon arcs ("Kiel" or "Maecker").[50]

2.3 OPTICALLY THIN EMISSION LINES INTEGRATED OVER WAVELENGTH PROFILES FROM A LAMINAR, HOMOGENEOUS PLASMA IN LOCAL THERMAL EQUILIBRIUM

2.3.1 Assumptions

A plasma is optically thin in a spectral region when radiative trapping is negligible, ie, when photons emitted anywhere along the line of sight have almost 100% chances of exiting the plasma without being absorbed. Everywhere along an optical path of length l, light is emitted and transmitted without significant radiative trapping.

Temperatures characterizing excited-state populations of all constituent atoms and ions have a common value because the homogeneous plasma is assumed to be in LTE. Excitation temperatures, ionization temperatures, and kinetic temperatures are thus all equal.

Spectral bandpass is assumed to be sufficiently broad to collect the light from all wavelengths into which an atomic or ionic transition is radiating. A pressure-broadened, dispersion-shaped line and its relation to entrance slit widths, both narrower and broader than the line's halfwidth (width a half-maximum intensity) are illustrated in Figure 2.2. The wide slit encompasses the preponderance of the profile area. Dispersion-shaped profiles have broad wings[1,2,28] so that slits have to extend outwards 10 halfwidths to reach line intensity levels of 1% of peak intensity. But, as indicated in Figure 2.2, very wide slit openings reduce S/B ratios and also invite problems with spectral interferences. A compromise, when recording an integrated line profile, is to compensate theoretically for far wing contributions and to set the slit opening equal to several line halfwidths. Attempting to fully capture the line wings within the bandpass $\Delta\lambda$ causes the line-minus-blank signal to behave approximately as $I_{tot}\Delta\lambda$ where I_{tot} is the integrated line intesity, as discussed in Section 2.3.2. A measurement of integrated intensity is made absolute, and the $\Delta\lambda$ dependence is nulled, by recording a standard, known intensity I_{std} using a regulated carbon arc[61] or a filament lamp[62] for forming the ratio $I_{tot}\Delta\lambda/I_{std}\Delta\lambda$.

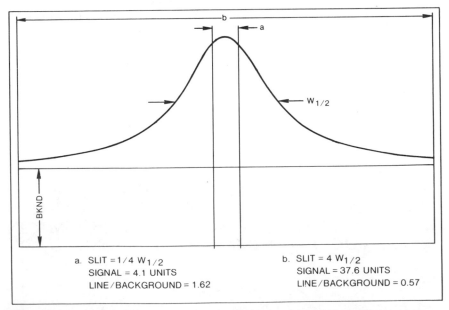

Figure 2.2. Pressure-broadened line profile, plus background, illustrating some consequences of broad and narrow slit widths. The line profile full width at half-maximum intensity is $w_{1/2}$. The wider slit provides more signal, but the reduced line-to-background ratio will reduce precision in the spectrochemical difference measurement of sample-minus-blank.

2.3.2   Radiation Intensity

Radiation intensity $I(\lambda)$ is the power $\Delta E/\Delta t$ passing through an area Δa within an element of solid angle $\Delta\Omega$ and within a wavelength interval $\Delta\lambda$[1-4,24]:

$$I(\lambda) \equiv \frac{\Delta E}{(\Delta t\,\Delta a\,\Delta\Omega\,\Delta\lambda)} \qquad [2.1]$$

$$\lim \Delta t, \Delta a, \Delta\Omega, \Delta\lambda \to 0$$

Radiation intensity is often expressed in units of $\text{ergs}/(\text{cm}^2\cdot\text{sr}\cdot\text{s}\cdot\text{cm})$ or $\text{W}/(\text{cm}^2\cdot\text{sr}\cdot\text{Å})$.

Radiation power $\xi(\lambda)$ from the volume ΔV is similarly defined as:

$$\xi(\lambda) \equiv \frac{\Delta E}{(\Delta t\,\Delta V\,\Delta\Omega\,\Delta\lambda)}$$

$$\lim \Delta t, \Delta V, \Delta\Omega, \Delta\lambda \to 0$$

Kirchhoff's law, applicable to plasmas in equilibrium, relates $\xi(\lambda)$, the absorption coefficient $K(\lambda)$, and the Planck function $B(\lambda, T)$ for blackbody radiation[1-4,24]:

$$\xi(\lambda) = K(\lambda)B(\lambda, T) \qquad [2.2]$$

and

$$B(\lambda, T) = \frac{2hc^2}{\lambda^5}\left[\exp\frac{hc}{\lambda kT} - 1\right]^{-1} \qquad [2.3]$$

where the Planck constant h, the Boltzmann constant k, and the light velocity c are given in Table 2.2. Planck function values for UV and visible wavelengths at 5000 and 10,000 K are presented in Figure 2.3.

Taking account of spontaneous emission and of absorption, but neglecting induced radiation, gives the radiation intensity from a homogeneous laminar source of thickness l[1-4,24]:

$$I(\lambda) = [1 - \exp(-K(\lambda)l)]B(\lambda, T) \qquad [2.4]$$

where $K(\lambda)l$, sometimes denoted as $\tau(\lambda)$, is the optical depth. In the presently considered optically thin case where reabsorption can be neglected, $\tau(\lambda) \to 0$:

$$I(\lambda) = \xi(\lambda) = K(\lambda)lB(\lambda, T) = \tau(\lambda)B(\lambda, T) \qquad [2.5]$$

2.3.3   Total Intensity of an Optically Thin Atomic or Ionic Line Emitted by a Plasma in Local Thermal Equilibrium

In the optically thin approximation of Equation 2.5, the intensity of an atomic line emitted from a homogeneous plasma of thickness l into the interval $\pm d\lambda/2$ about wavelength λ is[1-8]:

$$I(\lambda)d\lambda = \frac{h}{4\pi} A_{ul}n_u L(\lambda)d\lambda \qquad [2.6]$$

Table 2.2. Physical Constants and Conversion Factors

Quantity	Value	Units	
		SI	Cgs
Speed of 1-eV electron	5.931	10^5 m/s	10^7 cm/s
Temperature associated with 1 eV (E_0/k)	11,605.4	K	K
Energy of 1 eV per molecule (chem. scale)	23,055	10^{-3} kcal/mol^3	cal/mol
Wavenumber associated with 1 eV (s_0)	8,066.02	10^{-2} m^{-1}	cm^{-1}
Wavelength associated with 1 eV (λ_0)	12,397.7	10^{-10} m	10^{-8} cm
Energy of 1 eV	1.6020	10^{-17} J	10^{-12} erg
Velocity of light (c)	2.997925	10^8 m/s	10^{10} cm/s
First radiation constant ($8\pi hc$)	4.992579	10^{-24} J·m	10^{-15} erg·cm
Second radiation constant (hc/k)	1.438833	10^{-2} m·K	cm·K
Planck's constant (h)	6.626196	10^{-34} J·s	10^{-27} erg·s
Wein displacement law constant	0.289780	10^{-2} m·deg	cm·deg
Fine structure constant, $\alpha(2\pi e^2/hc)$	7.29735	10^{-3}	10^{-3}
Avogadro number (N_0)	6.023	10^{23} mol^{-1}	10^{23} mol^{-1}
Loschmidt number (n_0)	2.687	10^{25} m^{-3}	10^{19} cm^{-3}
Mass of unit atomic weight (chem. scale, M_0)	1.66026	10^{-27} kg	10^{-24} g
Electron charge (e)	1.6021917	10^{19} C	10^{20} emu
	4.803250	—	10^{-10} esu
Gas constant (R_0)	8.31434	10^3 J/mol K	10^7 ergs/mol K
Boltzmann constant (R_0/N_0, k)	1.380622	10^{23} J/K	10^{16} ergs/K
Bohr radius (a_0)	5.2917715	10^{-11} m	10^{-9} cm
Energy of 1 Rydberg	13.6048 eV	—	—
Wavelength units			
Ångstrom (Å)	—	10^{-10} m	10^{-8} cm
Micron (μm)	—	10^{-6} m	10^{-4} cm
Millimicron (mμ)	—	10^{-9} m	10^{-7} cm
Nanometer (nm)	—	10^{-9} m	10^{-7} cm

where $L(\lambda)$ is the line shape factor discussed in Section 2.4, which satisfies[1-4]:

$$\int_{-\infty}^{\infty} L(\lambda)\, d\lambda \equiv 1.0 \qquad [2.7]$$

and has dimensions of inverse wavelength. The Einstein coefficient A_{ul} (A value or transition probability) gives the probability for spontaneous transitions from upper state u to lower state l. Emitter number density n refers to the number density of atoms (or ions) in excited state u.

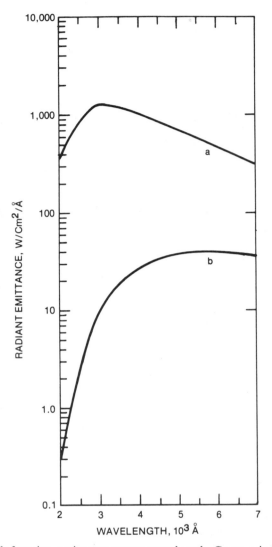

Figure 2.3. Planck function emittance versus wavelength. Curve a is for 10,000 K and curve b represents 5000 K. The Wein displacement of the wavelength of maximum emittance, λ_{max}, proportional to T^{-1}, can be noted.

Emitted photons have energy:

$$h\nu = \frac{hc}{\lambda} = E_{ul} \tag{2.8}$$

where c is the velocity of light and E_{ul} is the energy difference between the upper and lower state. For convenience, wavelengths and wavenumbers associated with 1 eV are listed in Table 2.2.

Integrating Equation 2.6 over wavelength corresponds to capturing all the energy contained within the hypothetical line profile illustrated in Figure 2.2 and gives the familiar expression for the total integrated intensity of the line:

$$I_{tot} = \frac{h}{4\pi} \frac{c}{\lambda} l A_{ul} N_u \qquad [2.9]$$

For plasmas in thermodynamic equilibrium, the mass action law takes an analogous form for dissociation $A_2 = A + A$, first ionization $A = A^+ + e$, and multiple ionization $A^+ = A^{2+} + e$.[1-8,29] For dissociation:

$$\frac{n_A^2}{n_{A_2}} = \frac{M_A^{3/2} v}{4\pi^{1/2} J_{A_2} (kT)^{1/2}} \frac{g_A^2}{g_{A_2}} \exp\left(\frac{-U_i}{kT}\right) \qquad [2.10]$$

where n_A is the atom number density, n_{A_2} is the molecular number density, J_{A_2} is the molecular moment of inertia, U_i is the dissociation energy, v is the molecular vibrational frequency, and g_A and g_{A_2} are the atomic and molecular ground-state degeneracies, respectively.

For the first ionization, the mass action law gives the Saha equation:

$$\frac{n_i n_e}{n_a} = \frac{2Z_i(T)}{Z_a} \left(\frac{2\pi mkT}{h^2}\right)^{3/2} \exp\left(\frac{-E_{ip}}{kT}\right) \qquad [2.11]$$

where n_i, n_a, and n_e are the number densities of first ions, atoms, and free electrons, respectively, and m is the electron mass. In multicomponent plasmas, n_e couples the Saha equations for first and multiple ionizations of all plasma components, requiring iterative solution. The atomic and ionic partition functions

$$Z_a = \sum_k g_k^a \exp\left(\frac{-E_k^a}{kT}\right) \qquad [2.12]$$

$$Z_i = \sum_j g_j^i \exp\left(\frac{-E_j^i}{kT}\right) \qquad [2.13]$$

are slowly varying functions of the temperature when T is less than 10,000 K. Inspection of tabulations for Z_a and Z_i[37,38] shows that the summations are dominated by the ground-state term g_1^a, or g_1^i when kT is less than 0.5 eV.

Degeneracy g of each state is:

$$g = (2J + 1) \qquad [2.14]$$

where

$$\mathbf{J = S + L} \qquad [2.15]$$

J, S, and L are the state's total, spin, and azimuthal quantum numbers, respectively. When a state corresponds to more than one electron in an excited orbit, S and L are themselves vector sums of the individual electronic spins and azimuthal quantum numbers. The $2J + 1$ refers to the multiplicity of quantized projections of J along the atom's (or ion's) z axis.[17-22]

Saha equilibrium for atoms (a) and the first ions (i) can be numerically expressed by:

$$\log_{10}\frac{m_i m_e}{m_a} = E_{ip}\theta - (\tfrac{3}{2})\log_{10}(\theta) + 20.94 + \log_{10}\left(\frac{2Z_i}{Z_a}\right) \qquad [2.16]$$

where $\theta = T/5040$ K and E_{ip} is the first ionization potential, expressed in electron volts, and n_j, n_e, and n_a are in cm^{-3}. This equation also applies to equilibrium between higher states of ionization when subscripts a and i are changed to z and $z + 1$ to represent the lower and higher ionization stages, respectively.[1-8]

Due to fluctuating plasma microfields, and an overall decrease in atomic and ionic potentials due to the plasma environment, the ionization potential E_{ip} of the atom is smaller than the ionization potential E_{ip} for an isolated atom of the same element. On average, each charged particle attracts a local excess of particles of opposite sign, causing the mutual Coulomb interaction potential to decline faster than $1/r$. This Debye shielding[4,63] causes each member of the electron–ion pair created at ionization to be attracted to local particles of opposite sign, thereby lowering the ionization potential. The lowering ΔE_{ip} is given approximately by[64,65]:

$$\Delta E_{ip} = \frac{-(z + 1)e^2}{(kT/4\pi e^2)^{1/2}}\left(n_e + \sum_j Z_j^2 n_j\right)^{1/2} \qquad [2.17]$$

where z is the stage of ionization for which the energy lowering is being computed (for the neutral atom $z = 0$, while for first ions $z = 1$), the summation is over the j components of the plasma having charge z_j and number density n_j, and e is the electron charge.

A ready approximation for ΔE_{ip} is:

$$\Delta E_{ip} \simeq 7.0 \times 10^{-7} n_e^{1/3}(z + 1)^{2/3} \qquad [2.18]$$

where, as before, z is the net charge of the lower ionization stage and ΔE_{ip} is expressed in electron volts when n_e is in cm^{-3}. Lowering of ionization potentials in laboratory plasmas changes partition functions typically by not much more than 1%.[1,64,65]

Additional conditions must also be satisfied, specifically:

(a) Mass conservation:

$$n_q + n_q^+ + n_q^{++} + \cdots = \alpha_q(n_{tot} - n_e) \qquad [2.19]$$

$q = 1, 2, \ldots, r$, and α_q is relative concentration of the qth component;

(b) Macroscopic charge neutrality:

$$n_e = \sum_j^r Z_j n_j \qquad [2.20]$$

(c) Since departures from ideal gas behavior, including those from shielded Coulomb interactions between charged constituents, are negligible when kT exceeds 0.5 eV, Dalton's law becomes a good approximation:

$$n_{tot} = \frac{P}{RT} \qquad [2.21]$$

where, because of velocity-induced head pressure, P is not precisely uniform throughout an ICP flow field.[66]

It is appropriate to pause here to compare the number of applicable equations with the number of unknowns in an r-component plasma. In the most simple case of a one-component plasma of a monatomic gas, suppose that we truncate solving for ion densities beyond first ions. The four unknowns are: n_a, n_i (where $n_i \equiv n^+$), n_e, and T. Applicable equations (2.11, 2.18, and 2.19) are only three, thus necessitating iterative solution. When a plasma having two or more components and a known composition is involved, Equation 2.17 applies, but this still leaves one excess unknown. For diatomic gases, one unknown and Equation 2.10 are added to the set. When higher degrees of ionization are included, one equation similar to Equation 2.11 is added for each higher stage of ionization, but there remains the single excess unknown and the need for iterative solutions. Computer programs[64] have been developed for this purpose, generally using P, T, and plasma chemical composition as input parameters. Pertinent number densities for various species in an atmospheric argon discharge are given in Table 2.3.

For argon at temperatures less than 25,000 K, it is usually adequate to consider $Ar \rightleftarrows Ar^+ + e$, $Ar^+ \rightleftarrows Ar^{2+} + e$, $e + Ar \rightleftarrows Ar^-$, because concentrations of higher ionization stages are negligible. For a diatomic gas such as N_2, the reactions $N_2 \rightleftarrows N + N$ and $N_2 + e \rightleftarrows N_2^-$, should also be taken into account. Such a program for argon[3] leads to the $I_{tot}(T)$ behavior illustrated in Figure 2.4 for Ar I 430.0 nm and Ar II 480.6 nm. Competition between Saha and Boltzmann equilibrium causes the intensity of Ar I 430.0 nm to peak around 15,500 K. Such maxima are the basis of "norm temperature" determinations.[2,25]

More importantly, Figure 2.4 illustrates an inherent photometric limitation of LTE plasmas:

1. Line intensities usually peak at temperatures equal to approximately 10% of E_{ip}/k (eg, for argon $E_{ip}/k = 182,850$ K).
2. Because line intensities are upper bounded in this way, but electron number densities and the continuum emissions, proportional to n_e^2, increase monotonically with temperature, S/B ratios in LTE plasmas are inherently limited by Saha–Boltzmann statistics.

If a line is intense enough to be interesting spectrochemically or diagnostically, it originates from a dipole, rather than a quadrupole, transition and obeys the dipole selection rules.[17–23] For a many-electron atom, where L and S are

Table 2.3. Equilibrium Number Densities in Argon Plasma at
$P = 1$ atm

T(K)	Densities $(cm^{-3})^a$				
	Ar	Ar$^+$	Ar^{2+}	n_e	n_{tot}
4,000	2.835 (18)	1.272 (10)	—	1.272 (10)	1.835 (18)
4,500	1.631 (18)	1.673 (11)	—	1.673 (11)	1.631 (18)
5,000	1.468 (18)	1.318 (12)	—	1.468 (18)	1.468 (18)
5,500	1.334 (18)	7.158 (12)	—	7.158 (12)	1.334 (18)
6,000	1.223 (18)	2.938 (13)	—	2.938 (13)	1.223 (18)
6,500	1.129 (18)	9.727 (13)	—	9.727 (13)	1.129 (18)
7,000	1.048 (18)	2.719 (14)	—	2.719 (14)	1.048 (18)
7,500	9.773 (17)	6.637 (14)	—	6.637 (14)	9.786 (17)
8,000	9.146 (17)	1.450 (15)	—	1.450 (15)	9.175 (17)
8,500	8.577 (17)	2.893 (15)	2.586 (5)	2.893 (15)	8.635 (17)
9,000	8.949 (17)	5.344 (15)	2.336 (6)	5.344 (15)	8.156 (17)
9,500	7.542 (17)	9.242 (15)	1.686 (7)	9.242 (15)	7.727 (17)
10,000	7.040 (17)	1.508 (16)	1.005 (8)	1.508 (16)	7.342 (17)
10,500	6.526 (17)	2.339 (16)	5.084 (8)	2.339 (16)	6.994 (17)
11,000	5.986 (17)	3.462 (16)	2.230 (9)	3.462 (16)	6.679 (17)
11,500	5.411 (17)	4.905 (16)	8.638 (9)	4.905 (16)	6.392 (17)
12,000	4.798 (17)	6.665 (16)	2.998 (10)	6.665 (16)	6.131 (17)
12,500	4.153 (17)	8.696 (16)	9.442 (10)	8.696 (16)	5.893 (17)
13,000	3.493 (17)	1.090 (17)	2.726 (11)	1.090 (17)	5.674 (17)
13,500	2.884 (17)	1.313 (17)	7.284 (11)	1.313 (17)	5.471 (17)
14,000	2.235 (17)	1.524 (17)	1.814 (12)	1.524 (17)	5.284 (17)
14,500	1.699 (17)	1.704 (17)	4.242 (12)	1.704 (17)	5.108 (17)
15,000	1.253 (17)	1.844 (17)	9.371 (12)	8.844 (17)	4.942 (17)
15,500	9.046 (16)	1.940 (17)	1.966 (13)	1.940 (17)	4.786 (17)
16,000	6.441 (16)	1.996 (17)	3.938 (13)	1.997 (17)	4.637 (17)

^{a} The applicable power of 10 is given in parentheses.

good quantum numbers for Russell–Saunders coupling, the selection rules are[17-23]: (a) parity must change in the transition; (b) total orbital quantum number L must change by ± 1 (lines tend to be stronger when ΔL and Δn change in the same direction); (c) total spin quantum number S must remain unchanged, and (d) the magnetic quantum number is unchanged if emitted light is polarized parallel to the z direction and must change by ± 1 if the light is polarized perpendicular to the z direction.

With respect to the transition probability in Equation 2.6, it should be appreciated that except for hydrogen and ionized helium lines, which can be calculated exactly, very few A_{ul} values are known to better than 10% accuracy.[32,33] Reliability of 20% is regarded as better than average, and discordances of 50% between authors are not rare.[32,33] Although errors in A_{ul} data are substantial, they are not necessarily the limiting factors in spectroscopic diagnostic studies, since absolute and relative photometry, including systematic error in calibration as well as error from interferences and from unsuspected

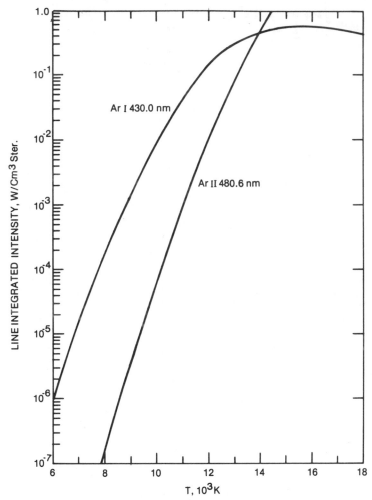

Figure 2.4. Integrated intensities of an argon atomic and first ionic line as functions of temperature. The emission is assumed to be optically thin and is proportional to the excited-state number densities. At 10,000 K, errors of $\pm 10\%$ in temperature translate into uncertainties of ± 1 order of magnitude in the Ar II 480.6 nm intensity and more than factor of ± 5 in the atom line intensity. The maximum in the integrated atom line intensity occurs at that temperature where the rate at which Boltzmann statistics populate the upper state is canceled by the rate at which Saha equilibrium depopulates the atomic ionization stage.

inhomogeneities and demixing,[50] can collectively contribute uncertainty of 10% to 20% even in careful experiments. Moreover, as indicated in Figure 2.1, plasma perturbations affect A_{ul} values. Quantitatively, this perturbation is not well understood,[1,5,6] but it is a less serious fundamental problem in $n = 10^{15}$ to 10^{16} cm^{-3} plasmas such as the ICP than in cases where $n_e = 20^{17}$ cm^{-3}.

Experimental and theoretical methods for finding A_{ul} values, along with assessments of strengths and weaknesses of the respective methods, are reviewed in the NBS critical compilations of A_{ul} data for the 20 lightest elements.[32] Bibliographies, covering A_{ul} findings for elements of all weights, are published regularly.[12,33] Extensive tabulations[34] of A_{ul} for medium-weight (transition metal) and heavier elements measured in free-burning arcs are convenient first approximations but should be checked for consistency with more recent results based on refined techniques.[12,32,33]

The probability for spontaneous transitions, A_{ul}, is expressed in reciprocal seconds. In the simplest case of only a single allowed channel for radiative decay, A_{ul} approximately equals the reciprocal of the upper state mean lifetime. More generally, the lifetime τ_u of excited state u is related to the A_{ul} values for spontaneous transition to states j by:

$$\tau_u^{-1} = \sum_j A_{uj} + Q \qquad [2.22]$$

where Q is for the net rates of collisional population and depopulation.

Conversion between A_{ul} and the absorption oscillator strength f_{lu} and the line strength S for dipole transitions is facilitated by use of Table 2.4. For example, the hydrogen beta line, H I 486.1 nm, for transition from principal quantum number n = 3 to n = 2, has $A_{ul} = 8.419 \times 10^6$ s^{-1}, $f_{lu} = 0.1193$, and $S = 15.27$, values typical of prominent visible nonresonance lines.[32,33] Corresponding values[32] for the ionic resonance line Ca II 393.3 nm are $A_{ul} = 1.50 \times 10^8$ s^{-1}, $f_{lu} = 0.69$, and $S = 18.0$. The absorption oscillator strength f_{lu} gives the equivalent number of classical electron oscillators represented by the transition.[22,23] It is related to the (negative) emission oscillator strength f_{ul} via:

$$g_u f_{ul} = -g_l f_{lu} \qquad [2.23]$$

and the $g_l f_{lu}$ value for a whole multiplet equals the sum of the $g_l f_{lu}$ of its individual lines. To help assess f values, it is useful to test[23,24,68] for conformity with the f-sum rules. These are:

(a) The Kuhn–Thomas–Reiche f-sum rule for transitions to and from a particular state u[23,24]:

$$\overset{i<u}{\underset{i}{\sum}} f_{ui} + \overset{j>u}{\underset{j}{\sum}} f_{uj} = z \qquad [2.24]$$

where the first sum is for allowed transitions to states i less energetic than u (and is negative) and the second sum extends to allowed transitions from all states j more energetic than u (including transitions from the continuum)[23,24] and z is the number of optical electrons (for strong transitions, $z = 1$ is usual),

(b) the Wigner–Kirkwood rule for a one electron jump.[23,24] If $l \rightarrow l - 1$;

$$\sum_i f_{ji} = \frac{l(2l-1)}{3(2l+1)} \qquad [2.24a]$$

and if $l \rightarrow l + 1$

$$\sum_i f_{ji} = \frac{(l+1)(2l+3)}{3(2l+1)} \qquad [2.24b]$$

Table 2.4. Conversion Factors[a,b] for Optical Transition Parameters

	A_{ul}	f_{lu}	S	
A_{ul}	1	$\dfrac{6.670 \times 10^{15}}{\lambda^2} \dfrac{g_l}{g_u}$	$\dfrac{2.026 \times 10^{18}}{g_u \lambda^3}$	
f_{lu}	$1.4992 \times 10^{16} \lambda^2 \dfrac{g_u}{g_l}$	1	$\dfrac{303.7}{g_l \lambda}$	
S	$4.935 \times 10^{19} g_u \lambda^3$	$3.292 \times 10^{-3} g_l \lambda$	1	

[a] The factor in each box converts by multiplication the quantity above into the one at its left.
[b] The line strength S is given in atomic units. For electric dipole transitions $a_0 e^2 = 6.459 \times 10^{-36}$ cm$^2 \cdot$esu^2. Wavelength is given in Ångstrom units, and g_l and g_u are the statistical weights of the lower and upper states, respectively.

where l is the orbital quantum number of the jumping electron. For example, for series where $p \to ns$, $f = -1/9$, whereas $p \to nd$, $f = 10/9$.

Dipole transition line strengths S are frequently given in atomic $|\chi|^2/a_0^2$ units.[24,32,35,36] Line strengths are symmetrical, $S_{ul} = S_{lu}$. For theoretical calculations, S tends to be preferred over f or A values because its units are computationally convenient. The line strength S is factored into three parts:

$$S = \sigma^2 S(M)\, S(L) \qquad [2.25]$$

where σ^2 depends on radial wave functions:

$$\sigma^2 = \frac{1}{4l^2 - 1} \left(\int_0^\infty R_i R_f r\, dr \right)^2 \qquad [2.26]$$

where l is the greater of the orbital quantum numbers involved in the transition, and R_i and R_f are the initial and final radial wave functions of the optical electron. The relative multiplet strength is represented by S(M), and S(L) gives the relative strength of a particular line within the multiplet. Tables based on LS coupling[2,24] permit rapid computation of S(M) and S(L). For J–J and other types of coupling, other matrices are available.[20,21] Numerical computations of σ in the scaled hydrogenic ("Coulomb") approximation can be executed quickly using published tables.[35,36] One first computes the effective principal quantum number n* of the upper state and of the lower state[1-3];

$$n^* = z \left(\frac{R}{E_{ip} - E_{ex}} \right)^{1/2} \qquad [2.27]$$

where z is the degree of ionization (1 for neutral, 2 for singly ionized, etc.), E_{ex} is the state excitation energy, and R is the Rydberg energy. Then, referring to tables[35,36] of the functions $F(n_l^*, l)$ and $I(n_{l-1}^*, n_l^*, l)$, one computes:

$$\sigma = \frac{1}{z} F(n_l^*, l) I(n_{l-1}^*, n_l^*, l) \qquad [2.28]$$

where z is the degree of ionization. The function $F(n_i^*, l)$ is a comparatively slowly varying function of n_i^*, while $I(n_{i-1}^*, n_i^*, l)$ is a sensitive function of $(n_{i-1}^* - n_i^*)$. Near the nodes in $I(n_{i-1}^*, n_i^*, l)$, the method loses reliability.[35,36] But for strong isolated lines, where σ is large, configurations are unmixed, and single electron jumps are involved. This approximation seems generally to be as reliable as most experiments or more sophisticated theoretical computations.[33]

2.3.4 Continuum Emission from an Optically Thin, Homogeneous, Plasma in Local Thermal Equilibrium

For well-controlled spectroscopic plasmas, it is sometimes feasible to utilize the continuum for diagnostics, either by line/continuum ratios[1-5] or by measuring the continuum variation with wavelength.[1-5] The latter method provides, in principle, a direct means for assessing the kinetic temperature of the free electrons. Spectrochemically, continuum is largely responsible for limiting sensitivity, because sample-minus-blank measurements become critical difference measurements as analyte concentrations decline. Efforts toward reducing continuum while maintaining or increasing analyte emission must deal with two facts. First, at UV–visible wavelengths, most continuum arises from free–bound transitions, ie, radiative recombination. Second, this continuum intensity is proportional to the product of ion and electron number densities, that is, n_e^2, and is only weakly dependent on plasma temperature per se.[1-8] Because n_e^2 is a more sensitive function of T than analytical atomic or first ionic lines, increasing temperature beyond analyte norm temperatures is counterproductive from the standpoint of S/B ratios. Switching to plasma gases of higher E_{ip} may destabilize the plasma because n_e may not couple the electrical energy source and the plasma strongly enough to offset conductive heat losses. Empirical S/B optimization via energy injection and gas flow refinements has in effect channeled present spectrochemical light source development toward non-LTE plasmas.[44]

For order-of-magnitude comparisons among the absorption cross sections $K(\lambda)$ of isolated lines (bound–bound transitions) $A_u \rightleftarrows A_l + h\nu$, (bound–free) photorecombination, $A^+ + e \rightleftarrows A + h\nu$ and (free–free) continuum due to electron retardation in ionic and atomic fields, $A^+ + e \rightleftarrows A^+ + e + h\nu$ and $A + e \rightleftarrows A + e + h\nu$, respectively, consider the optical cross sections $\sigma(\lambda)$ per absorbing particle:

$$n\sigma(\lambda) = K(\lambda) \qquad [2.29]$$

where n is the number density of absorbers at λ. The mean free path $l(\lambda)$ of a photon in the plasma is of order:

$$l(\lambda) \simeq K(\lambda)^{-1} = (n\sigma(\lambda))^{-1} \qquad [2.30]$$

For prominent UV and visible lines, $\sigma(\lambda)$ is of order 10^{-9} to 10^{-8} cm^{-2}, depending on the square of wavelength, the atomic structure, and the line's broadening.[3,4,24] The values of $\sigma(\lambda)$ for these lines are many orders of magnitude larger than the area $\pi(a_0)^2 = 8.8 \times 10^{-17}$ cm^2 of the first Bohr orbit. In a 5500 K plasma, where n_{tot} is $0.05 \times$ Loschmidt number $= 1.3 \times 10^{18}$ cm^{-3}, an argon resonance line would have $l(\lambda)$ of order 10^{-3} nm, while the resonance line of a trace analyte present at 10^{-9} relative partial pressure would be of order of 1 to 100 cm or more. Thus, even the strongest radiation from the trace analytes has an excellent chance of escaping the plasma and is optically thin even if some argon lines are optically thick. Cross sections for bound–free absorption (the photoelectric effect) are of order 10^{-20} to 10^{-17} cm^2, and those for free–free (bremsstrahlung) absorption are typically a further order of magnitude smaller.[1,2,4,24]

Changes from free–free to free–bound behavior, and from free–bound to bound–bound, are expected from theory[1,2,21,27] to be smooth (Figure 2.1). The classically derived Kramer formula gives the continuum absorption cross section $\sigma(v, n)$ for capture of a photon of frequency v by a hydrogenlike atom in its n quantum state with remainder charge z:

$$\sigma(v, \text{n}) = 7.9 \times 10^{-18} \frac{\text{n}}{z^2} \left(\frac{v_m}{v}\right)^3 \qquad [2.31]$$

where $\sigma(v, \text{n})$ is expressed in square centimeters and v_m is the minimum frequency a photon can have to remove an electron from the n level.[1–4] A characteristic of the cross section is the $(v_m/v)^3$ dependence, which strongly increases $\sigma(v, \text{n})$ at the absorption edge. The formula is improved by insertion of quantum mechanical correction terms g, the Gaunt factors, which are tabulated[1–3] and vary little from unity for hydrogenlike atoms at conditions encountered in spectrochemical plasmas.

For multielectron atoms and ions, the situation is more complicated, especially for photoionization originating from ground states.[3,4] But even where there are multiple electrons, states with large n (Rydberg states), whose large orbits induce hydrogenic behavior, tend to dominate radiative recombination. For example, overlapping of the recombination spectra for the densely spaced high n levels causes the argon IR continuum to be regular and featureless compared to UV-visible continuum.

The total absorption coefficient $K(v)$ for low-energy photons ($hv \ll E_{ip}$) resulting from both free–free and free–bound interactions, is approximated for hydrogenlike systems by the Kramer–Unsold expression[1–3]:

$$K(v) = \frac{16\pi^2 e^6 Z^2 kTn}{3^{3/2} h^4 Cv^3} \exp\left(\frac{-E_{ip} + hv}{kT}\right) \qquad [2.32]$$

where the relative contributions of free–bound to free–free are in the proportion[1,2]:

$$\left[\exp\left(\frac{hv}{kT}\right) - 1\right] : 1 \qquad [2.33]$$

Several authors[2,3,69,70] have proposed expressions for computing $K(v)$ for more complex atoms and ions. Compilations of continuum absorption coefficients for argon, from 200 to 700 nm, are given by Dresvin.[3a] The customary way of computing absorption in more complex spectra is to introduce a multiplicative correction factor $\xi(T_e, v)$ to make Equation 2.32 more exact.

At higher pressures and lower temperatures than are usual for spectrochemistry, the continuum for photoattachment, $A + e \rightleftarrows A^- + hv$, can also become important.[1-4,30,31] For example, in hydrogen plasmas at high pressure and $T < 5000$ K, the H^- is the strongest contributor to the visible continuum.

2.3.5 Factors Tending to Invalidate the Simple Model

Expressions stated above lose reliability when:

1. The plasma is inhomogeneous and no compensation is made for demixing of constituents of different mass,[2,32,50,51] hot spots,[26] or cool boundary layers.[1,32]
2. The observed lines are not optically thin.
3. The plasma is not in LTE.
4. Instrumental bandpasses do not capture the emission in the line wings.

Experimental tests for conformity with the modeled assumptions are difficult to do well. For example, effects from inhomogeneities can erroneously suggest departures from LTE.[26] Fortunately, there is a sufficient body of experience with 1-eV argon plasmas to relate n_e to LTE[3,64,65] with some allowance for plasma size and thermal gradients.[3,25]

Optical thickness can be assessed in several ways. The simplest is illustrated by Figure 2.5. If any line, b, nearby in wavelength, originating in the same part of the plasma as the desired line a, is markedly more intense than line a, then line a cannot have a large optical depth. In the illustration, line b appears about fourfold more intense than line a, therefore the peak intensity of line a cannot exceed $B(\lambda, T)/4$ and $(1 - e^{-\tau_p}) \leq 1/4$. Consequently the peak optical depth for line a is $\tau_0 < 0.29$, hence not more than approximately 15% of the line's total (integrated) energy can be radiatively trapped.

Another simple method to test for self-absorption is to place a mirror of known reflectivity behind the plasma and check whether the consequent doubling of the optical path length doubles the intensity of the desired line. Another test is to compare observed relative line strengths within multiplets with reliable gA data. Thermalization over atomic fine structure should obtain, so that the test is largely independent of LTE considerations. The presence of weaker multiplet members appearing consistently stronger, relative to the stronger members, than predicted by reliable relative gA values[32] indicates an optical depth problem. An analogous method is intramultiplet comparison of halfwidths. Within multiplets, lines should have essentially equal widths. Thus, if stronger lines systematically appear broader, radiative trapping is indicated.[1,27] While this latter test is easily performed, it is comparatively insensitive.

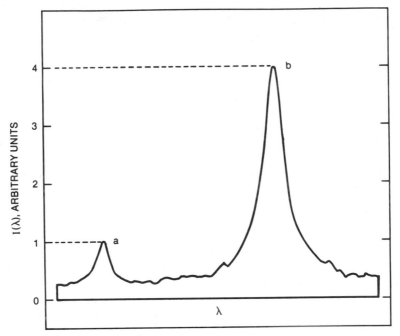

Figure 2.5. Presence of a more intense line has optical depth implications. Because the maximum intensity of *b* is four fold larger than that of *a*, yet cannot exceed the local Planck intensity, the peak optical depth of *a* must be less than 0.3 and radiative trapping must reduce the integrated intensity of *a* by less than 15%.

2.4   OPTICALLY THIN, WELL-RESOLVED LINES EMITTED FROM HOMOGENEOUS PLASMAS IN LOCAL THERMAL EQUILIBRIUM

Integration over line profiles lent simplicity to the preceding section on total line intensity. The shape and width of a resolved line profile are affected by instrumental broadening and various plasma broadening processes; these subjects are dealt with in the following two sections.

2.4.1   Slit Width Effects

In practice, line profiles are partially resolved (when the slit width *ws* is considerably less than the line halfwidth $w_{1/2}$), or partially integrated (when the slit opening *ws* considerably exceeds the line halfwidth $w_{1/2}$), as illustrated schematically in Figure 2.2. For a pressure-broadened line having a pure dispersion shape, the fraction of the line energy in the profile wings captured by the detector is[1,2]:

$$\frac{\int_{-ws/2}^{ws/2} I(\lambda)\, d\lambda}{\int_{-\infty}^{\infty} I(\lambda)\, d\lambda} = \left(\frac{2}{\pi}\right)\left[\tan^{-1}\left(\frac{ws}{w_{1/2}}\right)\right] \qquad [2.34]$$

If the slits are opened gradually to admit more of the line wings, background, noise-in-signal, and wavelength interferences also increase, and sample-minus-blank measurements increasingly involve dealing with small differences between two large signals. For example, in Figure 2.2b, the gross signal level for $ws/w_{1/2} = 4$ is a factor of 4.8 larger than when $ws/w_{1/2} = 0.25$, but the corresponding line-to-background ratios are 0.57 and 1.62. In principle, maximum L/B or S/B ratio is achieved when only the intensity at the line peak is sampled. But if the spectrometer has inadequate resolution,[71] slit width reduction does not necessarily diminish the image width but does decrease signal throughput. As a compromise between S/B, signal level, and costs, spectrometers often have $ws \simeq w_{1/2}$, where $w_{1/2}$ for typical analyte lines and spectrochemical source conditions is 5 to 10×10^{-3} nm.

A common ambiguity in the literature is use of the term "halfwidth," $w_{1/2}$, to describe either the full line width at half-maximum intensity (sometimes called full-$\frac{1}{2}$-width) or half this quantity (termed the half-$\frac{1}{2}$-width). Here, *halfwidth* means full-$\frac{1}{2}$-width unless explicitly denoted "half-halfwidth."

Unless ws is less than $(\frac{1}{3})w_{1/2}$, qualitative and quantitative broadening information are lost in the convolution of the instrumental and line profiles. Indications of reabsorption and self-absorption, contained in the hypothetical line shapes of Figure 2.6, are easily lost due to smoothing over by the slit. Accurate deconvolution of recorded profiles to recover emission line widths for plasma diagnostics is difficult unless ws is less than $(\frac{1}{3})\, w_{1/2}$.[2,24,28]

2.4.2 Effects of Various Line Broadening Processes

The shape function $L(\lambda)$ of the radiated line in Equation 2.7 depends on the atomic structure of the emitter, which determines susceptibility to pressure broadening and depends on the electron number density, temperature, and composition of the plasma. Numerically, the term is large. For example, suppose a line has a triangular profile of height h and base $b = 0.01$ nm. For the profile area $\frac{1}{2}bh = w_{1/2}h$ to satisfy the requirement for $\int^{\infty} L(\lambda)\, d\lambda = 1.0$, the peak profile value $L(\lambda_0) = h$ at the line center must be $w_{1/2}^{-1} = 2 \times 10^9$ cm^{-1}. For actual line profiles, which have a Voigt shape, area a is given by[2,24]:

$$a = phw_{1/2} = phb \qquad [2.35]$$

where $w_{1/2}$ is the halfwidth, which is also denoted by b in Equation 2.36, where h is the height of the central intensity and $1.064 < p < 1.571$, depending on the

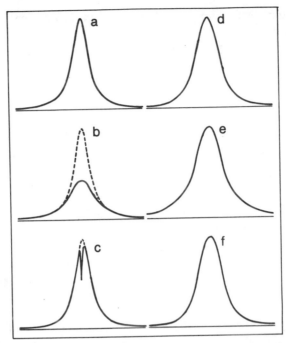

Figure 2.6. How the smoothing effect of convolution with instrumental profiles can mask line profile deformity from self-absorption and reabsorption. Profile *a* is a purely Lorentzian (dispersion) shape corresponding to a purely pressure-broadened optically thin line. Profile *b* represents the same line, but emitted with peak optical depth of 2.0. The dashed curve indicates the extent of self-absorption. Profile *c* has undergone boundary layer reabsorption. The absorption dip appears blue-displaced due to the red shift of the emission line (which originated in a higher n_e region). Profiles *d*, *e*, and *f* are generated from *a*, *b*, and *c*, respectively, by convolution with a Gaussian slit function of half-$\frac{1}{2}$-width equal to the half-$\frac{1}{2}$-width of the dispersion profile *a*. Profiles *d*, *e*, and *f* are normalized to the same peak intensity as profile *a* to help illustrate that on the basis of shape alone, it is very difficult to detect self-absorption or reabsorption when line and instrumental halfwidths are comparable.

relative proportion of damping (dispersion) and Gaussian components in the Voigt profile.[2,24,72]

Optical depth depends directly on $L(\lambda)$, via Equation 2.7. Lines arising from deep-lying states are well shielded from plasma broadening perturbations, and accordingly, resonance lines have particularly large $L(\lambda)$ values at their peaks. These lines also tend to have the large A values and involve highly populated states, further promoting high optical depth and self-absorption.

Because narrowness of emission lines contributes to photometric detectability and selectivity, it is advantageous if a non-LTE plasma can attain a higher than LTE value for the ratio of emitting state density to free electron number density.

Lines emitted by isolated atoms possess natural widths because the probability of spontaneous decay makes the lifetime in an excited level slightly uncertain and the uncertainty principle makes the emitting energy slightly uncertain.[1-4,27,28] Hence, line frequency is slightly uncertain. For lines of spectrochemical interest, natural halfwidths are of order 10^{-5} nm, which is completely negligible compared to the broadening imposed by spectrochemical plasma environments.

Two major kinds of broadening occur in the plasma:

1. Doppler broadening, due to the random thermal motion of the emitters.
2. Pressure broadening, due to perturbation of the emitter by charged particles (Stark broadening), by neutral particles with induced electric fields (van der Waals broadening), and by atoms or ions of the same kind as the emitter (resonance broadening).

   Doppler broadening imparts a Gaussian line shape. Pressure broadening produces a damping (dispersion, Lorentzian) line shape. The line image in the focal plane results from convolution of the line and instrumental profiles. When the instrumental halfwidth is small compared to the line halfwidth, it may often be simulated by a Gaussian shape without introducing significant error.[2,24,32] The profiles of the resultant emission line and recorded line then would have Voigt shapes.

2.4.3 Estimates of Major Causes of Broadening

Tabulations[2,24] of the Voigt shape are listed in terms of the ratio d/g of the relative Gaussian-to-Lorentzian halfwidths contributing to the total line width. The Gaussian term has half-$(1/e)$-width g in the intensity distribution function $\exp(-\lambda^2/g^2)$, where λ is displacement from line center. The Lorentzian power distribution of half-$\frac{1}{2}$-width d is the expression $1/(1 + \lambda^2/d^2)$. Partial widths d from the various pressure broadening mechanisms combine as $d = \sum d_i$. The total Gaussian width, from Doppler and instrumental effects, is obtained from $g = [\sum g_i^2]^{1/2}$. To within an uncertainty of $\pm 0.8\%$, the total width b of the Voigt profile is[24]:

$$b \simeq (d^2 + 2.80g^2)^{1/2} + d \qquad [2.36]$$

Except for resonance lines of argon or other rare gases, Stark broadening is the dominant pressure broadening mechanism when the degree of ionization exceeds 0.1%.[2,28,30] Stark broadening is due to time- and space-varying microfields, primarily from impacting electrons; but ions typically contribute a few percent to the Stark widths and also impart asymmetry to line profiles.[2,27] Degenerate emitters such as atomic hydrogen are subject[1-4,28,30,73] to the linear Stark effect, where $w_{1/2}$ (Stark) in proportion to $n_e^{2/3}$. Broadening profiles from linear Stark effect do not have the Lorentzian shape; rather, each line, such as the well-known Balmer series H_α ($\lambda = 656.2$ nm), H_β ($\lambda = 486.1$ nm), H_γ ($\lambda = 434.0$ nm) has its own characteristic profile.[1,27,74,75] Because it is double peaked and comparatively strong, H_β is particularly useful for comparing observed profiles to theoretical line shapes. Fitting via dimensionless log $I(\lambda)$ versus λ plots[74] helps in discriminating H_β line wings from background. Under ideal conditions, accuracy of n_e data obtained from H_β is seldom as good as 10%.[27,28,74] Additionally, comparison of the H_β halfwidth and red–blue peak separation constitutes a ready test for plasma homogeneity.[75] When fitting to theoretical Stark shapes, red and blue wings should be averaged because the H_β profile is asymmetric.[74,75] The H_β halfwidth is approximately 0.22 nm at $n_e = 10^{15}$ cm^{-3} and 0.99 nm at $n_e = 10^{16}$ cm^{-3}. Doppler and instrumental broadening tend to obscure profile details for $n_e < 10^{15}$ cm^{-3}. Use of H_α for n_e determinations is risky because this line is, for several reasons, prone to radiative trapping.

Quadratic Stark effect imparts a Lorentzian shape to nonhydrogenic lines, with a halfwidth proportional to n_e, regardless of whether the line is from an atomic or ionic emitter. However, theoretical treatments for atomic lines differ substantially from those ionic lines.[1,27,28,72,75,77] Extensive theoretical and experimental data for the Stark broadening parameters have been compiled.[28] Tabulated electron impact broadening parameters and parameters for broadening by ions for lines of lighter atoms[1,27] and ions[27,77] are combined to give the total Stark broadening parameter $w_{1/2}$[1,2]:

$$w_{1/2} = 2w\{1 + 1.75\alpha(1 - 0.75R)\} \qquad [2.37]$$

where w is the electron impact broadening parameter, customarily given as a half-$\frac{1}{2}$-width, α is the parameter for quasi-static broadening by ions and $R = (\frac{3}{4}\pi n_e)^{1/3}/\rho_D$ is the Debye length[1,27,73] and:

$$\rho_D = [kT/8\pi n_e e^2]^{1/2} \qquad [2.38]$$

Semiempirical formulas[78] based on Coulomb approximation wave functions and semiempirical Gaunt factors permit rapid hand computation of Stark broadening parameters for first ions.

Experimental and theoretical Stark broadening parameters agree, on average, to $\pm 20\%$.[27,28,77] Systematic trends[28,79–81] in the broadening of lines from atoms and ions of homologous structure are useful for extrapolating and assessing broadening parameter data. Generally, all lines in a multiplet are expected to have the same Stark broadening.[1,27,28,82] The Stark broadening parameters have only a slight temperature dependence.[1,27,28] Other things being equal, more Stark broadening is observed for lines having longer wavelengths. This is because of the relation:

$$\Delta\lambda = \frac{\Delta v}{v}\lambda \qquad [2.39]$$

Equation 2.39 suggest that similar relative energy perturbations $\Delta v/v$ cause broadening proportional to wavelength. Lines whose upper states have deep-lying, shielded orbits are much less susceptible to Stark broadening than those from states near the ionization limit. For example, the O I line widths, measured in 1-eV plasmas, at $n_e = 10^{16}$ cm^{-3}, are[28]:

λ (nm)	n*	$w_{1/2}$ (nm)
130.35	1.8	0.00014
394.7	3.2	0.015
725.4	3.9	0.16
604.6	4.9	0.4

Such data also illustrate that a line broadening, mostly associated with perturbations of the upper state, is sensitive to n*. Perturbations of the lower state rarely contribute more than a few percent to the Stark width.[1,27,28]

Resonance broadening affects both resonance lines and lines whose lower states are resonance states.[1,27,83] It is a quantum mechanical perturbation whose strength is proportional to r^{-3}. It contributes measurably to the broadening of Ar I resonance lines in spectrochemical atmospheric discharges.[83] For deep-lying states, both Stark and resonance broadening should be considered when argon is the major plasma constituent.[27,83] Due to their low partial pressures, analytes should have negligible resonance broadening.

Close encounters between the emitter and neutral particles generate[1,27,28] induced dipole fields that are proportional to r^{-6}. Parameters for van der Waals broadening of medium and heavier elements by argon are not well quantified.[1,27,28] However, because analyte lines usually involve well-shielded states of small effective principal quantum number n* = $Z (R/(E_{ip} - E))^{1/2}$, where R is the Rydberg constant, Z is the net atomic charge, and E is the energy of the state, and particle densities in spectrochemical discharges are factors of 20 to 40 smaller than at STP, this type of pressure broadening is usually negligible in spectrochemical plasmas.

Doppler broadening produces a Gaussian profile of full-$\frac{1}{2}$-width[1-4,28]:

$$w_{1/2} = 7.16 \times 10^{-7} \lambda \left(\frac{T}{M}\right)^{1/2} \qquad [2.40]$$

where M is the atomic weight, T is in Kelvins, and λ is the line wavelength in nanometers. At 10,000 K, when the most probable thermal velocity for hydrogen is 13 km/s, 500-nm lines of H and Fe would have $w_{1/2}$ of 0.036 and 0.005 nm, respectively.

Unresolved isotope splitting in some heavy elements, such as mercury, can cause their line profiles to appear anomalously broad.[84]

2.4.4 Estimates of Line Shifts

Line shifts from plasma Stark effect are proportional to n_e.[1,27,28] The shift, d, is usually toward high wavelengths, but blue shifts are not rare.[1,27,28] The shift is less than or equal to 0.6 $w_{1/2}$.[1,27,28] The critical compilations and computations of Stark broadening parameters[28] list shift-to-width ratios $d/w_{1/2}$, but as noted

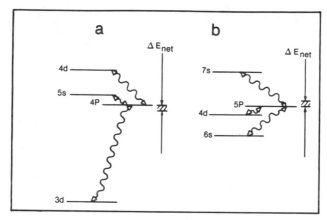

Figure 2.7. Hypothetical atomic level structures conducive to red or blue line shift by plasma Stark effect. Perturbation by time- and space-varying plasma microfields collisionally couple, and mutually repel, states between which dipole transitions are allowed. (*a*) The preponderance of interaction is with more energetic states; hence the time-averaged energy of the emitting state declines and the line shifts toward the red. (*b*) The converse occurs.

earlier, these seldom include data for medium-weight and heavier elements. Hypothetical atomic structures conducive to red and blue Stark shifts are shown schematically in Figure 2.7. Transient coupling between the emitting state and nearby states accessible by dipole transitions causes the interacting states to mutually repel each other in energy. In the illustrated case, Figure 2.7*a*, most perturbing states are at higher energy, depressing the energy of the emitting state. Since the lower state of the line is little perturbed, the net result is a less energetic photon, that is, a red shift. Because the energy density of states increases as the ionization limit is approached, the red shift is the more common pattern. Occasionally, however, more or more strongly coupled arrangements of perturbing states lie below the emitting state,[1,27,28] elevating its energy and producing a blue shift (Figure 2.7*b*).

2.5   EMISSION FROM INHOMOGENEOUS PLASMAS OR FROM PLASMAS OF NONLAMINAR GEOMETRY

2.5.1   Introduction

Homogeneous, laminar plasmas are simple to model but elusive in practice.[3,4] Cool boundary layers are inevitable in steady-state plasmas.[85] Within the body of the plasma, thermal gradients tend to demix constituents of disparate molecular weights.[50,86] Forced convective cooling sets up turbulence,[3,87] whose spectroscopic consequences can include systematic as well as random error. Side-on scanning of cylindrically zoned plasmas can be interpreted using the Abel inversion.[2,3,25] Plasmas with noncircular cross sections can be treated by other transformations[3] or can be probed by laser fluorescence techniques,[88,89] which do not require Abel inversion.

2.5.2 Boundary Layers

Luminous plasmas are generally sheathed in cooler boundary layers that strongly absorb atomic resonance lines.[32,90] Boundary layers in flowing argon plasmas tend to be thin because shear is set up by the marked viscosity differences accompanying the thermal gradients.[3,91] Steady-state laboratory plasmas possess boundary layers, since these constitute a transition zone between plasma and free stream conditions. Boundary layers can be avoided by using shock tube plasmas during microsecond initial sampling periods prior to growth of laminar boundary layers[4,92] and by hot argon flushing of certain end-on-viewed cascade arcs.[32]

Spectroscopically, consequences of boundary layers should be less important for chemical analyses than for plasma diagnostics. The boundary layer strongly reabsorbs atomic resonance radiation. From Equation 2.23, good line-emitting transitions have inverse transitions, which are good line absorbers.[21-23] Saha–Boltzmann equilibrium, however, effectively limits absorbing populations to atomic ground states. The reabsorption profiles that boundary layers superimpose on emergent resonance emission lines can substantially reduce intensities and modulate profile shapes, as illustrated in Figure 2.6. For resonance lines, boundary layers need not be thick to powerfully affect radiative transport. For example, for a 200-nm resonance line ($E_u = 6.1$ eV), the ratio of absorbing to emitting populations is 10^9-fold larger in a 2370-K (2.03-eV) absorbing plasma than in a 7100-K (6.1-eV) emitting plasma that it might enclose.

Analytically, the consequences of reabsorption in boundary layers are mainly a reduction in signal strength for atomic resonance lines, which reduces detectability. At high analyte concentration, reabsorption helps maintain linearity between analyte concentration and emission signal, since it removes the optically thick central portion of line profiles and leaves optically thin wings essentially unaffected because Stark and Doppler widths are small in the cool boundary layer.

Diagnostics of plasmas by the popular atomic-to-ionic line resonance intensity method[1-7] can be biased if boundary layer reabsorption is unrecognized due, most typically, to instrumental profiles washing out reabsorption "dips" (profile e in Figure 2.6). Since the ionic line of the pair is unaffected by reabsorption, the results skew towards erroneously high "Saha temperatures." The extent of atomic resonance reabsorption in plasma boundary layers is difficult to assess quantitatively, rendering these lines unsuitable for plasma diagnostic studies.

2.5.3 Cylindrically Symmetrical Plasmas

Side-on viewing of plasmas having cylindrical symmetry, illustrated schematically in Figure 2.8, poses no basic spectrochemical difficulty since the operator empirically selects the chord and observation height providing optimal signal-to-noise ratios and repeatability. Emission data for spatially mapping conditions in a plasma cross section are sampled by moving the line of sight in small, known increments from one side of the plasma column through to the other. Abel inversion is used to determine source conditions in the annular zones. Computer codes for performing Abel inversion are available.[2,93] For relevant ICP studies, the reader is referred to the chapter appendix.

To illustrate how ignoring Abel inversion generates bias, a hypothetical three-zoned plasma is considered (Figure 2.8). Viewed by a thin sampling cone crossing the center of the plasma, the zones have equal thicknesses, and the average of their temperatures is 10,000 K. Suppose two spectral lines with E_u of 6.0 and 9.0 eV are used to estimate excitation temperature by the two-line relative intensity method.[1,4,25,26] If the plasma is homogeneous and at 10,000 K, the intensity ratio of the 6.0-eV line to the 9.0-eV line is predicted to be a factor of 1.37 larger than the value expected from the three actual zones: 9000 K, 10,000 K, and 11,000 K. In this case, $\pm 10\%$ temperature inhomogeneities induce 37% errors in intensity ratio.

2.5.4 Inhomogeneous Plasmas

Numerical techniques have been developed to map conditions within cross sections of inhomogeneous plasmas using spectroscopic data collected side-on.[3]

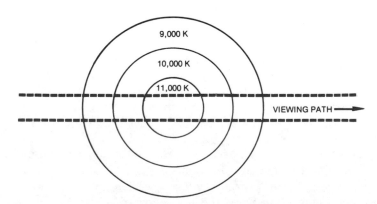

Figure 2.8. Section of a hypothetical light source with cylindrically symmetrical zones viewed side-on. Suppose two lines are in a spectrum and have excitation energies of 6.0 and 9.0 eV, and one records the intensity ratio of the former relative to the latter. Compared to a similarly viewed homogeneous 10,000 K plasma of equal cross section, the intensity ratio in the zoned plasma is smaller by a factor of 0.73.

An alternative is to use laser-excited fluorescence, where spatial resolution is excellent because the unit volume defined by the intersection of probe and collecting beams is minute and precisely positioned.[88,89]

The plasma diagnostician must be alert for repeatable inhomogeneities such as "hot spots" in the line of sight.[26] In analogy to arguments in connection with Figure 2.8, "hot-spot" inhomogeneities favor, in a nonlinear way, emission features that are difficult to excite. A pattern wherein diagnostic features of higher effective excitation energy yield higher apparent temperatures cannot per se be interpreted unambiguously, because such a pattern can be symptomatic of either a spatial inhomogeneity or a recombining non-LTE plasma.[26,94] Discrimination between these two possibilities is not easy.[26] As discussed in Section 2.7, the consensus is that a group of diverse tests is a more promising strategy than attempting a single precise measurement of one kind.[1-4,95,96]

2.6 OPTICALLY THICK LINES EMITTED BY A PLASMA IN LOCAL THERMAL EQUILIBRIUM

Resonance lines of analytes, matrix or plasma gases can become optically thick in spectrochemical plasmas. Analytical concern arises at optical depths $\tau(\lambda)$ when $(1 - \exp(-\tau(\lambda)) \simeq \tau(\lambda)$ ceases to be an acceptable approximation for use in interpolating between samples and standards. It may be unsound to use the optically thin approximation on a line of peak optical depth as low as $\tau(\lambda_0) = 0.1$. This, of course, depends on analytical accuracies required, on the optical depth of the local continuum (depending in turn on plasma composition), on other error sources, and on the concentration differences between sample and standards. For analyte lines, the operational remedies for self-absorption are simple: (a) dilute the sample to reduce $\tau(\lambda_0)$, (b) switch to weaker analyte lines, and (c) construct a nonlinear response curve based on several standard concentrations. Plots of log (I) versus log (ρ), where I is the line (total) intensity and ρ is the analyte concentrations, are linear when ρ is small but assume weaker-than-linear dependence as concentrations increase. Details of this "curve of growth"[29-31] depend on the line broadening parameters and on conditions in the plasma.

Due to the comparatively weak dependence of $I(\lambda)\,d\lambda$ on T in Equation 2.6, a few percent of radiative trapping might be acceptable for lines used in plasma thermometry. However, the higher the optical depth, the more the radiation becomes increasingly characteristic of the portion of the plasma closest to the detector. Optically thick lines, $\tau(\lambda_0) \gg 1.0$, preferentially convey information about the skin rather than the bulk plasma.

Resonance lines of argon and of any easily ionized matrix constituents present in relative abundances of parts per thousand, influence radiative cooling, internal energy transfer, and plasma stability.[3,4,97] Excitation conditions are sensitive to radiative transfer rates.[98] In instances of plasma conditions that are not robust against variation in matrix concentrations, radiative transfer processes, rather than the weak electron donor role of easily ionized elements, are likely to be responsible. The problem is accommodated analytically by matching matrices for sample and standard.[99]

Argon resonance lines attain optical depths several orders of magnitude larger than those of the more familiar visible Ar I lines (Table 2.5). Optical depths at line center $\tau(\lambda_0)$ are shown in Table 2.5 for Ar I 104.81 and 430.02 nm in 1.0-cm-thick atmospheric argon plasmas at 8000 K, 10,000 K, and 12,000 K. These lines are typical of atomic argon resonance and visible lines, respectively. For these computations, tabulated Ar I number densities and electron number densities of Table 2.3, are used. Stark,[27] resonance,[83] and Doppler broadening, but not van der Waals[2] broadening, are taken into account. Several features of Table 2.5 are noteworthy.

1.   Maximum optical depth of Ar I 430.0 nm increases only threefold in response to the 50% temperature increase from 8000 to 12,000 K. These temperatures are far below the norm temperature for Ar I 430.0 nm shown in Figure 2.4. Stark broadening, proportional to the n_e, increasingly spreads the line energy over wavelength as temperatures rise. The line is optically thin under all the tabulated conditions, so that to a good approximation, the central profile intensity equals the optical depth multiplied by the Planck function $B(\lambda, T)$. The function $B(\lambda, T)$ is strongly dependent on temperature, so that spectroscopic signals from the central portion of Ar I 430.0 nm increase faster than do the tabulated optical depths. But signal-to-background ratios cannot be improved simply by increasing temperature. On the contrary, the corresponding 40-fold increase is in n_e (from 1.45×10^{15} cm^{-3} to 6.66×10^{16} cm^{-3}) responsible for broadening Ar I 430.0 nm also increases the continuum (proportional to $n_e^2/T^{1/2}$) by more than three orders of magnitude. This well illustrates the inherent limitations of thermal plasmas in promoting good signal-to-noise ratios for inherently weak lines.

2.   Under all tabulated conditions, the Ar I 104.8 nm resonance line has a large optical depth at its center. This causes the line to saturate (at local $B(\lambda, T)$) outward, to approximately 100 halfwidths. The line, therefore, radiates over approximately 1.0 nm as a blackbody. Extrapolation of Figure 2.3, at 10,000 K indicates that this amounts to more than 50 W of radiative power. There are three other Ar I resonance lines,[16,32] so that power radiated from the Ar I by the Ar I resonance lines can readily reach several hundred watts.

3.   Between the highest and lowest temperature tabulated, the total dispersion partial width, $w_{1/2}$, goes through a minimum. At the lower temperature, resonance and van der Waals broadening dominate. At the hotter conditions, Stark broadening dominates.[27,83] Because the resonance line emission and absorption profiles narrow within intermediate plasma zones, radiative transport from hot input zones to the relatively cool analytical zones is markedly strengthened.

Table 2.5. Optical Depths of Two Atomic Argon Lines in 1-cm-Thick Argon Plasma at Atmospheric Pressure and Three Temperatures[a]

Line (nm)	T (K)	n_e (cm^{-3})	n_{ArI} (cm^{-3})	n_{up} (cm^{-3})	w_D (nm)	w_S (nm)	w_R (nm)	$w_{1/2}$ (nm)	$L(\lambda_0)$ (cm^{-1})	$\tau(\lambda_0)$
104.81	8,000	1.45×10^{15}	9.15×10^{17}	9.71×10^{10}	1.06×10^{-3}	1.60×10^{-5}	8.7×10^{-4}	1.32×10^{-3}	5.18×10^{9}	1.16×10^{5}
	10,000	1.49×10^{16}	7.04×10^{17}	2.31×10^{12}	1.19×10^{-3}	1.64×10^{-4}	6.72×10^{-4}	1.22×10^{-3}	5.82×10^{9}	1.01×10^{5}
	12,000	6.52×10^{16}	4.81×10^{17}	1.55×10^{13}	1.30×10^{-3}	7.17×10^{-4}	4.60×10^{-4}	1.51×10^{-3}	4.52×10^{9}	5.32×10^{4}
430.02	8,000	1.45×10^{15}	9.15×10^{17}	3.33×10^{9}	4.35×10^{-3}	5.22×10^{-3}	0	6.24×10^{-3}	1.10×10^{9}	4.23×10^{-4}
	10,000	1.49×10^{16}	7.04×10^{17}	1.72×10^{11}	4.87×10^{-3}	5.35×10^{-2}	0	5.37×10^{-2}	1.19×10^{8}	1.00×10^{-3}
	12,000	6.52×10^{16}	4.81×10^{17}	1.93×10^{12}	5.33×10^{-3}	2.35×10^{-1}	0	2.36×10^{-1}	2.70×10^{7}	1.43×10^{-3}

[a] n_{ArI} and n_{up} are the argon atom and emitting state number densities, respectively. Full halfwidths w_D, w_S, w_R, and $w_{1/2}$ are for Doppler, Stark, resonance, and total broadening, respectively.

Radiation of power from hot plasma zones and its interception in cooler plasma margins tends to reduce radial temperature gradients.[3,4,100] This radiative transport partially offsets the gradient-steepening influence of shear in forced-flow plasmas. Radiative coupling between remote portions of the plasma relies primarily on the Ar I resonance line wings.[1,27,29–31] These wings, due to Lorentzian portions of the profiles, are sensitive to electron number densities,[1–4] so that ionization of easily ionized elements can change spatial energy distributions by modulating argon–argon energy transport.[29–31,94] Radiation at the resonance line center λ_0 can diffuse at speeds not dissimilar to plasma flow velocities,[101] leading to long "confinement times" for λ_0 photons and enabling their advective transport to regions where no electrical or other energy input is expected.[101] For example, if conditions give a mean free path of order 10^{-6} cm at λ_0 for an Ar I resonance line, and its radiative lifetime is 10^{-8} s,[32] then λ_0 photons will propagate outward at less than 100 cm s^{-1}. The trapped resonance–radiation field, coupling ground and first excited states, helps to establish LTE inside the high-temperature zones.[1,3,4] The net flux of resonance radiation energy into cooler zones is hard to collisionally equilibrate in these zones due to low n_e and may contribute to non-LTE behavior at typical observation heights.[94,101,102–104]

Resonance lines of easily ionized elements such as sodium can constitute channels for significant radiative transfer, as might be suspected from watching a spectrochemical plasma take on the conspicuous D-line color upon aspiration of marine water sample. For example, in a 10,000 K argon plasma, sodium D-lines become optically thick at 10^{-4} partial pressure. Even though most sodium is ionized at these temperatures, and Stark broadening is almost two orders of magnitude larger than for the Ar I resonance lines, the 2.1-eV excitation potential of sodium D assures a high number density of emitting states. In cool plasma margins, sodium is mainly nonionized and resonance states are radiatively pumped by D-line flux from hotter regions. Only 3.0 eV is required to ionize Na I from its resonance state, so that radiative transfer can affect the degree of ionization in low-energy-density zones. This, in turn, can effect local absorption rates for Ar I resonance lines and rates of electrical coupling to the power supply.

Optically thick Na (or Li or K or Ca) lines tend to radiatively cool a plasma, whereas Ar I resonance radiation tends to spatially redistribute energy within the plasma. Widths of the sodium D lines are due primarily to Stark broadening. Reabsorption in cooler margins of the plasma occurs in low n_e regions, so that reabsorption profiles are much narrower than emission profiles, permitting most of the emission line energy to escape. But the reabsorption profile width of the Ar I resonance line has a greater width in cooler, denser layers due to resonance broadening than does the emission profile whose width results from Stark-plus-resonance broadening in the high-n_e, but low-n_A, energy input zones.

Computation of radiative transfer in inhomogeneous LTE plasmas is demanding because local absorption and emission profiles depend on kinetic temperatures, as well as densities of ground states and of resonance states and of free electrons.[29-31] For non-LTE situations, such as with astrophysical plasmas, iterative approximations are used to deal with the coupling between collisional and radiative rates.[29-31]

2.7 THEORETICAL CRITERIA AND EXPERIMENTAL TESTS FOR LOCAL THERMODYNAMIC EQUILIBRIUM

2.7.1 Introduction

Presumption of local thermodynamic equilibrium enables populations of internal and kinetic degrees of freedom to be simply computed and interrelated by the Saha–Boltzmann, Maxwell, and Planck equilibrium distribution functions. When these relationships are invalid, quantitatively reliable descriptions of emitting state population densities becomes a formidable, if not intractable, problem. So, for providing an intellectually appealing, unified framework, and for computational necessity, the equilibrium concept of temperature is one spectroscopists are loathe to abandon.

Faith in applying LTE axioms to atmospheric pressure discharges long appeared to be justified. Quantitative emission spectrometry was for decades synonomous with use of conventional carbon arcs, and it was documented that their emission conformed tolerably to the LTE assumption.[105] Against this background, it was perplexing when spectroscopic diagnostics of early atmospheric-pressure argon plasmas sometimes gave results incompatible with LTE.[106,107] Subsequent investigators have commented that reports of non-LTE behavior often resulted, in reality, from unrecognized inhomogeneities, self-absorption, faulty calibrations, or other experimental artifacts. Conflicting spectrochemical interpretations during the period 1950–1960 seem to have imparted a lingering mystification to the topic.[102-104]

However, the extensive body of theoretical[108-110] and more sophisticated experimental data[111-133] now available is in substantial collective agreement on the conditions required for LTE, as well as criteria for other, less complete, regimes of equilibration.[3] Retrospectively, differences in equilibration between carbon arcs and atmospheric pressure argon plasmas can be ascribed to: differences between respective resonance energies, ionization potentials, and consequences of forced versus free convective cooling.[3]

2.7.2 Thermal Equilibria: CTE, LTE, PLTE, and non-LTE

A plasma is in complete thermal equilibrium (CTE) if the energy distributions of all degrees of freedom—including dissociation and ionization, atomic, ionic, and molecular excitation, kinetic energies (of atoms, ions, electrons) of all species in all stages of ionization—are described by the same temperature, and this temperature also characterizes the plasma radiation field and the thermodynamic and transport properties. Equilibration with the radiation field occurs only for a cavity filled with blackbody radiation.[1–4,110] Within the cavity:

number of emissions = number of absorptions

number of collisions of second kind = number of exciting collisions   [2.41]

Power requirements for a blackbody plasma, having a temperature of several thousand degrees with even a small sampling window in the walls, would be formidable, and, aside from blackbody pyrometry, no diagnostic or spectrochemical information would be conveyed.

If the plasma radiation were optically thin, and if plasma gradients were not so steep that temperatures change perceptibly within an electron mean free path, then approximate *equilibrium can occur locally (LTE)* even in the absence of equilibrated cavity walls. Since emission and absorption are negligible in the optically thin approximation, the energy balance of Equation 2.41 is replaced by:

$$\begin{array}{c} \text{number of emissions} \\ plus \\ \text{number of collisions} \\ \text{of second kind} \end{array} = \begin{array}{c} \text{number of absorptions} \\ plus \\ \text{number of exciting} \\ \text{collisions} \end{array} \qquad [2.42]$$

However, as discussed in Section 2.6, plasmas hot enough to give useful intensity to analytical or diagnostic lines tend to possess optically thick atomic resonance lines, so that the predicted optically thin condition is not strictly met. For radiative relaxation to cause less than 10% departures from LTE, collision rates must exceed radiative decay rates tenfold.[3,4,108–110] The impacting electrons, because of their small mass and high speed, are the primary providers of the inelastic collisions for equilibrating the internal degrees of freedom in atoms and ions.[3,4,108–110] Three independent estimates of the requisite electron number densities for collisional domination of excited-state population dynamics concur that[110]:

$$n_e \geq D \times 10^{12} \times T_e^{1/2} E_{pq}^3 \qquad [2.43]$$

when n_e is in cm^{-3}, electron temperature T_e is in K, the energy separation E_{pq} between bound states p and q (same ionization stage) is in eV, and D is a constant that has been variously computed as 0.92, 1.3, and 1.6.[110]

As shown in Table 2.3, argon is only slightly ionized in 1-eV plasmas; hence the energy gap most difficult to collisionally equilibrate is the 11.6-eV resonance potential. Equation 2.43 predicts that an n_e of 1.6 to 1.7 × 10^{17} cm^{-3} is needed to fully thermalize this Ar I internal degree of freedom. This theoretical estimate is in gross agreement with results from seven experiments to test for attainment of LTE in atmospheric-pressure argon spectrographic arcs constricted to diameters of 3 to 5 mm.[111–113] Minimal n_e values for thermalizing argon were found experimentally to be 1.0 × 10^{16} to 20 × 10^{17} cm^{-3}, with a geometric mean value of 7×10^{16} cm^{-3}. In a free-burning argon arc, having less steep thermal gradients and higher optical depths to promote LTE at lower electron number densities,[94,110] apparent temperatures $T_{electron}$, from bremsstrahlung, T_{Saha}, from Ar II intensities, and T_{exc}, from Ar II line intensities, converged at $n_e \simeq 1.2$ to 3.0×10^{15} cm^{-3}.

Energy separations between resonance and metastable states are of order 0.1 eV in argon.[16] Application of Equation 2.43 predicts that the Ar I metastable and resonance states thermalize at electron densities as low as 10^{11} cm^{-3}. So, while in low-density, nonionized plasmas, metastable states have long lifetimes because radiative decay is forbidden, in spectrochemical plasmas collisional coupling to energetically close resonance states is strong enough to provide channels for prompt radiative decay to ground state.[40]

Figure 2.9 indicates some spectroscopic consequences of Equation 2.43. Energy levels are drawn approximately to scale. Hydrogen is a chief source of diagnostic lines, atomic argon is the usual ICP gas, atomic manganese is a typical transition metal, and atomic ruthenium is a typical heavy metal.[16] Within each element, the resonance and ground-state populations are the most difficult pair to collisionally equilibrate. Regarding the E_{pq}^3 scaling of Equation 2.43, E_{pq}^3 values, denoted in the figure as ΔE, are shown explicitly for the low-lying atomic levels. For hydrogen, and argon, electron number densities need to be three orders of magnitude larger to equilibrate between ground and resonance states than to equilibrate between states with $n^* > 3$. Stepwise equilibration, through excitation or cascading, rapidly couples excited states to the resonance state.[1–4] It is the collisional coupling to ground state that is tenuous. Elements such as Mn and Ru, due to their smaller resonance energies, require electron number densities two to three orders of magnitude *smaller* than needed to attain LTE in argon.

Mean energy spacings between emission line upper states and first ionization potentials are typically of the order of 0.5 eV,[13–15] corresponding to thermalizing electron number densities of order 2×10^{13} cm^{-3}. At the 10^{15} to 10^{16} cm^{-3} electron number densities prevalent in spectrochemical plasmas, excited states differing slightly in energy and communicating by dipole transitions thermalize rapidly. There exist correspondences between electron impact cross sections and optical cross sections,[21–24] where gf values gauge the latter.[21–24] *Partial local thermodynamic equilibrium* (PLTE) is achieved when n^* is high enough that n_e effects thermalization, allowing population of two excited states, m, n (of the same spectrum) to be related via:

$$\frac{n_m}{n_n} \simeq \frac{g_m}{g_n} \exp\left(\frac{E_{mn}}{kT_{exc}}\right) \qquad [2.44]$$

where E_{mn} is the energy separation and T_{exc} is the excitation temperature, where $T_{exc} \to T_e$, the free-electron kinetic temperature, in the limit $E_m \to E_{ip}$ of excited states approaching the ionization potential.[3,4,110] Coupling of high n^* states and the continuum is strengthened by the trends for collision cross sections to

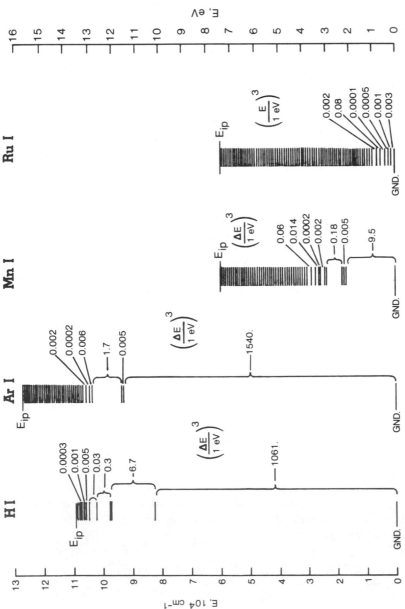

Figure 2.9. Schematic energy level diagrams of light, medium-weight, and heavy elements, indicating relative difficulty in collisionally equilibrating excited states with ground state. The depicted level structures are for atoms, but similar considerations hold for ions. Separations, ΔE, between levels are drawn to scale. Thermalization (equilibration) between levels of energy separation ΔE is driven primarily by inelastic electron impacts, cross sections for which scale approximately as $(\Delta E/kT)^3$. The $(\Delta E/kT)^3$ values corresponding to the energy gaps between depicted levels are listed to the right of each elements level diagram. For Ar I, the 11.6-eV resonance energy is a thousandfold greater impediment to LTE than is any other feature in the atom's structure.

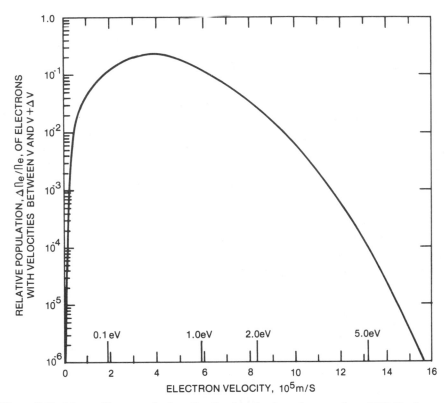

Figure 2.10. Maxwellian population distribution for free electrons in a 5000 K plasma. Practically all free electrons have energies in excess of typical atomic fine structure splitting, approximately 0.1 eV, but only a small minority possess the ArI 11.6-eV activation energy.

change as $(n^*)^4$, while optical transition probabilities, radiatively connecting large n^* to deep-lying states, decrease.

For large energy differences between excited states, stepwise collisional excitation and relaxation predominate over single, large energy transitions.[1–4] This difference in rates is shown in Figure 2.10. The Maxwellian distribution at spectrochemical temperatures contains a much larger percentage of free electrons with energies exceeding 0.5 eV than electron having energies exceeding 3 eV.

Simplified models for excitation in PLTE use two- or three-level hypothetical atoms to solve consistent rate equations. One objective of such models is the computation of coefficient b[114,115]:

$$b_i = \frac{n_i'}{n_i} \qquad [2.45]$$

where n_i for state i is the equilibrium (LTE) population number density and n_i' is the actual population density for that level.[114,115] For the atomic ground state, if overpopulation occurs, $b_1 > 1.0$, the PLTE plasma is termed "ionizing," while $b_1 < 1.0$ for ground states denotes a "recombining" PLTE plasma. Such situations might arise, respectively, in regions of plasmas that were net exporters or importers of resonance radiation. While simplified two- or three-level models are computationally tractable, their ability to simulate real atoms has shortcomings. For instance, rates of three-body recombination, which rely on capture into highly excited states, can err by an order of magnitude in such simple models.[1,3]

In the presence of electric fields, electrons and atoms can obey separate Maxwellian distributions, having temperatures T_e and T_a that may differ by as much as a few percent[3,4,116]:

$$\frac{T_e - T_a}{T_e} = \frac{m_a}{4m_e}\left(\frac{e\lambda_e E^2}{\frac{3}{2}kT_e}\right) \ll 1 \qquad [2.46]$$

where λ_e is the electron mean free path. Modest deviations between T_e and T_a do not propagate strongly into spectroscopic variables because dependence on T_a is only linear, via the ideal gas approximation used, and is employed only to obtain total particle density.

In non-LTE plasmas, energy distributions for kinetic degrees of freedom are non-Maxwellian.[3,4,110,117,118] Kinetic degrees of freedom equilibrate much faster than internal degrees of freedom.[1–8,108–110] If the free electron and atom kinetic energy distribution is not Maxwellian, then *no other* degrees of freedom are equilibrated according to the formulae developed for LTE situations. Specifically:

1. Excited states are not populated in accordance with the Boltzmann statistics, and excitation temperature T_{exc} is *undefined*.
2. Ion/atom equilibration is not governed by the Saha equations, and $T_{ionization}$ is *undefined*.
3. Detailed balancing, whereby in LTE the rate for *each* process is balanced by the rate of the inverse of that specific process (eg, per each pair of states, radiative excitation rates would equal radiative decay rates), ceases to hold.
4. In the absence of detailed balancing, the source function:

$$\frac{\xi(\lambda)}{K(\lambda)} \equiv S(\lambda) \qquad [2.47]$$

in the equation of radiative transfer:

$$\frac{dI(\lambda)}{d\tau(\lambda)} = -I(\lambda) + S(\lambda) \qquad [2.48]$$

where $I(\lambda)$, $\tau(\lambda)$, $K(\lambda)$ are defined by Equations 2.1 to 2.5, no longer satisfies Kirchhoff's law, so that:

$$S(\lambda) \neq B(\lambda, T)$$

and the Planck function is no longer appropriate for computing resonance line intensity changes in emitting and absorbing layers.

If kinetic energy is not Maxwellian in distribution, it is not possible to compare temperatures characterizing internal degrees of freedom or the radiation field, because the latter are not defined and there exists, for example, no Saha temperature.

No standard method exists for describing non-LTE plasmas.[1-8,29-31, 108-110,117-119] Interest in absorption and emission in stellar atmospheres has fostered astrophysical models concerned with kinetic energy and wavelength regimes akin to those in spectrochemical light sources.[29-31,115,124] Self-consistent solutions are sought for systems of coupled rate equations. At one atmosphere and 1 eV this involves the following as leading processes:

Impact ionization and three-body recombination:

$$M^* + e + E_{kin} \rightleftarrows M_0^+ + 2e \qquad [2.49]$$

Resonance photon emission and absorption:

$$M_0 + h\nu \rightleftarrows M_1 \qquad [2.50]$$

Electron impact excitation and cascading:

$$M^* + e + E_{kin} \rightleftarrows M^{**} + e \qquad [2.51]$$

where M^* (or M^{**}), M_0^+, M_0, and M_1 are an excited atomic state, a ground-state ion, a ground-state atom, and an atom in its resonance state, respectively.

Cross sections for inelastic collisions at 1-eV encounter energies are imprecisely known due to the paucity of low-energy-beam experiments.[121-124] It is usually known, however, which competing processes dominate. For instance, ionization-recombination channels in Equation 2.49 compete with auto-ionization-dielectronic recombination and photoionization-radiative recombination.[2-4,121-123] But in spectrochemical regimes (1 eV, $n_e = 10^{15}$ to 10^{16} cm^{-3}) these reactions are unimportant compared to Equation 2.49.[3,4] The computational expedient of separating the collisional rates from radiation flux rates[30,31,117] is not justified in atmospheric argon plasmas, where the flux in central portions of resonance line profiles can profoundly affect population dynamics. Importance of radiative transfer in atmospheric argon plasmas can be appreciated with data from 10,000 K, 12,000 K, and 14,000 K, showing the total wavelength integrated intensities to be 9, 180, and 950 W/cm^2 sr, respectively.

2.7.3 Factors Conducive to Local Thermal Equilibrium or Inhibiting Its Establishment

Factors favoring establishment of LTE all act to increase collisional rates relative to the rates of energy flux into or out of the plasma.

1.   Increasing the rates of elastic and inelastic collisions between particles promotes LTE. Since $v_e/v_M \simeq (m_M/m_e)^{1/2}$, $v_M \simeq 270$ for M = argon, even a 0.005 degree of ionization strongly affects elastic collision rates. The low mass of free electrons make them even more important for inelastic collision rates.[1,3,110,121] Other things being equal, increasing total pressure or total density increases collision frequencies.

2.   Decreasing contact and diffusive heat exchange between plasma and its surroundings helps foster LTE.[3,25,113] Hence, larger plasma size (ie, increased volume/area) promotes LTE. Forced convective–conductive cooling steepens gradients, thereby accelerating energy import or export rates and driving plasmas further from LTE.

3.   Increasing the coefficient of optical absorption for short-path (large-$K(\lambda)$) photons and weakening the absorption and radiation of long-path photons (eg, photons not reabsorbed within the plasma volume effect heat exchange with the environment) promotes LTE.[3]

4.   Reduction in the electric field strengths helps foster LTE.[1,3,4,110]

2.7.4 Tests for Conformity with an Assumption of Local Thermal Equilibrium

One of three methods is customarily used to test plasma sources for conformity with the assumption of LTE behavior.

The least difficult approach is to compare measured electron number density with the electron number density predicted by Equation 2.43, the minimum required to collisionally equilibrate resonance and ground states of the major plasma constituent. For this purpose, it is more advisable to obtain n_e from Stark broadening of H_β, where a few percent partial pressure doping with hydrogen, or hydrogen from aspirated aqueous samples, furnishes requisite H_β brightness, than to deduce n_e values from ionic-to-atomic line intensity ratios. Stark broadening theory does not invoke LTE to relate line halfwidth with n_e,[1,27] whereas use of the Saha equation presumes the LTE for which the test is being conducted.[1-4] The latter objection is more than a formality, since flowing argon plasmas often give higher ionic-to-atomic line ratios than one would expect from an LTE plasma with a particular Boltzmann slope temperature. Equation 2.43 is a minimal criterion for LTE.[110] The well-studied, constricted argon arcs[111-113] provide good basis for comparison with other plasmas.

A second class of tests for conformity with LTE, more theoretical than practical in appeal, consists of using a measured temperature as approximation and inquiring whether the coupled rate equations give stable and self-consistent state populations.[1,110,115] This approach has been instructive for hydrogen and helium plasmas, but adoption of simple (few levels) models for these atoms cause nontrivial loss of accuracy.[1,110,115] For complex atoms such as argon, lack of reliable rate coefficients precludes quantification.[121-124]

Comparison of experimental temperatures associated with various degrees of freedom is a third way of testing for conformity with LTE. Table 2.6 outlines the correspondence between internal degrees of freedom and spectroscopic variables.[2,26,45,95,128,132] In principle, temperatures characteristic of all degrees of freedom can be determined spectroscopically.[2] The radiation temperature T_{rad} specifying $B(\lambda, T)$ can be obtained from a line reversal measurement[133] on some optically thick feature, such as H_α. Kinetic temperature T_{kin} for the Maxwellian distribution of atoms and ions can be recovered from the Doppler widths of emission profiles via Equation 2.40.[2,130] Wide bandpass measurements of the continuum convey information about T_{kin} for free electrons, as in Equation 2.32.[1,2,3] Stark width measurements[132] or application of the Inglis–Teller formula[134] for merging of hydrogen or other series limits yields electron number density, which at known pressure can be inverted to estimate the Saha temperature T_{Saha}.[1,2,25,26,130,131] Boltzmann or excitation temperatures T_{exc} are extracted from absolute line intensity[1,2,25,26,131] measurements via Equation 2.6, from relative intensity measurements of several lines per spectrum by the "Boltzmann slope" method,[1,2,25,26,29] or from the Milne–Fowler "norm temperature" technique implied by Figure 2.4.[2,25,26,29] Unless one possesses vacuum UV capabilities and special boundary-layer-free sampling techniques,[83] it is not feasible to directly test whether the ground-to-resonance state population ratio agrees with the other measured distribution temperatures, but this ratio is the one most likely to depart from LTE in the Ar plasma.

Realization of meaningful accuracy in such a program of redundant temperature determinations can entail more vexing problems than are foreseen at the outset. Even in relatively homogeneous plasmas of simple geometry, designed for quantitative spectrometry,[111-113] systematic difficulties with absolute intensity calibrations and uncertainties in tabulated A_{ul} values and broadening parameters, and noise, tend to limit the reliability of temperature determination to a few percent.[1,2,25,32] In spectrochemical plasmas, excitation conditions and elemental abundance gradients compound the uncertainties. Unrecognized "hot spots" (Figure 2.8) can simulate patterns of disagreement expected from recombining PLTE plasmas.[26,94] Furthermore, each kind of temperature measurement is susceptible to particular sources of repeatable error. For example, Doppler width determinations are readily biased by competing Stark, resonance, and instrumental broadening.[1-3,27] Wide-band continuum measurements to establish electron kinetic temperatures must contend with the n_e^2 (vs $T^{1/2}$) sensitivity of the continuum, as well as with structure from negative hydrogen ions, interference from CN band and other impurities, and error in tabulated argon recombination spectral structure.[1-3,25] At usual spectrochemical n_e, Stark width measurements become critically dependent on slit

Table 2.6. Emitters, Degrees of Freedom and Associated Spectroscopic and Diagnostic Features

Emitter	Degree of freedom	Spectroscopic observable[a]	Comments on plasma diagnostics
Argon atom	Excitation	Nonresonance lines	Prominent in hotter plasma zones.
		Resonance lines (VUV)	Give blackbody intensity, and T_λ.
	Ionization	Free-bound, structured, continuum	Line/continuum ratios give T_{exc}, and broad band continuum gives n_e.
	(Kinetic)	(Doppler widths)	(Doppler broadening often swamped by
Metal atom	Excitation Ionization	Many resonance and nonresonance lines	Detailed Boltzmann slope data on thermalized populations; also norm temperature behavior per line.
Metal ion	Excitation Ionization	Ionic resonance lines	Ionic lines characterize hotter zones and are not reabsorbed in boundary layers; ion/atom line ratios are sensitive indicators of source Saha and excitation temperatures.
Hydrogen atom	Excitation Ionization	Balmer line intensity Balmer line width Merging of series limit	Hydrogen gA values are known precisely and b_i values for hydrogenic models are available; Balmer line width and series limit merging are useful n_e diagnostic at low n_e.
Free electrons	Kinetic, and indirectly, ionization	Stark widths and shifts continuum emissions	Stark broadening does not involve LTE assumptions; continuum emissions depend strongly (n_e^2) on electron densities.
Molecules	Electronic Rotational Vibrational Dissociation	Lines (IR), bands	Molecules, associated with cooler zones, and often with entrained air, characteristically emit in plasma margins.

[a] Figure 2.1 schematically illustrates some of these spectroscopic features.

deconvolution procedures.[28] But spectrochemical n_e ranges tend to exceed those where the Inglis–Teller formula gives its optimal precision.[2,3] Absolute and relative intensity measurements are prone to error from self-absorption and, for neutral resonance lines, boundary layer reabsorption.[1-8,32] Ratios of ionic to atomic lines, moreover, need scrutiny to ensure that lines from the two ionization stages arise from the same emitting zone.[25,26] Due to the shallowness of the intensity maximum in norm temperature determinations experiments seeking precision should range over both sides of the maximum,[2,25] which may exceed regimes of stable plasma operation.

Appreciation of these difficulties suggests that it is a sounder strategy to attempt to measure several kinds of temperature with modest precision than to strive for the utmost precision in one or two types of temperature measurement.

APPENDIX: BRIEF SURVEY OF THE LITERATURE

Several texts rigorously treat emission from low-energy plasmas.[1-8] Atomic and ionic lines are identified in the MIT tables,[9] and molecular lines are identified by Pearse and Gaydon.[10] Compilations of classifications are periodically published by the National Bureau of Standards[11] and the International Atomic Energy Agency.[12] Classifications of astrophysically prominent atomic and ionic lines are presented in Moore's visible[13] and ultraviolet tables.[14] These data are complimented by Striganov's and Zaidel's classifications of lines prominent in laboratory plasmas.[15] Atomic and ionic energy levels are provided by Moore's three-volume set.[16] The structure of spectra, selection rules, interval and line strength trends, and multiplet and supermultiplet relationships are examined in depth by Condon and Shortley[17] and more briefly by Johnson[18] and Candler.[19] Elementary quantum mechanical treatments of emission, absorption and scattering are given by Levison and Nikitin,[20] Slater,[21] Schiff,[22] and Bethe and Salpeter.[23] Sum rules are derived in Bethe and Salpeter[23] and are also summarized by Allen.[24] Besides the comprehensive works on plasma spectroscopy,[1-8] spectroscopic diagnosis of plasma temperature is reviewed by Tourin[25] and by Hefferlin.[26] Line broadening theory, is developed, with extensive predictions, by Griem.[27]

Theoretical and experimental Stark broadening data are critically reviewed by Konjevic and co-workers.[28] Radiative transfer in LTE and non-LTE plasmas is developed in several astrophysical works.[29-31] Critical compilations of atomic and ionic transition probabilities,[32] with periodic updatings,[33] are available from NBS. Semiquantitative line strengths for medium-weight elements are tabulated by Corliss and Bozman.[34] Line strengths may be computed in the scaled hydrogenic and LS coupling approximations by using tables of radial wave function integrals[35,36] and intra- and intermultiplet relative line strengths tabulations, respectively.[24] Partition functions and thermodynamic properties of argon plasmas are listed by Drellishak, Knopp, and Campbell.[37]

Viscosity, thermal, and electrical conductivity data of hot argon are reviewed by Dresvin.[3a] Partition functions of many elements are computed by Drawin and Felenbok.[38]

Spectroscopy and diagnostics of low-power argon plasmas are now largely within the purview of analytical chemistry, and much of the reported research is ICP oriented.[39–49] But the ICP was preceded by numerous types of low-power, flowing argon spectroscopic light sources whose popularity spanned three decades. Electron number densities and kinetic temperatures of these plasmas were similiar to those in the ICP. Wall-stabilized (Kiel-type) argon arcs were responsible for many of the better line broadening and line strength determinations[28,32,33] reported in the *Physical Review* and the *Journal of Quantitative Spectroscopy and Radiative Transfer.*[50,51] High-enthalpy argon plasma jets, used to simulate reentry conditions, had plume regions resembling of those in the ICP.[26] Water-stabilized carbon arcs with argon gas for CN band suppression (Margoshes arcs),[52–53] and the dc plasma jets that evolved from them were characterized by excitation conditions resembling those in energy input or analytical zones of the ICP.[54–56] Small, flowing, argon plasmas are all subject to radiative and diffusive energy loss mechanisms, which encourage non-LTE, and all also have roughly comparable electron impact rates, tending toward establishment of LTE. Review of the literature on these antecedent plasmas may promote better understanding of ICP.

REFERENCES

1. H. R. Griem, "Plasma Spectroscopy." McGraw-Hill, New York, 1964.
2. W. Lochte-Holtgreven, Ed., "Plasma Diagnostics." North-Holland, Amsterdam, 1968.
3. (a) S. V. Dresvin, "Physics and Technology of Low-Temperature Plasmas" (translation from Russian). Iowa State University Press, Ames, 1977. (b) H. W. Emmons, Arc Measurement of High-Temperature Gas Transport Properties, *Phys. Fluids* 10, 1125–1136 (1967).
4. Y. B. Zel'Dovitch and Y. P. Raizer, "Physics of Shock Waves and High-Temperature Hydrodynamics Phenomena," Vol. I. Academic Press, New York, 1966.
5. J. Cooper, *Rep. Prog. Phys.* 24, Part 1, 35 (1966).
6. M. P. Freeman, in "Progress in High Temperature Physics and Chemistry," Vol. II, C. A. Rouse, Ed. Pergamon Press, Oxford, 1967.
7. R. H. Huddlestone and S. L. Leonard, Eds., "Plasma Diagnostic Techniques." Academic Press, New York, 1965.
8. R. D. Cowan, "The Theory of Atomic Structure and Spectra." University of California Press, Berkeley, 1978.
9. G. R. Harrison, "Wavelength Tables with Intensities in Arc, Spark or Discharge Tube," MIT Press, Cambridge, Mass., 1969.
10. R. W. B. Pearse and A. G. Gaydon, "The Identification of Molecular Spectra." Chapman & Hall, London, 1965.
11. (a) L. Hagan and W. C. Martin, Bibliography on Atomic Energy Levels and Spectra, July 1968 through June 1971, NBS Special Publication No. 363, Government Printing Office, Washington, D.C., 1972; (b) Supplement 1, July 1971 through June 1975 (1977).
12. J. A. Hughes, Ed., *International Bulletin on Atomic and Molecular Data for Fusion.* International Atomic Energy Agency, Vienna, published monthly.
13. C. E. Moore, A Multiplet Table of Astrophysical Interest, Contributions from the Princeton University Observatory No. 20, and NBS Technical Note No. 36, Government Printing Office, Washington, D.C., 1959.

14. C. E. Moore, An Ultraviolet Multiplet Table, NBS Circular No. 488, Government Printing Office, Washington, D.C., 1950.
15. A. R. Striganov and N. S. Sventitskii, "Tables of Spectral Lines of Neutral and Ionized Atoms" (translation from Russian). IFI/Plenum, New York, 1968. See also: A. N. Zaidel, V. K. Prokofjev, S. M. Raiskli, V. A. Slavnyi, and E. A. Shreider, "Tables of Spectral Lines." IFI/Plenum, New York, 1970.
16. C. E. Moore, "Atomic Energy Levels," Vols. I, II, III, NBS Circular No. 467, Government Printing Office, Washington, D.C., 1949.
17. E. U. Condon and A. H. Shortley, "The Theory of Atomic Spectra." Cambridge University Press, Cambridge, 1963.
18. R. C. Johnson, "Atomic Spectra." Methuen & Co., London, 1961.
19. C. Candler, "Atomic Spectra and the Vector Model." Hilger and Watts, London, 1964.
20. I. B. Levison and A. A. Nikitin, "Handbook for Theoretical Computation of Line Intensities in Atom Spectra" (translation from Russian). Israel Program for Scientific Translations, Jerusalem, 1965.
21. J. C. Slater, "Quantum Theory of Atomic Structure," Vols. I and II. McGraw-Hill, New York, 1960.
22. L. I. Schiff, "Quantum Mechanics." Prentice-Hall, Englewood Cliffs, N.J., 1962.
23. (a) H. A. Bethe and E. E. Salpeter, "Quantum Mechanics of One- and Two-Electron Atoms." Springer, Berlin, 1957. (b) A. Dalgarno, in "Atomic Physics," B. Benderson, V. W. Cohen, and F.M.J. Pichanick, Eds. Plenum Press, New York, 1963.
24. (a) C. W. Allen, "Astrophysical Quantities, 2nd ed. Athlone, London, 1964. (b) L. Goldberg, Relative Multiplet Strengths in LS Coupling, Astrophys. J. 82, 1-24 (1935).
25. R. H. Tourin, "Spectroscopic Gas Temperature Measurement." Elsevier, Amsterdam, 1966.
26. (a) R. Hefferlin, in "Progress in High Temperature Physics and Chemistry," Vol. II, C. A. Rouse, Ed. Pergamon Press, Oxford, 1968. (b) G. M. Giannini, The plasma jet, Sci. Am. 80-88 (August 1957).
27. H. R. Griem, "Spectral Line Broadening by Plasmas." Academic Press, New York, 1974.
28. (a) N. Konjevic and J. R. Roberts, A Critical Review of the Stark Widths of Spectral Lines from Nonhydrogenic Atoms, J. Phys. Chem. Ref. Data 5, 209-257 (1976). (b) N. Konjevic and W. L. Weise, A Critical Review of the Stark Widths of Spectral Lines from Nonhydrogenic Ions, J. Phys. Chem. Ref. Data 5, 258-299 (1976).
29. V. V. Ivanov, "Transfer of Radiation in Spectral Lines," Chap. 1, in NBS Special Publication No. 385, Government Printing Office, Washington, D.C., 1973.
30. J. T. Jefferies, "Spectral Line Formation." Blaisdel, Waltham, Mass., 1968.
31. (a) D. Mihalas, "Stellar Atmospheres." W. H. Freeman, New York, 1970. See also: L. H. Aller, "Astrophysics—The Atmosphere of the Sun and the Stars." Roland Press, New York, 1953. (b) L. H. Aller, Quantitative Analysis of Normal Stellar Spectra, Chap. 4 in "Stellar Atmospheres," J. L. Greenstein, Ed. University of Chicago Press, Chicago, 1960.
32. (a) W. L. Wiese, M. W. Smith, and B. M. Glennon, "Atomic Transition Probabilities," Vol. I, NSRDC-NBS 4. Government Printing Office, Washington, D.C., 1967. (b) W. L. Wiese, M. W. Smith, and B. M. Glennon, "Atomic Transition Probabilities," Vol. II, NSRDC-NBS 4. Government Printing Office, Washington, D.C., 1969.
33. (a) B. M. Glennon and W. L. Wiese, Bibliography on Atomic Transition Probabilities, 1914 through October 1977, NBS Special Publication No. 505, Government Printing Office, Washington, D.C., 1978; (b) Supplement 1, November 1977 through March 1980 (1980).
34. C. H. Corliss and W. R. Bozman, U.S. Natl. Bur. Stand. Monogr. 53 (1962).
35. G. K. Oertel and L. P. Shomo, Tables for the Calculation of Radial Multipole Matrix Elements by the Coulomb Approximation, Astrophys. J. Suppl. 16, 175-218 (1968).
36. D. R. Bates and A. Damgaard, Phil. Trans. R. Soc. London, A242, 101 (1949).
37. K. S. Drellishak, C. F. Knopp, and A. B. Campbell, Partition Functions and Thermodynamic Properties of Argon Plasmas, AEDC-TDR-63-146. Arnold A.F. Development Center, Dayton, Ohio, 1963.
38. H. W. Drawin and P. Felenbok "Data for Plasmas in Local Thermodynamic Equilibrium." Gauthiers-Villars, Paris, 1965.
39. S. Greenfield, I.L.I. Jones, and C. T. Berry, High-Pressure Plasmas as Spectroscopic Emission Sources, Analyst 89, 713-772 (1964).
40. (a) R. H. Wendt and V. A. Fassel, Induction-Coupled Plasma Spectrometric Excitation Source, Anal. Chem. 37, 920-922, 1965. (b) G. W. Dickinson and V. A. Fassel, Emission Spectrometric Detection of the Elements at Nanogram per Milliliter Level Using Induction-Coupled Plasma Excitation, Anal. Chem. 41, 1021-1024 (1969).

41. (a) P.W.J.M. Boumans and F. J. DeBoer, An Experimental Study of a 1-kW, 50-MHz RF Inductively Coupled Plasma with a Pneumatic Nebulizer, and a Discussion of Experimental Evidence for a Nonthermal Mechanism, *Spectrochim Acta* 32B, 365-395 (1977). (b) P.W.J.M. Boumans, Conversion of "Tables of Spectral Line Intensities" for NBS Copper Arc into Table for Inductively Coupled Organ Plasmas, *Spectrochim. Acta* 36B, 169-203 (1981).

42. (a) G. F. Larson, V. A. Fassel, R. H. Scott, and R. N. Kniseley, Inductively Coupled Plasma-Optical Emission Analytical Spectrometry. A Study of Some Interelement Effects, *Anal. Chem.* 47, 238-243 (1975). (b) J.A.C. Broekaert, F. Leis, and K. Laqua, Some Aspects of Matrix Effects Caused by Sodium Tetraborate in the Analysis of Rare Earth Minerals with the Aid of Inductively Coupled Plasma Emission Spectroscopy, *Spectrochim. Acta* 34B, 167-175 (1979). (c) L. M. Faires, C. T. Apel, and T. M. Niemczyk, Intra-Alkali Matrix Effects in the Inductively Coupled Plasma, *Appl. Spectrosc.* 36, 558-565 (1983).

43. (a) R. K. Winge, V. A. Fassel, V. J. Peterson, and M. A. Floyd, "Inductively Coupled Plasma-Atomic Emission Spectroscopy: An Atlas of Spectral Information." Elsevier, New York, 1985. (b) P.W.J.M. Boumans, "Line Coincidence Tables for Inductively Coupled Plasma Atomic Emission Spectrometry," Vols. 1 and 2. Pergamon Press, Oxford, 1980. (c) M. L. Parsons, A. R. Forster, and D. L. Anderson, "An Atlas of Spectral Interferences in ICP Spectroscopy." Plenum Press, New York, 1980.

44. H. Falk, E. Hoffman, I. Jaeckel, and C. Ludke, Atomic Emission Trace Analysis by Nonthermal Excitation, *Spectrochim. Acta* 34B, 333-339.

45. N. Furuta and G. Horlick, Spatial Characterization of Analyte Emission and Excitation Temperature in an Inductively Coupled Plasma, *Spectrochim. Acta* 37B, 53-64 (1982).

46. R. S. Houk, H. J. Svec, and V. A. Fassel, Mass Spectrometric Evidence for Suprathermal Ionization in an Inductively Coupled Organ Plasma, *Appl. Spectrosc.* 35, 380-384 (1981).

47. V. A. Fassel, Quantitative Elemental Analysis by Plasma Emission Spectroscopy, *Science*, 202, 183-191 (1978).

48. R. M. Barnes, Ed. "Developments in Atomic Plasma Spectrochemical Analysis." Heydon, London, 1981.

49. J. P. Robin, ICP-AES at the Beginning of the Eighties, *Prog. Anal. Atom. Spectrosc.* 5, 79-110 (1982).

50. (a) K. P. Nick, J. Richter, and V. Helbig, Non-LTE Diagnostic of an Argon Arc Plasma, *J. Quant. Spectrosc. Radiat. Transfer* 32, 1-8 (1984). (b) H. Maecker, *Ann. Phys.* 18, 441 (1956).

51. J. M. Bridges and W. L. Wiese, Experimental Determination of Transition Probabilities and Stark widths of S I and S II Lines, *Phys. Rev.* 159, 31-38 (1967).

52. M. Margoshes and B. F. Scribner, Simple Arc Devices for Spectral Excitation in Controlled Atmospheres, *Appl. Spectrosc.* 17, 142-144 (1963).

53. C. D. Curtis, *Nature* 196, 1087 (1962).

54. V. M. Goldfarb, Recombination of Electrons and Ions in an Atmospheric Argon Plasma, in "Developments in Atomic Plasma Spectrochemical Analysis," R. M. Barnes, Ed. Heydon, London, 1981.

55. S. V. Desai and W. H. Corcoran, *J. Quant. Spectrosc. Radiat. Transfer* 9, 1371-1386 (1969).

56. M. H. Abdallah, J. M. Mermet, and C. Trassy, *Anal. Chim. Acta* 87, 329 (1979).

57. E. Michaud and J. M. Mermet, Iron Spectrum in the 200-300 nm Range Emitted by an Inductively Coupled Argon Plasma, *Spectrochim. Acta* 37B, 145-164 (1982).

58. B. DiBartolo, Ed., "Spectroscopy of the Excited State." Plenum Press (in cooperation with NATO Scientific Affairs Division), New York, 1976.

59. A. Montaser and V. A. Fassel, Electron Number Density Measurements in Ar and Ar-N_2 Inductively Coupled Plasmas, *Appl. Spectrosc.* 35, 613-617 (1981).

60. (a) R.W.P. McWhirter, in "Plasma Diagnostic Techniques," R. H. Huddlestone and S. L. Leonard, Eds. Academic Press, New York, 1965. (b) R.W.P. McWhirter and A. G. Hearn, *Proc. Phys. Soc., London* 82, 641 (1963).

61. A. T. Hattenburg, Spectral Radiance of a Low Current Graphite Arc, *Appl. Optics* 6, 95-100 (1967).

62. H. J. Kostkowski, Absolute Standards of Spectral Radiance, *J. Opt. Soc. Am.* 57, 1416 (1967).

63. N. A. Krall and A. W. Trivlepiece, "Principles of Plasma Physics." McGraw-Hill, New York, 1973.

64. C. A. Rouse, Ionization-Equilibrium Equation of State—III. Results with Debye-Hückel Corrections and Planck's Partition Function, *Astrophys. J.* 136, 636-670 (1962).

65. H. R. Griem, High-Density Corrections in Plasma Spectroscopy, *Phys. Rev.* 128, 997–1003 (1962).
66. R. M. Rotty, "Introduction to Gas Dynamics." John Wiley & Sons, New York, 1973.
67. C. R. Yorkley and J. B. Shumaker, Computer for the Abel Inversion, *Rev. Sci. Instr.* 34, 551–557 (1963).
68. M. H. Miller and R. D. Bengtson, Oscillator Strength Trends in Group IV Homologous Ions, *Astrophys. J.* 235, 294–297 (1980).
69. (a) D. Schluter, Die Berechnung der Übergangswahrscheinlichkeiten von Seriengrenzkontinia mil Anwendung auf die schweren E Delegase, *Z. Astrophys.* 61, 67–76 (1965). (b) E. Schulz-Gulde, The Continuous Emission of Argon in the Visible Range, *Z. Phys.* 230, 449–459 (1970).
70. R. L. Taylor and G. Caledonia, Experimental Determination of the Cross Sections for Neutral Brennsstrahlung of Ne, Ar and Xe, Research Report No. 311. Avco Everett Research Laboratory, Everett, Mass., 1968.
71. R. A. Sawyer, "Experimental Spectroscopy." Dover Publications, New York, 1963.
72. (a) J. T. Davies and J. M. Vaughan, A New Tabulation of the Voigt Profile, *Astrophys. J.* 137, 1302–1305 (1963). (b) M. L. Claude, Tables for Direct Determination of Spectral Line Parameters from Spectrometric Data—Voigt Broadening, *J. Quant. Spectrosc. Radiat. Transfer* 32, 17–48 (1984). (c) A. F. Jones and D. L. Misell, The Problem of Error in Deconvolution, *J. Phys. A, Gen. Phys.* 3, 462–472 (1970). (d) R. J. Exton, Widths of Optically Thick Lines, *J. Quant. Spectrosc. Radiat. Transfer* 11, 1377–1383 (1970).
73. (a) J. R. Fuhr, W. L. Wiese, and L. J. Roszman, Bibliography on Atomic Line Shapes and Shifts (1889 through March 1972). NBS Special Publication No. 366, Government Printing Office, Washington, D.C., 1972; (b) Suppl. 1, April 1972 through June 1973 (1974).
74. P. Kepple and H. R. Griem, Improved Stark Profile Calculations for the Hydrogen Balmer Lines, *Phys. Rev.* 173, 317 (1968).
75. (a) W. L. Wiese, D. E. Kelleher, and V. Helbig, Variations in Balmer-Line Stark Profiles with Ion-Atom Reduced Mass, *Phys. Rev.* A11, 1854–1863 (1975). (b) R. A. Hill, J. B. Gerardo, and P. C. Kepple, Stark Broadening of H_β, H_γ and H_δ: A Comparison of Theory and Experiment, *Phys. Rev.* A3, 855–861 (1971).
76. H. R. Griem, M. Baranger, A. C. Kolb, and G. Oerel, Stark Broadening of Neutral Helium Lines in a Plasma, *Phys. Rev.* 125, 177 (1962).
77. W. W. Jones, Comparison of Measured and Computed Stark Parameters for Singly Ionized Atoms, *Phys. Rev.* A7, 1826–1832 (1973).
78. H. R. Griem, Semiempirical Formulas for the Electron Impact Widths and Shifts of Isolated Ion Lines in Plasmas, *Phys. Rev.* 165, 258–266 (1968).
79. (a) W. L. Wiese and N. Konjevic, Regularities and Similarities in Plasma Broadened Spectral Line Widths, *J. Quant. Spectrosc. Radiat. Transfer* 28, 185–198 (1982). (b) P. E. Oettinger and J. Cooper, Stark Broadening of NaI 5682 and NaI 4978 Lines, *J. Quant. Spectrosc. Radiat. Transfer* 9, 591–596 (1969).
80. J. Puric, I. Lakicevic, and V. Glavonjic, Stark Width and Shift Dependence on the Ionization Potential, *Phys. Lett.* 76A, 128–130 (1980).
81. C. D. Curtis, Cyanogen Band Suppression in Direct Current Spectrographic Analyses, *Nature* 196, 1087–1088 (1962).
82. W. L. Wiese, in "Progress in Atomic Spectroscopy," W. Hanley and H. Kleinpoppen, Eds. Plenum Press, New York, 1978.
83. D. P. Aeshliman, R. A. Hill, and L. Evans, Collisional Broadening and Shift of Neutral Argon Spectral Lines, *Phys. Rev.* A14, 1421–1427 (1976).
84. R. W. Wood, *Phil. Mag.* 8, 205 (1929).
85. H. Rouse, Ed., "Advanced Mechanics of Fluids." John Wiley & Sons, New York, 1959.
86. W. L. Wiese, in "Physics of Ionized Gases," M. J. Kurepa, Ed. Institute of Physics, Beograd, 1972.
87. J. R. Welty, C. E. Wicks, and R. E. Wilson, "Fundamentals of Momentum, Heat and Mass Transfer." John Wiley & Sons, New York, 1976.
88. H. Uchida, M. A. Kosinski, and J. D. Winefordner, Laser Excited Atomic and Ionic Fluorescence in an Inductively Coupled Plasma, *Spectrochim. Acta* 38B, 5–13 (1980).
89. H. Uchida, M. A. Kosinski, N. Omenetto, and J. D. Winefordner, Studies on Lifetime Measurements and Collisional Processes in an Inductively Coupled Argon Plasma Using Laser Induced Fluorescence, *Spectrochim. Acta* 39B, 63–68 (1984).

90. D. G. Samaras, "Applications of Ion Flow Dynamics." Prentice-Hall, Englewood Cliffs, N.J., 1962.
91. R. H. Sabersky and A. J. Acosta, "Fluid Flow." Macmillan Co., New York, 1964.
92. I. I. Glass and G. Hall, Shock Tubes, Section 18 of "Handbook of Supersonic Aerodynamics," NAVORD Report No. 1438, Vol. 6 (1959).
93. (a) V. J. Friedrich, Zur Auswertung seitlicher Beobachtunger an zylindricken Bogen, *Ann. Phys.* 3, 327–333 (1959). (b) G. J. Cremers and R. C. Birkebak, Application of the Abel Integral Equation to Spectrographic Data, *Appl. Opt.* 5, 1057–1064 (1986).
94. M. H. Miller, D. Eastwood, and M. S. Hendrick, Excitation of Analytes and Enhancement of Emission Intensities in a dc Plasma Jet: A Critical Review Leading to Proposed Mechanistic Models, *Spectrochim. Acta* 39B, 13–56 (1984).
95. P. J. Dickerman, Ed., "Optical Spectrometric Measurements of High Temperatures." University of Chicago Press, Chicago, 1961.
96. R. M. Barnes and R. G. Schleicher, Temperature and Velocity Distribution in an Inductively Coupled Plasma, *Spectrochim. Acta* 36B, 81–101 (1981).
97. (a) A. C. G. Mitchell and M. W. Zemansky, "Resonance Radiation and Excited Atoms." Cambridge University Press, Cambridge, 1961. (b) T. Holstein, Imprisonment of Resonance Radiation in Gases—II, *Phys. Rev.* 83, 1159–1167 (1951).
98. B. Wende, Collisional Redistribution of Resonance Radiation and Related Phenomena, Chap. 8 in "Spectral Line Shapes," B. Wende, Ed. Walter DeGruyter, Berlin, 1981.
99. G. N. Coleman, W. P. Braun, and A. M. Allen, Characterization of an Improved dc Plasma Excitation Source, *Appl. Spectrosc.* 34, 24–30 (1980).
100. D. H. Sampson, "Radiate Contributions to Energy and Momentum Transfer in a Gas." Interscience, New York, 1965.
101. (a) J. W. Mills and G. M. Hieftje, A Detailed Consideration of Resonance Radiation Trapping in the Argon Inductively Coupled Plasma, *Spectrochim. Acta* 39B, 859 (1984). (b) Ref. 126.
102. G. M. Hieftje, G. D. Rayson, and J. W. Olesik, a Steady-State Approach to Excitation Mechanisms in the ICP, *Spectrochim. Acta* 40B, 167–176 (1985).
103. M. W. Blades and G. Horlick, Interferences from Easily Ionizable Element Matrices in Inductively Coupled Plasma Emission Spectroscopy—A Spatial Study, *Spectrochim. Acta* 36B, 881–900 (1981).
104. T. Fujimoto, Kinetics of Ionization-Recombination of a Plasma and Population Density of Excited Ions, *J. Phys. Soc. Japan.* (a) I. Equilibrium Plasma, 47, 265–272 (1979); (b) II. Ionizing Plasma, 47, 273–281 (1979); (c) III. Recombining Plasma-High-Temperature Case, 49, 1561–1568 (1980); (d) IV. Recombining Plasma-Low-Temperature Case, 49, 1569–1576 (1980).
105. P.W.J.M. Boumans, "Theory of Spectrochemical Excitation." Plenum Press, New York, 1966.
106. O. P. Bochkova and E. Y. Shreyder, "Spectroscopic Analysis of Gas Mixtures" (translation from Russian). Academic Press, New York, 1965.
107. R. C. Miller and R. J. Ayen, Temperature Profiles and Energy Balances for an Inductively Coupled Plasma Torch, *J. Appl. Phys.* 40, 5260–5273 (1969).
108. H. R. Griem, Validity of Local Thermodynamic Equilibrium in Plasma Spectroscopy, *Phys. Rev.* 131, 1170–1176 (1963).
109. R. Wilson, *J. Quant. Spectrosc. Radiat. Transfer* 2, 477 (1962).
110. (a) D. H. Sampson, Criteria for the Validity of Local Thermodynamic Equilibrium, TIS Report No. R64SD90 (1964). (b) J. J. Oxenius, *J. Quant. Spectrosc. Radiat. Transfer* 7, 837 (1967).
111. R. J. Rosado, Thesis (unpublished), University of Eindhoven, 1981.
112. D. Garz, *Naturforschung* 28a, 1459 (1973).
113. J. B. Schumaker and C. H. Popenue, *J. Res. U.S. Natl. Bur. Stand.* 76A, 71 (1972).
114. R. J. Lovett, A Rate Model of Inductively Coupled Argon Plasma Analyte Spectra, *Spectrochim. Acta* 37B, 969–985 (1982).
115. I.J.M.M. Raaijmakers, P.W.J.M. Boumans, B. Van Der Sijde, and D. C. Schram, A Theoretical Study and Experimental Investigation of non-LTE Phenomena in an Inductively Coupled Plasma—I. Characterization of the Discharge, *Spectrochim. Acta* 38B, 697–706 (1983).
116. A. Scheeline and M. K. Zoellner, Thompson Scattering as a Diagnostic of Atmospheric Discharges, *Appl. Spectrosc.* 38, 245–258 (1984).
117. (a) B. L. Caughlin and M. W. Blades, An Evaluation of Ion-Atom Emission Intensity Ratios and Local Thermodynamic Equilibrium in an Argon Inductively Coupled Plasma, *Spectrochim. Acta* 39B, 1583–1602 (1984). (b) Y. Noriji, K. Tanabe, H. Uchida, H. Haraguchi, K. Fuwa, and J. D. Winefordner, Comparison of Spatial Distributions of Argon Species Number

Densities with Calcium Atom and Ion in an Inductively Coupled Argon Plasma, *Spectrochim. Acta* 38B, 61–74 (1983). (c) R. S. Houk and J. A. Olivares, General Calculations of Vertically Resolved Emission Profiles for Analyte Elements in Inductively Coupled Plasmas, *Spectrochim. Acta* 39B, 575–587 (1984).

118. (a) W. W. Blades and G. M. Hieftje, On the Significance of Radiation Trapping in the Inductively Coupled Plasma, *Spectrochim. Acta* 37B, 191–197 (1982). (b) J. M. Mills and G. M. Hieftje, A Detailed Consideration of Resonance Radiation Trapping in the Argon Inductively Coupled Plasma, *Spectrochim. Acta* 39B, 859–866 (1984).

119. J. Jarosz, J. M. Mermet, and J. Robin, A Spectrometric Study of a 40-MHz Inductively Coupled Plasma—III. Temperature and Electron Number Density, *Spectrochim. Acta* 33B, 55–78 (1978).

120. H. Van Regemorter, in "Atoms in Astrophysical Plasmas in Atomic and Molecular Collision Theory," F. A. Gianturco, Ed. Plenum Press, New York, 1980.

121. (a) B. Bederson, Ed., "Atomic Physics." Plenum Press, New York, 1969. (b) L. B. Loeb, "Basic Processes of Gaseous Electronics." University of California Press, Berkeley, 1955.

122. J. O. Hirschfelder, Ed., "Intermolecular Forces." Interscience, New York, 1967. See also: S. C. Brown, "Basic Data of Plasma Physics." Technology Press and John Wiley, New York, 1959.

123. D. R. Bates and B. Bederson, Eds., "Advances in Atomic and Molecular Physics," Vol. II. Academic Press, New York, 1975.

124. B. L. Moiseiwitsch and S. J. Smith, Electron Impact Excitation of Atoms, NSRDS-NBS No. 25. Government Printing Office, Washington, D.C., 1968.

125. R. L. Taylor and G. Caledonia, Experimental Determination of the Cross Sections for Neutral Bremsstrahlung of Ne, Ar and Xe, Research Report No. 311. Avco Everett Research Laboratory, Everett, Mass., 1968.

126. T. Holstein, Imprisonment of Resonance Radiation in Gases, *Phys. Rev.* 72, 1212–1233 (1947).

127. L. DeGalan, Some Considerations on the Excitation Mechanism in the Inductively Coupled Argon Plasma, *Spectrochim. Acta* 39B, 537–550 (1984).

128. D. J. Kalnicky, R. N. Kniseley, and V. A. Fassel, Inductively Coupled Plasma Optical Emission Spectroscopy. Excitation Temperatures Experienced by Analyte Species, *Spectrochim. Acta* 30B, 511–525 (1975).

129. M. H. Miller, R. A. Roig, and R. D. Bengtson, Absolute Transition Probabilities of Phosphorus, *Phys. Rev.* 4, 1709–1722 (1971).

130. (a) F. Bastien, Spectroscopic Diagnostics in Gas Discharges, in "Electrical Breakdown and Discharges in Gases," E. E. Kunhardt and L. H. Leussen, Eds. Plenum Press, New York, 1983, pp. 267–291. (b) H.A.C. Human and R. H. Scott, The Shapes of Spectral Lines Emitted by an Inductively Coupled Plasma, *Spectrochim. Acta* 31B, 459–473 (1976).

131. J. S. Leonard, *J. Quant. Spectrosc. Radiat. Transfer* 12, 619 (1972).

132. (a) M. W. Blades and N. Lee, A Spatial Study of Electron Density and Analyte Emission in a dc Argon Plasma, *Spectrochim. Acta* 39B, 879–890 (1984). (b) A. T. Zander and M. H. Miller, Electron Density Profiles in the DCP Source, *Spectrochim. Acta* 40B, 1023–1037 (1985).

133. W.R.S. Garton, *J. Sci. Instr.* 36, 1 (1959).

134. A. Montaser, V. A. Fassel, and G. Larson, Electron Number Densities in Analytical Inductively Coupled Plasma as Determined via Series Limit Line Merging, *Appl. Spectrosc.* 35, 385–389 (1981).

LIST OF SYMBOLS

Å	ångstrom unit = 0.1 nm
A_{ul}	transition probability for spontaneous transition from upper state u to lower state l
a_0	radius of the first Bohr orbit
α	parameter for quasi-static broadening by ions
α_q	fractional partial pressure of species q
$B(\lambda, T)$	blackbody distribution function

b	profile halfwidth for a Voigt-shaped line
b	ratio of population to LTE population for state i
c	light velocity in vacuum
d	halfwidth of the Lorentzian component partial width for a Voigt-shaped line profile
E_{ex}	excitation energy
E_{ip}	ionization potential
E_{ul}	energy difference between states u and l
$\xi(\lambda, T)$	radiation power density
$\xi(T_e, v)$	emitter-specific correction factor to the Kramer–Unsold formula for the continuum
f_{lu}	absorption oscillator strength for transitions from lower state l to upper state u
g	the $1/e$ width for the Gaussian component of a Voigt-shaped line profile
g	the degeneracy $(2J + 1)$ of a state
h	Planck constant
$I(\lambda)$	radiation intensity at wavelength λ
I	total (wavelength-integrated) intensity of a spectral line
J	total angular momentum of a state
J_{A_2}	molecular moment of inertia for species A_2
$K(\lambda)$	optical absorption coefficient at wavelength λ
k	Boltzmann constant
L	azimuthal quantum number of an atom or ion in a specified state
$L(\lambda)$	line profile shape function
l	length of the optical path in the light source
l	azimuthal quantum number for a single electron
λ	wavelength
λ_0	wavelength at line center
m_e	mass of electron
m_a	mass of atom
n*	effective principal quantum number
n_A	number density of species A
n_a	number density of atoms
n_e	number density of electrons
n_i	number density of ions
n_{tot}	total number density
n_u	number density of atoms (or ions) in the (upper) state u
P	pressure
p	a constant for relating halfwidths and areas of Voigt profiles
Q	net collisional population or depopulation rate for a particular state
R	Rydberg constant
R	gas constant, R_u/M, where R_u is the universal gas constant and M is the species molecular mass
r	radial distance from atomic or molecular center of mass

ρ	analyte concentration
S	dipole transition line strength
$\mathbf{S}$	total spin of an atom or ion
$S(\lambda)$	source function at wavelength λ
$\sigma(\lambda)$	optical cross section at wavelength λ
σ^2	radial portion of the atomic wave function in the Coulomb approximation
T	temperature
T_a	kinetic temperature of atoms
T_e	kinetic temperature of electrons
T_{exc}	excitation or Boltzmann temperature
t	time
$\tau(\lambda)$	optical depth at wavelength λ
$\tau(\lambda_0)$	optical depth at line center
τ_u	radiative lifetime of (upper) state u
θ	$T/5040\text{K}$
v	frequency
v_n	minimum photon frequency to photoionize out of state of principal quantum number n
$w_{1/2}$	line width at half-maximum intensity (eg, "halfwidth" of line)
ws	halfwidth of instrument profile

3

Instrumentation for Optical Emission Spectrometry

A. STRASHEIM

Department of Chemistry
University of Pretoria
Hillcrest, Pretoria
South Africa

3.1   INTRODUCTION

Bunsen and Kirchhoff discovered two new elements, cesium and rubidium, using a spectroscope.[1] Rapid advances in instrumentation, engineering, and electronics during the nineteenth and present centuries, have allowed the basic theories of Planck and Bohr to be assessed. This allowed the practical development of optical spectroscopy as an analytical technique. Basically, the technique, as used today, is still similar to that of Bunsen and Kirchhoff, ie, the sample to be analyzed is introduced into an excitation source, the radiation then is dispersed by suitable means, and finally, the resulting spectra are evaluated via a photographic plate or an electronic signal.

This chapter describes the basic equipment necessary for spectrometry with the inductively coupled plasma (ICP), a high temperature atomization-excitation source in emission spectrochemical analysis. Certain optical phenomena, optical elements, and electronic techniques related to the construction and the performance of such instruments are also presented. Detailed discussions of instrumentation for optical emission spectrometry are given in the books cited in the Bibliography at the end of the chapter.

3.2   BASIC CONCEPTS AND DESCRIPTION OF DISPERSIVE SPECTROMETERS

In general, in any spectrometric system for sequential or simultaneous multiple-element measurements, one desires to: (a) have adequate wavelength selection, and (b) collect as much light as possible from a specific area of the radiative source being monitored. The system to achieve these goals consists of: (a) an entrance slit providing a narrow optical band of radiation that has the dimensions of the slit, (b) a collimator to produce a parallel beam of light, (c) a dispersing element, (d) a focusing element reforming the specific dispersed narrow bands of light, and (e) one or more exit slits to isolate the desired spectral band or bands.

The dispersive element mostly used in the optical train of the ICP–AES instrument is the grating. In certain high resolution systems, a prism also is an integral element in the optical train. Although a detailed knowledge of physical optics is not a necessity for the practicing analytical spectroscopist, a reasonable understanding of the theories of refraction and diffraction is desirable. The following discussions on refraction and diffraction are based on the wave nature of light.

3.2.1   Refraction

Partial reflection occurs when a wave is incident upon the boundary surface between two media in which the velocity of propagation of the waves differs. A part of the wave, however, passes into the second medium with a sudden change of direction. This phenomenon is called refraction.

Consider a series of plane waves incident upon a plane boundary surface XY between two media A and B. From the geometrical presentation in Figure 3.1, the extreme lateral point of the wavefront corresponding to I_1 reaches the surface XY at A_1 while the other extreme point corresponding to I_2 will only

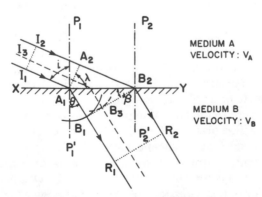

Figure 3.1. Wave theory of refraction.

have reached A_2. The point A_1 becomes a center of a new wavelet, which spreads into the medium B. If the velocities of propagation in the two media are V_A and V_B ($V_A > V_B$), the spherical front of the wavelet from A_1 will have reached the position B_1 by the time that the extreme point corresponding to I_2 has reached the boundary XY at B_2. If $V_A/V_B = n$, then

$$A_1B_1 = \frac{1}{n} \times A_2B_2 \qquad [3.1]$$

where A_1B_1 and A_2B_2 are the distances traveled by the radiation between the points A and B. An intermediate point, I_3 on the wavefront $I_1I_3I_2$, strikes the boundary XY before I_2. As I_3 travels further, it will reach the point B_3 at the same time as I_1 and I_2 reach the points B_1 and B_2, respectively. The refraction wavefront thus will be the common tangent $B_2B_3B_1$ to all the wavelets and will move in the direction of R_1R_2, as shown in Figure 3.1. Perpendiculars P_1P_1' and P_2P_2', and geometrical considerations, lead to Equation 3.2, which expresses the relationship between the incident angle i and the refracted angle θ:

$$\frac{\sin i}{\sin \theta} = \frac{A_2B_2}{B_1A_1} = n = \frac{V_A}{V_B} \qquad [3.2]$$

The ratio n, called the refractive index, is equal to the ratio of the velocities of propagation of the waves in the respective media. This relation is called Snell's law. The angles i and θ are always measured from the normals to the boundary surfaces. Since the paths of light rays are reversible, it makes no difference whether the ray passes from air or vacuum into the transparent medium or from the latter to air or vacuum.

For a prism, where the faces are not parallel, the beam undergoes two deviations, as indicated in Figure 3.2a. The ray entering the prism at the angle i is refracted through an angle r inside the prism and an angle r' on emerging, with

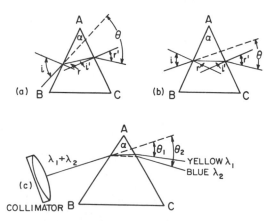

Figure 3.2. Refraction by a prism with an apical angle of α and a base length b. (a) Baseline case. (b) Minimum deviation. (c) Two light beams passing through a prism. Blue wavelengths undergo a greater dispersion than yellow.

a total deviation of θ. Most prism systems are used with the light beam traversing parallel to the base, BC. For this condition, the angle θ is a minimum deviation (Figure 3.2b) and the angular deviation of the emerging ray is a minimum. For most prism materials, such as quartz and fused silica, the change of the refractive index with wavelength is not uniform, increasing rapidly as one goes to shorter wavelengths. Consequently, the various wavelengths emerge from the second face of the prism at different angles (Figure 3.2c). The emerging light can be focused to form a spectrum.

Because the refractive index has a nonlinear wavelength dependency, the dispersion of a prism spectrometer is not linear with wavelength. For photographic prism instruments, several empirical formulas have been developed for the purpose of interpolation of wavelengths. The most used formula, the Hartmann equation, is given by:

$$\lambda = \lambda_0 + \frac{c}{d - d_0} \qquad [3.3]$$

where λ_0, c, and d_0 are constants. The values of these constants can be determined by starting with three known spectral lines A, B, C, measuring the distances on the focal plane A–B and A–C, and then solving the three simultaneous equations. This formula can be used to determine the wavelengths of unknown lines located between the known lines A and C. A new set of constants is needed for each other range. The birefringence, double refraction, of quartz is overcome by joining a 30° right-handed prism to a similar left-handed prism.

3.2.2   Diffraction

The diffraction of electromagnetic radiation is a phenomenon analogous to that shown by water waves, where straight wavefronts are incident on a barrier with a gap in it (Figure 3.3). The wave does not move in a straight line beyond the

Figure 3.3. Circular wave emanation from an aperture in a barrier. (Blackie and Son Ltd, with permission.)

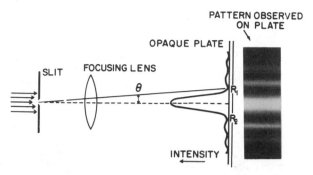

Figure 3.4. Fraunhofer diffraction pattern of a single slit, illuminated by monochromatic light. (Addison-Wesley Publishing Company Inc., with permission).

gap, but the motion of the water gives rise to a new wave center from which the wave spreads in circles behind the barrier. According to Huygens, each point on a wavefront is potentially a new source of waves. In the case of light, the spherical wave can be considered to be much diminished, spreading from a slit of infinitesimal width. Should a very large number of wavelets be produced side by side, as in a wide gap, the wavelets will interfere, giving a plane wavefront and apparent rectilinear propagation.

A diffraction pattern of a narrow slit, illuminated by monochromatic light, is illustrated in Figure 3.4. As pictured, the narrow slit gives no single sharp image, but several diffused, spaced ones. The various points across the slit become sources of secondary wavelets, and rays can propagate from each of them in all directions. If the amplitudes of the wavelets are in phase, reinforcement occurs as shown. For rays that leave the center at an angle θ (Figure 3.4) and terminate at R_1 or R_2 with a phase difference of $\lambda/2$, destructive interference occurs. Should there be two slits, each will produce a similar interference pattern as depicted in Figure 3.5a. In addition, the two parallel beams of light interfering with each other will result in an interference pattern as depicted in Figure 3.5b. For three equidistant slits the interference pattern is as shown in Figure 3.5c. The main feature of the three-slit interference pattern is that in comparison with the other two interference patterns, the pattern is more sharply defined. For 5, 6, and 20 slits, the interference patterns become progressively sharper. Thus, as the number of slits is increased, narrower and brighter interference lines are produced on a darker background. The optical element, having such a number of slits very close together at regular intervals in an opaque screen, is called a diffraction grating, and the distance between successive slit centers is called the grating constant. With the many slits of a diffraction grating, the secondary maxima, shown in Figure 3.6, are weakened greatly by multiple interference, and consequently are seldom detected. In modern practice, the grooves are ruled into a metal surface coated on a blank of quartz or a ceramic.

The formula describing the diffraction from a grating can be obtained by considering a parallel beam of *monochromatic* radiation striking a reflection

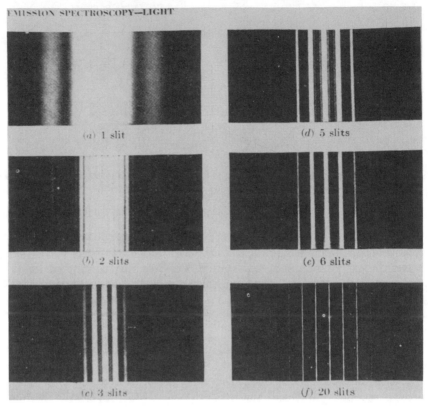

Figure 3.5. Fraunhofer diffraction patterns for gratings with a number of slits. (McGraw-Hill Book Co., with permission.)

plane grating perpendicularly. Consider a small portion of the radiation striking two grooves within the surface XY separated by a distance d (Figure 3.7). These two rays are diffracted and move away from the grating in parallel paths at an angle θ, measured with respect to the grating normal, N. The perpendicular BC on the ray R_1 cuts a segment AC on this ray. The two rays R_1 and R_2 are in phase and interfere constructively if the distance AC is equal to λ. The

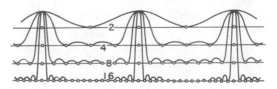

Figure 3.6. Variation of the maxima of the intensity values with the number of slits. (Blackie and Son Ltd, with permission.)

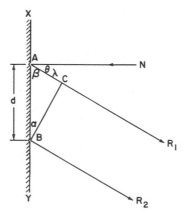

Figure 3.7. Diffraction by a plane reflection grating beam normal to surface.

geometrical relation between λ, θ, and d (groove spacing) is:

$$\sin \alpha = \frac{\lambda}{d}$$

but $\alpha + \beta = \beta + \theta = 90°$. Thus

$$\sin \theta = \frac{\lambda}{d}$$

or

$$\lambda = d \sin \theta \qquad [3.4]$$

or in general

$$m\lambda = d \sin \theta$$

where $m = 1$, 2, 3 and is called the order of diffraction, although the term "spectrum order" is generally used.

If the incident radiation is directed to the grating at an angle other than normal (Figure 3.8), the path difference between the two incident rays I_1 and I_2 at an angle i to the normal and the diffraction rays R_1 and R_2 at an angle θ will be $BD + BC$. Constructive interference occurs when $BD + BC = \lambda$. The general grating equation thus can be written:

$$m\lambda = d(\sin i \pm \sin \theta) \qquad [3.5]$$

where

$$BD = d \sin i \quad \text{and} \quad BC = d \sin \theta.$$

For incident and diffracted rays lying on the same side of the normal, Equation 3.5 is positive. For rays lying on the opposite side of the normal, the negative sign should be used.

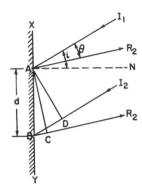

Figure 3.8. The general geometrical presentation for diffraction by a plane reflection grating.

3.2.3 Dispersion and Resolving Power

3.2.3.1 Prism

Dispersion of radiation within a spectrometer depends on the entrance slit width and the capability of the dispersive element to separate a band of radiation into the different frequencies present in the wavefront. As described, either refraction by a prism or diffraction by a grating is the process generally used. The dispersion of a prism is expressed by the derivative $d\theta/d\lambda$ (Figure 3.2c) and is a measure of the angular separation, $d\theta$, of two light rays differing in wavelength by an amount $d\lambda$. This derivative is not a constant, but increases from the infrared to the ultraviolet end of the spectrum. Quartz is mainly used for the manufacture of optical prism and lenses needed for ICP studies. The angular dispersion, $d\theta/d\lambda$ may be expressed as follows:

$$\frac{d\theta}{d\lambda} = \frac{d\theta}{dn} \cdot \frac{dn}{d\lambda} \qquad [3.6]$$

where $d\theta/dn$ is a geometric factor that includes the effect of prism shape and angle of incidence, while $dn/d\lambda$ is a factor dependent on the prism material characteristics. The useful transmission ranges of materials used for prisms, lenses, windows, and coatings are listed in Table 3.1.

As mentioned previously, most prism systems are used symmetrically at the angle of minimum deviation (θ in Figure 3.2). According to Snell's law, under these conditions, the refractive index n is related to θ and α (Figure 3.2), as shown in the following equations.

$$n = \frac{\sin[(\theta + \alpha)/2]}{\sin \alpha/2} \qquad [3.7a]$$

$$\frac{d\theta}{dn} = \frac{2 \sin(\alpha/2)}{1 - n^2 \sin^2 \alpha/2} \qquad [3.7b]$$

Table 3.1. Useful Transmission Ranges[a] for Selected
Optical Materials.

Material	Wavelength range	
	Low (nm)	High (μm)
Lithium fluoride	104	7
Calcium fluoride	125	9.0
Magnesium fluoride	110	7.5
Fused silica	170	3.6
Sodium chloride	200	25
Natural quartz	205	3.6
Glass	360	3.5
UV glass	185	—

[a] Limits are for wavelengths at which the optical transmission falls
to 60%, for a material thickness of 1 cm.
Source: Harper & Row, with permission.

Substituting Equation 3.6 in Equation 3.7b, we have:

$$\frac{d\theta}{d\lambda} = \frac{2 \sin \alpha/2}{1 - n^2 \sin^2 \alpha/2} \cdot \frac{dn}{d\lambda} \qquad [3.7c]$$

Equation 3.7c indicates that in addition to the prism material characteristics, resolution increases with larger apex angles.

The use of a prism at minimum deviation also has the following two advantages. First, because astigmatism is eliminated, optimum resolution is obtained. Second, internal reflections are minimized at the far prism face, thus reducing stray light.

3.2.3.2 Diffraction Grating

Angular dispersion of a grating is readily obtained by differentiating Equation 3.5:

$$\frac{d\theta}{d\lambda} = \frac{m}{d \cos \theta} \qquad [3.8]$$

Generally, the separation of adjacent wavelengths increases with m, the order number, and, also, as the grating constant d is decreased. Furthermore, angular dispersion also remains nearly constant when the wavelength is varied, as the range of θ is small for most spectrometers.

The term "angular dispersion" is seldom used in practice. The preferred term is the "reciprocal linear dispersion," indicating the number of nanometers per millimeter at the focus or in the focal plane of the instrument. This expression is related to the angular dispersion by the expression $1/f \times d\lambda/d\theta$, where f is the focal length of the spectrometer. Thus except for the echelle type, grating instruments have nearly constant reciprocal linear dispersion over their usable spectral range. For prism and echelle instruments, the reciprocal linear dispersion is related to a specific wavelength. In brochures of such instruments covering the wavelength range 200 to 500 nm, reciprocal linear dispersion values would be given at, say, 200, 300, and 400 nm.

The theoretical resolving power of a grating is given by the expression:

$$R = nm \qquad\qquad [3.9]$$

where n is the total number of illuminated grooves. By substituting the value of m from Equation 3.5 into Equation 3.9, we have:

$$R = \frac{W}{\lambda(\sin i \pm \sin \theta)} \qquad\qquad [3.10]$$

where W is the width of the ruled area ($W = nd$). Accordingly, maximum resolution for a particular line is achieved for a grating having a large ruled area. Furthermore, large incidence and diffraction angles are advantageous if this can be accommodated in the spectrometer.

3.2.4   Overlapping of Orders, Types of Grating, and Grating Errors

Because the use and performance of a diffraction grating depend on such properties as spectrum order used and the type and quality of the ruling, these properties warrant consideration.

3.2.4.1   Overlapping of Orders

The sensitive lines of interest to the spectrochemist cover the wavelength range 130 to 900 nm. The overlapping of spectral orders thus can be potentially troublesome. This means that wavelengths of 200, 300, and 400 nm in the first order will coincide with the wavelengths 400, 600, and 800 nm in the second and 600, 900 and 1200 nm in the third order, as shown by Equation 3.5. Spectral lines present in the first and second orders are usually monitored in analytical ICP–AES.

3.2.4.2   Type and Quality of the Rulings

Gratings vary widely in the percentage of incident light they transform into diffracted radiation. Generally specular reflection increases with the number of rulings per millimeter. Gratings can now be ruled with grooves shaped to direct up to 90% of the incident radiation in a specific order.

a.   Echellette Gratings.   Consider a very small portion of the ruled area of a modern grating, as illustrated in Figure 3.9. This echellette grating normally has a broad face (*AB*) at an angle θ relative to the normal of the grating and a

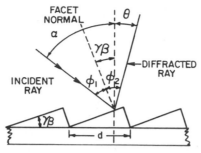

Figure 3.9. Section of the face of a plane echelette grating, $\gamma\beta$ blaze angle, α incident angle, θ diffraction angle.

narrow face *(BC)*. The "steps" between the faces of an echellette grating have heights equal to 0.5 to 5 wavelengths. According to the ordinary laws of reflection, the incident ray is reflected with a similar angle to the normal of the reflecting surface. Should the angle of the broad face *(AB)* be changed so that diffraction occurs at an angle close to the normal reflection from the facet surface, high-efficiency diffraction is effected. Referring to Figure 3.9, this condition is met when the angles ϕ_1 and ϕ_2 are nearly equal. High grating efficiency is thus obtained when:

$$\alpha\text{-}\gamma\beta = \gamma\beta - \theta \qquad \text{or} \qquad \theta = 2\gamma\beta\text{-}\alpha$$

Recall that θ is negative if the incident rays are on the opposite side of the perpendicular. Thus, if $\gamma\beta = 10°$ and $\alpha = 5°$, the maximum efficiency of diffraction occurs at an angle of 15°. The specific wavelength can be determined if the grating constant is known.

The wavelength for which first-order diffraction and specular reflection coincide is called the "blaze wavelength." The main advantage of an echellette grating is that light appears in the first two orders, centered around $\lambda_{\gamma\beta}$ in the first order and $\lambda_{\gamma\beta/2}$ in the second order.

b. Echelle Gratings. Echelle gratings,[2] which have "steps" equal to 10 to 200 wavelengths, have good efficiency in high orders. They are normally employed in the Littrow mount using the "short side" of the grooves for diffraction, as illustrated in Figure 3.10. The entering and leaving of the incident and the

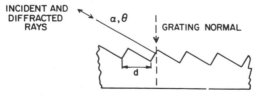

Figure 3.10. Section of a face of a plain echelle grating, α and θ incident and diffracted angles, respectively.

diffracted radiation, nearly parallel to the wider facets, is a unique feature of the echelle grating. Echelle gratings are designed to be used at blaze angles greater than 45°. By offering good spectral efficiency in high orders, the echelle makes possible very high resolution. Typically, the resolution afforded by an echelle grating is a factor of 10 higher than that of conventional gratings. A fore-prism or grating as order-sorter is an essential feature for use of an echelle grating.

c. **Concave Gratings.** Most direct-reading spectrometers in use today have a concave grating as the dispersive element. The dominance of the concave grating in such systems can be ascribed to the capability of this grating to act as both collimating and focusing medium, thus reducing the cost of the spectrometer. Further, limited size enhances a very important feature, ie, stability of a direct-reading spectrometer.

In Section 3.2.2 we developed the diffraction formulas for plane gratings. The grating equation, Equation 3.5, is also valid for the concave grating by virtual cancellation of two factors. The effect of the ruling of the grooves in the concave grating, with a constant period along the cord, is canceled by the departure of the concave grating from tangency with the Rowland circle. Further discussion of this subject is presented in Grove's, "Analytical Emission Spectrometry," cited in the Bibliography.

3.2.4.3   Grating Errors and the Testing of Concave Gratings

Even with a perfect blank, all gratings fall short of perfection. Errors in periodic spacing, groove parallelism, groove form, and so on have a marked effect on the resolution and efficiency of a ruled diffraction grating. Accordingly, such errors must be kept to a minimum. The well-known "Rowland ghosts," easily recognized as faint lines lying symmetrically around an overexposed spectral line, are caused by a periodic error of the screw of the ruling engine. Effectively, a coarse grating is superimposed on a fine grating. An annoying feature about a ghost is that, in higher orders, the intensity rises as the square of the spectral order.

"Lyman ghosts" are the result of multiple periodic errors in the ruling of the grating. Lyman ghosts occur at greater distances from the parent line than the Rowland ghosts. The former ghosts are readily detected using a mercury lamp and viewing the first order ultraviolet spectra. Appearance of any visible light indicates Lyman ghosts. For modern gratings, Rowland and Lyman ghosts should not exceed 0.1% of the intensity of the parent line present in the first order.

Should the spacings of the grooves change from one side of the grating to the other, which is known as "error of run," the resolving power of the grating will be impaired. By masking off portions of the ruled area and scanning across a narrow line, such as Cd 228.8 nm, with the appropriate widths of entrance and exit slits, this defect can be evaluated by measuring the half-intensity bandwidths of the scans (Section 3.2.8.2).

Lack of parallelism of the grooves of a grating also affects the resolving power, due to the change in the width of the spacings from top to bottom. This defect usually can be evaluated by decreasing the length of the slits and observing the Cd line scans as previously described. Care should be exercised that the entrance slit is critically parallel to the grating lines.

Defects in the shape of ruled grooves cause a serious deterioration in the resolution of the grating. This defect occurs when the grooves are cut to different depths in certain portions of the metal film, thus causing light to be less efficiently diffracted from the effected areas of the grating. The presence of defective groove shapes can be evaluated by photographing a few strong lines radiated by a mercury lamp, on a photographic plate placed about one-third of the way from the focal plane of the grating. Out-of-focus and much-widened line images are obtained. For a good grating, the pattern shows a uniform channeled structure. For an inferior grating, the target pattern shows areas of high and low blackening, and the whole pattern is irregular in shape. For plane gratings, a good quality focusing lens is needed.

An interesting new development[3] to improve the line shape of an echelle grating comes from the use of a diamond stylus for ruling that has an angle of $90°$ or less. Ruling the master with such a tool leaves a narrow unruled strip at the edge of the grooves. By ruling deep grooves, so that the width of the reflecting facet is greater than $d \cos \theta$, the blunt edges of the master grating are hidden in the steep corners of the grooves of the replica. This procedure enables the preparation of replica echelle gratings having a perfectly sharp groove edge and a guaranteed full width of the reflecting facet.

In addition to the errors described above for all gratings, a concave grating, except when illuminated with parallel light, can introduce aberrations, such as astigmatism and coma. Astigmatism is of greater concern because it produces images that are elongated and reduced in intensity. In quantitative analysis, this aberration distorts the relation between element concentration and line intensity.

The advent of lasers and the availability of photoresist materials with high resolution properties have allowed the production of gratings free of most of the errors previously described. The solubilities of photoresist coatings are enhanced when the coatings are exposed to light, thus producing a positive working mode of the photoresist.[4] If a blank coated with a thin layer of photoresist is irradiated with two laser beams at an angle, thus producing interference fringes along the photoresist surface, the irradiated areas can be dissolved by a suitable etching liquid. After drying, the blank can be vacuum-coated with aluminum to produce a very usable grating, known as a holographic grating. By proper choice of the recording medium, blazed holographic gratings can be made. The four main advantages of holographic gratings, compared to ruled gratings, are: (a) they are free from periodic error, and thus, free of ghosts, (b) they give a very low level of stray light, (c) they allow the recording of interference fringes on concave blanks, and (d) the preparation of large holographic gratings with minimal aberration and up to 6600 grooves/mm can be achieved by use of the frequency-doubled line at 257 nm from the argon laser. Although the transmission efficiency of the holographic grating is generally less than that of a ruled grating, the important feature for ICP studies is that the signal-to-noise ratios of spectral lines from a holographic grating is significantly superior.[5,6]

3.2.5 Commonly Used Grating Spectrometers

3.2.5.1 Spectrometers with the Paschen–Runge Mount

The Paschen-Runge mount[7] is presently the most commonly used optical configuration for direct-reading spectrometers. In this mount, the grating and the slits are in fixed positions, with exit slits mounted along a portion of the Rowland circle (Figure 3.11). A perspective drawing of a slit frame with a few mounted slits also is shown in Figure 3.11. For minimum astigmatism within the wavelength range 200 to 450 nm, an angle of incidence of approximately 27° is

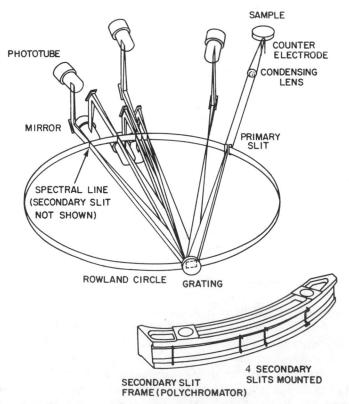

Figure 3.11. Diagram of a direct-reading spectrometer based on a Paschen–Runge mount and a secondary slit frame with a few mounted exit slits. (Applied Research Laboratories, with permission.)

used. Modern instruments mostly have a 1-m focal length and a reciprocal linear dispersion of 0.3 to 0.4 nm/mm, and thus require a grating that has 2400 grooves/mm. High dispersion is preferred because the ICP emits complex line-rich spectra. Considering stability and cost, it seems that for a Paschen–Runge mount, a maximum dispersion around 0.3 nm/mm in the first order has to be accepted for a 1-m spectrometer. With such instruments, normally an entrance slit of 20 μm and exit slits of 40 μm are used.

The basic design of a Paschen–Runge spectrometer can also be adapted for sequential measurements. In this instance, the exit slits on the Rowland circle are fixed at selected wavelengths, but the optics are so arranged that all the radiation passing through the exit slits for the spectral lines is directed by mirrors to a single photomultiplier tube. The radiation passing through the exit slits then can be sequentially determined by opening shutters in the optical trains, one after the other. An advantage of this system is that additional exit slits can be easily added, and that by coupling a stepping motor to the adjusting screw of the entrance slit, measurements of spectral background can be made at specific wavelength positions.[8] Such a feature makes this arrangement suitable for the analysis of materials using the ICP technique.

Another approach to the use of a Paschen–Runge mount for sequential spectrometry is shown in Figure 3.12. In this monochromator, an array of 255 fixed secondary slits, with a regular interslit spacing of 2 mm, is mounted in the slit frame instead of the normal individual slits. This arrangement allows a spectral section scan of ±1 mm across all exit slits for a movement of ±1 mm of the entrance slit. If the photomultiplier can be moved to a position behind each exit slit, full wavelength coverage is obtained by breaking the spectrum into 255 adjacent 2-mm sections. To set the monochromator to any desired wavelength, two movements are necessary; that is, the carriage must be set to the specific exit slit, and the entrance slit must be accurately moved to the desired spectral line.

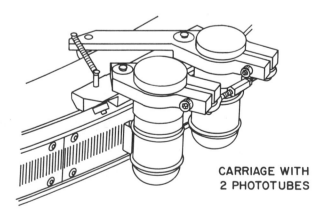

CARRIAGE WITH
2 PHOTOTUBES

Figure 3.12. A monochromator based on the Paschen–Runge mounting. (Applied Research Laboratories, with permission.)

The first movement need not be exact, but the entrance slit must be precisely positioned within the range of ± 1 mm. A position accuracy of 1 μm on an instrument of 1-m focal length can be achieved. This corresponds to a setting accuracy of 0.0005 nm in the second order for a grating having 1080 grooves/mm.[9] Filters, placed in the primary beam, significantly reduce stray light and overlapping orders that can interfere with specific spectral lines.

3.2.5.2 Echelle Spectrometers

Figure 3.13 illustrates an echelle grating spectrometer, an instrument highly efficient in separating the complex line-rich spectra from an ICP. Light from the entrance slit is directed by the collimating mirror to the echelle grating, and the diffracted beam then is reflected to the prism lens that acts as an order-sorter and focusing lens, forming a two-dimensional spectral pattern in the focal plane of the system. An aperture plate, containing preset apertures, is mounted at the focal plane. A mask plane, with slits and mirrors at appropriate positions, is located above the aperture plate. This assembly allows only the light from the selected wavelengths to reach the corresponding photomultipliers that are mounted in the photomultiplier rack. Dispersion and bandpass values of a commercial instrument using a 79 groove/mm echelle, with a blaze angle of $63°26'$ and having a ghost intensity of less than 0.005% of the parent line at the 42nd order, are given in Table 3.2. In an echelle spectrometer, the entrance slit is

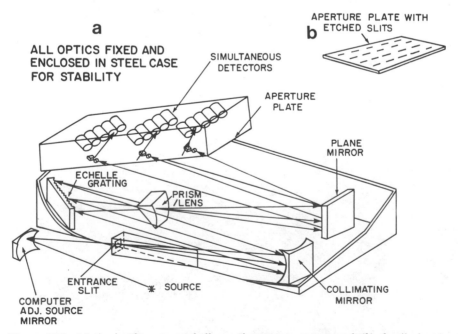

Figure 3.13. (*a*) A simultaneous echelle grating spectrometer and (*b*) detail showing aperture plate with etched slits. (Leeman Laboratories, Inc., with permission.)

Table 3.2. Reciprocal Linear Dispersion and Resolution of an Echelle
Spectrometer[a]

Wavelength (nm)	Dispersion (nm/mm)	Bandpass (nm)
200	0.083	0.008
400	0.137	0.016
600	0.205	0.021

[a] Effective aperture of spectrometer is $f/8$. The width of the entrance slit is 80 μm, and the width of the exit slit is 40 μm.
Source: Leeman Laboratories, Inc., with permission.

normally larger than the exit slits, whereas in the Paschen–Runge instrument, the entrance slit is smaller than the exit slits. Although the echelle grating provides resolution superior to that of a conventional grating, some resolution is sacrificed by using relatively large slits to gain high energy throughput and, most importantly, stability. Direct comparison[2] of conventional and echelle spectrometers only on optical data, as given in Table 3.3, is not an optimal method. Normally, the echelle spectrometers have two dispersive systems with the same or different f-numbers. Furthermore, the echelle systems using crossed dispersion have to limit the slit height to avoid interference between the orders, as reduced height affects the luminosity of the instrument. The main advantages of the echelle systems are superior resolution and dispersion, as indicated in Table 3.3, parameters of value for very complex spectra. Echelle systems also generally use larger entrance and exit slits than those used in conventional spectrometers. Experience has proved that for most analytical problems, using ICP radiation, conventional spectrometers having reciprocal linear dispersions between 0.3 and 0.4 nm/mm suffice. Normally, gratings with high groove numbers and

Table 3.3. Characteristics of Conventional and Echelle
Spectrometers

	Spectrometer type	
	Conventional	Echelle
Focal length (m)	1.0	0.5
Grating width (mm)	102	128
Groove density (lines/mm)	1800[a]	79
Angle of diffraction	—	63°26'
Order at 300 nm	1	75
Resolution at 300 nm	183600	763000
Reciprocal linear dispersion at 300 nm (nm/mm)	0.55	0.15
f-Number	$f/10.2$	$f/8.8$

[a] Special holographic gratings with nearly equal intensities are presently available, allowing the use of such gratings in the second order giving a reciprocal linear dispersion of 0.27 nm/mm.

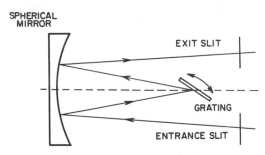

Figure 3.14. The Ebert mount.

instruments with focal lengths of 1 m are chosen. A grating having 1080 grooves/ mm has, however, been successfully used in an instrument with a 1-m focal length to separate close-lying spectral lines using up to the fourth order.

It is important to note that the efficiency of a grating, the percentage of incident light that is diffracted into the required order at a particular wavelength, changes within any spectral order and between the orders.[10] Thus, due to sag in the wings of the echelle efficiency profiles, spectral lines occurring near the ends of an order decline in intensity by a factor of 2 to 3. Relatedly, the widths of the efficiency profiles reduce as order increases (wavelength decreases).

3.2.5.3 Spectrometers with Ebert and Czerny–Turner Mounts

In an Ebert monochromator (Figure 3.14), the entrance and exit slits are on either side of the grating, and a single concave spherical mirror is used as a collimating and focusing element. Rays entering the entrance slit, which is in the focal plane of the mirror, strike the upper half of the collimating mirror and are reflected to the grating. After diffraction by the grating, the light beam strikes the bottom half of the mirror and is focused in the exit slit. Because the two reflections occur off-axis, there are no aberrations. Furthermore, as the entrance and the diffracted beams use different portions of the mirror, no serious scattering of light occurs. Wavelength scanning and selection of a specific line are accomplished by pivoting the grating about the monochromator axis (dashed line).

The Czerny-Turner mount (Figure 3.15) incorporates two smaller concave mirrors instead of a single mirror as used in the Ebert mount. The optical characteristics of the Ebert mount are similar to those of the Czerny–Turner mount. Mounts of the latter type are ideally adapted for linear wavelength or linear wavenumber presentation. Changes from one wavelength to another can easily be made, as most modern spectrometers are controlled by either a microprocessor or a computer. Accordingly, only the drive mechanism for linear wavelength presentation is described here.

A schematic diagram of a sine-bar grating drive for linear wavelength scan is shown in Figure 3.16*A*, while change of the diffraction grating angle, relative to the position of the lead screw, is depicted in Figure 3.16*B*. The drive mechanism

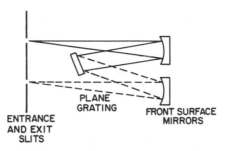

Figure 3.15. The Czerny–Turner mount.

is a precision-ground lead screw directly attached to the shaft of the stepping motor, which has a large number of steps per revolution. The number of steps needed depends on the focal length of the instrument and the resolution desired. For a typical spectrometer, a stepper motor having 10,000 steps per revolution is used. From Figure 3.16B, we see that if the sine bar, attached perpendicularly to the grating holder, has moved through an angle θ, the bar would have moved to B, a distance A–B. In this instance, the following trigonometric relations hold.

$$X = \frac{\sin \theta}{CB} \qquad [3.11]$$

Since $CB = CD$ = constant (length of sine bar),

$$X \alpha \sin \theta \qquad [3.12]$$

The lead screw thus advances a distance proportional to $\sin \theta$ and from Equation 3.12, any change in $\sin \theta$ is linearly related to a change in wavelength.

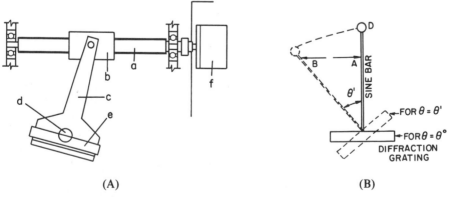

(A)   (B)

Figure 3.16 (A) Sine-bar grating drive for a linear scan: a, lead screw; b, lead screw nut; c, sine bar; d, grating pivot; e, grating; f, stepping motor. (B) Change of grating angle with position of the lead screw.

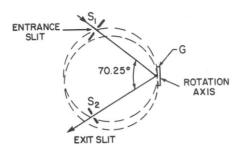

Figure 3.17. The Seya–Namioka mount: rotation axis is perpendicular to paper surface.

3.2.5.4   Spectrometers with Seya–Namioka Mount

The simplest scanning mechanism is achieved by rotating a grating about a vertical axis through the center of the grating. In the Seya–Namioka mount,[11] this is accomplished by using a concave grating and keeping the angle between the incidence and diffracted beams at a constant angle of approximately 70° (Figure 3.17). As the angle θ, through which the grating is turned, increases, the slits move from the Rowland circle and the wavelength at the exit slit increases. The main advantage of this mounting is that an acceptable, though curved, image is obtained over a wide spectrum from the vacuum ultraviolet to the visible region. Furthermore, for linear wavelength presentation, a sine drive scanning mechanism is used. The embodiment of the instrument also is reasonably compact.

3.2.5.5   Double Monochromators

Because of the complex spectra emitted by an ICP, especially for the transition elements, lanthanides, platinum group metals, and uranium, the use of a high-resolution monochromator is essential. A very elegant method for reducing stray light and also obtaining high-resolution spectra involves the use of an echelle spectrometer equipped with a predisperser.[12] This double monochromator (Figure 3.18) consists of a 1.5-m Ebert–Fastie monochromator equipped with an echelle, and a 0.4-m Seya–Namioka monochromator as predisperser. As the two instruments are operated in tandem, the exit slit of the predisperser is the entrance slit of the monochromator. Accordingly, the two monochromators are arranged with their slits parallel to each other. The predisperser is equipped with a replica grating, 1200 grooves/mm, which provides for a reciprocal linear dispersion at the entrance slit of the monochromator of approximately 1.5 nm/ mm at 190 nm to 1 nm/mm at 600 nm. The reciprocal linear dispersion and the theoretical resolving power of the 1.5-m monochromator having an echelle grating with 316 grooves/mm and a blaze of 63°26′ are listed in Table 3.4 for an entrance slit width of 60 μm. These data[12] indicate that in the wavelength range where most ICP spectral lines are located, this instrument has a reciprocal linear

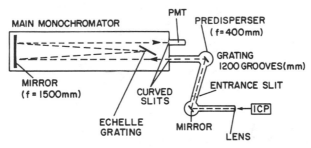

Figure 3.18. Schematic diagram of an echelle monochromator with disperser in parallel-slit arrangement. (From Reference 12, with permission).

dispersion that is approximately 10 times better than most monochromators used for ICP spectrometry.

3.2.5.6 Universal Spectrometers

The Paschen–Runge optical configuration provides a number of options for improving the versatility of such instruments. Two possible systems[13,14] using gratings with a curvature of 1.5 m are shown in Figure 3.19. In the system depicted in Figure 3.19a, the incoming beam is split into two beams by a semitransparent mirror, BS2. Each of the two identical holographic gratings, 2400 grooves/mm, gives a spectral coverage of 120 to 500 nm with a reciprocal linear dispersion of approximately 0.27 nm/mm. The two spectral ranges allow a number of spectral lines to be measured in the one range, while in the other range, background measurements close to the individual spectral lines are made. Two lines lying close together can also be measured, one in the one range and the other in the second range. To allow versatility, each exit slit–photomultiplier pair is mounted in a separate carriage, which can be moved to a specific position along the Rowland circle on a carriage supporting ring. The position of the detection system is controlled by a rotating arm that can be coupled rigidly to

Table 3.4. Certain Optical Properties of a 1.5-m Echelle Grating Monochromator Using a 0.4-m Seya–Namioka Monochromator as Predisperser

Order	Central wavelength (nm)	Reciprocal linear dispersion (nm/mm)	Theoretical resolving power
30	186	0.029	1,950,000
25	223	0.035	1,630,000
20	279	0.044	1,300,000
15	372	0.059	1,040,000
10	558	0.088	650,000

Source: Reference 12, with permission.

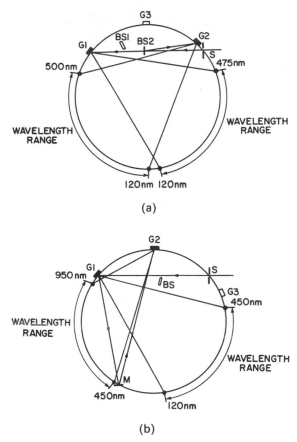

(a)

(b)

Figure 3.19. Optical layouts for the versatile direct-reading emission spectrometer: (*a*) with two similar spectral wavelength coverages and (*b*) using the direct reflected beam to extend the wavelength range of the instrument.

any carriage. The position of this arm is determined by making use of an incremental angle encoder, which can provide a linear resolution of 1.8 μm at the focal plane. Accurate alignment of the spectral line images, from an entrance slit 20 μm wide, within the 30-μm apertures of the exit slits is easily accomplished. Once the carriage is in position, it is rigidly clamped to the supporting ring by means of a special controllable magnet. The rotating arm is then uncoupled and coupled to another carriage. The system also can be used as a scanning monochromator in one range and as a direct-reading spectrometer in the other range. Background corrections can be made either by rotating a refractor plate behind the entrance slit or by moving the entrance slit.

In the second optical arrangement (Figure 3.19*b*), a mirror is placed slightly off the Rowland circle[15] to direct the normal reflected beam of the first grating to a second grating for diffraction. This allows the observation of a wavelength range of 120 to 950 nm.

3.2.6   Background Correction Devices

In the discussion of grating optical systems, background correction was briefly mentioned. The importance of background correction in ICP spectrometry warrants further discussion of the devices used for this purpose.

3.2.6.1   Entrance Slit Translation

A background correcting system based on the translation of the entrance slit[16] is illustrated in Figure 3.20. The system is operated as follows. After completion of the normal integration, the program activates the stepping motor, which translates the entrance slit to a predetermined position. This translation of the entrance slit changes the angle of incidence to the grating, thus effectively shifting the spectral line off the exit slit to allow measurement of background on one side of the line. After integration of the background signal, the motor will turn in the opposite direction to translate the entrance slit for measuring background on the other side of the spectral line. The exit slit is subsequently returned to its original position. This system allows background measurements to be made on both sides of the spectral line for each element at predetermined positions.

3.2.6.2 Refractor Plate Deflection

A simultaneous system for line and background measurements, described by Brinkman and Sacks,[17] is based on the subtraction of the background radiation from a nearby wavelength passing through the upper half of a specific exit slit from that of the spectral wavelength passing through the lower half. The wavelength displacement is accomplished by means of a refractor plate. This system requires a minimum of two detectors for the desired measurements.

When a stable source, such as the ICP, is studied, the measurement of the background and spectral line intensities can be done sequentially. The advantage of such an approach is that only one detector per spectral line is needed. Nordmeyer[18] placed a diffractor plate at the exit slit, while Visser and co-workers[19] used a refractor plate behind the entrance slit. Optical and

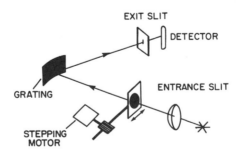

Figure 3.20. A background correcting system based on the translation of the entrance slit.

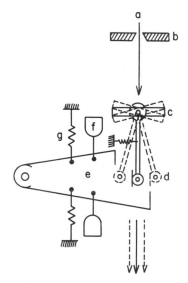

Figure 3.21. The use of a refractor plate behind the entrance slit for background correction: *a*, light beam; *b*, slit; *c*, refractor quartz plate; *d*, fixed positions of cam plate; *e*, cam plate; *f*, pull solenoids; *g*, balanced springs. (From Reference 18, with permission.)

mechanical details of the latter system are shown in Figure 3.21. The quartz refractor plate, 4.7 mm thick, is rigidly attached to an arm that is connected to a cam plate. The plate can be moved to three fixed positions by means of two balanced springs and pull solenoids. With this device, sequential measurements are possible on both sides of the spectral line, and on the spectral line. Rotating the quartz plate by 15° to either side of the spectral line shifts the wavelength by 0.1 nm. When a refractor plate is used, refocusing of the spectrometer is necessary. In modern spectrometers, the refractor plate allows measurements at variable background locations.

3.2.6.3   Direct Scanning

Computer-controlled scanning monochromators[20] can be programmed to measure background in the vicinity of spectral lines at predetermined positions on either side of the line.

3.2.6.4   Simultaneous Measurements

As discussed in Section 3.2.5.6, a spectrometer with two similar spectral ranges allows the simultaneous recording of background and line intensities. For such instruments, background measurements can be made only at one side of a spectral line. Alternatively, a multiple-exit-slit system[21] may be used for background correction.

3.2.7   Focusing Devices

It is general practice in ICP spectrometry to observe a specific portion of the plasma for analytical purposes. This requires a spherical lens or mirror system to select the desired portion of the plasma. Detailed discussions of focusing devices may be found in the second edition of "Fundamentals of Optics," by Jenkins and White (see Bibliography). Briefly, the action of both lenses and mirror is described by the thin-lens equation:

$$\frac{1}{p} + \frac{1}{q} = \frac{1}{f}$$ [3.13]

where p and q are the source-to-lens and lens-to-image distances, respectively. To provide a uniform illumination of the slit, especially for spectrography, a lens arrangement (Figure 3.22) is used. In this instance, an image of the ICP is made on a diaphragm that is focused on the collimating lens or mirror. In many modern applications, front surface mirrors are preferred to quartz lenses because of chromatic aberration, that is, the variation of the focal length of a quartz lens with wavelength. The equation for correcting the focal length of quartz lenses with wavelength is given by:

$$F_2 = F_1\left(\frac{n_1 - 1}{n_2 - 1}\right)$$ [3.14]

where F_1 is the focal length at wavelength 1, and n_1 and n_2 are the refractive indices of quartz at wavelengths 1 and 2, respectively. Concave off-axis mirrors[22] also are used to transfer ICP radiation to the spectrometer. Rotation of the mirror along its axis enables the observation of various portions of the ICP without the physical translation of the ICP, as required with lens focusing devices.

3.2.8   Figures of Merit for Grating Spectrometers

To evaluate a monochromator as the dispersing device for ICP–AES, the experimenter is interested in the dispersion, resolving power, resolution, speed, and stray light of such an instrument. Grating efficiency in the different orders is also of significance, since certain line interferences can be overcome if a grating can be efficiently used in a higher order.

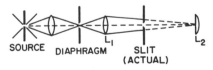

Figure 3.22. Lens arrangement for a uniform illumination of the spectrometer slit.

3.2.8.1 Dispersion

For ICP spectrometry, the practice of using spectrometers having reciprocal linear dispersions adequate for earlier emission sources with simpler spectra is slowly disappearing, and spectrometers with dispersions of 0.3 nm/mm or less are being adopted. As an example, consider the relative positions of two adjacent exit slits of a simultaneous instrument (Figure 3.23*A*). With exit slits cut into 3-mm base assemblies, the minimum distance on the Rowland circle between the centers of the two exit slits is about 4 mm for an instrument having a focal length of 1 m and a reciprocal linear dispersion of 0.55 nm/mm. Such a system, with a grating having 1800 grooves/mm, an entrance slit of 25 μm, and exit slits of 70 μm, will allow two lines 0.1 nm apart to be effectively separated, as illustrated in Figure 3.23*C*. The scan shows that the two lines, Fe 375.823 nm and Ti 375.929 nm, being 0.106 nm apart, are fully resolved. For stability, the image of the entrance slit ideally should be within the exit slit, as shown in Figure 3.23*B*.

3.2.8.2 Resolving Power and Resolution

The expressions related to the theoretical resolving power for a prism and a grating have been given in Equations 3.7c and 3.9, respectively. A working

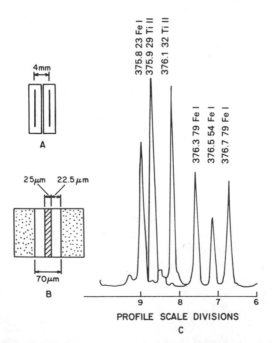

Figure 3.23. (*A*) Minimum distance between two slits. (*B*) Ideal positioning of the image of the entrance slit within the aperture of the exit slit. (*C*) A spectral scan in the vicinity of the line Fe I 376.379 nm.

definition of the resolving power is given by the expression:

$$R = \frac{\bar{\lambda}}{\Delta\lambda}$$ [3.15]

where $\Delta\lambda$ is the wavelength separation of two lines that can just be distinguished and $\bar{\lambda}$ is the mean wavelength of the pair. Thus, for the Hg 313.1 nm doublet, which is separated 0.0287 nm, we need only a resolving power of approximately 11,000.

The resolution of a monochromator is the smallest wavelength interval that can be resolved. This wavelength interval depends not only on the resolving power of the instrument but also on the slit width, slit height, and the perfection of the optical system. Resolution usually is reported as the half-intensity bandwidth, illustrated in Figure 3.24. When the image of a monochromatic beam of light from the entrance slit is passed across the exit slit, as in a scanning monochromator, the intensity of the radiation will increase until the image and the exit slit coincide. Further scanning results in a decrease of intensity at the

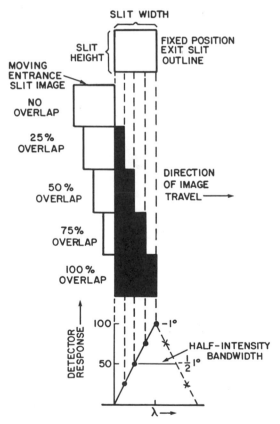

Figure 3.24. The dependence of the intensity of the radiation passing the exit slit versus successive images. (Harper & Row, with permission.)

A. STRASHEIM

exit slit. If the slit width is decreased, the width of the detector response curve also decreases. Due to aberrations present in all optical systems, the image is somewhat widened and the distribution of intensity with wavelength is more Gaussian than angular. Accordingly, the exit slit of a monochromator not only should be set accurately on the focal plane, but its shape should be made to match the image of the entrance slit or the resolution is sacrificed.

For measurement of the bandpass of a monochromator, the Hg 313.1 nm line or the Cd 288.8 nm line from a hollow cathode lamp can be used, as they have a half-intensity bandwidth much less than the bandpasses of most monochromators. Recorder scans (Figure 3.25) of the Hg 313.1 nm doublet in the first and second orders, using a 1-m Czerny-Turner monochromator, allow the separation of the two lines, measured at halfwidth, to be converted to wavelength units. For the first and second orders, the line separations are 6 and 11.8 mm,

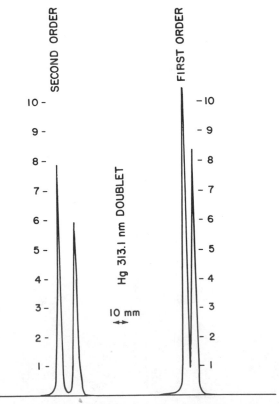

Figure 3.25. Scans of the Hg 313.183 and 313.154 nm doublet in first and second orders using a Czerny-Turner monochromator, 1 m focal length, 1800 grooves/mm grating, reciprocal linear dispersion (first order) 0.59 nm/mm, 15 μm × 2 mm slit dimensions, 10 mm corresponds to 0.05 and 0.025 nm in first and second order, respectively. Blaze wavelength of grating is at approximately 250 nm in the second order.

respectively, while the widths of lines measured are 2.25 and 2 mm at half-intensity. As these differences correspond to a wavelength separation of 0.0287 nm, the measured half-intensity bandwidths for the first and second orders thus are 0.011 and 0.0049 nm, respectively. The approximate half-intensity band-width of a monochromator may be calculated by multiplying the reciprocal linear dispersion by the slit width. To determine whether an instrument operates according to specification, certain manufacturers use the resolution factor, R.F. This factor is the ratio T/p, expressed as a percentage, where T is the mean height of the two peaks measured from the trough between the two peaks, and p is the mean height of the two peaks measured from the zero of the trace. Using the recorder scans of the Hg doublet (Figure 3.25), we obtain R.F. values of 91 and 98%, respectively, for first and second orders. This result indicates a satisfactory alignment of the monochromator, because the accepted R.F. value by the manufacturer for the specific type of instrument for this line in the second order is 95%. For a similar tracing recorded for a 3100 groove/mm grating on the same instrument (Figure 3.26), an R.F. value of 92% is observed. Because the previous experiment, using the 1800 groove/mm grating, provided satisfactory

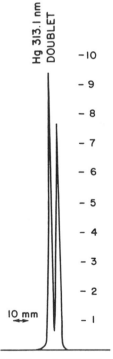

Figure 3.26. Scans of the Hg 313.183 and 313.154 nm doublet in first order using a Czerny–Turner monochromator, 1 m focal length, 3100 grooves/mm grating, reciprocal linear dispersion 0.3 nm/mm, 15 μm × 2 mm slit dimensions, 10 mm corresponds to 0.05 nm.

A. STRASHEIM

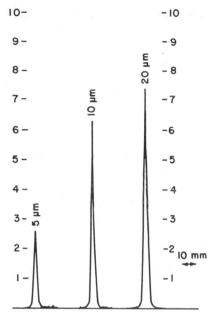

Figure 3.27. The dependence of the response of monochromator with a 1800 groove/mm grating on the width of the entrance slit for the Cd 228.8 nm excited in an Ar ICP, 10 mm correspond to 0.05 nm. The concentration of Cd is 10 μg/mL.

R.F. values, the entrance and exit must be parallel. Thus, the grooves of the grating having 3100 grooves/mm are not parallel to the axis of rotation.

Although the use of narrow slits enhances the resolution, a significant reduction in response is observed if very narrow slits are used (Figure 3.27). A typical slit width for ICP–AES is approximately 20 μm.

3.2.8.3 Speed

Light-gathering ability of an optical system is described by the f-number, which is given by the equation:

$$f = \frac{F_1}{A} \qquad [3.16]$$

where F_1 is the focal length of the collimator lens and A is the diameter of the limiting aperture.

In monochromators, the grating is often used as the aperture stop, as its ruled area is rectangular. This defining aperture must be equated to an equal circular area from which the diameter A can be calculated using the formula:

$$A = \left(\frac{4WH}{\pi}\right)^{1/2} \qquad [3.17]$$

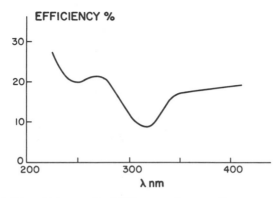

Figure 3.28. The efficiency of a 3600 groove/mm grating versus wavelength.

where H is the height and W is the width of the aperture. A monochromator having a low f-number possesses a high speed, or a high optical throughput.

3.2.8.4 Grating Efficiency

The efficiency of a grating is defined as the ratio of the intensity of the diffracted light for a particular wavelength in a specific order, to the intensity of the undispersed light at that wavelength that would be reflected by an aluminized mirror. Normally, specular reflection increases with the number of rulings. To increase the efficiency of a grating in a specific wavelength region, specific blaze angles are normally chosen. In selecting the blaze wavelength, it must be remembered that the efficiency of a grating falls off on either side of the blaze wavelength. A useful rule of thumb for predicting the range in which the efficiency should be better than 50% is given by Equation 3.18, where λ_b is the blaze wavelength in the first order:

$$\left(\frac{2}{2n+1}\right)\lambda_b < \lambda < \left(\frac{2}{2n-1}\right)\lambda_b. \qquad [3.18]$$

As most spectral lines used in ICP spectrometry are located below 300 nm, a blaze angle at 250 nm covering the range 165 to 370 nm is suggested for monochromators. The efficiency curve for a grating having 3600 grooves/mm is shown in Figure 3.28.

The blaze of a grating has a significant effect on the grating efficiency. Nearly equal efficiencies in the first and second orders are observed (Figure 3.25), for the grating having 1800 grooves/mm and a blaze wavelength at approximately 250 nm in the second order. For the other grating tested, also having a symmetrical profile, but having 3100 grooves/mm and a blaze wavelength at 225 to 250 nm, the efficiency in the first order for Hg 256.2 nm, approaches 92%, while in the second order, the efficiency is quite low (Figure 3.29).

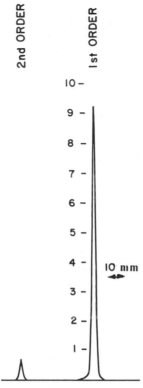

Figure 3.29. Illustration of the efficiency of a modern grating in first and second orders; 10 mm corresponds to 0.50 nm and 0.025 nm in first and second orders, respectively. Blaze wavelength of grating is at approximately 225 to 250 nm in the first order.

3.2.8.5 Instrument Stray Light

Due to the efficient excitation of certain elements in the ICP, stray radiation within spectrometers was a serious problem in the early days of this technique. The main interferences came from Ca, Mg, and Fe.[23] Stray light equivalents (mg/L) due to Ca radiation in a 1-m Paschen–Runge spectrometer with a reciprocal linear dispersion of 0.55 nm/mm are listed in Table 3.5. These results indicate that all the exit slits of this particular instrument are affected by the stray light originating from the Ca radiation. The main sources of stray light are reflections from interior surfaces, filters, lenses, prisms, etc. Light scattered by imperfections, dirt, or scratches associated with the optical components also can contribute to stray light. Several ways to reduce stray light in spectrometers are listed in Table 3.6. The percentage of stray light can be determined by measuring the transmission of certain optical filters just below their cutoff wavelengths. The wavelengths at which the measurements can be made, typical filters, and sources used in measuring stray light are listed in Table 3.7. For further discussions of

Table 3.5. Stray Light Equivalents from Calcium.[a]

Element	Wavelength (nm)	Ca, 50 μg/mL	Ca, 500 μg/mL
Zn	213.9	0.008	0.032
Pb	220.4	0.094	0.563
Fe	259.9	−0.002	0.009
Sn	303.4	0.000	0.648
Al	396.1	0.156	1.146
Ba	455.4	0.001	0.005

[a] Measurements were made at wavelengths corresponding to spectral lines for the elements indicated. Bandpass = 0.038 nm.

Table 3.6. Methods to Reduce Stray Light in Spectrometers

1. Use holographic grating.
2. Paint the interior of the instrument with flat black paint.
3. Insert baffles to obstruct radiation from all except the key direction.
4. Enclose the photomultiplier in a suitable housing so that only light passing through an exit slit can reach the photocathode.
5. Use isolation filters, or a calcium rejection filter for solutions with high calcium concentrations.
6. Use solar blind photomultipliers for recording the characteristic radiation of spectral lines below 300 nm.

Table 3.7. Filters and Sources used in the Determination of Stray Light

Wavelength (nm)	Filter type	Source
280	Glass	Deuterium lamp
320	Corning 4303	Deuterium lamp
440	Corning 9863	Tungsten lamp
650	Chance OB10	Tungsten lamp

stray light from holographic and ruled gratings, the reader is referred to other publications.[24,25]

3.3 DETECTION SYSTEMS

The photographic plate has played a very important role in the development of atomic emission spectroscopy. For many decades, the photographic emulsion served as a composite detector–transducer–amplifier–recorder of the radiation dispersed along the focal plane of a spectrograph. Its main advantages, ie, multiradiation detection and permanency of record, have not been surpassed by modern detectors. The major limitations of the photographic emulsion are the dynamic range, which typically is limited to about two decades of response, and the speed of analysis. The main detectors used in AES today, photomultipliers, vidicons, image dissectors, and photodiode arrays, are discussed next.

3.3.1 Photomultipliers

In principle, the operation of a photomultiplier is similar to that of a vacuum photocell (Figure 3.30). Under certain conditions, the absorbed photon energy may cause the escape of an electron from the cathode. If the emitted electron moves away from the cathode to the anode under the influence of an applied voltage, a current is produced depending on the intensity of the incident radiation. Normally, the cathodes of photoemissive devices require minimal energies for electron emission. The sensitivity of a photoemissive tube can be increased by a factor of 10^6 to 10^8 by accelerating the electrons from the cathode to a set of electrodes, dynodes, to cause the secondary emission of electrons (Figure 3.31). Usually, seven or more dynodes are used in photomultiplier tubes, which are typically operated at voltages in the range from 700 to 2000 V. The typical response time of a photomultiplier is in the nanosecond range. As the wavelength range of interest for recording ICP radiation stretches from W 174.2 to Cs 852.1 nm, a few typical response curves covering this range are illustrated in Figure 3.32. The basic corresponding characteristics are summarized in Table 3.8. Photocathode and window materials normally are selected on the basis of the desired spectral range for measurements. Because many spectral lines used in ICP–AES occur below 250 nm, the importance of the solar blind photomultiplier, which produces no response for radiation above 300 nm, must be emphasized. Quantum efficiencies of photomultipliers at peak wavelength sensitivity vary from 40% for the solar blind tube to 1% for the tube sensitive to near-infrared radiation. For the spectral responses shown, the anode dark current varies from 1 to 50 nA, while current amplification varies from 1.0×10^5 to 3.8×10^6. Because a photomultiplier tube exhibits a high input impedance, the current output can be easily amplified and converted to a voltage signal for measurement.

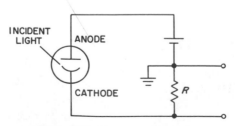

Figure 3.30. Vacuum phototube.

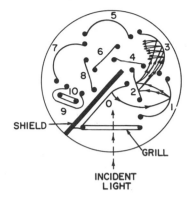

Figure 3.31. Electron photomultiplier tube.

In the absence of illumination, the photomultiplier exhibits a residual current, known as the dark current. The most important component of the dark current is the thermionic emission of electrons from the photocathode and the dynodes. The thermionic dark current can be reduced significantly by cooling most photomultipliers to approximately $-40°C$. The remainder of the dark current is due to such factors as radioactive substances occurring naturally in the envelope materials. A typical value of the dark current for photomultipliers is around 5 nA.

The signal from the photomultiplier tube is also affected by other noise sources, the most important of which are shot noise and Johnson noise. Mostly, the magnitude of the signal generated by a photomultiplier is large compared

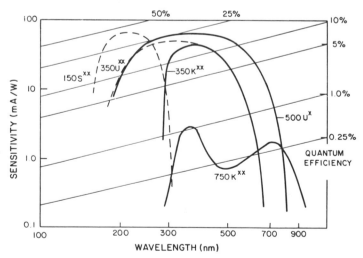

Figure 3.32. Response curves for some typical photoemissive surfaces used for recording ICP spectra: x, semitransparent photocathode; xx, opaque photocathode, reflection mode. (Hamamatsu Corporation, with permission.)

Table 3.8. Characteristics of a Few Typical Photomultiplier Tubes Used for Recording ICP Spectra

Type no.	Remarks	Spectral response			Photo-cathode material	Window material	Current amplification typical	Anode dark current after 5 s typical (nA)
		Curve code	Range (nm)	Peak wavelength (nm)				
Side-On Type, 28 mm diameter								
931 A	S-4 response general purpose	350 K (S-4)	300–650	400	Sb-Cs	Boro-silicate	2.5×10^6	50
1 P21	Medium gain and very low dark current variant of 931A	350 K (S-4)	300–650	400	Sb-Cs	Boro-silicate	3.8×10^6	2
1 P28	S-5 response, general purpose	350 U (S-5)	185–650	340	Sb-Cs	UV glass	2.5×10^6	30
R406	For red to 1R detection near S-1 response	750 K (S-1)	400–1100	730	Ag-O-Cs	Boro-silicate	2.5×10^5	
Side-On Type 13 mm diameter								
R427	Solar blind response	250 S	160–320	200	Cs-Te	Fused silicon	3.3×10^6	1
Head-On Type, 13 mm diameter								
R1463	High gain wide spectral response. Multi alkali photocathode.	500 U	185–850	420	Multi alkali	UV glass	1.0×10^5	10

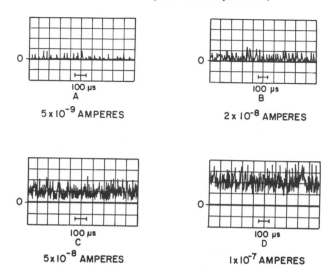

Figure 3.33. Simulated oscilloscope display of the output of a photomultiplier showing discrete current pulses, with increasing dc photo current levels from *A* to *D*. (From Reference 26, with permission.)

with that of the noise. Yet at low levels of light, the magnitude of noise becomes a greater fraction of the signal and the uncertainty of measurement increases. In such cases, photon counting[26–30] is very effective in achieving high S/N ratios. The oscilloscopic displays of photomultiplier output (Figure 3.33) at various light intensities show that at low light levels, the discrete current pulses are easily resolved. Furthermore, as the light level increases, the current pulses begin to overlap and pile up as a result of the finite frequency response of the measurement system. Using a photon counting system (Figure 3.34), the current

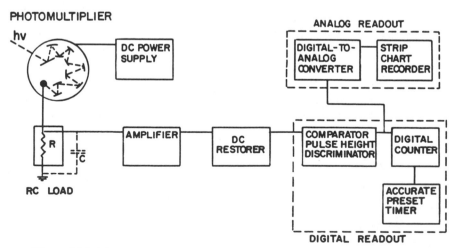

Figure 3.34. Photon counting system. (From Reference 26, with permission.)

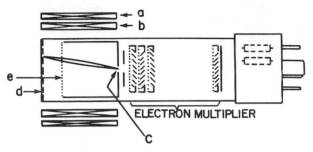

Figure 3.35. Image dissector tube: *a*, focusing coil; *b*, deflecting coil; *c*, aperture (8 × 100 µm); *d*, photocathode; *e*, Accelerating electrode (mesh). (From Reference 31, with permission.)

pulses, proportional to the number of photons striking the photocathode, can be counted. Compared to other measuring systems, the advantages[26–30] of the photon counting system can be summarized as follows: direct digital processing of the inherently discrete spectral information, decrease of the dark current effect by orders of magnitude, improvement in the signal-to-noise ratio, satisfactory sensitivity at very low light levels, and improved precision of analytical results.

3.3.2   *Television-type Detectors*

3.3.2.1   Image Dissector Tubes

An image dissector tube[31] (Figure 3.35) consists of a photocathode, electron-optics, and an electron multiplier. Between the electron multiplier and the electron-optics is an aperture. The optical spectrum is focused on the photocathode. By choosing suitable values of the electrostatic accelerating field (*e*) and of the axial magnetic focusing field (*a*), the photoelectrons emitted from the photocathode are focused to form an electron image on the screen containing the aperture (*c*). The photoelectrons from any chosen spot on the photocathode can be directed into the electron multiplier by adjusting the current to the two orthogonal deflection field coils (*b*). The intensity is finally measured by photon counting. When dispersed radiation from an echelle spectrometer is focused on the photocathode of an image dissector, spectra similar to that shown in Figure 3.36 are obtained . The spectral range from 200 to 800 nm is segmented in 91 spectral orders. The reciprocal dispersion varies from 0.16 nm/mm at 200 nm to 0.63 nm/mm at 800 nm.[31] Compared to other imaging detectors, the image dissector exhibits the best sensitivity, and its dynamic range spans six to seven orders of magnitude. In contrast to other TV-type detectors, the image dissector is not an integrating device, its main disadvantage seems to be the extensive computer control needed for operation.

WAVELENGTH

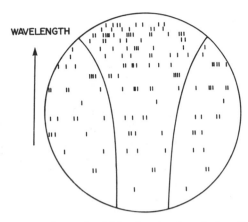

Figure 3.36. Echelle spectra focused within the circular photocathode area, 40 mm in diameter, of an image dissector tube. The spectra in the central area within the two curves represent spectra with the highest intensity. (From Reference 31, with permission.)

3.3.2.2 Photodiode Arrays

Self-scanning, linear silicon photodiode arrays have been used successfully in ICP studies.[32] Arrays having 256, 512, and 1024 elements are commercially available. The detector diodes are on a 25.4-μm spacing, which results in a density of 39.4 diodes/mm. Hence, the lengths of the 256, 512, and 1024 arrays are 6.5, 13.0, and 26.02 mm, respectively. A 1024-element array is shown in Figure 3.37a. The photo sensitive elements are along the center of the middle rectangle. The latest arrays have a height of 2 mm. Representative spectral response curves for photodiode array detectors appear in Figure 3.37b.

The signals from a photodiode array spectrometer, for a sample and a blank, are illustrated in Figure 3.38. A simplified schematic of a complete integrated circuit necessary for scanning a linear photodiode is shown in Figure 3.39. Each photodiode, paralleled by a capacitor, operates in a change storage mode; hence the diode is inherently an integrating-type detector. For effective use of a photodiode array, a computer data acquisition system must be used, and the array must be cooled, usually with a Peltier cooler, to reduce the dark current.[33] Because the electronic background on any of the arrays is exactly repeatable from scan to scan, it can be removed by subtraction. Figure 3.40 compares the integration performances of a cooled and an uncooled array. Furthermore, the spectral background can be automatically subtracted (Figure 3.41). One of the key features of an array is the ease with which the integration time can be extended to increase signal-to-noise ratios. The main advantage of the photodiode array, versus a photomultiplier tube, is that the photodiode array measures spectra simultaneously. However, the wavelength coverage of a diode array detector for moderate- and high-resolution studies is quite limited, roughly 10 nm, at best. The diode array is especially useful for the recording of

(a)

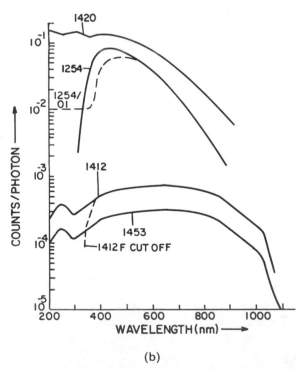

(b)

Figure 3.37. (*a*) A 1024-element linear array. (Photo from Reticon Corporation, with permission.) (*b*) Representative spectral response curves for models 1200 and 1400 series photodiode array detectors. (EG&G Princeton Applied Research, with permission.)

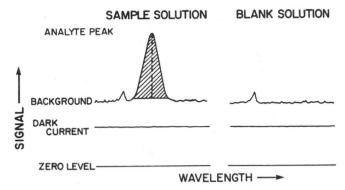

Figure 3.38. Signals from a photodiode array spectrometer for a sample and for a blank solution.

time-resolved spectra, the measurement of transient events, and spectral diagnostics studies of the ICP.[34]

3.3.2.3 Vidicon Detection Systems

A vidicon, a semiconductor photosensitive detector, is similar in operation to a photodiode array with two important differences. In contrast to a linear array, vidicons are usually configured with a two-dimensional array of 517 × 512 individual sensors in a 12.5-cm² area. Second, these photosensors are not accessed by wire circuitry, as in the photodiode array, but by a scanning electron beam. Figure 3.42a is a line diagram of a vidicon. Like photodiode arrays, vidicons can be operated, in a charge storage mode and thus can act as integrating devices. The electronics needed for vidicons are more complex than those required by the diode array systems. In contrast with the photodiode, it is possible to randomly address various sections of the vidicon because of the use of a scanning electron beam.

Compared to a photodiode array, vidicons suffer from two undesirable characteristics: bloom and lag. Bloom, the effect of a high light intensity falling on a specific sensor area causing a false reading in the immediate adjacent sensor, can be overcome by programming. The incomplete recharging of a

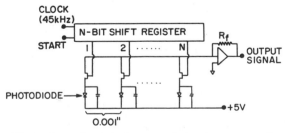

Figure 3.39. Simplified schematic of the integrated circuit necessary for scanning a linear photodiode.

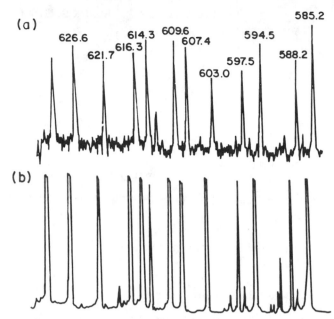

Figure 3.40. Comparison of the integration performance of (*a*) an uncooled array and (*b*) an array cooled to $-15°C$. The two spectra have been computer scaled to the same amplitude to facilitate comparison. (From Reference 33, with permission.)

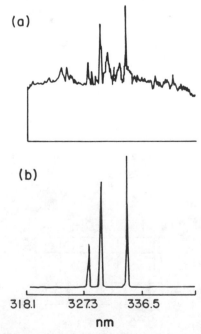

Figure 3.41. Spectra illustrating the power of electronic background subtraction. (*a*) Zn dc arc emission spectra. (*b*) Same as (*a*) with electronic background subtraction. (From Reference 33, with permission.)

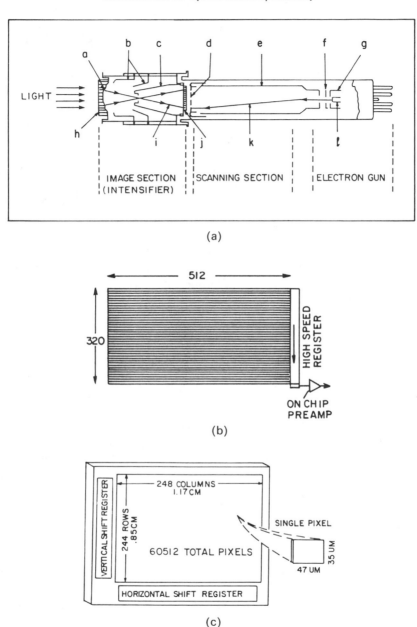

(a)

(b)

(c)

Figure 3.42 (*a*) A cross section of a silicon-intensified target (SIT) vidicon detector: *a*, photocathode; *b*, image focus grids; *c*, anode; *d*, field mesh grid; *e*, gun focusing grid; *f*, accelerating grid; *g*, control grid; *h*, higher optic face plate; *i*, high-velocity photoelectron beam imaged on target; *j*, silicon target; *k*, low-velocity scanning beam; *l*, cathode. (Radio Corporation of America, with permission.) (*b*) The pixel layout of a charge-coupled device (CCD) having an array of 512 × 320 or 163,840 sensor elements. (Radio Corporation of America, with permission.) (*c*) Dimensions and layout of a charge-injection device (CID) having 60,512 pixels in 244 rows and 248 columns. (From Reference 32, with permission.)

sensor, during a single scan of the electron beam, is called lag. Lag can be overcome by repeat recharging, but additional time is required.

A silicon-intensified target (SIT) vidicon coupled to a relatively high resolution monochromator[35] having a reciprocal linear dispersion of 0.4 nm/mm, and a 5-nm spectral window covering 500 electronic channels, gave ICP detection limits of nanograms per milliliter for a number of elements. By cooling the SIT vidicon to $-20°C$, detection limits for a 1-kW argon ICP, based on a total measurement time of 10 s, were not substantially worse than those for a photomultiplier, except for Mn, Co, Cd, and Zn, which have spectral lines lower than 260 nm.[36] Recently, a vidicon detector has been coupled to a direct-reading spectrometer having multiple entrance slits.[37] Optical fibers are used to direct radiation from the source to the entrance slit. The dispersed spectral bands from the grating are then focused on the vidicon detector. In such a spectrometer, spectral lines may be moved across the detector target to avoid potential spectral interferences.[37] Since the input ends of the optical fibers can be moved to different positions relative to the source, individual observation heights may be selected for the elements. Vidicon detectors are also especially useful in diagnostic studies of the ICP.[38]

3.3.2.4   Charge-Coupled and Charge-Injection Detectors

Denton and associates[32] have been the chief promoters of charge-coupled and charge-injection detectors (CCD and CID) for atomic spectrometry. In both devices, the photon-generated charge is collected and stored in metal oxide semiconductor (MOS) capacitors, thus allowing precise pixel addressing with minimal lag. Both devices are less expensive and easier to cool than the vacuum tube imaging detectors. Yet, they are two dimensional detectors with random or pseudorandom pixel-addressing capabilities. A CCD (Figure 3.42b) can be thought of as a large number of photodetecting analog shift registers. After the photon-generated charges have been stored, they are clocked horizontally, row by row, through a high-speed shift register to a preamplifier. The resulting signal is stored by a computer. For the CID (Figure 3.42c) the light strikes a row–column structure of discrete pixels, each of which is composed of a pair of silicon-type MOS capacitors. The photon-generated charge, stored in the one capacitor, may be transferred to another capacitor during the readout of each pixel, thus providing a nondestructive readout. For the destructive readout mode, the stored charge is injected into the substrate. By mixing destructive and nondestructive readouts, blooming, which is a problem in vidicon detectors, can be eliminated, especially when complex spectra are being monitored.

Compared to the vidicon and the photodiode arrays, CCD and CID are smaller. The CCD and CID used in atomic spectrometry have dimensions of 1.41×1.73 cm, with 163,840 pixels, and 0.9×1.1 cm, with 60,512 pixels, respectively. Detectors with larger numbers of pixels will soon be available. The maximum spectral response of the CCD and CID detectors are located at 500 to 800 nm, and the response falls off quickly in the ultraviolet. These detectors have been used to monitor direct-current plasmas, but not yet an ICP.

3.4 READOUT DEVICES

The electrical signal from the detector must be processed by either an analog or digital electronic circuit before it is measured by the readout device. Most modern spectrometers are interfaced to dedicated micro- or minicomputers, which control the operating conditions, data acquisition, and the data treatment, including corrections for spectral interferences and background. For a comprehensive discussion of this subject, the reader is referred to the Bibliography for the books edited by Elving and authored by Malmstadt, Enke, and Crouch, respectively.

3.5 FOURIER TRANSFORM SPECTROMETRY

3.5.1 General Principles

The modern direct reader, based on the exit slit–photomultiplier combination, is a powerful multiple-channel measuring system. However, even at best, only a very small fraction of the spectrochemical information present in the emitted radiation is measured. This limitation can be overcome by utilizing an electronic image detector, which can provide, over a moderate wavelength range, a multiple-channel detection system having continuous wavelength access. The Michelson interferometer, using a periodic mirror scan mechanism, provides a large, continuous-wavelength coverage, after Fourier transformation of the data. The main advantages of the Fourier transform spectrometer over conventional dispersive instruments are: higher energy throughput (because no entrance slit is required), achievement of higher optical resolution, and the ability to monitor all spectral information simultaneously for an extended period.

The basic components of a Michelson interferometer are shown in Figure 3.43. Radiation from the source S falls on a beamsplitter O, where it is either reflected (at the back surface) toward mirror M_1 or transmitted to mirror M_2. The reflected beams from these two mirrors are recombined at O and emerge as a single beam. The detector E will sense either darkness or light depending on whether the optical path differences of the two light beams OM_1O and OM_2O are exactly either $\lambda/2$ or λ, respectively. If provision is made for uniform translation of M_1, the detector output, as a function of time, is similar to that shown in Figure 3.43. The frequency of the detector is determined by the translation velocity of M_1 and the wavelength of the monochromatic radiation. The amplitude of the detector signal is proportional to the intensity of the incident monochromatic radiation. If the incident radiation is polychromatic, each frequency component will be transformed so that a detector output wave of unique frequency is produced for each component. The variation of the output signal with time, the interferogram, has a form as shown in Figure 3.44.

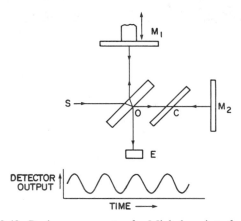

Figure 3.43. Basic components of a Michelson interferometer.

To transfer the interferogram to a spectrum, complex calculations are required. The operation involves the translation of an intensity–distance plot into an intensity–frequency plot (Figure 3.45). The exact mathematical equations require that the mirror be moved an infinite distance and that an infinite frequency range be observed. Certain restrictions, such as summation instead of integration of the data over a finite distance interval, must be accepted. Furthermore, in the computation, an appropriate function, called an apodizing function, should be included to control the rate of truncation of the interferogram. Suitable apodizing functions are the boxcar and the triangle (Figure 3.45c). Such a computation would improve the resolution further and effect the line shape and fine structure at the base of a line. The Fourier transforms of a monochromatic interferogram are graphically illustrated in Figure 3.45d. The results clearly show the effect of mirror distance movement on resolution and the effect of certain apodizing functions.

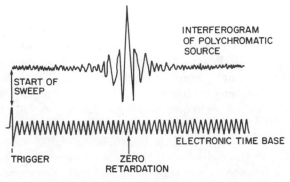

Figure 3.44. Interferogram of a polychromatic source. (Harper & Row, with permission.)

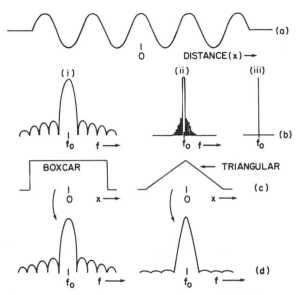

Figure 3.45. The Fourier transform of a monochromatic interferogram. (*a*) Interferogram: only a short section of the actual infinite wave is shown, when the interferometer mirror is moved a path difference $-x$ to a path difference $+x$. (*b*) Fourier transform for mirror movement corresponding to different path difference: (i) $x = 16\lambda$, (ii) $x = 128\lambda$, (iii) $x = \infty$. (*c*) Boxcar and triangular apodizing functions. (*d*) The effect of applying the apodizing functions in (*c*) to the Fourier transfer with a path difference of $x = 16\lambda$. (Addison-Wesley Publishing Company Inc., with permission.)

3.5.2 Inductively Coupled Plasma–Atomic Emission Fourier Transform Spectrometry

Most of the investigations of atomic emission–Fourier transform spectrometry have been conducted by Horlick and associates.[39-47] The schematics of typical Fourier transform spectrometers used by these investigators for monitoring an ICP are shown in Figures 3.46 and 3.47. The spectrometer shown in Figure 3.46 has three optical inputs incorporated into a single optical axis: a He–Ne laser, a white light (tungsten bulb) source, and the spectral source of interest. The laser system generates clock pulses for the analog-to-digital converter (ADC), and thus, controls the digitation sequence and the velocity of the moving mirror. The main function of the white-light channel is to ensure the coherent addition of the repetitive scans. The mirror drive system is driven by a magnetic coil assembly from a loudspeaker, thus allowing the critical and precise translation of the moving mirror.

In a recent series of articles,[43-47] Horlick and associates critically evaluated ICP–FTS spectrometry for simultaneous multiple-element analysis. In their final design,[45] these investigators coupled a slew-scanning monochromator with wide entrance and exit slits to a Michelson interferometer. Simultaneous

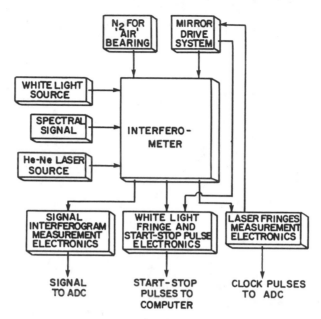

Figure 3.46. Block diagram of a Michelson interferometer system. (From Reference 39, with permission.)

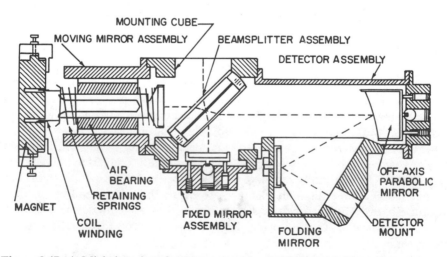

Figure 3.47. A Michelson interferometer capable of Fourier transform measurements. (From Reference 39, with permission.)

multiple-element measurements could be conducted over a 4-nm spectral window, but the monochromator provided a flexible method for selecting different windows. Because the monochromator is used, the effect of the strong emission of other elements on the standard deviation of the baseline of the spectrum was reduced such that the detection limits found with the FTS system approached those obtained from the conventional ICP-AES spectrometers.[45]

For ultra-high-resolution studies, a Fourier transform spectrometer recently developed at Los Alamos National Laboratory[48-50] operates in the wavelength region from 200 nm to 2 μm. This instrument has a resolution of 0.0025 cm^{-1} (0.000015 nm) at 250 nm, a wavenumber accuracy of about 0.0001 cm^{-1}, and an intensity accuracy of about 0.1%. Thus, interferometry has a unique place in high-resolution studies of the ICP, especially in fundamental studies of the plasma. For further discussion of ICP-Fourier transform spectrometry, the reader is referred to Chapter 7 of this book.

3.6   THE FUTURE

In the immediate future, robots will be used in routine applications of the ICP techniques, especially for sample preparation and sample introduction. With reference to the basic dispersing module, it seems that the main problems of slew scan systems (ie, the accurate positioning of the instrument at a desired wavelength and temperature drift) have been overcome. The main advantages of the scanning monochromators (ie, the selection of any wavelength or wavelength region, which allows easy background correction, the study of interference effects, and the ease of setting up a new program) will make the monochromator the first choice for research in ICP for many years to come.

For routine applications, simultaneous instruments with resolution and dispersion higher than currently used will soon be considered essential. Many instruments, specially designed for measuring the complex spectra from an ICP source, have been developed. With reference to Fourier transform spectrometry, one manufacturer, Applied Research Laboratories, has recently reported on the performance of a prototype FTS-ICP-AES instrument. Also, the combination of an array-type detector with a direct-reading spectrometer equipped with a slit mask (Figure 3.48) may prove a viable alternative to the current fixed-slit, direct-reading spectrometers. In this spectrometer, the light is first dispersed by a low-resolution, concave-grating monochromator. A mask allows selection of the desired spectral region before the separated beams are recombined to pass into a high-resolution echelle spectrometer. The high-resolution spectral information in multiple orders is then focused on the linear diode array detector. The instrument has a resolution of 0.0015, 0.0022, and 0.0030 nm at 200, 300, and 400 nm, respectively.

As far as detectors are concerned, the photomultiplier, with its unique characteristics, will remain the detector of choice for the immediate future. The

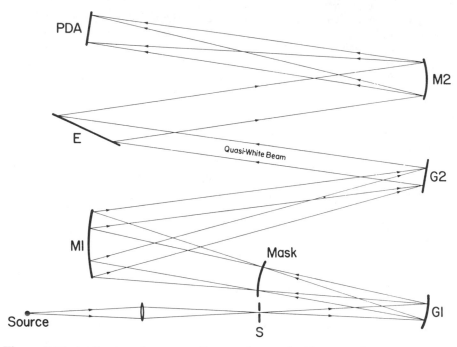

Figure 3.48. A direct-reader spectrometer using a double monochromator and a photodiode array detector. G1 and G2 are concave and plane grating, respectively having the same groove density (580 g/mm); M1 and M2 are concave mirrors; E is an echelle grating having 31.6 grooves/mm; and PDA is a photodiode array detector. (Photochemical Research Associates, with permission.)

photodiode array detectors will soon possess higher sensitivity, especially for the ultraviolet and vacuum ultraviolet regions. The spectral responses of other detectors, such as charge-coupled and charge-injection devices, will be enhanced in the ultraviolet, and their size will be increased to allow their widespread use in analytical plasma spectrometry.

ACKNOWLEDGMENTS

The author expresses his sincere gratitude to Mrs. J. Turnbull for the line drawings, to Mrs. A. Kok for the typing, and to Mrs. I. Strasheim for correcting and editing the manuscript.

BIBLIOGRAPHY

Barnes, R. M., "Emission Spectroscopy." Dowden, Hutchison, and Ross, Stroudsburg, Pa., 1976.
Elving, P. J., Ed., "Treatise on Analytical Chemistry," Part I, "Theory and Practice," Vol. 4, Section E, Principles of Instrumentation for Analysis, Chaps. 1-9. John Wiley & Sons, New York, 1984.

Grimsehl, E., "A Textbook of Physics," Vol. IV, "Optics." Blackie and Son Ltd., London, 1933.

Grove, E. L., "Analytical Emission Spectroscopy," Part 1. Marcel Dekker, New York, 1971.

Hutley, M. C., "Diffraction Grating." Academic Press, New York, 1982.

Jenkins, F. A., and H. E. White, "Fundamentals of Optics," 2nd ed. McGraw-Hill, New York, 1950.

Klein, M. V., "Optics." John Wiley & Sons, New York, 1970.

Malmstadt, H. V., C. G. Enke, and S. R. Crouch, "Electronics and Instrumentation for Scientists." Benjamin-Cummings Publishing, Reading, Mass., 1981.

Mann, C. K., T. J. Vickers, and W. M. Gulick, "Instrumental Analysis." Harper & Row, New York, 1974.

Sawyer, R. A., "Experimental Spectroscopy," 3rd ed. Dover Publications, New York, 1963.

Slavin, M., "Emission Spectrochemical Analysis." Wiley-Interscience, New York, 1971.

Strobel, H., "Chemical Instrumentation: A Systematic Approach," 2nd ed. Addison-Wesley, Reading, Mass., 1973.

REFERENCES

1. G. Kirchhoff and R. Bunsen, Chemical Analysis by Spectrum-Observation, *Phil. Mag.* 20(4), 89–98, 106–109 (1980). Reprinted in "Emission Spectroscopy," R. M. Barnes, ed., Dowden, Hutchinson, and Ross, Stroudsburg, Pa., 1976, pp. 28–41.

2. P. N. Keliher and C. C. Wohlers, Echelle Grating Spectrometers in Analytical Spectrometry, *Anal. Chem.* 48, 333–349 (1976).

3. S. Engman and P. Lindblom, A New Technique for Ruling Echelle Gratings, *Spectrochim. Acta* 38B, Suppl., 389 (1983).

4. J.L.R. Williams, M. F. Molaire, in "Kirk-Othmer Encyclopedia of Chemical Technology," Vol. 17, John Wiley & Sons, New York, 1982, p. 680.

5. N. M. Walters, A. Strasheim, and A. R. Oakes, The Influence of Dispersion and Stray Light on the Analysis of Geological Samples by Inductively Coupled Plasma Atomic Emission Spectroscopy (ICP-AES). *Spectrochim. Acta* 38B, 959–965 (1983).

6. J. W. McLaren and J. M. Mermet, Influence of the Dispersive System in Inductively Coupled Plasma Atomic Emission Spectrometry, *Spectrochim. Acta* 39B, 1307–1322 (1984).

7. C. R. Runge and F. Paschen, *Abh. K. Akad. Wiss. Berlin*, 1 (1902).

8. A. Strasheim, C. Schildhauer, and R. G. Böhmer, A Method to Adjust and Evaluate the Positioning of Exit Slits of Certain Direct-Reading Spectrometers, *Spectrochim. Acta* 39B, 1037–1044 (1984).

9. D. F. Sermin, The Importance of Making Measurements at "Line Peak" Positions When Using a Computer Controlled Monochromator with an ICP Source, *Spectrochim. Acta* 38B, Suppl., 301 (1983).

10. A. T. Zander, M. H. Miller, M. S. Hendrick, and D. Eastwood, Spectral Efficiency of the Spectraspan—III. Echelle Grating Spectrometer, *Appl. Spectrosc.* 39, 1–5 (1985).

11. R. M. Barnes and R. F. Jarrell, in Analytical Spectroscopy Series, Part I, E. L. Grove, Ed., Marcel Dekker, New York, 1971, p. 278.

12. P.W.J.M. Boumans and J.J.A.M. Vrakking, High-Resolution Spectroscopy Using an Echelle Spectrometer with Predisperser—I. Characteristics of the Instrument and Approach for Measuring Physical Line Widths in an Inductively Coupled Plasma, *Spectrochim. Acta* 39B, 1239–1260 (1984).

13. A. Strasheim, M. E. Thain, N. M. Walters, C. Claase, H.G.C. Human, and W. P. Ferreira, New Versatile Computer-Controlled Direct Reading Emission Spectrometer, *Spectrochim. Acta* 38B, 921–936 (1983).

14. A. Strasheim and M. E. Thain, Variable Optical Systems for a Universal Spectrometer, *Spectrochim. Acta* 39B, 119–123 (1984).

15. M. W. McDowell and H. K. Bouwer, Optimization of a Dual Mode Rowland Mount Spectrometer Used in the 120–950 nm Wavelength Range, *Spectrochim. Acta* 38B, 1311–1317 (1983).

16. N. W. Walters, A. Strasheim, and A. R. Oakes, The Influence of Dispersion and Stray Light on the Analysis of Geological Samples by Inductively-Coupled Plasma Atomic Emission Spectroscopy (ICP-AES), *Spectrochim. Acta* 38B, 959–965 (1983).

17. D. W. Brinkman and R. D. Sacks, Simple, Inexpensive Monochromator Modification Permitting Dual-Channel Operation, *Anal. Chem.* 47, 1723-1725 (1975).
18. M. Nordmeyer, Eine einfache Messanordnung zur photoelektrischen Spektrometrie möglichst kleiner Konzentrationen, *Spectrochim. Acta* 27B, 377-383 (1972).
19. K. Visser, F. M. Hamm, and P. B. Zeeman, Quantitative Discrimination Between Spectral Background and Ionization Interference Caused by the Sample Matrix in Flame Emission Analyses, *Appl. Spectrosc.* 30, 620-625 (1976).
20. M. A. Floyd, V. A. Fassel, R. K. Winge, J. M. Katzenburger, and A. P. D'Silva, Inductively Coupled Plasma-Atomic Emission Spectroscopy: A Computer-Controlled, Scanning Monochromator System for the Rapid Sequential Determination of the Elements, *Anal. Chem.* 52, 431-438 (1980).
21. A. De Wit and R. J. Neugebauer, Multiple Exit Slit System for Background Correction in Emission Spectroscopy, *Spectrochim. Acta* 39B, 939-940 (1984).
22. R. N. Savage and G. M. Hieftje, Development and Characterization of a Miniature Inductively Coupled Plasma Source for Atomic Emission Spectrometry, *Anal. Chem.* 51, 408-413 (1979).
23. G. F. Larson, V. A. Fassel, R. K. Winge, and R. N. Kniseley, Ultratrace Analysis by Optical Emission Spectroscopy: The Stray Light Problem, *Appl. Spectrosc.* 30, 384-391 (1976).
24. W. Kaye, Stray Radiation from Holographic Gratings, *Anal. Chem.* 55, 2018-2025 (1983).
25. W. Kaye, Stray Radiation from Ruled Gratings, *Anal. Chem.* 55, 2022-2025 (1983).
26. M. L. Franklin, G. Horlick, and H. V. Malmstadt, Basic and Practical Considerations in Utilizing Photon Counting for Quantitative Spectrochemical Methods, *Anal. Chem.* 41, 2-10 (1969).
27. J. D. Ingle, Jr., and S. R. Crouch, Critical Comparison of Photon Counting and Direct Current Measurement Techniques for Quantitative Spectrometric Methods, *Anal. Chem.* 44, 785-794 (1972).
28. E. J. Darland, G. E. Leroi, and G. E. Enke, Pulse (Photon) Counting: Determination of Optimum Measurement System Parameters, *Anal. Chem.* 51, 240-245 (1979).
29. T. M. Niemczyk, D. G. Ettinger, and S. G. Barnhart, Optimization of Parameters in Photon Counting Experiments, *Anal. Chem.* 51, 2001-2004 (1979).
30. E. J. Darland, G. E. Leroi, and G. C. Enke, Maximum Efficiency Pulse Counting in Computerized Instrumentation, *Anal. Chem.* 52, 714-723 (1980).
31. A. Danielsson and P. Lindblom, The IDES System—An Image Dissector Echelle Spectrometer System for Spectrochemical Analysis, *Appl. Spectrosc.* 30, 151-155 (1976).
32. See, for example: (a) F. Grabau and Y. Talmi, Inductively Coupled Plasma Atomic Emission Spectroscopy with Multichannel Array Detectors, (b) G. R. Sims and M. B. Denton, Multielement Emission Spectrometry Using a Charge-Injection Device Detector, and (c) M. B. Denton, H. A. Lewis, and G. R. Sims, Charge-Injection and Charge-Coupled Devices in "Practical Chemical Analysis: Operation Characteristics and Considerations," in "Multichannel Image Detectors," Vol. II, Y. Talmi, Ed. ACS Symposium Series No. 236, American Chemical Society, Washington, D.C., 1983, and references therein.
33. G. Horlick, Characteristics of Photodiode Arrays for Spectrochemical Measurements, *Appl. Spectrosc.* 30, 113-123 (1976).
34. See, for example: S. W. McGeorge and E. P. Salin, Image Sensor Applications in Analytical Atomic Spectroscopy, *Prog. Anal. Spectrosc.* 7, 387-410 (1981), and references therein.
35. N. Furuta, C. W. McLeod, H. Haraguchi, and K. Fuwa, Evaluation of a Silicon-Intensified Target Image Detector for Inductively Coupled Plasma Emission Spectrometer, *Appl. Spectrosc.* 34, 211-216 (1980).
36. J.A.C. Broekaert, F. Leis, and K. Laqua, Evaluation of a Flexible Sequential ICP-Spectrometer Using SIT as Radiation Detector, *Spectrochim. Acta* 38B, Suppl., 44 (1983).
37. K. W. Busch, Multiple Entrance Aperture Dispersion Optical Spectrometer, U.S. Patent 4,494,872 (1985).
38. J. W. Olesik and G. M. Hieftje, Optical Imaging Spectrometers, *Anal. Chem.* 57, 2049-2055 (1985), and references therein.
39. G. Horlick and K. W. Yuen, A Modular Michelson Interferometer for Fourier Transform Spectrochemical Measurements from the Mid-Infrared to the Ultraviolet, *Appl. Spectrosc.* 32, 38-46 (1978).
40. G. Horlick, R. H. Hall, and W. K. Yuen, in "Atomic Emission Spectrochemical Measurements with Fourier Transform Infrared Spectroscopy," Vol. III, J. R. Ferraro, L. J. Basile, and A. Mantz, Eds. Academic Press, New York, 1982, pp. 37-38.

41. G. Horlick, Detection and Measurement of Radiation: Rewards and Penalties of New Approaches, *Spectrochim. Acta* 38B, Suppl. 391 (1983).
42. E. A. Stubley and G. Horlick, Some Near-IR Spectral Emission Characteristics of the Inductively Coupled Plasma, *Appl. Spectrosc.* 38, 162–168 (1984).
43. E. A. Stubley and G. Horlick, A Fourier Transform Spectrometer for UV and Visible Measurements of Atomic Emission Sources, *Appl. Spectrosc.* 39, 800–804 (1985).
44. E. A. Stubley and G. Horlick, Measurement of Inductively Coupled Plasma Emission Spectra Using a Fourier Transform Spectrometer, *Appl. Spectrosc.* 39, 805–810 (1985).
45. E. A. Stubley and G. Horlick, A Windowed Slew-Scanning Fourier Transform Spectrometer for Inductively Coupled Plasma Emission Spectrometry, *Appl. Spectrosc.* 39, 811–817 (1985).
46. R.C.L. Ng and Gary Horlick, A Real-Time Correlation-Based Data Processing System for Interferometric Signals, *Appl. Spectrosc.* 39, 841–847 (1985).
47. R.C.L. Ng and Gary Horlick, A Real-Time Correlation-Based Data Processing System for Interferometric Signals, *Appl. Spectrosc.* 39, 841-847 (1985).
48. L. M. Faires, B. A. Palmer, and R. E. Engleman, Jr. (a) Temperature Determination in the Inductively Coupled Plasma Using a Fourier Transform Spectrometer, *Spectrochim. Acta* 39B, 819-828 (1984); (b) ICP Argon Emission in the Near Infrared by High-Resolution Fourier Transform Spectrometry, *Spectrochim. Acta* 40B, 545-551 (1985).
49. L. M. Faires, B. A. Palmer, and J. W. Brault, Line Width and Line Shape Analysis in the Inductively Coupled Plasma by High Resolution Fourier Transform Spectrometry, *Spectrochim. Acta* 40B, 135-143 (1985).
50. L.M.H. Faires, Fourier Transform and Polychromator Studies of the Inductively Coupled Plasma, Los Alamos National Laboratory, LA-9888-T, Los Alamos, October 1984.

4

Common Radio Frequency Generators, Torches, and Sample Introduction Systems

STANLEY GREENFIELD

Loughborough University of Technology
Department of Chemistry
Loughborough, Leicestershire, England

4.1 INTRODUCTION

One of the principal interferences in analytical atomic spectrometry arises from the formation of refractory compounds. Thus, when calcium and aluminum are introduced together into an air–acetylene flame, the calcium emission is depressed by the formation of a very stable compound. Appreciation of this type of phenomenon led to a search for sources with higher gas temperatures. A hotter source should dissociate refractory compounds more effectively, and an added bonus would probably be higher excitation, ionization, and electron temperatures. The annular inductively coupled plasma was found to be such a source.[1,2]

A possible definition of a plasma is a gas ionized sufficiently for its properties to depend significantly on the ionization. Macroscopically, the gas remains electrically neutral, and it becomes a good conductor of electricity.

When a gas is seeded with a few electrons and is caused to flow through a tube held within coils through which a high-frequency current passes, the rapid changes in the magnetic field induce eddy currents in the gas. Resistance to the eddy current flow produces Joule heating. Once ionizing temperatures have been reached, the process is self-sustaining, and an inductively coupled plasma (ICP) is formed almost instantly.

The first inductively coupled plasma was probably produced by Hittorf,[3] who discharged a Leyden jar through a coil surrounding a glass tube filled with air at less than atmospheric pressure. Discrete bursts of damped oscillations with frequencies in the Megahertz range were used to generate a plasma that filled the discharge tube. Shortly thereafter, J. J. Thomson[4] grounded a similar coil and formed a ring discharge. A grounded-coil circuit was used to excite various gases in an examination of gas spectra with a direct-vision spectroscope.[4] All these discharges were generated in low-pressure gases.

There are important differences between the gas and the excitation temperatures of low-pressure plasmas and of atmospheric-pressure plasmas. In a low-pressure plasma, a free electron has a long mean free path. Thus, if it is subjected to an electric or electromagnetic field, it is accelerated for a comparatively long time and can acquire a high energy. On collision with an atom, the kinetic energy of an accelerated electron can be used to excite an atom, which can emit its characteristic radiation. Thus, a low pressure plasma can be used to produce atomic radiation. Because of the low frequency of collisions, however, most of the atoms will have low kinetic energy, so that the plasma gas temperature is low.

In a high-pressure plasma, the frequency of collisions is high, and small quantities of kinetic energy are transferred frequently. The high frequency of energy transfer leads to a more uniform distribution of energy among the particles and provides a higher gas temperature than that in a low-pressure plasma. As stated earlier, this high gas temperature is attractive for spectrochemical analysis, and therefore, high-pressure plasmas are to be preferred for this purpose even though they require much more power.

The first serious experiments with high-pressure plasmas were those of Babat,[5] who operated high-powered (tens of kilowatts) thermal induction plasmas in closed tubes for possible industrial applications. Inductively coupled plasmas, generated in flowing gas streams in open tubes, were first described by Reed[6,7] for crystal growing. Greenfield and associates[1] used a modified Reed torch to generate an annular, or tunnel, ICP for spectrochemical analysis. An aerosol stream from aqueous solutions, slurries, or powders was passed through the tunnel of plasma, thus producing a long tailflame, which was used as a spectroscopic source.

Shortly thereafter, Fassel and co-workers[8] generated a spheroidal ICP in a laminar flow torch. In contrast to the annular plasma, sample aerosol flowed around the plasma in the spheroidal ICP.[8,9] Subsequently, Fassel and associates reported on the properties of an annular plasma,[2] which is the preferred system today. Many spectroscopists have taken up the ideas of the early workers in this field, as summarized in review articles.[10–13]

4.2   RADIO FREQUENCY GENERATORS

The radio frequency (RF) generators used to supply power to ICPs are oscillators that basically generate an alternating current at a desired frequency. The basic circuit for such an oscillator is relatively simple, consisting of a capacitor and an inductor (coil) in parallel. This circuit is called a "tuned circuit" or "tank circuit." When the capacitor is discharged through the inductor, the subsequent collapse of the magnetic field causes charge buildup on the capacitor. This charge is opposite in polarity to the original charge. Were it not for the resistance of the circuit, the oscillatory process would continue indefinitely. In fact, the oscillation will gradually decrease unless enough electrical energy is transferred into the tuned circuit. The additional energy has to be supplied at just the right moment. A convenient way of sustaining oscillation is to connect the tank to the grid of a vacuum tube (thermionic valve), thus amplifying the oscillating voltage. When a small portion of the amplified voltage is fed back in the correct phase, enough electrical energy will be fed into the tank circuit to overcome the resistance losses. The major differences between the various oscillators lies in the feedback process.

The two general types of oscillator most used in plasma spectrometry are referred to as "free running" and as "crystal controlled". In the first category, the basic frequency of oscillation is fixed by the values of the components in the tank circuit; these, in turn, are modified by any changes in the plasma impedance and in the coupling of the plasma to the load (induction) coil. In the second category, a piezoelectric crystal is used to control the feedback and to maintain constant frequency of operation. Oscillators of both categories are used extensively and produce excellent results.

The majority of instrument manufacturers use generators with nominal powers of 2 kW and a frequency of 27.12 MHz. A representative cross section of generators that have been, or are, used in plasma spectrometry is given in Table 4.1. There is some evidence[14] that higher frequency of oscillation leads to lower excitation and ionization temperatures, lower electron number densities, and lower continuum background, together with improved background stability. Improved stability, along with lower spectral background, leads to superior detection limits that are clearly noticeable as the frequency is increased from 5 to 40 MHz. The high-frequency field must never be allowed to radiate beyond the confines of the generator case, torch enclosure, and the associated electronics. In other words, *the equipment must be adequately shielded, not only to comply with national and international regulations on RF exposures, but to avoid interference with other electronic equipment.*

Table 4.1. RF Plasma Generators for Use in ICP Spectrometry[a]

Type and manufacturer	Model number, generators or spectrometer	Type of oscillator[b]	Nominal power output (kW)	Frequency (MHz)	Ref.
Free-running generators					
Lepel	T-5-3-MC-J-S	TATG	5	3.4	8
Lepel	T-2-5-1-MC2-J-B	TATG	2.5	23–48	2,9
Radyne	RD150	Colpitts	15	7	25,57
Radyne	H3OP	Colpitts	5	36	50
Radyne	SC15	Colpitts	2.5	36	1,26
Radyne	R50/P	Colpitts (MBT)	5.0	27.1	43
Radyne	R15/P	Colpitts (MBT)	1.5	27.1	—
Philips	PV8490	Colpitts	0.7–2.0	50	—
Linn	FS-4	Huth–Kuhn	2.5	27.1	18
Linn	FS-4	Huth–Kuhn	4.0	27.1	18
Linn	FS-4	Huth–Kuhn	7.0	27.1	18
Linn	FS-4	Huth–Kuhn	10.0	27.1	18
STEL	—	TATG	6.6	5.4	—
Leeman Laboratories	103001	Huth–Kuhn	2.5	27.1	—
ESI-RC Durr	3364	Tuned line	3.3	64	19
ESI-RC Durr	4364	Tuned line	4.3	64	19
Crystal-controlled generators					
Perkin–Elmer	ICP-6500	Pierce	2	27.1	—
Plasma Therm	HFP-2500D	Pierce	2.5[c]	27.1	30
International Plasma Corp.	120–127	—	2	27.1	35
Applied Research Laboratories (ARL)	—	—	1.5	27.1	—
ARL–Henry Radio	2500 PGC-27	—	2.5	27.1	—
Labtest Aus.	Plasmaspan	—	2	27.1	—
Solid-state generators					
RF Plasma Products	ICP1500S	—	1.5	40.7	—

[a] For information on the spectrometers incorporating these generators, see the 1984 issue of Annual Review of Atomic Spectroscopy.
[b] TATG: tuned anode tuned grid; MBT: magnetic beamed triode.
[c] Plasma Therm manufactures generators with power rating of up to 5 kW.

4.2.1.   Free-Running Generators

An electronic oscillator consists of a tank circuit and an amplifier with a feedback circuit. All oscillators are very much alike. The simplest form of these devices is the Armstrong oscillator, and its basic principles underlie those of most oscillators.

In the Armstrong oscillator, a feedback coil, in the anode circuit of the amplifier tube, is wound adjacent to the coil in the tank circuit, so that when an anode current flows through this feedback coil, the rapidly expanding magnetic field induces a voltage in the tank circuit coil. The voltage surge is enough to start oscillation. The tube is biased well below cutoff, ie, nonconducting, for most of the cycle of operation (class C operation), the high negative bias being maintained by a resistor and a capacitor in the grid circuit. Only when the capacitor in the tank circuit is at its maximum positive charge will it counterbalance the charge on the grid capacitor and allow an anode current to flow. This pulse operation maintains the oscillatory process.

The Hartley oscillator is more frequency stable than the Armstrong oscillator. Here, the feedback coil is made part of the coil, which is in the tank circuit. This single coil is tapped so that the cathode current flows through the lower part of the coil, inducing a "kick" voltage in the grid portion. The feedback voltage can be altered by moving the tap location on the single coil.

The Colpitts oscillator, used extensively in ICP spectrometry, has greater frequency stability than either the Armstrong or the Hartley device. This oscillator is very similar to the Hartley type, but the feedback coupling is between a pair of capacitors that are parallel with the inductor in the tank circuit. To ensure that the feedback is in correct phase, the anode and the grid of the amplifier tube are connected across the inductor.

Another version of oscillator used in ICP spectrometry is the tuned-anode, tuned-grid (TATG) oscillator. As the name implies, both the input and output circuits are tuned, and feedback is accomplished from anode-to-grid circuits via the interelectrode capacitance of the tube. Block diagrams of two typical commercially available free-running generators are shown in Figure 4.1. (A) with a conventional triode tube and (B)[15] with a magnetic beamed triode (MBT) in a TATG circuit. In a MBT, electrons emitted from the cathode of a vacuum tube are focused by a magnetic field into a rectangular beam that is collected by the anode. This beam is controlled by a gate that functions similarly to the grid of a conventional triode. The combined effect of the magnetic and electrostatic fields, which are also present, causes the emitted electrons to spiral toward the anode. According to Whittle and Behenna,[15] the gates of these tubes operate at much lower temperatures than the grids of conventional vacuum tubes. This low gate dissipation enables the use of less expensive materials for construction. Other properties of the MBT are low grid currents, high grid resistance, low cathode heating power, low drive power, and slightly higher anode and overall efficiency than conventional triodes.

4.2.1.1 Contactor

All generators have a system of relays and contacts for switching on and switching off the RF power.

4.2.1.2 Regulator

The regulator provides for control and stabilization of output power. In earlier generators, variable transformers or saturable reactors were used to control power. Thyristors are used to control power in some of the Radyne generators

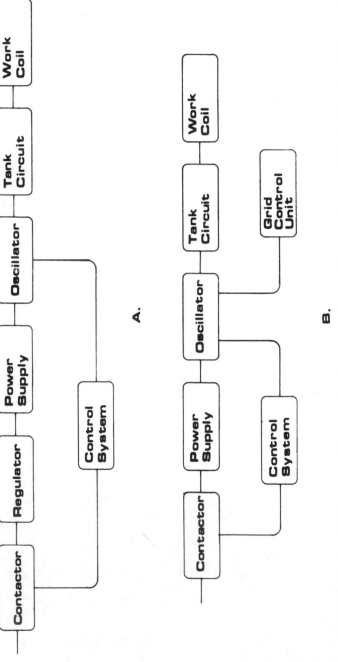

Figure 4.1. Free-Running RF Generator: (*A*) conventional vacuum tube and (*B*) magnetic beamed triode. (Courtesy of Radyne Ltd.).

that have conventional vacuum tube arrangements. In these cases, the thyristors, which are semiconductor devices, are used as switches in the ac line to control the mains voltage supply to the transformer–rectifier system. The thyristors, which have very fast response, are either on or off according to the level of a control signal supplied to the gate of the device. Once they have been "fired," they can be turned off only by the natural decay of current to zero or by application of a reverse voltage. The control signal is supplied by a feedback amplifier, and the method of control is called "phase control." For phase control, the firing point of every half-cycle is maintained in relation to the zero point to allow a set fraction of the half-cycle to flow. Some Radyne generators use MBT tubes that have their own system of power control and stabilization, as explained below. The author is indebted to Radyne Ltd. for the above description of power regulation by thyristors.

The Philips generator employs a serial stabilizer to control the anode voltage of a Colpitts oscillator. The serial stabilizer is placed between the rectifier, with its smoothing filter, and the oscillator. The voltage drop across a Zener diode in the serial stabilizer is fed to a comparator. A signal from the comparator then controls a servo-driven variable transformer that feeds the rectifier. This system achieves a voltage stability of better than $\pm 0.05\%$. (Courtesy of Philips Scientific and Analytical Equipment.)

4.2.1.3 Power Supply

A three-phase system of silicon semiconductor rectifiers invariably is used to convert the ac voltage from the secondary winding of the input transformer to dc. To reduce objectionable audio noise, a mains filter unit is also needed (except if an MBT is used) to reduce the ripple.

4.2.1.4 Oscillator

Vacuum triodes normally are used, with air- or water-cooled anodes, in Colpitts or TATG circuits. A simplified equivalent circuit of the Colpitts oscillator used by Philips[16] is shown in Figure 4.2.

The power triode is represented by a voltage source E and an associated internal resistance r_p. This tube delivers anode current pulses i_p. Capacitors of fixed values C_1 and C_2 make up the oscillating circuit. The voltage feedback V_g to the grid is taken from capacitor C_2. The inductance is composed of two parts: a fixed inductance L, which forms part of the generator itself, and the remainder L_t, which together with a series resistance R_t, represents the association of the load coil with the plasma.

A form of power stabilization associated with the free-running generator is typified by the Philips version of the Colpitts oscillator. In this version, no matching circuit is required because the load coil and plasma form an integral part of the tank circuit. The generator automatically corrects for small changes in power to the plasma, assuming the ratio L_t/L is less than 2, by an opposing compensatory change in the coupling factor k.[17]

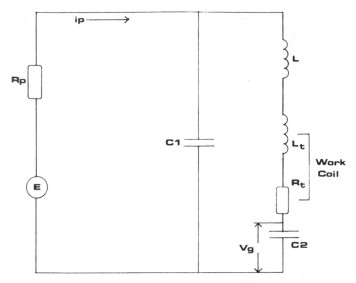

Figure 4.2. Simplified equivalent circuit for a Colpitts oscillator. (Redrawn from Reference 16, courtesy of Philips Scientific and Analytical Equipment.)

A detailed derivation of the equations governing this method of power stabilization is given in a Philips Technical Review,[17]* but the essential features are given below by courtesy of Philips Scientific and Analytical Equipment.

The load coil and plasma (Figure 4.2) are represented by the equivalent circuit of a transformer with the coil as a primary and the plasma as a single-turn secondary, which can be represented by a conducting cylinder of diameter d. The value of d will vary directly with the power dissipated in the plasma. The coupling between the primary and secondary increases with plasma diameter, and thus, k increases with d. From general transformer equations, one can also show that as k increases, the inductance L_t will decrease and the resistance R_t will rise. If k decreases, then k_t will increase, and R_t will fall.

Another standard parameter of the load coil, governing plasma power, is the quality factor Q:

$$Q = \frac{w_c L_t}{R_t} \qquad [4.1]$$

where w_c is the fundamental frequency of the oscillator. The factor Q will vary inversely with the plasma power. A plot of Q versus L_t/L at various values of plasma power will give a family of characteristic curves with maxima for Q at $L_t/L = 2$.

If we assume that a change in the plasma occurs, causing a decrease in the coupling factor k, L_t will increase and the plasma power will tend to rise to oppose the decrease in k. These changes will be followed by a rapid rise in Q so that the generator operating point returns to the same characteristic curve at the original power setting, but at a slightly higher value of Q, thus achieving automatic stabilization. A similar stabilizing action occurs for increases in the

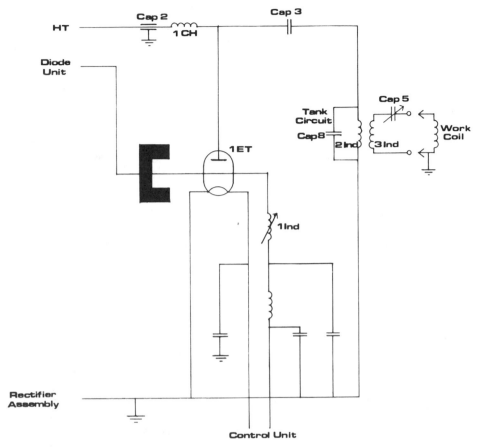

Figure 4.3. Tuned-anode, tuned-gate oscillator with a magnetic beamed triode. (Courtesy of Radyne Ltd.)

coupling factor. Plasma operation under conditions where L_t/L is greater than 2 is unstable, and no power stabilizing action occurs.

The oscillator circuit of the Radyne R50P generator is shown in Figure 4.3. and the following description of its operation is by courtesy of Radyne Ltd. High voltage is applied to the MBT, 1ET, via a feed-through capacitor Cap 2 and anode choke 1CH. These components ensure that RF energy cannot be fed back to the power supply. The MBT provides power to the tank capacitor Cap 8, and tank inductance 2Ind. The blocking capacitor Cap 3 isolates the tank circuit from any dc voltage. A coupling coil 3Ind and tuning capacitor Cap 5 provide the required current in the load coil for plasma applications.

The oscillator circuit is a tuned-anode, tuned-gate arrangement with a common filament. Feedback is provided by 1Ind and is preset by adjusting the spacing between turns. The dc bias voltage is obtained by resistor combinations on the line to the rectifier assembly and the control unit.

4.2.1.5 Tank Circuit

The frequency of oscillation is determined by the tank circuit, which is a capacitance–inductance network. In the Huth–Kuhn circuit, used in the Linn generator,[18] the tank circuit takes the form of a cavity in a TATG circuit. A cavity may be regarded as a capacitor with widely separated plates joined by a very large number of straight wire inductors in parallel.

4.2.1.6 Load Coil

The load coil, or work coil, its size, and the number of coil turns necessary to maintain a stable plasma within a torch are discussed in Section 4.2.2.9.

4.2.1.7 Control System

All generators have various relay protection circuits and instrumentation necessary to provide safe monitored operation.

4.2.1.8 Grid Control Unit

An MBT free-running generator (Figure 4.1B) does not have a regulator section because power regulation and stabilization are performed in the grid control unit. Regulation is achieved by varying the voltage on the grid of the MBT via a clipping diode. A control potentiometer is set to a particular voltage level that is amplified and appears on the diode. When the RF voltage on the grid of the MBT rises above the amplifier voltage, the diode conducts, thus limiting any further rise. A change in tank circuit voltage causes a change in anode current that is monitored by a sensing resistor and then applied to the input of a comparator. The comparator gives an output proportional to the error between the required voltage (from the potentiometer) and the feedback voltage. The foregoing description is by courtesy of Radyne Ltd.

In another type of free-running generator, described by Trassy and Mermet[19] (Figure 4.4), the tank circuit does not set the frequency, but, to a first approximation, the diameter, length, and separation of its transmission lines determine the oscillation frequency. Interelectrode capacitance of the valve should also affect the oscillator frequency. The resistor–capacitor combination, connected to the grid, provides automatic bias control to maintain a steady output amplitude. The coupling line (5) connected to a tuning capacitor (2) and inductor (1) picks up power from the plate lines (3). The frequency variation[18] is less than 10^{-3}. The dynamic adjustment of impedance makes it unnecessary to manually adjust the generator when starting the plasma. Because the system is entirely electronic, it provides extremely fast adjustment of the impedance.

4.2.2 Crystal-Controlled Generators

Crystals, such as quartz and Rochelle salts, compressed between two metal plates generate a voltage difference across the plates. Conversely, if an ac voltage

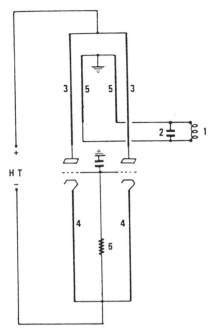

Figure 4.4. Tuned-line oscillator: 1, inductor; 2, variable capacitor; 3, anode line; 4, cathode line; 5, coupling lines, 6, polarizing resistor. (From Reference 19, with permission.)

is applied to the two plates, the crystal will expand and contract with the changing polarity. These interactions are known as the piezoelectric effect. Maximal expansion and contraction are attained if the frequency of the applied voltage is the same as the natural mechanical frequency of the crystal. Furthermore, this natural frequency is remarkably constant, which is why these crystals are used in RF generators, where they act as parallel-resonant circuits when placed between two plates. The frequency at which a crystal vibrates is inversely related to the thickness of the crystal. Thus, for all other factors equal, a thick crystal vibrates at a lower frequency than a thin crystal.

4.2.2.1 Block Diagram

A typical layout for a crystal-controlled RF generator is shown in Figure 4.5.

4.2.2.2 Oscillator Stage

The oscillator stage is similar to the Colpitts and TATG circuits previously discussed, except that the frequency-determining part of the system is replaced with a piezoelectric crystal, thus imparting great stability to the oscillation frequency. Solid-state components or a vacuum tube are used in the oscillator circuit of a crystal-controlled RF generator.

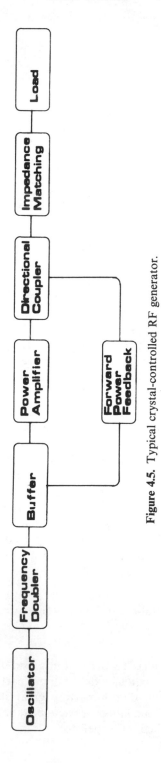

Figure 4.5. Typical crystal-controlled RF generator.

4.2.2.3 Frequency Doubler

The crystals that are used in the oscillator generally have a natural frequency of approximately 10 MHz. Crystals that operate at very high frequencies are impractical to manufacture. Therefore, the desired frequencies are obtained by frequency multiplication. The frequency doubler circuit is somewhat similar to an amplifier that normally operates at a fixed input frequency. The difference is an anode-tuned circuit that is tuned to operate at twice the frequency of the input. The frequency doubler tube generates current pulses at the same frequency as the input signal, energizing the anode-tuned circuit, and thus causing it to oscillate at twice the input signal at the grid. The current pulses arrive at the same time during alternate cycles of the doubled frequency, thus ensuring continuation of oscillation.

4.2.2.4 Buffer Stage and Drive Control

High power output from a crystal-controlled oscillator is achieved through the use of a number of amplification stages. The first stage, known as the buffer stage, is used to amplify the RF signal from the oscillator to drive the power amplifier and to isolate the oscillator for improved stability. As shown in Figure 4.5, automatic power control can be achieved by driving the buffer stage with low-level signals.

4.2.2.5 Power Amplifier Section

The tank circuit of the power amplifier is usually tuned to the frequency of the grid signal. A high-voltage RF signal, developed across the tank circuit, appears on the load coil that couples the RF field to the plasma.

4.2.2.6 Directional Coupler

A coupling circuit transfers energy from the power amplifier section through the coaxial transmission linematching network to the load or work coil (Figure 4.5) and provides proper signal for the feedback circuit. The directional coupler circuit generally consists of a pickup coil and a variable capacitor in series. Sensors are added to measure the forward and reflected powers, the latter occurring if the line does not terminate in the characteristic impedance. In such a case, not all the RF energy will be absorbed at the load and that part of the energy will be reflected back to the power amplifier.

4.2.2.7 Forward Power Feedback Circuit

The feedback circuit senses changes in the forward power by probes that sample a small portion of the electromagnetic field of the coaxial transmission line. Compensation for changes in forward power is achieved by an arrangement of diodes in a feedback network. The feedback signal is transferred to the grid of the tube in the buffer section, thus altering the drive control of this first amplification stage.

4.2.2.8 Impedance Matching Network

The impedance of the gas at the point of plasma initiation is very different from that which exists after formation of the plasma discharge. For the maximum power transfer efficiency, the impedance of the load must be matched with the output impedance of the power amplifier, thus requiring use of a matching network.[9] Such a network is not needed for free-running generators, where a change of impedance in the plasma results in a slight change in frequency of the tank circuit with no loss of power in the plasma. One form of matching network is a variable capacitor placed in series with the load coil. Other forms of matching network have been suggested that take account of phase changes in an attempt to compensate for the sudden rise in the impedance on ignition.[20] Automatic impedance matching networks facilitate plasma formation and stabilization, especially when molecular gases are used in the ICP. Of some interest is the work on this topic of Barnes and his associates.[21,22]

4.2.2.9 Induction Coil Configuration

The essential factors governing the design of the load coil for generating a plasma are very similar to those important to induction heating of a metal object. The heating occurs in the vicinity of the turns of the coil encircling the object to be heated. Thus, a plasma is mainly confined within the load coil. Roughly, the longer coil produces a longer plasma, and, if the plasma torch is as wide as possible to fit within the coil, a larger diameter coil gives a larger diameter plasma. The coupling depends on the distance of the plasma from the coil; thus a small torch in a large coil will lead to inefficient coupling. It should be noted that RF currents tend to flow only on the outer surface of a conductor. This phenomenon, known as the skin effect, is discussed separately in Section 4.3.

   In coils with more than one turn, the adjacent coils must not touch (to prevent discharges between coil windings), but the coils should be wound as close as possible, to provide a uniform field. Current flowing in one coil turn should flow in the same direction in an adjacent coil turn so that the magnetic fields from each will not be in opposition. Such opposition of magnetic fields causes phase cancellations and a greatly reduced heating effect. Coils are wound in copper tubing having either a round or square cross section. Square-section tubing enables closer tube spacing, a more compact coil, and higher coupling. A flow of cooling water, at constant temperature, through the copper tubing limits the heating and stabilizes the surface resistance. Silver-plated copper coils are often used to reduce the surface resistance and to prevent the corrosion that usually occurs on the surfaces of plain copper coils. For AES, the lower turn of the coil usually is grounded. The upper turn of the coil adjacent to the plasma is often protected by a quartz bonnet to prevent plasma flashover damage to the coil.

4.2.2.10   Radio Frequency Shielding and Grounding

Shielding should prevent RF radiation outside the generator, matching network, transmission line, and the plasma enclosure. This means that the oscillator compartment and that of the load coil and torch should be constructed to approach the containment of a Faraday cage. Doors of the enclosures should be capable of being completely closed, and the windows, except for the aperture facing the spectrometer, should be covered with a fine mesh screen. All such screening should be grounded at one point, as should all grounds from ancillary equipment.

To avoid undesirable noise on the signal, RF filters should be on all input leads for the generator and on any other electronic equipment associated with the spectrometer.

4.2.3   Solid-State Generators

Solid-state generators based on solid state-static inverter circuits have been available for some years. An inverter is the reverse of a rectifier or converter: it takes direct current and converts it to an alternating current of the desired frequency. The following description of its operation is by courtesy of Radyne Ltd.

4.2.3.1   Block Diagram

A general schematic of a typical solid-state inverter is shown in Figure 4.6. In the ac–dc–ac inverter, which is generally favored by Radyne for their solid-state generator, alternating current of one frequency can be converted to ac of another frequency.

4.2.3.2   Converter

A typical three-phase bridge rectifier in which the diodes are replaced by thyristors comprises the converter. The firing order of these thyristors is arranged to give six pulse rectification.

4.2.3.3   Medium-Frequency Filter

The dc power is fed to a water-cooled dc choke that acts as a constant current source for the inverter and as a filter for mains-frequency currents from the inverter and for medium-frequency currents from the mains.

4.2.3.4   Inverter

The parallel-tuned bridge inverter uses four thyristors that are arranged in diagonally opposite pairs, which, when fired together, produce a square wave current. When this square wave is fed into a parallel-tuned circuit, at a

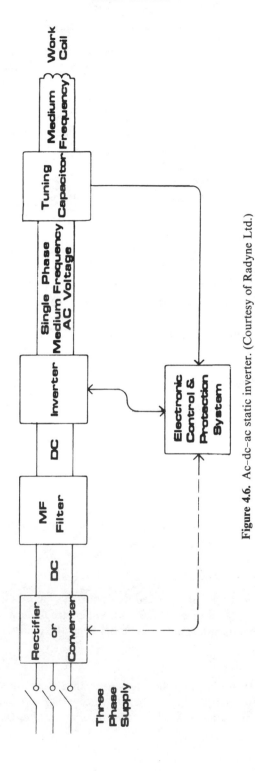

Figure 4.6. Ac–dc–ac static inverter. (Courtesy of Radyne Ltd.)

reasonable Q load, the output is modified into a voltage sine wave across the parallel-tuned circuit.

4.2.3.5 Tuning Capacitor

Together with the work coil, the tuning or power-factor-correction capacitors form the resonant circuit load.

4.2.3.6 Electronic Controls

The control system monitors the dc current and the medium frequency voltage, and continuously adjusts the operation of the rectifier and inverter to maintain constant output. This type of solid-state generator is essentially a free-running generator and will maintain a constant power output by means of a frequency swing against load inductance changes.

Solid-state ICP systems were introduced into the marketplace by RF Plasma Products in 1984. The 1001S series solid-state generator is a solid-state analog of the crystal-controlled oscillator shown in Figure 4.5. It is a single phase system with provisions for computer control. The generator can be run either off its own oscillator or off an external oscillator and can also feed another generator with an oscillating signal, if required. Such a solid-state generator is applicable to low-gas-flow torches. Recently, Applied Research Laboratories introduced a commercial low-gas-flow, low-power ICP–AES.

4.2.4   Power Measurements

A knowledge of the actual power in the plasma, although desirable for the comparison of experimental data from research workers, is not essential for normal plasma spectrometry. Most RF generators have some form of power meter. Crystal-controlled generators have reflected power meters that give the power into the impedance matching circuit but not necessarily the power into the plasma. Free-running generators should have anode voltage, anode current, and grid current meters from which it is possible to determine the power in the plasma from previous calibration. This calibration may be derived by the use of a water-cooled load cylinder placed in the load coil.[23] The method in use[24,25] is to operate plasmas at various recorded values of anode voltage, anode current, and grid current. The torch is then replaced by a glass tube containing iron and copper rods through which water is circulating. The ratio of iron to copper is adjusted, as is the distance the load is inserted into the work coil, until similar values of the previously recorded meter readings are obtained. At this point, the inlet and outlet temperatures of the circulating water are measured and the power in kilowatts computed from the following equation:

$$W = 4.18 \times F \times \Delta T \qquad [4.2]$$

where W is the power in kilowatts, F is the flow rate in milliliters per second, and ΔT is the temperature rise in degrees Celsius. This approach has been confirmed by a direct calorimetric assessment of the power in the plasma[26] in which the heat from a plasma was absorbed by water flowing through a silica condenser confining the discharge. The flow rate and temperature of the gas issuing from the condenser were measured, as was the temperature rise of the cooling water. The temperature of the torch body was measured to calculate the heat lost by radiation. An estimate of the radiation from the fireball was also made. The foregoing information provided the approximate power consumed by the plasma, which was similar to the value found by the dummy-load procedure.[24] With free-running generators of the Colpitts type, approximately 40 to 45% of the input power reaches the plasma. Bogdain[27] estimated that a Huth–Kuhn oscillator transfers 43% of the input power into the plasma. The power efficiency of the Philips Colpitts oscillator was determined to be 50% of the plate power.[28] Other power measurements using the Huth–Kuhn oscillator have been described.[29]

4.2.5   Power Stability

The intensities of spectral lines from elements introduced into an ICP and the spectral background from the plasma are strongly dependent on the power in the plasma. These generalized observations have implications both for accuracy of determination and for achievable detection limits. Short-term fluctuations in power are therefore clearly undesirable.

Short-term fluctuations attributed[30] to variations in line voltage, for both free-running and crystal-controlled oscillators, require that some form of feedback control be incorporated in generators for use in spectrometry, giving regulation of forward power to better than 1%. Stabilization of the dc voltage on free running oscillators is variously quoted[31] as being ± 0.05 to 0.5% for a 10% mains variation. However, a high degree of regulation is of little avail if the noise introduced by, for instance, the nebulizer is orders of magnitude greater. Anderson and co-workers[32] implied that changes of a few watts in power could produce changes in the final result of tens of parts per billion in concentration.

4.3   TORCHES

A single silica tube, placed inside the load coil of an RF generator, is sufficient to generate an ICP when the laminar flow of argon within the tube is seeded with a few electrons from a Tesla coil. The resulting plasma takes the form of a prolate spheroid and is self-sustaining, running as long as a high-frequency current is supplied to the load coil and gas is supplied to the tube. As the RF current is increased, a set of conditions is reached where cooling of the silica tube is necessary to avoid its melting. These conditions depend on the diameter of the tube, the gas flow, and the power in the plasma. Cooling may be achieved either by water or by a tangential or laminar flow of gas. With water cooling, the plasma tube becomes the inner wall of a condenser. With gas cooling, another tube is placed coaxially around the plasma tube, extending beyond its end, and a gas flows through the annulus formed, either tangentially or in a laminar fashion.

Reed[33] thought that it was necessary to create a region of low pressure along the axis of the tubes by causing the gas to enter tangentially and thus create a vortex that pulls the plasma in the opposite direction of the gas flow. Otherwise, he argued, the plasma would be blown from the region of the magnetic field and extinguished. However, the first plasma torch described by Wendt and Fassel[8] had a laminar flow of gas, thus establishing that the plasma is not readily extinguished at modest flow rates. Sample solutions are commonly converted to aerosols by a nebulizer, and the aerosol is introduced into the plasma via a narrow-bore tube placed along the central axis of the torch.

In the early days of plasma spectrometry, it was usual to refer to the outer gas stream as the coolant gas, the intermediate stream as the plasma gas, and the inner stream carrying the aerosol as the injector or nebulizer gas. Clearly, such a sharp demarcation between the functions of the two outer gases is not justified because, from what has already been written, either gas flow will independently sustain a plasma. Indeed, today, the Fassel torch, although retaining three concentric tubes, is frequently operated with only an argon gas entry (sometimes called the plasma stream) to the outer tube and an injector gas stream (sometimes referred to as the carrier). The formation of a toroidal, or tunnel, plasma rather than the spheroidal plasma, which is initially formed, is of utmost importance to proper generation of the ICP source for analytical spectrometry. When an aerosol flows around a spheroidal plasma, the tailflame produced is short, wide, and diffuse. In contrast, the tailflame from an aerosol flowing through an annular or tunnel plasma is long, narrow, and sharply bounded. For an annular ICP, the emission per unit volume is therefore high, and excellent detectability and sensitivity are obtained when the tailflame is used as a spectroscopic source. Because the tailflame, unlike the fireball within the core, emits relatively little continuum radiation, the spectral background is low. Little mixing occurs between the central gas stream carrying the sample aerosol and the surrounding gas flows, so that the atoms and ions from the sample are confined to the narrow tailflame. The cooler surroundings of the tailflame contain relatively few sample atoms, and, consequently, the tailflame behaves like an optically thin source that exhibits little self-absorption. A further consequence of this confinement is that because most of the power in the plasma is dissipated in the outer layers whose electrical conductivity is unaltered by the presence of sample atoms and ions in the tunnel, the power dissipated in the plasma is substantially independent of the nature and concentration of the sample. This feature provides for good stability. These properties, among others, make an annular plasma an excellent spectroscopic source and the preferred source in use today.

In general, extremely high temperature gradients are observed in plasmas. Relatedly, plasmas are dense systems in which gases are expanding and accelerating outward in a direction perpendicular to the gas flow. Consequently, injection of an aerosol into a plasma is very difficult. This task is easier if the spheroidal plasma is converted into an annular ICP. In principle, the frequency of the current should influence the ease with which an annular plasma is formed.[2,34,35] In high-frequency induction heating of a conductor, the magnetic field falls to a value $1/e$ of that at the surface over a distance known as the skin depth. Most of the power absorbed by the conductor is dissipated in this skin-depth layer. The skin depth is inversely proportional to the square root of the frequency. At higher frequencies, most of the power is dissipated nearer the surface, thus facilitating the formation of a central, cooler tunnel. At the frequencies and powers normally used for emission spectrometry,[10,24] the torch design and the gas flow rates have greater importance in the formation of the axial channel than the frequency.

The difficulty of injecting an aerosol into a plasma has been explained in terms of magnetohydrodynamic thrust.[36,37] Measurements of the radial distribution of the magnetic field in the plasma, by means of magnetic probes, have confirmed the reduction in field in the axial channel as expected from the skin-depth effect. Similarly, the magnetic pressure, which is the force exerted on electrically charged particles by the magnetic field, exhibits a drop toward the center, leading to an inward radial flow of plasma by "magnetic pumping." As a result of magnetic pumping, the kinetic pressure at the plasma axis is increased, causing an axial flow of plasma to the top and bottom of the discharge. In this way, vortex rings are formed, superimposed on the main axial flow and the thermal expansion of the gas. This is analogous to the convection currents in a hot water tank. For the injector flow to penetrate the plasma, its velocity must exceed the magnetohydrodynamic thrust velocity. Once this is achieved, an annular or tunnel plasma is formed and the injector stream passes through the plasma, forming the long narrow tail. The best way to achieve an annular plasma is to use an aerosol injector tube with a narrow-bore tip. The tip of the injector tube should be placed close to the base of the plasma, without melting the tip. The penetration is helped, in the case of large diameter torches, if the base of the plasma is flattened by the intermediate gas stream, although penetration is also achieved in smaller diameter torches without any intermediate gas.

4.3.1   The Greenfield Torch

The design of the Greenfield torch, the developmental stages of which are shown in Figure 4.7, is based on the early work of Reed.[6,7,38] Reed's observations[33,38] indicate that he obtained an annular plasma only at frequencies of approximately 100 MHz, a frequency that he wished to avoid. It is difficult to see how he could have failed to produce an annular ICP with the configuration shown in

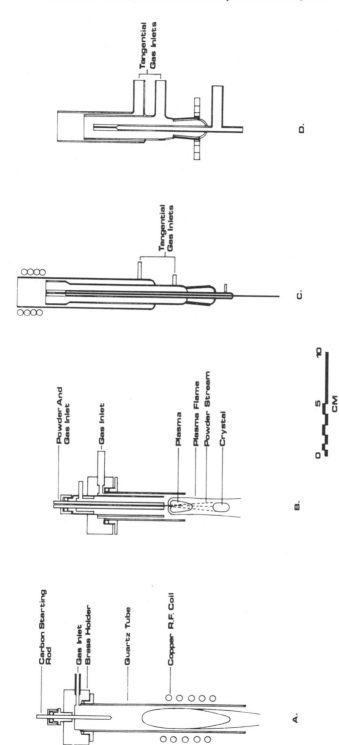

Figure 4.7. The plasma torches of Reed (*A*, *B*) and Greenfield (*C*, *D*). (From references 6, 7, and 39, with permission.)

Figure 4.7*B, unless the gas velocity along the axis were insufficient to penetrate the plasma.* However, Greenfield and associates,[39,40] with much experience in introducing aerosols into dc plasma jets, deliberately set out to produce an annular ICP (Figure 4.7*C*).

The Greenfield torch consists of three concentric tubes. The two outer tubes are made from silica and are used to contain the plasma. The inner tube is borosilicate glass and is used to inject an aerosol through the plasma. The torch is placed concentrically within the load coil of the RF generator. Argon gas is fed tangentially into the intermediate tube, and once the plasma has formed, argon, nitrogen, air, or another gas is introduced into the outer tube. A hole is then punched through the flattened base of the plasma by the introduction, through the injector, of an aerosol in argon gas. In Figure 4.7*C*, a fine capillary is seen in the injector tube; this is an integral nebulizer and is used to nebulize solutions, slurries, and powders into the plasma. This total consumption nebulizer was abandoned for an external nebulizer, mainly because of its delicacy of construction.

The final conception of the Greenfield torch (Figure 4.7*D*) is a straightforward design that can be constructed by a glass blower of average ability. The integral nebulizer has been replaced by a wide bore glass tube terminating in a 2-mm-bore which has two purposes. The capillary injector creates a narrow stream of aerosol at sufficient velocity to penetrate the plasma and allows the positioning of the capillary tip away from the heat of the ICP, where salts are likely to deposit. However, the end of the injector should be as close to the base of the plasma as possible. One can achieve this by deliberately allowing the tip of the injector tube to melt, cutting off the melted portion with a diamond wheel, and repeating the process until melting no longer occurs. Alternatively, the injector tube can be cut so that the tip is 35 to 40 mm from the top of the torch, with the actual length depending on the applied RF power.

Although similar detection limits are obtained with the Greenfield and Fassel torches, the former torch requires higher RF powers and higher gas flows than the smaller diameter Fassel torch. Typical gas flows of argon are[41]: intermediate gas, 0 to 9 L/min; outer gas, 16 to 40 L/min; and injection gas, 1 to 2.7 L/min, at powers of 1.4 to 4.0 kW in the plasma. When nitrogen is used as the outer gas (coolant), and argon for the other flows, the values are[41]: intermediate, 4.5 to 32 L/min; outer, 7 to 40 L/min; and injector 1.5 to 2.8, L/min, at powers in the plasma of 0.8 to 2.6 kW. The analytical properties of the plasma, such as attainable detection limits and freedom from interference, depend greatly on gas flows, forward power, and the observation height in the ICP. The ease of starting a plasma, plasma stability, and freedom from overheating are functions of the gas velocities, which, in turn, depend on the torch dimensions and concentricity. The values for these variables should be determined by a multiple-variable optimization method, such as simplex optimization[42,43] or another method,[41] even if later one uses compromise values for a multichannel spectrometer. When nitrogen gas is used as the outer flow (coolant), the spectral background continuum is significantly lower than the argon continuum, and the plasma is closer to local thermodynamic equilibrium than is the case with argon.[41,44]

The Greenfield torch is rugged, fairly tolerant of imperfections in construction, and able to function under extreme conditions. It will, for instance, function quite well with saturated solutions of many elements and will tolerate the ingress of air through the nebulizer without the plasma being extinguished.

4.3.2   The Fassel Torch

The developmental stages of the Fassel torch are shown in Figure 4.8. The torches in Figure 4.8A and 4.8B produced spheroidal ICPs with the aerosol flowing around the plasma. The good detection limits obtained with these torches are attributable to the use of an ultrasonic nebulizer that transports more analyte to the plasma than pneumatic nebulizers. An annular ICP, generated in a later torch[2] (Figure 4.8C), provided substantial improvement in detection limits. The design on which the present-day Fassel torch is based is shown in Figure 4.8D. In its original form, it had three concentric silica tubes, with only one tangential gas entry to the annulus between the outer and the intermediate tubes. Now it has two tangential gas entries to the outer and intermediate tubes. Aerosol is introduced through the small-diameter tube along the torch axis. Argon gas is used almost exclusively in the gas streams, although helium and molecular gases have been used in the ICP, as discussed in Chapter 15. Typical argon flow rates for the outer gas and for the injector gas are 10 to 20 L/min and 0.5 to 1 L/min, respectively. The power used is typically in the range 1 to 1.7 kW.

In most respects, the Fassel torch gives an analytical performance similar to that of the larger Greenfield torch, and because it requires much less gas and power, it is commonly used by most instrument manufacturers. Compared to the Greenfield torch, the Fassel torch requires greater accuracy in its construction and has the disadvantages of less tolerance to solutions of high solute contents, and a greater tendency to be extinguished by ingress of air into the nebulizer, especially when operating at low power levels.

4.3.3   Gas Flow Systems for Plasma Torches

The emission from an ICP depends on a number of factors: the power in the plasma, the viewing or observation height, and the gas flows, particularly the nebulizer or injection gas flow. Therefore, a high degree of control of these gas flows is essential. Above all, there must be no interaction between the gas flows brought about by taking each flow from a common main source. The low gas flow to the nebulizer is significantly perturbed by such interactions when it is taken from the gas line that carries the relatively high flow of outer gas. The best method of supply is to take each gas flow from a separate cylinder, fitted with a

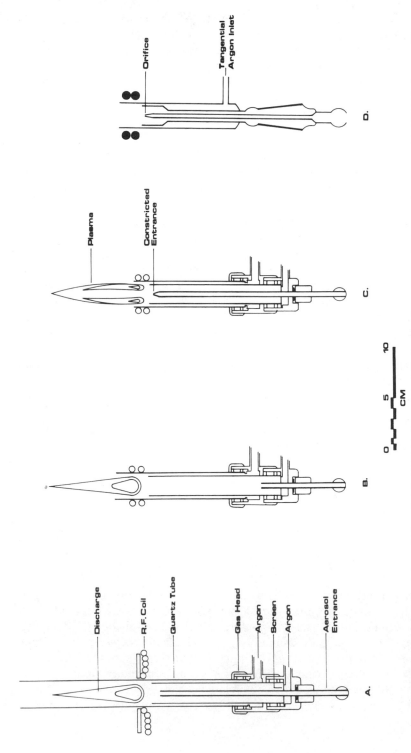

Figure 4.8. The plasma torches of Fassel: (*A*) 1965, Reference 8. (*B*) 1968, Reference 9. (*C*) 1969, Reference 2. (*D*) 1974, Reference 35. (From References 2, 8, 9 and 35, with permission).

two-stage regulator. If the argon is supplied by a liquid argon tank, a constant mass flow controller on each gas line is highly desirable. The common method of passing the gas from the two-stage regulator through a needle valve and float flowmeter (rotameter) is inadequate. A better practice is to pass the gas from the regulators through filters and then through restrictors, taking the upstream pressure as a measure of flow.

By far the best method of precise control of gas flow is to use mass controllers and flowmeters. The most accurate and precise mass flow controller, and the most expensive device, uses a heating element in the sensor tube carrying the flow. This device contains temperature-measuring elements equidistant from the heater. With no flow, there is no temperature difference between the elements. Increasing flow develops a temperature difference that is measured by a bridge circuit. The output from this circuit is amplified and used to control the flow. An inexpensive type of mass flow controller incorporates a spring-loaded diaphragm that actuates a control valve. A change in the inlet or outlet pressures will alter the balance of forces, causing the diaphragm to open or close the valve. Ideally, all the gas flow lines should be fitted with a mass flow controllers. When this is not possible, a mass flow controller preceded by a particulate filter should be used for the injection gas.

4.4   SAMPLE INTRODUCTION SYSTEMS

4.4.1   Classification

A variety of devices (Table 4.2) transport gas, liquid, and solid samples into ICPs. Liquids are generally converted into aerosols before introduction into the ICP. Viscous liquids generally require dilutions before conversion into aerosols. Electrothermal vaporizers can nebulize viscous liquids directly. The liquid is placed onto graphite yarn,[45] a graphite rod,[46] or a tantalum filament,[47] and the solvent is evaporated by electrical heating. The solute then is volatilized into the injection gas stream. A hydride generator—using, for example, sodium borohydride solution—may be used with solutions containing hydride-forming elements to introduce gaseous hydrides into the injection gas stream of the ICP.[48]

Many ways have been tried, with varying degrees of success, to introduce solids into plasmas in a quantitative manner. At the time of writing, a *generally routinely applicable* method does not exist. However, commercial systems are offered by Allied Analytical Systems and Applied Research Laboratories to introduce aerosols of conductive solids directly into ICPs. Reed[7] injected powders into plasmas, but he was not concerned with the quantitative aspects of solids delivery. Greenfield[1,39] used a pneumatic nebulizer, placed just below the plasma, to introduce powders from a rotating hopper. Although qualitatively successful, this method suffered from an inherent inability to maintain a

Table 4.2. Methods for the Introduction of Samples[a] into an Annular Inductively Coupled Plasma

Sample type	Method	Ref.
Liquid	Pneumatic nebulizer	60
	Ultrasonic nebulizer	72
	Electrothermal vaporizer	45–47
	Hydride generator	48
Solid	Pneumatic nebulizer	39
	Swirl cup	49, 50
	Fluidized bed	50
	Electrothermal vaporizer	45–47
	Direct insertion	51
	Arc or spark chamber	52,53
	Direct laser ablation	54
	Laser volatilization of powders on tape	55
	Spark volatilization of powders on tape	56
	Ultrasonic nebulization	9
	Slurry nebulization	1,39,57
Gas	Direct introduction along with the injector gas	See Chapter13

[a] See Chapters 11 to 13.

constant, uniform feed due to voiding, impacting, and agglomeration of the powders tested. Swirl cups,[49,50] in which the injection gas stream passes through powdered solid agitated in a container, also cannot be regarded as a general solution to the problem of powder transport; but in selected cases with uniform matrices and fairly constant concentration of the element of interest, they may have some application. The same is true of fluidized beds,[50] electrothermal vaporizers,[45–47] direct insertion into the plasma on graphite rods,[51] and arc and spark chambers,[52,53] The problems associated with all these methods are numerous and varied. Swirl cups, fluidized beds, and spark chambers suffer from the differing particle sizes and varying densities of the constituents of the powdered sample, leading to segregation and inhomogeneity. Electrothermal vaporizers, and all methods in which small samples are manipulated, are inherently imprecise, and, because of inhomogeneity, the small samples may not be representative of the whole sample. Selective volatilization is also a problem for electrothermal vaporization and for direct insertion methods.

Laser ablation [54] volatilizes only a few micrograms of material. However, it is very rapid and the repetition rate can be as high as 10 pulses per second. Since a large area of the sample can be surveyed, and the results averaged, greater precision and accuracy can be obtained than are possible with single-shot experiments. This is especially true for powdered samples, where each firing can involve fresh material originating not only from the surface but from the depth of the sample, thus improving the averaged result. In principle, methods that involve the placing of powders onto tapes, followed by transporting them through an intermittent laser[55] or spark[56] discharge, offer the most promise to date.

Ultrasonic nebulizers have been used to produce aerosols of liquid metals.[9] Slurries have also been nebulized into plasmas.[1,39,57,58] The latter method is also prone to the problems of density settling and consequent segregation, and the density of the liquid should be near that of the solid.

Detailed discussions on the introduction of liquids, solids, and gases into ICPs are presented in Chapters 11 to 13.

4.4.2 Common Pneumatic Nebulizers

Pneumatic nebulizers for use with ICPs have two basic configurations. In the concentric type, the sample solution passes through a capillary surrounded by a high-velocity argon gas stream parallel to the capillary axis. The cross-flow nebulizer has a liquid-carrying capillary set at a right angle to the tube carrying the gas stream. In both configurations, a pressure differential created across the sample capillary draws the sample solution through the capillary according to Poiseuille's equation:

$$Q = \frac{\pi R^4 P}{8 \eta L} \qquad [4.3]$$

where Q is the rate of flow of the liquid, R the capillary radius, P the pressure differential, η the viscosity of the liquid, and L the length of the capillary. Poiseuille's equation assumes that the velocity of the liquid at the capillary wall is zero, whereas the liquid slips over it. In this case, the R^4 term is replaced by $(R^4 + 4\eta R^3/B)$, where B is the coefficient of sticking friction of the liquid on the wall.

The mechanism of droplet formation following the aspiration of the sample is discussed in Chapter 11. Nukiyama and Tanasawa[59] gave an empirical relationship between the aerosol droplet diameter and the solution properties in conjunction with nebulizer parameters:

$$d_0 = \frac{585}{v(\sigma/\rho)^{1/2}} + 597 \left[\frac{\eta}{(\sigma\rho)^{1/2}}\right]^{0.45} \left(\frac{1000 Q_{\text{liq}}}{Q_{\text{gas}}}\right)^{1.5} \qquad [4.4]$$

where d_0 is the mean droplet diameter (μm), v is the velocity of gas (m/s: or, more strictly, the difference between gas and liquid velocities), σ is the surface tension (dyne·cm), ρ is the density of the liquid (g/mL), η is the viscosity (poises), and Q_{liq} and Q_{gas} are the volume flows of the liquid and gas. This empirical formula was derived from experiments in which the solution properties were within the following limits: $30 < \sigma < 73$; $0.01 < \eta < 0.3$; $0.8 < \rho < 1.2$. Extrapolation outside these ranges can be misleading. The Nukiyama–Tanasawa equation is also strictly valid for air as the gas, although using argon, which has nearly the same velocity of sound, should not introduce any great error. Other restrictions on the use of Equation 4.4 are given in Chapter 11.

The Babington nebulizer is a variant of the cross-flow nebulizer. In the original device, the solution is pumped through a glass tube that terminates in a hollow sphere, where it emerges from a small hole in the top of the sphere to form a thin film over the outside surface. Gas is forced through a small horizontal slot in the sphere, thus rupturing the film and producing an aerosol. Several variations of the Babington design are described in Section 4.4.2.4.

4.4.2.1   Cross-Flow Nebulizers

The first cross-flow nebulizer designed to produce aerosols for introduction into an ICP was described by Kniseley and associates.[60] This nebulizer has two adjustable capillary tubes set at right angles to each other in a polytetrafluoroethylene (PTFE) body. A vertical tube is for liquid uptake, and the other tube is for the injector gas stream. The nebulizer is designed to operate at a gas flow of approximately one liter per minute of argon, and at that flow, the rate of liquid uptake is approximately 3 mL/min. The positions of the capillary tips can be adjusted to achieve maximum performance. Unfortunately, this adjustability also allows the needles to creep, thus affecting the long-term stability of the nebulizer. For this reason, the fixed cross-flow nebulizer with nonadjustable capillaries and with an impact bead[61] is often used. In both designs, the shapes of the capillary tips and the positions of the tips, relative to each other, have profound effects on the performance of the nebulizer. Generally, the cross-flow nebulizers are less prone than concentric nebulizers to salt buildup at the tip, when spraying solutions having high solute concentrations.

4.4.2.2   Concentric Nebulizers

The concentric nebulizers are constructed from many materials and are used for a variety of purposes. A glass version,[62] suitable for use with an ICP, can be fabricated easily by a competent glass blower. A commercial version of the glass, concentric nebulizer, having a small venturi end, is produced by Meinhard Associates, Incorporated. The Meinhard nebulizers, which currently are the nebulizers most widely used in ICP spectrometry, are supplied in two configurations, with a range of operating specifications and a choice of three nozzles. The input pressure at one liter per minute of argon ranges from 70 to 350 kPa, while the liquid uptake rate ranges from 0.5 to 4 mL/min. For lower argon flow rates, in the range from 0.5 to 0.8 L/min, the solution uptake rate of less than 0.5 to 1 mL/min can be obtained. While this low gas flow rate may be suitable for the small Fassel torch, it is not enough to form an annular plasma in the large Greenfield torch, which calls for an auxiliary injection gas. The Meinhard concentric nebulizer exhibits good long-term stability, but because of the very small annular gap (10 to 35 μm), it is very prone to blockage by particles in the analyte solution and by solutions having high solute concentrations.

4.4.2.3   High-Pressure Nebulizers

Earlier in this chapter, the poor long-term stability in cross-flow nebulizers was attributed to the relative movement of the adjustable glass needles. The inner tube of the concentric nebulizer is also known to vibrate, resulting in lower precision than would otherwise be obtainable. To eliminate this problem, Anderson, Kaiser, and Meddings[32] constructed a cross-flow nebulizer from thick-walled glass capillary tubes that were fixed into nonadjustable alignment. Relatedly, precise gas control was achieved by operating the nebulizer at pressures much higher than are commonly used with the Meinhard nebulizer. A patented version of this device, known as the MAK nebulizer, is illustrated in Figure 4.9, along with an integral spray chamber using the essential baffles.

The MAK nebulizer operates at approximately 1400 kPa, with an argon flow rate of 500 mL/min. Such a low flow needs an auxiliary injection gas to penetrate a plasma generated in a Greenfield torch. The nebulizer exhibits good long-term stability,[32] a precision better than 0.5% relative standard deviation, and an excellent resistance to salt buildup at the tip of the nebulizer. Sodium chloride solutions with concentrations up to 30% can be used in the MAK nebulizer. Strasheim and Olsen,[63] have compared the MAK nebulizer with nebulizers of other types.

4.4.2.4   Babington Nebulizers

The interesting feature of the Babington nebulizer is that the liquid emerges through a relatively large slot or hole in a surface, and therefore, the nebulizer is not prone to blocking by particulates or solutions of high solute content. Its simplest form is described in Section 4.4.2. The formation of a liquid film depends more on the surface tension of the liquid than it does on viscosity. Fry and Denton[64] used a modified Babington[65] system to determine metals in

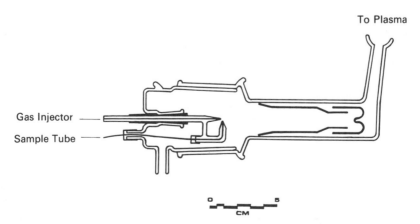

Figure 4.9. The MAK nebulizer. (Sherritt Gordon Mines, Ltd., with permission.)

tomato sauce, evaporated milk, and whole blood by flame atomic absorption spectrometry. Babington nebulizers, in their original form, and when modified for use in atomic absorption spectrometry, use gas flows and liquid uptake rates that are far too high for ICP usage, although other versions suitable for ICP spectrometry have been evaluated.

In a design originated by Garbarino and Taylor,[66] a tube having a hemispherical end with a small hole in the side was used to deliver argon, while the liquid sample was introduced from a separate tube positioned vertically above the gas tube. With a gas orifice of 0.1 mm, the gas flow rate and the solution delivery rates were 600 and 3 to 5 mL/min, respectively. Suddendorf and Boyer[67] patented a nebulizer in which the liquid flows down a V-groove having the gas orifice somewhere near the top of the groove. However, the latter nebulizer is undoubtedly more difficult to make and does not appear to offer any advantages over the former system. Wolcott and Sobel[68] evaluated a simplified version of the design by Suddendorf and Boyer[67] to achieve a gas flow rate of 0.4 to 1.2 L/min.

The GMK Babington nebulizer (Figure 4.10) is available commercially. The gas injector nozzle is constructed from a piece of 6-mm capillary bore glass tubing with a V-groove cut into the face and through the capillary. The sample tube is tapered down to 1-mm bore. A glass impactor bead further fragments the droplets issuing from the nebulizer. The argon flow rate for this nebulizer ranges from 0.8 to 1.2 L/min and is obtained by use of different gas injector nozzles. The operating pressure is 280 kPa. The GMK nebulizers are said by the manufacturers to exceed or equal the performance of the cross-flow and concentric nebulizers with respect to detection limits, precision, memory effects, sample analysis time, particle-handling capacity, salt effects, and acid effects. The

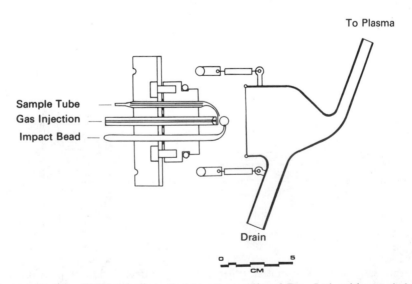

Figure 4.10. The GMK nebulizer. (Labtam International Pty., Ltd., with permission.)

author's experience would tend to confirm this assertion. Similar to other Babington nebulizers, the GMK nebulizers can handle solutions containing particulates and saturated solutions without any nebulizer blockage. It should be reemphasized that saturated solutions can be used in the plasmas generated in a large Greenfield torch operated at moderate powers[69] but not in a Fassel torch operated at the normal power levels.

An improved version of the GMK nebulizer has recently been introduced into the marketplace. The new GMK nebulizer obviates leakage and ensures precise positioning of the components by the judicious use of precision-ground tubing and the use of a PTFE support rod for the glass impact bead. The sample tube is a continuous length of PTFE tubing from a peristaltic pump through to the V-groove. The sequence of events that has been devised for the new nebulizer is as follows.

Prior to the analysis of a new sample, and commencing at the onset of the normal flush period, a two-way valve is actuated to divert the argon injector gas away from the injector tube while still maintaining flow through the spray chamber to the plasma. At the instant the gas is diverted, there is approximately double the normal gas flow in the chamber, and through the plasma. This has the effect of sweeping the spray chamber completely free of the previous aerosol. A needle valve sets the final flow rate for the diverted gas stream. The nebulizer aerosol flow commences to decay at this instant, and when the decay is complete, the sample pump is switched into high-speed operation. The pump operates in the high-speed mode long enough to ensure that the new sample has reached the V-groove and that it has completely flooded and preconditioned the surface. The whole system then reverts to normal operation, with the gas once again being fed to the gas injector nozzle of the nebulizer, with the pump operating at the normal speed.

The foregoing operating system is said by the manufacturer to be sufficient to ensure that crystals do not form at the gas–liquid interface and to reduce the memory effect dramatically.

4.4.3   Ultrasonic Nebulizers

Another method of producing an aerosol from a liquid is to use ultrasound to break the liquid mass into small particles. In this method, an ultrasonic generator drives a piezoelectric crystal at a frequency between 200 kHz and 10 MHz. The longitudinal wave, which is propagated perpendicularly from the surface of the crystal toward the liquid–air interface, produces a pressure that breaks the surface into an aerosol. The wavelength of the surface wave is given by[9]:

$$\lambda = \left(\frac{8\pi\sigma}{\rho f^2}\right)^{1/3} \qquad\qquad [4.5]$$

where λ is wavelength, σ is surface tension, ρ is liquid density, and f is ultrasonic frequency. The average droplet diameter is given as:

$$D = 0.34\lambda \qquad\qquad [4.6]$$

There are basically two types of ultrasonic nebulizer. In the first type, the sample solution is contained in a small glass cell with a base made from a thin membrane of Mylar, which is transparent to ultrasonic waves. The cell is usually coupled to the transducer by a column of water. Either the ultrasound is focused onto the surface of the liquid by a methylmethacrylate (Plexiglas) lens,[8] for greater efficiency in the transport of energy, or a waveguide is used to achieve the same purpose.[70] A vertical spray chamber encloses the cell, with entries and exits for the solution drain, the carrier gas, and the aerosol. In the second type of ultrasonic nebulizer,[71] the transducer is bonded to a chemical-resistant plate and placed vertically in a horizontal spray chamber. The sample solution is fed continuously onto the transducer plate, with the excess falling by gravity into the chamber, down to the drain.

Because a constant load is maintained on the transducer of the first type of nebulizer, the stability of conversion of acoustic energy and the coupling efficiency are probably greater than those achieved in the second type. While the second type of nebulizer has the disadvantage of requiring water cooling, it exhibits less memory effect and allows a rapid change of samples. Since for both types of ultrasonic nebulizer the production of aerosol is very large and is independent of the gas flow, more analyte can be transported to the ICP at a slower flow rate of injection gas than can be achieved with a pneumatic nebulizer, thus giving longer residence times for analytes in the plasma and subsequently a greater number of emitting species. Detection limits with ultrasonic nebulizers are an order of magnitude better than with pneumatic devices, *if the matrix is not complicated*. If there are interferences, such as background shifts or spectral coincidences, these effects will also be enhanced to similar extent. Likewise, salt buildup on the injector tube orifice will be increased when solutions having high salt concentrations are nebulized. This latter effect is overcome[72] by the addition of a laminar flow of gas surrounding the aerosol stream. Interferences, such as chemical, which are particle-size dependent, are likely to be reduced because the aqueous droplets produced by a 1-MHz ultrasonic nebulizer can be smaller than 0.4 μm.[70] However, an increase in "aerosol ionic redistribution interference" may occur.[73]

4.4.4   Spray Chambers and Desolvation Systems

Aerosol transport efficiency is defined as the percentage of the mass of nebulized solution that actually reaches the plasma. For this percentage to be high, and for rapid desolvation, volatilization, and atomization of the aerosol droplets when they reach the plasma, a nebulizer must produce droplets less than 10 μm in diameter. Unfortunately, nebulizers, particularly pneumatic nebulizers, produce

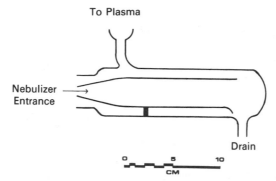

Figure 4.11. The Scott spray chamber. (From Reference 35, with permission).

aerosols that are highly polydispersive, with droplet diameters up to 100 μm. These large droplets must be removed by means of a spray chamber. Such a spray chamber, as described by Scott and co-workers,[35] is shown in Figure 4.11.

A barrel-type spray chamber removes the larger droplets by turbulent deposition on the inner walls of the chamber, or by gravitational action, as discussed in Chapter 11. Relatedly, the inner concentric tube reduces the random fluctuations in signal intensity, much of which comes from flicker noise in the plasma resulting from aerosol density changes in the spray chamber. This type of noise is greatly influenced by the shape and positioning of the baffles and aerosol exit tubes.[74] Minor pressure changes in the chamber can greatly affect the plasma, and the fall of drops in a poorly designed drain can cause severe noise spikes in the recorded signal from the plasma emission. Drains should be designed so that the liquid drains away smoothly through devices such as sand beds, fritted glass discs, or capillaries. Also, liquid buildup in the chamber should be avoided, for this causes the pressure to change, resulting in a corresponding change in aerosol velocity and a drift in the signal.

In plasma spectrometry, the time required for the signal from the plasma to rise to a steady state and then return to background level after completion of the aspiration of the sample depends heavily on features of the spray chamber, such as shape, size, and position of baffles or other internal parts. Other factors affecting the rise and decay of signals are the gas velocity and the element being nebulized. Spray chambers and nebulizer systems for ICP spectrometry are much slower than those used in atomic absorption spectrometry: they require minutes rather than seconds. The transport efficiency of sample introduction systems is less than 3% in ICP spectrometry, compared to 15% or less for flame atomic absorption spectrometry. Some controversy continues over whether the removal of large droplets in barrel-type spray chambers is due to gravitational[75] or inertial[76] processes, but general agreement exists that the droplet size distribution plays a large part in transport efficiency, freedom from interferences, and measurement precision. Impact surfaces, or beads, may help to reduce the number of large droplets by fragmenting them into smaller ones, thus increasing the transport efficiency.

Cyclone chambers[41,77] also have been used as spray chambers in ICP spectrometry. These chambers are conical, and the aerosol enters through a tangential inlet on the periphery of the vessel to follow a downward spiral motion. This motion leads to the development of a centrifugal force, which acting on the droplets, throws the droplets to the wall. At the bottom of the chamber, the aerosol changes direction and moves toward the top of the vessel in an even tighter spiral that is concentric with the original path. The large particles that are thrown against the wall leave by the drain tube at the base, and the small droplets pass through the outlet tube, which protrudes a short way into the vessel from the top. Cyclone spray chambers are not as commonly used as the barrel type, but they are very efficient and merit more attention than they have thus far received. More detailed discussions of the cyclone chamber are presented in Chapter 11.

One great disadvantage of high-efficiency pneumatic and of ultrasonic nebulizers particularly is that they convey such a greatly increased amount of solvent to the plasma that low-power, small-torch plasmas are extinguished. Therefore, desolvation of the aerosol becomes necessary. Desolvation is achieved by heating the aerosol sufficiently to evaporate the solvent from the aerosol. Heating is accomplished in either a heated spray chamber[78] or in a separately heated tube.[2,79] The resulting vapor and dried particles are then passed through a condenser. The condenser removes most of the solvent from the vapor, and some of the dried particles are also lost. The liquid solvent passes out through the drain tube at the bottom of the condenser, and the dry particles are transported to the plasma through a side tube. The operation of removing the solvent from the aerosol can result in the introduction of what are known as "desolvation interferences." Boumans and de Boer[80] have suggested that one cause of this type of interference is the differing volatility between the analyte and the matrix. Certainly, this effect is not present in simple solutions of analytes at low concentrations.

4.5   THE FUTURE

The present RF generators perform their tasks well, and any imprecision that they may introduce to ICP-AES is smaller than that which is introduced by sample introduction systems, gas controller, and similar adjuncts to the method. Therefore, any changes that manufacturers may make to common generators can reasonably be thought of as largely cosmetic. More solid-state RF generators may be appearing for use in ICP-AES, especially if they can be made small and compact. If miniature torches are demonstrated to be useful, more low-power, solid-state generators will make their appearance. On the other hand, with the advent of ICPs as both primary line sources and atomizers in atomic fluorescence spectrometry, generators capable of delivering 5 kW and driving two plasma torches may become more common.

As with generators, the torches in use today perform well, and the two most well known, the Greenfield and the Fassel torches, despite many claims of superior versions, have stood the test of time. Any notable advance being made in this direction is difficult to see, unless miniature torches make some headway. In the same way that small torches are more demanding than large torches, one might expect that miniature torches will be more critical in construction, more susceptible to matrix interferences, and more prone to extinguishing of their plasmas by air, by high-solute solutions, and by nonaqueous solvents than are small torches. On the other hand, one reason for thinking that miniature torches have a future is that they consume less argon and less power than their larger counterparts, which could make miniature torches attractive for field applications and for applications on board ships, etc. In this connection, one should not overlook the possibility of reducing the use of argon in larger torches by substituting nitrogen or air.

Despite the greatly increased knowledge now available on the topic, pneumatic nebulizers remain basically highly inefficient devices for which 1 to 3% of the sample uptake reaches the plasma and the rest ends up down the drain. Furthermore, any *major* improvement in their efficiency is very doubtful. Babington nebulizers may have some scope for improvement, and in some ways, they have greater possibilities than conventional nebulizers, not least of which is their ability to handle particulates and high-solute solutions. Ultrasonic nebulizers may be used more widely, and again, there is scope for further improvement. An order-of-magnitude improvement in detection limits makes worthwhile efforts to overcome the problems of memory effects and other interferences.

By far the greatest challenge to be met is that of finding new methods to introduce liquids, and more particularly, solids into inductively coupled plasmas. With man's armory of devices capable of hitting solid samples with a range of energetic projectiles, surely someone can devise a better and more general method for introducing material into a plasma, without the need for tedious chemical operations!

ACKNOWLEDGMENTS

The author is greatly indebted to the following for their generous gifts of time and information, which made the task of writing this chapter easier than it would otherwise have been.

Mr. D. Baker	Radyne, Ltd., Wokingham, UK
Dr. P.W.J.M. Boumans	Philips Research Laboratories
	Eindhoven, The Netherlands
Mr. I.Ll.W. Jones	Albright and Wilson, Ltd.
	Oldbury, UK

Mr. T. Knight	Labtam International Pty Ltd.
	Stafford, Australia
Mr. B. A. Ronksley	Philips Industries, S.A.
	Wavre, Belgium
Dr. M. W. Routh	ARL, Inc.
	Sunland, California
Mr. W. Slavin	Perkin-Elmer Corporation
	Norwalk, Connecticut
Mr. D. I. Spash	Radyne, Ltd., Wokingham, UK
Dr. D. A. Yates	Perkin-Elmer Corporation
	Norwalk, Connecticut

REFERENCES

1. S. Greenfield, I.Ll. Jones, and C. T. Berry, High-Pressure Plasmas as Spectroscopic Emission Sources, *Analyst* 89, 713–720 (1964).
2. G. W. Dickinson and V. A. Fassel, Emission Spectrometric Detection of the Elements at the Nanogram per Milliliter Level Using Induction-Coupled Plasma Excitation, *Anal. Chem.* 41, 1021–1024, (1969).
3. W. Hittorf, Über die Electrizitätsleitung der Gase. *Ann. Phys.* 21, 90–139 (1891).
4. J. J. Thomson, On the Discharge of Electricity Through Exhausted Tubes without Electrodes, *Phil. Mag.* 32, 321 and 445 (1891).
5. G. I. Babat, Electrodeless Discharges and Some Allied Problems, *J. Inst. Electr. Eng.* (*London*) 94, 27–37 (1947).
6. T. B. Reed, Induction-Coupled Plasma Torch, *J. Appl. Phys.* 32, 821–824 (1961).
7. T. B. Reed, Growth of Refractory Crystals Using the Induction Plasma torch, *J. Appl. Phys.* 32, 2534–2535 (1961).
8. R. H. Wendt and V. A. Fassel, Induction-Coupled Plasma Spectrometric Excitation Source, *Anal. Chem.* 37, 920–922 (1965).
9. V. A. Fassel and G. W. Dickinson, Continuous Ultrasonic Nebulization and Spectrographic Analysis of Molten Metals, *Anal. Chem.* 40, 247–249 (1968).
10. S. Greenfield, H. McD. McGeachin, and P. B. Smith, Plasma Emission Sources in Analytical Spectroscopy—III. *Talanta* 23, 1–14 (1976).
11. R. M. Barnes, Recent Advances in Emission Spectroscopy: Inductively Coupled Plasma Discharges for Spectrochemical Analysis, *CRC Crit. Rev. Anal. Chem.* 7, 203–296 (1978).
12. V. A. Fassel, Quantitative Elemental Analyses by Plasma Emission Spectroscopy, *Science* 202, 183–191 (1978).
13. J. P. Robin, ICP–AES at the Beginning of the Eighties, *Prog. Anal. Atom. Spectrosc.* 5, 79–110 (1982).
14. M. H. Abdalla, R. Diemiaszonek, J. Jarosz, J. M. Mermet, J. Robin, and C. Trassy,: Étude Spectrométrique d'un Plasma Induit par Haut Fréquence—I. *Anal. Chim. Acta* 84, 271–282 (1976).
15. R.W.H. Whittle and J. J. Behenna, Controlled RF Power Generation Using Magnetically Beamed Triodes, *Radio Electron. Eng.* 46, 605–613 (1976).
16. J. Schmitz, Paper read to the Groupement pour l'Avancement des Méthodes Spectroscopiques et Physico-Chimiques d'Analyse (GAMS), Paris, May 31, 1977. Philips Bulletin No. E.S. 39.
17. P.W.J.M. Boumans, F. J. de Boer, and J. W. de Rieter, A Stabilized RF Argon-Plasma Torch for Emission Spectroscopy, *Philips Tech. Rev.* 33, 50–59 (1973).
18. H. Linn, RF Generators for ICP Applications, *ICP Inf. Newsl.* 2, 51–60 (1976).
19. C. Trassy and J. M. Mermet, "Les Applications Analytiques des Plasmas Haute-Frequence," Technique et Documentation Lavoisier, Paris. 1984.
20. C. Allemand, Design of an ICP Discharge System for Residual Fuel Analysis for Marine Applications, *ICP Inf. Newsl.* 2, 1–26 (1976).

21. R. G. Schleicher and R. M. Barnes, Remote Coupling Unit for Radio Frequency Inductively Coupled Plasmas Discharges in Spectrochemical Analysis, *Anal. Chem.* 47, 724–728 (1975).
22. C. D. Allemand and R. M. Barnes, Design of a Fixed-Frequency Impendence Matching Network and Measurement of Plasma Impedence in an Inductively Coupled Plasma for Atomic Emission Spectroscopy, *Spectrochim. Acta* 33B, 513–534 (1978).
23. British Standard 1799:1952, Power Rating of Valve-Driven High-Frequency Induction Heating Equipment, British Standards Institution, London.
24. S. Greenfield, I.Ll. Jones, H. McD. McGeachin, and P. B. Smith, Automatic Multi-Sample Simultaneous Multi-Element Analysis with a High-Frequency Plasma Torch and Direct Reading Spectrometer, *Anal. Chim. Acta* 74, 225–245 (1975).
25. S. Greenfield, H. McD. McGeachin, and P. B. Smith, Nebulization Effects with Acid Solutions in ICP Spectrometry, *Anal. Chim. Acta* 84, 67–78 (1976).
26. S. Greenfield and H. McD. McGeachin, Calorimetric and Dimensional Studies on Inductively Coupled Plasmas, *Anal. Chim. Acta* 100, 101–119, (1978).
27. B. Bogdain, The Power Game, *ICP Inf. Newsl.* 2, 269–270 (1977).
28. R. Treptow, D. Gold, and A. El-Shamy, Electronically and Calorimetrically Measured Energy Balances of a Thermal Induction Plasma, *IEEE Trans. Plasma Sci.* 6, 121–129 (1978).
29. P.A.M. Ripson and L. De Galan, Empirical Power Balances for Conventional and Externally Cooled Inductively Coupled Argon Plasmas, *Spectrochim. Acta*, 38B, 707–726 (1983).
30. D. R. Demers and A. I. Priede, Ultratrace Analysis via Inductively Coupled Plasma Optical Emission Spectrometry: The Importance of Precision RF Power Regulation and of Monitoring Nebulizer Operation, *ICP Inf. Newsl.* 3, 221–228 (1977).
31. B. A. Ronksley, Philips Industries, Belgium, private communication.
32. H. Anderson, H. Kaiser, and B. Meddings, High Precision (< 0.50% RSD) in Routine Analysis by ICP Using a High-Pressure (200 psig) Cross-Flow Nebulizer, in "Developments in Atomic Plasma Spectrochemical Analysis," R. M. Barnes, Ed. London, Heyden, 1981, pp. 251–277.
33. T. B. Reed, Induction Plasma Torch with Means for Recirculating the Plasma, U.S. Patent 3,324,334 (1967).
34. P.W.J.M. Boumans and F. J. de Boer, Studies of Flame and Plasma Torch Emission for Simultaneous Multielement Analysis—I. Preliminary Investigations, *Spectrochim. Acta* 27B, 391–414 (1972).
35. R. H. Scott, V. A. Fassel, R. N. Kniseley, and D. E. Nixon, Inductively Coupled Plasma Optical Emission Analytical Spectrometry, a Compact Facility for Trace Analysis of Solutions, *Anal. Chem.* 46, 75–80 (1974).
36. J. D. Chase, Magnetic Pinch Effect in the Thermal RF Induction Plasma, *J. Appl. Phys.* 40, 318–325 (1969).
37. J. D. Chase, Theoretical and Experimental Investigations of Pressure and Flow in Induction Plasmas, *J. Appl. Phys.* 42, 4870–4879 (1971).
38. T. B. Reed, Plasma Torches, *Int. Sci. Technol.* June, 42–48 (1962).
39. S. Greenfield, I.Ll. Jones, C. T. Berry, and D. I. Spash, Improvements Relating to Spectroscopic Methods and Apparatus. British Patent 1,109,602 (1968).
40. S. Greenfield, I.Ll. Jones, and C. T. Berry, Plasma Light Source for Spectroscopic Investigation, U.S. Patent. 3,467,471 (1969).
41. S. Greenfield and D. T. Burns, A Comparison of Argon-Cooled and Nitrogen-Cooled Plasma Torches Under Optimized Conditions Based on the Concept of Intrinsic Merit, *Anal. Chim. Acta* 113, 205–220 (1980).
42. S. N. Deming and L. R. Parker, Jr., A Review of Simplex Optimization, *Anal. Chem.* 7, 187–202 (1978).
43. L. Ebdon, M. R. Cave, and D. J. Mowthorpe, Simplex Optimization of Inductively Coupled Plasmas, *Anal. Chim. Acta* 115, 179–187 (1980).
44. R. M. Barnes and G. A. Meyer, Low-Power Inductively Coupled Nitrogen Plasma Discharge for Spectro-Chemical Analysis, *Anal. Chem.* 52, 1523–1525 (1980).
45. R. L. Dahlquist, Method of Preparing Analyte Material for Spectrochemical Analysis, U.S. Patent. 3,832,060 (1974).
46. A. M. Gunn, D. L. Millard, and G. F. Kirkbright, Optical Emission Spectrometry with an Inductively Coupled Radiofrequency Argon Plasma Source and Sample Introduction with a Graphite Rod Electrothermal Vaporization Device, *Analyst* 103, 1066–1073 (1978).
47. D. E. Nixon, V. A. Fassel, and R. N. Kniseley, Inductively Coupled Plasma Optical Emission Analytical Spectroscopy: Tantalum Filament Vaporization of Microlite Samples, *Anal. Chem.* 46, 210–213 (1974).

48. M. Thompson, B. Pahlavanpour, S. J. Walton, and G. F. Kirkbright, Simultaneous Determination of Trace Concentrations of Arsenic, Antimony, Bismuth, Selenium and Tellurium in Aqueous Solution by Introduction of the Gaseous Hydrides into an Inductively Coupled Plasma Source for Emission Spectrometry—I. Preliminary Studies, *Analyst* 103, 568–579 (1978); II. Interference Studies, *Analyst* 103, 705–713 (1978).

49. H. C. Hoare and R. A. Mostyne, Emission Spectrometry of Solutions and Powders with a High-Frequency Plasma Source, *Anal. Chem.* 39, 1153–1155 (1967).

50. R. M. Dagnell, D. J. Smith, T. S. West, and S. Greenfield, Emission Spectroscopy of Trace Impurities in Powdered Samples with a High-Frequency Argon Plasma Torch, *Anal. Chim. Acta* 54, 397–406 (1971).

51. E. D. Salin and G. Horlick, Direct Sample Insertion Device for Inductively Coupled Plasma Emission Spectroscopy, *Anal. Chem.* 51, 2284–2286 (1979).

52. H.G.C. Human, R. H. Scott, A. R. Oakes, and C. D. West, The Use of a Spark as a Sampling-Nebulizing Device for Solid Samples in Atomic Absorption Atomic Fluorescence and Inductively Coupled Plasma Emission Spectrometry, *Analyst* 101, 265–271 (1976).

53. R. H. Scott, Spark Elutriation of Powders into an Inductively Coupled Plasma, *Spectrochim. Acta* 33B, 123–125 (1978).

54. M. Thompson, J. E. Goulter, and F. Sieper, Laser Ablation for the Introduction of Solid Samples into an Inductively Coupled Plasma for Atomic Emission Spectrometry, *Analyst* 106, 32–39 (1981).

55. F. N. Abercrombie, M. D. Silvester, A. D. Murry, and A. R. Barringer, A New Multielement Technique for the Collection and Analysis of Airborne Particulates in Air Quality Surveys, in "Applications of Inductively Coupled Plasmas to Emission Spectroscopy, Conference, Proceedings," R. M. Barnes, Ed. Franklin Institute Press, Philadelphia, 1978, pp. 121–145.

56. S. Greenfield and T. P. Sutton, Investigation into the Interfacing of a Danielsson Tape Machine with an ICP Polychromator for the Analysis of Powderable Samples, *ICP Inf. Newsl.* 7, 375–383 (1980).

57. A. E. Watson and P. J. Humphries-Cuff, The Direct Analysis of Solid Materials in the Form of Slurries by the Inductively Coupled Plasma Technique. MINTEK 50 Johannesburg, March 26–30, 1984.

58. L. Ebdon and M. R. Cave, A Study of Pneumatic Nebulization Systems for Inductively Coupled Plasmas Emission Spectrometry, *Analyst* 107, 172–178 (1982).

59. S. Nukiyama and Y. Tanasawa, Experiments on the Atomization of Liquids in an Air Stream (E. Hope, transl.). Defense Research Board Department of National Defense, Ottawa, Canada, 1950.

60. R. N. Kniseley, H. Amenson, C. C. Butler, and V. A. Fassel, An Improved Pneumatic Nebulizer for Use at Low Nebulizing Gas Flow, *Appl. Spectrosc.* 28, 285–286 (1974).

61. J. W. Novak, Jr., D. E. Lillie, A. W. Boorn, and R. F. Browner, Fixed Crossflow Nebulizer for Use with Inductively Coupled Plasmas and Flames, *Anal. Chem.* 52, 576–579 (1980).

62. R. H. Scott, Concentric Glass Nebulizer Fabrication Technique, *ICP Inf. Newsl.* 3, 425–427 (1978).

63. A. Strasheim and S. D. Olsen, Correlation of the Analytical Signal to the Characterized Nebulizer Spray, *Spectrochim. Acta* 38B, 973–975 (1983).

64. R. C. Fry and M. B. Denton, High Solids Samples Introduction for Flame Atomic Absorption Analysis, *Anal. Chem.* 49, 1413–1417 (1977).

65. R. S. Babington, Method of Atomizing Liquid in a Mono-Dispersed Spray, U.S. Patent 3,421,692 (1969).

66. J. R. Garbarino and H. E. Taylor, A Babington-Type Nebulizer for Use in the Analysis of Natural Water Samples by Inductively Coupled Plasma Spectrometry, *Appl. Spectrosc.* 34, 584–590 (1980).

67. R. F. Suddendorf and K. W. Boyer, Mechanical Device to Produce a Finely Dispersed Aerosol, U.S. Patent. 4,206,160 (1980).

68. J. F. Wolcott and C. Sobel, A Simple Nebulizer for an Inductively Coupled Plasma System, *Appl. Spectrosc.* 32, 591–593 (1978).

69. S. Greenfield and M. Thomsen, Atomizer, Source, Inductively Coupled Plasmas in Atomic Fluorescence Spectrometry (ASIA). Some Recent Work, *Spectrochim. Acta* 40B 1369–1377 (1985).

70. J. M. Mermet, and C. Trassy, Design of a New Ultrasonic Nebulizer for Routine Analysis in ICP-AES, in "Developments in Atomic Plasma Spectrochemical Analysis," R. M. Barnes, Ed. Heyden, London, 1981, pp. 245–250.

71. K. W. Olson, W. J. Haas, Jr., and V. A. Fassel, Multielement Detection Limits and Sample Nebulization Efficiencies of an Improved Ultrasonic Nebulizer and a Conventional Pneumatic Nebulizer in Inductively Coupled Plasma–Atomic Emission Spectrometry, *Anal. Chem.* 49, 632–637 (1977).
72. C. Trassy, Device to Produce an Aerosol Stream, French Patent 79-15575 (1979).
73. R. F. Browner, M. S. Black, and A. W. Boorn, Recent Studies with Sample Introduction into the RFICP, in "Developments in Atomic Plasma Spectrochemical Analysis," R. M. Barnes, Ed. Heyden, London, 1981.
74. P. Schutyser and E. Janssens, Evaluation of Spray Chambers for Use in Inductively Coupled Plasma–Atomic Emission Spectrometry, *Spectrochim. Acta* 34B, 443–449 (1979).
75. J. W. Novak, Jr., and R. F. Browner, Characterization of Droplet Sprays Produced by Pneumatic Nebulizer, *Anal. Chem.* 52, 792–796 (1980).
76. A. Gustavson, A Review of the Theory for, and Practical Aspects on Aerosol Chambers, in thesis, "Theoretical and Practical Aspects of Nebulizer Systems for Spectroscopy," Department of Analytical Chemistry, The Royal Institute of Technology, Stockholm, Sweden (1983).
77. S. Greenfield and P. B. Smith, The Determination of Trace Metals in Microlitre Samples by Plasma Torch Excitation, *Anal. Chim. Acta* 59, 341–348 (1972).
78. C. Veillon and M. Margoshes, A Pneumatic Nebulization System Producing Dry Aerosol for Spectroscopy, *Spectrochim Acta*, 23B, 503–555 (1968).
79. S. Greenfield, Plasma Sources in Spectroscopy, *Metron* 3, 224–230 (1971).
80. P.W.J.M. Boumans and F. J. de Boer, Studies of a Radio Frequency Inductively Coupled Argon Plasma for Optical Emission Spectrometry—III. Interference Effects Under Compromise Conditions for Simultaneous Multielement Analysis, *Spectrochim. Acta* 31B, 355–375 (1976).

5

Analytical Performance of Inductively Coupled Plasma–Atomic Emission Spectrometry

MICHAEL THOMPSON

Applied Geochemistry Research Group
Department of Geology
Imperial College of Science and Technology
London, England

5.1   INTRODUCTION

The introduction, in 1963, of the inductively coupled plasma (ICP) as a source for analytical atomic emission spectrometry (AES) constituted a revolutionary advance in that field. The performance characteristics of ICP–AES, namely its versatility, wide applicability, and ease of use, were almost unparalleled among methods of elemental analysis.[1,2] Many tasks, previously regarded by the analyst as difficult or time-consuming, could be undertaken with ease and rapidity by ICP–AES.

In principle, any element, other than constituents of the injector gas, can be determined by ICP–AES, and, in practice, all but a few elements can be determined by commercially available equipment. The accuracy, precision, and

© 1987 VCH Publishers, Inc.
Radiation Chemistry: Principles and Applications

sensitivity attainable by ICP–AES are suitable for all but the most stringent application requirements. Furthermore, the almost invariable presence of a dedicated computer in ICP–AES instrumentation means that despite the complexities of the method, new users find themselves rapidly producing good results. Many applications of ICP–AES for elemental analysis of a variety of samples are reviewed in Chapter 16.

The purpose of this chapter is to enable potential users of ICP–AES to compare the capabilities of the method with their specific requirements. The technical characteristics of foremost concern to practical analysts are accuracy (or absence of interference effects), precision (or absence of variability), detection limit, and calibration range. These are discussed here in detail. Tests of performance of another kind can also be applied to ICP–AES. For example, the ease of use and calibration, sample throughput rate, space and work-force requirements, and capital and operating costs are also very important to practical analysts, but such factors, which often are specific to the user, the task, and the environment, are not discussed in detail here.

A discussion of the performance characteristics of any method naturally tends to dwell longer on problems than on merits. This weighting toward problems may incline the beginner to feel that the problems predominate. However, in ICP–AES, there are no grounds at all for this feeling.[3] The commonplace precautions of analytical chemistry suffice for nearly all circumstances. Moreover, the method is genuinely a pleasure to use.

In the construction of ICP–AES equipment, many features that are critical to the performance are fixed at manufacture. Thus, the typical user has little control over the light-gathering power and resolution of the spectrometer, the size of the ICP torch, the frequency of the RF generator, etc. On the other hand, there are several operating parameters that are adjustable by the analyst: factors including the RF power supplied to the plasma, the gas flow rates for the torch, and the observation height in the plasma used for spectroscopic measurements. These adjustable parameters affect the performance of ICP–AES in complex, interrelated manners, and different spectral lines are affected in individually different ways. Thus, the selection of the best combination of these parameters is a critical prelude to good analysis, and one that might seem a daunting task to the beginner.

Modern commercial equipment, however, is usually thoroughly characterized by the manufacturer in relation to these parameters, and the user is invariably provided with recommended operating conditions. Often, the only real choice facing the user is that of the best spectral lines to use, and many instruments now provide a library of suitable lines, and their characteristics, as part of the software package supplied with the instrument.

Because of its flexibility, ICP–AES can be operated with a variety of different injection devices for the introduction of sample material, which can be a solid, liquid, or gas. This chapter is restricted to the most widely used and reliable method of sample introduction, namely the injection of the analyte as a nebulized solution.

5.2 OPERATING CONDITIONS AND ANALYTICAL PERFORMANCE

5.2.1 Influence of Operating Parameters

The intensity of an analyte line in ICP–AES is a complex function of several factors that are open for the operator to adjust. The variable parameters that are of principal concern are those that affect the "kitchen" area of an ICP–AES instrument, namely the RF power coupled into the plasma, the gas flow rates for the ICP torch, the observation height for spectroscopic measurements, and the solution uptake rate of the nebulizer. The most significant effects on an analyte signal intensity are produced by variations in forward power, observation height and injector gas flow rate. These three factors, however, interact in a complex fashion, and their combined effects are different for different spectral lines. It is the selection of an appropriate combination of these factors that is of critical concern in the use of ICP–AES.

The outer and the intermediate gas flows and the solution uptake rate have relatively small effects on spectral line intensities, and their influence can be understood and adjusted almost independently of the other factors.

5.2.1.1 Major Factors Affecting Line Intensities

The influences of the RF power coupled into the plasma, the injector gas flow rate, and the observation height on the intensities of various spectral lines are not yet fully understood. There is, as yet, no complete theoretical model of the ICP purely in terms of magnetohydrodynamics and analyte excitation mechanisms. However, a number of recent papers describing the spatial variations of emission within the plasma have elucidated some of the problems, as described in Chapter 8. These studies have accounted for many of the ICP–AES phenomena in terms of "soft" and "hard" spectral lines, which demonstrate rather different emission characteristics.[4]

Soft lines are excited mainly by collisional mechanisms and give their greatest intensities low in the plasma, ie, less than about 10 mm above the load coil. At greater observation heights in the plasma, increasing ionization of the analyte atoms reduces the neutral atom population, and a decrease in intensity results. Hard lines, with greater excitation potentials, are not excited at the temperatures prevailing at low levels in the plasma, and at greater observation heights they are excited by mechanisms that probably involve excited argon species. Eventually, however, the plasma gas cools to the extent that this type of excitation also diminishes, so that hard lines also exhibit a point of maximum intensity, although at higher points in the plasma.

One can conveniently display the effects of the three predominant parameters on graphs showing emission intensity as a function of observation height. A simplified summary of the effects of these parameters on both soft lines and hard

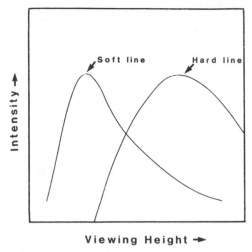

Figure 5.1. Typical behavior of emission intensity as a function of viewing (observation) height for a soft line and a hard line. The hard line peaks at a greater height in the plasma.

lines, under typical operating conditions, is provided by Figures 5.1 to 5.3. The peak intensity of a soft line typically occurs below 10 mm, in contrast to a hard-line peak at about 20 mm (Figure 5.1). Under conditions of constant RF power, the effect of variation in injector gas flow rate on the analyte intensity is shown in Figure 5.2. The main effect is a decrease in the peak intensity and a vertical shift upward of the peak position for an increase in the flow rate. Both soft lines and hard lines respond in this way, but soft lines are more strongly affected.

The effect of increasing RF power on analyte intensity, at a constant injector gas flow rate, is shown in Figure 5.3. The effect on the hard line is straightforward; the intensity increases with the RF power at all observation heights, and

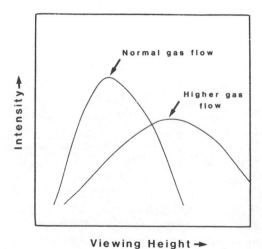

Figure 5.2. Typical effect of changes in the flow rate of the injector gas on the emission intensity at various viewing (observation) heights.

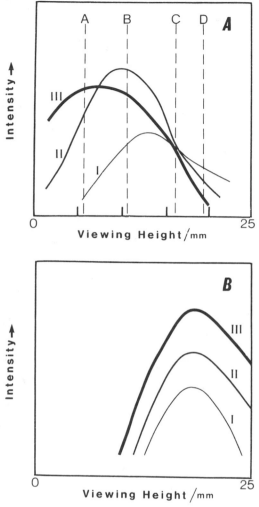

Figure 5.3. Typical behavior of emission intensity as a function of viewing (observation) height for different RF power levels supplied to the plasma. Curves I to III show the effect of increasing RF power on (*A*) a soft line and (*B*) on a hard line.

the height at which the peak intensity occurs does not vary much. Thus, at any point in the axial channel of the ICP, intensity increases with RF power. The soft line is affected in a more complex way. An initial increase in power both increases the peak intensity and shifts the peak to a lower position in the axial column. A further increase in power causes a further downward shift of the peak, but decreases the intensity. Thus, the relationship between intensity and power will vary with observation height. At point *A* in Figure 5.3*A*, intensity increases with RF power. At point *B* in Figure 5.3*A*, a steady increase in RF power causes intensity to rise to a maximum and then decline again. At point *D*, an increase in RF power causes a small decrease in intensity. At point *C*, increased power can be seen to have a very small effect.

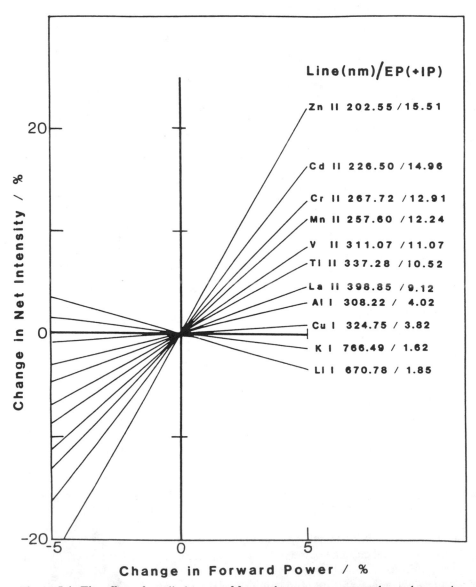

Figure 5.4. The effect of small changes of forward power, at commonly used operating conditions, on emission intensity for a range of spectral lines. The magnitude of the effect is strongly related to the excitation potential (EP) or, for ion lines, the excitation potential plus the ionization potential (EP + IP).

This last point is important in practical analysis because point C is close to the commonly used observation height. The finite size of the viewing "window," usually a square of about 4-mm sides, tends to reinforce this effect by averaging the intensity over a short range. Thus, under suitable operating conditions, the apparently anomalous situation arises that soft lines are less susceptible to power changes (and indeed to injector gas flow rate changes) than are hard lines, despite their generally greater susceptibility.

This effect can be clearly seen in Figure 5.4, which shows the influence of small RF power variations on line intensities in an ICP–AES system operating under conditions used for routine analysis.[5] The soft lines (eg, Li) have small susceptibility to RF power, whereas the graphs for hard lines (eg, Zn) have much larger slopes. The slopes of these graphs are strongly related to the excitation potentials of the spectral lines. Small variations in injector gas flow produce similar effects that are likewise related to excitation potentials.

5.2.1.2 Minor Factors Affecting Line Intensity

With pneumatic nebulizers, the injector gas flow rate has a much greater effect than the sample uptake rate. In instruments designed for practical analysis, the nebulizer gas comprises the whole of the injector gas, and there is little scope for varying the injector gas flow rate.

The solution uptake rate can be varied with some benefit however, and the results, on first glance, are surprising. Increasing the solution uptake beyond the "free-uptake" level by pumping produces a slight decrease in signal intensity and an increase in noise. Decreasing the solution flow rate below the free-uptake level, creating a "starved" nebulizer, produces an increase in signal up to a maximum level of sometimes 1.5 times the original intensity. At this stage, the noise is reduced, so that the signal-to-noise (S/N) ratio is improved. At uptake rates lower than that corresponding to the maximum S/N, the intensity falls sharply.

The use of a starved nebulizer is especially advantageous when solution volumes are limited. Satisfactory results can be obtained by these means with as little as 0.5 mL of solution, without resorting to special equipment. Starvation can be induced simply by reducing the pump rate, but this tends to increase pulsation in the liquid flow, and hence produces more noisy signals. Better results are achieved by using lengths of flexible capillary uptake tube to control the flow. A 300-mm length of polyethylene tubing of 0.6-mm bore is used routinely in this laboratory to give sample flow rates of about 0.6 mL/min for aqueous solutions in a Meinhard nebulizer. Careful control of viscosity is needed, however.

Adjustment of outer and intermediate gas flows in the ICP torch plays only a minor role in practical analysis. A zero intermediate gas flow is commonly used for aqueous solutions. Low levels of gas flow, up to about 1 L/min, are used with organic solvents. The main effect of this practice is to lift the whole plasma, relative to the torch, to help prevent the deposition of carbon on the injector tip.

Outer gas flow is often set at a level that reflects economy in the use of argon rather than the maximization of line intensities. If the outer gas flow is reduced below certain limits, spectral interference from molecular species, such as NO, can occur.

5.2.2 Optimization Strategies

5.2.2.1 The Value of Optimization

In ICP–AES, the response, ie, the signal intensity of an analyte line, is a complex function of several operational variables. The most useful combination of these variables is, in many cases, the set that maximizes that response. For a system of n such independently adjustable variables, this point in n-space provides the maximum extension of the analytical response in the $(n + 1)$th dimension. There are many ways of finding or closely approaching this optimum point. Until recently, trial-and-error procedures have predominated in the search for response optima in analytical methods. The shortcomings of such an approach are illustrated in Figure 5.5A, which shows a set of hypothetical response contours for two variables, x and y. A naive trial-and-error search might involve two univariate scans at intermediate values of x and y. If the x-scan were conducted first, the response curve shown in Figure 5.5B would be produced. If the y variable were then investigated at the x value corresponding to the maximum (R), a second response curve (Figure 5.5C), would be produced, with a higher maximum (Q). The experimenter might assume that the point Q was the optimum, which clearly is an incorrect conclusion.

The surest method of finding an optimum is to conduct a complete factorial experiment, but that is too time-consuming when many variables are involved. For example, if six variables were to be investigated and the range of each variable were divided into five steps, a total of 5^6 (15,625) experimental points would have to be evaluated for response. Several methods that are much more efficient than this have been devised. A method that is both fairly efficient and easy to understand and execute is the simplex method.[6–10]

5.2.2.2 Sequential Simplex Optimization

A simplex is a geometrical figure in n-space that has the smallest number of vertices (ie, $n + 1$). Thus, in 2-space, where there are two different variables, the simplex is a triangle (Figure 5.5A). In three dimensions, the simplex is a tetrahedron. The optimization procedure is easy to visualize only in 2-space, but simple mathematics allows us to extend the method to spaces of higher dimensionality.

To locate an optimum on the contours shown in the previous example, the procedure is initiated by drawing the simplex, an equilateral triangle ABC, on the graph in an arbitrary position. The responses for values of x and y represented by each vertex of the simplex are evaluated, and the lowest point is

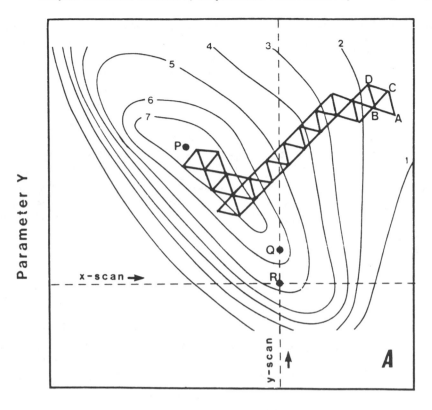

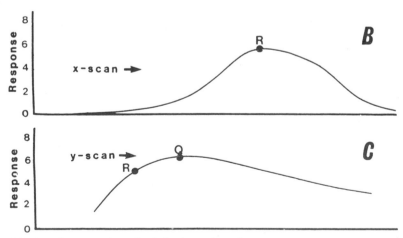

Figure 5.5. (*A*) Contours 1 to 7 of the intensity of a response due to two adjustable parameters. Two successive univariate scans (*B* and *C*) fail to find the true optimum (*P*). The simplex method closely approaches *P*.

identified: A in Figure 5.5A. The simplex is then "flipped over" about line BC to form a new simplex BCD for which the process is repeated. As a result, the simplex climbs the hill, although not necessarily in the shortest direction, until it is close to the peak, which it then circles indefinitely. If the initially selected simplex is too small, an unduly large number of steps would be required to locate the optimum. If the initial simplex were larger, a correspondingly greater uncertainty in locating the optimum would result. An occasional deviation in the procedure is needed when the newly created point in the simplex is again the lowest in the response. This can happen when the simplex straddles a ridge or is close to the optimum. In this case, the next lowest point is reflected through its opposite side, and the simplex proceeds in a new direction. When an optimum set of parameters has been tentatively identified, it must be checked by performing short univariate searches in each parameter around the optimum point.

Several variants on the simplex method[6-10] have been formulated. Perhaps the most useful is the Nelder–Mead version that allows the simplex to change shape and size, depending on the rate of increase in the response (Figure 5.6). When the optimum is approached, the simplex becomes smaller, allowing a convergence on the required point.

A further factor that has to be taken into account is that of boundary conditions, as in Figure 5.5A, where data are obtainable only within certain ranges of each parameter. If a simplex tried to "step outside" the permitted zone, an unobtainable vertex would be specified. This is avoided by giving arbitrary low responses to such vertices, which has the effect of reflecting the simplex away from the boundary.

5.2.2.3   Problems with Optimization

A problem immediately encountered in the optimization of an analyte response is the question of which response is the correct one to optimize. For instance, it

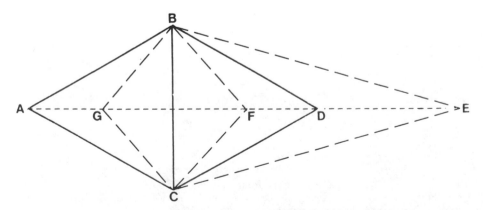

Figure 5.6. A modified simplex method. By reflection (D), expansion (E), or contraction (F, G), the simplex moves away from the lowest response at A toward the maximum.

is possible to maximize line intensity, net signal, signal-to-noise ratio, signal-to-background ratio, signal-to-background-noise ratio, signal-to-interference-effect ratio, and many others. Naturally, the intended use of the analytical method usually suggests a choice. For example, optimized signal-to-noise ratio will tend to give the best precision for high concentrations of analyte, whereas signal-to-background or signal-to-background noise will optimize for lowest detection limit, in the absence of interferences. It is noteworthy that optimizations reported to date for argon ICPs have not been concerned with the unit cost of analysis, interference effects, or freedom from instrumental drift.

One should also note that optimizing one of the above-mentioned ratios may produce a false sense of security concerning the stability of the net signal, which is the actual quantity that is related to analyte concentration. For example, in Figure 5.7, the signal-to-background ratio is maximized at point *A*. The net signal, however, is still strongly dependent on RF power. It is maximized and least dependent on small power variations only at a higher power (point *B*). Because power drifts probably make a significant contribution to signal instability, it might well be thought preferable to optimize on net signal to minimize drift problems.

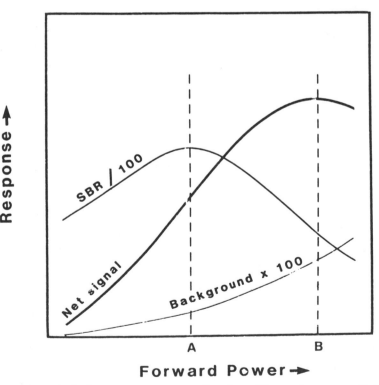

Figure 5.7. Hypothetical response curves as a function of forward power applied to the ICP. Net signal and signal-to-background ratio (SBR) reach maxima at different power levels.

Having settled on a suitable response to optimize, the analyst who wants to optimize for simultaneous analysis now faces a serious obstacle in that each analyte will have a unique set of optimized parameters. A compromise of some kind is therefore required.

In practice, this problem is often solved in a crude way relating to the performance of a key element. For instance, in the analysis of soils for environmental purposes, lead is perhaps the most important single element. Good detection limits for many metals are easy to obtain under a wide range of operating conditions, but the detection limit for lead, at its best, is not as low as desirable (in relation to median levels in uncontaminated soil). A workable compromise would amount to the optimization of the instrument for low detection limits for lead, and the determination of all the other elements under suboptimal conditions.

A more objective way to optimize for simultaneous analysis would involve the optimization of a combined response consisting of a weighted mean of the individual element responses.[10] The weights could reflect the relative importance of each element, and for any element the difference between the actual and the required analytical performance.

5.2.2.4 The Practical Position

Rigorous optimization is hardly ever a major preoccupation of practical analysts using ICP–AES. Manufacturers will almost invariably provide "cookbook" instructions, which include recommended conditions for various types of analysis, and there is rarely any need to deviate from these. Some manufacturers have even prevented any adjustments by fixing such parameters as gas flow rates and observation height. In the author's opinion, this is the correct approach in task-oriented analysis. Complete stability in operating conditions, hence complete predictability of behavior, is more useful than complete optimization, as long as the response is not grossly suboptimal.

5.2.3 Commonly Used Operating Conditions

Operating conditions used in practice tend to fall into clusters, depending mainly on the type of ICP torch used. Larger torches, based on Greenfield's original design, need large gas flows to sustain the plasma, and correspondingly high levels of power can be coupled into the plasma.

Most manufacturers of ICP–AES equipment employ a smaller torch based on Fassel's design, which requires considerably lower gas flows. Consequently, the operating conditions reported for practical applications most often fall within the small ranges suitable for this type of torch (Table 5.1).

In the interest of cutting operating costs, much attention has been paid to reducing the various dimensions of the torch, as discussed in Chapter 14. This would enable less argon to be used and would facilitate the use of a solid-state

Table 5.1. Operating Conditions Commonly Used in ICP-AES for Nebulized Aqueous Solutions.

Torch:	Fassel type
Injector gas flow (Ar):	0.6 to 1 L/min
Intermediate gas flow (Ar):	0 to 0.5 L/min
Outer gas flow (Ar):	12 to 18 L/min
Forward RF power:	1.1 to 1.3 kW
Observation height:	Centered at 14 to 18 mm above load coil
Nebulizer uptake rate:	0.5 to 2.0 mL/min

RF generator because lower RF powers are required. It is, of course, desirable to maintain the excellent analytical performance of the plasma generated by the Fassel torch. Clearly, a different set of optimum conditions is appropriate for any low-gas-flow torches.

5.3   DETECTION LIMITS

5.3.1   The Concept and Definition of Detection Limits

5.3.1.1   Instrumental Detection Limits

Any line intensity used as a measure of analyte concentration is necessarily a difference between two quantities. The total light intensity passing through an exit slit of the spectrometer consists of photons originating from analyte atoms plus photons from other sources in the plasma. This total intensity is the quantity that is measured by the instrument. To obtain the net response from the analyte atoms alone, it is necessary to estimate the background response (ie, the intensity that would have been produced if no analyte atoms were present) and subtract this quantity from the total.

The manner in which the background intensity estimate is produced depends on the type of spectrometer used (Section 5.5.3). In scanning instruments, the background is estimated from values measured on either side of the analyte line within a few seconds of the analyte line measurement. In polychromators, the background is normally measured at the peak wavelength when a blank solution is nebulized at a separate time. Either way, there is uncertainty in the background estimate. When the signal from the analyte atoms is very low, and similar in magnitude to the uncertainty in the background level, the presence of analyte atoms cannot be definitely established.

The detection limit is defined by IUPAC as the concentration that produces a net line intensity equivalent to three times the standard deviation of the background signal. The standard deviation is often estimated from 11 consecutive integrations of the signal during the continuous nebulization of the blank solution. Given certain reservations, this is a useful definition, as the detection

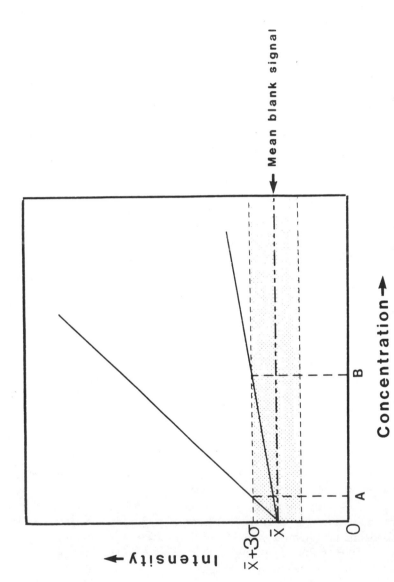

Figure 5.8. The detection limit refers to the concentration at which the analyte signal intensity is equal to the mean blank signal ($\bar{x}$) plus three times the standard deviation (σ) of a blank solution. For the same $\bar{x}$ and σ, a line of higher sensitivity (slope) will have a lower detection limit (A) than that of a less sensitive line (B).

limit can be estimated rapidly for a comparison between different instruments or between differing operating conditions on a single instrument. However, these detection limits are unrealistically low for application to most practical analysis and should be qualified as "instrumental detection limits" (IDL). The numerical value of IDL depends on both the value of the blank uncertainty and the sensitivity, ie, the slope of the calibration line (Figure 5.8). Instrumental factors that increase the sensitivity without affecting background noise improve the IDL. In contrast, factors that increase the uncertainty in the background estimate, such as spectral interference corrections, degrade the IDL. Values of IDL attainable[11] in conventional ICP–AES are given in Table 5.2.

For a given line, and under optimized operating conditions, the sensitivity can be increased only by increasing the rate at which analyte is injected into the plasma. This can be effected by a variety of methods, including: (a) using a more efficient nebulizer or utilizing an existing nebulizer more effectively, (b) nebulizing a preconcentrated solution, and (c) injecting analytes as hydrides or other volatile species. These methods are discussed in Chapters 11 and 13. A reduction of background noise is more difficult to achieve for the general user, but some improvement can be made by the use of a starved nebulizer, or, as discussed in Chapter 4, through the use of a high-pressure nebulizer.

5.3.1.2 Practical Detection Limits

Practical detection limits are greater than IDLs by a factor that typically ranges from 1.5 to 5, a factor that depends on the sample type and the conditions of the analysis. The final quantification in any analysis is the result of a blank subtraction, either explicit or implicit, to produce the net spectral line intensity. This doubles the variance, or increases the detection limit by a factor of $\sqrt{2}$.

A further reason for higher practical detection limits is illustrated in Figure 5.9. The background signal varies both randomly and systematically within the period between recalibration points (Figure 5.9). Integrations used for IDL estimation are spaced very closely in time and represent a minimum possible variance that is dependent mostly on random errors. Integrations spaced at random within the whole calibration period show a larger variance because of the systematic changes ("drift") also included, and this larger variance is closer to the practical situation. Thus, the practical detection limits vary with the interval between blank calibration checks. As a rule of thumb, the detection limits may be assumed to change by about 50% for each hour between calibration checks, although this number may vary widely between instruments.

Real samples usually contain concomitant elements, many of which give rise to spectral interference (ie, background enhancement or line overlap) that has to be corrected. As discussed in Section 5.5.3, the correction term obtained by "on peak" or "off-peak" methods has to be subtracted from the apparent analyte response to give the corrected analyte response. The variance in the correction terms is also added to the total uncertainty in estimating the background intensity. When spectral interferences are high, this factor may predominate

Table 5.2. ICP-AES Instrumental Detection Limits (μg/L) Giving the Most Sensitive Line (nm) for an Element, with Nebulization of Aqueous Solutions[a]

0.1–0.31	0.32–0.99	1–3.1	3.2–9.9	10–31	32–99	100–316
Be 313.04	Os 225.58	Ba 455.40	Ag 328.06	Al 309.27	As 193.69	Rb 780.02
Ca 393.36	Sr 407.77	Cd 214.43	B 249.77	Au 242.79	Bi 223.06	U 385.95
Mg 279.55		Eu 381.96	Co 238.89	Dy 353.17	C 193.09	
		Li 670.78	Cu 324.75	Er 337.27	Ce 413.76	
		Lu 261.54	Cr 205.55	Gd 342.24	Ga 294.36	
		Mn 257.61	Fe 238.20	Hf 277.33	Ge 209.42	
		Sc 361.38	Ho 345.60	Hg 194.22	In 230.60	
		Yb 328.93	I 178.28	Ir 224.26	K 766.49	
		Zn 213.85	La 394.91	Na 588.99	Nb 309.41	
			Mo 202.03	Ni 221.64	Nd 401.22	
			Re 197.31	Pt 214.42	P 178.29	
			Ti 334.94	Ru 240.27	Pb 220.35	
			V 309.31	Si 251.61	Pd 340.45	
			Y 371.03	Sn 189.98	Pr 390.84	
			Zr 343.82	Ta 226.23	Rh 233.47	
				Tb 350.91	S 180.73	
				W 207.91	Sb 206.83	
				Cs 825.11	Se 196.02	
					Sm 359.26	
					Te 214.28	
					Th 283.73	
					Tl 190.86	

[a] Practical detection limits may be higher, depending on circumstances.

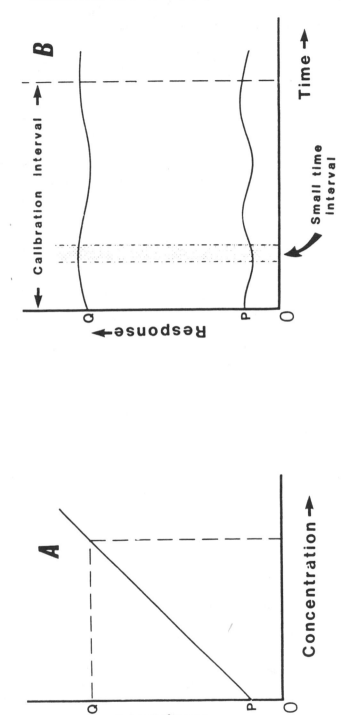

Figure 5.9. (*A*) A schematic calibration with two fixed points *P* and *Q*. (*B*) Variation in the position of *P* and *Q* on the response axis with the passage of time, showing exaggerated drifts. The standard deviation of a number of measurements of *P* and *Q* repeated within a small time interval (eg, 11 integrations within 2 min) would be much smaller than a standard deviation of results scattered within the calibration interval (eg, 1 to 2 hours). Precision and detection limits usually cited refer to the small interval.

among the error terms at low analyte concentration and become the main determinant of detection limit.

For samples that need chemical manipulations before the instrumental determination, errors in these various stages also introduce additional variance into the total error determining the practical detection limit.

5.3.1.3 Other Measures of Detectability

A less frequently used figure of merit, related to IDL, is the background equivalent concentration (BEC), the analyte concentration producing a net signal equal to the background intensity at the line. Because the background noise is often roughly 1% of the background intensity, the BEC is about 30 times the IDL. Variations in the signal-to-noise ratio from line to line, however, render the BEC less useful to the practical analyst than the IDL.

The main advantage of the BEC over IDL is that the BEC can be estimated with a much smaller relative error than IDL, from the same data. If a background intensity (I) is measured 11 times, giving a mean estimate $\bar{I}$ and a standard deviation estimate s, the BEC estimate is equal to $\bar{I}/k$, where k is the sensitivity. The standard error of $\bar{I}$ is $s/\sqrt{11}$, the relative standard error is $s/\sqrt{11}\,\bar{I}$. If $s/\bar{I}$ is 0.01, a typical level for ICP–AES, the RSD of $\bar{I}$ is approximately 0.003, or 0.3%. Hence, because the value of k is accurately known, the estimate of BEC has a %RSD of better than 1.

The standard error estimate of s is $s/\sqrt{22}$, so the RSD of the IDL is $1/\sqrt{22}$, ie, about 20%. Thus, estimates of IDL will be prone to large sampling variations, a fact that must be remembered when values of IDL are compared. Differences of 50% in IDL estimates should not be considered as significant. BEC is therefore capable of showing the effects of any instrumental modifications more clearly than IDL.

5.3.2 Some Practical Factors Affecting Detection Limits

5.3.2.1 Nebulizer Type

Pneumatic nebulizers, as a group, yield similar detection limits for a given analyte line and integration time. For strict comparison, each nebulizer, with all its operating parameters optimized, should be tested on a single instrument. This is seldom done in practice. An example, however, can be found in a paper of Walton and Goulter,[12] where a Babington-type nebulizer was compared with a concentric glass nebulizer. Detection limits given by the nebulizers for 11 elements were compared and found to be very similar. This similarity in detection limits between different pneumatic nebulizers implied that their absolute efficiencies (not relative efficiencies) were very similar. Indeed, the main difference in performance between these nebulizers lies not in their IDLs but in their degree of tolerance to high-solids solutions or the presence of suspended material. This last factor may well give rise to better practical detection limits in solution with high dissolved solids contents.

The rate at which analyte is injected into an ICP can be increased an order of magnitude over that of pneumatic nebulizers by the use of an ultrasonic nebulizer (Chapter 4). These devices simply produce fine aerosol more efficiently than pneumatic nebulizers, and thereby increase the rate of analyte injection into the plasma. However, the rate of solvent injection is increased by a similar factor, a circumstance that cools the plasma and strongly reduces the excitation efficiency. Improved detection limits for argon ICPs can therefore be obtained only if the excess solvent is largely removed from the injector gas flow by desolvation.[13] Again, improved detection limits can be achieved only when spectral interferents are at a low level. The contribution of spectral interference corrections to the total variance proportionately increases with the sensitivity enhancement produced by the higher rate of analyte injection. Thus, if this contribution is the dominant term in the total background uncertainty, no improvements in detection limits are effected by the use of an ultrasonic nebulizer. Hence, ultrasonic nebulization may provide improved detection limits for samples such as fresh waters, but such improvement may not be observed for complex matrices.

5.3.2.2   The Effect of Integration Time

Both sensitivity and blank noise level depend on the length of time over which the stabilized signal from the ICP is integrated. The sensitivity is proportional to the integration time, but the standard deviation of the blank noise should vary as the square root of the integration times, given that the noise is random and has a "white" spectrum. Under these conditions, the detection limit should vary inversely with the square root of the integration time; ie, an increase of four times in integration time would lower the detection limit by a factor of 2.

The limited data available suggest that this relationship is true for integration times in the normal range, 1 to 10 s, and for instrumental detection limits. The relationship may not be valid under practical conditions, however, where factors other than instrumental noise come into play. For instance, if the detection limit is dominated by drift or correction errors, no improvement may be apparent with increased integration times.

5.3.2.3   Sequential and Simultaneous Analysis

Little difference is noted between instrumental detection limits obtained by simultaneous and sequential instruments when the same lines and comparable integration times are used. The differences are of the same magnitude as the differences between instruments of the same quality from different manufacturers. Because operating conditions may be optimized for individual elements in sequential spectrometry, practical detection limits, in the absence of complications, should be slightly better than those achieved with simultaneous spectrometers.

5.3.2.4   Detection Limits in Strong Salt Solutions

Given identical operating conditions, detection limits in strong salt solutions tend to be inferior to those measured from pure aqueous solutions for two separate reasons. First, the sensitivity of the analyte may be lowered because of a matrix effect in the ICP due to the concomitant salt. Second, the baseline is destabilized because of increased noise and background shift[14-16] by the salt solution to an extent depending on the nebulizer type. Each of these factors degrades the IDL. Babington-type nebulizers probably are less prone to baseline destabilization by strong salt solutions than are concentric nebulizers, and therefore, their detection limits are expected to degrade to a smaller extent.

5.3.2.5   Detection Limits for Organic Solvents

Organic solvents can, under suitable conditions, be readily nebulized into an ICP. A high loading of solvent vapor tends to cool the plasma, resulting in analyte signal depression. Therefore, work with organic solvents tends toward the following conditions: (a) use of a solvent that is not too volatile, (b) use of restricted uptake rates, and (c) use of a forward power somewhat higher than that used with aqueous solutions, typically 1.5 to 1.8 kW. Organic solvents have been used principally in two areas: the analysis of lubricating and hydraulic oils for wear metals, and the analysis of solvent extracts from aqueous solutions following separation and preconcentration.

Detection limits for a range of analyte lines and organic solvents have been compared with those of aqueous solutions.[17-19] Generally, detection limits are, within a factor of 2, comparable to those for aqueous solutions, except for certain analytes in very volatile solvents, such as carbon tetrachloride, where somewhat inferior detection limits are obtained. The similarity between limits may be explained by the sensitivity changes brought on through a combination of increased analyte injection due to nebulizer effects, which cause small enhancements, and plasma cooling by the volatile solvents, which generally leads to signal depression. Detection limits have been reported[13] for 15 elements, for oils diluted 10 times with 4-methylbutan-2-one. Despite variations in viscosity of the solutions, the detection limits are close to those obtained at the same lines for aqueous solutions.

5.3.3   Comparison with Other Plasma–Source Methods

5.3.3.1   Inductively Coupled Plasma–Mass Spectrometry

Mass spectrometry using the ICP as an ion source is described in Chapter 10. The instrument is commercially available. Detection limits for most elements by this technique are one to three orders of magnitude superior to those of ICP-AES. At the present stage of development, however, ICP-MS cannot compete with ICP-AES in providing a rapid, precise method for routine analysis of a wide variety of samples.

5.3.3.2   Inductively Coupled Plasma–Atomic Fluorescence Spectrometry

The ICP–AFS technique, discussed in Chapter 9, is a variant of the basic ICP method that combines the best features of emission and absorption techniques while rejecting the less desirable properties of both. Commercial ICP–AFS instrumentation, suitable for routine use in the simultaneous determination of up to 12 elements, is available. The main benefits of the technique are the virtually complete absence of spectral overlaps and good baseline stability combined with long, linear calibration ranges. The main drawback focuses on detection limits, which for many elements are an order of magnitude inferior to those of ICP–AES, especially for the elements that form refractory oxides. This deficiency can be offset somewhat by the introduction of propane into the injector gas to reduce oxygen activity, and by using ultrasonic nebulization.

5.3.3.3   Emissions from Microwave-Induced Plasmas and Direct-Current Arc Plasmas

Comparative studies of microwave-induced plasmas and direct-current arc plasmas with ICP–AES as atomic emission sources have been reviewed.[20] The main limitation of both plasmas is related to the difficulty in injecting sample aerosol into the hottest part of the plasma. This results in a set of matrix, chemical, and ionization interferences familiar to analysts having experience with combustion flames. Thus, these methods are at their best with "low-matrix" samples, such as fresh waters. "High-matrix" samples need ionization buffering and frequent recalibration to achieve good results. Detection limits are somewhat inferior to those obtained by ICP–AES.

5.4   WORKING RANGES OF CONCENTRATION

5.4.1   Limits of Quantitative Determination

While the detection limit is clearly the lowest point at which an analytical signal is greater than the background noise level, most applications require a greater certainty for the quantification to be realistic. This can be achieved only at concentrations higher than the detection limit, and one such standard, the limit of quantitative determination (LQD), is arbitrarily defined as five times the detection limit. The practical detection limit is more appropriate in this context than the IDL, although this aspect is often unspecified.

In practical analysis, it is usually possible to relate the standard deviation (σ) of a concentration measurement (c) in a particular matrix by a linear relationship:

$$\sigma = \sigma_0 + kc \qquad [5.1]$$

where σ_0 is the standard deviation at zero concentration and k is a constant. In terms of relative standard deviation:

$$\text{RSD} = \frac{\sigma}{c} = \frac{\sigma_0}{c} + k \qquad [5.2]$$

The term k is thus the asymptotic RSD at high concentrations. The detection limit, c_L, is the concentration at which RSD $= \frac{1}{3}$. Thus,

$$\frac{\sigma_0}{c_L} + k = \frac{1}{3}, \qquad [5.3]$$

A combination of Equations 5.2 and 5.3 gives the RSD in terms of c', the concentration-normalized detection limit, ie, c/c_L:

$$\text{RSD} = \frac{[k(c'-1) + \frac{1}{3}]}{c'} \qquad [5.4]$$

At the LQD, c' is 5 by definition, and therefore from Equation 5.4 the RSD is $(4k + \frac{1}{3})/5$. Thus, at the LQD, the RSD cannot be less than about 0.066 (6.6%). For ICP-AES, the value of k in practical analysis usually is within the range 0.015 to 0.03, so that at the LQD, the RSD falls between 0.08 and 0.09. An RSD of 0.1 is achieved at a concentration corresponding to about four to five times the detection limit.

5.4.2   Upper Limit of Calibration

Calibration relationships of analyte concentration versus net intensity are essentially linear in ICP-AES over concentration ranges spanning three to six orders of magnitude. This is an important feature of ICP-AES and is essential for effective simultaneous analysis. Most ICP-AES users are aware that deviation from strict linearity can be detected even at low concentrations.[21] These deviations have no practical effect, usually, until a concentration of 10^3 or 10^4 times the detection limit is encountered. Beyond this point, nonlinear equations have to be fitted if the calibration is to be used at higher concentrations. Differences in the degree of line curvature for different elements may be partly attributed to the line broadening occurring in the plasma (Chapter 8).

5.5   ACCURACY AND INTERFERENCE EFFECTS

5.5.1   Introduction

Accuracy can be achieved by an analytical method in the absence of any uncontrolled interference effect. Interference effects are responses of the analytical sensor due to solution constituents other than analyte. The analytical sensors of present interest are the spectral line intensities measured in ICP-AES.

When ICP–AES was introduced into the laboratory, it was acclaimed to be virtually interference free. This claim has not survived its application to the analysis of real samples. It is now realized that significant interferences do occur, but in almost every field of application, they can be successfully overcome by the normal measures used by cautious analysts. In addition, interference effects in ICP–AES are smaller probably than in any other field of atomic spectrometry. Certainly, interference effects are of far greater concern and severity in microwave-induced plasmas and direct-current plasmas.

This relative freedom from interference effects in ICP–AES is the result of the high temperature to which the sample material is subjected on passing through the ICP (Chapter 8). Because of this high temperature, the volatilization and atomization of the sample material are thought to be virtually complete in the observation zone of the plasma.[22,23] Thus, there is little scope for chemical interference (ie, molecule formation), which reduces the atom population in the plasma. The problem of selective volatilization does not arise, and oxide formation does not occur to reduce sensitivity.

Two indirect effects contributing to interferences are associated with the nebulizer and the ICP torch. The first such effect is caused by aerosol ionic redistribution[24] in the nebulizer. The smallest droplets produced by the nebulizer have a composition different from that of the bulk solution. However, no practical problems relating to this phenomenon have been reported. The second interference is the occurrence of a memory effect in the plasma torch. If a concentrated potassium solution is nebulized into the plasma, followed by a blank solution, the baseline returns to the background level in less than a minute. Subsequent nebulization of a sodium solution, however, releases a burst of potassium into the plasma. This effect can survive the vigorous washing of the torch with concentrated acids. No study of this effect has been published, and it is not known how general the effect is.

5.5.2   Matrix Effects

5.5.2.1   Nebulizer-Related Matrix Effects

The operation of a nebulizer is affected by the physical properties of the liquid, in particular, by density, viscosity, and surface tension. The median droplet size produced by a pneumatic nebulizer is reported to be governed by the Nukiyama–Tanasawa equation (Chapter 11) as a function of these physical properties. The droplet size, in turn, affects the rate of analyte injection, hence the sensitivity of measurements. The effects of changes in physical properties of the sample solution on relative sensitivity for the analyte are most noticeable where concentrated solutions of mineral acids are used, especially sulfuric and phosphoric acids with their large changes in density and viscosity (Figure 5.10). The effects of mineral acid concentration have been described by Greenfield and his co-workers,[25] and the same effects can be seen in solutions containing viscous organic substances, such as oils or thickening agents. Important though they

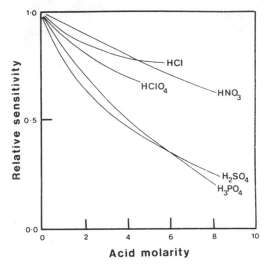

Figure 5.10. The effect of the concentration of several mineral acids on the sensitivity of an analyte. (Calculated from the data given in Reference 17.)

are, these nebulizer effects cause very little problem in practical analysis. The sample solutions and the standard calibration solutions usually can be adjusted to a constant matrix concentration so that the efficiency of the nebulizer operation, hence the instrumental sensitivity, are constant for both the samples and the standards.

When matrix matching is not possible, the method of standard additions should always prove helpful in ICP–AES because of the long linear calibration ranges. However, standard additions necessarily increase the analysis time by at least 100%. Internal standardization is not recommended in the general case, because of the considerable number of complications that may arise to distort the results in complex matrices. One should note that the use of a peristaltic pump does not overcome the effects of changing viscosity. While the pump delivers solution to the nebulizer tip at a constant rate, the nebulization itself is affected by the viscosity, and pumping cannot alleviate differences attributable to viscosity changes.

5.5.2.2 Plasma-Related Matrix Effects

a. The Nature and Cause of the Effects.   The matrix effects discussed previously are related to gross changes in the physical properties of the solution being nebulized. A separate set of matrix effects arises in the plasma itself. For any analyte line, the effect is strongly related to the line excitation potential, ie, the energy required to raise the analyte atom from its ground state, or a low-lying energy level, to the excited state. Analyte lines with high excitation potentials are much more susceptible to the effect than those with low excitation potentials.

The matrix effect is brought about by a wide range of substances.[26-31] Almost any matrix element can cause a change in the sensitivity of a susceptible spectral line (eg, Zn II 202.6 nm). The effect seems to be related to the ionization of the matrix element in the plasma, but it is not simply an instance of ionization buffering. Easily ionizable elements, such as sodium and potassium, cause smaller effects than some elements with higher ionization potentials, at least at observation heights 14 to 18 mm above the load coil. The effects are rather complicated and far from fully characterized. Nevertheless, a grasp of the broad aspects is important, as it bears upon the conduct of practical analysis.

For the analyst, the study of Thompson and Ramsey[26] on matrix effects caused by calcium (Figure 5.11) is illuminating. The effect of calcium up to 10,000 μg/mL on analyte sensitivity is shown for a selection of analytes with a range of excitation potentials, under commonly used operating conditions. As the concentration of the interfering element increases, the analyte sensitivity decreases, in a manner quantitatively similar to an exponential decay, towards

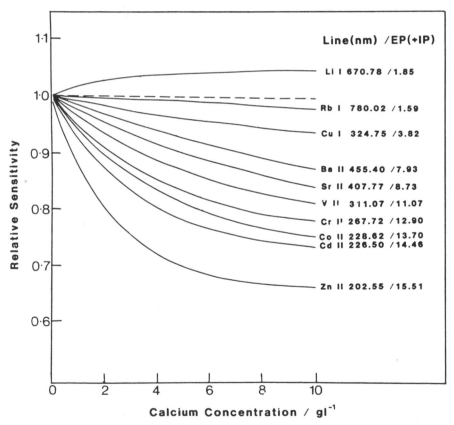

Figure 5.11. Matrix effects due to calcium, showing changes in net sensitivity for several analyte lines at commonly used operating conditions. Lines are affected according to their excitation potentials.

an asymptotic lower level, except for Li I 670.78 nm, which shows a slight enhancement. The effect is strongest for analyte lines with highest excitation potentials. The pattern is very similar to the effect of a reduction in the forward power supplied to the plasma. It may be related to the population of excited states derived from the Boltzmann distribution as a function of temperature. Furthermore, experiments with dual nebulizers, where the matrix element is introduced through a separate nebulizer, directly demonstrate that the effect is independent of nebulizer operation.

Spatial studies of the interference effect have been conducted by several research groups,[32-34] but the matrix elements studied have been restricted to alkali metals. In terms of mechanisms, Blades and Horlick[34] deduced that the matrix effects are not due to ionization suppression by an excess of electrons from the easily ionizable element. In the lower regions of the analyte channel (ie, up to about 10 mm above the load coil), where both atom lines and ion lines are enhanced by the matrix, the effect is ascribed to increased collisional excitation, as a result of the increased number of high-energy electrons. At higher regions in the axial channel, both atom and ion lines are suppressed in intensity in the presence of the interferent. This is thought to be due to ambipolar diffusion, a mechanism brought about by electrons diffusing radially out of the central channel (Chapter 8). This would tend to drag positive ions away from the center to cause a reduction in numbers and, therefore, in analyte emission intensity from this region. The changeover point from enhancement to suppression coincides for many, but not all, analyte lines at the observation height commonly used for routine analysis.

b. Implications in Practical Analysis.   The existence of substantial matrix effects has important ramifications in practical analysis. Clearly, standard and sample solutions should be matrix-matched as much as possible so that the sensitivity for any analyte line is constant throughout. In certain cases, this is easy to effect. In the analysis of low-alloy steels, for example, the matrix is a solution of iron of predictable concentration, which varies only slightly as the alloying element concentrations change. If the standard solutions for the alloying elements contain the appropriate concentrations of iron, the matrix effect will not be observed. In other cases, the major constituents of the samples might vary in concentration from sample to sample, as in soil analysis, and no such constant matrix can be devised for the calibration solutions.

When a constant matrix cannot be assumed, several courses of action are open to the analyst. The method of standard additions is widely applicable to systems affected by matrix effects and is very effective in ICP–AES. However, the method undoubtedly requires considerably more time for each sample. Another valuable technique is chemical separation of trace analytes from the matrix. Separation can be combined with preconcentration to provide a useful lowering of detection limits. Again, however, a considerable increase in the time spent on each sample is required. Several other methods are suggested by Thompson and Ramsey,[26] including automatic matrix matching and mathematical techniques.

Calibration strategies may have to be carefully designed to avoid problems of varying sensitivity due to matrix effects. Segregation of the trace constituents into a single solution, to avoid spectral interference from the much higher concentrations of matrix elements, is a method often used. This practice, which can result in systematic errors of 5 to 20% in trace element concentrations, explains the bias found by some practitioners between results obtained by ICP–AES and atomic absorption spectrometry on the same solutions. Relatedly, a simple serial dilution of a concentrated multiple-element standard gives rise to serious inaccuracies in calibration at the low concentration range.

5.5.3   Spectral Interferences

5.5.3.1   Introduction

Spectral interferences, discussed in detail in Chapter 6, result from the inability of a spectrometer to resolve a spectral line emitted by a specific analyte from light emitted by other atoms or ions. Atomic spectra originating from ICPs are complex because of the high temperatures of the ICP. Broad-band spectra are relatively weak, however, because of the low concentrations of molecules. Complex spectra are most troublesome when produced by the major constituents of a sample, as spectral lines from other analytes tend to be overlapped by lines from the major elements. Notable examples of elements that produce complex line spectra fall mainly among the transition elements (eg, Fe, Ti, and Mn, and U, the lanthanides, and the noble metals).

To some extent, spectral complexity can be overcome by the use of high-resolution spectrometers.[35] High resolution cannot overcome exact line coincidences, or complete broad-band overlap. Where these features cannot be avoided by selecting an alternative line, correction procedures must be used. Stray light is generally very small in most spectrometers built for ICP–AES, but recombination continuum produces a broad-band background enhancement or shift.

5.5.3.2   Correction Procedures

a. Types of Spectral Interference. Three types of spectral interference that commonly accompany an analyte peak are coincident line overlap, wing overlap from a much more intense line nearby, and broad-band background enhancement or background shift. These are illustrated in Figure 5.12, along with a no-interference case, and in each diagram, the response produced in the absence of analyte is also shown. The peak difference between each pair of spectra is the analyte intensity required for the measurement. Interferences of the types illustrated cause an additive (translational) interference effect. In absolute terms, the bias caused by the interferent is independent of the analyte concentration. Therefore, it is relatively much more important at low analyte levels. Methods available for correcting these translational interferences depend on whether a

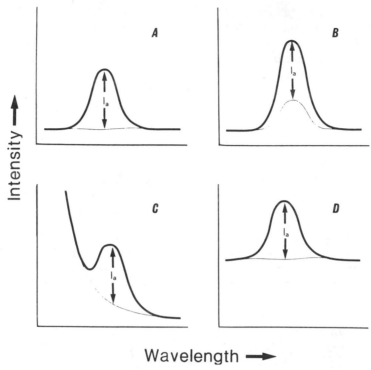

Intensity

Wavelength ⟶

Figure 5.12. Spectral line intensities under different circumstances showing the gross signal (bold line), the background signal (fine line) and the analyte peak net intensity, I_a. (A) Spectral line with no interference. (B) Analyte peak with a small coincident interferent peak. (C) Analyte peak with a wing overlap from a strong interferent peak. (D) Analyte peak with broad background enhancement, or background shift.

monochromator or polychromator is used. The method of standard additions is completely ineffective for the correction of background enhancement. For instrumental methods of background correction, the reader is referred to Chapter 3.

b. Monochromator Systems.   When complex sample matrices are injected into the ICP, the test solution may contain substances that cause background enhancement. Aluminum causes enhancement in the spectral region from 190 to 220 nm. Calcium and magnesium behave similarly. If the enhanced background beneath the analyte peak is flat, as shown in Figure 5.12D, the gross analyte intensity can be corrected by subtracting the enhanced background by the method of "off-peak" correction. In fact, the method of correcting background enhancement interference due to broad-band spectra is no different from the normal process of peak location and quantification. When a scanning monochromator is used, a series of short integrations is made over a spectral region of about 0.05 nm in the vicinity of the analyte line (Figure 5.13). After ensuring that

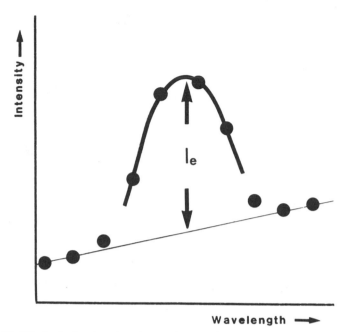

Figure 5.13. Method of estimating the analyte peak net intensity, I_e, from scan data.

a genuine peak is present, the maximum is located by fitting a mathematical function to the highest points. The background intensity is estimated by interpolation of the outlying observations, and the intensity difference at the peak wavelength is measured. This technique requires that the background emission of the ICP be relatively flat under the peak, an assumption that can be tested by running matrix-matched solutions not containing the analyte.

If the background is structured, or is the result of a wing overlap, the off-peak method may give incorrect results because the background at the peak cannot be estimated accurately from the outlying points (Figures 5.12C and 5.12D). In these cases, the uncertainty in the background intensity estimation may be the principal cause of poor detection limits. With a sequential ICP–AES, however, an alternative line often can be utilized. It must also be remembered that exact spectral coincidences cannot be corrected at all by the off-peak method.

c. Polychromator Systems. In certain polychromator systems, the background can be estimated only by measurements on a separate blank solution. The integrated blank signal on a particular channel is then subtracted from all measurements of the sample solutions made at that channel. The blank measurement should be made a short time before that of the sample so that drifts in the blank reading do not affect the accuracy of the measurements.

Background enhancements due to interfering concomitant elements are, however, not automatically corrected as they are in sequential spectrometers.

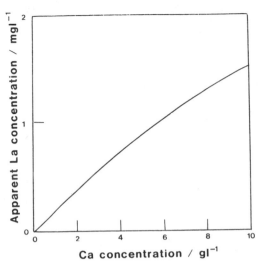

Figure 5.14. The relationship between spectral interference (apparent lanthanum concentration at La 398.8 nm) and the concentration of calcium.

Matrix matching of the standard solutions to the sample solutions would eliminate the problem, if the matching were sufficiently close, but this works only if the matrix is almost constant from sample to sample. This is not usually the case in analysis of natural materials.

Polychromators therefore employ the "on-peak" correction method. This method requires an estimation of the enhancement obtained from a simultaneous determination of the interfering elements on other channels. The relationship between interferent concentration and response at the analyte channel is determined in advance (Figure 5.14). This "pseudocalibration" is used to estimate the interference. Such curves are often nonlinear and are difficult to use as a simple "correction factor." In complex samples, there is often more than one matrix constituent that contributes to the correction term to be subtracted from the apparent analyte signal. These correction terms are assumed to be additive. All types of translational interference can be corrected by the "on-peak" method, including coincidence, wing line overlap, and background enhancement.

Many polychromators have a scanning facility on the entrance slit, which can subtract backgrounds in the same manner as do monochromators. As discussed in Chapter 6, it is desirable to have the capability of selecting different background wavelengths for channels on polychromators.

5.5.3.3 Problems in Correcting for Spectral Interference

All corrections for background, regardless of origin, are estimates and are, therefore, subject to uncertainty. In the absence of spectral interference, the uncertainty is due to noise and, more importantly, drifts in the background. This uncertainty determines the magnitude of the detection limit. In the presence of interference, the increased uncertainty degrades the detection limit of the analyte. As an example,[36] Mo 281.6 nm has an instrumental detection limit equivalent to about 0.6 μg/g in soils. The correction term for Mg (at typical concentrations of 3%, by weight) is equivalent to 18 μg/g of Mo, and the estimated uncertainty at three standard deviations is $18 \times 0.1 = 1.8$ μ/g, assuming a 10% relative uncertainty. The detection limit for Mo in soils with 3% Mg would be $[(1.80)^2 + (0.6)^2]^{1/2} = 1.9$ μg/g, an increase exceeding 300%. The effect of aluminum is more drastic. The appropriate correction term is 90 μg/g. The detection limit is increased to $[(0.6)^2 + (1.8)^2 + (9.0)^2]^{1/2} = 9.2$ μg/g, which is quite useless for soil analysis.

5.6   PRECISION

5.6.1   The Evaluation of Precision in Inductively Coupled Plasma–Atomic Emission Spectrometry

Precision is a measure of repeatability, and, like detection limit, its magnitude depends on the time scale of measurement. If repeated integrations are made over the shortest possible time scale, with unbroken nebulization of the solution, the highest precision is obtained. If the repeated integrations are spaced over the normally much longer time interval between calibration adjustments, a more realistic value of precision is obtained (Figure 5.9). In practical analysis, additional sources of variance become important: mainly those related to spectral interference corrections and to the sample preparation procedure.

According to Equation 5.4, the RSD of a determination approaches the value k asymptotically as c' increases. The value of k can be estimated from repeated measurements by using analyte solutions in which c' exceeds 1000, although the estimates will be subject to high sampling variance if a small number of integrations is used.

For 5-s integrations with continuous nebulization, the value of k lies within the range 0.003 to 0.006, corresponding to 0.3 to 0.6% RSD for most analytes measured with modern ICP–AES instruments. Under the same conditions, the k value for medium-term precision is higher at 0.01 to 0.02, and for practical analysis, precisions of 0.015 to 0.03 are obtained. These values represent relative standard deviations at high analyte concentrations. At lower concentrations ($1 < c' < 1000$), the RSD can be obtained from Equation 5.4 with the insertion of appropriate values of c' and k. Values of c' can be obtained from tables of instrumental detection limits, with suitable adjustments if practical detection limits are required. This simple equation provides a very adequate model for estimation of precision in ICP–AES at various analyte levels.

5.6.1.1   The Effect of Integration Time on Precision

For an analyte that is present at high concentration relative to the detection limit, an increase in integration time should produce an improved precision. If the signal noise is purely random, ie, having a "white" spectrum, the RSD should be inversely proportional to the square root of the integration time. However, because these conditions are not true of ICP–AES signals, no such relationship has been observed,[37] and the RSD is largely independent of integration times.

5.6.1.2   The Effect of Calibration Interval on Precision

The signal stability of ICP–AES is remarkably good, unless an actual instrumental fault occurs, such as a nebulizer blockage. For instance, Ramsey and Thompson[5] reported RSDs for 24 analyte lines measured over a 4-hour period that were only two to three times larger than those determined over a short interval. These data indicate that calibration intervals of up to an hour would give results that are not perceptibly worse than those from the analysis of samples closely bracketed by standard solutions. However, regular checks within the longer integration intervals may be necessary to guard against the possibility of nebulizer blockage or other drifts in practical analysis.

5.6.2   Sources of Variation in Inductively Coupled Plasma–Atomic Emission Spectrometry

Variations in sensitivity in an ICP–AES are due largely to small uncontrolled changes in the operating conditions within and between measurement periods. If all the operating parameters were held absolutely constant, only the small random variation due to electronic noise and shot noise would be observed. These random effects, however, are estimated to contribute less than 5% of the observed, short-term RSD for uninterrupted nebulization. The remaining variations are therefore attributed to the operating parameters, particularly those relating to the "kitchen" area of the ICP–AES.

Forward power is a parameter that is stabilized to better than 0.2% in many instruments, but its variations can strongly affect signal intensity. The effect varies from line to line,[5] however, depending largely on the excitation energy (Figure 5.4). For example, under conditions of practical analysis, Li I 670.78 nm (EP = 1.85 eV) is almost unaffected by small forward power fluctuations, whereas for Zn II 202.55 nm (EP + IP = 15.51 eV) a 1% increase in forward power brings about an increase of approximately 5% in the net line intensity. The influence of injector gas flow rate is similar, but opposite, to that of forward power variations. An increase of 1% in the injector gas flow rate causes a decrease in the zinc line intensity of about 4%. Variations in nebulizer uptake rate also affect line intensities, but in a completely different manner. In contrast to the parameters previously discussed, spectral lines with low excitation energies are the most strongly affected.

5.6.3   Chemometric Improvements to Precision

As previously seen, the extent to which a line is affected by parameter variations under conditions of routine analysis depends on its excitation energy. For example, two lines from different analytes, but with closely matched excitation energies, would be affected similarly by variations in operating conditions. Therefore, one such analyte could act as an internal standard[38,39] for the other, and by simple ratioing, a marked improvement in precision should be obtained. The corrected net intensity for the analyte R_{corr} is related to the observed net intensity R by the relationship:

$$R_{corr} = R\left(\frac{R_s^o}{R_s}\right) \qquad [5.6]$$

where R_s is the observed net intensity of the internal standard line, and R_s^o is its original intensity determined at calibration time. Close matching[40,41] of excitation energies is not a practicable proposition in routine multiple-element analysis, however. Several workers have recommended the use of an internal standard in ICP–AES, although, in most cases, the practice is not properly justified, either theoretically or by a soundly designed experiment.

Two closely related methods of overcoming the need for internal standards with matched excitation energies have been reported recently.[5,42,43] The method that is easier to explain is the parameter-related internal standard method (PRISM) of Ramsey and Thompson.[5] In PRISM, an internal standard line (A), which is strongly affected by changes in an operating parameter, eg, Zn II 202.55 nm, affected by forward power, is chosen to represent fluctuations of that, or other parameters with similar effects. The corrected net intensity for any analyte is then given by:

$$R_{corr} = \frac{R}{[1 + k_A(R_A/R_A^o - 1)]} \qquad [5.7]$$

The constant, k_A takes account of the different response of the analyte and the internal standard to variations in the parameter. If the effects of two independent parameters, such as forward power and nebulizer uptake rate, are considered, two internal standards (A and B) are required. The correction equation then expands into the form:

$$R_{corr} = \frac{R}{[1 + k_A(R_A/R_A^o - 1)][1 + k_B(R_B/R_B^o - 1)]} \qquad [5.8]$$

Using two internal standards, Ramsey and Thompson reported a substantial improvement in medium-term precision, measured over a 4.4-hour period for 24 analyte lines, covering a wavelength range 178 to 766 nm.

Lorber and co-workers[42,43] were first to publish on this topic, and put forward their "generalized internal reference method" (GIRM), which is similar to, but more generalized than PRISM. In terms similar to Equation 5.8, the GIRM correction is of the form:

$$R_{corr} = \frac{R}{\Pi[1 + f_i\,(R_i/R_i^o - 1)]} \qquad [5.9]$$

where Π indicates the product of several similar terms and f_i is a functional relationship replacing the constant of proportionality k in PRISM. In GIRM, it is not necessary for the effects of the parameters to be uncorrelated, and the equations are solved by a numerical method.[42,43] Also, GIRM, which requires considerable computing power, provides a substantial improvement in precision, probably better than PRISM, although the two methods have not been directly compared.

In principle, for simultaneous multiple-element analysis at least, these chemometric methods can provide a substantial improvement in precision without the need for expensive instrumental changes. Neither method has been implemented for routine analysis.

5.7 FUTURE DIRECTIONS

ICP–AES has arrived at the stage of development where further conventional improvements in hardware are capable of effecting only relatively minor improvements in technical performance. For example, an increased resolution would overcome some of the remaining problems of spectral interference but there would still remain an irreducible minimum. Again, improved instrumental stability could improve precision, but not beyond the level set by electronic noise and shot noise. Nonetheless, these improvements are likely to appear in conventional instrumentation, and analysis will benefit from them. Sample introduction procedures and nebulizers, in particular, will receive much attention.

In contrast to purely technical considerations, the economics of ICP–AES are far from optimized. While the method is cost effective where argon is plentiful, in countries where argon is scarce, ICP–AES may be prohibitively expensive solely for this reason. Researchers and manufacturers are responding to this shortcoming by developing low-gas-flow torches with performance characteristics similar to the Fassel torch. Such torches can operate on lower argon gas flow and can be powered by solid-state generators that provide the smaller RF power requirement (Chapter 14). The capital costs of such instruments could even be reduced, to make ICP–AES as available in the future as atomic absorption spectrometry is today. Alternatively, mixed-gas and molecular gas ICPs may be used to reduce the operating costs (Chapter 15).

A strategy for further improvement in instrumentation consists of chemometric improvements to signals from existing equipment. The Myers–Tracy correction, GIRM, and PRISM, although not yet widely implemented by manufacturers, are capable of providing real-time corrections for both noise and drift in analyte sensitivity to levels approaching the irreducible minimum variance. A similar technique can be used to correct background variation. Each of these methods requires at least one reference channel, which greatly increases the cost and complexity of a scanning system. However, GIRM or PRISM, implemented on a monochromator, would require no physical change to the instrument and very little else beyond a trivial amount of computing time. We are likely to see further developments in data improvement along these chemometric lines, both by research groups and in commercial equipment. The recent review article by Boumans[44] addresses the use of chemometrics for improving specificity of ICP-AES through line selection and correction of spectral interferences. Wirsz and Blades[45] have also discussed the application of pattern recognition and factor analysis for qualitative and quantitative analysis of complex emission spectra of the ICP in simultaneous multielement analysis.

The introduction of sample into the ICP in the form of a solution will remain the predominant technique; however automatic solution manipulation before the nebulizer is a field open for considerable development and is likely to transform current practice. Continuous solution flow, with computer-controlled switching, is a more flexible system than the discrete solution handling by mechanical-style robotics. Techniques such as dilution, preconcentration, extraction, calibration, standard addition, and matrix matching can all be currently accomplished by flow injection systems. With full interactive control, a smart system seems to allow the construction of virtually autonomous analytical machines, capable of diagnosing any analytical problem and, apart from mechanical or electronic failures, providing the necessary remedial activity.

Such an autonomous system, perhaps assisted by chemometric improvement of the quality of results, will produce data that are the result of extensive manipulations. Even if the instrument operator understands the general principles involved in the manipulation and calculation, the person may be unable to say what has happened for a particular sample or analyte. The system would thus approach the black-box status. Under these circumstances, an independent and carefully designed protocol for data quality control is an essential adjunct to the analysis, and we will soon see such packages incorporated into commercial software designed for ICP-AES instruments.

Perhaps the most revolutionary changes in ICP-AES are likely to develop from the use of Fourier transform spectrometry to produce a whole emission spectrum from a sample at a very high resolution. This would enable the simultaneous use of many lines for measurement of the concentration of a single analyte, and many other lines for interference correction or chemometric data quality improvement. In principle, this approach could provide substantial improvement in performance, given the necessary expert system software to extract the required information from the plethora of data. Exactly how this could be done would depend critically on the methods by which Fourier transform is implemented in future commercial instruments.

REFERENCES

1. S. Greenfield, I. L. Jones, and C. T. Berry, High Pressure Plasmas as Spectroscopic Emission Sources, *Analyst* 89, 713–720 (1964).
2. R. H. Wendt and V. A. Fassel, Induction-Coupled Plasma Spectrometric Excitation Source, *Anal. Chem.* 37, 920–922 (1965).
3. M. Thompson, The Capabilities of Inductively Coupled Plasma Atomic Emission Spectrometry, *Analyst* 110, 443–449 (1985).
4. M. W. Blades and G. Horlick, The Vertical Spatial Characteristics of Analyte Emission in the Inductively Coupled Plasma, *Spectrochim. Acta* 36B, 861–880 (1981).
5. M. H. Ramsey and M. Thompson, Correlated Variance in Simultaneous Inductively Coupled Plasma Atomic Emission Spectrometry: Its Causes and Correction by a Parameter-Related Internal Standard Method, *Analyst* 110, 519–530 (1985).
6. W. Spendley, G. R. Hext, and F. R. Himsworth, Sequential Application of Simplex Designs in Optimization and Evolutionary Operation, *Technometrics* 4, 441–460 (1962).
7. S. L. Morgan and S.N. Deming, Simplex Optimization of Analytical Chemical Methods, *Anal. Chem.* 46, 1170–1181 (1974).
8. S. N. Deming and S. L. Morgan, Teaching the Fundamentals of Experimental Design, *Anal. Chim. Acta* 150, 183–198 (1983).
9. L. Ebdon, M. R. Cave, and D. J. Mowthorpe, Simplex Optimization of Inductively Coupled Plasmas, *Anal. Chim. Acta* 115, 179–187 (1980).
10. J. J. Leary, A. E. Brookes, A. F. Dorrzapf, Jr., and D. W. Golightly, An Objective Function for Optimization Techniques in Simultaneous Multiple-Element Analysis by Inductively Coupled Plasma Spectrometry, *Appl. Spectrosc.* 36, 37–40 (1982).
11. R. K. Winge, V. J. Peterson, and V. A. Fassel, Inductively Coupled Plasma Atomic Emission Spectrometry: Prominent Lines, *Appl. Spectrosc.* 33, 206–219 (1979).
12. S. J. Walton and J. E. Goulter, Performance of a Commercial Maximum Dissolved Solids Nebulizer for Inductively Coupled Plasma Spectrometry, *Analyst* 110, 531–534 (1985).
13. K. W. Olson, W. J. Haas, and V. A. Fassel, Multielement Detection Limits and Sample Nebulization Efficiencies of an Improved Ultrasonic Nebulizer, *Anal. Chem.* 49, 632–637 (1977).
14. G. F. Larson, V. A. Fassel, R. K. Winge, and R. N. Kniseley, Ultratrace Analyses by Optical Emission Spectroscopy: The Stray Light Problem, *Appl. Spectrosc.* 30, 384–391 (1976).
15. V. A. Fassel, J. M. Katzenburger, and R. K. Winge, Effectiveness of Interference Filters for Reduction of Stray Light Effects in Atomic Emission Spectrometry, *Appl. Spectrosc.* 33, 1–5 (1979).
16. G. F. Larson and V. A. Fassel, Line Broadening and Radiative Recombination Background Interferences in Inductively Coupled PLasma–Atomic Emission Spectroscopy, *Appl. Spectrosc.* 33, 592–599 (1979).
17. A. W. Boorn and R. F. Browner, Effects of Organic Solvents in Inductively Coupled Plasma Atomic Emission Spectrometry, *Anal. Chem.* 54, 1402–1410 (1982).
18. V. A. Fassel, V. J. Peterson, F. N. Abercrombie, and R. N. Kniseley, Simultaneous Determination of Wear Metals in Lubricating Oils by Inductively Coupled Plasma Atomic Emission Spectrometry, *Anal. Chem.* 48, 516–519 (1976).
19. P. Barrett and E. Pruszkowska, Use of Organic Solvents for Inductively Coupled Plasma Analyses, *Anal. Chem.* 56, 1927–1930 (1984).
20. M. Thompson and J. N. Walsh, "A Handbook of Inductively Coupled Plasma Spectrometry." Blackie, Glasgow, (1983), pp. 242–246.
21. M. Thompson and J. N. Walsh, "A Handbook of Inductively Coupled Plasma Spectrometry." Blackie, Glasgow, (1983), pp. 24–25.
22. G. F. Larson, V. A. Fassel, R. H. Scott, and R. N. Kniseley, Inductively Coupled Plasma–Optical Emission Analytical Spectrometry. A Study of Some Interelement Effects, *Anal. Chem.* 47, 238–243 (1975).
23. V. A. Fassel, Quantitative Elemental Analysis by Plasma Emission Spectroscopy, *Science* 202, 183–190 (1978).
24. J. A. Borowiec, A. W. Boorne, J. H. Pillard, M. S. Cresser, and R. F. Browner, Interference Effects from Aerosol Ionic Redistribution in Analytical Atomic Spectrometry, *Anal. Chem.* 52, 1054–1059 (1980).
25. S. Greenfield, H. McD. McGeachin, and P. B. Smith, Nebulization Effects with Acid Solutions in ICP Spectrometry, *Anal. Chim. Acta* 84, 67–78 (1976).

26. M. Thompson and M. H. Ramsey, Matrix Effects Due to Calcium in Inductively Coupled Plasma-Atomic Emission Spectrometry, *Analyst* 110, 1413-1422 (1985).
27. S. R. Koirtyohann, J. S. Jones, C. P. Jester, and D. A. Yates, Use of Spatial Emission Profiles and a Nomenclature System as Aids in Interpreting Matrix Effects in the Low-Power Argon Inductively Coupled Plasma, *Spectrochim. Acta* 36B, 49-59 (1981).
28. W. H. Gunter, K. Visser, and P. B. Zeeman, Some Aspects of Matrix Interference Caused by Elements of Low Ionization Potential in Inductively Coupled Plasma-Atomic Emission Spectrometry, *Spectrochim. Acta* 37B, 571-581 (1982).
29. W. H. Gunter, K. Visser, and P. B. Zeeman, Ionization Interferences under Various Operating Conditions in a 9, 27, and 50 MHz Inductively Coupled Plasma, and a Study of Shifts in Level Populations of Calcium through Simultaneous Absorption-Emission Measurements in a 9 MHz Inductively Coupled Plasma, *Spectrochim. Acta* 40B, 617-629 (1985).
30. F.J.M.J. Maessen, J. Balke, and J.L.M. de Boer, Preservation of Accuracy and Precision in the Analytical Practice of Low-Power Inductively Coupled Plasma-Atomic Emission Spectrometry, *Spectrochim. Acta* 37B, 517-526 (1982).
31. S. S. Que Hee, T. J. Macdonald, and J. R. Boyle, Effects of Acid Type and Concentration on the Determination of 34 Elements by Simultaneous Inductively Coupled Plasma Atomic Emission Spectrometry, *Anal. Chem.* 57, 1242-1252 (1985).
32. H. Kawaguchi, T. Ito, K. Ota, and A. Mizuike, Effects of Matrix on Spatial Profiles of Emission from an Inductively Coupled Plasma, *Spectrochim. Acta* 35B, 199-206 (1980).
33. L. M. Faires, C. T. Apel, and T. M. Niemczyk, Intra-alkali Matrix Effects in the Inductively Coupled Plasma, *Appl. Spectrosc.* 37, 558-563 (1983).
34. M. W. Blades, and G. Horlick, Interference from Easily Ionizable Element Matrices in Inductively Coupled Plasma Emission Spectrometry—A Spatial Study, *Spectrochim. Acta* 36B, 881-900 (1981).
35. P.W.J.M. Boumans and J.J.A.M. Vrakking, Spectral Interferences in Inductively Coupled Plasma-Atomic Emission Spectrometry—I. A Theoretical and Experimental Study of the Effect of Spectral Bandwidth on Selectivity, Limits of Determination, Limits of Detection, and Detection Power, *Spectrochim. Acta* 40B, 1085-1125 (1985).
36. M. Thompson and J. N. Walsh, "A Handbook of Inductively Coupled Plasma Spectrometry." Blackie, Glasgow, (1983), pp. 124-128.
37. R. M. Belchamber and G. Horlick, Effect of Signal Integration Period on Measurement Precision in Inductively Coupled Plasma Emission Spectrometry, *Spectrochim. Acta* 37B, 71-74 (1982).
38. R. M. Belchamber and G. Horlick, Correlation Study of Internal Standardization in ICP Atomic Emission Spectrometry, *Spectrochim. Acta* 37B, 1037-1046 (1982).
39. S. A. Myers and D. H. Tracy, Improved Performance Using Internal Standardization in Inductively Coupled Plasma Spectroscopy, *Spectrochim. Acta* 38B, 1227-1253 (1983).
40. W. B. Barnett, V. A. Fassel, and R. N. Kniseley, Theoretical Principles of Internal Standardization in Analytical Atomic Emission Spectroscopy, *Spectrochim. Acta* 23B, 643-664 (1968).
41. W. B. Barnett, V. A. Fassel, and R. N. Kniseley, An Experimental Study of Internal Standardization in Analytical Emission Spectroscopy, *Spectrochim. Acta* 25B, 139-161 (1970).
42. A. Lorber and Z. Goldbart, Generalized Internal Reference Method for Simultaneous Multichannel Analysis, *Anal. Chem.* 56, 37-42 (1984).
43. A. Lorber, Z. Goldbart, and M. Eldan, Correction for Drift by Internal Reference Methods in Inductively Coupled Plasma Simultaneous Multielement Analysis, *Anal. Chem.* 56, 43-48 (1984).
44. P.W.J.M. Boumans, A Century of Spectral Interferences in Atomic Emission Spectroscopy—Can We Master Them with Modern Apparatus and Approaches? Proceedings of the 24th Colloquium Spectroscopicum Internationale, Garmisch-Partenkirchen, 1985, published in *Fres. Zeit. Anal. Chem.* 324, 397-425 (1986).
45. D. F. Wirsz and M. W. Blades, Application of Pattern Recognition and Factor Analysis to Inductively Coupled Plasma Optical Emission Spectra, *Anal. Chem.* 58, 51-57 (1986).

6

The Problem of Spectral Interferences and Line Selection in Plasma Emission Spectrometry

ANDREW ZANDER

North American Instrument Division
Perkin-Elmer Corporation
Norwalk, Connecticut

6.1   INTRODUCTION

The most significant impediment to the effective use of inductively coupled plasma–atomic emission spectrometry (ICP–AES) is spectral interference. Thus, a thorough knowledge of the spectral emission characteristics of ICPs, particularly the argon ICP, is a prerequisite to the most efficient use of the ICP in analytical atomic emission spectrometry. The utility of the ICP derives from its excitation capability, which provides high signal-to-background (S/B) and signal-to-noise (S/N) ratios for many spectral lines. Because nearly every species that exists in or is injected into the plasma is caused to emit rather profusely, a considerable amount of spectral "clutter" is generated. As a result, the choice of a spectral line for determining a trace element becomes dependent on the relative freedom from other spectral features at that wavelength. No spectral line is ever completely free of other spectral signals, but many are less affected than others. However, with an adequate knowledge of the argon spectrum, and through the use of proper dispersing devices, compilations of wavelengths useful for quantitation at the trace and ultratrace levels may be prepared.

This chapter classifies various spectral interferences, discusses the selection of prominent lines, and considers the practical resolution required for minimal spectral interferences. A discussion of the methods for identification of spectral interferences is followed by a review of correction techniques. Because this chapter is mainly concerned with the Ar ICP, the reader is referred to Chapter 15, which treats ICP discharges in other gases. The applications of high-resolution AES are reviewed in Chapter 7.

6.2   CLASSIFICATION OF SPECTRAL INTERFERENCES

Spectral interferences are observed in the ICP, as in every emission source. Compared to other sources, such as flames, arcs, or sparks, spectral interferences are more important in the ICP because emission lines that might be expected to be weak or nonexistent frequently are quite intense.

All emission spectral interferences have their origins in either the inherent argon spectrum (line and continuum) or spectral features (also line and continuum) generated from molecular and atomic species entrained in or injected into the argon plasma. The severity of the interference depends on the wavelength proximity of the nonanalyte feature to the analyte line, on its spectral distribution, and on the intensity and stability of the interfering signal. The alleviation, minimization, or correction for the interference depends greatly on these aspects.

Spectral interferences fall into a few broad classes. Some spectral features fall into more than one class, and consequently, have to be treated with multiple approaches. The principal classes are:

1. *Overlap from spectral lines.* Spectral lines, both atomic and molecular in origin, may directly overlap the spectral line of interest. Unidentified or unrecorded lines are a particular nuisance. Partial overlap of analyte lines with line wings of other lines, resulting from line broadening and instrument broadening, is common. Because of the high temperature of the ICP, significant Doppler broadening occurs. Relatedly, collisional processes, such as resonance and Stark effects, contribute to the broadening of spectral lines.

2. *Variation of the background continuum.* The background continuum from the Ar ICP is affected by the solvent and matrix elements.

3. *Stray light.* Stray light is radiation at wavelengths outside the instrumental bandpass that reaches the detector. It results from the intense emission of easily excited species or those in high concentration in the ICP, and from the imperfections of the optical system.

6.2.1   The Discharge Spectrum

The emission from a pure argon ICP at atmospheric pressure consists essentially of Ar atom lines superimposed on a background continuum (Figure 6.1). The continuum intensity, which is proportional to the square of the electron number density n_e, is due predominantly to ion–electron recombination processes at the series limit and, to some extent, to brehmsstrahlung, ie, free–free transitions of electrons in the field of ions. Variations of the forward power and of the injector gas flow rate, both of which affect the n_e value, change the intensity of the background continuum. Furthermore, the injection of hydrogen into the plasma, through aspiration of water or other solvent, also alters the continuum intensity by causing variations in n_e. The introduction of analyte or matrix elements, particularly those that are easily ionized, has no appreciable influence on the continuum intensity[1] because of the very high n_e prevailing in the plasma. This background continuum is encountered from the vacuum-ultraviolet (VUV) to the near-infrared (NIR) regions.

All recorded transitions for neutral atom argon (Ar I) have been observed in the emission spectrum of the Ar ICP,[2–4] but no observation of transitions from argon ions (Ar II) has been reported. Nearly 200 Ar lines can be observed, with none occurring between 200 and 300 nm. The most intense emissions occur around 430 nm. Some of the transitions are more prone to broadening than others; for example, transitions to the 4p level are usually broader than those to the 4s level.[4,5] This is quite important, since line widths of a few tenths of a nanometer result from this broadening.

Atomic transitions of hydrogen, carbon, and silicon can often be observed in the argon ICP spectrum. The hydrogen lines, some of which are almost 1 nm wide, are the result of the dissociation of water molecules originating from entrainment of the atmosphere or directly from injection of water solvent. A torch having an extended outer tube or a purged system eliminates entrained species. The C I lines at 193.09 and 247.68 nm arise from low-molecular-weight impurities, such as CO and CO_2, in the commercial argon. The use of the boil-off from a tank of liquid argon eliminates carbon-containing species. Silicon lines result from the volatilization of the quartz tubes in the torch assembly, and the intensities of these emissions usually diminish with the aging of the quartz torch.

6.2.2   The Concomitant Spectrum

Besides argon, the predominant species are H_2O, from entrained or injected water, and N_2, from entrainment or dissolved gas in the injected solvent. After dissociation in the hot plasma, these species generate OH, N_2^+, NH, and NO, which are excited to produce rich molecular rotational–vibrational emission manifolds, "molecular bands." The bands are spread throughout the VUV to

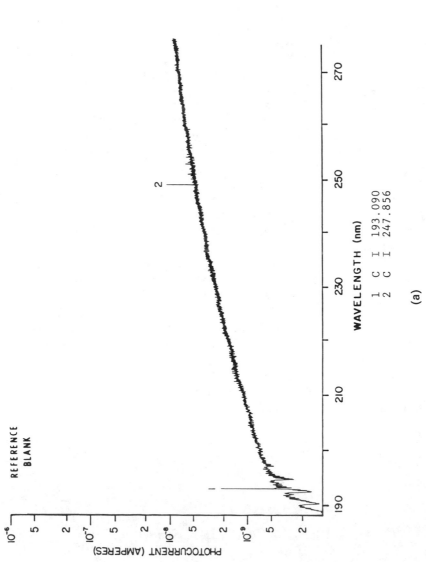

Figure 6.1. (a) The blank emission spectrum of an Ar ICP showing the principal features of the Ar spectrum. Numbers locate Ar marker wavelengths. (From Reference 2, with permission.)

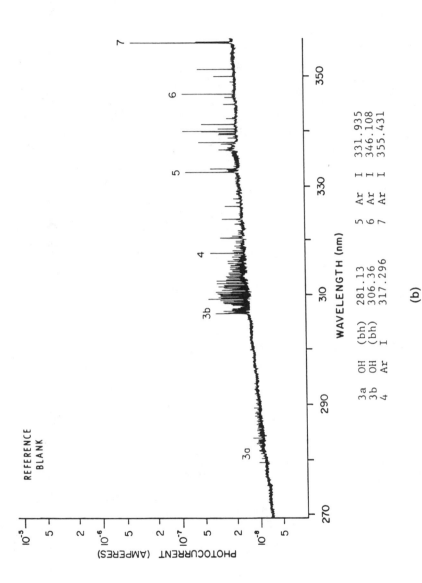

Figure 6.1. (b) The blank emission spectrum of an Ar ICP showing the principal features of the Ar spectrum. Numbers locate Ar marker wavelengths. (From Reference 2, with permission.)

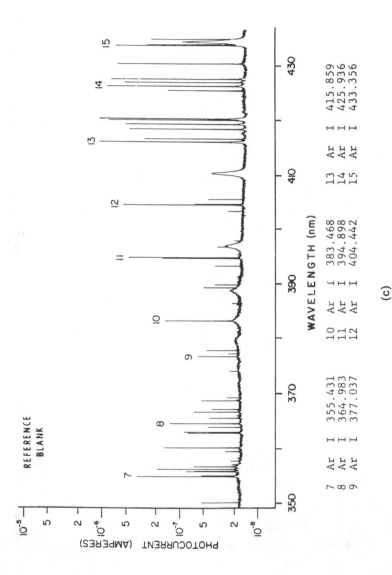

Figure 6.1. (c) The blank emission spectrum of an Ar ICP showing the principal features of the Ar spectrum. Numbers locate Ar marker wavelengths. (From Reference 2, with permission.)

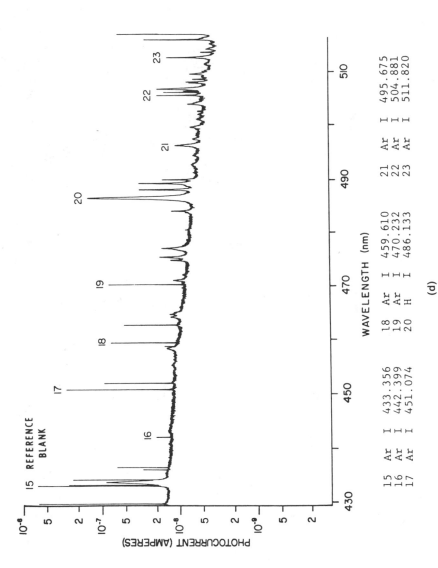

(d)

15	Ar	I	433.356	18	Ar	I	459.610	21	Ar	I	495.675
16	Ar	I	442.399	19	Ar	I	470.232	22	Ar	I	504.881
17	Ar	I	451.074	20	H	I	486.133	23	Ar	I	511.820

Figure 6.1. (d) The blank emission spectrum of an Ar ICP showing the principal features of the Ar spectrum. Numbers locate Ar marker wavelengths. (From Reference 2, with permission.)

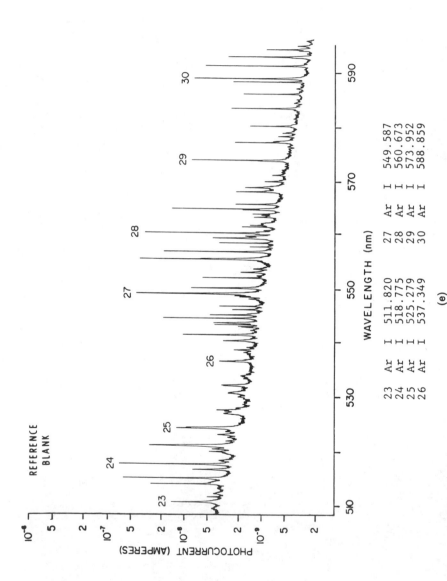

Figure 6.1. (e) The blank emission spectrum of an Ar ICP showing the principal features of the Ar spectrum. Numbers locate Ar marker wavelengths. (From Reference 2, with permission.)

NIR region. The ultraviolet and low-visible portions of the spectrum, illustrating certain bands, are shown in Figure 6.2.

Below 200 nm, the only features evident are the molecular oxygen absorption bands (Figure 6.3), which arise from oxygen in the optical path between the plasma and the detector. Evacuation or purging of the optical path with a nonabsorbing gas, such as Ar or He, eliminates these features.

The OH radical is the prominent emitter in dry, unsheathed or wet Ar ICPs. In a pure Ar plasma, without water injection, the tailflame of the plasma shows strong emission from the OH (0,0) band around 306.4 nm, which results from atmospheric water entrainment. When water is introduced, most of the OH band features can be observed easily at commonly used observation heights. The bands appear in the regions 281.0 to 294.5 nm and 306.0 to 324.5 nm. Because of the numerous features of these band spectra, the use of analyte lines in these regions usually requires high resolution. The relative intensities of lines of

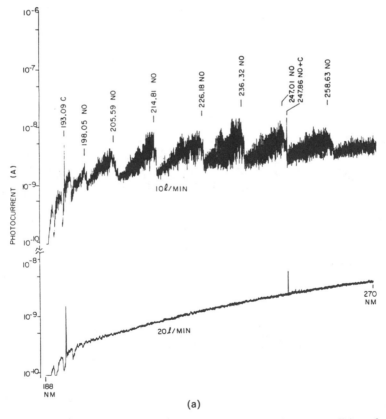

(a)

Figure 6.2. (a) Background spectrum of the Ar ICP operating at conditions favoring molecular emission (upper) and conditions minimizing molecular emission (lower). (From Reference 2, with permission.)

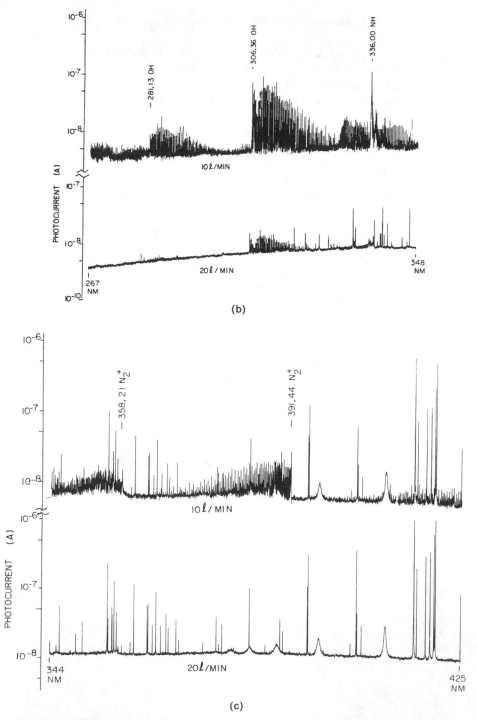

Figure 6.2. (b) and (c) Background spectrum of the Ar ICP operating at conditions favoring molecular emission (upper) and conditions minimizing molecular emission (lower). (From Reference 2, with permission.)

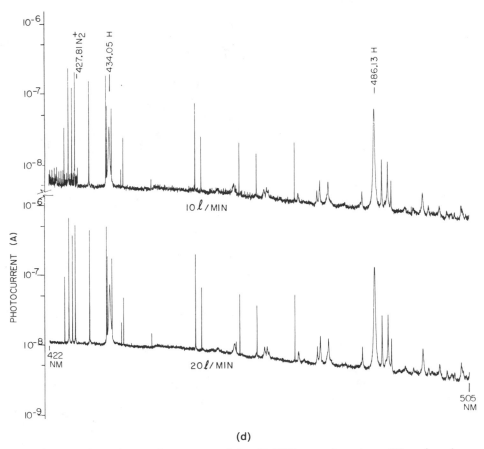

(d)

Figure 6.2. (d) Background spectrum of the Ar ICP operating at conditions favoring molecular emission (upper) and conditions minimizing molecular emission (lower). (From Reference 2, with permission.)

the (0,0) band between 307 and 309 nm have been used to measure OH rotational temperatures in the plasma[6,7] (Chapter 8).

The most prominent NH emission occurs around 336 nm, with the band sequence stretching from 302.2 to 380.4 nm. The band head at 336.0 nm is observable in high-power and in low-power plasmas using reduced gas flow. At lower outer flow rates, entrainment of ambient air is increased, resulting in a pronounced band head at 324 nm (Figure 6.2).

Features of the main system of N_2^+, principally around the (0,0) and (1,1) bands at 391.4 and 388.4 nm, are possible to observe. Normal precautions to avoid entrainment reduce these features to a manageable level.

Band emission from NO is commonly observed in the Ar ICP. The spectral structure of the more prominent gamma system extends from 195.6 to 345.9 nm (Figure 6.4). The most intense portions of the system lie between 200 and 280 nm, with the 237.0, 247.9, and 272.2 nm band heads being the strongest. The

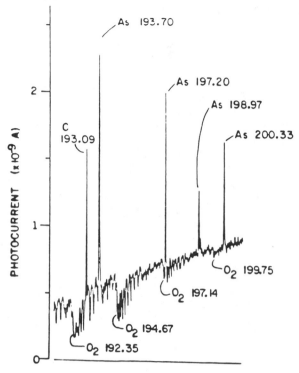

Figure 6.3. Short section of the background spectrum of the Ar ICP showing molecular oxygen absorption bands. (From Reference 2, with permission.)

(1,0) subsystem between 210 and 216 nm is suspected of causing interference on Zn I 213.86 nm.[8] The signal-to-background ratio of the NO–gamma system is relatively unaffected by the forward power or the observation height, but is reduced at higher outer flow rates[2] (Figure 6.2). Enclosing the plasma to control its immediate surroundings minimizes, but does not eliminate, the NO spectrum.

Molecular spectra arising from the sample matrix, other than the solvent, are not frequently observed in ICP spectra. Typical examples are as follows. An elevation of the background at Li I 670.78 nm, when high Ti concentrations are present, is the result of TiO emission.[2] A similar situation arises at the Pb I 405.78 nm line when matrix concentrations of B exceed 0.1%. Elements with rather stable oxides, such as yttrium, scandium, gadolinium, samarium, lutecium, and zirconium, produce monoxide spectra in the ICP. Oxide emissions are observed most easily in the preheating zone of the plasma, where desolvation and decomposition processes are incomplete; and in the plume region far above the observation height, where atmospheric entrainment is greatest. Even in plasma zones that do not favor oxide emission, small levels of molecular emission occur, usually as a minor increase in the background continuum.

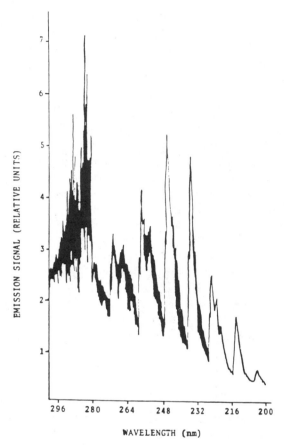

Figure 6.4. Short section of the background spectrum of the Ar ICP showing molecular emission features of the OH (band at left) and NO (bands at center and to right) species. (From Reference 3, with permission.)

Increasing forward power or argon flow rate, or the use of an extended torch, tends to reduce monoxide emission.

In spite of the predictability of the molecular spectra of more common species, or the relative infrequency of occurrence of diatomic molecule spectra of other matrix constituents, there is always a reasonable chance of finding spectral features of nearly every species existent in the ICP. Many of these features initially appear as unrecognizable structure on the analyte spectrum. Upon closer inspection, they are seen to be a myriad of low-intensity, smeared-out lines forming a pseudocontinuum and contributing an offset to the background spectrum. At the commonly used observation height, and in particular, for ultratrace determinations, the presence of this pseudocontinuum is a less important factor than the stability of the pseudocontinuum signal. The text by Pearse and Gaydon,[9] is an excellent reference on molecular spectral features, even though it does not treat the ICP discharge.

6.2.3   Spectral Line Interferences

Continual publication of reports of observed spectral interferences of various types, and the recent availability of atlases of spectral line interferences and line coincidence tables,[2,4,10] attest to the fact that interpretation of spectra generated by an ICP must be approached cautiously.

The severity of interference for two spectral lines is expressed in terms of the Rayleigh criterion. For two lines of equal intensity, the lines are resolved, based on the Rayleigh criterion, if the depth of the valley between the peaks is at least 19% of the peak height. Direct overlap occurs if the Rayleigh criterion is not met, and no amount of resolving power can alleviate this type of interference. In a closely similar situation, there may actually be some wavelength separation between two lines, but the dispersing apparatus may not be adequate for the isolation of the analyte line. Overlap again occurs, but it is instrument dependent. Then there is the case of unidentified, unlisted lines. The severity of their interference is more a matter of lack of predictability than presentation of a new type of interference necessitating different correction techniques. Given the prolific production of spectral lines from an ICP, and the diversity of sample matrix types encountered, it is unlikely that any spectral atlas will ever be complete enough to minimize the occurrence of unrecorded lines. Finally, partial overlap by a line wing can occur due to the broadening of the interfering line.

6.2.3.1   The Effect of Spectral Bandpass on Line Overlap

True overlap of lines occurs frequently with especially line-rich spectra, such as those produced from uranium, iron, or lanthanide matrices. Total overlap occurs when the lines have no wavelength separation. Examples are the Fe 367.007 nm overlap of U II 367.007 nm[5] and the Hf 273.88 nm overlap of Mn 273.88 nm.[2] The interfering line may arise not only from matrix species, but also from the argon and concomitant spectra. For example, La II 394.910 nm is nearly overlapped by Ar I 394.915 nm. If the Rayleigh criterion is not met, even if it is the result of unequal intensities of the lines, line coincidence is considered to have occurred. This is quite frequently the case for closely spaced lines. Examples of this type are the Zn I 213.856 nm overlap with Cu I 213.853 nm and the Zn II 202.551 nm overlap with Cu II 202.547 nm.

If the wavelength separation between two lines exceeds the Rayleigh criterion, isolation of the analyte line may be provided if an optical system with adequate dispersion is used. A common means of rating the dispersing ability or resolving power of a spectrometer is through the calculation of its spectral bandpass (SBP). High resolution instruments provide an SBP between 0.001 and 0.01 nm. Medium resolution spectrometers have an SBP in the range of 0.01 to 0.10 nm. The narrower the SBP, the more likely it is that closely spaced lines are separated. For example, P I 213.618 nm is easily isolated from Cu II 213.598 nm with a spectrometer bandpass of 0.004 nm; while these lines are not resolved at all if the bandpass is 0.02 nm. Clearly, the isolation of the analyte line is instrument dependent. The line coincidence tables of Boumans[10] indicate the magnitude of interference effect to be expected at particular values of SBP. The practical resolution needed for minimization of spectral interference is addressed in Section 6.3.

A spectrometer with a significantly narrow SBP is a definite asset for doing ICP analyses. There is, however, a limit to the effectiveness of even the best optical instrument. This limit is determined by the broadening of the lines in the plasma.

6.2.3.2   The Effect of Line Broadening on Line Overlap

The subject of line broadening is discussed in detail in Chapters 7 and 8, and elsewhere.[5,11] Briefly, the major causes of line broadening are natural broadening, self-absorption broadening, Doppler broadening, resonance and Lorentz broadening, and Stark broadening (Table 6.1). For the vast majority of lines used in ICP–AES, natural broadening contributes approximately 0.00001 nm to the half-intensity bandwidth, and is, therefore, insignificant. For an ICP at 5000 to 6000 K, Doppler widths can be on the order of 0.001 to 0.01 nm, depending on the element. The broadening produces a Gaussian line profile, the wings of which decrease in intensity rapidly. Thus, Doppler broadening does not contribute significantly to background shifts at wavelengths away from the line center of the potentially interfering line.

When an element, monitored at a resonance line, is present in high concentration in the plasma, resonance broadening can occur. The width of the line profile broadens in proportion to the number of collisions between excited and ground state atoms. Resonance broadening leads to profiles with significantly extended and intense line wings. Prominent examples of this are the Ca I 393.37 nm and Ca II 396.85 nm lines at a Ca concentration of 1 mg/mL[11] (Figure 6.5). The wings of these lines can be expected to produce measurable interference, in the form of a nonlinear elevation of the background, as much as 10 nm away from line center. Thus, the determination of Al at either its 394.4 or 396.2 nm line would be seriously impaired in a Ca matrix. Relatedly, the determination of

---

Table 6.1. Causes of Line Broadening

1. Interaction with the radiation field
    A. Natural broadening
    B. Self-absorption broadening
2. Random motion of the atoms: Doppler broadening
3. Collisional or pressure broadening
    A. With neutral particles
        i. Like atoms: resonance (Holtzmark) broadening
        ii. Unlike atoms: Lorentz broadening
    B. With charged particles: Stark broadening

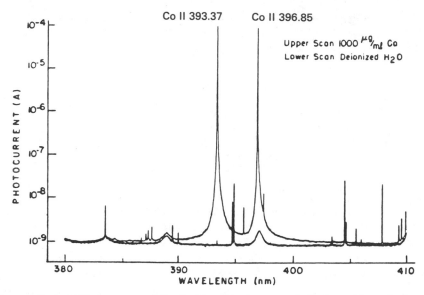

Figure 6.5. Ar ICP spectrum of Ca doublet showing extended wings. (From Reference 11, with permission.)

boron in gallium arsenide at 249.771 nm is impaired by the same mechanism[5] (Figure 6.6).

The collision of trace element atoms with neutral atoms of argon provides the most usual cause of broadening in the ICP,[5] the so-called Lorentz broadening, but its effect is less than that of Doppler broadening. For such broadening, the wings of the line profiles are enhanced even at a substantial distance away from the line center. As a result, although the effect on the full width at half-maximum is relatively weak, it will provide additional background enhancement to nearby analyte lines.

Collisions of analyte species with charged particles, electrons, and ions lead to Stark broadening. The perturbation of hydrogenlike orbitals by the charged particles arises from the linear Stark effect and leads to very broad, symmetrical lines. The perturbation of more complex orbitals is due to the quadratic Stark effect and results in less widely broadened, asymmetric lines. The line broadening that occurs depends on the number densities of the charged particles.[12,13] Extensive tabulations of Stark broadening parameters for hydrogen, and some neutral atom and singly ionized lines, are now available.[12]

As far as spectral interference is concerned, the Stark effect is more significant for certain lines of argon (Figure 6.7) and for certain lines of hydrogen. A spectrum of Mg that shows substantial resonance broadening, and also broadening effects for some weak lines, appears in Figure 6.8. The series of triplets in the 257-to-274 nm region shows increased broadening and decreased intensity for higher members of the series, a behavior due to the linear Stark

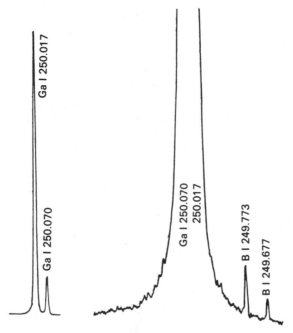

Figure 6.6. Influence of the resonance broadening of Ga on the determination of B in GaAs. (From Reference 5, with permission.)

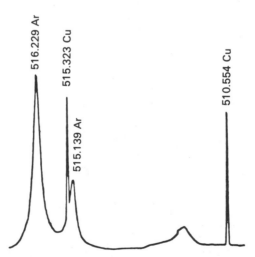

Figure 6.7. Overlap of Stark-broadened wings of Ar I 515.139 nm with Cu 515.323 nm. (From Reference 5, with permission.)

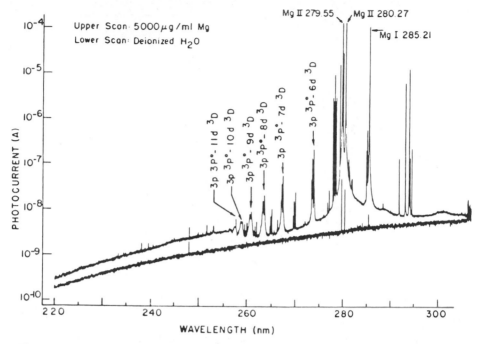

Figure 6.8. Portion of the Mg spectrum showing Stark broadening of strong and weak lines. (From Reference 11, with permission.)

effect on the hydrogenic d orbitals.[11] The importance to analytical determinations is that spectral line backgrounds are produced at large distances from the centers of the broadened lines. Relatedly, if the weak lines are not diffuse or broad, or if they have not been listed in wavelength tables, detection and correction of such interferences are difficult.

6.2.4   Continuum Radiation

Throughout an ICP spectrum there is broadband background beneath the line radiation that adds an offset to the signal measured. Thus, it is necessary to compensate for this signal offset, usually through the use of sample blanks. However, the continuum presents a spectral interference when the level and the stability of the background offset vary with the sample. Some form of dynamic background compensation must be applied then.

The continuum originates from a number of sources, including the electrons, Ar, and matrix species, both atomic and molecular. As discussed in Section 6.2.1, any operational parameter that alters n_e, such as the forward power or the gas flow, also changes the Ar continuum intensity. Many molecular spectra, of low intensity and reasonably broadened, result in smeared-out features that resemble low-intensity pseudocontinuum. This pseudocontinuum is difficult to characterize, but the sample blank suffices to compensate for it in many cases. However, with particularly complicated or highly concentrated matrices, the offset contribution from this source may become less controllable. Relatedly, the resonance-broadened weak lines of the Mg spectrum (Figure 6.8) merge into a continuum for the highest members of the series.[11]

The Mg spectrum shows another source of continuum radiation, which is the result of recombinations of Mg ions with free electrons to produce radiation over an exceptionally wide range. This leads to a background shift that requires compensation. Because the final state in the recombination process is not necessarily the ground state,[11] the recombination continuum is considered to be a spectral interference, as it cannot be predicted from information in the traditional wavelength tables or from the ionization potential. However, the recombination continuum is proportional to the analyte concentration, and its effect can be reasonably modeled.

6.2.5   Stray Light

Stray light is radiation that reaches a detector unintentionally. As discussed in Chapter 3, stray light occurs in a number of ways. It can be line "ghosts" (Rowland or Lyman ghosts) from periodic errors in grating spacing. It can appear as satellites, "grass," or near scatter (within a few bandpasses of a line), generated from regular or irregular grating defects. Far scatter, which might be generated many bandpasses away from an affected line, can result from localized grating imperfections or roughness of the groove shape. No dispersing device is perfectly constructed; and under intense illumination, even slight errors in design or construction may lead to unintended instrumental beam aberrations, reflections, or scattering of radiation that eventually reaches a detector.

Methods for producing diffraction gratings have been consistently improved over a period of many years. Ruled gratings and their replications are now of an exceptionally high quality, with very low levels of stray light. The advent of interferographically produced gratings (holographic gratings) has provided further advances in grating quality leading to exceptionally low levels of stray and scattered light. However, even with the best gratings, other components of a monochromator contribute to the stray light. For example, instrumental line broadening and stray light can result from imperfections in slits and slit mechanisms, design flaws leading to optical aberrations, and simple optical misalignment. Degraded antireflection coatings on optics, dust on optical surfaces, and particles in the optical path also can scatter radiation.

When trace levels of an analyte are being determined by ICP–AES, the desired line frequently has a significantly lower intensity than spectral lines from concomitants. If the analyte line lies near a strong concomitant line, the potential for stray-light interference is increased. Because of the intense lines, the effects of diffraction defects such as "grass" and ghosts are more prominent.[8] Background shifts, not necessarily proportional to the concentration of the concomitant causing them, can add to the analyte signal. Grating ghosts, which can overlap the analyte line, cannot be predicted and tabulated, and thus, they can be particularly troublesome. As higher orders of diffraction are used to enhance dispersion, the grating ghosts become even more prominent. Ghost intensities are proportional to the square of the order used.[14]

Exceptionally intense concomitant lines can affect analyte lines very many bandpasses away from them. For example, the "far-scatter" effect of the Ca 393.37 and 396.85 nm lines on the Zn 213.68 nm line is documented.[8] The far-scatter effect has been attributed to grating blank and grating groove shape roughness[15,16] and can be quite troublesome because wavelength cutoff filters, such as borosilicate glass to reduce visible far scatter from the ultraviolet region, are not always effective.

Grating and optical component imperfections may be the primary source of stray radiation, but instrumental imperfections also contribute. Under intense illumination, interior surfaces used for baffling and supports may reflect or scatter the optical beam. In polychromators, in particular, the secondary transfer optics at the exit slits can cause signals to appear at adjacent detectors. The more closely spaced the detectors, the more likely this is to occur. This was a problem with one early version of the detector housing of a SpectraMetrics SpectraSpan III echelle spectrometer, which used 20 exit slits in a close-packed, two-dimensional array. Light-tight tubes and periscopes from the exit slits to the photomultiplier tube (PMT) detectors eliminated the scattered light. The principal point is that spectrometer design plays a significant role in stray and scattered light reduction.

Minimization of stray light will be discussed here, although it is a hardware matter rather than a question of methodology, as are the other spectral interference types. Proper design and construction of a spectrometer are paramount when stray light reduction is a primary goal. High-quality optical components, assembled, aligned, and tested under high illumination, can usually assure a minimum of instrumental sources of stray light. Specific approaches include the use of detectors with selected spectral response. For example, solar blind PMTs do not respond to radiation above 350 nm, thus minimizing sources of far scatter. Narrow-band rejection filters can effectively block the radiation from particularly intense concomitant lines.[17] However, the transmission of the desired radiation is reduced. Double monochromators have proved to be quite effective for minimizing stray light.[16] The prism predispersion of some types of echelle spectrometer are effective in a similar manner. If it is possible to monitor a line known to be generating stray radiation, as with a polychromator, correction coefficients can be empirically determined. They can be used then to adjust analyte signals for the stray light. Since stray radiation is predominantly broadband, ghosts being the exception, wavelength modulation, synchronous scanning, or linear scanning across the analyte line can be used to provide dynamic correction.

6.3 SELECTION OF PROMINENT LINES

The relative intensities of the spectral lines of an element depend strongly on the type of excitation source and the operating conditions. The Ar ICP, operated at atmospheric pressure, generates elemental spectra.[2,4,10] with characteristic features different from those of other sources.

When spectral interferences have to be taken into account, the selection of the most desirable line depends on the sample type. At the present time, the considerable variety of sample types does not allow the selection of a single set of lines to cover all analyses for all elements, in spite of the spectral reproducibility and low-level interelement interferences of the ICP.

6.3.1 Criteria for Line Selection

Generally agreed upon criteria for selecting a wavelength for analytical purposes are:

Prominence of the line
Freedom from spectral interferences
Sufficient range of linearity
Practical utility in the specific situation

6.3.1.1 Line Prominence

Prominent spectral lines are intense lines that stand out[2] from the surrounding spectral features as the concentration is progressively reduced. Such lines were initially referred to as "persistent" lines.[18] For the ICP, the sensitivity and the signal-to-background ratio (S/B) are used to compare the prominences of lines. The sensitivity relates to the limiting detectability of an element at a specific line. On a relative basis, a specific atomic concentration of an element provides a certain normalized signal level. Since it is assumed that the ratio of concentration to signal is constant (ie, the emission intensity is proportional to the number of emitters at a constant temperature), a reduced concentration provides a linearly decreased signal level. A more sensitive line will be observable at a lower concentration; ie, its limiting detectability is lower. In modern terms, the sensitivity is defined as the inverse of the ratio given above (ie, the ratio of signal level to concentration). This is the same as the slope of the curve of growth of an element. In contrast to relative intensity, the use of the S/B ratio does not require normalization of all line intensities to 1.0.

Also included within the same context, the detection limit follows from the considerations of sensitivity and S/B. The important connection is the observation that, in a majority of cases, the limiting noise in ICP determinations is a predictable fraction of the background signal level.[19-22] If detection limits are defined as concentrations giving net signals equivalent to three times the standard deviation of the limiting noise, then:

$$C_{DL} = \frac{3\sigma_L}{S} \qquad [6.1]$$

where C_{DL} is the detection limit in units of concentration, σ_L is the standard deviation of the limiting noise, and S is the sensitivity. The standard deviation of the background, σ_L, is approximately 1% of the background signal level. Therefore, the detection limit can be given by:

$$C_{DL} = \frac{0.03X_B}{S} \qquad [6.2]$$

where X_B is the background signal level at the analytical line. In terms of the net signal to background ratio, the detection limit is given by:

$$C_{DL} = \frac{0.03C}{(S/B)} \qquad [6.3]$$

where C is the analyte concentration yielding the observed S/B. A listing of lines with intensities (sensitivities), or S/B's and detection limits constitutes a list of prominent lines.

6.3.1.2 Freedom from Spectral Interference

Freedom from spectral interference has proved to be a difficult measure to quantitate. A compilation of a composite list of lines by wavelength order is one means by which this factor has been addressed. The proximity of lines to each other in such a list is an indication of potential spectral interference.

A semiquantitative approach to listing spectral interferences has been proferred.[10,23] A measure of the interference level that might occur is given in the form of the critical concentration ratio (CCR). The CCR is the concentration ratio of interfering element to analyte at which the ratio of signal intensities of the two species is equal to unity. The CCR is a function of the intensity ratio of the two potentially interfering lines, the wavelength separation between them, and the spectral bandpass of the dispersing device used. That is, the CCR is instrument dependent. High values of CCR are indicative of lesser interference.

The computation of CCR values takes into account an assumed Voigt profile line shape overlapping a rectangular spectral window where the analyte line occurs. When the separation between lines is less than twice the spectral bandpass, line overlap is serious enough to warrant consideration of a different analyte line. Of course, the intensity of the interfering line complicates this generalization, and the CCR value is meant to emphasize the significance of the interference. For the situation in which lines are separated by more than twice the spectral bandpass, values of CCR of 100, or less, indicate line wing interference of such significance that the determination is adversely affected, also warranting consideration of a different analyte line.

While there is merit to using CCR values for predicting spectral interference, the CCRs are still empirically based. In that regard, their use requires a reasonable knowledge of the composition of the samples to which they are to be applied. That is, the analyst must still run preliminary tests to verify the predicted interferences. In the jargon of computer science, CCR values are a good first approximation to an "expert system" for atomic emission spectrometry. Much wider utility of CCRs for ICP methods development awaits a future generation of analytical spectrometric systems.

6.3.1.3 Sufficient Range of Linearity

As discussed in Chapter 5, the linear dynamic range of most prominent spectral lines in ICP–AES ranges between three and five orders of magnitude.

6.3.1.4 Practical Utility

The most subjective criterion for line selection is the practical utility of the line in the specific analytical situation at hand. Following line choice based on signal level (intensity or sensitivity) or S/B, detection limit, and proximity of matrix constituent lines or a CCR value for matrix constituents, an experimental profile of the spectral region surrounding the line is obtained. Standards closely approximating the samples to be analyzed are used. Observable spectral and background interferences, their levels, and their reproducibility are documented. Of great importance when interferences are observed is the identification of wavelength locations nearby the analyte line, which can be used for applying a procedure for correction of the interferences. Wavelength listings provide essentially no assistance for this step of line selection in methods development.

Winge and associates[2] provide superimposed profiles of prominent lines and those of abundant elements present in diverse matrices, such as water, sediments, the crust of the earth, and biological materials. These coincidence profiles were usually obtained for the four most prominent spectral lines of each of 70 elements. Interference solutions used in generating the coincidence profiles contained aluminum, calcium, chromium, copper, iron, magnesium, manganese, nickel, titanium, and vanadium, each within the concentration range of 200 to 1000 μg/mL.

The coincidence profile[2] for Cr II 205.55 nm shows four separate plots (Figure 6.9). The plots on the left-hand side of the figure have a linear intensity scale, while those on the right use a logarithmic intensity scale. For clarity, the top plot shows interference profiles of five elements (Al, Ca, Cr, Cu, Fe) and those on the bottom show interferences due to five elements (Mg, Mn, Ni, Ti, V). Such profiles indicate likely interferences and the relative magnitude of the spectral interferences from major elements. Although these profiles vary from one instrument to another, a cautious analyst may yet use these data for estimating errors resulting from spectral interferences. In general, spectral interferences causing errors greater than 5% in the analyte concentration are considered to be significant. For example, the recombination continuum from Al and the line

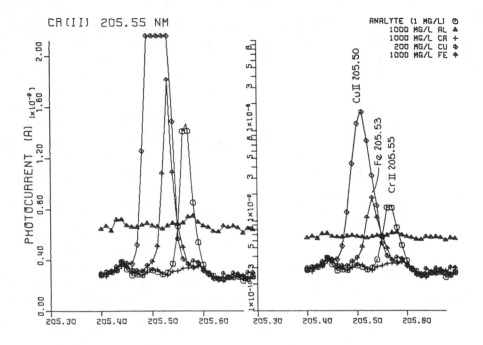

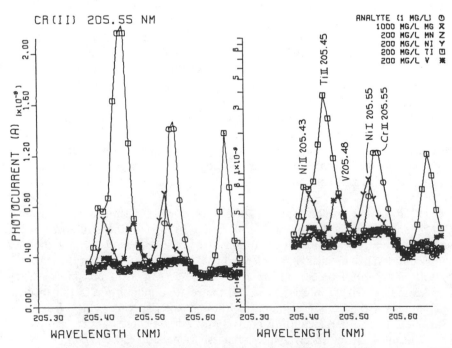

Figure 6.9. Coincidence profiles from an Ar ICP for Cr II 205.50 nm in the presence of Al, Ca, Cu, Fe, Mg, Mn, Ni, Ti, and V. (From Reference 2, with permission.)

overlap of Ni 205.55 nm produce significant interference at CCRs of 1000 and 200, respectively. If the error in analyte concentration is less than 5%, the interference is considered marginal. Typical examples of marginal interference on Cr II 205.55 nm are Cu 205.50 and Fe 205.53 nm, at the concentrations shown in Figure 6.9. No apparent interference is observed here on Cr II 205.55 nm in the presence of calcium, magnesium, manganese, and vandium.

6.3.2 Wavelength Tables

The comprehensive listing of spectral lines with accompanying information about the suitability of the line for particular analytical purposes has become a popular endeavor. Existing wavelength tables have been invaluable sources of information for the identification of spectral lines and for interpretation of spectra for a variety of excitation sources.[9,18,24-31] The use of these tables with ICP sources is limited because of the unique character of ICP spectra. However, compilations of spectral lines observed in the ICP are now available,[2,4,10,22,32-41] but they are not equally extensive or complete.

Certainly, line lists and coincidence lists will continue to be published and refined. Consequently, the tabulations cited here may well be viewed as necessary for the proper conduct of analytical spectrochemistry with an argon ICP; but they are not likely to be sufficient, except for sample types of the simplest matrix. A recent list of prominent lines[2] for an Ar ICP is given in the Appendix of this book.

6.4 PRACTICAL RESOLUTION FOR MINIMUM SPECTRAL INTERFERENCE

A good first approximation to the optical performance necessary to minimize the occurrence of spectral interferences is that the experimentally obtainable spectral bandpass (SBP) be as close to the physical line width as possible. In view of the narrow width of the atomic lines used in ICP-AES (on the order of 0.001 to 0.005 nm), instrumentation available for routine photometric quantitation that can supply such performance is at the state-of-the-art of development. This discussion is centered on conventional diffraction-based dispersion devices. Other techniques, such as Fourier transform spectrometry, are discussed in Chapter 7.

Typical values of the reciprocal linear dispersion (R_d) for spectrometers used in ICP-AES range from below 0.1 nm/mm for the highest performance systems to around 1.0 nm/mm. Echelle grating systems that operate in very high orders of diffraction have wavelength-dependent values of R_d between 0.02 and 0.1

nm/mm. The monochromator bandpass, SBP, approaches a minimum value as the slit width is reduced to a minimum. As the slit width is decreased, then, the diffraction-limited bandpass, $\Delta\lambda_d$, will become a larger fraction of the total bandpass:

$$\Delta\lambda_d = \frac{R_d\lambda f}{A} \qquad [6.4]$$

where λ, f, and A are wavelength, monochromator focal length, and effective aperture, respectively. The value $\Delta\lambda_d$ is sometimes referred to as the theoretical resolution of a spectrometer.[42,43] It is important to realize that theoretical resolution is not attainable. Optical imperfections in an instrument system, such as spherical aberration, coma, and astigmatism, and mechanical imperfections such as inadequate design, ragged or uneven slit edges, degraded optical elements, and misalignment can cause the experimentally observed bandpass to be larger than the calculated diffraction-limited value. This additional "aberration-limited" bandpass term, $\Delta\lambda_a$, which can be obtained experimentally only, must also be included. In low-resolution spectrometers, $\Delta\lambda_a$ can be as large as the instrumental spectral bandpass; but in systems of reasonable quality, it should be definitely less than the diffraction-limited bandpass.

Also, the physical line width of an analyte in the ICP is dominated by Doppler broadening.[5,43] Other types of broadening exert much less influence on the line profile. Within a kinetic temperature range of 5000 to 6000 K, the line widths of analytes are between 0.0014 and 0.0060 nm, depending on the atomic mass of the element and the wavelength.[43]

The profiles of atomic lines in ICP spectra are recorded as the convolution of various instrumental and spectral functions, at least one of which is determined by the identity of the constituents being monitored. Thus, there is a separate, best choice of spectral bandpass for each and every interference situation that arises. The choice depends on the physical separation between the analyte line and its interfering spectral feature. Some generalization, however, is possible.

The effect of spectral bandpass on signal level is different for line and continuum radiation. The intensity of a monochromatic line is directly proportional to the bandpass, while that of continuum radiation is proportional to the square of the bandpass. In the case of line emission superimposed on a continuum background, a narrower bandpass results in a higher signal-to-background ratio. Reducing the bandpass beyond a certain point, though, decreases the line intensity enough to adversely affect the signal-to-noise ratio and the detectability of the line.

Elimination of spectral interference is paralleled by a reduction in the capability of quantitatively monitoring the analyte. The detection limit in ICP–AES can be given by:

$$C_{\text{DL}} = \frac{0.03\text{RSD}_B C}{(S/B)} \qquad [6.5]$$

where RSD_B is the relative standard deviation of the background signal level in percent and S/B is the observed signal-to-background ratio provided by concentration, C. Since line and continuum radiation are each functions of the bandpass, the S/B is a function of spectral bandpass. So also is the RSD_B, being described earlier as a nearly constant fraction of the background signal level. Therefore, the detection limit is dependent on the bandpass (Figure 6.10). In the situation in which the bandpass is much narrower than the line width, the spectral line acts as continuum radiation.[45] In this circumstance, the S/B is effectively constant with changes in bandpass. A reduction of bandpass in that case leads to an increase of RSD_B as the background signal level drops. Eventually, the overall signal level becomes so low that shot noise dominates the detection limit. As the bandpass is widened to values equal to or greater than the line width, the line acts less and less as continuum radiation and the S/B becomes dependent on the bandpass. In that region, as the S/B begins to decrease with increased bandpass (since the background signal increases faster with bandpass than does the line signal), the loss of S/B is approximately balanced by a decrease of the controlling noise level. That is, RSD_B improves because the background signal level rises faster than the line signal. Eventually, the detection limit again degrades as the S/B continues to decline and the RSD_B is less well approximated as a constant fraction of the background level and soon becomes dominated by signal flicker noise. Typical medium-to-high resolution ICP-AES spectrometers, having spectral bandpasses of 0.02 to 0.03 nm, operate in this region.[45]

The optimum bandpass for best detection limit is approximately in the region in which the analyte line is close to behaving as continuum radiation: that is, when the bandpass is approximately equal to the physical line width. The spectral bandpass for minimum spectral interference, then, is approximately equal to the separation between interfering, equal-intensity, spectral lines. Whether such a bandpass actually is effective depends heavily on the relative strength of the lines and the exact causes of the interference. In general, a ratio of spectral bandpass to line width of approximately 1, and probably up to 3, provides the best compromise.[43,45]

Presented in terms of the resolution, R, where:

$$R = \frac{\lambda}{\Delta\lambda} \qquad [6.6]$$

an *experimentally attainable* resolution of between 50,000 and 150,000 is necessary. For operation in the wavelength range between about 190 and 850 nm, this converts to spectral bandpasses between 0.001 and 0.005 nm. It should be realized that, traditionally, optical spectrometers with this performance have been the very best and finest systems assembled. This is not meant to imply that analysis by ICP-AES using a spectrometer of this type of performance completely alleviates the effect of spectral interferences. It only suggests that, in cases in which spectral interferences occur, they can be minimized.

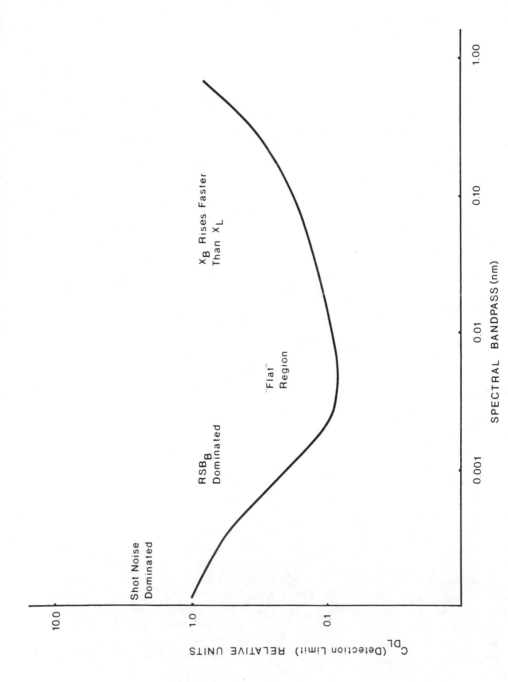

Figure 6.10. Effect of spectral bandpass on detection limits.

6.5 IDENTIFICATION OF SPECTRAL INTERFERENCES

Once it is possible to measure a spectral signal from the ICP, one must determine what portion of the signal is due just to the analyte. Analyte line suitability, ensuring element selectivity in trace determination, begins with a requirement for knowledge of the major and minor constituents of the sample. An order-of-magnitude estimation of the analyte concentration is also helpful. This information can be used with wavelength tabulations for the compilation of candidate lines. Constituent line proximities or coincidences and overlaps are usually identified in a direct manner in this step. Next, wavelength profiles of the candidate lines in the sample are acquired, along with equivalent profiles of the lines in single and multiple element standards. Furthermore, wavelength profiles at the candidate analyte lines are obtained with samples of only the potentially interfering species. From this data set, unanticipated line overlaps and the character of background shifts can be observed. A desirable line is free of overlaps, has minimal background shifts, and has near it wavelength locations suitable for the application of correction techniques.

Without the capability of wavelength profiling, either photoelectrically or photographically, determination of the suitability of a line becomes dependent on the characterization of the signals obtained from the types of samples and standards mentioned above. This situation implies, for the majority of cases, that no form of correction at a close-lying wavelength is possible. Consequently, the exact identification of the origin of an interference is less a concern than the characterization of it. The procedures for modeling of that sort are well documented in earlier texts, such as those on spectrophotometric analysis.

With the expanded wavelength tables now available, unanticipated direct-line overlaps are rare. In spite of this, if a line overlap is still suspected, it is easily observable in an appropriate reagent blank, although its origin may be difficult to identify. Tabulations of molecular lines are useful in identifying undocumented lines. Relatedly, grating ghosts are frequently located at calculable distances away from strong lines. Higher order lines of intense interfering lines should be considered, as even these are now being documented.[35]

All other spectral interferences are manifested as some form of background shift. Depending on the effective spectral bandpass, the shift may appear structured, curved, sloped, or flat. For the order listed, the severity of interference decreases from the structured to the flat background shift. Structured background shifts commonly originate from low-intensity, smeared-out molecular features, or they can be from broad or diffuse grating ghosts, or from low-intensity lines of a line-rich concomitant. Identification of the latter is straightforward. Structured background shifts due to stray light, such as Rowland or Lyman ghosts from the grating, can be identified through a comparison with a different spectrometer known to be more free of these effects.

Curved and sloped background shifts are somewhat more manageable. This is because they are probably caused by either stray light or the line wing broadening of a concomitant. They show no maxima within the spectral window of interest and are generally smoothly changing functions of wavelength. It is important to be able to unravel the contributions from either of these sources, because stray radiation is amenable to elimination with instrumental modifications while line broadening effects are not. An appropriate choice of cutoff filters can be used to block stray radiation from intense interfering lines. Transmission, interference, or rejection filters of various types, broadband and narrow-band, are available. The presence and absence of the filters should result in discernible signal-level changes while monitoring the analyte wavelength suspected of having interference. Care must be taken to ensure that the filters function properly, not inadvertently passing radiation that supposedly had been eliminated.

Another approach to verify the existence of the stray radiation, as opposed to the effect of line broadening, is illustrated by the background shift of Al 396.2 nm caused by Ca 396.8 nm.[11] The center of the Ca 396.8 nm line showed curve-of-growth curvature above a certain concentration, whereas its resonance-broadened line wing retained a linear plot. The nonlinear dependence of the background shift at Al 396.2 nm (without Al present) on Ca concentration provided proof that the origin of the background shift was predominantly stray radiation from the line center, in combination with existing instrumental broadening.

In minimizing or rejecting stray radiation, one should also note that the dispersed radiation, after passing an exit slit, can be scattered and reflected from secondary optics that precede the photomultiplier tube. This effect is particularly troublesome in polychromators, which necessarily use exit slits that are quite close together. Scanning a totally blocked analyte exit slit, while providing a strong signal from a suspected interfering line, results in a signal if secondary scattering exists.

It must be remembered that stray radiation effects are not only the result of particularly intense lines, such as those of calcium and magnesium. Less intense, but more numerous molecular lines or bands, such as those from NH, OH, and N_2^+ lines, contribute to stray radiation. Organic matrices also produce prominent Swan bands from C_2.

If, after adequate rejection of stray light, there still exist background shifts due to the presence of concomitants, the shift is attributable to a real modification of the background that is inherently under the analyte line. No guidelines to adequately identify the cause of such background shifts are available. Such background shifts have to be modeled to allow the proper correction of the analyte signal.

Simple, flat background shifts can occur for many reasons. Contributions of stray light to the shifts can be identified as already described. A modification of the excitation capability of the ICP, with a consequent alteration of the electron number density, changes the continuum background. However, such background shifts are small compared to the accompanying changes in intensities of analyte lines.

6.6   CORRECTION TECHNIQUES

6.6.1   Direct Line Interference

When the desired analyte line is directly overlapped with a background spectrum line, the interfering line is more a nuisance than a real hindrance to any determination. An example is the OH molecular line interference with Al I 308.2 nm. The OH signal, which is relatively constant, can be subtracted out with the blank signal, but it adds background noise that can interfere with determinations at very low analyte levels.

However, line overlap can cause a greater problem when it originates from a concomitant. A typical example is the coincidence of Cu I 213.859 nm with Zn I 213.856 nm. In such a case, one should use an alternative line. For either of these elements, this is easy to do, though less so with Zn, because each has other sensitive spectral lines that can be used. The coincidence of As I 228.812 nm on the Cd I 228.802 nm line is more problematical because Cd has fewer sensitive lines. The obvious solution is to use a higher resolution spectrometer, as in this case, the lines are farther apart than their physical widths, assuming minimal line broadening. Once resolved, interference is still possible through line–wing overlap or stray light. If the interference from these causes is too severe, another line must be chosen. If, however, the interference is not overwhelming, some form of correction technique might be applicable.

6.6.2   Background Correction Approaches

The selection of a particular background correction technique to alleviate or minimize spectral interference depends on the relative spectral distribution of the interfering feature (Figure 6.11). It has less to do with the origin of the feature, except in the case of stray light, which can be handled through instrumental modifications.

6.6.2.1   Simple Flat Background

In Figure 6.11A the analyte signal is offset by some constant amount at all locations. Correction involves measuring the simple flat background at some point away from the analyte peak and subtracting that value from the analyte-plus-background signal. A single off-peak point is sufficient.

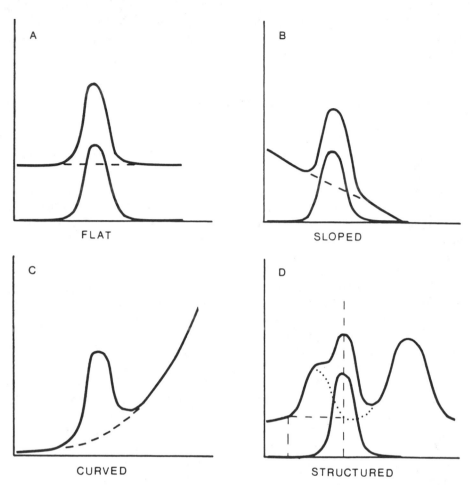

Figure 6.11. General cases of spectral distribution of background emission: (*A*) flat background, (*B*) sloped, linear background, (*C*) simple curved background, and (*D*) moderately structured background in which maxima or minima exist.

6.6.2.2 Sloped Background

The background in Figure 6.11*B* is not constant with wavelength and changes in an approximately linear manner. With a high-resolution spectrometer, the background offset from widely broadened line wings appears as the dashed line, and the use of a single off-peak measurement provides erroneous results. The wavelength window over the analyte peak can be scanned to construct the tangent baseline beneath the peak. The analyte signal is represented by the distance from the peak to the artificial baseline. Alternatively, background measurements at approximately equal intervals off the peak on either side of it can be taken. The average of these signals, or their weighted average, if unequal distances are used, is subtracted from the peak signal.

6.6.2.3   Simple Curved Background

The case of a simple curved background is really one of the more difficult situations to treat (Figure 6.11C). For example, background interference may be the result of proximity to a strong concomitant line. Two off-peak measurements will give only an approximate correction in this case. Multiple off-peak measurements can be attempted with the aim of generating a more accurate artificial background, a task that is time-consuming and not always effective. Scanning over the analyte peak might provide a somewhat better interpolation. With greater storage capabilities available in the control computers on many ICP–AES systems, it is possible to scan a matrix sample without any analyte, store the spectrum, and then subtract the spectrum point by point from an equivalent sample spectrum. The effectiveness of this approach depends on the availability of a matrix containing no analyte and the reproducible generation of the background. An inappropriate background spectrum removes either too little or too much of the background contribution. For trace determinations, this error may not be tolerable.

6.6.2.4   Moderately Structured Background

A background with low-level maxima or minima is common for ICP–AES (Figure 6.11D). Numerous low-intensity, blended lines of various origins litter the baseline beneath many analyte lines. This case can be handled by a dual-wavelength technique. Two wavelengths are chosen for which the background signals are identical, but for which the analyte signal is not the same. The difference in signals at the two wavelengths is a measure of the analyte alone. The difference is not a function of the background intensity as long as the shape of the background remains the same. Of the possible choices for wavelengths, the ones providing the largest difference signal are best.

Such an approach has been implemented in certain ICP–AES systems through the use of a procedure that automatically switches between two selectable wavelength settings, continually computing the difference signal. Of course, wavelength resettability affects the attainable signal precision. Also, this approach does not work for either sloped or curved background, unless a sacrifice of the analyte signal level is acceptable, in which case the off-peak wavelength is taken to the high-background side of the peak. For trace determinations, this sacrifice frequently is not acceptable.

6.6.2.5   Derivative Spectrometry, Wavelength Modulation, and Repetitive Scanning

The technique that works well for any background shape much broader than the analyte line is derivative spectrometry. In this approach, the rate of change of spectral intensity with respect to wavelength is recorded, ie, $dI/d\lambda$. Narrow features, such as lines, which have a high rate of change with wavelength, are emphasized compared to broader features and the background. A particular advantage is that the exact shape of the background is less important, as long as the background is definitely much broader than the analyte line. That is, as long as there is no discernible structure near the analyte line, the derivative approach works. Obviously, it will apply to the flat background because the derivative of a line of zero slope is zero. For a sloped background, only a second derivative technique is rigorously correct, ie, $d^2I/d\lambda^2$. For a curved or moderately structured background, the derivative technique can be only approximate, because the background generates finite derivative signals of its own. The background structure can be taken advantage of, however, because derivative signals from the background have zero crossings of the signal. Analyte measurements made at background zero crossings are free of background interference.[44] The tedious implementation of this rather effective approach has prevented it from being more frequently used.

Derivative techniques have been applied to atomic emission spectrometry,[46–50] with devices that modulate the radiation in the spectrometer. Because the most useful information relative to performing a background correction is at the analyte peak and at location(s) somewhere well off the peak, these versions of derivative spectrometry have been modified to enhance the magnitude of the corrected signal.

Alterations to the waveform used to scan over the peak are not frequently symmetric in time with respect to the points at which measurements are taken. That is, the same amount of signal-retrieval time is not spent at each point interrogated. One optimization of wavelength modulation[51] indicated that 91% of the signal acquisition time should be spent at the analyte peak, while background information can be obtained in the other 9% nearly anywhere off the peak. These modifications to the derivative technique are considered to be variations of multiple-wavelength-based corrections, discussed in Section 6.6.2.4. The principal function performed in them all is that of acquisition of reproducible data for the background and using such data to correct the analyte signal in a time interval so short that there is no adverse impact on the precision of the determination.

6.7   CONCLUSIONS AND FUTURE DIRECTIONS

A variety of spectral interferences can affect the analyte spectra from the ICP. A number of tools for dealing with these interferences are available to the spectrochemist. It must be remembered that each new sample type has to be treated as a unique case, with its own set of specific interference effects to be deciphered, no matter how similar it is to other sample types. The analyst is wise to remain properly skeptical when evaluating the merit of any ICP spectrum.

With the growing use of on-line computers to control chemical instrumentation, interest is increasing in having the computer systems take over the tedious work of deconvolving the spectral interferences encountered. Commentary is beginning to appear in the literature on various aspects and attempts at this endeavor. Predominantly, what has been accomplished is the characterization of and access to the data sets and wavelength tabulations central to interference minimization. The computer is fast in organizing and sorting data in model sets and sample sets. At the more advanced levels of computerization, the computer applies preset instructions for the manipulation of data to arrive at an analytical result. This, too, is only an acceleration of the actions of the operator. Worse, it is still fraught with all the error generated by the analyst, and no new information is provided. The generation of new information from no more data than now are being gathered is the key to the use of so-called artificial intelligence (AI) machines.

The application of AI—that is, truly "smart" spectrometers, capable of conducting the analyst's work of unraveling spectral interferences on analyte signals—awaits a new age of techniques and instrumentation. The approaches to extract more useful information from the signals presently generated in an ICP-AES system, such as forms of pattern recognition, rank annihilation, and varieties of chemometric conversions and inversions, are only now being investigated. One should note that atomic spectroscopy is not yet one of the leading subtopical development areas popular with chemometricians. The application of this technology will surely require modes of operation not now familiar to practicing spectrochemists. More important, the instrumentation able to generate the data streams required by AI approaches is not likely to be forthcoming rapidly.

In the interim, computerization of ICP-AES systems will at least provide more efficient and convenient use of reference data sets and wavelength listings and scans. The near-term result should be higher quality of quantitative results and gradually increased levels of information output from the presently accepted techniques in ICP-AES.

REFERENCES

1. A. Batal, J. Jarosz, and J. M. Mermet, A Spectrometric Study of a 40 MHz ICP—VI. Argon Continuum in the Visible Region of the Spectrum, *Spectrochim. Acta* 36B, 983–992 (1981).
2. R. K. Winge, V. A. Fassel, V. J. Peterson, and W. A. Floyd, "Inductively Coupled Plasma—Atomic Emission Spectroscopy: An Altas of Spectral Information," 1st ed. Elsevier, New York, 1985.
3. R. D. Reeves, S. Nikdel, and J. D. Winefordner, Molecular Emission Spectra in the RF ICAP, *Appl. Spectrosc.* 34, 477–483 (1980).
4. A. R. Forster, T. A. Anderson, and M. L. Parsons, ICP Spectra—I. Background Emission, *Appl. Spectrosc.* 36, 499–503 (1982).
5. J. M. Mermet and C. Trassy, A Spectrometric Study of a 40 MHz ICP—V. Discussion of Spectral Interferences and Line Intensities, *Spectrochim. Acta* 36B, 269–292 (1981).
6. G. R. Kornblum and L. DeGalan, Arrangement for Measuring Spatial Distributions in an Argon Induction Coupled Radio Frequency Plasma, *Spectrochim. Acta* 29B, 249–261 (1974).
7. G. R. Kornblum and L. DeGalan, Spatial Distribution of the Temperature and the Number Densities of Electrons and Atomic and Ionic Species in an Inductively Coupled RF Argon Plasma, *Spectrochim. Acta* 32B, 71–96 (1977).

8. G. F. Larson, V. A. Fassel, R. K. Winge, and R. N. Kniseley, Ultratrace Analyses by OES: The Stray Light Problem, *Appl. Spectrosc.* 30, 384–391 (1976).

9. R.W.B. Pearse and A. G. Gaydon, "The Identification of Molecular Spectra," 4th ed. John Wiley & Sons, New York, 1976.

10. P.W.J.M. Boumans, "Line Coincidence Tables for Inductively Coupled Plasma Atomic Emission Spectrometry." Vols. 1 and 2. 1st ed. 1980; 2nd ed., 1984. Pergamon Press, Elmsford, New York.

11. G. F. Larson and V. A. Fassel, Line Broadening and Radiative Recombination Background Interferences in ICP—AES, *Appl. Spectrosc.* 33, 592–599 (1979).

12. H. R. Griem, "Plasma Spectroscopy." McGraw-Hill, New York, 1964.

13. D. J. Kalnicky, V. A. Fassel, and R. N. Kniseley, Excitation Temperatures and Electron Number Densities Experienced by Analyte Species in Inductively Coupled Plasma with and without the Presence of an Easily Ionized Element, *Appl. Spectrosc.* 31, 137–150 (1977).

14. G. R. Harrison, The Production of Diffraction Gratings—II. The Design of Echelle Gratings and Spectrographs, *J. Opt. Soc. Am.* 39, 522–528 (1949).

15. G. W. Stroke, Diffraction Gratings, in "Handbuch der Physik," Vol. 29, S. Flugge, Ed. Springer-Verlag, Berlin, 1967.

16. R. M. Barnes and R. F. Jarrell, Gratings and Grating Instruments, in "Analytical Emission Spectroscopy," Vol. 1, Part 1, E. L. Grove, Ed. Marcel Dekker, New York, 1971.

17. V. A. Fassel, J. M. Katzenberger, and R. K. Winge, Effectiveness of Interference Filters for Reduction of Stray Light Effects in AES, *Appl. Spectrosc.* 33, 1–5 (1979).

18. W. F. Meggers, C. H. Corliss, and B. F. Scribner, Tables of Spectral Intensities: Part I—Arranged by Elements; Part II—Arranged by Wavelengths, NBS Monograph No. 145. Government Printing Office, Washington, D.C., 1975.

19. P.W.J.M. Boumans and F. J. DeBoer, Studies of an Inductively Coupled High-Frequency Argon Plasma for Optical Emission Spectrometry—II. Compromise Conditions for Simultaneous Multielement Analysis, *Spectrochim. Acta* 30B, 309–334 (1975).

20. P.W.J.M. Boumans, F. J. DeBoer, F. J. Dahmen, H. Hoelzel, and A. Meyer, A Comparative Investigation of Some Analytical Performance Characteristics of an Inductively Coupled RF Plasma and a Capacitively Coupled Microwave Plasma for Solution Analysis by Emission Spectrometry, *Spectrochim. Acta* 30B, 449–469 (1975).

21. P.W.J.M. Boumans, Derivation of and Comments on the Relationship Between the RSD of a Net Line Signal and the Concentration of the Analyte, *ICP Inf. Newsl.* 4, 232 (1978).

22. (a) R. K. Winge, V. J. Peterson, and V. A. Fassel, Inductively Coupled Plasma—Atomic Emission Spectrometry: Prominent Lines, *Appl. Spectrosc.* 33, 206–209 (1979).
    (b) R. K. Winge, V. A. Fassel, V. J. Peterson, and M. A. Floyd, Inductively Coupled Plasma-Atomic Emission Spectrometry: On the Selection of Analytical Lines, Line Coincidence Tables, and Wavelength Tables, *Appl. Spectrosc.* 36, 210–221 (1982).

23. P.W.J.M. Boumans, ICP—DC Arc in a New Jacket? *Spectrochim. Acta* 35B, 57–71 (1980).

24. G. R. Harrison, "Massachusetts Institute of Technology Wavelength Tables." MIT Press, Cambridge, Mass., 1969.

25. F. M. Phelps, "Massachusetts Institute of Technology Wavelength Tables," Vol. 2, "Wavelengths by Element." MIT Press, Cambridge, Mass., 1982.

26. R. J. Kelly, "A Table of Emission Lines in the Vacuum Ultraviolet for All Elements (6 Angstroms to 2000 Angstroms)," UCRL 5612. University of California Lawrence Radiation Laboratory, Livermore, 1959.

27. R. J. Kelly and L. J. Palumbo, "Atomic and Ionic Emission Lines Below 2000 Angstroms—Hydrogen Through Krypton," NRL Report No. 7599. Naval Research Laboratory, Washington, D.C., 1973.

28. J. Kroonen and D. Vader, "Line Interference in Emission Spectrographic Analysis." Elsevier, Amsterdam, 1963.

29. J. Kuba, L. Kucera, F. Pizak, M. Dvorak, and J. Mraz, "Coincidence Tables for Atomic Spectroscopy." Elsevier, Amsterdam, 1965.

30. M. L. Parsons, B. W. Smith, and G. E. Bentley, "Handbook of Flame Spectroscopy." Plenum Press, New York, 1975.

31. A. N. Zaidel, V. K. Prokof'ev, S. M. Raiskii, V. Slavnyi, and E. A. Shreider. "Tables of Spectrum Lines." IFI/Plenum, New York, 1970.

32. P.W.J.M. Boumans and M. Bosveld, A Tentative Listing of the Sensitivities and Detection Limits of the Most Sensitive ICP Lines as Derived from the Fitting of Experimental Data for an

Argon ICP to the Intensities Tabulated for the NBS Copper Arc, *Spectrochim. Acta*, 34B, 59–72 (1979).

33. I. B. Brenner and H. Eldad, A Spectral Line Atlas for Multitrace and Minor Element Analysis of Geological and Ore Mineral Samples by ICP-AES, *ICP Inf. Newl.* 10(6), 451–508 (1984).

34. T. A. Anderson, A. R. Forster, and M. L. Parsons, ICP Emission Spectra—II. Alkaline Earth Elements, *Appl. Spectrosc.* 36, 504–509 (1982).

35. T. A. Anderson and M. L. Parsons, ICP Emission Spectra—III. The Spectra of the Group IIIA Elements and Spectral Interferences Due to Group IIA and IIIA Elements, *Appl. Spectrosc.* 38, 625–634 (1984).

36. E. Michaud and J. M. Mermet, Iron Spectrum in the 200-300 nm Range Emitted by an ICP, *Spectrochim. Acta* 37B, 145–164 (1982).

37. M. L. Parsons, A. R. Forster, and D. Anderson, "An Atlas of Spectral Interferences in ICP Spectroscopy." Plenum Press, New York, 1980.

38. R. K. Winger, V. J. Peterson, and V. A. Fassel, "Inductively Coupled Plasma Atomic Emission Spectroscopy: Prominent Lines." EPA-600/4-79-017, National Technical Information Service, Springfield, Va., 1979.

39. (a) R. C. Fry, S. J. Northway, R. M. Brown, and S. K. Hughes, Atomic Fluorine Spectra in the Argon Inductively Coupled Plasma, *Anal. Chem.* 52, 1716 (1980).
    (b) S. K. Hughes, R. M. Brown, and R. C. Fry, Photodiode Array Studies of Near Infrared and Red Atomic Emissions of C, H, N, and O in the Argon Inductively Coupled Plasma, *Appl. Spectrosc.* 35, 396 (1981).
    (c) S. K. Hughes and R. C. Fry, Nonresonant Atomic Emissions of Bromine and Chlorine in the Argon Inductively Coupled Plasma, *Anal. Chem.* 53, 1111 (1981).
    (d) S. K. Hughes and R. C. Fry, Near Infrared Atomic Emissions of S and C in the Argon Inductively Coupled Plasma, *Appl. Spectrosc.* 35, 493 (1981).

40. C. C. Wohlers, Experimentally Obtained Wavelength Tables for the ICP, *ICP Inf. Newsl.* 10(8), 593–688 (1985).

41. D. D. Nygaard and D. S. Leighty, Inductively Coupled Plasma Emission Lines in the Vacuum Ultraviolet, *Appl. Spectrosc.* 39, 968–976 (1985).

42. P.W.J.M. Boumans and J.J.A.M. Vrakking, High-Resolution Spectroscopy Using an Echelle Spectrometer with Predisperser—I. Characteristics of the Instrument and Approach for Measuring Physical Line Widths in an ICP, *Spectrochim. Acta* 39B, 1239–1260 (1984).

43. J. W. McLaren and J. M. Mermet, Influence of the Dispersive System in ICP-AES, *Spectrochim. Acta* 39B, 1307–1322 (1984).

44. T. C. O'Haver, Chap. 2 in "Trace Analysis: Spectroscopic Methods for Elements," J. D. Winefordner, Ed. John Wiley & Sons, New York, 1976.

45. P.W.J.M. Boumans and J.J.A.M. Vrakking, High-Resolution Spectroscopy Using an Echelle Spectrometer with Predisperser—II. Analytical Optimization for ICP-AES, *Spectrochim. Acta* 39B, 1261–1290 (1984).

46. W. Snelleman, T. C. Rains, K. W. Yee, H. D. Cook, and O. Menis, Flame Emission Spectrometry with Repetitive Optical Scanning in the Derivative Mode, *Anal. Chem.* 42, 394–398 (1970).

47. G. M. Hieftje and R. J. Sydor, Application of a Wavelength Modulation Device to Problems Concerning Spectrometer Misalignment, *Appl. Spectrosc.* 26, 624–631 (1972).

48. M. S. Epstein and T. C. O'Haver, Improvements in Repetitive Scanning Techniques for Reducing Spectral Interferences in Flame Emission Spectrometry, *Spectrochim. Acta*, 30B, 135–146 (1975).

49. R. K. Skogerboe, P. J. Lamothe, G. J. Bastiaans, S. Freeland, and G. N. Coleman, A Dynamic Background Correction System for Direct-Reading Spectrometry, *Appl. Spectrosc.* 30, 495–500 (1976).

50. S. R. Koirtyohann, E. D. Glass, D. A. Yates, E. Hinderberger, and F. E. Lichte, Effect of Modulation Waveform on the Utility of Emission Background Corrections Obtained with an Oscillating Refractor Plate, *Anal. Chem.* 49, 1121–1126 (1977).

51. L. R. Layman, A Programmable Wavelength Modulator for Background Correction in Simultaneous Multielement Emission Spectrometry. Paper No. 188, presented at the FACSS Meeting, Philadelphia, 1979.

7

High-Resolution Plasma Spectrometry

M. C. EDELSON

Ames Laboratory
U.S. Department of Energy
Iowa State University
Ames, Iowa

7.1   INTRODUCTION

In this chapter, we consider the topic of high-resolution plasma spectrometry (HRPS). Several different experimental approaches to HRPS are reviewed, and the benefits derived from increased spectral resolution are discussed. Results from the scientific literature are presented, but no attempt is made to provide an encyclopedic review. The discussion is limited to plasma emission spectrometry and, in particular, to inductively coupled plasma–atomic emission spectrometry (ICP–AES).

To a laser spectroscopist, the term "high-resolution spectroscopy" is applied to studies that exploit precise wavelength, line width, or temporal measurements to investigate primary physical processes. Tunable lasers, atomic or molecular beams, and high-vacuum conditions allow measurements with kilohertz to megahertz resolution.[1] It is important to realize that 1 MHz at 400 nm is equivalent to 0.533 fm or $5.33 \times 10.^{-7}$ nm!

This chapter is directed toward the chemist who is concerned with the spectroscopic determination of the concentrations of elements that often are at the part-per-million (μg/g) or part-per-billion (ng/g) level. The atmospheric pressure plasma, eg, the ICP, is widely recognized as being useful for this purpose, but the physical characteristics of this source preclude ordinary spectroscopic measurements at the high resolution previously cited.

By "high-resolution plasma spectrometry," we refer to studies that feature optical detection systems capable of resolving the width of spectral lines emitted by atoms or ions produced in an atmospheric-pressure plasma. These line widths are typically in the 2 to 6 pm range, depending on the atomic weight of the emitter, and they are 100 to 100,000 times greater than the line widths observed in certain high-resolution laser studies.

The spectroscopic apparatus needed to resolve plasma line widths has been commercially available for many years. However, there are few high-resolution ICP spectrometers in operation today. It is noteworthy that manufacturers have decided not to offer high-resolution monochromators with their ICP sources. Nevertheless, commercial plasma instruments have achieved great acceptance by analytical chemists and have been successfully applied to many difficult analytical problems. Certain applications of plasma emission spectrometry, such as isotopic analysis, cannot be accomplished without resorting to high-resolution instruments. Other applications, such as the determination of trace impurities in complex matrices, can be performed more efficiently with a high-resolution instrument. These applications are described in this chapter.

7.2   THE PROFILE OF EMISSION LINES

7.2.1   Source Line Broadening

To properly discuss the subject of HRPS, we must first explore the intrinsic limitations of resolution that are imposed by the plasma source. This discussion focuses on the ICP; however many of the arguments apply to the direct-current plasma (DCP) with little modification.

An ICP spectrometer system consists of a well-regulated plasma source, a device for converting a sample to an aerosol for introduction into the plasma, an optical or a mass spectrometer for selectively detecting species atomized, ionized, and excited by the plasma, and an electronic signal-processing apparatus for measuring the current from the radiation or mass detector. Because this chapter is devoted to optical spectroscopy, no extensive reference to the allied technique of ICP-mass spectrometry is made.

The light emitted by atoms and ions from materials introduced into an ICP consists of the superposed line spectra of all emitting species. Individual spectral lines are subjected to several broadening mechanisms that are related to the environment of the emitting species. The complex environment of emitting atoms and ions can lead to complex line shapes. Human and Scott[2] utilized a Fabry–Perot interferometer to examine the shapes of lines emitted by an ICP operated at commonly used conditions. The spectral line shape was determined to be closely approximated by a Voigt profile,[3] which is the convolution of the Gaussian and Lorentzian line profiles.

Inductively coupled plasmas are very "hot" sources that have estimated temperatures of 6500 K in the observation zone.[4] Atoms and ions in the plasma are in violent thermal motion, and their thermal velocity, through the Doppler effect, modifies the wavelength of radiation emitted. The Doppler effect leads to a Gaussian line shape,[5] with DL, the full width at half-maximum (FWHM) intensity of the line given by:

$$DL = 7.16 \times 10^{-7} \times L \times \left(\frac{T}{M}\right)^{1/2} \qquad [7.1]$$

where L is the line wavelength, T is the temperature of the emitting atom or ion (K), and M is the mass of the emitting species (amu).

Emitting atoms and ions in plasmas operated at atmospheric pressure suffer numerous collisions with other plasma constituents. The line broadening caused by collisions is not easily calculated, but such broadening is known to lead to a Lorentzian line shape.[5] The degree of mixing of Gaussian and Lorentzian character in the Voigt profile is described by the Voigt parameter a, which is defined in Equation 7.2.

$$a = \frac{HWMH(L)}{HWMH(G)} \times (\ln 2)^{1/2} \qquad [7.2]$$

where $HWHM(L)$ and $HWHM(G)$ are the halfwidths at half-maximum height due to Lorentzian broadening (eg, collisional broadening), and Gaussian broadening (eg, thermal broadening), respectively.

Human and Scott[2] found that the Voigt profile of lines emitted by ICP species was dominated by the Gaussian component, as confirmed by recent investigations.[6-9] Thus, the Voigt parameter for Fe lines has been shown[6] to be approximately 0.10. The presence of a Lorentzian component in the spectral line shape is most noticeable in the wings of the line, as illustrated by the Voigt profile in Figure 7.1. A good approximation to the FWHM of an ICP spectral line can be calculated from Equation 7.1. The contribution from collisional broadening to the FWHM of the spectral line equals only about 10% of the total line width and therefore may be neglected in this simple model. Equation 7.1 also can be used to estimate the plasma temperature from measured line widths (the FWHM of the Gaussian component should be used).

7.2.2 Instrumental Line Broadening

Once the shape and width of lines emitted by species introduced into the analytical plasma are approximately known, the selection of a spectroscopic

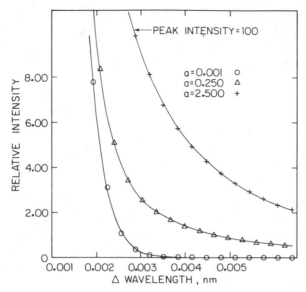

Figure 7.1. Comparison of Voigt line shapes far from the line peak. Three lines with different Voigt parameters are displayed. Each line has a FWHM of 2.5 pm and a peak intensity of 100. The line with $a = 0.001$ is essentially Gaussian. (1 nm = 1000 pm).

instrument for measurement is possible. The resolving power (R_p) of a spectrometer is defined as the ratio of the line wavelength to the minimum resolvable line width (FWHM) at that wavelength:

$$R_p = \frac{L}{DL} \qquad [7.3]$$

Each spectroscopic device has an associated slit function that must be convoluted with the plasma line shape to determine the experimental line shape.[3] The slit function is determined by measuring the line profile of an extremely sharp line, such as that produced by a low-power, continuous-wave laser. Direct measurement of the true spectral line shape is not possible; there is always some instrumental broadening. The effect of instrumental broadening on the effective width of the emission lines does not depend linearly on the instrument resolving power (Figure 7.2), and there is increasingly little benefit for R_p exceeding 300,000. Recent work[10] indicates that the optimum R_p for spectral lines emitted by a plasma is about 150,000, with conventional diffraction grating or echelle spectrometers. Such a resolving power provides a reasonable photon flux at the detector and only moderately broadens the line being measured. The commonly used monochromators provided with commercial sequential ICP systems (typically with a focal length of 1 m, a 2400 groove/mm grating used in first order, and fixed 25-μm slits) usually operate at a resolving power of approximately 20,000 and severely broaden the lines emitted by the ICP.

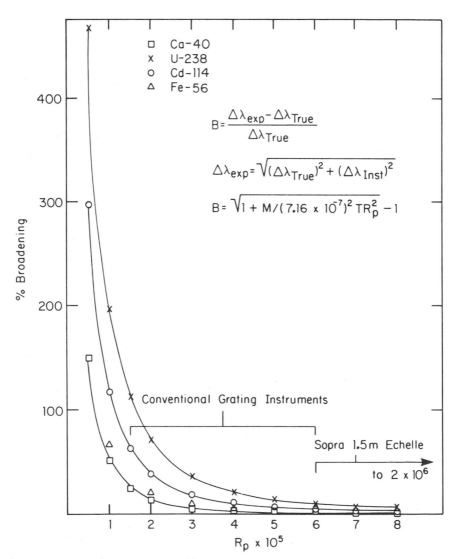

Figure 7.2. The percent line broadening (*B*) of thermally broadened lines (assumed to be Gaussian) by the monochromator. The effect of increasing instrumental resolving power (*R$_p$*) on *B* is shown to rapidly diminish when *R$_p$* exceeds 300,000. The temperature of the plasma is assumed to be 6000 K.

7.3   INSTRUMENTATION FOR HIGH-RESOLUTION SPECTROMETRY

The techniques used to achieve high-resolution detection of ICP emission lines, briefly described in this section, include Fabry–Perot interferometry, echelle and grating spectrometry, and Fourier transform spectrometry.

7.3.1   Fabry–Perot Interferometry

The Fabry–Perot interferometer (FPI), which can operate at high spectral resolving power (> 1,000,000), is basically a simple device. It consists of a pair of precisely orientable, high-reflectivity mirrors. The FPI achieves high spectral resolution through the principle of interference. The fundamental principles governing the performance of the FPI are well documented.[11] It is noteworthy that standard treatments on the theory of the FPI do not often discuss how recent advances in technology have benefited modern Fabry–Perot interfero-metry. Developments in thin-film production, microprocessors, and transducers have greatly improved the analytical utility of Fabry–Perot interferometry.

The resolution of the FPI is given by:

$$DL = \frac{FSR}{F} \qquad [7.4]$$

where FSR and F are the free spectral range and the finesse, respectively. The FSR, which is the wavelength range accessible in any one order of interference, is inversely proportional to the mirror spacing. The flatness and reflectivity of the FPI mirrors chiefly determine the instrument quality factor known as finesse, F. FPI plates with F between 25 and 50 are commercially available. The resolving power of the FPI is given by:

$$R_p = \frac{L}{DL} = \frac{L \times F}{FSR} \qquad [7.5]$$

To achieve high resolving power with the FPI, the ratio F/FSR must be maximized. This goal is realized when very-high-reflectivity mirror coatings[11] are used, but such coatings are useful only over a small wavelength range and thus, many different mirror sets are required to cover a useful spectral region. Currently, one FPI mirror set is available that can be used throughout the visible and near-IR spectral region,[12] with F approximately 30.

The FPI offers excellent spectral resolution but very poor spectral selectivity. Thus, an FPI mirror set coated for operation between 350 and 450 nm will transmit all lines in that region without discrimination, ie, many different orders of interference will be transmitted simultaneously. To achieve wavelength selection within the FPI-mirror transmission range,[11] the use of optical glass filters or of a monochromator is often required. This practice is similar to the use of glass filters with monochromators as "order-sorters." To permit the unambiguous interpretation of FPI spectra, it is often necessary to set the monochromator bandpass equal to the FSR of the interferometer. This requirement is particularly important for line-rich sources such as the ICP. The FPI-monochromator combination is often the least expensive route to HRPS and is particularly attractive for an instrument dedicated to operation in a limited wavelength range. Applications of the FPI to HRPS include the pioneering ICP line shape measurement studies of Human and Scott,[2] the first ICP isotope determination by Edelson and Fassel,[13] and a recent study of the wavelength shift of ICP lines by Kato and co-workers.[14]

7.3.2 Echelle Spectrometry

The echelle, originally proposed by Harrison,[15] is a coarsely ruled grating that is used in high orders ($n > 80$) and at large angles of incidence. The theoretical resolving power of an echelle, or a grating, is:

$$R_p = n \times N_r \qquad [7.6]$$

where n is the order of diffraction, and N_r is the total number of grooves on the illuminated area of the echelle or grating. The operating principles and the analytical applications of the echelle have been thoroughly described,[16] and the echelle now is being used in several commercial plasma spectrometers. In a recent series of articles, Boumans and Vrakking[17-19] have evaluated a new echelle spectrometer designed for HRPS (Figure 7.3). In contrast to the earlier commercial versions of the echelle spectrometer, where a prism is used to isolate the orders, Boumans and Vrakking used a grating predisperser prior to an Ebert

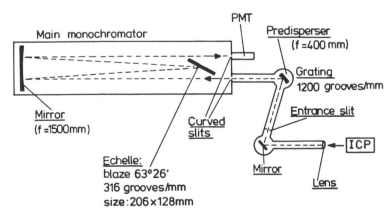

Figure 7.3. Schematic diagram of the Boumans–Vrakking echelle monochromator with predisperser in parallel slit arrangement. (From Reference 17, with permission.)

Figure 7.4. The DPS 1500 High-Resolution Echelle Spectrometer. (Reproduced with permission from SOPRA, 68, rue Pierre-Joigneaux, F 92270 Bois-Colombes, France.)

monochromator equipped with an echelle grating. The theoretical resolving power of this instrument exceeds 1,000,000, and a commercial version (Figure 7.4) is now available.

Optimization of an ICP spectrometer involves the consideration of several factors, including resolution and powers of detection.[17–19] The detection limit, C_L, is defined[18] as:

$$c_L = \frac{0.01 \times k \times (\text{RSD})_B \times c_o}{\text{SBR}} \qquad [7.7]$$

where k is 2 (a somewhat arbitrary choice), $(\text{RSD})_B$ is the percent relative standard deviation of the background, SBR is the signal-to-background ratio, and c_o is the analyte concentration at which the SBR is measured.[18] While the SBR for a Mn ion line remains relatively constant over a wide range of slit widths, as shown in Figure 7.5, the RSD of the background decreases dramatically as the slit width is increased. The best powers of detection are achieved at rather wide slits. At these slit widths, the practical resolving power of the echelle spectrometer is reduced to that of a medium-resolution monochromator, which is adequate for single-element trace determinations in aqueous solutions. Detection limits obtained with wide slits are comparable to or better than those obtained with a conventional medium-resolution instrument.[18,20] In the presence of a complex matrix, the increase in the photon flux at the detector permits efficient operation with small slits. Here, the detection limit advantage of the echelle over a medium-resolution monochromator is approximately twofold, even when there is no line from the matrix in the close vicinity of the analyte line.

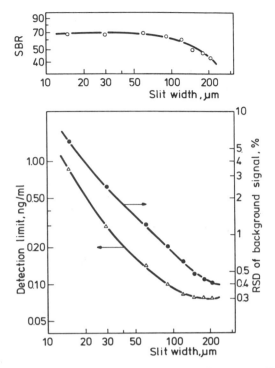

Figure 7.5. Double logarithmic plots of SBR, $(RSB)_b$ and c_l versus slit width for Mn II 257.610 nm. (From Reference 18, with permission.)

Based on these studies, practical resolving powers between 120,000 and 200,000 are required for high-resolution ICP spectroscopy in the visible, whereas powers of 100,000 to 140,000 are necessary in the low ultraviolet region.

7.3.3 Grating Spectrometry

Equation 7.6 suggests that high resolving power can be achieved by operating in high orders of diffraction or by utilizing finely ruled diffraction gratings. Modern techniques of holography permit the economic fabrication of large (eg, ≥ 100 mm wide), finely ruled gratings with groove densities of 2400 to 3600 lines per millimeter. Special gratings can be prepared with substantially greater groove densities. However, the construction of a monochromator capable of routine operation at high resolving power is nontrivial. One high-resolution grating instrument is commercially available (the THR1500 monochromator, by Jobin-Yvon) that can, in first order, resolve the Doppler width of ICP lines. This monochromator has a 1.5-meter focal length and can be used either in a single-pass or a double-pass mode. Resolving powers exceeding 300,000, with a 2400 groove/mm grating, or exceeding 450,000, with a 3600 groove/mm grating, are

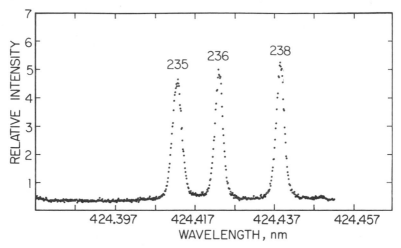

Figure 7.6. Resolution of U II 424.437 nm isotopic splitting with the THR1500 monochromator. Total U concentration is about 0.6 mg/mL, and the signal integration time was 1 s/point. The spectrum is not smoothed. The distance between the ^{235}U and ^{236}U components is about 10 pm. The ^{238}U line width (FWHM) is about 2 pm, and broadening of the ^{235}U component, due to unresolved hyperfine splitting, is evident. (M. C. Edelson and V. A. Fassel, Ames Laboratory, unpublished results.)

achieved for the double-pass mode, if room temperature is regulated within $\pm 0.6°C$. An example of the working resolution of this instrument, used to monitor the isotopic splitting of the U II 424.437 nm line in an ICP is given in Figure 7.6. Despite the lack of thermostatting on the spectrometer, a resolving power close to 300,000 with the 2400 groove/mm grating in the double-pass mode is observed by the author when this spectrometer is operated in a laboratory having poor temperature control ($\pm 4°C$).

Grating instruments have several advantages relative to echelle spectrometers and Fabry–Perot interferometers. First, in contrast to echelle spectrometers and Fabry–Perot interferometers, the effective free spectral range of a grating used in low order is very large. Order-sorting problems in grating monochromators are very easily controlled with simple glass bandpass or interference filters. These problems can become acute with Fabry–Perot interferometers and can require the use of medium-focal-length monochromators for order sorting when using Fabry–Perot interferometers with line-rich sources, such as the ICP.[13] Second, the efficiency of a grating varies slowly with wavelength so that, in contrast to echelle instruments, it is usually possible to operate in one order over the entire spectral region of interest. To measure widely separated lines with useful efficiency, echelle instruments must be operated in many different orders. This requires complex wavelength scanning programs for computer-controlled echelle monochromators.[17] Relatedly, the mirror coatings of Fabry–Perot interferometer plates must be optimized for high reflectivity to achieve high finesse, and these coatings are often useful only over a small spectral range. To achieve spectral coverage over the range 170 to 700 nm, a conventional FPI could require more than a dozen mirror sets. Third, the resolution of a grating monochromator is relatively constant across its working range, but the practical resolution of an echelle monochromator can vary considerably with wavelength. Diffraction efficiency in an echelle monochromator is maintained by keeping the angle of incidence of light impinging on the echelle close to the blaze angle. This requires that the diffraction order be varied as the wavelength is changed, and as the order of diffraction is lowered, the resolution of the echelle is decreased.

In contrast, the inherently higher theoretical resolving power of the echelle, relative to the diffraction grating, allows the instrument maker to achieve high resolving power with a relatively compact instrument. The echelle also is a very efficient device when used close to its optimum blaze angle,[18] and detection limits obtained with echelle plasma spectrometers are comparable to those achieved by grating spectrometers.

7.3.4 Fourier Transform Spectrometry

Many of the inherent difficulties associated with the application of the Fabry–Perot interferometer to HRPS do not pertain to the Michelson interferometer (Figure 7.7). Michelson interferometry is well known to infrared spectroscopists,[21] but until recently, very few applications of this technique in ultraviolet and visible spectroscopy had been reported. The problems of utilizing the Michelson interferometer in the ultraviolet and visible spectral regions are mainly mechanical, as described by Nordstrom.[22] In the last 10 to 20

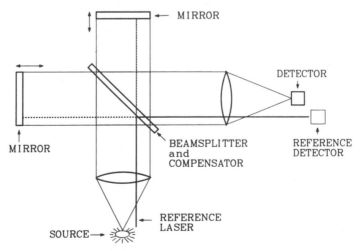

Figure 7.7. Basic Michelson interferometer design. (From Reference 25, with permission.)

years, these problems have been solved as a result of major advances in laser and computer technology.

In contrast to the Fabry–Perot interferometer, which produces an intensity-versus-wavelength display when the distance between its mirrors is varied and its central spot transmission through a pinhole is observed, the Michelson interferometer generates an interferogram. The interferogram must be mathematically transformed to produce the conventional spectrum of intensity versus wavelength. This operation is accomplished by the Fourier transform; hence, the name Fourier transform spectrometry has been attached to this application of Michelson interferometry. The first application of Fourier transform spectrometry (FTS) to analytical atomic emission spectrometry was accomplished by Horlick and his co-workers. It is important to note that these investigators have not critically discussed the FTS instrumental resolution in their most recent series of articles.[23] The intent of the articles seems to have been to demonstrate that a small, inexpensive FTS instrument can be used to perform the majority of the common analyses now handled by medium-resolution, sequentially scanned monochromators and direct-reading spectrometers. Because the FTS described by Horlick[23] was neither designed nor intended for HRPS, we shall not describe the instrument in detail here. These studies have touched on some of the general limitations of FTS that are pertinent to the application of FTS to HRPS, and, where appropriate, we shall refer to these results in the text. Recently, Faires and co-workers at Los Alamos National Laboratory utilized the Kitt Peak National Observatory Fourier transform spectrometer[24] to demonstrate the feasibility of FTS for HRPS–ICP spectrometry.[25]

The interferometric techniques offer two major advantages, relative to conventional spectrometry, as performed with scanning diffraction grating (or echelle) instruments. These are the multiplex or Fellgett advantage and the throughput or Jacquinot advantage. To appreciate the multiplex or Fellgett advantage, one should note that the interferometer samples the entire spectrum simultaneously, whereas, at any given time, a scanning monochromator monitors only the narrow bandwidth of the spectrum that is defined by the exit slit. If the spectral noise is random and is independent of the intensity of the optical signal, the FTS instrument derives an important signal-to-noise benefit[22] given by Equation 7.8:

$$\frac{SNR_{FTS}}{SNR_{grating}} = \left(\frac{\text{spectral range [nm]}}{\text{bandpass [nm]}}\right)^{1/2} \qquad [7.8]$$

The SNR advantage is approximately 385 for a bandpass of 2 pm and a range of 300 nm. However, ultraviolet–visible detectors exhibit complex noise characteristics, such as signal-dependent shot noise, and thus, in contrast to the thermal detectors used in infrared spectrometry, there is no multiplex or Fellgett advantage. Faires[25] has pointed out that the ICP–FTS technique offers an important operational benefit relative to conventional scanning spectrometry that may compensate for the loss of the multiplex advantage. Conventional ICP scanning spectrometers operate in a slew-scan manner to minimize the time required for analysis. Direct-reading ICP spectrometers are limited to a fixed number of recording channels by geometrical constraints. In each of the cases above, it is not practical to record extensive portions of the spectrum emitted by the ICP. Such recordings are convenient with the FTS instrument. The digitized interferogram may be stored on a mass storage device, such as magnetic tape, to allow the selection of lines customized to the individual solution matrix after the spectrum has been recorded. This procedure should permit the selection of relatively interference-free lines for elemental concentration determinations and should, therefore, result in a more accurate assay of solution constituents. An unexpected solution constituent that interferes with the signal from an analyte can more easily be corrected for in this instance than is possible in conventional ICP–AES, where the analyst is locked into a fixed set of spectral lines at the time of sample measurement.

The Jacquinot advantage in FTS arises because the interferometer is a slitless device and can utilize a much greater portion of the emitted radiant flux of the ICP than can a grating monochromator. High resolving power in a grating monochromator is achieved with narrow slits (eg, 10 to 100 μm). Since the ICP is an extended source, the use of narrow slits strongly attenuates the photon flux that reaches the detector. The Jacquinot advantage is estimated to exceed 200 in the infrared region of the spectrum[22] and is operative in the visible and ultraviolet portions of the spectrum as well. The signal-to-noise gain due to the increased throughput is proportional to the square root of the throughput gain above and therefore is approximately 14.

There is a multiplex disadvantage associated with the use of Fourier transform–ICP spectrometry.[22–25] The background signal in an ICP spectrum is provided by unresolved line wings, argon emission lines, lines from other plasma constituents, etc. In grating spectrometers the contribution to the background is localized to the vicinity of the noise source. Thus, for example, if one wishes to determine the concentration of Al in a matrix that contains a high concentration of U, it would be prudent to select for measurements an Al line that is far removed from intense U lines. Close to an intense U line, the average background signal is larger than the average background signal far from such a line. Since the Al detection limit depends inversely on the signal fluctuations of the background and those fluctuations increase with the average background signal, the detection limit for Al is degraded in the vicinity of a strong matrix line. In ICP–FTS, the noise (Figure 7.8), which is related to the total concentration of all solution components, is spread evenly throughout the entire emission spectrum. In complex, concentrated matrices, the average noise in an ICP–FTS spectrum may be greater than the noise in a relatively line-free region of an ICP spectrum recorded with a diffraction instrument. In this event, the ICP–FTS instrument could exhibit powers of detection inferior to those from an ICP monitored by a grating instrument. Faires[25] found comparable detection limits for ICP–FTS and ICP–grating instruments for analytes contained in a multiple-element matrix.

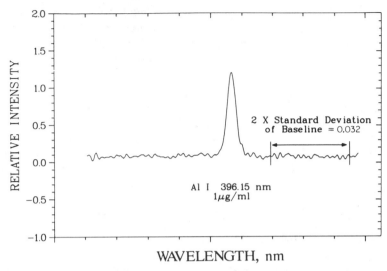

WAVELENGTH, nm

Figure 7.8. The determination of $(RSD)_B$ for the ICP–FTS is performed on a structure-free portion of the recorded spectrum. Since the spectral noise is averaged over the spectrum, this determination suffices for all lines in the spectrum. (Graph courtesy of L.M.H. Faires, Los Alamos National Laboratory.)

Stubley and Horlick[23a–c] coupled a low-resolution monochromator to a FTS to prevent strong matrix emission lines far from analyte wavelengths from contributing to the background noise. This achieved improved detection limits at the expense of wavelength coverage. Without the monochromator, the ICP–FTS detection limits were inferior to those obtained with an ICP–AES instrument using a diffraction grating spectrometer. Stubley and Horlick[23a–c] also investigated the noise distribution in ICP–FTS spectra and reported that the background noise depended on the concentration of all sample constituents but was not evenly distributed throughout the spectrum. In common with conventional spectrometers, the background noise increased in the vicinity of a strong emission line.

Baudais and Buijs[26] discussed the relative efficiencies of a grating instrument and a commercial Michelson interferometer used in the visible and ultraviolet regions. When the effects of the modulation and transfer efficiencies on the throughput of the FTS were factored into the calculation of relative efficiency,[26] the FTS instrument was found to be less efficient than a grating instrument used for low-resolution ultraviolet spectral measurements. However, the efficiency of the FTS compared favorably with that of a high-resolution monochromator used for ultraviolet spectrometry.

The resolution of a Michelson interferometer is determined by the path difference X between the interfering beams[25]:

$$DL = \frac{1}{(2 \times X)} \tag{7.9}$$

Table 7.1. Specification of Los Alamos Fourier Transform Spectrometer

Optical path difference:	2.5 meters (5.0 meters, double pass)
Spectral range:	200 to 20,000 nm
Resolution:	80 fm (at 632.8 nm) (40 fm, double pass)
Detectors:	Photomultiplier tubes, Si PIN diodes, InSb, Si(As)
Intensity accuracy:	0.1%
Wavelength accuracy:	4 fm (at 632.8 nm)
Position accuracy:	1 millifringe rms at 632.8 nm
Scan time (full resolution):	4 min
Optical throughput:	88.9-mm-diameter working aperture

Source: Adapted from information provided by L.M.H. Faires, Los Alamos National Laboratory.

The maximum resolving power of the Kitt Peak Fourier transform spectrometer is approximately 1,000,000 at 300 nm,[25] while the maximum resolving power of the proposed Los Alamos National Laboratory (LANL) Fourier transform spectrometer is 30,000,000. The characteristics of the LANL FTS are given in Table 7.1, and the optical layout is illustrated in Figure 7.9. The resolution of these spectrometers, which are specialized, one-of-a-kind instruments, exceeds that of any existing grating monochromator. A commercial Fourier transform spectrometer (Bomem, Inc., Quebec, Canada) that can be used for visible and ultraviolet spectrometry is reported to achieve resolving powers greater than 1,000,000 for infrared wavelengths.

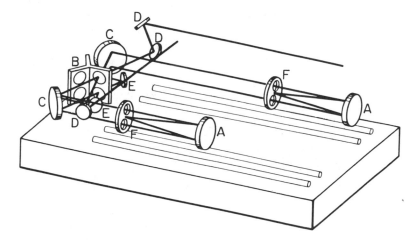

Figure 7.9. Design of proposed Los Alamos National Laboratory Fourier transform spectrometer. Folded Michelson interferometer design uses cat's-eye retroreflectors to help minimize alignment errors. A = cat's-eye reflectors, B = turret mounted beamsplitters, C = folding mirrors, D = collimating optics, E = camera mirrors and F = secondary mirrors. (Diagram courtesy of L.M.H. Faires, Los Alamos National Laboratory.)

7.4 APPLICATIONS OF HIGH-RESOLUTION PLASMA SPECTROMETRY

7.4.1 Isotopic Discrimination in Actinide Spectrometry

7.4.1.1 Introduction

Based on the quantum theoretical description of atomic structure, the position of an electron relative to the atomic nucleus is described by a probability function. Since there is a finite probability of finding certain electrons close to the atomic nucleus, or even within the atomic nucleus itself, it should not be difficult to imagine that the electron energy can be slightly perturbed by changes in nuclear volume, spin, and quadrupole moment that can occur with a change in the number of neutrons in the nucleus. The effect of these perturbations on the electronic energy is quite small, but can be observed with sufficient instrumental resolution.[5] The magnitude of these changes, the isotopic shifts, depends critically on the electronic states involved and on the nuclear mass effect.[5] These isotopic shifts have been used to study various isotopes over the past 65 years.[28]

To be suitable for monitoring isotopic composition, an optical transition should be intense and relatively free from spectral interference, and it should possess resolvable isotopic structure. The electronic spectra of the actinides, which possess easily deformable nuclei and large numbers of electrons, are among the most complex of all atomic spectra. Relatedly, many spectral lines of the actinides exhibit resolvable isotopic shifts. Atomic emission spectrometry for the analytical measurement of the isotopic composition of actinides has been used over the past 30 years.[29] To a large degree, such applications of optical spectroscopy were discontinued in favor of the mass spectrometric determination of isotopic composition. This shift in analysis methods was induced by the great precision attainable with mass spectrometry and by the relative inconvenience of the optical emission techniques available for use in the 1950s and 1960s. The advent of the atmospheric-pressure ICP revitalized the analysis of materials by optical emission spectrometry. The first use of the ICP for study of individual isotopes was reported in 1981.[13]

Because the spectral lines emitted by the ICP are strongly Doppler and pressure broadened, the isotopic splitting of optical transitions is difficult to resolve. The advantages of using the ICP for isotopic analysis lie in its convenience, speed, and ability to be automated. The precision of ICP–AES isotope determinations will not rival that obtainable by mass spectrometry, but many analytical applications do not require ultrahigh precision and the ICP approach may be justifiable in terms of efficient manpower and resource utilization. In contrast to mass spectrometers, ICP–AES does not count sample ions (or atoms) directly, but counts the photons emitted by plasmas species. Thus, highly radioactive analytes can be studied by ICP–AES without contamination of the spectrometer.

7.4.1.2 Uranium Isotope Analysis

The emission spectrum of uranium is very complex and consists of thousands of resolvable lines.[30] Detailed spectral atlases of the U emission spectrum produced by classical sources, such as the hollow cathode lamp,[30,31] are invaluable to high-resolution spectrometric studies of uranium excited in an ICP. Extensive compilations of the isotopic splitting of U lines also have been published.[32]

The first U isotopic assay of HRPS was conducted with a FPI–monochromator pair,[13] but the apparatus was eventually replaced with a high-resolution grating monochromator to increase wavelength coverage. The resolving power of the monochromator was approximately 300,000. The latter ICP apparatus has been used to perform $^{235}U/^{238}U$ assays (Figure 7.10) of U dissolved in dilute nitric acid and dissolved in more complex matrices. Several intense lines with resolvable isotopic shifts are emitted by U in the ICP, but $^{235}U/^{238}U$ assays at low concentrations of ^{235}U are made difficult by interferences between the ^{235}U component chosen for study and other ^{238}U lines.[33] The U 424.437 nm line has been successfully used (Figure 7.11) to perform such isotopic assays at ratios below 1%. The ^{236}U component of the 424.437 nm line is interfered with by a weak ^{238}U line, whereas the 468.91 nm line is not so affected (Figure 7.12).

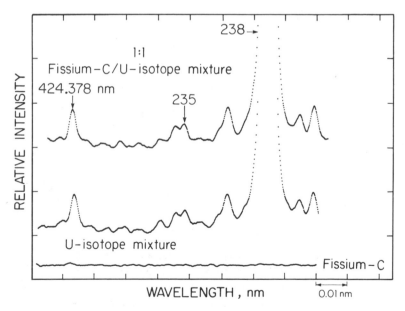

Figure 7.10. Measurement of ^{235}U (0.84%, relative to ^{238}U) by ICP–AES using the U II 424.437 nm line. The determination of ^{235}U is not interfered with by components of "fissium-C," a simulated fuel dissolver matrix that contained no Pu. The upper spectrum is from a 1 : 1 mixture of U + fissium-C; the middle spectrum is from 1 : 1 U + nitric acid; the bottom spectrum is from undiluted fissium-C. The ^{235}U and ^{238}U components of U II 424.437 nm are indicated in the figure along with another U line at 424.378 nm. (M. Edelson and V. A. Fassel, Ames Laboratory, unpublished results.)

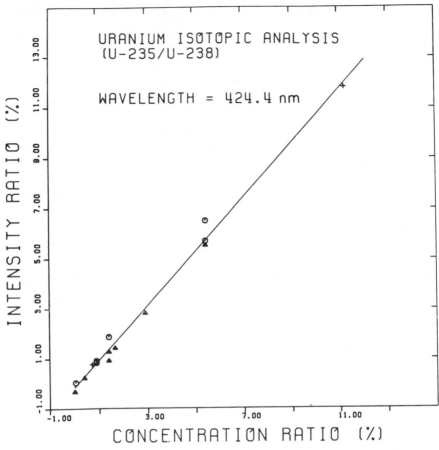

Figure 7.11. A ^{235}U–^{238}U determination by ICP-AES using the 424.437 nm line. The ^{238}U (99.93% isotopic purity) spectrum is stripped from the spectra of mixtures prior to ^{235}U determination. The three symbol types refer to results from three experimental runs performed over the course of several months. The same ^{238}U reference spectrum is used throughout. (M. C. Edelson and V. A. Fassel, Ames Laboratory, unpublished results.)

7.4.1.3 Plutonium Isotope Analysis

The Ames ICP group has recently evaluated, using the experimental facility shown in Figure 7.13, the application of HRPS to elemental and isotopic assays of plutonium.[34] The safe study of trace amounts of Pu requires the confinement of Pu particulates and the prevention of their dispersal into the atmosphere. Detection limits measured with this apparatus for Pu are listed in Table 7.2. A recent publication from the Argonne National Laboratory[35] provided a wealth of precise wavelength and isotope shift data on Pu emission lines and greatly aided the ICP work.

Whereas the most abundant isotope of U is an even–even isotope (^{238}U), the major isotope of Pu (^{239}Pu) has an odd atomic weight, and its spectral lines are

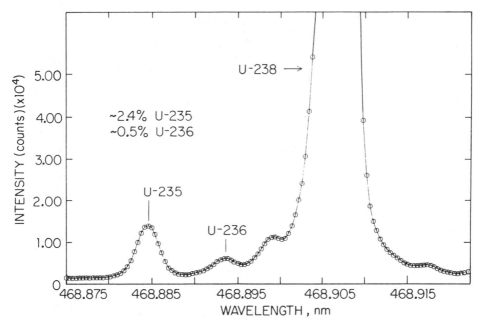

Figure 7.12. Determination of ^{235}U and ^{236}U using U II 468.91 nm emission line. The signal integration time was 0.25 s and measurements were made every 0.4 pm. Data were smoothed once with a 7-pt Savitsky–Golay filter. (M. C. Edelson and V. A. Fassel, Ames Laboratory, unpublished results.)

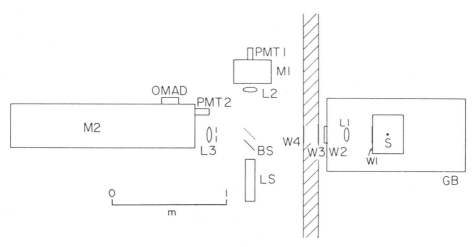

Figure 7.13. Arrangement of equipment for ICP spectrometry of Pu: BS = fused silica beamsplitter, GB = stainless steel glovebox, L = fused silica lens, LS = He–Ne alignment laser, M1 = 0.32-m monochromator, M2 = 1.5-m monochromator, PMT = photomultiplier tube, OMAD = optical multichannel analyzer detector, S = ICP source, W = fused silica window. The plasma image on the entrance slit of M2 is demagnified by a factor of 3. (From Reference 34, with permission.)

Table 7.2. Detection Limits of Prominent Plutonium ICP Emission Lines

Wavelength (nm)[a]	Detection limit (ng/mL)[b]	^{239}Pu–^{240}Pu isotope splitting (pm)[c]	^{239}Pu hyperfine splitting (pm)[d]	Transition assignment (wavenumbers)[e]
453.61461	15	−1.8	5.1, 0.95	22038-0
363.22100	25	0.4	(2.8)	27523-0
358.58665	45	0.2^	(~3)	27879-0
340.10960	50	−0.9	(3.0)	29393-0
340.10960	50	−0.1^	(3.0)	33363-3969
450.49165	52	−1.4	3.3	24206-2014
300.05716	54	−0.6^	—	35332-2014
381.01876	54	−0.4^	(1.8)	28252-2014
397.54246	56	−0.4	(1.2)	27162-2014
397.54246	56	−1.2	1.9	30865-5717
299.40455	57	−0.4^	⟨6.2⟩	33389-0
390.72145	58	−4.0	⟨4.8⟩	34296-8709
296.46435	61	0.3^	—	33721-0
402.15421	66	−7.6	1.2	33057-8198
449.37819	66	−2.6^	0.8	30956-8198
297.24995	68	—	—	43339-9707
346.51002	69	−0.6^	—	32820-3969
427.33368	72	−0.2^	2.0	32032-8638
295.00552	76	0.2^	—	35902-2014
319.84467	81	−0.1^	—	31255-0
299.64065	81	−0.6^	—	39081-5717
447.27884	83	−2.7	3.3	24366-2014
435.27081	97	−2.5	—	31677-8709
476.71698	118	−0.5^	1.1	24206-3235
298.02268	122	—	—	43252-9707

[a] For the ^{240}Pu isotope; values from Reference 35.
[b] Measured on a 0.01 mg/mL ^{242}Pu solution; data obtained with the THR1500 monochromator used in the double-pass mode with PMT detection. The monochromator had a 2400 groove/mm holographic grating, and the slits were 0.030 mm.
[c] Isotope shift originally given in mK (1 mK = 0.001 wavenumber); carets ^ indicate stated as uncertain in Reference 35.
[d] No parentheses, values from References 36 and 37; values in parentheses were measured from ICP spectra of 0.1 mg/mL solutions of ^{239}Pu (99.9% isotopic purity); angle brackets ⟨ ⟩ indicate the measured width (FWHM) of unresolved hyperfine splitting; dash—indicates not measured.
[e] Energy levels connected by listed Pu spectral line.
Source: From material in Reference 34 except as otherwise indicated.

split by hyperfine effects.[36–38] The efficient isotopic assay of the principal isotopes of Pu by HRPS requires an intense Pu line with resolvable ^{239}Pu–^{240}Pu splitting and negligible ^{239}Pu hyperfine splitting. Plutonium lines evaluated for use as Pu isotope monitors are listed in Table 7.3. Examples of resolved Pu isotope splitting and resolved ^{239}Pu hyperfine structure are shown in Figures 7.14 and 7.15.

7.4.2 Improved Selectivity in Elemental Determinations

The major benefits from increased spectral selectivity are most evident when assaying dilute components dissolved in a complex, line-rich matrix. Broekaert

Table 7.3. Plutonium Isotopic Shifts[a]

| Wavelength (nm)[b] | $^{239}Pu-^{240}Pu$ isotope shift (pm)[c] | | $^{240}Pu-^{242}Pu$ isotope shift (pm) | Intensity | |
	Ref. 34	Ref. 35		$^{242}Pu^{d}$	Ref. 35
381.48833	+2.5	(+3.33)	4.40	80000	1000
385.68484	−4.8	(−4.52)	5.86	37000	300
390.72145	−4.7	(−4.00)	5.04	260000	3000
391.34803	−5.3	(−5.12)	6.64	90000	1000
393.55516	−6.8	(−7.00)	9.2	23000	30
397.22030	−4.4	(−5.03)	6.46	183000	3000
398.98770	^	(−10.6)	^	^	300
402.15421	−7.9	(−7.60)	9.65	153000	3000
412.35012	−7.6	(−8.60)	^	^	300
419.00625	−4.2	(−4.49)	5.72	82000	10000
437.99035	+6.0	(+5.62)	7.89	101000	10000
439.64458	−5.9	(−5.74)	7.35	183000	10000

[a] Caret ^ indicates interference with neighboring lines; deconvolution required.
[b] ^{240}Pu wavelengths from Reference 35.
[c] Disagreements between values from References 34 and 35 may be due to differing measurement strategies. It is assumed that the isotopic shifts listed in Reference 35 were measured from the ^{240}Pu peaks to the centroid of the corresponding ^{239}Pu hyperfine multiplets. The Pu isotopic shifts from Reference 34 were the intervals between the ^{240}Pu peaks and the apparent ^{239}Pu peaks without allowing for the effects of hyperfine splitting.
[d] Intensity of line counts from OMA ^{242}Pu spectrum.
Source: From material in Reference 34 except as otherwise indicated.

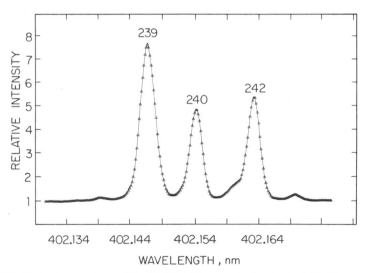

Figure 7.14. Isotopic structure of Pu II 402.145 nm emission in the ICP. The concentrations of individual isotopes are approximately 0.2 mg/mL. Measurements were made every 0.2 pm, and the integration time was 0.25 s/point. The spectrum was smoothed once with a 7-pt Savitsky–Golay filter. (From Reference 34, with permission.)

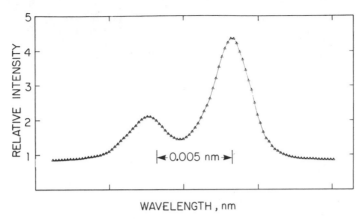

WAVELENGTH , nm

Figure 7.15. Hyperfine splitting of Pu II 453.614 nm emission in the ICP measured with 0.1 mg/mL ^{239}Pu solution. Measurement conditions are the same as those listed for Figure 7.14. (From Reference 34, with permission.)

and co-workers[40] found that the determination of the lanthanides in mineralogical samples was aided by the use of high-resolution spectrometry. A 3.4-m spectrograph, equipped with a 1180 groove/mm grating operated in second order, was used to achieve a theoretical resolving power of 460,000.

Boumans and Vrakking[18] studied the effect of high spectral resolution in the determination of analytes in a line-rich matrix, such as cobalt and nickel at 1 mg/mL. The detection powers were improved by a factor of 3 to 4 with a high resolution echelle spectrometer, as compared to results obtained with a medium-resolution instrument. The gain in detection power was accompanied by a corresponding increase in accuracy.

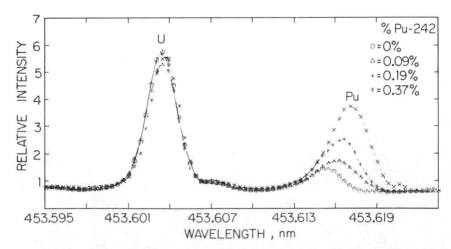

WAVELENGTH , nm

Figure 7.16. Spectrum of ^{242}Pu in a concentrated U matrix. The Pu is detectable at U/Pu > 1000. Scan conditions are the same as those listed for Figure 7.12. (From Reference 41, with permission.)

Edelson and co-workers[41] examined the use of HRPS for the detection of Pu in a U matrix (Figure 7.16). The measurement of Pu to concentrations below 0.1%, relative to U, in the line-rich U matrix was possible.

7.5 ALLIED TECHNIQUES

High-resolution spectrometery was applied to ICP–AES to determine whether increased resolution would lead to improved analytical performance. It is useful now to consider whether such benefits are shared by allied measurement techniques, such as ICP–mass spectrometry.

7.5.1 Inductively Coupled Plasma–Mass Spectrometry

The merits of an argon ICP as an ion source for analytical mass spectrometry[42,43] are discussed in Chapter 10. ICP–MS is well suited for isotopic assay work and has been used to obtain elemental assays in very complex matrices.[44] As an example of the potential simplification afforded by ICP–MS, relative to ICP–AES, one may consider the determination of trace impurities in a U matrix. The U emission spectrum has many thousands of spectral lines, whereas the ICP–MS spectrum of U consists of only a few lines that are attributable to U isotopes, U oxides, doubly charged U ions, and normal solution components at low masses. Impurity elements having M less than 100 should not be interfered with by any U species. ICP–MS can be utilized with flow-injection techniques for the analysis of microliter quantities of solution[45]; it offers detection limits that are superior to those obtained with ICP–AES,[45] and the quadrupole mass spectrometer normally supplied with ICP–MS instruments can be operated in a manner analogous to that of a very fast, sequentially scanned ICP–AES spectrometer for multiple-element assays.

Recent reports indicate that ICP–MS instruments suffer from greater matrix sensitivity than do ICP–AES instruments,[44–45] a problem that may be attributed to the sampling method used to collect ions from the ICP. Present ICP–MS instruments are first- or second-generation devices, and future technical advances may overcome the currently reported deficiencies.

The ICP–MS technique derives its excellent sensitivity by coupling ion counting electronics with an intense ion source. This means that ICP–MS, like all mass spectrometry techniques, cannot offer the quality of sample–instrument isolation that is provided by optical assay methods, such as ICP–AES. In optical spectrometry, sample information is encoded onto a set of photons that can be remotely detected, without any detector contamination, an important consideration for many nuclear analysis tasks. For nonnuclear applications, the advantages of ICP–MS over ICP–AES for determining analytes in complex matrices are already substantial and may increase as ICP–MS matures.

7.5.2   Atomic Fluorescence Spectrometry

The potential of an ICP as an atom cell for atomic fluorescence spectrometry (AFS), with a variety of primary excitation sources, is discussed in Chapter 9. Compared to ICP-AES, the chief advantage of ICP-AFS[46–48] is the great spectral selectivity of the technique, especially when narrow-line tunable dye lasers are used to excite the analyte. However, such lasers are costly and generally require operation with many dyes and frequency-doubling crystals to cover the necessary spectral range for multiple-element analysis. Although recent results[49] have demonstrated that laser-excited atomic fluorescence (LEAF) with ICP provides detection limits that surpass those obtained by ICP-AES, there has been no successful attempt to capitalize on the potentially high-resolution isotopic measurement possibilities offered by the narrow-line laser sources in ICP-AFS.

7.6   FUTURE TECHNIQUES AND INSTRUMENTATION FOR HIGH-RESOLUTION PLASMA SPECTROMETRY

7.6.1   Breaking the Doppler Limit

As we have seen, the line widths from plasma sources are largely determined by Doppler broadening. Regardless of the resolving power of the monochromator or interferometer used for line detection, recording of spectral information beyond the source-imposed limit would not be possible. Two benefits would accrue if this effect of the source on the line width were substantially reduced. First, the plasma emission spectrometric analysis of complex materials would be enormously simplified because of the reduction of spectral interference. Second, the applications of HRPS to isotopic analysis could be extended to additional elements, such as cesium and neodymium, and especially to those that do not possess large isotopic shifts.

Substantial reduction of Doppler broadening by reducing the temperature of the ICP is not possible without simultaneously deteriorating the relative freedom of the ICP from chemical matrix effects. However, other spectroscopic techniques can, in principle, effect such a reduction.

7.6.2   Modern Spectroscopic Measurements

The application of high-resolution lasers to spectroscopy has resulted in a rethinking of the "natural" limitations on attainable spectral resolution.[50] Narrow-line laser sources can be utilized to eliminate environmental spectral line broadening. For example Brown and co-workers[51] have used laser excitation to obtain high-resolution spectra from dilute solutions of complex polyaromatic hydrocarbons. These same solutions, probed with wide-bandwidth sources, yielded broad spectral features caused by environmental inhomogeneous broadening. Atomic spectroscopists in the prelaser era utilized beam methods to reduce Doppler broadening.[52] The experimental geometry was carefully controlled, with the exciting light source and the emitted photon sensor orthogonal to each other and to the atomic beam being studied. An inherent difficulty with these experiments was that only a minuscule fraction of the sample was available for study, which made the technique very inefficient and the observation of the resulting fluorescence signal very difficult. The use of narrow-band laser sources has made the use of the atomic beam method more feasible, so that real-time observation of rare, and often, rapidly decaying atomic nuclei[53] now is possible. Such studies have provided new insights into nuclear properties through the systematic measurement of hyperfine and isotope effects.

In 1974, the first application of Doppler-free, two-photon spectroscopy was reported.[54] This method involves the use of counterpropagation laser beams, each beam contributing one photon to an eventual two-photon absorption transition. Since an atom interacting with the first photon experiences a Doppler effect opposite in sense to the same atom's interaction with the second photon, the effect of Doppler broadening on the two-photon transition was canceled. This technique resulted in spectroscopic measurements of unparalleled resolution. The highest resolution spectrometer alone cannot penetrate the Doppler barrier. It will be necessary to utilize atomic beams or the selectivity offered by laser excitation to remove this impediment to the further development of HRPS.

ACKNOWLEDGMENTS

The author acknowledges Professor V. A. Fassel, Ames Laboratory, and Mr. Joseph Goleb, Office of Safeguards and Security of the U.S. Department of Energy, for recognizing the potential high-resolution plasma spectrometry. The following kindly made the results of their research available to the author in advance of publication: Dr. L.M.H. Faires (Los Alamos National Laboratory), Dr. P.W.J.M. Boumans (Philips Research Laboratory), and Dr. J. W. McLaren (National Research Council of Canada).

This work was supported by the Office of Safeguards and Security of the U.S. Department of Energy under contract W-7405-ENG-82. Commercial products are discussed in this article for descriptive reasons only; no endorsement is implied or intended.

REFERENCES

1. K. Shimoda, Ed., "High-Resolution Laser Spectroscopy." Springer-Verlag, Berlin, 1976.
2. H.G.C. Human and R. H. Scott, The Shapes of Spectral Lines Emitted by an Inductively Coupled Plasma, *Spectrochim. Acta* 31B, 459–473 (1976).

3. J. F. Kielkopf, New Approximation to the Voigt Function with Applications to Spectral-Line Profile Analysis," *J. Opt. Soc. Am.* 63, 987-995 (1973).
4. L. M. Faires, B. A. Palmer, and J. W. Brault, Line Width and Line Shape Analysis in the Inductively Coupled Plasma by High-Resolution Fourier Transform Spectrometry," *Spectrochim. Acta* 40B, 135-143 (1985).
5. R. D. Cowan, "Theory of Atomic Structure and Spectra." University of California Press, Berkeley, 1981.
6. T. Hasegawa and H. Haraguchi, Physical Line Widths of Atoms and Ions in an Inductively Coupled Argon Plasma and Hollow Cathode Lamps as Measured by an Echelle Monochromator with Wavelength Modulation, *Spectrochim. Acta* 40B, 123-133 (1985).
7. H. Kawaguchi, Y. Oshio, and A. Mizuike, Interferometric Measurements of Spectral Line Widths Emitted by an Inductively Coupled Plasma, *Spectrochim. Acta* 37B, 809-816 (1982).
8. A. Batal and J. M. Mermet, Calculation of Some Line Profiles in ICP-AES Assuming a van der Waals Potential, *Spectrochim. Acta* 36B, 993-1003 (1981).
9. M. C. Edelson and V. A. Fassel, Feasibility Evaluation of Remote, On-Line Plasma Atomic Emission Spectroscopy for the Direct Multielement and Multi-Isotopic Analysis of Dissolver Tank Solutions, *Proc. 3rd ESARDA Symp. Safeguards Nucl. Mater. Management*, Karlsruhe, May 6-8, 1981, pp. 97-103.
10. J. W. McLaren and J. M. Mermet, Influence of the Dispersive System in Inductively Coupled Plasma Atomic Emission Spectrometry, *Spectrochim. Acta* 39B, 1307-1322 (1984).
11. M. V. Klein, "Optics." John Wiley & Sons, New York, 1970, pp. 205-215.
12. Burleigh Instruments, Inc., Burleigh Park, Fishers, NY 14453.
13. M. C. Edelson and V. A. Fassel, Isotopic Abundance Determinations by Inductively Coupled Plasma Atomic Emission Spectrometry, *Anal. Chem.* 53, 2345-2347 (1981).
14. K. Kato, H. Fukushima, and T. Nakajima, Observation of Spectral Line Profiles Emitted by an Inductively Coupled Plasma—I. On the Wavelength Shift of Spectral Lines, *Spectrochim. Acta* 39B, 979-991 (1984).
15. G. R. Harrison, The Production of Diffraction Gratings—II. The Design of Echelle Gratings and Spectrographs, *J. Opt. Soc. Am.* 39, 522-528 (1949).
16. (a) R. K. Skogerboe and I. T. Urasa, Evaluation of the Analytical Capabilities of a DC Plasma-Echelle Spectrometer System, *Appl. Spectrosc.* 32, 527-532 (1978). (b) L. A. Fernando, Figures of Merit for an ICP-Echelle Spectrometer System, *Spectrochim. Acta* 37B, 859-868 (1982).
17. P.W.J.M. Boumans and J.J.A.M. Vrakking, High-Resolution Spectroscopy Using an Echelle Spectrometer with Predisperser—I. Characteristics of the Instrument and Approach for Measuring Physical Line Widths in an Inductively Coupled Plasma, *Spectrochim. Acta* 39B, 1239-1260 (1984).
18. P.W.J.M. Boumans and J.J.A.M. Vrakking, High-Resolution Spectroscopy Using an Echelle Spectrometer with Predisperser—II. Analytical Optimization for Inductively Coupled Plasma Atomic Emission Spectrometry, *Spectrochim. Acta* 39B, 1261-1290 (1984).
19. P.W.J.M. Boumans and J.J.A.M. Vrakking, High-Resolution Spectroscopy Using an Echelle Spectrometer with Predisperser—III. A Study of Line Wings as a Major Contribution to the Background in Line-Rich Spectra Emitted by an Inductively Coupled Plasma, *Spectrochim. Acta* 39B, 1291-1305 (1984).
20. R. K. Winge, V. J. Peterson, and V. A. Fassel, Inductively Coupled Plasma-Atomic Emission Spectrometry: Prominent Lines, *Appl. Spectrosc.* 33, 206-219 (1979).
21. For example, H. Sakai, High Resolving Power Fourier Spectroscopy, in "Spectrometric Techniques," Vol. 1, G. A. Vanasse, Ed. Academic Press, New York, 1977.
22. R. J. Nordstrom, Aspects of Fourier Transform Visible/UV Spectroscopy, in "Fourier, Hadamard, and Hilbert Transforms in Chemistry," A. G. Marshall, Ed. Plenum Press, New York, 1982.
23. (a) E. A. Stubley and G. Horlick, A Fourier Transform Spectrometer for UV and Visible Measurements of Atomic Emission Sources, *Appl. Spectrosc.* 39, 800-804 (1985). (b) E. A. Stubley and G. Horlick, Measurement of Inductively Coupled Plasma Emission Spectra Using a Fourier Transform Spectrometer, *Appl. Spectrosc.* 39, 805-810 (1985). (c) E. A. Stubley and G. Horlick, A Windowed Slew-Scanning Fourier Transform Spectrometer for Inductively Coupled Plasma Emission Spectrometry, *Appl. Spectrosc.* 39, 811-817 (1985). (d) R.C.L. Ng and G. Horlick, Correlation-Based Data Processing for an Inductively Coupled Plasma/Fourier Transform Spectrometer System, *Appl. Spectrosc.* 39, 834-840 (1985). (e) R.C.L. Ng and

G. Horlick, A Real-Time Correlation-Based Data Processing System for Interferometric Signals, *Appl. Spectrosc.* 39, 841-847 (1985).

24. J. W. Brault, Rapid-Scan High-Resolution Fourier Spectrometer for the Visible, *J. Opt. Soc. Am.* 66, 1081 (1976) (conference presentation abstract).

25. L.M.H. Faires, Fourier Transform and Polychromator Studies of the Inductively Coupled Plasma, Los Alamos National Laboratory (Report No. LA-9888-T, Los Alamos, October 1984.

26. F. L. Baudais and H. Buijs, Electronic Fourier Transform Spectroscopy, *Am. Lab.* 31-41 (February 1985).

27. L.M.H. Faires, Los Alamos National Laboratory, private communication.

28. (a) K. Heilig, Bibliography on Experimental Optical Isotope Shifts, 1918 through October 1976, *Spectrochim. Acta* 32B, 1-57 (1977). (b) W. H. King, "Isotope Shifts in Atomic Spectra." Plenum Press, New York, 1984.

29. L. E. Burkhart, G. L. Stukenbrocker, and S. Adams, Isotope Shifts in Uranium Spectra, *Phys. Rev.* 75, 83-85 (1949).

30. B. A. Palmer, R. A. Keller, and R. Engleman, Jr., An Atlas of Uranium Emission Intensities in a Hollow Cathode Discharge, Los Alamos National Laboratory Report No. LA-8251-MS, Los Alamos, July 1980.

31. D. W. Steinhaus, M. V. Phillips, J. B. Moody, L. J. Radziemski, Jr., K. J. Fisher, and D. R. Hahn, The Emission Spectrum of Uranium Between 19,080 and 30,261 cm^{-1}, Los Alamos Scientific Laboratory Report No. LA-4944, Los Alamos, August 1972.

32. R. Engleman, Jr., and B. A. Palmer, Precision Isotope Shifts for the Heavy Elements—I. Neutral Uranium in the Visible and Near Infrared, *J. Opt. Soc. Am.* 70, 308-317 (1980).

33. G. Rossi and M. Mol, Isotopic Analysis of Uranium by an Optical Spectral Method—III. Determination of U235/U238 Ratios with a Hollow Cathode Source and a Direct Reading Attachment, *Spectrochim. Acta* 24B, 389-398 (1969).

34. M. C. Edelson, E. L. DeKalb, R. K. Winge, and V. A. Fassel, Analytical Atomic Spectroscopy of Plutonium—I. High-Resolution Spectra of Plutonium Emitted in an Inductively Coupled Plasma, *Spectrochim. Acta* 41B, 475-486 (1986).

35. J. Blaise, M. Fred, and R. G. Gutmacher, The Atomic Spectrum of Plutonium, Argonne National Laboratory Report No. ANL-83-95, Argonne, Ill., 1984.

36. C. Bauche-Arnoult, S. Gerstenkorn, J. Verges, and F. S. Tompkins, Extended Experimental and Theoretical Analysis of the Hyperfine Structure in the Ground Multiplets of Pu I and Pu II, *J. Opt. Soc. Am.* 63, 1199-1203 (1973).

37. L. A. Korostyleva and A. R. Striganov, Hyperfine and Isotope Structure in the Plutonium Spectrum and Its Classification, *Opt. Spectrosc.* 20, 309-312 (1966).

38. S. Gerstenkorn, Étude du Plutonium par Spectroscopie à Haute Resolution—I. Contribution à la Classification du Spectre d'Arc, *Ann. Phys. (Paris)* 7, 367-404 (1962).

39. M. C. Edelson, E. L. DeKalb, R. K. Winge, and V. A. Fassel, Atlas of Atomic Spectral Lines of Plutonium Emitted by an Inductively Coupled Plasma, Ames Laboratory Report No. IS-4883, Ames, IA, September 1986.

40. J.A.C. Broekaert, F. Leis, and K. Laqua, Application of an Inductively Coupled Plasma to the Emission Spectroscopic Determination of Rare Earths in Mineralogical Samples, *Spectrochim. Acta* 34B, 73-84 (1979).

41. M. C. Edelson, E. L. DeKalb, R. K. Winge, and V. A. Fassel, Analytical Atomic Spectroscopy of Plutonium—II. Direct Determination of Plutonium Contained in a Concentrated Uranium Matrix by High-Resolution Inductively Coupled Plasma Atomic Emission Spectrometry, *Appl. Spectrosc.* (in preparation).

42. R. S. Houk, V. A. Fassel, G. D. Flesch, H. J. Svec, A. L. Gray, and C. E. Taylor, Inductively Coupled Argon Plasma as an Ion Source for Mass Spectrometric Determination of Trace Elements, *Anal. Chem.* 52, 2283-2289 (1980).

43. A. L. Gray and A. R. Date, Inductively Coupled Plasma Source Mass Spectrometry Using Continuum Flow Ion Extraction, *Analyst* 108, 1033-1050 (1983).

44. R. M. Brown, S. E. Long, and C. J. Pickford, Applications of ICP-MS in the Nuclear Industry, Paper No. 3, presented at the Karlsruhe International Conference on Analytical Chemistry in Nuclear Technology, Karlsruhe, June 3-6, 1985.

45. J. A. Olivares, Continuum Flow Sampling Mass Spectrometer for Elemental Analysis with an Inductively Coupled Plasma Ion Source. Doctoral thesis, Iowa State University, Ames, 1985.

46. A. Montaser and V. A. Fassel, Inductively Coupled Plasmas as Atomization Cells for Atomic Fluorescence Spectrometry, *Anal. Chem.* 48, 1490-1499 (1976).

47. D. R. Demers and C. D. Allemand, Atomic Fluorescence Spectrometry with an Inductively Coupled Plasma as Atomization Cell and Pulsed Hollow Cathode Lamps for Excitation, *Anal. Chem.* 53, 1915–1921 (1981).

48. N. Omenetto, H.G.C. Human, P. Cavalli, and G. Rossi, Laser-Excited Atomic and Ionic Nonresonance Fluorescence Detection Limits for Several Elements in an Argon Inductively Coupled Plasma, *Spectrochim. Acta* 39B, 115–117 (1984).

49. H.G.C. Human, N. Omenetto, P. Cavalli, and G. Rossi, Laser Excited Analytical Atomic and Ionic Fluorescence in Flames, Furnaces and Inductively Coupled Plasma—II. Fluorescence Characteristics and Detection Limits for Fourteen Elements, *Spectrochim. Acta* 39B, 1345–1363 (1984).

50. A. L. Schawlow, Spectroscopy in a New Light, *Science* 217, 9–16 (1982).

51. J. C. Brown, M. C. Edelson, and G. J. Small, Fluorescence Line Narrowing Spectrometry in Organic Glasses Containing ppb Levels of Polycyclic Aromatic Hydrocarbons, *Anal. Chem.* 50, 1394–1397 (1978).

52. H. G. Kuhn, "Atomic Spectra," 2nd ed. Longmans Green, London, 1969.

53. H. A. Schuessler, Laser Spectroscopy On-Line with Nuclear Accelerators, *Phys. Today* 34, 48–55 (1981).

54. (a) F. Biraben, B. Cagnac, and G. Grynberg, Experimental Evidence of Two-Photon Transition Without Doppler Broadening, *Phys. Rev. Lett.* 32, 643–645 (1974). (b) M. D. Levenson and N. Bloembergen, Observation of Two-Photon Absorption without Doppler Broadening on the 3S–5S Transition in Sodium Vapor, *Phys. Rev. Lett.* 32, 645–648 (1974).

8

Fundamental Properties of Inductively Coupled Plasmas

TETSUYA HASEGAWA AND HIROKI HARAGUCHI

Department of Chemistry
Faculty of Science
The University of Tokyo
Bunkyo-ku, Tokyo, Japan

8.1 INTRODUCTION

The fundamental properties of inductively coupled plasmas (ICP) are of importance for the characterization of plasmas and for their efficient use for analytical purposes. As is well known, the plasma temperatures, electron number densities, atom and ion emission intensities, number densities of analyte and argon species, and spectral line widths are generally considered to be the fundamental parameters of plasmas. Because the ICP is an inhomogeneous plasma, the spatial distributions or differences of fundamental properties should be examined to characterize the physical and spectral features of the ICP.

The ICP has a doughnut, or toroidal, structure that allows the efficient introduction of cold sample aerosols into the central channel of the plasma. The doughnut, or annular, structure of the ICP provides several unique analytical capabilities, such as low detection limits, wide linear dynamic ranges, and relative freedom from chemical and physical interferences and from self-absorption. Relatedly, the high temperatures and large electron number densities of the plasma provide higher emission intensities for ionic species than for neutral-atom species. However, ionic and neutral atom species are not necessarily in local thermodynamic equilibrium (LTE). Such non-LTE phenomena have

Radiation Chemistry: Principles and Applications

been major subjects of studies on the characterization and on excitation mechanisms of the ICP. From experimental results that are discussed later and certain theoretical considerations following Mermet's proposal of the Penning ionization processes with argon-metastable atoms, various excitation mechanisms have been proposed for the analyte species. One should note that this proposed mechanism does not successfully provide the total picture of the excitation processes prevailing in the plasma.

Recently, the authors presented a collisional–radiative model for theoretical consideration of excitation in the argon ICP. This and other models are described in this chapter. The following sections describe the methods used for measuring various plasma parameters, along with the relevant data. The composite knowledge obtained from such measurements facilitates the discussion of excitation mechanisms.

8.2 TEMPERATURE MEASUREMENTS

Species in a plasma, such as neutral atoms, ions, molecules, and electrons, are distributed over many energy states. The states of the distribution are often defined by different temperatures, depending on the species used for the temperature measurements. The theory and techniques used for temperature measurement in an ICP are described in this section, followed by an account of Abel inversion and a discussion of experimental results.

8.2.1 Plasma Temperatures and Measurement Methods

8.2.1.1 Excitation Temperature

For a system in thermal equilibrium (Section 8.7.1), at temperature T, the population density of atomic level p follows a Boltzmann distribution:[1,2]

$$n(\mathrm{p}) = n_a \left[\frac{g_p}{Z_a(T)} \right] \exp\left(\frac{-E_p}{kT} \right) \qquad [8.1]$$

where n_a is the total concentration of atom a, g_p is the statistical weight of the level p, $Z_a(T)$ is the partition function of atom a, E_p is the excitation energy of the level p, and k is the Boltzmann constant. The temperature governing this energy population is defined as excitation temperature, T_{exc}. When the radiation source is optically thin, the observed emission intensity I_{pq} (ergs/srs cm^2) of a transition from a higher level p to lower level q is expressed as:

$$I_{pq} = \frac{(l/4\pi)n(\mathrm{p})A_{pq}hc}{\lambda_{pq}} \qquad [8.2]$$

$$= \left(\frac{l}{4\pi} \right) n_a \left[\frac{g_p}{Z_a(T)} \right] A_{pq} \left(\frac{hc}{\lambda_{pq}} \right) \exp\left(\frac{-E_p}{kT} \right) \qquad [8.3]$$

where l is the path length of the source, A_{pq} is the transition probability for spontaneous emission, h is Planck's constant, c is the velocity of light, and λ_{pq} is the wavelength of the emission line. If n_a and the *absolute transition probability* are known, T_{exc} can be determined from the absolute intensity of an emission line according to Equation 8.3. Therefore, in the case of an argon ICP, this method is only applicable to Ar, whose n_a can be estimated from the ideal gas law.

Instead of the absolute method, the relative intensities of two or more lines are most often used in the measurement of T_{exc}. The logarithmic form of Equation 8.3 is:

$$\ln\left(\frac{I_{pq}\lambda_{pq}}{(g_p A_{pq})}\right) = \frac{-E_p}{kT} + \ln\left[\frac{n_a lhc}{(4\pi Z_a(T))}\right] \qquad [8.4]$$

Thus, T_{exc} is derived from the slope of the straight line $(-1/kT)$ fitted to a plot of the left-hand side of Equation 8.4 against E_p, where the *relative transition probabilities* are required. The selection of a set of spectral lines should be based on three major factors.[3] First, reliable transition probabilities must be available.[4,5] Second, the spectral lines should be close together to avoid the calibration of the detector systems. Third, excitation energies of the upper levels should be greatly different to enhance the precision of temperature measurements.

The emission methods described above require knowledge of the absolute or relative values of the transition probabilities. However, it should be noted that the error in T_{exc} associated with these methods is mainly caused by the inaccuracy in the transition probabilities (A values).

Another method for T_{exc} measurement is an absorption method. The absorbance of radiation emitted from a narrow-line source is expressed as:[6]

$$A = \frac{(\lambda^4 g_p/4\pi^2 cg_q)n(q)lA_{pq}V(a,0)(\pi \ln 2)^{1/2}}{\Delta\lambda_D} \qquad [8.5]$$

where V is the Voigt function, and a and $\Delta\lambda_D$ are the a parameter and Doppler width of an absorption line, respectively (Section 8.5). The ratio of the emission intensity and absorbance (I/A) gives T_{exc} without knowing the A value, as follows:[7]

$$\frac{I}{A} = \frac{\Delta\lambda_D(\pi/\ln 2)^{1/2}hc^2}{(\lambda^5 V(a,0))\exp(-E_p/kT)} \qquad [8.6]$$

Therefore, this method is free from the error caused by the uncertainty of the A value, but the data for the line width are necessary.

8.2.1.2 Ionization Temperature

If neutral atoms and ions are collisionally equilibrated, their concentration ratio is given by the Saha formula:[1,2]

$$\left(\frac{n_i n_e}{n_a}\right) = \left(\frac{2\pi m_e kT}{h^2}\right)^{3/2}\left(\frac{2Z_i}{Z_a}\right)\exp\left(\frac{-E_i}{kT}\right) \qquad [8.7]$$

where n_i is the concentration of ions, n_e is the electron number density, m_e is the mass of the electron, and E_i is the ionization potential. Combining Equation 8.3 with Equation 8.7 results in the following relationship:

$$\frac{n_e(I_{kl}^+ A_{pq}\lambda_{kl}^+)}{I_{pq}A_{kl}^+\lambda_{pq}} = (2g_k/g_p)(2\pi m_e kT/h)^{3/2}\exp(-(E_i + E_k - E_p)/kT) \quad [8.8]$$

where the superscript + denotes the ion. According to Equation 8.8, ionization temperature T_{ion}, which defines the ratio of atom to ion, is determined from the spectral line intensity ratio of atom to ion and the electron number density.

Recently Houk and co-workers[8,9] reported the ICP–mass spectrometric (MS) method for temperature measurement. The ICP–MS approach was used to directly measure the concentration of singly and doubly charged ions for calculating T_{ion} from the Saha equation.

8.2.1.3   Rotational Temperature

The emission intensities of the rotational lines emitted from a diatomic molecule are given by:[2,10]

$$\ln\left[\frac{I\lambda^4(K+1)}{(K+1)^2 - 1)}\right] = D - \frac{BhcK(K+1)}{kT} \quad [8.9]$$

where I is the intensity of the line, K is the rotational quantum number, D is a constant, and B is the rotational constant of the upper vibrational level. A plot of this function yields two parallel straight lines from which the rotational temperature T_{rot} is calculated in a manner similar to that for T_{exc}. The rotational temperature T_{rot} is generally recognized as the gas kinetic temperature due to the rapid exchange between rotational and kinetic energy of the molecule.[7]

8.2.1.4   Doppler Temperature

The kinetic motions of species in a high-temperature medium are subject to the Maxwell–Boltzmann law.[11] The temperature that governs the velocity distribution is defined as electron temperature in the case of an electron, as described later, and as Doppler temperature, T_D, for heavy particles. The relative motions of atoms in the plasma cause line broadening, known as Doppler broadening (Section 8.5.1). The resulting line width $\Delta\lambda_D$ is related to T_D, as shown in Equation 8.10:

$$\Delta\lambda_D = 2(2R\ln 2)^{1/2}\left(\frac{\lambda_0}{c}\right)\left(\frac{T_D}{M}\right)^{1/2} \quad [8.10]$$

where R is the gas constant, λ_0 is the wavelength at the center of the line, and M is the atomic weight. Since the widths of spectral lines emitted from an ICP are, in general, less than 10 pm, they can be determined only by high-resolution optics.[12–15] In addition, the observed emission profiles are broadened by other causes such as Lorentz, Stark, and instrumental effects. Therefore, the deconvolution of these broadenings is necessary for precise evaluation of each line width.[13,14] Line width measurements are described in detail in Section 8.5.

8.2.1.5 Electron Temperature

The free-bound transition of an electron, accompanied by the emission of continuum radiation, is called radiative recombination (Equation 8.49 in Section 8.7.2.5). The intensity of the continuum depends on the electron number density and the kinetic temperature of electrons, ie, electron temperature, T_e. If it is assumed that the plasma, as a whole, is electrically neutral, the emission intensity observed within the narrow interval of wavelength $\Delta\lambda$ is:

$$I_{cont} = \left(\frac{l}{4\pi}\right)g(\lambda, T_e)n_e n_{Ar^+}\,\Delta\lambda = \left(\frac{l}{4\pi}\right)g(\lambda, T_e)n_e^2 \qquad [8.11]$$

where $g(\lambda, T_e)$ is a function of λ and T_e,[16,18] and n_{Ar^+} is the number density of argon ions. According to this equation, T_e can be derived from the absolute emission intensity and electron number density.

The slope of $\ln(I_{cont})$ versus λ is also used for the estimation of T_e in a relative way, without knowing n_e.[19]

$$\frac{d\ln(I_{cont})}{d\lambda} = \frac{\partial\ln(g(\lambda, T_e))}{\partial\lambda} \qquad [8.12]$$

The combination of Equations 8.11, 8.2, and 8.7 (the Saha equation), applied to argon, gives another expression for T_e:[20–23]

$$\frac{I_{cont}}{I_{line}} = \left(\frac{g(\lambda, T)\lambda\Delta\lambda}{A_{pq}hc}\right)\frac{(2\pi m_e kT)^{3/2}}{h^3(2g_i/g_p)}\exp\left(-\frac{E_i - E_p}{kT}\right) \qquad [8.13]$$

where g_i is the statistical weight of the ground state of the argon ion. The electron temperature T_e can be derived from the emission intensity ratio of an Ar atomic line and the adjacent continuum with a precision better than that of other methods. However, it should be considered that the Saha equation, assumed for Ar ion in this method, is not always valid in the Ar ICP,[18] as described in Section 8.7.

8.2.2 Abel Inversion

The study of spatial distributions of various plasma parameters, both in vertical and radial directions, documents that the ICP is not homogeneous and has a spatial structure. Spectroscopic information, such as emission and absorption intensities, when observed with side-on projection, is obtained as the integrated intensity over the entire depth of the light source. The procedure to transform the side-on data, that is, lateral profile, into the radial profile is the Abel inversion.[24–28]

If the light source is assumed to be cylindrically symmetrical, the lateral intensity in the direction of the x-axis of observation is expressed as:

$$I(x) = 2 \int_x^{R_0} \frac{i(r)r}{(x^2 - r^2)^{1/2}} \, dr \qquad [8.14]$$

where R_0 is the radius of the source, and $i(r)$ is the radial intensity at radius r. Then, $i(r)$ is given by the Abel equation:

$$i(r) = -\frac{1}{\pi} \int_r^{R_0} \frac{dI(x)/dx}{(x^2 - r^2)^{1/2}} \, dx \qquad [8.15]$$

The numerical solution of Equation 8.15 is usually performed in the following sequence.[26] With respect to the central axis of the ICP, the observed lateral intensity data along the x-axis are actually asymmetric. The symmetric intensity data are first calculated by averaging the intensity data from both sides of the central vertical axis. Then, the averaged lateral profile is fitted to a nth-order polynomial expression by least-squares regression. The radial profile is finally obtained by integrating the first derivative of the polynomial expression, in accordance with Equation 8.15.

Certain features of the intensity profiles, useful in the characterization of the ICP, can be masked if symmetry is imposed by averaging the left- and the right-hand sides of the lateral profiles.[28] With little modification, the procedure above can be extended to the asymmetric Abel inversion, which keeps asymmetry in the radial data.[28]

8.2.3 Literature Values for Temperatures in the Inductively Coupled Plasma

The reported values for various plasma temperatures are summarized in Table 8.1, along with the measurement methods and the measured species. The excitation temperature T_{exc} has been measured by many workers using the relative emission method for fewer than 10 thermometric species. Because of the availability of reliable transition probabilities, iron has frequently been used as the thermometric species. Jarosz and co-workers[4] compared the reliabilities of the reported A values for argon, iron, titanium, and magnesium by means of the F-test, and selected the best data set. Kalnicky and co-workers[3] measured the radial profile of T_{exc} for Fe I at three observation heights, with and without the presence of an easily ionizable element (6900 $\mu g/mL$ of Na). Their observations showed that the large amount of sodium did not change T_{exc} under the commonly used operating conditions. Cesium and phosphate, which cause severe interferences in combustion flames and the dc arc, have minor influences on T_{exc}.[37]

The radial distributions of T_{exc} reported by Kalnicky and co-workers[32] are shown in Figure 8.1. It should be pointed out that at the observation height of 15 mm, the temperature at the central channel is relatively low compared with that

Table 8.1. Literature Values for Various Temperatures of ICP

Temperatures (K)	Species and measurement method	Frequency (MHz)	Power (kW)	Observation height (mm)	Ref.
Excitation temperature					
5200–5600	Ar, Fe; slope method	27	1.0	15	3
4800	Ar, Fe, Ti, V; slope method	40	1.3	2	4
2300–4000	Ar, Ca, Mg; absorption method	50	0.5	3.75–30	7
5000–7000	Ar; slope method	5.4	6	0–12	10
4380	Ar; slope method	50	0.27	0.5	19
4500–5100	Ar, Cd, Fe, Ti; slope method	40	1.5	5	20
5900–7000	Ar, H; slope method	40	1.5	2	21
6200	Fe; slope method	31	2.5	15	29
4000–4700	Ar, Zn; slope method	50	2	6	30
6200–6300	Ar, Fe, Ti; slope method	5.4	6	0–10	31
5700–6400	Fe; slope method	27	1.0–1.2	15	32
6313–14,723	H; slope method	9	12	10	33
4900–5400	Fe; slope method	50	0.4–0.7	20	34
5000–7000	Ar, Ti; slope method	144	0.05		35
4800–5000	Ar, Fe, Ti; slope method	40	1.8		36
3800	Mg; slope method	50	0.53	7.5	37
4500–5100	Ar, Fe, Ti, V; slope method	40	0.5–4		38
6900–8500	Fe; slope method		1.2	10–30	39
7500–8300	Ar; absolute method	26.5	0.75	−7–12	40
6200–8500	Ar; absolute method	26.5	0.5–0.75	−7–18	41
5600–6700	Fe; slope method	27	1.3	0–35	42
7000	Ar; slope method	27	1.5	15	43
4500	Fe; slope method	27	1.25	14	44
4000–6700	Fe; slope method	8–56	1.1–1.5		45
3500–3800	Fe; slope method	27	1.25	10–25	46
4300–5200	Fe; slope method	27	1.8	0–20	47

(*Continued*)

Table 8.1. (*Continued*)

Temperatures (K)	Species and measurement method	Frequency (MHz)	Power (kW)	Observation height (mm)	Ref.
4700–9000	Fe; slope method			20	48
4800	Fe; slope method	40	1.6		49
5900	Sr; slope method	9.1	9	4	50
7000	Ar; slope method	27	1.5	15	51
12,670	H; slope method		1.5		52
4400–5200	Co, Fe, Ni, V; slope method	27	1.1		53
5800	Fe; slope method		1.25	15	54
Ionization temperature					
6700–7400	Ar, Mg, V; Saha method	40	1.3	2	4
5300	Mg; Saha method	50	0.5	7.5	7
7400–8200	Ba, Sr, Cd/I; mass spectrometric method	27	1		8
8000	Cd/I; mass spectrometric method	27	1.2		9
3500	Ti; Saha method	144	0.05		35
7350	Ar; Saha method	40	1.8		36
6200	Mg; Saha method	50	0.53	7.5	37
7700–8400	Ba, Ca, Cd, Fe, Mg, Ti, Zn; Saha method		1.2	10–30	39
9000	Ar; Saha method	27	1.5	15	51

Rotational temperature

Temperature	Species/Method				Ref.
1600	OH band	50	0.5	12	7
6100	C$_2$ band	5.4	6	0–3	10
2200	CN band	50	0.27	0.5	19
4500–5000	C$_2$, N$_2^+$ band	40	1.5	5	20
3000	OH band	50	2	6	30
4500–5000	N$_2^+$ band	40	0.5–4		38
2800–4700	OH band	27	1.3	0–35	42
4500	N$_2^+$	40	1.6		49
5000	BO band	6.3	12		55
8290–6140	N$_2^+$ band	9.2		7–41	56

Doppler temperature

Temperature	Species/Method				Ref.
5200–6900	Ar, Ca, Sr	27	1	10–25	12
3300–7900	34 lines	27	1.0	15	13
6310	Average of 81 Fe lines	27	1.1	11	15
2070	Ar	50	0.27	0.5	10

Electron temperature

Temperature	Method				Ref.
7000–8400	Continuum method	27	1.1	5–25	18
11,400	Continuum method	50	0.27	0.5	19
8000–10,000	Line/continuum ratio method	40	0.9–2.0	5	20
12,000	Line/continuum ratio method	40	1.5	2	21
23,000–52,000	Electric probe method	144	0.05		35
5480	Continuum method	27	1.25	14–16	57
7200	Ar line; slope method			4	58
7700	Line/continuum ratio method	27	1.25	10	59

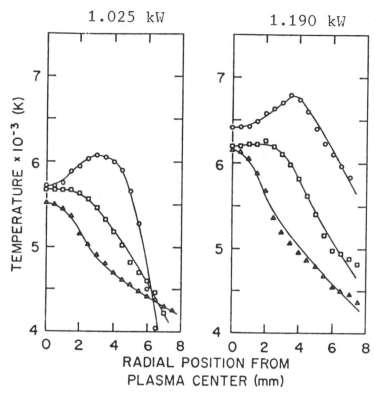

Figure 8.1. Radial excitation temperature distributions for two power levels and three observation heights using Fe I: —○—, 15 mm; —□—, 20 mm; —△—, 25 mm (From Reference 32, with permission.)

of the surrounding zone. This structure of the ICP, called a doughnut or annular shape, is more evident at the lower region of the plasma, but it is no longer seen in the plasma plume, where the central region is rather hot; ie, the radial temperature profile shows a bell shape. The spatial characteristics of the temperature can be explained by the gas flow and heat conduction, as follows.[18] In the off-axis region, the temperature decreases monotonically with distance from the hottest region inside the load coil due to the heat conduction to the surrounding atmosphere. In the central core, however, the temperature inside the load coil is significantly low because the cool injector gas flows very fast (> 50 m/s).[40,41] At observation heights close to the commonly-used observation height, the temperature increases[47] because of the gas mixing between the inner core and the hot surrounding zone. In the plasma plume, where the mixing is nearly completed, the temperature decreases again. Recently, by using image detectors, such as photodiode arrays,[47,54] the spatial characteristics of T_{exc} have been studied in detail, thus confirming the complicated behavior of T_{exc} discussed above.

The experimental conditions greatly influence the plasma parameters.[3,7,14,32,37,40,41,51,54] For example, Capelle and associates[45] found that temperature and electron number density decreased with increasing generator frequency. Even when these factors are taken into account, the literature values for T_{exc} in Table 8.1 are widely different, ranging from 3000 to 7000 K, possibly due to the uncertainty of the A values employed. The absorption method may enable one to circumvent this problem.[7,37] However, since the hollow cathode lamp, commonly used as a primary radiation source, is not intense enough to provide sufficient absorption signal, the absorption measurements are limited to temperature determinations involving resonance radiations of alkaline earth elements.

It is also known that T_{exc} increases along with the excitation energy of the upper level of the transition.[39,48] Alder and co-workers[39] measured the relative intensities of Fe I for 20 different transitions and divided the data into three groups with increasing excitation energy, resulting in three discrete values of T_{exc}. Furthermore, the extended observation by Kornblum and Smeyers-Verbeke[48] showed the continuous increase of T_{exc} with respect to the excitation energy. These results indicate the departure of the energy populations from the Boltzmann distribution.

The species available for T_{ion} measurements in the ICP are restricted to those having sufficiently intense atomic and ionic lines. Alder and co-workers[39] measured T_{ion} for seven elements along with T_{exc} for Fe I, and found that T_{ion} values were equivalent to T_{exc} estimated from high energy levels. Their data are in good agreement with T_{ion} values for barium and strontium obtained by the mass spectrometric method.[8,9] It is clearly seen from Table 8.1 that T_{ion} is generally higher than T_{exc}.

Rotational temperature, T_{rot}, shows a wide variation depending on the molecular species.[7,10,19,20,30,38,42,49,55,56] The T_{rot} values measured from N_2^+ and C_2 provide relatively high temperatures (4500 to 5000 K),[20,38,49,56] while those from other species are below 3000 K. At present, these differences cannot be interpreted.

The data for T_D are scarce[12,13,15,19] because of the difficulty of measuring spectral line widths and the associated deconvolution procedure. However, except for the value by Kleinmann and Cajko (2070 K),[19] other T_D values determined by Fabry–Perot interferometry,[12] echelle spectrometry,[13] and Fourier transform spectrometry[15] are quite consistently within the range 4000 to 6000 K.

Electron temperature T_e is certainly the most important temperature for characterizing the ICP. This is because the real significance of T_{exc} and T_{ion} as "plasma temperatures" is lost in the non-LTE plasma.[18,23] However, the determination of T_e from the background emission of Ar is still difficult, and, in addition, requires complicated theoretical calculations for the continuum radiation.[19–21,57,58] Batal and co-workers[20,21] reported the radial profiles of T_e, which were deduced from the intensity ratio of an Ar I line to continuum in the visible region of the ICP spectrum. The observed T_e, ranging from 8000 to

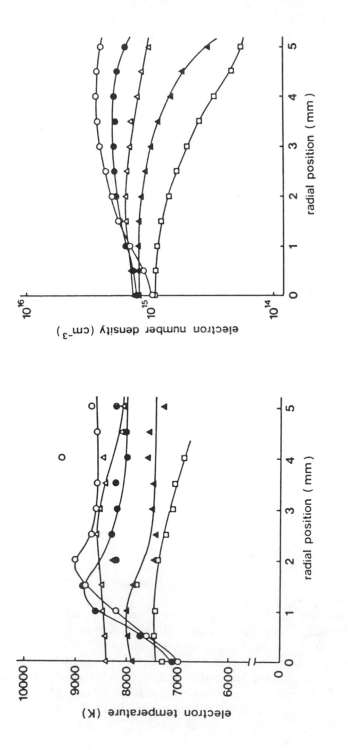

Figure 8.2. (*A*) Radial distributions of electron temperature at various observation heights: 5 mm, $-\bigcirc-$; 10 mm, $-\bullet-$; 15 mm, $-\triangle-$; 20 mm, $-\blacktriangle-$; 25 mm, $-\square-$. (*B*) Radial distributions of electron number density at various observation heights: 5 mm, $-\bigcirc-$; 10 mm, $-\bullet-$; 15 mm, $-\triangle-$; 20 mm, $-\blacktriangle-$; 25 mm, $-\square-$.

10,000 K, showed considerably higher values than T_{exc} and T_{rot} estimated by using several species. Hasegawa and Haraguchi[18] determined the spatial distribution of T_e by fitting the theoretical curve to the observed continuum profile. The results are shown in Figure 8.2, along with n_e profiles. The electron temperature T_e, at 15 mm above the load coil in the plasma center, was 8400 K, which agrees well with that by Batal and co-workers[20,21] within experimental error. However, curve fitting to a gray body gave a significantly low value, 5480 K.[57] It seems that more direct techniques, such as the Thomson scattering method,[60a] should be examined in T_e measurements. Recently, Huang and co-workers reported theoretical considerations[60b] and experimental results[60c] for measurements of electron temperatures by the Thomson scattering method. The T_e values of 11,600 and 9200 K were measured at the respective observation heights of 10 and 15 mm, where electron number densities were determined to be 3.3×10^{15} and 7.3×10^{14} cm^{-3}, respectively. These values seem to be a little high compared with those obtained by Ar continuum methods.

The literature data for temperatures in Table 8.1 can be summarized in the statement:[7]

$$T_{rot} < T_D \simeq T_{exc}(E_p) < T_{ion} < T_e \qquad [8.16]$$

The fact that the various temperatures disagree with each other, especially that higher energy species provide higher temperatures, suggests deviation from local thermodynamic equilibrium, which is discussed in detail in Section 8.7.1.

8.3 ELECTRON NUMBER DENSITY MEASUREMENTS

Because a laboratory plasma is a partially ionized gas, electron number density n_e is an important parameter indicating the degree of ionization. Contrary to temperature, the definition of n_e is simple, ie, the number of free electrons in a unit volume. Nevertheless, it is not so easy to determine a definite value of n_e in the ICP. This section presents several methods for n_e measurements and discusses the reliability and precision of literature values associated with individual techniques.

8.3.1 Theory and Methods of Measurement for Electron Number Density

8.3.1.1 Stark Width Method

Spectral lines from the plasma are broadened by the interaction between the emitting atoms and the local electric field generated by the surrounding ions and electrons. This phenomenon is called Stark broadening.[61-63] As discussed in Section 8.5, for the analyte elements introduced into the ICP, the principal cause of line broadening is the Doppler effect.[12,13,15] However, since atomic hydrogen is subject to an exceptionally large linear Stark effect, the Stark broadening is dominant for hydrogen lines.

With the aid of an adequate theory for line broadening, the hydrogen line profile is used to determine n_e in the plasma. According to the quasi-static theory,[64] the halfwidth of hydrogen line, $\Delta\lambda_H$, is proportional to $n_e^{2/3}$. Thus, n_e can be obtained from the following relationship:

$$n_e = C_H(n_e, T)\Delta\lambda_H^{3/2} \qquad [8.17]$$

where C_H is a constant that depends only weakly on n_e and temperature. In the measurement of Stark width, the H_β line (486.1 nm), which is the second line of the Balmer series, is commonly used for the following reasons: (a) reliable data for C_H are available,[60,61,63] (b) the H_β line is relatively free from spectral interference caused by the emission lines of plasma components,[65] and (c) the line profile is sufficiently intense and broadened. The precision of n_e associated with the Stark calculation from the H_β line is approximately 5%.[2] In addition, this method does not require the assumption of thermal equilibrium for the plasma. Consequently, the H_β Stark method is considered to be more reliable than other spectroscopic methods, as discussed later.

In the case of non-hydrogen-like atoms, which are subject to the quadratic Stark effect,[2] the broadening is mainly caused by electron impact. Thus, the halfwidth ($\Delta\lambda_S$) is proportional to n_e.

$$n_e = C_S(n_e, T)\Delta\lambda_S \qquad [8.18]$$

Argon atomic lines originating from higher energy levels are commonly used for the measurement because their Stark widths are relatively large compared to their Doppler widths. However, the data reported[61] for the proportionality constant C_S have some uncertainties, and corrections are necessary for the Doppler and Lorentz contributions to the total line width. The accuracy in n_e is unlikely to be better than 20 to 30%.

8.3.1.2 Saha Method

The Saha equation is also applied to the determination of n_e. When the temperature, which has been estimated by the proper method, usually T_{exc}, is substituted for T in Equation 8.8, n_e is calculated from the intensity ratio of atomic and ionic lines:

$$n_e = \frac{I_{pq}, A_{kl}^+\lambda_{pq}}{I_{kl}^+ A_{pq}\lambda_{kl}^+} \times \left[\left(\frac{2g_k}{g_p}\right)\left(\frac{2\pi m_e kT}{h}\right)^{3/2} \exp\left(-\frac{E_i + E_k - E_p}{kT}\right)\right] \qquad [8.19]$$

In this method, therefore, thermal equilibrium between atom and ion is assumed.

From a practical point of view, the precautions necessary for n_e measurements are similar to those for T_{exc} measurements,[3] ie, proximity of the wavelengths, availability of reliable transition probabilities, freedom from spectral interference, etc. In addition, the excitation energies for atomic and ionic lines should be closely matched to minimize the effect of temperature on the exponential term in Equation 8.19.

8.3.1.3 Continuum Method

Equation 8.11 demonstrates that n_e can be readily derived from the absolute intensity of continuum. If T_e is estimated by the slope or line-continuum ratio methods (see Section 8.2.1.5), n_e is given by:

$$n_e = \left[\frac{(4\pi/l)I_{cont}}{\Delta\lambda g(\lambda, T_e)}\right]^{1/2} \qquad [8.20]$$

For the continuum around 430 nm, $g(\lambda, T_e)$ is given by:[7,30]

$$g(\lambda, T_e) = 1.013 \times 10^{-29} T_e^{-1/2} \text{ (ergs/s cm}^{-3} \text{ } \mu\text{m)} \qquad [8.21]$$

Because n_e weakly depends on T_e ($n_e \propto T_e^{1/4}$), it can be obtained from an adequately chosen T_e.

The uncertainty of n_e measured by the technique above is mainly ascribed to the experimental error associated with the absolute intensity measurement and to the theoretical error associated with the calculation of the constant of proportionality.

8.3.1.4 Series Limit–Line Merging Method

The Stark splitting of spectral lines in a series increases with larger values of principal quantum number, n_p, while the energy difference between the adjacent states rapidly decreases. Therefore, as n_p increases, the broadening lines start to merge, and finally, they are smoothed into a continuum before the series limit. The n_p of the last discernible line in the series (n_m) is related to n_e by the Inglis–Teller equation:[66–68]

$$\log(n_s) = 23.23 - 7.5 \log(n_m) \qquad [8.22]$$

where n_s is the sum of n_e and the number density of the ion (cm^{-3}):

$$n_s = n_e + n_i \qquad [8.23]$$

Equation 8.23 is valid only for hydrogen and hydrogenlike atoms. For other species, Equation 8.20 approximately holds for atoms in higher energy states, because such atoms are hydrogenic.[2] Considering that other broadening processes cause the lowering of the apparent series limit and that the Abel inversion technique is hardly applicable, this method should be used for the estimation of n_e in the central channel of ICPs.

8.3.2 Literature Values for n_e

Reported n_e values for commonly used operating conditions for the ICP are summarized in Table 8.2. The n_e values measured by the Stark method, which is considered to be the most reliable method, show surprisingly wide variations;

Table 8.2. Literature Values for Electron Number Densities n_e in ICP

n_e(cm⁻³)	Species and measurement method	Frequency (MHz)	Power (kW)	Observation height (mm)	Ref.
2×10^{13}	Ca, Cd, Fe, Mg, Zn; Saha method	27	1	15	3
5×10^{14}	H$_\beta$; Stark method		1.3	2	4
5.5×10^{14}	H$_\beta$; Stark method	40	0.5	3.75–7.5	7
2×10^{15}–4×10^{15}	Continuum method	50	1		8,9
1×10^{15}	Line merging method	27	1.1	5–25	18
1×10^{15}–2×10^{15}	H$_\beta$; Stark method	40	1.5	5	20
2×10^{14}–5×10^{14}	H$_\beta$; Stark method	40	1.5	2	21
1×10^{15}	H$_\beta$; Stark method	50	2	6	30
6×10^{14}	Mg; Saha method				
8×10^{14}	Continuum method				
7×10^{15}–1×10^{16}	Ar; Stark method	5.4	6	0–10	31
5×10^{14}	Ar; Stark method	40	1.8		36
2×10^{14}–3×10^{15}	H$_\beta$; Stark method		1.2	10–30	39
4×10^{15}	H$_\beta$; Stark method	27	1.5	15	43
3×10^{13}–5×10^{14}	Ar; Stark method	8–56	1.1–1.5		45
3.4×10^{14}	Sr; Saha method	9.1	9	4	50
2×10^{15}	H$_\beta$; Stark method		2	15	54
3.5×10^{14}	H$_\beta$; Stark method			4	58
7×10^{14}	Al, Ca, H, K, Li, Mg; line merging method	27	1.2–1.8	15–20	67
4×10^{14}–1.4×10^{15}	Al; line merging method	27	1.0–2.0	15	68
1×10^{14}–4×10^{14}	H$_\beta$; Stark method	27	1.1	12–30	69
5×10^{14}–4×10^{15}	H$_\beta$; Stark method		1–2	4–20	70

the accepted n_e values range from 5×10^{14} to 5×10^{15} cm^{-3}. This inconsistency may be attributed to such factors as: (a) error introduced by the Abel inversion, (b) uncertainty associated with the correction of Doppler and instrumental broadenings, (c) misalignment of the optical system, (d) asymmetry of the plasma, and (e) differences in operating conditions. To improve the precision, Goode and Deavor[71] investigated the one-parameter least-squares fit to the Stark-broadened hydrogen line. A comparison of data obtained by the halfwidth method and the curve-fitting approach showed that the halfwidth method had a tendency to overestimate n_e values by a factor of 2 to 3.

The radial distributions of n_e, determined by the H$_\beta$ Stark and Saha methods,[3] are shown in Figure 8.3. Both n_e profiles obviously exhibit off-axis peaks similar to those observed for temperatures. More detailed maps of n_e were published by Blades and Caughlin[54,70] and Furuta and co-workers.[69] The n_e-values from the Saha equation (Figure 8.3) show small differences, which might be attributed to the uncertainty in the transition probabilities. However, the n_e profile measured by the Stark method provides n_e values that are 30 to 50

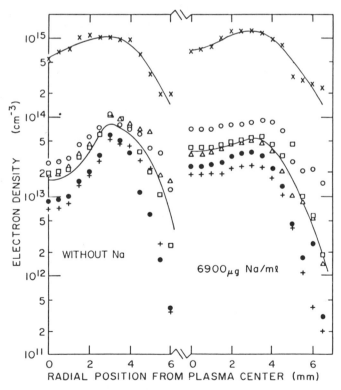

Figure 8.3. Radial electron number density distributions obtained from H$_\beta$ Stark broadening ($-\times-$) and Saha calculations: 10 μg/mL of Ca ($-\bigcirc-$), 150 μg/mL of Fe ($-\square-$), 10 μg/mL of Mg ($-\triangle-$), 10 μg/mL of Cd ($-+-$), 10 μg/mL of Zn ($-\bullet-$). (From Reference 3, with permission.)

times greater than the n_e from the Saha method, a difference that cannot be associated with the errors in A values and Abel inversion alone. The large discrepancy between data sets indicates deviation from LTE and thus reveals that the ICP is a non-LTE plasma.

Another important point seen from Figure 8.3 is the minor effect of an easily ionizable element (EIE) on n_e.[72] In general, if an EIE is introduced into a plasma, n_e increases and, as a result, the ionization equilibrium between atom and ion shifts toward the atom. This is known as ionization interference.[1,73] In the case of the ICP, n_e is on the order of 10^{15} cm^{-3}. From the number density measurements for analyte species, on the other hand, it is estimated that even a solution of 1000 μg/mL of the EIE releases only 10^{12} cm^{-3} of electrons in the plasma. Therefore, the large amount of the EIE has a negligible influence on n_e for commonly used operating conditions.

Only a few examples for n_e measured by the continuum method have been reported. Kornblum and de Galan[7] determined the radial profile of n_e from the absolute continuum intensity at 450 nm in low- and high-gas-flow ICPs. The n_e value at the center of the plasma with a low flow of injector gas is about 1×10^{15} cm^{-3}, which is consistent with the Stark n_e value.

The series limit–line merging method was applied to the determination of n_e in the ICP by Montaser and co-workers.[67,68] Values of n_e estimated from the line merging of six elements are comparable to or slightly lower than Stark n_e values, although the Inglis–Teller formula generally tends to give a high value.

8.4 SPATIAL DISTRIBUTIONS OF VARIOUS SPECIES

Plasma components move in the plasma because of the gas flows, and thermal and electric diffusions. Under the steady-state condition, which is approximately valid for the conventional ICP, the concentrations of species at any location can be regarded as constant. Therefore, the spatial distributions of various species may reflect the macroscopic and microscopic characteristics of the discharge. Before discussing plasma diagnostics, it is worthwhile to summarize studies on the spatial distributions of analyte and argon atoms.

8.4.1 Profiles for Excited Species

Because the number density of excited Ar atoms is proportional to the emission intensity of the corresponding transition (Equation 8.2), the relative distribution profiles of excited species can be readily obtained from relative intensity measurements.[25,26,28,30,44,46,47,50,69,74–92]

Horlick and co-workers.[47,74-78] have investigated spatial distributions, both experimentally and theoretically, using a self-scanning photodiode array. They first reported the vertical distributions of various analyte species in the central channel without Abel inversion, and noticed the following remarkable features. For "soft" atomic lines, whose excitation energies are relatively low or intermediate (< 5 eV), the maximum position of the vertical intensity profile was less than 15 mm above the load coil; the position shifted upward with higher excitation energy and depended on the plasma operating parameters, such as RF power and Ar gas flows. In contrast, other emission lines (ie, atomic lines with higher excitation energies and all ionic lines) called "hard lines" exhibited maxima at an observation height of approximately 15 mm, independent of the species and plasma conditions. Further studies[76] showed the linear relationship between vertical peak position for atomic lines and the norm temperature, defined as the temperature giving the maximum population of the energy state of interest. Thus, it was concluded that the analyte is excited by electron impact in the lower region of the plasma (thermal region) and by the Penning process in the upper (nonthermal) region.

The measurements were extended to the radial distributions with the Abel inversion technique.[47] As shown in Figure 8.4, the higher energy lines exhibited off-axis peaks in the lower region of the ICP, reflecting the temperature distribution described in Section 8.2

The application of a spectrograph to analyte distribution measurements makes it possible to obtain spatial profiles using a single exposure.[78,83,84]. Niebergall and others[83,84] evaluated the resulting spectrogram by using equidensitometry. Based on the emission profiles of N_2^+, they described the gas flow dynamics of the atmosphere surrounding the ICP. According to their observations, the atmospheric gases, entrained into the plasma mantle by turbulence, were directed inward and upward through the action of the forced aerosol flow.

Although relative freedom from matrix interferences caused by EIE is known as one of the prominent analytical advantages of ICP–AES,[85,86] such interferences could occur under particular experimental conditions. The effect of the matrix on the emission profiles of analytes has been the subject of several diagnostic studies.[46,50,75,77,87-90] In general, the emission profiles of ionic lines were not affected significantly, but those of atomic lines were enhanced by the presence of EIE. The enhancement effect for atomic lines increased with decreasing ionization potential. This pattern is easily understandable from the high ionization efficiency of an analyte in the ICP: namely, the analyte is highly ionized in the ICP so that the increase of n_e caused by EIE results in minor changes of the concentration of ions. Conversely, the slight shift of ionization equilibrium toward the atom results in the sensitive enhancement of atom lines.

The theoretical calculations for radial[91] and vertical[92] distributions of analytes were based on the Saha and Boltzmann formulas. These calculations require the use of experimental values for plasma parameters and the presence of LTE, but they make possible the prediction of reactions occurring in the plasma.

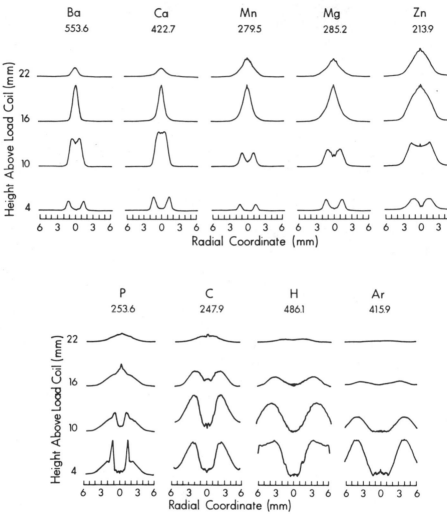

Figure 8.4. Radial intensity emission profiles for Ba, Ca, Mn, Mg, Zn, P, C, H, and Ar neutral atom lines. (From Reference 47, with permission.)

8.4.2 Profiles for Ground-State Species

Because the ground state is the most populated state, even at temperatures higher than 5000 K, information about its absolute density and spatial distribution provides the fundamental data that make possible a quantitative discussion of excitation mechanisms, and especially, reaction rates. These distributions have been measured by both absorption[7,29,37,51,58,93,94] and fluorescence[95] methods.

The absorption method is based on the principle that absorbance of resonance radiation is proportional to the population of the ground state in an absorption cell (Equation 8.5). If the a parameter and Doppler width of the absorption line profile are known, or adequately assumed, absolute number density can be estimated from this equation. Kornblum and de Galan[7,37] determined the absolute radial profiles of Ca I ground state by using a hollow cathode lamp (HCL) as a primary radiation source. The line parameters were derived from interferometry using a curve-fitting procedure. The plasma, operating with a low flow of injector gas, showed a parabolical distribution of calcium atoms, while with a high flow of injector gas, the plasma exhibited a toroidal distribution, suggesting a sideward diffusion of Ca atoms because of the high flow of injector gas.

To measure the radial distributions of both Ca atoms and ions, Nojiri and associates[51] used a second ICP in place of a hollow cathode lamp. The results from calculations of absolute number densities indicated that a 1000 μg/mL solution nebulized into the ICP produces 7×10^{11} cm^{-3} of Ca atoms and 1×10^{12} cm^{-3} of Ca ions in the central channel. These values are quite consistent with the theoretical values estimated from aspiration efficiency, gas flow, and solution uptake rate,[50] but they are about 50 times lower than the densities measured by Kornblum and de Galan.[7] Based on this work, Eckert[96] estimated that the total detected Ca particles transported in the axial channel lies between 11 and 55% of the injected Ca, assuming a nebulizer efficiency between 5 and 1%. Relatedly, Eckert noted that one Ca particle can be detected in the presence of 2×10^{14} Ar particles by ICP–AES.

The radial concentration profile measured by the absorption method suffers from the errors associated with the Abel inversion procedure. To avoid this problem, Walters and co-workers[93] used the saturated absorption method,[97] which allows the direct calculation of the spatially resolved concentration. The number of absorbing species, n, in the intersecting volume, is related to the emission intensities obtained with and without the perturbing saturation beam, I_s and I, respectively, according to the following simplified expression:[93]

$$\frac{\Delta n}{n} = \text{const} \times \log\!\left(\frac{I_s}{I}\right) \qquad [8.24]$$

where n is the total number of absorbing species along the path probing beam. This method, applied to several configurations of the ICP for the absorbing species Sr and Ba, showed the complicated behavior of $\Delta n/n$, depending on the experimental parameters.

Another technique for measurement of radial concentration profiles, with no need for an Abel inversion procedure, is the fluorescence method. The spatially resolved information on the concentration mapping of species can be obtained by observing the fluorescence radiation. Omenetto and co-workers[95] reported the relative spatial profiles of Ba atoms and ions measured by laser-excited fluorescence.

8.4.3   Profiles for Argon-Metastable Atoms

Since the proposal by Mermet,[98] metastable argon (Ar[m]) has been considered to be a major particle that governs analyte excitation in the ICP through the Penning ionization reaction, which is discussed in Section 8.7.2. This proposal is based on the assumption that Ar[m] would be significantly overpopulated because of its long lifetime.[99,100] The absolute number density of Ar[m] should be known to confirm this mechanism. However, the emission method is not applicable to this measurement because the radiation transition from Ar[m] to the ground state is optically forbidden. The absorption and fluorescence methods require intense primary radiation sources for argon, that is, sources that are difficult to produce.

Uchida and co-workers[101] overcame this problem by utilizing an atmospheric-pressure argon microwave-induced plasma (MIP) as a radiation source for absorption measurements. They measured the spatial profile of absorbance of Ar 811.5 nm, which is related to one of the metastable levels ($4s[3/2]_2^{\circ}$), along with T_{exc} (Ar I) and n_e. In the lower region of the plasma, the radial absorbance profile showed off-axis peaks around 4 mm from the center, while above the observation height of 15 mm, the absorption profile is rather flat or parabolical. Observation of the vertical profile along the central axis of the discharge, shows a maximum at 15 mm (ie, at the normal observation height for the analyte). This spatial characteristic of absorbance, corresponding to the density of Ar[m], suggests the inflow of Ar[m] produced in the surrounding hot zone to the inner core of the plasma.[72] In addition, a large amount of potassium (3%) does not essentially change the absorbance values of Ar[m], thus indicating the buffer action[72] of Ar[m] (see Section 8.7.2.1).

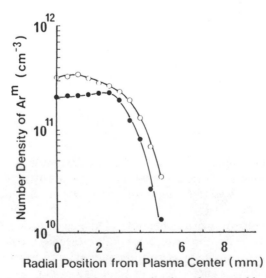

Figure 8.5. Radial distributions of number density of metastable argon atoms with (—●—) and without (—○—) carrier argon gas flow. (From Reference 51, with permission.)

Absolute densities of Ar^m were estimated by Nojiri and co-workers[51] from the absorbance values by assuming the optical parameters of the absorption line. The radial profile of Ar^m, with and without injector Ar gas flow, at the height of 15 mm is shown in Figure 8.5. The number density of Ar^m at the plasma center is approximately 10^{11} cm^{-3}, which is not overpopulated compared to the LTE value calculated from the experimental T_{exc} and n_e. Evidently, this result contradicts the basic assumption made for Penning ionization.

Using a similar experimental procedure, Blades and Hieftje[102] extended the observation to a number of other transitions, which correspond to the absorption from not only the metastable levels but also from adjacent radiative 4s levels. The concentration ratio of metastable to radiative levels was calculated to be 0.6, which also supports the nonoverpopulation of Ar^m.

8.5   SPECTRAL LINE PROFILES

The width and shape of a spectral line reflect the various processes occurring in the plasma. Sections 8.5.1.1 to 8.5.1.5 deal with the theoretical and experimental aspects of line broadening in the ICP.

8.5.1   Theory of Line Shape

According to the simplest classical model for atomic structure, line widths should be infinitely narrow. The actual spectral lines are, however, broadened by the following causes.[2,103]

8.5.1.1   Natural Broadening

The energy of an atomic level cannot be strictly determined because of an uncertainty ΔE due to the Heisenberg principle:

$$\tau \, \Delta E \sim \frac{h}{2\pi} \qquad [8.25]$$

where τ is the mean lifetime of the level. Therefore, the spectral line corresponding to any transition has a wavelength spread, $\Delta \lambda_N$. The broadened profile is found to be a Lorentzian function:[103]

$$P(\lambda) = \frac{\text{const}}{1 + [2(\lambda - \lambda_0)/\Delta \lambda_N]^2} \qquad [8.26]$$

where λ is the wavelength and λ_0 the wavelength at the intensity maximum. The wavelength spread $\Delta \lambda_N$ is usually less than 10^{-2} pm, and is negligibly small compared to the halfwidths of other line broadenings.

8.5.1.2   Doppler Broadening

The relative motion of emitting species to an observer causes an apparent shift in wavelength, known as the Doppler effect. Because the species in a hot medium follow the Maxwell–Boltzmann velocity distribution, atoms having different velocities emit radiations with a spread of wavelengths. In this case, the line profile is defined by the Gaussian distribution:

$$P(\lambda) = \text{const} \times \exp\left\{-\ln 2\left[\frac{2(\lambda - \lambda_0)}{\Delta\lambda_D}\right]^2\right\} \qquad [8.27]$$

where $\Delta\lambda_D$ is given in Equation 8.10.

8.5.1.3   Lorentz Broadening

Collisions with other species also cause line broadening, called Lorentz broadening. This phenomenon is interpreted as radiation damping, ie, the interruption of radiation through a collision in Lorentz's classical theory,[104] and as the perturbation of atomic structure in the quantum theory.[105–109] According to the impact theory, the perturbation shifts each energy level E_i by an amount ΔE_i, which depends on the distance between the two atoms, ρ. The frequency of emission from the level 2 to level 1 is given by:

$$v(\rho) = \frac{E_2(\rho) - E_1(\rho)}{h} \qquad [8.28]$$

The rapid change of ρ results in the spread of frequency, and thus wavelength. In the argon ICP, the argon ground-state atom is considered to be a main perturber because of its high number density.

8.5.1.4   Holtzmark Broadening (Resonance Broadening)

The Holtzmark type of line broadening[110] occurs when the excited atom, usually in the resonance state, X^*, collides with the identical atom in the ground state, X:

$$X^* + X \rightarrow X + X^* \qquad [8.29]$$

This broadening becomes important at high concentrations of the emitting and ground-state atoms.

8.5.1.5   Stark Broadening

In a strong electric field, a degenerate level splits into several components. This Stark effect, caused by charged particles in the plasma, broadens the spectral line. As described in Section 8.3.1.1, in all the atoms except hydrogen, which shows a linear Stark effect, the perturbation is quadratic; ie, the interaction energy is proportional to the square of the electric field. The broadenings discussed in Sections 8.5.1.3 to 8.5.1.5 are called "pressure broadening," and the line shapes are approximately expressed by the Lorentzian form.[108]

Perturbations by foreign particles cause not only line broadening but also line shift. This process is outside the realm of classical theory and requires a quantum mechanical treatment.[106,109] The statistical average of $v(\rho)$ (Equation 8.28) over ρ is usually reduced, compared to the central frequency of the unshifted line. The corresponding emission line, therefore, tends to be shifted to a longer wavelength. This line shift is observed for all the pressure effects (Sections 8.5.1.3 to 8.5.1.5) except for the linear Stark effect, which splits the energy levels symmetrically.[2]

Because a spectral line from the plasma is subject to all the broadening processes mentioned above, the line profile is not a simple Gaussian or Lorentzian function but a combination of both components, called the Voigt function:[103]

$$V(a, \omega) = \left(\frac{a}{\pi}\right) \int_{-\infty}^{+\infty} \frac{\exp(-t^2)}{a^2 + (\omega - t)^2} \, dt \qquad [8.30]$$

$$a = (\ln 2)^{1/2} \frac{\Delta\lambda_P}{\Delta\lambda_D} \qquad [8.31]$$

$$\Delta\lambda_P = \Delta\lambda_N + \Delta\lambda_L + \Delta\lambda_R + \Delta\lambda_S \qquad [8.32]$$

$$\omega = \frac{2(\ln 2)^{1/2}(\lambda - \lambda_0 - d)}{\Delta\lambda_D} \qquad [8.33]$$

where $\Delta\lambda_R$, $\Delta\lambda_S$, and $\Delta\lambda_P$ are the Holtzmark, Stark, and total pressure widths, respectively, d the line shift, and a the a parameter (a value) indicating the contributions of Gaussian and Lorentzian components.

Factors further complicating the actual line shape are hyperfine structure and self-absorption. A spectral line of an element that has several isotopes or a nonzero nuclear spin shows splitting, or fine structure, known as "hyperfine structure" (hfs).[103,111] In this case, the line profile is given by a summation of all the hyperfine components:

$$V_{sum} = \sum_i V_i(a_i, \omega_i) \qquad [8.34]$$

The emission line from an optically thick light source suffers from selective absorption by identical atoms, ie, self-absorption.[12,14,103] Because this effect preferentially occurs at the intensity maximum, where the absorption coefficient is large, it causes apparent broadening of the line profile. Particularly, when the light source has fringes that are cooler than the inner core, the high number density of absorbing atoms in the fringes may result in severe intensity depression at the line center. This phenomenon is called self-reversal.

8.5.2 Literature Values for Line Widths and Shifts

The effective line profile observed through a dispersive system is not a true physical profile, but it is broadened and distorted by imperfections of the instrument, such as finite resolving power, aberrations, and misalignment of the optical components. To minimize these instrumental effects, a high-resolution optical system is required for the measurements. The experimental data for line widths and shifts in the ICP obtained with various high-resolution systems are summarized in Table 8.3, along with theoretical values for pressure broadenings.

Human and Scott[12] used a Fabry–Perot interferometer to measure the line profiles of calcium, strontium, and argon emitted from the ICP. From the comparison of a theoretical Voigt function to the observed line profiles, the a parameter was found to be less than 0.4 under conditions of low optical density, (ie, low solution concentration) and low or intermediate observation height. This observation led the investigators to conclude that spectral lines from the ICP are predominantly Doppler broadened and are less subject to collisional broadening. Furthermore, the self-absorption or self-reversed profiles observed at higher solution concentrations were confirmed. The studies well explained the wide dynamic range experienced in ICP–AES.

The wavelength region observable by a Fabry–Perot interferometer is usually limited to wavelengths above 400 nm because of the difficulty in fabricating dielectric coatings for use in the ultraviolet region, where a number of prominent emission lines exist.[65] Kawaguchi and co-workers[112] extended observations to the ultraviolet region by using an aluminum coated etalon, and reported the halfwidths of 15 emission lines. The physical line widths from the ICP were corrected for instrumental broadening, which was estimated by measuring the emission lines from hollow cathode lamps. Assuming $T_D = 5000$ K, they obtained a values ranging from 0.2 to 0.7.

High-resolution monochromators that can cover wide wavelength ranges and are convenient to use have been applied to line width measurements.[13,14,113–116] Broekaert and co-workers[113] determined the line widths for the emission lines of 16 lanthanide elements using a 3.4-m Ebert mount monochromator. Certain line profiles indicated that the typical hfs, is probably due to nonzero nuclear spin. Larson and Fassel[114] discussed the spectral interferences caused by line broadening and radiative recombination. The wings of significantly intense emission lines (eg, ionic resonance lines of alkaline earths observed by a double monochromator) elevated the background even at wavelengths far from the line center. Pressure effects are responsible for broadening in the wing region, since the Gaussian function rapidly attenuates as the wavelength distance from the intensity maximum increases.

Hasegawa and Haraguchi[13] reported the use of an echelle monochromator (cross-dispersion type) with wavelength modulation that has some remarkable advantages over conventional monochromators.[117] They first estimated the true halfwidths of HCL lines by extrapolating the effective halfwidth to a slit width of 0 μm. Then, based on the assumption that the instrumental profile is expressed as a Voigt function, Gaussian and Lorentzian components were separately derived as a function of wavelength by a curve-fitting procedure for several HCL lines. Finally, the optical parameters of 24 ICP emission lines were deduced from the observed line profiles after the deconvolution of the instrument function. As is clearly seen from Table 8.3, a values were in the range from

Table 8.3. Literature Data for Line Widths and Shifts

Species	Wavelength (nm)	Concentration (μg/mL)	a	Line widths (pm)			T_D (K)	d (pm)	Ref.
				$\Delta\lambda_D$	$\Delta\lambda_P$	$\Delta\lambda_v$			
H I	486.133		3.1	24.5	90	96.6	5000[a]		118
Li I	460.286		2.0	8.8	21.2	24.6	5000[a]		118
Li I	610.364							2.76	120
Be I	234.861	10	0.19	4.2	0.93	4.7	5600		13
Be II	313.042	10	0.11	5.6	0.74	6.0	5500		13
B I	249.773	100	0.22	4.4	1.1	5.0	6400		14
Na I	588.995							0.31	120
Na I	589.592		12	2.1	30.5	30.7	5000[a]	0.24	118
Mg I	285.213	10	0.32	2.9	1.1	3.5	4800	0.28	120
			0.2	2.9	0.8	3.3	5000[a]	0.28	13,14
						3.4			119
		10				3.8			112
		0.1							116
Mg I	518.362					3.6		-0.27	120
Mg II	279.079	1000	0.29	3.0	1.1	3.5	5000[a]		119
			0.3	2.9	1.2	3.3	5000		13
Mg II	279.553	10	0.17	3.0	0.59	3.6	5500	0.27	13,14
		1				3.4			112
		0.1							116
Mg II	280.270	10	0.16	3.2	0.63	3.5	6100		13
			0.2	2.9	0.7	3.3	5000[a]		119
Al I	396.153	100	0.22	4.2	1.1	4.8	5900	0.67	13,14
		100				5.1			112
		100				5.4			116
Si I	251.612	200	0.30	2.3	0.82	2.8	4600		13
P I	213.620	1000	0.23	2.0	0.55	2.3	5300		13
		1000				2.8			112

(Continued)

Table 8.3. (*Continued*)

Species	Wavelength (nm)	Concentration (µg/mL)	Line widths (pm)				T_D (K)	d (pm)	Ref.
			a	$\Delta\lambda_D$	$\Delta\lambda_P$	$\Delta\lambda_V$			
Ar I	415.859		0.65	4.0	3.1	5.6	7300		13
			1.16	3.33	4.63	6.44	5000ᵃ	0.38	36
Ar I	425.936		0.27	3.4	1.1	4.0	5000ᵃ		118
Ar I	427.217			2.2			2070		19
				4.0			6900		12
Ar I	549.587		3.7	4.4	19.6	20.6	5000ᵃ		118
Ar I	696.543		1.06	5.6	7.12	10.3	5000ᵃ		36
Ca I	422.673	25	0.16	3.8	0.72	4.2	6200	0.22	13,14
		1	0.1	3.36	0.4	3.6	4960		12
			0.46	2.8	1.5	3.7	3800ᵇ		7
			0.4	3.4	1.7	4.4	5000ᵃ		119
		100				4.2			112
		0.5				4.6			116
Ca II	317.933	100	0.33	2.8	1.1	3.4	5800		13
Ca II	393.367	1	0.24	3.0	0.84	3.5	4500	0.13	13,14
			0.62	2.6	1.9	3.8	3800ᵇ		7
		1	0.4	3.2	1.6	4.1	5000ᵃ		119
		0.05				3.9			112
						4.4			116
Ca II	396.847	1	0.23	3.2	0.89	3.7	5200		13
Sc II	361.384	10				7.1ᶜ			14
						6.0			113
Ti II	334.941	10	0.09	2.6	0.30	2.8	5700		13
V II	309.311	10				6.7ᶜ			14
Cr I	357.869	20	0.36	2.1	0.88	2.6	3300	0.18	13,14
Cr II	205.552	1000	0.31	1.4	0.52	1.7	4700		13
Mn II	257.610	10				4.8ᶜ			14
		1				3.6			116

Line	λ (nm)								Ref.
Mn II	260.569	1	0.33	1.4	0.57	4.2	3600	0.17	116
Fe I	248.327	300	0.21	1.7	0.43	1.8	4800	0.03	13,14
Fe II	259.940	100	0.12	2.74	0.41	2.0	6402		13,14
Fe I	358.120	1000	0.11	2.81	0.36	2.98	6216		15
Fe I	371.994	1000	0.13	2.86	0.43	3.00	6390		15
Fe I	373.487	1000	0.10	2.82	0.35	3.10	6212		15
Fe I	373.713	1000				3.00			15
Co II	238.892	100				5.5c			14
Ni I	232.003	100	0.27	1.5	0.47	1.7	4500	0.20	13,14
Ni II	231.604	100	0.17	1.5	0.30	1.7	5000	0.06	13,14
Cu I	324.754	10				6.5c			14
Cu I	510.554					1.5		0.29	120
Cu I	521.820					1.5		1.14	120
Zn I	231.856	100	0.12	1.4	0.21	1.4	5500	0.09	13,14
Zn II	202.551	100	0.35	1.2	0.49	1.5	4200	0.04	112
Ga I	294.364	250	0.40	1.4	0.67	1.4	3900		13,14
Ge I	265.118	500	0.22	1.5	0.38	3.0	5800		14
As I	228.812	300	0.27	2.8	0.89	1.8	6300		13
Sr I	460.733	1000	0.25	2.54	0.76	3.3	5230	0.47	13
			0.7	2.5	2.2	3.0	5000a		13,14
						3.8			12
Sr II	407.771	1	0.28	2.1	0.69	2.5	4400	0.06	119
		100				3.1		0.14	120
		0.25				3.5			13,14
									112
									116
Sr II	421.552	1	0.27	2.3	0.74	2.7	5000		13
			0.6	2.3	1.8	3.3	5000a		119
				2.3			4880		12
Sr II	430.545							0.21	120
Y II	371.029	10				1.8		−0.85	120
						2.8			14
Zr II	339.198	10	0.15	1.8	0.33	2.0	5200		113
									13

(Continued)

Table 8.3. (Continued)

Species	Wavelength (nm)	Concentration (μg/mL)	a	Line widths (pm)			T_D (K)	d (pm)	Ref.
				$\Delta\lambda_D$	$\Delta\lambda_P$	$\Delta\lambda_V$			
Mo II	281.615	200				2.3			14
Rh I	343.489	100	0.14	1.6	0.26	1.7	4100		14
Pd II	229.651	1000				1.1			14
Ag I	328.068	10	0.26	1.9	0.58	2.2	6700		14
Cd I	228.802	1000	0.25	1.2	0.36	1.4	6000	0.16	13,14
		100				1.4			112
		0.5				1.4			116
Cd I	479.992							−0.25	120
Cd I	508.582							−0.31	120
Cd II	214.438	100	0.58	0.96	0.67	1.4	4400	0.09	13,14
Cd II	226.502	100				1.6			112
		0.5				2.2			116
Sn I	283.999	500				1.3			14
Sb I	217.581	1000	0.45	1.1	0.60	1.5	6100		14
Te I	214.281	1000				0.97			14
Ba I	553.548	500	0.34	3.0	1.2	3.7	7900		13
		1000				3.7			112
Ba II	455.403	10	0.59	1.7	1.2	2.4	3500	0.09	120
		10				3.4			13
									112
Ba II	493.409							0.15	120
								0.12	120
La II	398.852	100				15.6c			14
La II	408.672	10				4.8			113
		4.3				3.1			112
						3.4			116
Ce II	401.239					3.8			113
Ce II	413.765	50	0.23	1.9	0.53	2.2	6000		13
Pr II	390.844	100				14.5c			14

Pr II	422.533		4.3	113
Nd II	401.225	50	3.4	14
Nd II	417.732		2.8	113
Sm II	359.260	50	1.2	14
Sm II	392.827		2.3	113
Eu II	381.967	10	9.9[c]	14
Eu II	412.974		9.5	113
Gd II	342.247	50	2.0	14
Gd II	364.620		5.6	113
Tb II	350.917	10	9.8[c]	14
			12	113
Dy II	353.170	20	1.6	14
Dy II	400.045		3.1	113
Ho II	345.600	10	20.1[c]	14
			19	113
Er II	337.271	10	1.7	14
Er II	369.265		2.9	113
Tm II	313.126	50	1.6	14
Tm II	376.133		4.5	113
Yb II	328.937	10	2.3	14
Yb II	369.420		2.7	113
Lu II	261.542	10	20.1[c]	14
Lu II	347.248		11	113
W II	244.875	1000	0.63	14
Au I	242.795	200	1.6	14
Pb I	283.306	100	1.6	112
		50	2.4	116
Pb II	220.353	1000	1.7	14
		100	1.6	112
		50	1.4	116
Bi I	223.061	1000	1.4	14

[a] Assumed to be 5000 K.
[b] Assumed to be 3800 K.
[c] Not corrected for the instrument function.

0.2 to 0.3 for most of the lines, thus verifying the conclusion of Human and Scott.[12] Recently, this investigation was extended to prominent lines of 33 elements. A curve-fitting procedure was used to correct for spectral interferences.[14] The hfs observed for certain emission lines restricted the practical use of the curve-fitting procedure, especially for lines having unknown hfs.

Boumans and Vrakking[116] also used an echelle monochromator with a predisperser in a parallel-slit arrangement for line width measurements. Although they did not estimate the a parameters and Doppler widths, their results for the total halfwidths of 13 emission lines agreed well with those published by other investigators.[12-15,112]

Using a Fourier transform spectrometer, Faires and co-workers[15] measured the line widths of 81 Fe I lines in the spectral range 290 to 390 nm, originating from the usual analytical zone of an ICP. The observed line profiles were split by a curve-fitting procedure into Gaussian and Lorentzian components. The Doppler temperatures, mean value of 6310 K, calculated from Doppler widths, showed somewhat higher values than other published data presented in Section 8.2. This may be due to instrumental effects.

Theoretical calculations of line shape in the ICP have been dealt with in certain publications.[36,118,119] Batal and Mermet[119] calculated line profiles for calcium, magnesium, and strontium, assuming a van der Waals potential between the emitting atom and perturbing Ar ground state:

$$V(\rho) = \frac{-C_6}{\rho^6} \qquad [8.35]$$

where ρ is the distance between the emitting and perturbing atoms, $V(\rho)$ is an electronic interaction potential, and C_6 is a proportionality constant. The results for the Doppler, Lorentz, and Stark widths confirmed the previous observation that the main cause of line broadening in the ICP is the Doppler effect, and the emission lines originating from levels with higher quantum number were predominantly broadened by the Stark effect.

Only a few examples have been reported for line shift measurements. Kato and co-workers[120] observed the profiles of 16 emission lines with a pressure-scanning Fabry-Perot interferometer, and determined the wavelength shifts of 14 ICP lines compared to HCL lines used as the "wavelength reference". Most of the lines were shifted below 0.3 pm toward the longer wavelength (ie, red shift), except for the lines emitted from the higher energy levels. The decrease in the magnitude of the wavelength shift with increasing observation height might indicate the larger perturbation effect in the lower portion of the plasma. The shift phenomena were attributed to the Stark effect,[120] although theoretical calculations of line broadening,[118,119] predicted the major role of the Lorentzian effect by Ar atoms.

Hasegawa[14] measured the line shifts of 17 ICP lines of 10 elements compared to HCL lines by using a wavelength–modulation echelle monochromator. All the shifts were red shifts less than 0.5 pm, except for Al, and showed a tendency to be proportional to the total halfwidth of pressure broadenings ($\Delta\lambda_P$). This result suggests that the line shift and pressure broadening are caused by the same effect. Based on the assumption that Ar atoms are the main perturbers interacting with emitting species according to the Lennard–Jones expression $(V(\rho) = C_{12}\rho^{-12} - C_6\rho^{-6})$, the coefficients C_6 and C_{12} were calculated from the measured $\Delta\lambda_L$ and d with the aid of the Lindhelm–Foly theory.[109]

8.6   GAS FLOW DYNAMICS

Argon gas, flowing through an ICP torch, experiences electromagnetic heating by RF power before it reaches the upper region of the plasma, carrying a part of the heat. For any small zone in the plasma, there is an energy balance between the supplied energy and the heat transported away by the gas. Thus, the gas flow is one of the most fundamental factors affecting the conditions of the plasma.

Earlier work on gas flow dynamics in the ICP was aimed at improving the analytical performance, such as detectability, stability, and ease of ignition.[121–124] The most remarkable point concerning the gas flow in the pioneering work by Reed[125] was vortex stabilization, which was accomplished by the tangential introduction of the argon outer flow. The low-pressure zone created in the center of the discharge tube prevented the plasma from being extinguished and established the recirculation of some of the plasma. The further modified torch, consisting of three concentric quartz tubes, allowed the feeding of powders and gaseous samples efficiently into the central core of the plasma.[126] Since the first applications of the ICP to spectrochemical analysis made by Greenfield and co-workers[121] and Wendt and Fassel,[122] a number of modifications have been attempted with the objective of achieving superior analytical capabilities. However, Reed's basic design of the ICP has not been altered fundamentally up to now.

Barnes and co-workers[40,41,80,127–130] and Boulos,[131] who recognized the importance of gas flow dynamics as a fundamental property of the ICP, performed theoretical and experimental investigations of gas velocity distributions. They modified the mathematical model developed by Miller and Ayen[132] to account for the spatial characteristics of gas properties and major energy losses in high-temperature argon[128] and nitrogen[129,130] discharges. Their model, which incorporated a three-dimensional gas flow pattern, sample particle motion, and decomposition, yielded the radial and axial profiles of temperature and gas velocity by solving the energy and magnetic flux equations. The basic concepts and equations are presented in Reference 129. Although many assumptions were made to simplify calculations (eg, laminar gas flow in the outer torch tube), the results of computer simulation exhibited the evident toroidal configuration that depends on the RF power and gas flow rate.

Barnes and Nikdel[129] compared argon and nitrogen discharges for the purpose of evaluating their analytical utilities and found that the nitrogen discharge provided the higher efficiency for the decomposition of Al_2O_3 particles, in spite of a lower temperature. The model calculation was also applied

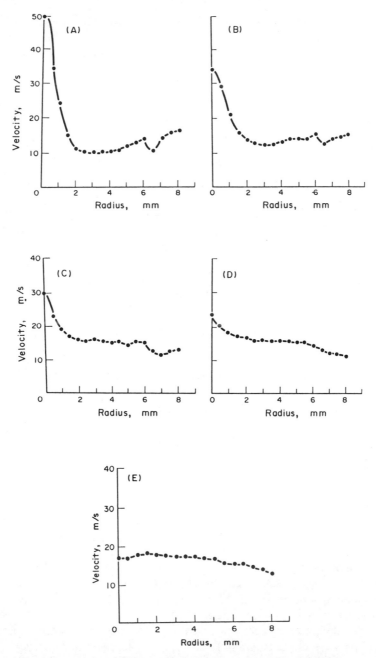

Figure 8.6. Radial velocity profiles for a 0.75-kW discharge with 14 L/min outer gas flow, 1.00 L/min aerosol flow. Axial positions are (*A*) −7 mm, (*B*) 0 mm, (*C*) 4 mm, (*D*) 6 mm, and (*E*) 12 mm from the top of the coil. (From Reference 41, with permission.)

to designing the plasma torch.[80] The prediction by computer simulation helped to design a torch capable of achieving high sample introduction efficiency, low gas consumption, and easy ignition.

Recently, Boulos[131] published a review paper dealing with the basic principles and the main design features of the ICP. Many diagnostic works to determine the characteristics of the electric and magnetic fields, and the temperature, velocity, and concentration distributions in the discharge region are summarized in detail and critically reviewed along with different plasma models.

These theoretical works were followed by experimental studies.[40,41] Barnes and Schleicher[40] measured radial velocity distributions at various operating conditions using an impact (Pitot) probe. The gas velocity was obtained from the static pressure by means of the Bernoulli equation. The flow model proposed, after adjustment of the empirical parameters, agreed well with the experimental results within an accuracy of 15% inside the core of the plasma and up to 15 mm above the load coil. Near the torch wall, however, a large discrepancy was noted because of the relatively high uncertainty associated with the pressure measurement.

Barnes and Genna[41] further improved the sensitivity and resolution of the velocity measurements by utilizing a smaller probe. The radial velocity profiles measured for a 0.75-kW discharge with 14 L/min outer gas flow are shown in Figure 8.6. Based on their measurements, the ICP was divided into three radial regions: (a) a high-velocity central flow region, approximately 1.5 mm from the central axis, (b) a bright luminous core extending from the edge of the high-velocity central flow region to approximately 2 mm from the torch wall, and (c) a torch wall region, within approximately 2 mm from the wall. In the high-velocity central flow region, formed by cool and fast injector flow, gas velocity rapidly decreased with increasing height up to 12 mm, above which the radial profile was rather flat. This result suggests the lateral diffusion of inner gas by turbulence. The luminous core region, maintaining relatively constant velocity, was less affected by the turbulence. This region was considered to be supported by the large flow of the outer gas. The radial profile in the wall region showed a more complex behavior (Figure 8.6), probably due to effects of both the lateral diffusion of argon gas and air entrainment into the plasma. These spatial characteristics of gas velocity closely match the distribution patterns of plasma temperatures and electron number density, which change from a toroidal to a bell-shaped configuration with increased observation height (see Sections 8.2 and 8.3).

8.7   EXCITATION MECHANISMS

The ultimate purpose of the diagnostic work on the ICP is certainly to clarify the excitation mechanisms for support argon gas and analytes introduced into the plasma. In the remainder of this chapter, we present general considerations on the deviation from LTE and briefly summarize the proposed excitation models, discussing whether the non-LTE phenomena would be satisfactorily interpreted.

8.7.1   Deviation from Local Thermodynamic Equilibrium

If every energy exchange process occurring in a system is exactly balanced by its reverse process, the system is said to be in the state of complete thermodynamic equilibrium (CTE).[1,2] From thermodynamic arguments, all energy distributions in CTE follow the Boltzmann, Saha, and Maxwell–Boltzmann expressions, as discussed in Section 8.2, and are described by only one temperature. The energy exchange processes mentioned above include not only collisional processes but also radiative processes, so that emitted photons must be completely absorbed in the CTE system.

In actual plasmas, however, the radiative processes are out of balance, since the complete absorption of photons requires high optical densities for all the radiative transitions. When collisional processes are dominant in a system, each point of the system can be described by an individual temperature, which is governed by the various distribution laws. This state is known as local thermodynamic equilibrium (LTE).[1,2] Therefore, the criterion for the presence of LTE is that collisional excitation and ionization processes exceed radiative processes. Mathematically, the criterion for LTE is:[61]

$$n_e \gg 1.6 \times 10^{12} \times T_e^{1/2}(\Delta E)^3 \ (\mathrm{cm}^{-3}) \qquad [8.36]$$

where T_e is the electron temperature (K) and ΔE is the energy difference (eV) between the states in question.

The conventional argon ICP is generally accepted not to be in LTE because various measured temperatures differ from each other (Section 8.2). Especially, from the analytical point of view, the relationship $T_{exc} < T_{ion}$ should be noted because it is equivalent to the statement of higher sensitivities of ionic lines, or overpopulation of ions.[7,72] In addition, increase of T_{exc} with excitation energy is also a remarkable phenomenon indicating deviation from Boltzmann's law. This is commonly represented by another expression that higher energy levels are overpopulated compared to lower levels. Next we discuss these two points as typical non-LTE phenomena to check the reliabilities of the plasma models.

8.7.2   Proposed Excitation Mechanisms

8.7.2.1   Penning Ionization Reaction

The process first postulated as the mechanism for analyte excitation[98] is a Penning reaction in which argon-metastable atoms ionize analyte atoms with their high excitation energies, 11.55 and 11.76 eV:

$$\mathrm{Ar^m + X \rightarrow Ar + X^+ + e^-} \qquad [8.37]$$

$$\mathrm{Ar^m + X \rightarrow Ar + X^{+*} + e^-} \qquad [8.38]$$

(The reverse processes are neglected on the assumption that Ar^m is fairly overpopulated.) Obviously, this mechanism provides the higher population of ions, but it cannot be applied to the ionic levels for which the ionization energy plus excitation energy is higher than the excitation energy of Ar^m. In this case, a two-step reaction, including electron collision, should be considered:

$$Ar^m + X \rightarrow Ar + X^+ + e^- \qquad [8.39]$$

$$X^+ + e^- \rightarrow X^{+*} + e^- \qquad [8.40]$$

The overpopulation of higher atomic levels can be interpreted in terms of direct excitation by Ar^m or by the recombination of the overpopulated ionic ground state with electrons in a three-body recombination.

$$Ar^m + X \rightarrow Ar + X^* \qquad [8.41]$$

$$X^+ + 2e^- \rightarrow X^* + e^- \qquad [8.42]$$

Boumans and de Boer[72] further postulated that Ar^m was overpopulated as a result of its inflow from the surrounding zone and played a dual role in the plasma: ie, an "ionizer" causing the Penning processes and an "ionizant" acting as follows:

$$Ar^m + e^- \rightarrow Ar^+ + 2e^- \qquad [8.43]$$

Their hypothesis allows a quantitative understanding of the high electron number density in the ICP and, thus, the relatively small ionization interference.

Until the absolute number density for Ar^m was measured by Uchida and co-workers[43,51,101] (Section 8.4.3), the dominance of these processes involving Ar^m had been believed. The measured number density of Ar^m, 2×10^{11} cm^{-3}, was not significantly different from the value predicted from the Boltzmann equation and ideal gas law, 3×10^{10} cm^{-3}, and from Saha and Boltzmann equations, 5×10^{12} cm^{-3}. The proximity of these values appears to be the most serious contradiction against the Penning mechanism.[113,114]

8.7.2.2 Charge Transfer Reaction

Recently, instead of Penning processes, charge transfer has attracted attention as a possible reaction for analyte excitation:[58,113]

$$Ar^+ + X \rightarrow Ar + X^+ \qquad [8.44]$$

$$Ar^+ + X \rightarrow Ar + X^{+*} \qquad [8.45]$$

where ionization (plus excitation) potential of X must be lower than that of Ar (15.76 eV). These processes allow one to explain the non-LTE phenomena of the ICP, similar to the Penning reactions. In addition, the larger population of argon ions, which is equal to n_e, relative to Ar^m favors this mechanism.

Nonlinear reactions, such as Penning ionization and charge transfer, tend to occur most frequently for an analyte whose total excitation energy is close to the threshold energy of Ar^m or Ar^+. Schram and co-workers[58] reported the exceptional overpopulation of Mg ionic levels having the total excitation energy of approximately 16 eV and attributed it to the pumping effect by charge transfer. The major shortcoming of this mechanism is, however, that the following reverse reaction was neglected:

$$Ar + X^+ \rightarrow Ar^+ + X \qquad [8.46]$$

That is, if Expression 8.46 occurs, the number density of analyte ions may be reduced through a collision with highly populated argon ground-state atoms, which is estimated to be of the order of 10^{18} cm^{-3}. As a result, the plasma may be approaching LTE.

As the foregoing discussion shows, the Penning and charge transfer processes are still hypothetical and should be assessed as competitive reactions with electron impact ionization (see Section 8.7.2.5).

8.7.2.3   Recombining Plasma Model

The proposal of a recombining plasma model as the excitation mechanism for support argon and analytes is based on the physical consideration of the kinetics of ionization–recombination of a plasma.[135-140] Using the collisional–radiative model developed by Bates and co-workers[135] (see Section 8.7.2.5), Fujimoto calculated the population density distributions and reaction rates for a hydrogen plasma under the following idealized conditions: (a) an ionization equilibrium plasma, in which the ionization and recombination rates are equivalent,[137] (b) a purely ionizing plasma, in which the number density of ions is zero,[138] and (c) a purely recombining plasma, in which the population of the neutral ground state is zero.[139,140] The actual plasma is, in general, not in one of these idealized situations but is a superimposition of the purely ionizing and purely recombining plasmas.

Boumans[141] noted that the recombining plasma was appropriate for describing the overpopulation of analyte ions in the nonthermal region of the ICP and proposed a mechanism in which Penning ionization, three-body recombination, and spontaneous emission were dominant. However, in his model, the state of the recombining plasma is established by the Penning process, which presumably should far exceed electron impact ionization. As discussed in Section 8.7.2.1, analyte excitation by such a mechanism is unlikely to prevail in the ICP.

Raaymakers and co-workers[142] pursued the discussion into further detail by considering the various plasma phases. These investigators realized that the n_e values alone sufficed to characterize the plasma discharge if the deviation from LTE was not too large. The features of the plasma were separated into regions having different n_e values. The analytical ICP was defined as in a state of partial LTE, in which Saha equilibrium exists between the ionic ground state and higher excited levels of neutral atoms, but the lower atomic levels deviate from Saha equilibrium.

Schram and associates[58] evaluated the features of the Ar ICP as an ionizing or recombining plasma by means of a gas dynamic approach. From the mass and energy balances of the plasma constituents, it was concluded that the axial convection flow of Ar ions and electrons, estimated to be several orders of magnitude larger than the ionization or recombination flows, reduced the population of neutral Ar atoms in the recombining zone normally used for analysis. The contribution of the sideward diffusion was significant in the active zone inside the load coil. Furthermore, the charge transfer reaction was suggested to play an important role in analyte excitation in the recombining zone, since the reduction of Ar ground state favors the charge transfer process rather than the reverse process.

Aeschbach[143] presented a two-temperature model in which the electron temperature was assumed to be significantly greater than the gas kinetic temperature (Doppler temperature) and calculated the radial profiles of the two temperatures and of n_e. The high electron number density was ascribed to ambipolar diffusion. The electron–ion pairs, produced in the hot surrounding zone, flow into the cool central core because of the extreme concentration gradient. Aeschbach did not describe this situation as a recombining plasma.

The recombining plasma model provides an explanation for the overpopulation of ions. Thus, we now consider the relative population densities of the energy levels in neutral atoms. The quasi-steady-state calculation for population densities by Fujimoto[139,140] predicted that in the recombining plasma, the excited states of neutral atoms were also underpopulated with respect to LTE values, and the extent of underpopulation increased with lower excitation energy. The excitation temperature derived from the slope of the Boltzmann plot, thus, decreases with the excitation energy. This is, however, obviously inconsistent with the experimental results. To explain this discrepancy, the aid of other mechanisms should be required.

8.7.2.4 Radiation Trapping Model

Based on the experimental result that metastable and resonant states of argon have almost the same number densities, as discussed in Section 8.4.3, Blades and Hieftje[102] proposed the following radiation trapping model. The metastable and resonant levels are rapidly mixed through mutual collisions leading to the "averaging" of the two states. The resonant states, which would ordinarily deactivate through spontaneous emission, maintain high populations due to radiation absorption. Namely, the ground-state atoms in the center of the light source substantially absorb or capture the photons arising from the surrounding zone. This absorption, which counteracts the radiative deactivation and enhances the population of the upper level of the transition, is known as radiation trapping. Assuming an infinite cylindrical configuration and a Doppler-broadened line profile,[144] Blades and Hieftje[102] estimated the apparent lifetime of the resonant state ($4s[3/2]_1^\circ$) to be 1.16×10^{-3} s, which is far greater than the normal[145,146] lifetime of 1×10^{-8} s.

Blades[147] combined the hypothesis of Boumans and de Boer[72] and the radiation trapping model to offer another excitation mechanism. According to his proposal, the populations of the resonant and metastable atoms in the axial channel are enhanced by radiation trapping to a degree of those in the hot outer zone. The overpopulated excited levels "assist" ionization of neutral argon, resulting in suprathermal population of electrons in the central core. The overpopulation of analyte ion is explained by the mass action effect of the ionization equilibrium on the excess population of electrons.

In a recent paper, Mills and Hieftje[148] revised the apparent lifetime of the argon resonant state by considering both Doppler and pressure broadenings of line profiles, to 1.6×10^{-6} s. This result suggests that the radiation trapping no longer contributes significantly to the population of argon, which contradicts the earlier argument.[102] If the recent value is correct, the argon resonance line should provide an emission intensity similar to that of the ionic resonance line of calcium observed when aspirating a solution of 1000 μg/mL Ca.[149] However, emission spectra of the ICP in the ultraviolet region, measured by Carr and Blades,[52] showed that the resonance line of argon, which was slightly self-absorbed, was not as intense as the emission lines originating from the gaseous impurities in the argon support gas. This experimental result clearly disagrees with the theoretical prediction by Mills and Hieftje.[148]

Based on observations by Carr and Blades,[52] it appears that resonance radiation of argon is almost completely trapped in the plasma core, and this effect has a large influence on the population distribution of argon. However, the nonoverpopulation of the resonant and metastable argon, confirmed experimentally, does not support the postulates of Blades,[147] ie, suprathermal population of electrons and "assisted" ionization caused by radiation trapping.

8.7.2.5 Collisional–Radiative Model

As a consequence of the discussion presented in the preceding section, it is realized that any excitation model must be refined to satisfactorily account for the non-LTE phenomena of the ICP. Recently, quantitative treatments that consider various reactions as the rate processes occurring in the plasma have been attempted.[14,18,23,150–155]

To appreciate the basic properties of the argon ICP, Hasegawa and Haraguchi[18,23] proposed the collisional–radiative model,[135,136] which incorporates the following six processes:[156–158]

1. Electron impact excitation and de-excitation:

$$\text{Ar(p)} + \text{e}^- \underset{k_{qp}}{\overset{k_{pq}}{\rightleftharpoons}} \text{Ar(q)} + \text{e}^- \qquad [8.47]$$

2. Electron impact ionization and three-body recombination:

$$\text{Ar(p)} + \text{e}^- \underset{k_{ip}}{\overset{k_{pi}}{\rightleftharpoons}} \text{Ar}^+ + 2\text{e}^- \qquad [8.48]$$

3. Spontaneous emission and induced absorption:

$$\text{Ar(q)} \underset{\rho_{pq} B_{pq}}{\overset{A_{qp}}{\rightleftharpoons}} \text{Ar(p)} + h\nu_{\text{line}} \qquad [8.49]$$

4. Radiative recombination:

$$Ar^+ + e^- \underset{\alpha_p}{\rightleftharpoons} Ar(p) + h\nu_{cont} \qquad [8.50]$$

where $Ar(p)$ is the neutral argon atom in the level p; A_{qp}, B_{pq}, and ρ_{pq} are the transition probability for spontaneous emission, Einstein's B coefficient for absorption, and radiation density, respectively, for the transition between p and q; and k or α is the reaction rate for each process as a function of T_e.[23] Besides these reactions, the diffusions of electrons and argon ions are also taken into consideration:[143,156]

5. Ambipolar diffusion:

$$\Gamma_a = D_a \nabla^2 n_e \qquad [8.51]$$

6. Convection:

$$\Gamma_c = -v\nabla n_e \qquad [8.52]$$

where Γ is the number of electron–Ar ion pairs flowing into a unit volume at the location (x,y,z) per second, D_a is the ambipolar diffusion coefficient,[159] and v is the gas velocity.[40,41] Therefore, in this model, T_e and n_e are the most important fundamental parameters for characterizing the plasma. If the spatial distributions of T_e and n_e are known, the number densities of argon energy states can be calculated for the steady-state approximation.[135,136] It is assumed in the steady-state condition that the number densities of all energy levels are time independent, ie, the derivatives of n(p)'s, relative to time, are zero. For the ground-state atom:

$$\frac{dn(1)}{dt} = -\left[\left(\sum_{q>1} k_{1q} + k_{1i}\right)n_e + \sum_{q>1}\rho_{1q}B_{1q}\right]n(1) + \sum_{q>1}(k_{q1}n_e + A_{q1})n(q)$$

$$+ k_{i1}n_e^3 + \alpha_1 n_e^2 - (D_a\nabla^2 n_e - v\nabla n_e) = 0 \qquad [8.53]$$

and for the excited-state atom:

$$\frac{dn(p)}{dt} = \sum_{q<p}(k_{qp}n_e + \rho_{qp}B_{qp})n(q)$$

$$- \left[\left(\sum_{q>p}k_{pq} + k_{pi}\right)n_e + \sum_{q<p}A_{pq} + \sum_{q>p}\rho_{pq}B_{pq}\right]n(p)$$

$$+ \sum_{q>p}(k_{qp}n_e + A_{qp})n(q) + k_{ip}n_e^3 + \alpha_p n_e^2 = 0 \qquad [8.54]$$

where $n(1)$ is the number densities of the ground-state atoms.

Table 8.4 gives the computed argon population densities for the observation heights of 15 and 25 mm at the plasma center.[18] Figure 8.7 demonstrates the plot of $\log[n(p)/g_p]$ versus excitation energy (Boltzmann plot) for the height of 15 mm, where $T_e = 8400$ K and $n_e = 1.4 \times 10^{15}$ cm^{-3} were used. It should be emphasized that the number densities of the higher excited levels are very close

Table 8.4. Population Densities n(p) of Argon Atomic Levels
Calculated by a Collisional-Radiative Model

Term	Excitation energy (eV)	Statistical weight, g_p	n(p)(cm^{-3}) Height 15 mm	n(p)(cm^{-3}) Height 25 mm
(3p)^6	0	1	4.0×10^{17}	2.2×10^{18}
4s[3/2]$_2$[a]	11.548	5	3.4×10^{11}	1.9×10^{11}
4s[3/2]$_1$	11.624	3	1.8×10^{11}	1.0×10^{11}
4s'[1/2]$_0$[a]	11.723	1	4.6×10^{10}	2.9×10^{10}
4s'[1/2]$_1$	11.828	3	1.2×10^{11}	7.4×10^{10}
4p	13.097	24	1.3×10^{11}	8.3×10^{10}
4p'	13.319	12	4.1×10^{10}	2.9×10^{10}
3d	14.008	40	5.9×10^{10}	3.4×10^{10}
5s	14.077	8	1.1×10^{10}	6.1×10^9
3d'	14.240	20	1.9×10^{10}	1.2×10^{10}
5s'	14.252	4	3.8×10^9	2.3×10^9
5p	14.509	24	1.7×10^{10}	9.3×10^9
5p'	14.690	12	6.2×10^9	3.5×10^9
4d	14.780	40	2.0×10^{10}	1.0×10^{10}
6s	14.843	8	3.6×10^9	1.9×10^9
4f	14.906	56	2.3×10^{10}	1.2×10^{10}
4d'	14.968	20	7.1×10^9	3.8×10^9
6s'	15.020	4	1.3×10^9	7.0×10^8
6p	15.028	24	8.4×10^9	4.2×10^9
4f'	15.083	28	8.4×10^9	4.5×10^9
5d	15.147	40	1.2×10^{10}	5.8×10^9
7s	15.183	8	2.3×10^9	1.1×10^9
6p'	15.205	12	3.1×10^9	1.6×10^9
5d'	15.317	20	4.4×10^9	2.2×10^9
7s'	15.359	4	8.2×10^8	4.2×10^8

[a] Metastable level.
Source: Reference 18.

to those in LTE, whereas the lower states, especially the ground states, are slightly overpopulated. Based on reaction rates, this result is explained in the following way, as shown in Figure 8.8.[18] For the higher excited levels, electron impact processes predominate the radiative processes, since the energy difference between two states is small near the ionization limit. This results in the approach to LTE, as indicated in Section 8.7.1. On the other hand, spontaneous emission between the low-lying states contributes significantly because of the relatively large energy differences. To balance the reaction rates of upward (electron impact excitation) and downward (electron impact de-excitation plus spontaneous emission) transitions, the number densities of lower states increase compared to the LTE values (solid line in Figure 8.7). Therefore, the argon ground state, which requires 11 eV for excitation to 4s levels, is expected to be greatly overpopulated. However, the argon ground-state atoms distribute widely in the plasma as support gas, so that the resonance radiation emitted from the surrounding region is almost completely absorbed in the central

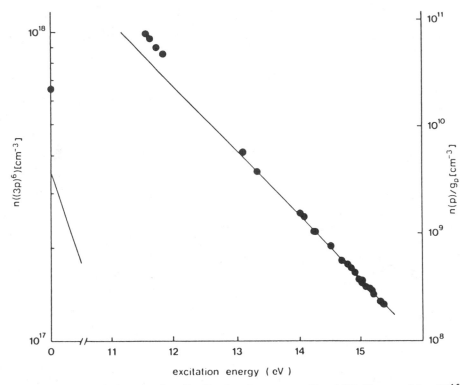

Figure 8.7. Population density distribution for argon: $T_e = 8400$ K, $n_e = 1.4 \times 10^{15}$ cm^{-3}, observation height = 15 mm, radial position = 0 mm. Line gives LTE value; points indicate calculated values. (From Reference 18, with permission.)

channel zone. This radiation trapping reconciles the effect of spontaneous emission, resulting in approach to LTE. In addition, the frequent interconversion between radiative and metastable 4s levels by electron collision "averages" the populations of both states. As a result of these phenomena, the extent of overpopulation for argon neutral states is small in comparison to the ideal case for an optically thin plasma.[23]

Apparently, *overpopulation of neutral atoms and lower energy levels* predicted by this model seem to contradict the experimental results, ie, *overpopulation of ions and higher energy levels*. However, this inconsistency is due to the difference of temperatures used for the description of LTE.[18,23] As shown in Figure 8.9, the former means the deviation from the LTE value (solid line) calculated from T_e, while the latter is based on the T_{exc} determined from the slope of the Boltzmann plot for lower levels. This T_{exc} value apparently is lower than T_e due to the effect of spontaneous emission. Underestimation of temperature obviously leads to the conclusion that ions and higher excited levels are overpopulated.

The importance of diffusion processes should be emphasized further.[18,23] At the normal observation height, the diffusion flow was estimated to be negligible compared to the ionization and recombination flow (about 1%). This

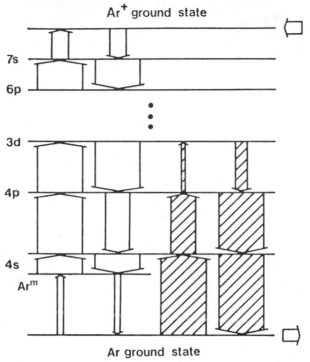

Figure 8.8. Schematic diagram of dominant electron flows between argon energy states in the normal analytical zone of the argon ICP. The widths of the arrows indicate the magnitudes of the electron flows. Open arrows denote the nonradiative processes and hatched arrows are the radiative processes.

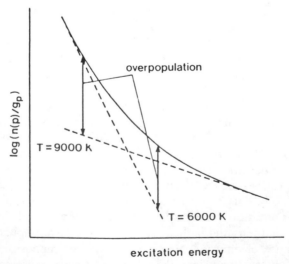

Figure 8.9. Schematic representation of population density distribution showing the overpopulations of upper or lower levels.

corresponds to the behavior predicted for an ionization–equilibrium plasma described by Fujimoto.[137] At the upper and the lower regions of the axial channel, however, there were large inflows of electrons and ions by convection and ambipolar diffusion, respectively.

The collisional–radiative model has also been applied to the elucidation of the excitation mechanism for analytes.[14,154,155] Lovett[154] postulated the excitation model that involved Penning ionization along with electron impact and radiative processes. However, the predicted population densities for alkaline earths demonstrated values very close to LTE and unfortunately failed to explain the non-LTE phenomena. Hieftje and co-workers[155] compared the spatial behavior of analyte emission intensity with the concentration product of reacting species in the steady-state collisional reactions. The correlation plots suggested that the excitation of neutral atoms occurred through an ion–electron recombination, whereas ion generation and excitation were mainly due to the Penning process.

According to the recent calculation for the excitation of magnesium and cadmium by a collisional–radiative model,[14] in which the Penning and charge transfer processes were taken into account, population distributions similar to that of argon are obtained. Figure 8.10 illustrates the population density

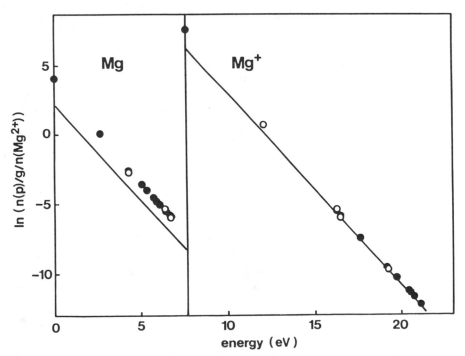

Figure 8.10. Population density distribution for magnesium: $T_e = 8400$ K, $n_e = 1.4 \times 10^{15}$ cm^{-3}, observation height 15 mm, radial position $= 0$ mm. Lines indicate LTE values; open points are measured values normalized to the calculated value for Mg$^+$(3p); solid points are calculated values. (From Reference 14, with permission.)

distributions of Mg neutral and ionic levels. In the case of analyte atoms, the solution of rate equations (Equations 8.53 and 8.54) is given as the concentration ratios to one standard level. The results in Figure 8.10 show the population relative to the ground state of Mg^{2+}. Obviously, neutral atoms in all energy levels and ions in the lower levels are overpopulated. Further comparison of reaction rates leads to the conclusion that the main mechanism for analyte ionization is the electron impact process, and the contributions of Penning ionization and charge transfer processes are less than 0.1 and 1%, respectively.

In summary, the collisional–radiative model, which can include other mechanisms proposed previously, seems to be most promising to elucidate the excitation mechanisms in the ICP, but it has not been verified yet. To prove the effectiveness of the collisional–radiative model, it is hoped that the experimental data will be collected for plasma properties, which can be predicted by theoretical calculations such as absolute number densities (see Section 8.4) and effective lifetime[160] of the energy levels.

8.8   CONCLUSIONS

As described in this chapter, the physical and spectroscopic properties of the ICP have been fairly elucidated. In the lower region of the plasma, the ICP has an annular structure; ie, temperatures and electron number densities of the central channel are lower than those of the off-axis zone. This probably is due to two effects; the skin conduction effect, in which the RF current tends to flow on the surface of a conductor,[86] and fast injector gas flow (see Section 8.6). With increasing observation height, however, the annular shape gradually changes to the "bell" shape as a result of the mixing of gases between the central core and surrounding region.

The axial temperature profile in the central channel becomes a maximum around 15 mm above the load coil, where all ionic and atomic lines with higher excitation energies are effectively excited. The "norm" temperatures for ionic lines, which provide the maximum populations of excited levels, are much higher than the usual ICP temperature, so that the ionic species show similar distribution patterns in the plasma, as described in Section 8.4. On the other hand, most of the atomic species, whose "norm" temperatures are within the range of the ICP temperature, reveal their individual behaviors. This is consistently described by the temperature distribution, especially below the usual observation height.

The spectral lines emitted from the ICP are dominantly broadened by the Doppler effect, except for hydrogen and argon lines, which are subject to a large Stark effect (see Section 8.5). Therefore, the shape is close to the purely Gaussian, and its line width is approximately expressed by the Doppler temperature. This means that the ICP emission line causes less interference than the Lorentz-broadened profile, because the Gaussian function rapidly decreases with the distance from the line center. In the case of extremely intense lines, such as ionic resonance lines of alkaline earth elements, however, the line wings broadened by the Lorentz effect severely enhance the background level, even in the wavelength region far away from the line center.[114]

The excitation mechanism in the ICP also has been discussed in detail. The most interesting characteristic of the ICP is the deviation from LTE, especially, the overpopulation of ions and higher excited atomic species, which is observed as the disagreement of temperatures determined by the different methods (see Sections 8.2.3 and 8.7.1). The nonlinear processes, such as Penning ionization and charge transfer, were first postulated as excitation mechanisms to interpret non-LTE phenomena. Recently, recombining plasma and radiation trapping have been introduced with the aid of physical models. However, these concepts can explain only one aspect of non-LTE phenomena with little experimental evidence. On the other hand, the collisional–radiative model can directly consider the rate processes occurring in the plasma and integrate all models described above. The calculated population densities show the overpopulation of atoms and lower ionic levels, which apparently contradict the experimental observations. This discrepancy can be attributed to the differences of temperature as the basis for LTE calculation, ie, excitation temperature (T_{exc}) and electron temperature (T_e). The collisional–radiative model may be most consistent, but it still needs experimental support. The electron temperature is the most important parameter determining the physical condition of the plasma as an excitation source. The precise measurement of T_e is desirable.

REFERENCES

1. P.W.J.M. Boumans, Excitation Spectal, in "Analytical Emission Spectrometry." E. L. Grove, Ed. Marcel Dekker, New York, 1972.
2. A. P. Thorne, "Spectrophysics." Chapman & Hall, London, 1974.
3. D. J. Kalnicky, V. A. Fassel, and R. N. Kniseley, Excitation Temperature and Electron Number Densities Experienced by Analyte Species in Inductively Coupled Plasmas With and Without the Presence of an Easily Ionized Element, *Appl. Spectrosc.* 31, 137-150 (1977).
4. J. Jarosz, J. M. Mermet, and J. P. Robin, A Spectrometric Study of a 40-MHz Inductively Coupled Plasma—III. Temperatures and Electron Number Density, *Spectrochim. Acta* 33B, 55-78 (1978).
5. M. Kubota, Y. Fujishiro, and R. Ishida, Selection of Spectral Lines and Transition Probability Data in Plasma Temperature Measurements by Photoelectric Scanning Spectrometry, *Spectrochim. Acta* 36B, 697-710 (1981).
6. G. R. Kirkbright and M. Sargent, "Atomic Absorption and Fluorescence Spectrometry." Academic Press, London, 1974.
7. G. R. Kornblum and L. de Galan, Spatial Distribution of the Temperature and the Number Densities of Electrons and Atomic and Ionic Species in an Inductively Coupled RF Argon Plasma, *Spectrochim. Acta* 32B, 71-96 (1977).
8. R. S. Houk, H. V. Svec, and V. A. Fassel, Mass Spectrometric Evidence for Superthermal Ionization in an Inductively Coupled Argon Plasma, *Appl. Spectrosc.* 35 380-384 (1981).
9. R. S. Houk, A. Montaser, and V. A. Fassel, Mass Spectra and Ionization Temperatures in an Argon–Nitrogen Inductively Coupled Plasma, *Appl. Spectrosc.* 37, 425-428 (1983).
10. J. F. Alder and J. M. Mermet, A Spectroscopic Study of Some Radio Frequency Mixed Gas Plasma, *Spectrochim. Acta* 28B, 421-433 (1973).
11. T.-Y. Wu, "Kinetic Equations on Gases and Plasmas." Addison-Wesley, Reading, Mass., 1966.

12. H.G.C. Human and R. H. Scott, The Shapes of Spectal Lines Emitted by an Inductively Coupled Plasma, *Spectrochim. Acta* 31B, 459–473 (1976).
13. T. Hasegawa and H. Haraguchi, Physical Line Widths of Atoms and Ions in an Inductively Coupled Plasma and Hollow Cathode Lamps as Measured by an Echelle Monochromator with Wavelength Modulation, *Spectrochim. Acta* 40B, 123–133 (1985).
14. T. Hasegawa, "Basic Studies on Spectroscopic and Physical Characteristics of Argon Plasma as an Excitation Source for Spectrochemical Analysis." Ph.D. thesis, University of Tokyo, 1985.
15. L. M. Faires, B. A. Palmer, and J. M. Brault, Line Width and Line Shape Analysis in the Inductively Coupled Plasma by High-Resolution Fourier Transform Spectrometer, *Spectrochim. Acta* 40B, 135–143 (1985).
16. L. M. Biberman and G. E. Norman, Emission of Recombination Radiation and Bremsstrahlung from a Plasma, *J. Quant. Spectrosc. Radiat. Transfer* 3, 221–245 (1963).
17. D. Hofsaess, Emission Continua of Rare Gas Plasmas, *J. Quant. Spectrosc. Radiat Transfer* 19, 339–352 (1978).
18. T. Hasegawa and H. Haraguchi, A Collisional-Radiative Model Including Radiation Trapping and Transport Phenomena for Diagnostics of an Inductively Coupled Plasma, *Spectrochim. Acta* 40B, 1505–1515 (1985).
19. I. Kleinmann and J. Cajko, Spectrophysical Properties of the Plasma of a High-Frequency Low-Power Discharge in Argon at Atmospheric Pressure, *Spectrochim. Acta* 25B, 657–668 (1970).
20. A. Batal, J. Jarosz, and J. M. Mermet, A Spectroscopic Study of a 40-MHz Inductively Coupled Plasma—VI. Argon Continuum in the Visible Region of the Spectrum, *Spectrochim. Acta* 36B, 983–992 (1981).
21. A. Batal, J. Jarosz, and J. M. Mermet, A Spectrometric Study of a 40-MHz Inductively Coupled Plasma—VII. Continuum of a Binary Mixture in the Visible Region of the Spectrum, *Spectrochim. Acta* 37B, 511–516 (1982).
22. G. J. Bastiaans and R. A. Mangold, The Calculation of the Electron Density and Temperature in Ar Spectroscopic Plasmas from Continuum and Line Spectra, *Spectrochim. Acta* 40B 885–892 (1985).
23. T. Hasegawa and H. Haraguchi, Appreciation of a Collisional-Radiative Model for the Description of an Inductively Coupled Plasma, *Spectrochim. Acta* 40B, 1067–1084 (1985).
24. A. Scheeline and J. P. Walters, Considerations for Implementing Spatially Resolved Spectrometry Using the Abel Inversion, *Anal. Chem.* 48, 1519–1530 (1976).
25. M. W. Blades and G. Horlick, Photodiode Array Measurement System for Implementing Abel Inversions on Emission from an Inductively Coupled Plasma, *Appl. Spectrosc.* 34, 696–699 (1980).
26. B. S. Choi and H. Kim, On Abel Inversions of Emission Data from an Inductively Coupled Plasma, *Appl. Spectrosc.* 36, 71–74 (1982).
27. M. M. Prost, Critical Mathematical Considerations Involving the Abel Integral Equation for Converting Side-on Experimental Data in Plasma Spectroscopy—I. A Study of Basic Analytical Solutions, *Spectrochim. Acta* 37B, 541–570 (1982).
28. M. W. Blades, Asymmetric Abel Inversions on Inductively Coupled Plasma Spatial Emission Profiles Collected from a Photodiode Array, *Appl. Spectrosc.* 37, 371–375 (1983).
29. W. B. Barnett, V. A. Fassel, and R. N. Kniseley, An Experimental Study of Internal Standardization in Analytical Emission Spectroscopy, *Spectrochim. Acta* 25B, 139–161 (1970).
30. G. R. Kornblum and L. de Galan, Arrangement for Measuring Spatial Distributions in an Argon Induction-Coupled RF Plasma, *Spectrochim. Acta* 29B, 249–261 (1974).
31. J. M. Mermet, Comparaison des températures et des densités électroniques mésurées sur le gaz plasmagène et dur des elements éxcités dans un plasma haute frèquence, *Spectrochim. Acta* 30B, 383–396 (1975).
32. D. J. Kalnicky, R. N. Kniseley, and V. A. Fassel, Inductively Coupled Plasma-Optical Emission Spectroscopy: Excitation Temperature Experienced by Analytical Species, *Spectrochim. Acta* 30B, 511–525 (1975).
33. K. Visser, F. M. Hamm, and P. B. Zeeman, Temperature Determination in an Inductively Coupled RF Plasma, *Appl. Spectrosc.* 30, 34–38 (1976).
34. P.W.J.M. Boumans and F. J. de Boer, Studies of a Radio Frequency Inductively Coupled Argon Plasma for Optical Emission Spectrometry—III. Interference Effects under Compromise Conditions for Simultaneous Multielement Analysis, *Spectrochim. Acta* 31B, 355–375 (1976).
35. P. E. Walters, T. L. Chester, and J. D. Winefordner, Measurement of Excitation, Ionization and

Electron Temperatures and Positive Ion Concentrations in a 144-MHz Inductively Coupled Radio Frequency Plasma, *Appl. Spectrosc.* 31, 1–8 (1977).

36. J. M. Mermet and T. C. Trassy, Étude de transferts d'excitation dans un plasma induit par haute fréquence entre gaz plasmagène et elements introduits, *Rev. Phys. Appl.* 12, 1219–1222 (1977).

37. G. R. Kornblum and L. de Galan, A Study of the Interference of Cesium and Phosphate in the Low-Power Inductively Coupled Radio Frequency Argon Plasma Using Spatially Resolved Emission and Absorption Measurements, *Spectrochim. Acta* 32B, 455–478 (1977).

38. M. H. Abdallah and J. M. Mermet, The Behavior of Nitrogen Excited in an Inductively Coupled Argon Plasma, *J. Quant. Spectrosc. Radiat. Transfer* 19, 83–91 (1978).

39. J. F. Alder, R. M. Bombelka, and G. F. Kirkbright, Electronic Excitation and Ionization Temperature Measurements in a High-Frequency Inductively Coupled Plasma Source and the Influence of Water Vapor on Plasma Parameters, *Spectrochim. Acta* 35B, 163–175 (1980).

40. R. M. Barnes and R. G. Schleicher, Temperature and Velocity Distributions in an Inductively Coupled Plasma, *Spectrochim. Acta* 36B, 81–101 (1981).

41. R. M. Barnes and J. L. Genna, Gas Flow Dynamics of an Inductively Coupled Plasma Discharge, *Spectrochim. Acta* 36B, 299–323 (1981).

42. H. Kawaguchi, T. Ito, and A. Mizuike, Axial Profiles of Excitation and Gas Temperatures in an Inductively Coupled Plasma, *Spectrochim. Acta* 36B, 615–623 (1981).

43. H. Uchida, K. Tanabe, Y. Nojiri, H. Haraguchi, and K. Fuwa, Spatial Distributions of Metastable Argon, Temperature and Electron Number Density in an Inductively Coupled Argon Plasma, *Spectrochim. Acta* 36B, 711–718 (1981).

44. J. P. Rybarczk, C. P. Jester, D. A. Yates, and S. R. Koirtyohann, Spatial Profiles of Interelement Effects in the Inductively Coupled Plasma, *Anal. Chem.* 54, 2162–2170 (1982).

45. B. Cappelle, J. M. Mermet, and J. P. Robin, Influence of the Generator Frequency on the Spectral Characteristics of Inductively Coupled Plasma, *Appl. Spectrosc.* 36, 102–106 (1982).

46. J. E. Roederer, G. J. Bastiaans, M. A. Fernandez, and K. J. Fredeen, Spatial Distribution of Interference Effects in ICP Emission Analysis, *Appl. Spectrosc.* 36, 383–389 (1982).

47. N. Furuta and G. Horlick, Spatial Characterization of Analyte Emission and Excitation Temperature in an Inductively Coupled Plasma, *Spectrochim. Acta* 37B, 53–64 (1982).

48. G. R. Kornblum and J. Smeyers-Verbeke, Some Comments upon the Behavior of the Excitation Temperature in the ICP, *Spectrochim. Acta* 37B, 83–87 (1982).

49. M. H. Abdallah and J. M. Mermet, Comparison of Temperatures in ICP and MIP with Ar and He as Plasma Gas, *Spectrochim. Acta* 37B, 391–397 (1982).

50. W. H. Gunter, K. Visser, and P. B. Zeeman, Some Aspects of Matrix Interference Caused by Elements of Low Ionization Potential in Inductively Coupled Plasma Atomic Emission Spectrometry, *Spectrochim. Acta* 37B, 571–581 (1982).

51. Y. Nojiri, K. Tanabe, H. Uchida, H. Haraguchi, K. Fuwa, and J. D. Winefordner, Comparison of Spatial Distributions of Argon Species Number Densities with Calcium Atom and Ion in an Inductively Coupled Argon Plasma, *Spectrochim. Acta* 38B, 61–74 (1983).

52. J. W. Carr and M. W. Blades, Emission Spectra of an Argon Inductively Coupled Plasma in the Vacuum Ultraviolet: Background Spectra from 85 to 200 nm, *Spectrochim. Acta* 39B, 567–574 (1984).

53. L. M. Faires, B. A. Palmer, R. Engleman, Jr., and T. M. Niemczyk, Temperature Determinations in the Inductively Coupled Plasma Using a Fourier Transform Spectrometer, *Spectrochim. Acta* 39B, 819–828 (1984).

54. M. W. Blades and B. L. Caughlin, Excitation Temperature and Electron Density in the Inductively Coupled Plasma—Aqueous vs. Organic Introduction, *Spectrochim. Acta* 40B, 579–591 (1985).

55. B. Talayrach, J. Besombes-Vailhe, and H. Triche, Spectrographic Analysis of Boron Monoxide Bands and Calculation of Rotation Temperature from One of These Bands during Titanium Diboride Formation in a Plasma Torch, *Analysis* 1, 135–139 (1972).

56. P. B. Zeeman, S. P. Terblanche, K. Visser, and F. H. Hamm, Temperature Determination on a 9.2-MHz Inductively Coupled Plasma Source using N_2^+ Bands as Monitor, *Appl. Spectrosc.* 32, 572–576 (1978).

57. D. H. Trassy and S. A. Myers, Absolute Spectral Radiance of 27-MHz Inductively Coupled Argon Plasma Background Emission, *Spectrochim. Acta* 37B, 1055–1068 (1982).

58. D. C. Schram, I.J.M.M. Raaymakers, B. van der Sijde, H.J.W. Schenkelaars, and P.W.J.M. Boumans, Approaches for Clarifying Excitation Mechanisms in Spectrochemical Sources, *Spectrochim. Acta* 38B, 1545–1557 (1983).

59. V. M. Goldfarb and H. V. Goldfarb, ICP-AES Analysis of Gases in Energy Technology and Influence of Molecular Additives on Argon ICP, *Spectrochim. Acta* 40B, 177-194 (1985).
60. (a) A. Scheeline and M. J. Zoellner, Thomson Scattering as a Diagnostic of Atmospheric Pressure Discharge, *Appl. Spectrosc.* 38, 245-258 (1984). (b) M. Huang and G. M Hieftje, Thomson Scattering from an ICP, *Spectrochim. Acta* 40B, 1387-1400 (1985). (c) M. Huang, K. A. Marshall, and G. M. Hieftje, Electron Temperature and Electron Number Densities Measured by Thomson Scattering in the Inductively Coupled Plasma, *Anal. Chem.* 58, 207-210 (1986).
61. H. R. Griem, "Plasma Spectroscopy." McGraw-Hill, New York, 1964.
62. W. L. Wiese, Line Broadening, in "Plasma Diagnostic Techniques," R. H. Huddlestone and S. L. Leonard, Eds. Academic Press, New York, 1965.
63. H. R. Griem, "Spectral Line Broadening by Plasmas." Academic Press, New York, 1974.
64. P. Kepple and H. R. Griem, Improved Stark Profile Calculations for the Hydrogen Lines H_α, H_β, H_γ and H_δ, *Phys. Rev.* 173, 317-325 (1968).
65. R. K. Winge, V. J. Peterson, and V. A. Fassel, Inductively Coupled Plasma-Atomic Emission Spectroscopy: Prominent Lines, *Appl. Spectrosc.* 33, 206-219 (1979).
66. D. R. Inglis and E. Teller, Ionic Depression of Series Limits on One-Electron Spectra, *Astrophys. J.* 90, 439-448 (1939).
67. A. Montaser, V. A. Fassel, and G. Larsen, Electron Number Densities in Analytical Inductively Coupled Plasmas as Determined via Series Limit Line Merging, *Appl. Spectrosc.* 35, 385-389 (1981).
68. A. Montaser and V. A. Fassel, Electron Number Density Measurements in Ar and Ar-N_2 Inductively Coupled Plasmas, *Appl. Spectrosc.* 36, 613-617 (1982).
69. N. Furuta, Y. Nojiri, and K. Fuwa, Spatial Profile Measurement of Electron Number Densities and Analyte Line Intensities in an Inductively Coupled Plasma, *Spectrochim. Acta* 40B, 423-434 (1985).
70. B. L. Caughlin and M. W. Blades, Spatial Profiles of Electron Density in the Inductively Coupled Plasma, *Spectrochim. Acta* 40B, 987-993 (1985).
71. S. R. Goodes and J. P. Deavor, Determination of Electron Density in an Atomic Plasma by Least-Squares Fit to the Stark Profile, *Spectrochim. Acta* 39B, 813-818 (1984).
72. P.W.J.M. Boumans and F. J. de Boer An Experimental Study of a 1-kW, 50-MHz RF Inductively Coupled Plasma with Pneumatic Nebulizer, and a Discussion of Experimental Evidence for a Nonthermal Mechanism, *Spectrochim. Acta* 32B, 365-395 (1977).
73. I. Rubeska, Chemical Interferences in the Vapor Phase, in "Flame Emission and Atomic Absorption Spectrometry," J. A. Dean and T. C. Raines, Eds. Marcel Dekker, New York, 1969.
74. T. E. Edmonds and G. Horlick, Spatial Profiles of Emission from an Inductively Coupled Plasma Source Using a Self-Scanning Photodiode Array, *Appl. Spectrosc.* 31, 536-541 (1977).
75. G. Horlick and M. W. Blades, Clarification of Some Analyte Emission Characteristics of the Inductively Coupled Plasma Using Emission Spatial Profiles, *Appl. Spectrosc.* 34, 229-233 (1980).
76. M. W. Blades and G. Horlick, The Vertical Spatial Characteristics of Analyte Emission in the Inductively Coupled Plasma, *Spectrochim. Acta* 36B, 861-880 (1981).
77. M. W. Blades and G. Horlick, Interference from Easily Ionizable Element Matrices in Inductively Coupled Plasma Emission Spectrometry—A Spatial Study, *Spectrochim. Acta* 36B, 881-900 (1981).
78. G. Horlick and N. Furuta, Spectrographic Observation of the Spatial Emission Structure of the Inductively Coupled Plasma, *Spectrochim. Acta* 37B, 999-1008 (1982).
79. P.W.J.M. Boumans and F. J. de Boer, Studies of Flame and Plasma Torch Emission for Simultaneous Multielement Analysis—I. Preliminary Investigations, *Spectrochim. Acta* 27B, 391-414 (1972).
80. C. D. Allemand and R. M. Barnes, A Study of Inductively Coupled Plasma Torch Configurations, *Appl. Spectrosc.* 31, 434-443 (1977).
81. J. Robin, Emission Spectrometry with the Aid of an Inductive Plasma Generator, *ICP In. Newsl.* 4, 495-509 (1979).
82. J. Mostaghimi, P. Proulx, M. I. Boulos, and R. M. Barnes, Computer Modeling of the Emission Patterns for a Spectrochemical ICP, *Spectrochim. Acta* 40B, 153-166 (1985).
83. K. Niebergall, H. Brauer, and K. Dittrich, Equidensitometry—A Method for Plasma Diagnostics—XIII. Investigations of the Spatial Emission Distribution of Different Spectral Lines in an ICP, *Spectrochim. Acta* 39B, 1225-1237 (1984).

84. K. Dittrich and K. Niebergall, Equidensitometry—A Method for Plasma Diagnostics in Atomic Spectroscopy, *Prog. Anal. Atom. Spectrosc.* 7, 315–372 (1984).
85. V. A. Fassel and R. N. Kniseley, Inductively Coupled Plasma-Optical Emission Spectroscopy, *Anal. Chem.* 46, 1110A–1120A (1974).
86. P.W.J.M. Boumans, Inductively Coupled Plasma–Atomic Emission Spectroscopy: Its Present and Future Position in Analytical Chemistry, *Z. Anal. Chem.* 229, 337–361 (1979).
87. G. F. Larson, V. A. Fassel, R. H. Scott, and R. N. Kniseley, Inductively Coupled Plasma-Optical Emission Analytical Spectrometry: A Study of Some Interelement Effects, *Anal. Chem.* 47, 238–243 (1975).
88. H. Kawaguchi, T. Ito, K. Ota, and A. Mizuike, Effect of Matrix on Spatial Profiles of Emission from an Inductively Coupled Plasma, *Spectrochim. Acta* 35B, 199–206 (1980).
89. S. R. Koirtyohann, J. S. Jones, C. P. Jester, and D. A. Yates, Use of Spatial Emission Profiles and a Nomenclature System as Aids in Interpreting Matrix Effects in the Low-Power Argon Inductively Coupled Plasma, *Spectrochim. Acta* 36B, 49–59 (1981).
90. R. Rezaaiyaan, J. W. Olesik, and G. M. Hieftje, Interferences in a Low-Flow, Low-Power Inductively Coupled Plasma, *Spectrochim. Acta* 40B, 73–83 (1985).
91. H. U. Eckert and A. Danielsson, An Equilibrium Model for the Radial Distribution of Analyte Lines in the ICP Discharge, *Spectrochim. Acta* 38B, 15–27 (1983).
92. R. S. Houk and J. A. Olivares, General Calculations of Vertically Resolved Emission Profiles for Analyte Elements in Inductively Coupled Plasmas, *Spectrochim. Acta* 39B, 575–587 (1984).
93. P. E. Walters, G. L. Long, and J. D. Winefordner, Spatially Resolved Concentration Studies of Ground State Atoms and Ions in an ICP: Saturated Absorption Spectroscopic Method, *Spectrochim. Acta* 39B, 69–76 (1984).
94. S. W. Downey, G. L. Keaton, and N. S. Nogar, Spatially Resolved, Intracavity Absorption for Inductively Coupled Plasma Diagnostics, *Spectrochim. Acta* 40B, 927–932 (1985).
95. N. Omenetto, S. Nikdel, R. D. Reeves, J. B. Bradshaw, J. N. Bower, and J. D. Winefordner, Relative Spatial Profiles of Barium Ion and Atom in the Argon Inductively Coupled Plasma as Obtained by Laser-Excited Fluorescence, *Spectrochim. Acta* 35B, 507–517 (1980).
96. H. U. Eckert, A Check on the Continuity of Analyte Transport in the Chemical ICP Torch, *Spectrochim. Acta* 40B, 145–151 (1985).
97. V. G. Muradov, Determination of Absolute Atomic Concentration by the Saturated Vapor Total Absorption Method, *Spectrochim. Acta* 38B, 1151–1156 (1983).
98. J. M. Mermet, Excitation Mechanisms of Elements Inducted into a HF (High-Frequency) Argon Plasma, *C.R. Acad. Sci., Ser. B* 281, 273–275 (1975).
99. A. H. Futch and F. A. Grant, Mean Life of the $^3P_2$ Metastable Argon Level, *Phys. Rev.* 104, 356–361 (1956).
100. N. E. Small-Warren and L.-Y. C. Chiu, Lifetime of the Metastable $^3P_2$ and $^3P_0$ States of Rare-Gas Atoms, *Phys. Rev.* A11, 1777–1783 (1975).
101. H. Uchida, K. Tanabe, Y. Nojiri, H. Haraguchi, and K. Fuwa, Measurement of Metastable Argon in an Inductively Coupled Argon Plasma by Atomic Absorption Spectroscopy, *Spectrochim. Acta* 35B, 881–883 (1980).
102. M. W. Blades and G. M. Hieftje, On the Significance of Radiation Trapping in the Inductively Coupled Plasma, *Spectrochim. Acta* 37B, 191–197 (1982).
103. A.C.G. Mitchell and M. W. Zemansky, "Resonance Radiation and Excited Atoms." University Press, Cambridge, 1934.
104. M. Baranger, Spectral Line Broadening in Plasmas in "Atomic and Molecular Processes." D. R. Bates, Ed. Academic Press, New York, 1962.
105. H. M. Foley, The Pressure Broadening of Spectral Lines, *Phys. Rev.* 69, 616–628 (1946).
106. S.-Y. Chen and M. Takeo, Broadening and Shift of Spectral Lines due to the Presence of Foreign Gases, *Rev. Mod. Phys.* 29, 20–73 (1957).
107. R. G. Breene, Jr., Line Shape, *Rev. Mod. Phys.* 29, 94–143 (1957).
108. M. Baranger, Simplified Quantum-Mechanical Theory of Pressure Broadening, *Phys. Rev.* 111, 481–493 (1958).
109. W. R. Hindmarsh, A. D. Petford, and G. Smith, Interpretation of Collision Broadening and Shift on Atomic Spectra, *Proc. R. Soc. London* A297, 296–304 (1967).
110. I. I. Sobelman, L. A. Vainshtein, and E. A. Yukov, "Excitation of Atoms and Broadening of Spectral Lines." Springer-Verlag, Berlin, 1981.
111. G. Herzberg, "Atomic Spectra and Atomic Structure." Dover Publications, New York, 1944.
112. H. Kawaguchi, Y. Oshio, and A. Mizuike, Interferometric Measurements of Spectral Line Widths Emitted by an Inductively Coupled Plasma, *Spectrochim. Acta* 37B, 809–816 (1982).

113. J.A.C. Broekaert, F. Leis, and K. Laqua, Application of an Inductively Coupled Plasma to the Emission Spectroscopic Determination of Rare Earths in Mineralogical Samples, *Spectrochim. Acta* 34B, 73–84 (1979).

114. G. F. Larson and V. A. Fassel, Line Broadening and Radiative Recombination Background Interferences in Inductively Coupled Plasma–Atomic Emission Spectroscopy, *Appl. Spectrosc.* 33, 592–599 (1979).

115. T. Hasegawa, H. Haraguchi, and K. Fuwa, Conventional Measurements of Spectral Line Profiles from Hollow Cathode Lamps and Inductively Coupled Argon Plasma by a Wavelength-Modulation Echelle Monochromator, *Chem. Lett.* 1983, 397–400.

116. P.W.J.M. Boumans and J.J.A.M. Vrakking, High-Resolution Spectroscopy Using an Echelle Spectrometer with Predisperser—I. Characteristics of the Instrument and Approach for Measuring Physical Line Widths in an Inductively Coupled Plasma, *Spectrochim. Acta* 39B, 1239–1260 (1984).

117. P. N. Keliher and C. C. Wohlers, Echelle Grating Spectrometers in Analytical Spectrometry, *Anal. Chem.* 48, 333A–340A (1976).

118. J. M. Mermet and C. Trassy, A Spectrometric Study of a 40-MHz Inductively Coupled Plasma—V. Discussion on Spectral Interferences and Line Intensities, *Spectrochim. Acta* 36B, 269–292 (1981).

119. A. Batal and J. M. Mermet, Calculation of Some Line Profiles in ICP–AES Assuming a van der Waals Potential, *Spectrochim. Acta* 36B, 993–1003 (1981).

120. K. Kato, H. Fukushima, and T. Nakajima, Observation of Spectral Line Profiles Emitted by an Inductively Coupled Plasma—I. On the Wavelength Shift of Spectral Lines, *Spectrochim. Acta* 39B, 979–991 (1984).

121. S. Greenfield, I. L. Jones, and C. T. Berry, High-Pressure Plasmas as Spectroscopic Emission Sources, *Analyst* 89, 713–720 (1964).

122. R. H. Wendt and V. A. Fassel, Induction-Coupled Plasma Spectrometric Excitation Source, *Anal. Chem.* 37, 920–922 (1965).

123. R. N. Savage and G. M. Hieftje, Development and Characterization of a Miniature Inductively Coupled Plasma Source for Atomic Emission Spectrometry, *Anal. Chem.* 51, 408–413 (1979).

124. E. Sexton, R. N. Savage, and G. M. Hieftje, Hydrodynamic Flow Patterns as a Simple Aid to Effective Inductively Coupled Plasma Torch Design, *Appl. Spectrosc.* 33, 643–646 (1979).

125. T. B. Reed, Induction-Coupled Plasma Torch, *J. Appl. Phys.* 32, 821–824 (1961).

126. T. B. Reed, Growth of Refractory Crystals Using the Induction Plasma Torch, *J. Appl. Phys.* 32, 2534–2535 (1961).

127. R. M. Barnes and S. Nikdel, Mixing of Nitrogen in a Flow Argon Inductively Coupled Plasma Discharge, *Appl. Spectrosc.* 29, 477–481 (1975).

128. R. M. Barnes and R. G. Schleicher, Computer Simulation of RF Induction-Heated Argon Plasma Discharges at Atmospheric Pressure for Spectrochemical Analysis—I. Preliminary Investigations, *Spectrochim. Acta* 30B, 109–134 (1975).

129. R. M. Barnes and S. Nikdel, Temperature and Velocity Profiles and Energy Balances for an Inductively Coupled Plasma Discharge in Nitrogen, *J. Appl. Phys.* 47, 3929–3934 (1976); Computer Simulation of Inductively Coupled Plasma Discharge for Spectrochemical Analysis—II. Comparison of Temperature and Velocity Profiles, and Particle Decomposition for Inductively Coupled Plasma Discharges in Argon and Nitrogen, *Appl. Spectrosc.* 30, 310–320 (1976).

130. R. M. Barnes, N. Kovacic, and G. A. Meyer, Computer Simulation of a Nitrogen ICP Discharge, *Spectrochim. Acta* 40B, 907–918 (1985).

131. M. I. Boulos, The Inductively Coupled RF (Radio Frequency) Plasma, *Pure Appl. Chem.* 57, 1321–1352 (1985).

132. R. C. Miller and R. J. Ayen, Temperature Profiles and Energy Balances for an Inductively Coupled Plasma Torch, *J. Appl. Phys.* 40, 5260–5273 (1969).

133. J. P. Robin, ICP–AES at the Beginning of the Eighties, *Prog. Anal. At. Spectrosc.* 5, 79–110 (1982).

134. L. de Galan, Some Considerations on the Excitation Mechanism in the Inductively Coupled Argon Plasma, *Spectrochim. Acta* 39B, 537–550 (1984).

135. D. R. Bates, A. E. Kingston, and R.W.P. McWhirter, Recombination between Electrons and Atomic Ions—I. Optically Thin Plasmas, *Proc. R. Soc. London* A267, 297–312 (1962).

136. R.W.P. McWhirter and A. G. Hearn, A Calculation of the Instantaneous Population Densities of the Excited Levels of Hydrogenlike Ions in a Plasma, *Proc. Phys. Soc.* 82, 641–654 (1963).

137. T. Fujimoto, Kinetics of Ionization-Recombination of a Plasma and Population Density of Excited Ions.—I. Equilibrium Plasma, *J. Phys. Soc. Japan* 47, 265–272 (1979).

138. T. Fujimoto, Kinetics of Ionization–Recombination of a Plasma and Population Density of Excited Ions.—II. Ionizing Plasma, *J. Phys. Soc. Japan* 47, 273–281 (1979).
139. T. Fujimoto, Kinetics of Ionization–Recombination of a Plasma and Population Density of Excited Ions.—III. Recombining Plasma—High-Temperature Case, *J. Phys. Soc. Japan* 49, 1561–1568 (1980).
140. T. Fujimoto, Kinetics of Ionization–Recombination of a Plasma and Population Density of Excited Ions—IV. Recombining Plasma—Low-Temperature Case, *J. Phys. Soc. Japan* 49, 1569–1576 (1980).
141. P.W.J.M. Boumans, Comment on a Proposed Excited Mechanism in Argon ICPs, *Spectrochim. Acta* 37B, 75–82 (1982).
142. I.J.M.M. Raaymakers, P.W.J.M. Boumans, B. van der Sijde, and D. C. Schram, A Theoretical Study and Experimental Investigation of non-LTE Phenomena in an Inductively-Coupled Argon Plasma—I. Characterization of the Discharge, *Spectrochim. Acta* 38B, 697–706 (1983).
143. F. Aeschbach, Evaluation eines Elektronendiffusionsmodelles zur Berechnung von nicht Gleichwichts-Electronenkonzentrationen im induktiv gekoppelten Argonplasma für die spektrochemische Analyse, *Spectrochim. Acta* 37B, 987–998 (1982).
144. T. Holstein, Imprisonment of Resonance Radiation in Gases—II, *Phys. Rev.* 83, 1159–1168 (1951).
145. W. L. Wiese, M. W. Smith, and B. M. Miles, "Atomic Transition Probabilities." NSRDS–NBS 22. National Bureau of Standards, Washington, D.C., 1969.
146. K. Katsonis and H. W. Drawin, Transition Probabilities for Argon (I), *J. Quant. Spectrosc. Radiat. Transfer* 23, 1–55 (1980).
147. M. W. Blades, Some Considerations Regarding Temperature, Electron Density and Ionization in the Argon Inductively Coupled Plasma, *Spectrochim. Acta* 37B, 869–879 (1982).
148. J. M. Mills and G. M. Hieftje, A Detailed Consideration of Resonance Radiation Trapping in the Argon Inductively Coupled Plasma, *Spectrochim. Acta* 39B, 859–866 (1984).
149. T. Hasegawa, unpublished data.
150. R. J. Ginnaris and F. P. Incropera, Radiative and Collisional Effects in a Cylindrically Confined Plasma—I. Optically Thin Considerations, *J. Quant. Spectrosc. Radiat. Transfer* 13, 167–181 (1973).
151. R. J. Ginnaris and F. P. Incropera, Radiative and Collisional Effects in a Cylindrically Confined Plasma—II. Absorption Effects, *J. Quant. Spectrosc. Radiat. Transfer* 13, 183–195 (1973).
152. B. van der Sijde, H.M.J.Willems, and D. C. Schram, A Numerical Collisional Radiative Model for the Argon Neutral System, *Proc. 16th ICPIG*, Düsseldorf, 1983.
153. T. Hasegawa, K. Fuwa, and H. Haraguchi, A Steady-State Approximation of Population Density Distributions of Argon Atoms in an Inductively Coupled Plasma as an Excitation Source for Atomic Spectrometry, *Chem. Lett.* 1984, 2027–2030.
154. R. J. Lovett, A Rate Model of Inductively Coupled Argon Plasma Analyte Spectra, *Spectrochim. Acta* 37B, 969–985 (1982).
155. G. M. Hieftje, G. D. Rayson, and J. W. Olsik, A Steady-State Approach to Excitation Mechanisms in the ICP, *Spectrochim. Acta* 40B, 167–176 (1985).
156. D. R. Bates, Ed., "Atomic and Molecular Processes." Academic Press, New York, 1964.
157. E. W. McDaniel, "Collision Phenomena in Ionized Gases." John Wiley & Sons, New York, 1964.
158. J. B. Hasted, "Physics of Atomic Collisions." 2nd ed. Butterworths, London, 1972.
159. R. S. Devoto, Transport Coefficients of Partially Ionized Argon, *Phys. Fluids* 10, 354–364 (1967).
160. H. Uchida, M. A. Kosinski, N. Omenetto, and J. D. Winefordner, Studies on Lifetime Measurements and Collisional Processes in an Inductively Coupled Argon Plasma Using Laser-Induced Fluorescence, *Spectrochim. Acta* 39B, 63–68 (1984).

LIST OF SYMBOLS

A	absorbance
A_{pq}	transition probability of transition from p to q
a	a parameter

a_i	a parameter of hyperfine component i
B	rotational constant
B_{qp}	Einstein's B coefficient for transition from q to p
$C_H(n_e, T)$	function of n_e and T; see Equation 8.17
$C_S(n_e, T)$	function of n_e and T; see Equation 8.18
C_6, C_{12}	constant; see Equation 8.35
c	velocity of light
D	constant; see Equation 8.9
D_a	ambipolar diffusion coefficient
d	line shift
E_i	ionization potential
E_p	excitation energy of level p
g_i	statistical weight of ionic ground state
g_p	statistical weight of level p
$g(\lambda, T_e)$	function of λ and T_e; see Equation 8.21
h	Planck's constant
I_{cont}	emission intensity of continuum band
I_{line}	emission intensity of atomic line
I_{pq}	emission intensity of transition from p to q
I_s	emission intensity with perturbing saturation beam
$I(x)$	lateral intensity at distance x
$i(r)$	radial intensity at radius r
K	rotational quantum number
k	Boltzmann's constant
k_{ip}	rate coefficient for three-body recombination to level p
k_{pi}	rate coefficient for electron impact ionization of level p
$k_{pq}(p < q)$	rate coefficient for electron impact excitation from p to q
$k_{qp}(p < q)$	rate coefficient for electron impact de-excitation from q to p
l	path length of light source
M	atomic weight
m_e	mass of electron
n	total number of absorbing species along probe beam path
n_{Ar}^+	number density of argon ion
n_a	total concentration of atom
n_e	electron number density
n_i	total concentration of ion
n_m	principal quantum number of last discernible line
n_p	principal quantum number
$n(p)$	number density of atom in level p
n_s	sum of n_e and number density of ion
$n(1)$	number density of neutral ground state
$P(\lambda)$	line profile
R	gas constant
R_0	radius of plasma source
r	distance from plasma center

T	temperature
T_e	electron temperature
T_{exc}	excitation temperature
T_{ion}	ionization temperature
T_{rot}	rotational temperature
t	time
$V(a, \omega)$	Voigt function
$V_i(a_i, \omega_i)$	Voigt function of hyperfine component i
V_{sum}	overall line profile
$V(\rho)$	electronic interaction potential between two atoms
v	axial gas velocity
x	distance from plasma center along x-axis
$Z_a(T)$	partition function of atom
$Z_i(T)$	partition function of ion
α_p	rate coefficient for radiative recombination to level p
ΔE	uncertainty of energy
Δn	number of absorbing species in intersecting volume
$\Delta\lambda$	wavelength interval
$\Delta\lambda_D$	Doppler width
$\Delta\lambda_H$	Stark width of hydrogen line
$\Delta\lambda_L$	Lorentz width
$\Delta\lambda_N$	natural width
$\Delta\lambda_P$	total pressure width
$\Delta\lambda_R$	Holtzmark (resonance) width
$\Delta\lambda_S$	Stark width
$\Delta\lambda_V$	total halfwidth of Voigt profile
Γ_a	amount of inflow to unit volume by ambipolar diffusion
Γ_c	amount of inflow to unit volume by convection
λ	wavelength
λ_{pq}	wavelength of transition from p to q
λ_0	wavelength of center of line
v	frequency of emission line
ρ	distance between two atoms
ρ_{pq}	radiation density at wavelength λ_{pq}
τ	mean lifetime of energy level
ω	dimensionless parameter; see Equation 8.33
ω_i	dimensionless parameter of hyperfine component i

9

Atomic Fluorescence Spectrometry with the Inductively Coupled Plasma

NICOLÒ OMENETTO

CCR, European Community Center
Chemistry Division
Ispra, Italy

JAMES D. WINEFORDNER

Department of Chemistry
University of Florida
Gainesville, Florida

9.1 INTRODUCTION

Atomic fluorescence spectrometry (AFS) is based on the radiational activation of atoms and ions produced in a suitable atomizer (ionizer) and the subsequent measurement of the resulting radiational deactivation, called fluorescence. Atomic fluorescence spectrometry has been of considerable interest to researchers in atomic spectrometry because of its use for both analytical and diagnostic purposes. Unfortunately, the analytical applications of AFS have suffered from the lack of commercial instrumentation until the recent marketing of the Baird multiple-element, hollow cathode lamp-excited inductively coupled plasma system.

This chapter is concerned strictly with the use of the inductively coupled plasma (ICP) as a cell and as a source for AFS. Many of the major references[1-38] concerning the ICP in analytical AFS are categorized in Table 9.1, along with several reviews and diagnostical studies. For more detailed discussions of the fundamental aspects of AFS, the reader is referred to previous reviews.[34,35]

The extreme versatility of the ICP in AFS should be stressed. The ICP is both an excellent atomization and ionization cell, and an intense source of line and continuum radiation. Consequently, the ICP has been used as a cell for excitation of atoms (ions) by means of electrodeless discharge lamps (EDLs),[1] hollow cathode lamps (HCLs),[12-19] ICPs,[9-11] and dye lasers,[22-31] and as a source of excitation for atoms (ions) in flames,[2,3-6,8] furnaces,[37] and ICPs.[9-11]

Two basic approaches to ICP-AFS are commonly used. In conventional AFS (CAFS), the sample is introduced into the cell [ICP, electrothermal atomizer (ETA), flame] and is excited by means of a suitable source (HCL, EDL, dye laser, or a second ICP containing a high concentration of the desired element). This process is called ICP-CAFS, and is the most commonly used ICP-AFS technique. In the other situation, the sample is introduced into the primary source ICP, and a constant concentration of 1 to 100 μg/mL of an analyte standard is introduced into the atomization cell, which is either an ICP, a flame, or a furnace. The analyte emission from the primary source ICP excites the fluorescence of the analyte in the cell with extremely high spectral selectivity. Resolution of the order of 0.002 nm, the halfwidth of absorption lines in the cell, is commonly obtained. This system is called ICP-RMAFS, ie, ICP-resonance monochromator atomic fluorescence spectrometry. Actually, this system could also be called atomic emission spectrometry with AF detection. The ICP-CAFS method results in detection limits of less than 1 ng/mL, depending on the combination of primary source of excitation, the cell, and the atom line used. The upper concentrations, where linearity has not degraded by more than 5%, are less than 1 μg/mL. The ICP-RMAFS method provides detection limits and upper concentrations about 1000 times higher than those of CAFS. The original analytical application of the resonance monochromator concept was by Sullivan and Walsh.[32]

The ICP has two characteristics that result in its usefulness in ICP-CAFS. First, the ICP is an efficient atomizer and ionizer of virtually all elements and has a low spectral background compared to other thermal sources of about the same temperature. Second, it has a remarkable freedom from chemical interferences. Similarly, the ICP has two main characteristics that result in its usefulness in ICP-RMAFS. First, it produces intense, un-self-reversed atomic and ionic lines for virtually all elements introduced at concentrations of less than 10 to 20 mg/mL. Second, if a suitable line is available in the vicinity of the analyte line, the two-line, single-source method[35] may be used to correct the primary source scattering.

Table 9.1. Experimental Components Used and Types of Study Performed with ICP-Atomic Fluorescence Spectrometry[1-38]

System[a]	Components				Studies				
	Source[b]	Cell[c]	Dispersing device	Electronics	Exp.[d]	Fund.[e]	Appl.[f]	Rev.[g]	Ref.
EDL-ICP-CAFS	EDL	ICP	Grating	Chopper-lock-in	X	—	—	—	1
ICP-FL-CAFS	ICP	FL	?	?	X	X	—	—	2
ICP-FL-AFS	ICP	FL	Grating	Chopper-lock-in	X	X	—	—	3, 33
ICP-FL-AFS	ICP	FL	Grating	Chopper-lock-in	X	—	X	—	4-6
ICP-ETA-AFS	ICP	ETA	Grating	Chopper-lock-in	X	—	—	—	7
ICP-FL-AFS	ICP	FL	Grating	Chopper-lock-in	X	—	—	—	8
ICP-ICP-AFS	ICP	ICP	Grating	Chopper-lock-in	X	—	—	—	9-13
HCL-ICP-CAFS	HCL	ICP	Filter	Gated detector	X	—	X	—	12-14, 17
HCL-ICP-CAFS	HCL	ICP	Filter	Gated detector	X	—	—	—	15, 16, 18-21
L-ICP-CAFS	ADL	ICP	Grating	Lock-in	X	—	—	—	23
L-ICP-CAFS	FDL/NDL	ICP	Grating	Gated detector	X	—	X	—	24
L-ICP-CAFS	EDL	ICP	Grating	Gated detector	X	—	—	—	25, 27
L-ICP-CAFS	NDL	ICP	Grating	Gated detector	X	—	—	—	26
L-ICP-CAFS	EDL	ICP	Grating	Gated detector	—	X	—	—	28
L-ICP-CAFS	EXDL	ICP	Grating	Gated detector	X	X	—	—	29, 31
L-ICP-CAFS	EXDL	ICP	Grating	Gated detector	X	X	—	—	30
ICP-AFS					X	—	—	X	34-36, 38
ICP-AFS					—	—	—	X	37

[a] System symbols: see text

[b] The analyte emission from the primary saura ICP excites the fluorescence of the analyte in the cell with extremely high spectral selectivity. The original analytical application of the resonance monochromator concept was by Sullivan and Walsh.[32] Source Symbols: ICP = inductively coupled plasma, EDL = electrodeless discharge lamp, HCL = hollow cathode lamp (pulsed), FDL = pulsed flashlamp dye laser, NDL = pulsed nitrogen dye laser, ADL = continuous-wave argon ion dye laser, EXDL = pulsed excimer dye laser.

[c] Cell Symbols: FL = flame, ICP = inductively coupled plasma; ETA = electrothermal atomizer.

[d] Paper dealt with experimental aspects, such as calibration curves, analytical figures of merit, and optimization.

[e] Paper dealt with fundamental aspects, such as signal strength, shapes of calibration curves, and scatter curves.

[f] Paper dealt with real applications.

[g] Review paper, even if not specifically related to ICP only.

It should be noted that the ICP is particularly suited to ionic emission and ionic fluorescence, whereas flame cells are primarily useful for atomic fluorescence. Also, HCL and EDL sources are primarily atom emitters, while tunable dye lasers are excellent sources for exciting both atom and ion lines.

References 33 to 38 are reviews of atomic fluorescence spectrometry, and the reader is referred to them for different approaches, more details in some cases, and a historical perspective of AFS.

9.2   THEORY OF ATOMIC FLUORESCENCE

9.2.1   Types of Atomic Fluorescence

The types of atomic fluorescence resulting with single-wavelength excitation, lasers and nonlaser sources, and dual-wavelength excitation are described in Figure 9.1.

9.2.2   Fluorescence Radiance Expressions

The limiting fluorescence expressions[39] for two-level atoms, or ions, for high-, low-, or intermediate-intensity line or continuum sources and for high-, low-, or

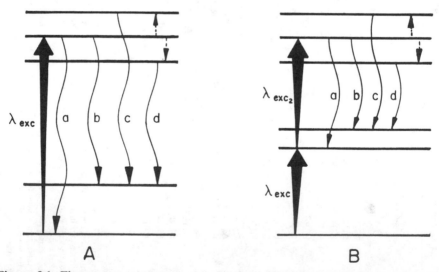

Figure 9.1. Fluorescence processes. (*A*) Single-wavelength excitation: *a*, resonance fluorescence (same wavelengths for excitation and fluorescence); *b*, direct line fluorescence (same upper levels for excitation and fluorescence: if $\lambda_{exc} < \lambda_{fl}$, the fluorescence is called Stokes direct line fluorescence; if $\lambda_{exc} > \lambda_{fl}$, the fluorescence is called anti-Stokes direct line fluorescence); *c, d*, stepwise line fluorescence (different upper levels for excitation and fluorescence: if $\lambda_{exc} < \lambda_{fl}$, the fluorescence is called Stokes stepwise line fluorescence; if $\lambda_{exc} > \lambda_{fl}$, the fluorescence is called anti-Stokes stepwise line fluorescence). (*B*) Dual-wavelength excitation: *a* to *d* correspond to these processes in (*A*).

Table 9.2A. Fluorescence Radiance Expressions for Dominant Cases of Several Source–Cell Combinations:[34,39] Assumptions and Symbols Given in Tables 9.2B and 9.2C

Source of excitation	Cell	Kind and intensity of source	Optical density of absorber	Pertinent expression	Expression no.
Any	Any	Line; high or low	High or low	$B_F = \left(\dfrac{1}{4\pi}\right) Y E_L k_m L \left(\dfrac{E_L^s}{E_L + E_L^s}\right)\left\{1 - \dfrac{1}{2}\bar{k}\left(\dfrac{E_L^s}{E_L + E_L^s}\right)\left(l + L\dfrac{k_m}{k}\right) + \cdots\right\}$	[9.1]
Any	Any	Continuum; high or low	High or low	$B_F = \left(\dfrac{1}{4\pi}\right) Y E_{Cv} L \int k(v)\, dv \left(\dfrac{E_{Cv}^s}{E_{Cv} + E_{Cv}^s}\right)\left\{1 - \dfrac{1}{2}\bar{k}(l + L)\left(\dfrac{E_{Cv}^s}{E_{Cv} + E_{Cv}^s}\right) + \cdots\right\}$	[9.2]
ICP	Flame, ETA ICP	Line; Low	Low	$B_F = \left(\dfrac{1}{4\pi}\right) Y E_L k_m L$	[9.3]
			High	$B_F = \left(\dfrac{1}{4\pi}\right) Y E_L \dfrac{2\sqrt{a}}{(\sqrt{\pi} k_0 l)^{1/2}}$	[9.4]
			Intermediate	$B_F = \left(\dfrac{1}{4\pi}\right) Y E_L k_m L \left\{1 - \dfrac{1}{2}\bar{k}\left(l + L\dfrac{k_m}{k}\right) + \cdots\right\}$	[9.5]
ICP	Flame, ETA ICP	Continuum; low	Low	$B_F = \left(\dfrac{1}{4\pi}\right) Y E_{Cv} L \int k(v)\, dv$	[9.6]
			High	$B_F = \left(\dfrac{1}{4\pi}\right) Y E_{Cv} L a \delta v_{\text{eff}} \left(\dfrac{4L}{\pi l}\right)^{1/2}$	[9.7]
			Intermediate	$B_F = \left(\dfrac{1}{4\pi}\right) Y E_{Cv} L \int k(v)\, dv \{1 - \tfrac{1}{2}\bar{k}(l + L) + \cdots\}$	[9.8]

(*Continued*)

Table 9.2A. (*Continued*)

Source of excitation	Cell	Kind and intensity of source	Optical/density of absorber	Pertinent expression	Expression no.
HCL or EDL	ICP	Line/Low	Low High Intermediate	Same as Eqn. 9.3 Same as Eqn. 9.4 Same as Eqn. 9.5	
Laser	ICP	Line/Low	Low High Intermediate	Same as Eqn. 9.3 Same as Eqn. 9.4 Same as Eqn. 9.5	
Laser	ICP	Continuum/Low	Low High Intermediate	Same as Eqn. 9.6 Same as Eqn. 9.7 Same as Eqn. 9.8	
Laser	ICP	Line/High	Low, Intermediate High[a]	$B_F = \left(\dfrac{1}{4\pi}\right) Y E_L^s k_m L = \left(\dfrac{1}{4\pi}\right) A h\nu_0 L \left(\dfrac{n_T}{2}\right)$	[9.9]
Laser	ICP	Continuum/High	Low, Intermediate High[a]	$B_F = \left(\dfrac{1}{4\pi}\right) Y E_{c\nu}^s L \displaystyle\int k(\nu)\, d\nu = \left(\dfrac{1}{4\pi}\right) h\nu_0 A \left(\dfrac{n_T}{2}\right)$	[9.10]
ICP	ICP, Flame	Continuum/Low	Low[c]/Low[c,e]	$B_F = B_\nu^p(\nu_D, T) k_{0,s} L_s k_{0,c} L_c \delta\nu_{eff}$	[9.11]
		Continuum/Low	High[d]/Low[d,e]	$B_F = B_\nu^p(\nu_0, T) k_{0,c} L_c \delta\nu_{eff}$	[9.12]

[a] If saturation is achieved, self-absorption does not enter in.
[b] This is the RM case. The basic relationship is as follows.

$$B_F = \left\{\int B_\nu^p(\nu_0, T)(1 - e^{-k_s(\nu)L_s})\, d\nu_s\right\}\left\{\int\int [1 - \exp(-k_f(\nu)L_c)]\, d\nu_c\right\}\left\{\frac{\int\int [1 - \exp(-k_f(\nu)L_c)]\, d\nu_c}{l_c \int k_f(\nu)\, d\nu_c}\right\} Y$$

[c] Low/low means low optical density in source and in cell. The absorption profile and coefficient are assumed to be identical in the source and cell.
[d] High/low means high optical density in source and low optical density in cell. Here the peak absorption coefficient in the source is $k_{0,s}$ and in the cell is $k_{0,c}$.
[e] No high optical density cases for the cell are given, since a rather low concentration, 100 μg/mL or below, is introduced into the cell for all **RM** cases.

where the first term is the emission spectral radiance from the primary source ICP near the analyte spectral line, the second term is the equivalent line width (total absorption) of the analyte line in the cell, the third term is the correction for self-absorption of the fluorescence radiation from the cell, and the Y is the fluorescence quantum efficiency.

Table 9.2B. Assumptions Necessary for Table 9.2A

1. All expressions are given for two-level atoms. Although expressions could be given for multilevel atoms, the basic relationships of the fluorescence radiance with analyte number density, source irradiance, and quantum efficiency would not change. In addition, the B_F expressions for multilevel atoms are much more complicated, since they depend in a complex way not only on the radiational rate constants but also on the nonradiational rate constants as well as cell temperature.
2. All expressions assume steady-state conditions, namely, that the source on-time is sufficiently long for steady-state conditions (rate of activation = rate of deactivation) to occur.
3. All expressions are given for radiative excitation only (ie, thermal and chemical excitation processes are assumed to be negligible).
4. Prefilter or postfilter effects are assumed to be absent; namely, the cell is fully illuminated and the resulting fluorescence across the entire illuminated cell is imaged on the spectral measurement system. In other words, only self-absorption needs to be considered.
5. Significant excited-state chemical reactions or ionization processes are assumed not to occur.
6. The absorption coefficient for the excitation process is assumed to be identical to the absorption coefficient for the fluorescence process. Also, the damping constants for the excitation and fluorescence processes are equal.
7. The concentration (number density) of species in the lower level is assumed to be essentially the same as the total concentration of species in all levels (prior to excitation).
8. The statistical weights of the upper and lower levels involved in the absorption and fluorescence transitions are assumed to be equal.

intermediate absorber concentrations are given in Table 9.2A. All assumptions necessary for the validity of the expressions, as well as definitions and units of all symbols, are given in Tables 9.2B and 9.2C. The types of source (ICP, EDL, HCL, or dye laser) and the types of cell (ICP, flame, or ETA) that are applicable to each expression are indicated in Table 9.2A. It is important to note that when HCLs, EDLs, an ICP, or a low-intensity laser is used to excite atomic fluorescence in an ICP cell, such as in the HCL–ICP–AFS system by Baird, only Expressions 3 to 5 of Table 9.2A are applicable for predicting fluorescence radiance.

9.2.3 Signal-to-Noise Considerations

Two major sources of noise are shot and flicker noises. The shot noises are given by the general form:

$$N_{\text{shot}} = (2S_x \, t_m)^{1/2} \qquad [9.13]$$

where S_x is the spectral source of the shot noise (fluorescence, emission, scatter, etc.) and t_m is the measurement time in seconds. The flicker noises are given by the general form:

$$N_{\text{flicker}} = \xi S_x t_m \qquad [9.14]$$

where ξ is the fractional portion of the signal S_x attributed to the flicker noise. No attempt is made to give specific noise expressions or to evaluate them for a

Table 9.2C. List of Symbols Used in Tables 9.2A and 9.2B

Symbol	Definition
a	classical damping constant, $\sqrt{\ln 2}(\delta v_c/\delta v_D)$, dimensionless
A	Einstein coefficient of spontaneous emission, s^{-1}
B	Einstein coefficient of induced absorption (emission), m^3 Hz/J·s
B_F	fluorescence radiance, J/s m^2 sr
$B_v^P(v_0, T)$	Planck blackbody spectral radiance evaluated at frequency v_0 and temperature T, J/s m^2 sr Hz
c	speed of light in vacuum, m/s
E_{Cv}	spectral irradiance of continuum source, J/s m^2
E_{Cv}^s	saturation spectral irradiance, J/s m^2
	$$\frac{1}{2}\left(\frac{cA}{BY}\right) = \left(\frac{4\pi h v_0^3}{c^2}\right)\frac{1}{Y}$$
E_L	irradiance of narrow line source, J/s m^2
E_L^s	saturation irradiance of line source, J/s m^2
	$$\frac{1}{2}\left(\frac{cA}{BY}\right)\delta v_{\text{eff}}$$
h	Planck's constant, J·s
$k(v)$	atomic absorption coefficient at frequency v, m^{-1}
k_c	cell analyte absorption coefficient, m^{-1}
k_m	atomic (average) absorption coefficient at frequency v_0 as measured by a line source, m^{-1}
k_m	$k_0 V(a, \bar{\omega})$
k_0	atomic absorption coefficient at frequency v_0, m^{-1}
	$$\frac{2}{\delta v_0}\left(\frac{\ln 2}{\pi}\right)^{1/2}\left(\frac{\lambda_0^2}{8\pi}\right)n_T A$$
$k_s(v)$	source analyte absorption coefficient, m^{-1}
$\bar{k}$	$\dfrac{\int k^2(v)\,dv}{\int k(v)\,dv}$, m^{-1}
$\int k(v)\,dv$	integrated absorption coefficient, m^{-1}·Hz
	$$\frac{\pi k_0 \delta v_D}{2\sqrt{\ln 2}} = \left(\frac{h v_0}{c}\right)n_T B$$
l	fluorescence path length, m
l_c	cell fluorescence path length, m
L	excitation path length, m
L_c	cell absorption path length, m
L_s	source path length, m
n_T	total concentration (number density) of absorbers, m^{-3}
$V(a, \bar{\omega})$	Voigt factor, dimensionless
	$$\frac{a}{\pi}\int_{-\infty}^{+\infty}\frac{e^{-y^2}\,dy}{a^2 + (\bar{\omega} - y)^2}$$
y	integration variable, dimensionless
Y	quantum (power) efficiency of fluorescence process, ie, number of photons per second emitted to number of photons per second absorbed, dimensionless
$\delta v_{C\text{ or }L}$	collisional (or Lorentzian) halfwidth of absorption line, Hz

Table 9.2C. (*Continued*)

Symbol	Definition
$\delta v_{D \text{ or } G}$	Doppler (or Gaussian) halfwidth of absorption line, Hz
δv_{eff}	effective halfwidth of absorption line, Hz
	$\dfrac{\pi \delta v_L}{2}$ Hz for a Lorentzian-shaped line
	$\dfrac{\sqrt{\pi} \delta v_G}{2\sqrt{\ln 2}}$ Hz for a Gaussian-shaped line
δv_s	source line width, Hz
λ_0	wavelength peak of absorption (fluorescence) line, m
v_0	frequency peak of absorption (fluorescence), line, Hz
$\bar{\omega}$	$\sqrt{\ln 2} \dfrac{\delta v_s}{\delta v_D}$, dimensionless

particular situation. The reader is referred to Epstein and Winefordner[40] and to Omenetto and Winefordner[34,38] for further discussion of the expressions and calculations.

The signal-to-noise ratio S/N in AFS[34] is given by:

$$S/N = \frac{B_F \Omega_F A_0 T_o K_0 t_m}{[N_{FS}^2 + N_{ES}^2 + N_{I_fS}^2 + N_{I_eS}^2 + N_{BS}^2 + N_{SS}^2 + (N_{FF} + N_{SF} + N_{I_fF})^2 \atop + (N_{EF} + N_{I_eF} + N_{BF})^2 + N_E^2 + N_D^2]} \quad [9.15]$$

where B_F = fluorescence radiance, W/m^{2}sr

Ω_F = solid angle of fluorescence collected, sr

A_0 = limiting area of aperture, eg, slit area of monochromator, m^2

T_o = transmittance of optical train, dimensionless

K_0 = conversion factor to account for electronic measurement system, electrical output/J

t_m = measurement time, s

N_{FS} = analyte fluorescence shot noise, same units as numerator of K_0

N_{ES} = analyte emission shot noise, same units as N_{FS}

N_{I_fS} = interferent fluorescence shot noise, same units as N_{FS}

N_{I_eS} = interferent emission shot noise, same units as N_{FS}

N_{SS} = source scatter shot noise, same units as N_{FS}

N_{BS} = background emission shot noise, same units as N_{FS}

N_{FF} = analyte fluorescence flicker noise, same units as N_{FS}

N_{EF} = analyte emission flicker noise, same units as N_{FS}

N_{SF} = source scatter flicker noise, same units as N_{FS}

N_{I_fF} = interferent fluorescence flicker noise, same units as N_{FS}

N_{I_eF} = interferent emission flicker noise, same units as N_{FS}

N_{BF} = background emission flicker noise, same units as N_{FS}

N_E = electronic measurement noise, same units as N_{FS}

N_D = detector noise, same units as N_{FS}

Table 9.3A. Means of Increasing Signal-to-Noise Ratios Through Reduction of Various Noise Sources in Atomic Fluorescence Spectrometry

Means of minimizing noise (see Table 9.3B)	Noise source													
	N_{FS}	N_{ES}	N_{IrS}	N_{IeS}	N_{SS}	N_{BS}	N_{FF}	N_{EF}	N_{IrF}	N_{IeF}	N_{SF}	N_{BF}	N_E	N_D
1. Reduce slit width	0	0	0/+	0/+	0	+	0	0	0/+	0/+	0	+	+/0/−	+/0/−
2. Pulsed source gated detector	+	+	+	+	+	+	0(−)	+	0(−)	+	0(−)	+	+	+
3. Amplitude modulation	0	0	0	0	0	0⁻	0	+	0	+	0	+	+	+
4. Wavelength modulation	0	0	0⁻	0⁻	0	0⁻	0⁻	+	0⁻	+	0⁻	+	+	+
5. Improve nebulizer stability	0	0	0	0	0	0	+	+	+	+	+	+	0	0
6. Improve source stability	0	0	0	0	0	0	+	+	+	+	+	+	0	0
7. Increase measurement time	+	+	+	+	+	+	0	0	0	0	0	0	+	+

Explanation of symbols:
0: no effect on S/N
+: increased S/N
−: decreased S/N
0/+: no effect if interferent is a line overlapping analyte fluorescence line or is within spectral measurement interval, or + increased S/N if interferent is a band overlapping analyte fluorescence line.
+/0/−: S/N may increase, not change, or decrease depending on spectrometer slit width, w, or aperture. As w increases, S/N is enhanced and may reach a plateau or a peak depending on the other noise sources.
Other comments: For the case of shot noises with pulsed source–gated detector, the pulsed source intensity is assumed to exceed the hypothetical continuous (cw) intensity.
0(−): no improvement in S/N, unless pulsing the source seriously degrades its stability.
0⁻: amplitude modulation actually causes a decrease in S/N by a factor of $\sqrt{2}$ since signal is decreased by 2 and noise by $\sqrt{2}$. Also, S/N may degrade if interferent emission or fluorescence and background emission are spectrally broad so that during wavelength modulation, the analyte fluorescence signal is smaller, but the interferent emission and fluorescence and the background emission signals do not change. Also, improved stability is assumed not to change the average signal levels for each process.
Source: References 34 and 40.

Table 9.3B. Means of Minimizing Noise Sources in AFS Systems

AFS system	Means of minimizing noise[a]
Pulsed laser–ICP–CAFS	1, 2, 5, 6, 7
ICP–FL–CAFS/RMAFS	1, 3, 4, 5, 7
ICP–ETA–CAFS/RMAFS	1, 3, 4, 7, and use of platform, matrix modifiers, pyrolyte graphite, good temperature control, and fast temperature ranging
ICP–ICP–CAFS/RMAFS	1, 3, 4, 5, 7
HCL–ICP–CAFS	2, 5, 7

[a] Numbers correspond to items in column 1 of Table 9.3A.

Detection limits for various atomic methods have been calculated for typical analytical conditions.[34,41] In a classic paper, Alkemade[42] discussed the factors and expressions that control the intrinsic detection limits of elements in atomic spectrometry. He concluded that only laser-enhanced atomic (ionic) fluorescence and ionization approaches are capable of single-atom (ion) detection.

The effects of various parameter changes on the signal-to-noise ratios for individual noise types in atomic fluorescence spectrometry are given in Table 9.3A. Table 9.3B lists the means of minimizing noise sources in AFS systems.

9.3   INSTRUMENTS

9.3.1   The Early Systems

The use of an ICP as an atomization cell for AFS was first reported by Montaser and Fassel,[1] who used electrodeless discharge lamps to excite atomic fluorescence from cadmium, mercury, and zinc in plasmas generated in conventional and extended torches operated at 27 MHz. These investigators studied the effects of RF power, observation height, and gas flow rates on the fluorescence signal and signal-to-noise ratios for these three elements and found superior detection limits for the AFS approach over conventional AES. In contrast to the commonly used operating conditions for AES (Table 5.1 in Chapter 5), the optimum conditions for AFS measurements were approximately 0.7 kW forward power, 45 to 65 mm observation height, and 2.4 L/min Ar injector gas flow. For these operating conditions, no significant matrix interference or scattering was observed for the elements studied. As discussed below, similar operating conditions were adopted for a commercial AFS system manufactured by Baird. However, it should be noted that this commercial ICP operates at 40 MHz. The uses of ICP–AFS, prior to 1981, for analysis of metals have been summarized by Omenetto and associates.[43] The ICP has been also used as a source of excitation for atoms (ions) in flame,[2, 3–6, 8] furnaces,[37, 44] and ICPs[9–11] (Figure 9.2A).

The first use of an ICP as a primary excitation source in atomic absorption studies is credited to Greenfield and associates.[45] Hussein and Nickless[2] first

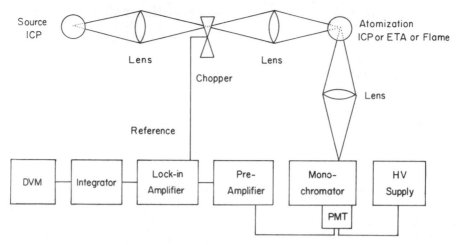

Figure 9.2. (*A*) Block diagram of ICP–excited flame or ETA or ICP–atomic fluorescence spectrometer.

used an ICP as a primary excitation source to excite atomic fluorescence, but the detection limits measured in the flame atomization cell were not impressive in this initial study.

9.3.2 Hollow Cathode Lamp Excited ICP–AFS: The Baird Plasma AFS System

A schematic diagram of the Baird Plasma AFS System[12-21] is shown in Figure 9.2*B*. In this system, 12 pulsed HCLs, arranged circularly around the ICP, are used to excite atomic fluorescence at observation heights optimal for each element. The fluorescence radiation passes through lens–filter combinations before striking the photomultiplier tubes. For the determinations of sulfur and phosphorus in the vacuum UV,[16] argon-flushed modules have been used. To improve detection limits of the elements, especially for refractory elements, an ultrasonic nebulizer has replaced the pneumatic nebulizer used in the initial system. The pulsing of the HCLs is under the control of a microcomputer.

Two points concerning the torch used in this commercial system should be particularly noted. First, the torch has an extended outer tube, and the observation height is located at 1 to 10 cm above the edge of the outer tube. As discussed later, the use of such a large observation height may enhance matrix effects and oxide formation. Second, the graphite tip of the injector tube is located inside the load coil and within the body of the plasma. The presence of the graphite tip within the plasma seems to enhance the stiffness of the discharge, especially at large observation heights.

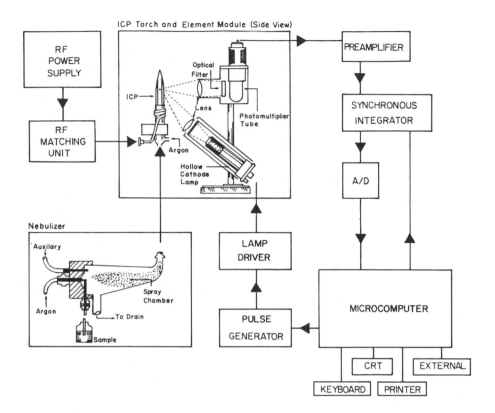

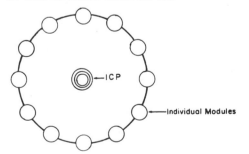

Figure 9.2. (*B*) Block diagram of the Baird Plasma AFS System.

9.3.3   The Inductively Coupled Plasma as a Source of Excitation in Conventional and Resonance Monochromator Atomic Fluorescence Spectrometry

Two modes of operation may be used for an ICP as an excitation source. In both modes, the optical configuration of the systems are essentially the same (see Figure 2A). The atomization cell can be an ICP, a flame, or an ETA. In the conventional fluorescence mode, ICP–CAFS, a selected analyte in solution at a high concentration in the range of 10 to 20 mg/mL is introduced into the ICP source to produce intense atomic and ionic lines of the analyte. These spectral lines are then amplitude modulated prior to excitation of the analyte within the cell. The resulting amplitude-modulated atomic–ionic fluorescence is measured by a photomultiplier-phase-lock-in amplifier system. Because flames and ETAs are primarily atom reservoirs, only fluorescence of atomic species is measured. On the other hand, with an ICP cell, either atomic or ionic fluorescence is measured, depending on sensitivity, detectability, and linear dynamic range.

If the sample is introduced into the source ICP, and a constant concentration of analyte, corresponding to the upper region of linearity of the calibration curve, is introduced into the cell (ICP, flame, ETA), the instrumental system is referred to as an ICP–RMAFS. Again, flames and ETAs are excellent atomizers for nonrefractory elements. The graphite ETA is an efficient atomizer of certain

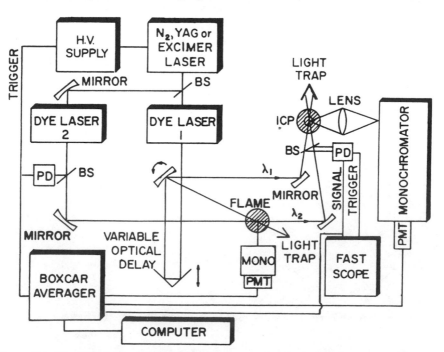

Figure 9.3. Block diagram of an ICP–AFS system using dual-pulsed dye laser. Redrawn from Ref. 30.

refractory elements, as long as carbide formation is minimized, but it exhibits both poor precision (5 to 10% RSD) and matrix errors, unless suitable atomization surfaces, heating rates, and matrix modifiers are used. On the other hand, the ICP cell has much greater freedom from matrix interferences, at the expense of higher background emission.

9.3.4 Dye-Laser-Excited ICP–AFS

A single dye laser can be used to excite atomic fluorescence from an ICP.[22–29] Recently, Omenetto and co-workers[30] demonstrated the advantages of a second dye laser for dual-wavelength excitation (Figure 9.3). The extremely high spectral selectivity achieved, even by single dye-laser-excited ICP–AFS, and the excellent detectability demonstrated, should make this type of AFS measurement one of the most powerful analytical techniques presently available.

9.4   ANALYTICAL FIGURES OF MERIT

9.4.1   Detection Limits

Detection limits for a number of elements obtained by ICP-excited flame CAFS, ICP-excited flame RMAFS, ICP-excited ETA–CAFS, ICP-excited ICP–CAFS, and ICP-excited ICP-RMAFS are listed in Table 9.4. In Table 9.5, the best ICP-AFS results are compared with detection limits from ICP-AES, where pneumatic nebulization and electrothermal vaporization (ETV) of analytes are used to introduce sample into the ICP. For comparison, detection limits are also listed for flame-AAS and ETA-AAS. One should note that the ICP-ETA-CAFS approach gave the lowest detection limits for several elements (Ag, Cd, Cu, Mg, and Zn) and that ICP-ICP-CAFS provides detection limits similar to those achieved for ICP-AES. The HCL-ICP-CAFS and ICP-flame-CAFS gave detection limits for nonrefractory elements that are comparable to or better than those for ICP-AES. For the refractory elements, the opposite trend is observed because the atomization efficiencies in a plasma generated in an extended torch and in the flame are poor, and also, atomic lines, rather than ionic lines, are used.

9.4.2   Spectral Selectivity

Possible interferences in the AFS methods are listed in Table 9.6. The cause of the interferences, the effect on the analytical signals, and the means of minimization are also given. The effect of a variety of interferences on the noise and the signal-to-noise ratio were summarized in Table 9.3. In general, vaporization-atomization interferences are minimal in ICP cells, compared to flames and ETAs.

Table 9.4. Detection Limits for ICP–Atomic Fluorescence Spectrometry with Flame, ETA, and ICP Atomization Cells

	Detection limits (ng/mL)						
	ICP–Flame				ICP–ETA	ICP–ICP	
	CAFS			RMAFS	CAFS	CAFS	RMAFS
Element	Ref. 43	Ref. 8	Ref. 3	Ref. 33	Ref. 44	Ref. 11	Ref. 11
Ag	3	8	—	8000/2000	0.04	—	—
Al	—	—	1000	—	—	10	—
As	—	—	5000	—	—	—	—
B	—	—	—	—	—	10	—
Ba	—	—	—	—	—	0.9	1000
Be	—	10	—	—	—	—	—
Ca	—	20	4	5000/1000	—	0.4	1000
Cd	0.7	6	0.8	1000/700	0.01	—	—
Co	—	20	11	10000/5000	—	—	—
Cr	—	30	2	15000/7000	20	10	2000
Cu	4	8	2	10000/2000	0.02	0.4	3000
Fe	—	30	6	6000/6000	—	—	—
Hf	—	—	—	—	—	30	—
Ho	—	—	—	—	—	10	—
K	—	—	—	—	—	100	—
Mg	0.5	4	0.09	2000/400	0.001	0.2	3000
Mn	5	8	2	2000/900	—	—	—
Mo	—	—	400	—	—	—	—
Na	—	—	—	—		1	20000
Ni	60	50	—	20000/3000	—	—	—
P	—	—	—	—	—	80	—
Pb	—	40	800	—	—	—	—
Pd	4000	4000	—	60000/15000	—	—	—
Pt	—	—	—	—	—	30	—
Sb	—	800	—	—	—	—	—
Sc	—	20	—	—	—	—	—
Se	—	400	—	—	—	—	—
Si	—	—		—	—	7	—
Sn	—	—	—	—	—	20	—
Sn	—	1000	—	—	—	—	—
Sr	—	—	—	—	—	0.2	—
Te	—	100	—	—	—	—	—
Th	—	—	—	—	—	100	—
Tl	—	90	—	—	—	—	—
V	—	—	400	—	—	40	—
Y	—	—	—	—	—	20	—
Yb	—	—	—	—	—	10	—
Zn	0.7	6	0.5	500/700	0.003	2	4000
Zr	—	—	—	—	—	10	—

Provided that saturation broadening is not important, spectral selectivity is greatly enhanced by the use of dye laser excitation, in which a single, narrow spectral line selectively excites a given atom or ion in the ICP. The prominent excitation and fluorescence lines used in laser-excited ICP–AFS are given in Table 9.7, along with the type of species, atomic (I) and ionic (II), used in such studies.

9.4.3   Advantages and Limitations of Laser Excitation

Characteristics of major excitation sources for atomic fluorescence studies are presented in Table 9.8. The capital cost and the operating costs of laser systems, especially those suitable for frequency doubling, are high compared to nonlaser excitation sources. However, lasers are less stable and more difficult to use than sources such as the HCL and EDL, but the intensity of laser radiation is 10^6 to 10^{12} times higher than the intensity of conventional sources. The steady-state spectral (pseudocontinuum source) irradiance and irradiance (line source) required to optically saturate hypothetical two-level atomic or ionic lines at 200, 300, 400, 500, 600, 700, and 800 nm are listed in Table 9.9 in terms of the fluorescence quantum efficiency, Y. The steady-state saturation spectral irradiance is the value of the source spectral irradiance E_λ^s, or source irradiance E^s needed to give a fluorescence radiance equal to 50% of the maximum possible value. Clearly, the nonlaser sources will not optically saturate any level involved in transitions between 200 and 800 nm, except possibly at 800 nm if $Y = 1$. On the other hand, nearly all pulsed dye-lasers will cause saturation in the visible (fundamental) region of 360 to 800 nm, as long as Y exceeds 0.01. However, in the UV (doubled) region of 210 to 360 nm, only the Nd:YAG dye laser and the excimer dye laser will saturate transitions, as long as Y exceeds 0.05.

9.4.4   Precision

The precision of analysis for AFS depends greatly on the source–cell combination. With dye laser excitation, the %RSD is seldom less than 5. With ICP or HCL excitation of an ICP, or ICP excitation of a flame, the %RSD is less than 1. With ICP-excited ETA–AFS, the precision is 5 to 10% RSD, characteristic of ETAs.

9.4.5   Linearity of Analytical Calibration Curves

The linearity of analytical calibration curves, linear dynamic range (LDR) in ICP–AFS is generally four to seven orders of magnitude and is comparable to ICP–AES and ICP–MS (Chapters 5 and 10). Typical LDRs are summarized in Section 9.8.

Table 9.5. Detection Limits (ng/mL) for Aqueous Solutions by Atomic Emission, Atomic Absorption, and Atomic Fluorescence Spectrometries

| Element | ICP-AES[a] | | AAS | | HCL-ICP-CAFS[d] | ICP-FL-CAFS[e] | ICP-AFS | | | |
	Pneum.	ETV	Flame[b]	ETA[c]			ICP-ETA-CAFS[f]	ICP-ICP-AFS[g]	ICP-CAFS[h]
Ag	2	30	2	0.05	1	3	0.04		1
Al	30	60	30	0.4	40	1000		10	0.4
As	300		300	0.4		5000			
Au	10		10	0.2	20				11
B	500		500	50				10	4
Ba	20	70	20	0.8				0.9	0.7
Be	1		1	0.04	0.8[i]	10			
Ca	1	60	1	0.03	0.4	4		0.4	0.007
Cd	1	6	0.2	0.5	0.7	0.7	0.01		
Co	5	9	5	0.2	3	11			
Cr	6	4	6	0.1	8	2	20	10	
Cu	3	10	3	0.1	1	2	0.02	0.4	
Fe	6	2	6	0.1	10	6			
Ga	80		80	1					
Hf	2000		2000					30	1
Ho	40		40					10	16
In	40		40	1	20				
Ir	500		500					58	
Mg	0.3	0.1	0.3	0.01	0.5	0.09	0.001	0.2	0.05
Mn	2	0.4	2	0.03	3	2			
Mo	20		20	0.4	100	400			5
Na	0.2		0.2	0.01	0.3			1	
Nb	2		2000					11	
Ni	10	90	10	0.5	5	50			
P	40000		40000	200				80	

Pb	10	40	10	0.1	70	40			1
Pd	10		10	0.6	20	4000		30	4
Pt	100		100	4	60				
Rh	5		5		3				
Ru	80	160	80						34
Sb	40		40	0.5		800			
Sc	50		50			20			
Se	500		500	1	100	400			1
Si	250	80	250	1	200			7	1
Sm	1000		1000					20	3
Sn	100		100	1	200[i]	1000			1
Sr	2		2	0.1	2			0.2	
Ta	2000		2000					20	
Th								100	
Ti	80		80	2	300[i]				1
Te	30	86	30	1		100			
Tl	20		20	1	20	90			
V	70		70	1	100[i]	400		40	3
W	1000		1000		700[i]				
Y	200		200		500			20	0.6
Zn	0.8	6	0.8	0.01	0.4	0.5	0.003	2	
Zr	1		1000					10	7

[a] Values for pneumatic nebulizer/ICP-AES are from Perkin-Elmer literature on ICP/6500 System. Values for ETV-AES are from A. Aziz et al, *Spectrochim. Acta* 37B, 369 (1982) and H. M. Swaldan and G. D. Christian, *Anal. Chem.* 56, 120 (1984).
[b] Values are from Varian literature on AA-1275 and AA-1475 atomic absorption spectrophotometer; best value selected for either C_2H_2–air or C_2H_2–N_2O flame.
[c] Values are from Varian literature on AA-75 atomic absorption spectrophotometer with a GTA95 graphite furnace.
[d] From References 16 and 18.
[e] From References 3, 8, 33, and 43.
[f] From References 37 and 44.
[g] From Reference 11.
[h] From References 25, 27, 30, and 31.
[i] Propane added.

Table 9.6. Possible Interferences in Atomic Fluorescence Spectrometry, Effect on Analytical Signal, and Means of Minimization

Interference type	Cause of interference	Effect on analytical signal	Means of minimization of interference
Spectral			
Not Source Induced	Atomic–molecular species from matrix are thermally excited and emit at fluorescence wavelength	None	Improve spectral discrimination to reduce noise contribution.
Cell-Induced	Atomic–molecular species from matrix have overlapping absorption line with analyte	Decrease in signal	Unlikely with most line (ICP, HCL, EDL) sources; very unlikely with laser sources; must be considered with xenon arc excitation.
	Atomic–molecular species from matrix have overlapping absorption and fluorescence lines with analyte	Increase in signal	Very unlikely, only with xenon arc excitation must it be considered; even here emission monochromator can discriminate.
	Incandescence due to heating of particles by laser	Increase in signal	Laser power density must be high to observe this interference; also unlikely in ICP because of few particles at observation height.
Source-Induced	Source scatter from unvaporized matrix particles	Increase in signal	Many means[38] exist to minimize this interference: two-line correction, continuum correction, Zeeman correction, nonresonance fluorescence, intermodulation, or second-harmonic methods.
Chemical			
Matrix Related	Compound formation	Decrease in signal	Primarily in flames and furnaces; use higher temperature source, matrix modifiers, etc.

Matrix-Related	Matrix vaporization; these effects decrease solute vaporization and atomization efficiencies	Decrease in signal	Primarily in flames; neither compound formation nor matrix vaporization is significant in ICP cells if conventional torches and normal observation heights are used. In flames, use combinations of standard addition methods, matrix-matched standards, internal standards, and releasing agents. In furnaces, use same approach as in flames and combinations of special graphite surface such as pyrolytic or glassy or refractory carbides, matrix modifiers, L'vov platform or L'vov probe and rapid heating rate: $> 10^5$ K/s
Physical	Decreased aspiration and solvent vaporization efficiencies; high salt contents can partially clog nebulizer	Decrease in signal	Likely in all nebulizer-based methods. Minimize by use of combinations of matrix-matched standards, internal standards, ultrasonic nebulizers, and high-solids nebulizers.
	Losses of analyte during drying or ashing steps in ETA or ETV	Decrease in signal	Use matrix modifiers or matrix-matched standards
Ionization of Matrix Species	Shifts of analyte ionization equilibria	Increase or decrease in signal depending on type of atomization cell	Use ionization buffer in flame; minimal significance in furnaces. In ICP, ionizers affect atom and ion signals; minimize by choice of observation height, argon flow rate, and RF power.

Table 9.7. Excitation and Fluorescence Lines used in Laser Atomic Fluorescence Spectrometry with ICPs

Element (and line type)	λ_{ex} (nm)	λ_{fl} (nm)	Ref.
Ag(I)	328.0	338.3	31
Al(I)	394.4	396.1	27
Au(I)	267.6	267.6	31
B(I)	249.7	249.8	27
Ba(I)	553.5	553.5	25
Ba(II)	455.4	614.2	27
		455.4	25
	455.4	389.2	30
	416.6		
Ca(I)	422.7	422.7	25
Ca(II)	393.4	393.4	25
	396.8	373.7	30
	370.6		
Ga(I)	287.4	294.4	27
		294.3	
	403.3	417.2	25
Hf(II)	263.97	303.1	31
Ir(I)	285.0	292.7	31
Mg(II)	279.5	279.1	31
	279.8		
Mo(I)	313.3	317.0	27
			31
Nb(II)	292.8	269.7	31
Pb(I)	283.3	405.8	27
	405.8	283.3	25
Pd(I)	324.3	340.5	31
Pt(I)	265.9	271.9	31
Ru(I)	287.5	366.3	31
Sc(II)	364.3	364.3	25
Si(I)	288.2	251.4	27
		251.6	
		251.9	
Sn(I)	300.9	317.5	27
Sr(I)	460.7	460.7	25
Sr(II)	407.8	460.7	25
	407.8	416.2	30
	430.5		
Ta(II)	268.5	276.2	31
Ti(II)	307.8	316.2	27
		316.8	
V(I)	411.2	411.2	25
V(II)	268.8	290.9	27
	390.3	290.9	25
Y(I)	508.7	376.0	27
			25
Zr(II)	310.7	257.0	27
	431.7	349.6	31
			25

Table 9.8. Characteristics of Excitation Sources in Atomic Fluorescence Spectrometry

Source	Peak spectral irradiance (J/s cm² nm)		Pulse width (ns)	Optical saturation		Wavelength range		Ease of use[a]		Stability[b]	Spectral interference[c,d]	Cost[e]	
	360–1000 nm	210–360 nm		VIS	UV	VIS	UV	VIS	UV			Initial	Operation
Nd-YAG dye	10^{10}–10^{11}	10^{8}–10^{10}	10–20	Yes	Yes(?)	Yes	≳ 200 nm	F	V	G	V	H	M
Excimer-dye	10^{9}–10^{10}	10^{7}–10^{9}	8–15	Yes	Yes(?)	Yes	≳ 220 nm	F	V	G	V	H	M
N$_2$-dye	10^{8}–10^{9}	10^{6}–10^{7}	3–10	Yes	No(?)	Yes	≳ 220 nm	F	V	G	V	M	L
Xenon arc	10^{-1}–10^{1}	10^{-3}–10^{-1}	Cw	No	No	Yes	Yes	E	E	E	P,F	L	L
EDL	10^{-2}–10^{-5}	10^{-1}–10^{-5}	Cw	No	No	Yes	Yes	E	E	E	F	L	L
HCL	10^{-2}–10^{-6}	10^{-2}–10^{-6}	Cw or pulsed	No	No	Yes	Yes	E	E	E	F	L	L
ICP	10^{0}–10^{-5}	10^{0}–10^{-5}	Cw	No	No	Yes	Yes	E	E	E	F	M	M

[a] E = easy, F = fairly difficult, V = very difficult.
[b] G = good, E = excellent.
[c] Spectral interferences refer to those expected when source is used to excite atom or ion lines in flame or ICP.
[d] V = very few, F = few, P = possible.
[e] H = high, M = moderate, L = low.

Table 9.9. Saturation Steady-State Spectral
Irradiance and Irradiance for a Two-Level^a Atom

λ_0 (nm)	E_λ^s (J/s cm^2 nm)	E^s (J/s cm^2)^b
200	$2.5 \times 10^5/Y$	$1.2 \times 10^3/Y$
300	$3.0 \times 10^4/Y$	$1.5 \times 10^2/Y$
400	$7.0 \times 10^3/Y$	$3.5 \times 10^1/Y$
500	$2.5 \times 10^3/Y$	$1.2 \times 10^1/Y$
600	$1.0 \times 10^3/Y$	$5.0 \times 10^0/Y$
700	$2.0 \times 10^2/Y$	$1.0 \times 10^0/Y$

^{a} The statistical weights of the upper and lower levels are
assumed to be equal; Y is the fluorescence quantum
efficiency.
^{b} $E^s = E_\lambda^s \, \delta\lambda_{eff}$ where $\delta\lambda_{eff}$ (0.005 nm) is the effective
absorption line width.
Source: Reference 34.

9.5   ANALYTICAL STUDIES

9.5.1   *ICP–Flame–CAFS and ICP–ETA–CAFS*

The ICP is an extremely versatile source because even high concentrations (1 to
2%) of analyte in the plasma, generated in a conventional torch, result in
spectral lines remarkably free from self-absorption at the observation heights
commonly used in analytical measurements.[33] The excellent detectability of
ICP–flame–CAFS and ICP–ETA–CAFS has already been documented (Tables
9.4 and 9.5). However, these detection limits are difficult to achieve on a
simultaneous basis because nebulization of a solution containing 1 to 2% by
weight of each of more than 10 elements into the ICP would be impractical.
Sequential introduction of solutions (1 to 2%) of single-element standards into
the ICP source allows multiple-element determinations on a sequential basis.
Alternatively, a multiple-element solution, containing 0.1% of each analyte, may
be used for simultaneous determinations. The spectral selectivity of this approach is quite high. However, scatter from unvaporized particles in the flame or
ETA is a problem that must be minimized (Table 9.6). The simplest and the best-
known method[35] to correct for the scatter is the two-line method, which
assumes that scattering is the same in the vicinity of the fluorescence wavelength.
Thus, one simply needs a source with a line, which does not excite fluorescence,
near the fluorescence line. The scatter signal from such a line is simply
subtracted from the measured analyte signal (fluorescence plus scatter).
    A second method of scatter correction is based on the different behaviors of
fluorescence and scatter when the concentration of standard analyte in the
source ICP is increased. As the concentration of analyte in the ICP is increased,
self-absorption eventually predominates, and the ICP radiations follow a
square-root dependence with concentration. The scatter signal follows the

source behavior. However, when self-absorption occurs in the source, the peak spectral emission intensity does not change; ie, the increased emission is solely in the emission line wings, which accounts for the well-known square-root dependence of the emission radiance with analyte concentration.[46] Thus, the fluorescence signal now shows no variation at all with increase in standard analyte concentration in the ICP source. By simple mathematical expressions, scatter can be corrected:

$$S_1 = x + s \qquad [9.16]$$

$$S_2 = \alpha x + \beta s \qquad [9.17]$$

where x is the fluorescence signal, s is the sample scatter signal, S_1 is the total observed intensity from cell at a lower analyte concentration in the ICP, S_2 is the total observed intensity from cell at a higher analyte concentration in the ICP, α is the ratio of analyte fluorescence at higher and lower concentration in the ICP, and β is the ratio of scatter intensity at higher and lower concentrations in the ICP. Such an approach has been used to determine iron in fly ash.[4] Several other approaches to correct for scatter in AFS are available.[35]

9.5.2   ICP–ICP–CAFS

Similar to the techniques discussed in Section 9.5.1, detection limits for both nonrefractory and refractory elements are good (Tables 9.4 and 9.5). However, minimal spectral interferences[11,34] are observed in ICP–ICP–CAFS compared to ICP–AES (Figure 9.4). The upper recording of Figure 9.4 shows the AES spectrum of an ICP for a solution containing 1000 μg/mL of calcium and 50 μg/mL of aluminum. The Al lines lie in the valley between the two collisionally broadened Ca peaks. For this AES spectrum, one must resort to background correction. In the lower spectrum, the same solution is measured by AFS while introducing a 2% Al solution into the source ICP. The lack of any background near the Al lines is evident. Of course, the broad, intense Ca emission increased the noise in the region of the Ca peaks in the AFS spectrum.

In the cell ICP, molecular fluorescence of metal oxides, metal hydroxides, etc, is not efficiently excited by the source ICP, thus resulting in no noticeable background fluorescence, as sometimes occurs with flame and ETA atomizers. Enhanced background noise can be reduced by an increase in integration time.

Spectral overlap of the fluorescence lines of two elements is rare with ICP excitation sources and is even less frequent with laser excitation sources, especially when monochromators are used. Because interference filters are used in the commercial HCL–ICP–AFS system (Section 9.3.2), spectral overlaps are more prevalent, but still not very significant. For a spectral interference to occur in AES, it is only necessary for the interfering radiation to fall within the spectrometer bandpass. However, spectral interference in AFS occurs only if: (a) the interfering element has a significant absorption cross section over the spectral profile of the emission source, (b) the population of the interfering

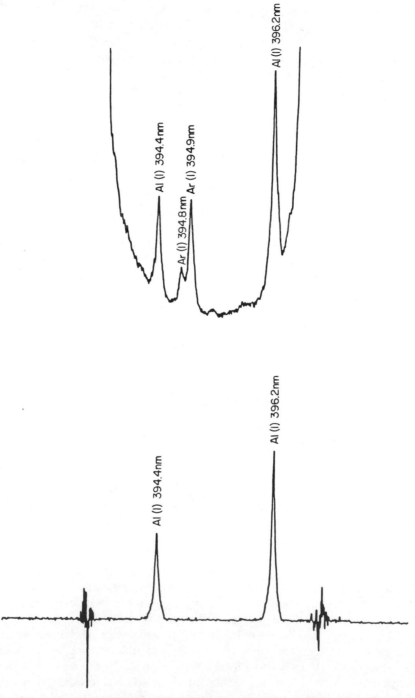

Figure 9.4. Recorder tracings showing the interference of Ca on Al for ICP–AES (upper) and ICP–ICP–AES (lower). In both cases, 50 μg/mL of Al and 1000 μg/mL of Ca are nebulized. The spectral bandpass is 0.044 nm and the instrumental time constant is 100 ms. (From Reference 11, with permission.)

species in the lower level is significant, and (c) the fluorescence quantum efficiency of the interfering species is high for a fluorescence line within the spectrometer bandpass.

9.5.3   ICP–Flame–RMAFS, ICP–ETA–RMAFS, and ICP–ICP–RMAFS

Because no pronounced self-absorption occurs in the emission profile of the ICP,[33] as discussed in Chapter 8, the linearity of analytical calibration curves can be extended to very high concentrations by simply switching the aspiration of the sample solution from the cell to the source ICP. By doing so, the calibration curve linearity can be extended to percentage concentration levels, making such an approach the analytical technique with the greatest linearity. In other words, when the conventional AFS analytical calibration curve begins to bend toward the abscissa, the sample solution is nebulized into the ICP source and a pure analyte standard, about 100 μg/mL, is introduced into the cell, which can be an ICP, a flame, or an ETA. As a result, the cell acts as a resonance monochromator. Obviously, the detection limits obtained by the RM approach are about 1000 times inferior to those by the CAFS approach.

To enhance the detecting power of RMAFS-based method, Cavalli, and co-workers[44] used a modified Varian CRA 90 ETA (Figure 9.5), for their ICP-ETA-RMAFS system. The good precision of ICP-ETA-RMAFS for

Figure 9.5. Modified Varian CRA 90 electrothermal atomizer for use in ICP-ETA-CAFS and ICP-ETA-RMAFS.[37,44]

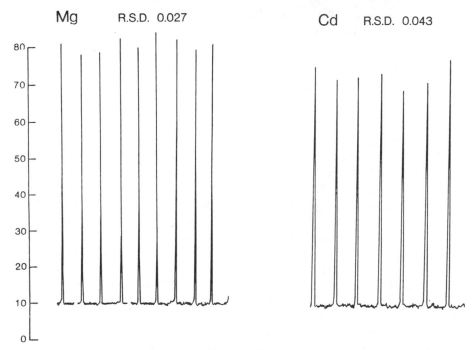

Figure 9.6. Recorder tracings for either 5 μL of 10 μg/mL Mg or Cd in the ETA and 3 μg/mL of Mg or Cd in the ICP for ICP-ETA-RMAFS.[37,44]

magnesium and cadmium measurements are apparent from Figure 9.6. With the ETA cell, the detection limits for the RM approach are more than 10 times better than with the ICP or flame cells. For comparison, the excellent sensitivity of the ICP-ETA-CAFS is shown in Figure 9.7.

9.5.4 ICP–AFS: The Baird Plasma AFS System

The detection limits for nonrefractory metals, measured on the Baird Plasma AFS System, are comparable to those achieved by ICP-AES. For alkali metals, superior detection limits are obtained with the AFS system. However, for the refractory elements (V, Al, Zr, W, B, etc), the detection limits are 10 to 100 times inferior to those from ICP-AES. This is because (a) hollow cathode lamps primarily emit atomic lines, while ionic species are often more predominant in the ICP, and (b) measurements are made at observation heights of 1 to 10 cm above the top of a torch having a long, extended outer tube. Because such a plasma has a lower temperature, lower electron number density, and more entrained air, compared to a conventional ICP, its characteristics resemble those of a hot combustion flame.

Demers[16] has described several means of improving the detection limits of refractory elements. One method is to introduce propane at 30 mL/min into the spray chamber or into the injector tube to produce a long, blue, pencil-like

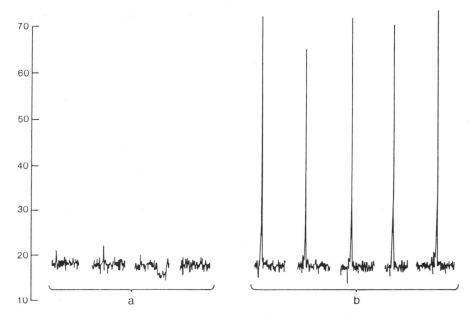

Figure 9.7. Recorder tracings of (*a*) water blank in ETA with 10,000 μg/mL Cd in ICP source and (*b*) 5 μL of 1 ng/mL Cd in ICP–ETA–CAFS.[37,44]

plasma, which resembles a flame. In general, the detection limits have improved by factors of 5 to 10 for certain elements (Table 9.10). Alternatively, organometallics in dioxane, rather than in water, were introduced into the ICP, producing a highly luminous, yellow discharge, which could be transformed back to a tall, green plasma by introducing sufficient O_2 to oxidize the carbon-containing species (Table 9.10). Unfortunately, in the two cases above, the extended torch was used, and because HCLs are not sufficiently intense, measurement at observation heights of 15 to 20 mm could not be conducted. Obviously, refractory oxide formation occurs to a lesser extent at 15 to 20 mm, but the high background emission from the plasma saturates the photomultiplier, because interference filters are used for spectral isolation.

Other improvements to the Baird Plasma AFS System include the use of an ultrasonic nebulizer,[19] which results in a gain in detectability of a factor of 5 to 10, and the utilization of a graphite tip on the injector tube for the determination of aluminum, boron, beryllium, and silicon.[16] Demers[47] also evaluated EDLs as sources for the Baird Plasma AFS System and found higher pulsed intensities for the EDLs than for the pulsed HCLs for most elements.

9.5.5 Laser-Excited ICP–AFS

Without a doubt, the most powerful single-element, atomic fluorescence spectrometric system consists of a dual dye laser pumped by the same source, a

Table 9.10. HCL–ICP–AFS Detection Limits of Elements in Water with and Without Propane and in Organic Solvents

| Element | λ_{fl} (nm) | HCL–ICP–AFS detection limit (ng/mL) | | Organic solution | ICP–AES detection limit in MIBK[a] (ng/mL) |
| | | Water solution | | | |
		No propane	With propane	Dioxane	
Ag	328.1	3	—	3	3
Al	Σ396.2	50	20	30	15
B	249.8	ND[b]	500	200	—
Ba	553.5	500	500	80	—
Be	—	3	0.8	—	—
Ca	422.7	0.4	—	1	0.06
Cr	Σ357.9	40	6	20	6
Cu	Σ324.9	3	—	5	1
Fe	Σ248.3	10	—	20	6
Ge	—	ND	200	—	—
K	766.5	2	—	10	—
Mg	285.2	0.5	—	1	1
Mo	Σ313.5	ND	200	50	—
Na	Σ589.0	1	—	5	—
Ni	Σ232.0	7	—	15	15
P	Σ178.3	20000	—	20000	—
Pb	364.0	100	—	300	50
S	Σ180.7	1000	—	2000	—
Si	Σ251.7	300	—	700	10
Sn	Σ284.0	8000	300	200	5
Ta	—	ND	2000	—	—
Ti	Σ335.5	ND	400	80	5
V	Σ318.4	ND	300	100	5
W	—	ND	2000	—	—
Yb	—	300	20	—	—
Zn	213.9	0.5	—	3	6

[a] MIBK = Methyl isobutyl ketone.
[b] ND = Not detectable.
Source: References 16 and 18.

conventional ICP torch, a medium-resolution spectrometer, and a photomultiplier tube–boxcar averager (Figure 9.3). Such a system has an extremely high spectral selectivity, high detection powers, although not approaching single atom per cubic centimeter detection limits, and good precision and accuracy. Unfortunately, this system currently costs approximately $200,000, is complex, and is useful only for single-element determinations.

9.5.6 ICP-Excited Molecular Fluorescence

In a novel article, Tallant[48] reports the use of an ICP with various metal salts aspirated to excite fluorescence of molecules in the condensed phase. To our knowledge, this approach has not been pursued further.

9.5.7   Matrix Effects in the ICP–AFS Systems

With the conventional torch[1] operated at 27 MHz, matrix interferences such as ionization and vaporization are identical to those in ICP-AES. With the extended torch, the outer tube of which is 4 cm longer, the temperature of the plasma in the observation region is lower[26] than that at the commonly used observation height of 15 to 20 mm (with the conventional short torch). As a result, matrix interferences are expected to be more severe with the extended torch. Demers and Jansen[16,19] have found that such interferences (eg, P on Ca, Na on Ca) are significant only if the forward power is low or if the interferent-to-analyte concentration ratio is high. Demers[16] suggested that the rate of temperature decrease and the quenching due to air entrainment are less in the extended torch than in a conventional torch, but no direct experimental proof has been given. Krupa and co-workers,[11] using an ICP-ICP-CAFS system, also found no sulfate or phosphate interference on calcium, but did find an enhancement of 300% for Ca I fluorescence and a depression of 50% for Ca II fluorescence, when a 1 μg/mL Ca solution contained 10,000 μg/mL Na.

9.6   APPLICATIONS

Several applications of ICP–AFS methods, including analyses of reference standards, are summarized in Table 9.11. Excellent correlation is apparent in all determinations.

9.7   DIAGNOSTICAL STUDIES OF THE INDUCTIVELY COUPLED PLASMA

Atomic fluorescence excited by pulsed dye laser can be used to obtain spatial fluorescence profiles in the ICP without the need for an inversion technique. The first report of such profiles was by Omenetto and co-workers,[49] who studied spatial profiles of Ba I and Ba II. Determinations of the lifetimes, τ, of excited atoms and ions in the ICP are possible by use of a short-pulse dye laser and measurement of the fluorescent decays. Because $Y = \tau/\tau_{sp}$, where $\tau_{sp} =$ the spontaneous (radiative) lifetime available in the literature, measurements of quantum efficiencies are also possible. Uchida and associates[50,51] were the first to measure fluorescence lifetimes of excited atoms and ions in ICPs.

Because this chapter deals primarily with analytical studies, these subjects are not treated in detail. For further discussions of diagnostic studies, the reader is referred to the literature for argon-metastable measurements,[52,53] concentration profiles of nonmetals obtained by two-photon excitation,[30] the effect of RF power on fluorescence,[54,55] and the use of laser-excited fluorescence for spatial concentration profiles.[49,56,57]

Table 9.11. Applications of ICP-AFS

Method	Cell	Element	Concentration[a]	Sample	Ref.
ICP-FL-CAFS	H_2/air	Cd	1.6 ± 0.5 μg Cd (1.45 ± 0.06)	Fly ash	38
		Zn	211 μg/g Zn (211 ± 20)		
		Fe	$6.0 \pm 0.2\%$ Fe (6.2%)		
ICP-FL-CAFS	Sep C_2H_2/air[b]	Cd	No certified value	Sediments	4
ICP-FL-CAFS/RMAFS	Sep C_2H_2/air[b]	Pd	250 μg/ml (265 by AAF)	Nuclear wastes	6
ICP-ICP-CAFS	ICP	Cu, Cr, Zn	Within 1% relative	High-carbon steel	11
ICP-ICP-RMAFS	ICP	Fe	Within 1% relative	High-carbon steel	11
HCL-ICP-CAFS	ICP	Au	(95.7% fire assay)	Gold alloy	13
		Pd	2.88% (2.84% fire assay)	Gold alloy	
		Ag	1.01% (1.02% fire assay)	Gold alloy	
		Pt	0.33% (0.31% fire assay)	Gold alloy	
		Ag	70.7% (70 ± 1 nominal)	Brazing alloy	13
		Cu	$28.7 \pm 0.1\%$ (28 ± 1 nominal)	Brazing alloy	
		Ni	$0.62 \pm 0.02\%$ (0.75 ± 0.25 nominal)	Brazing alloy	
		Cd	$< 0.0001\%$ ($< 0.002\%$ nominal)	Brazing alloy	
		Pb	$< 0.0001\%$ ($< 0.002\%$ nominal)	Brazing alloy	
		Zn	$0.005 \pm 0.001\%$ ($< 0.002\%$ nominal)	Brazing alloy	
HCL-ICP-CAFS	ICP	Al	0.64% (0.51% accepted)	Iron ore	14
			5.35% (5.14% accepted)	Slag	
		Co	0.025% (0.027% accepted)	Low-alloy steel	
			0.14% (0.10% accepted)	Stainless steel	

Method		Element	Value	Sample	Ref.
		Cr	12.53 % (12.82 % accepted)	Stainless steel	
		Cu	1.31 % (1.31 % accepted)	Low-alloy steel	
		Cu	0.084 % (0.080 % accepted)	Stainless steel	
		Mo	0.103 % (0.100 % accepted)	Low-alloy steel	
		Mo	0.030 % (0.028 % accepted)	Low-alloy steel	
		Mn	2.38 % (2.38 % accepted)	Stainless steel	
		Mn	0.45 % (0.42 % accepted)	Slag	
		Ni	1.47 % (1.50 % accepted)	Low-alloy steel	
		Ni	0.035 % (0.034 % accepted)	Low-alloy steel	
		Si	12.0 % (12.2 % accepted)	Stainless steel	
		Si	3.5 % (3.8 % accepted)	Iron ore	
		Si	0.50 % (0.49 % accepted)	Low-alloy steel	
HCL–ICP–CAFS	ICP	Na	87 ± 4 μg/mL (89 ± 3 μg/mL)	Fuel oil (NBS SRM 1634a)	19
		Ni	30 ± 1 μg/mL (29 ± 1 μg/mL)	Fuel oil (NBS SRM 1634a)	
		V	55 ± 3 μg/mL (56 ± 2 μg/mL)	Fuel oil (NBS SRM 1634a)	
		Zn	2.7 ± 0.1 μg/mL (2.7 ± 0.2 μg/mL)	Fuel oil (NBS SRM 1634a)	
HCL–ICP–CAFS	ICP	Na	2094 ± 21 μg/mL (2077 ± 78 AAS)	Blood	19
		K	1711 ± 25 μg/mL (1710 ± 10 AAS)	Blood	
		Fe	496 ± 7 μg/mL (463 ± 36 AAS)	Blood	
		Ca	65.7 ± 0.4 μg/mL (57 ± 4 AAS)	Blood	
		Mg	40.4 ± 0.6 μg/mL (33.7 ± 2 AAS)	Blood	
		Zn	5.3 ± 0.2 μg/mL (5.2 ± 0.8 AAS)	Blood	
		Cu	1.30 ± 0.05 μg/mL (1.05 ± 0.02 AAS)	Blood	

[a] Parenthetical value is reference value from other method(s).
[b] Separated C_2H_2/air flame (use of Ar sheath).

9.8   CONCLUSIONS AND FUTURE DIRECTIONS

A comparison of the capabilities of atomic fluorescence methods with other atomic spectrometric methods is given in Table 9.12. Three conclusions are apparent. First, LEIS provides the best detection limits, but the technique is subject to matrix interferences. Second, the linear dynamic ranges of the ICP-based techniques are excellent, especially for the IIAFS, which has an LDR of four to eight orders of magnitude. Third, compared to ICP-based methods commercially available, the detection limits for ICP-MS are superior to all others. The chief limitation of ICP-AFS is its inferior detectability for the refractory elements. However, the ICP-AFS provides the greatest spectral selectivity compared to ICP-AES and ICP-MS. As discussed in Chapter 10, the determinations of certain elements by ICP-MS is not feasible because of mass spectral interferences. In this context, the future of ICP-AFS seems bright.

The Baird Plasma AFS System provides high detection powers for the non-refractory elements, large dynamic ranges, excellent precision, few spectral or

Table 9.12. Comparison of Atomic Fluorescence Spectrometry with Other Atomic Spectrometric Methods

Method[a]	Source	Cell	LOD (ng/mL)[b]	LDR[c]	Spectral inter-ferences[d]	Matrix inter-ferences[e]
LEAFS	Laser	FL	$1–10^3$	$10^3–10^6$	V	S
		ETA	$10^{-3}–10^3$	$10^2–10^5$	V	M
		ICP	$10^{-1}–10^3$	$10^3–10^6$	V	V
HCAFS	HCL	ICP	$1–10^3$	$10^2–10^4$	F	F
XAFS	Xenon arc	FL	$1–10^3$	$10^2–10^4$	F	S
IIAFS	ICP	ICP	$1–10^2$	$10^4–10^8$ (RM)	V	V
IFAFS	ICP	FL	$1–10^3$	$10^4–10^7$ (RM)	V	S–V
LEIS	Laser	FL	$10^{-3}–10^2$	$10^3–10^6$	F	S
		ETA	$10^{-6}–10^1$	$10^3–10^6$	V	M
AAS	HCL	FL	$1–10^2$	$10^1–10^2$	F	S
	HCL	ETA	$10^{-3}–10^2$	$10^1–10^2$	F	M
AES		ICP	$10^{-2}–10^2$	$10^4–10^5$	M	V
		FL	$10^{-1}–10^4$	$10^2–10^5$	F	S
ICP–MS		ICP	$10^{-3}–10^1$	$10^3–10^5$	S	S–V

[a] LEAFS = laser-excited atomic fluorescence spectrometry, HCAFS = hollow cathode excited atomic fluorescence spectrometry, XAFS = xenon arc excited atomic fluorescence spectrometry, IIAFS = ICP-excited ICP atomic fluorescence spectrometry, IFAFS = ICP excited flame atomic fluorescence spectrometry, LEIS = laser enhanced ionization spectrometry, AAS = atomic absorption spectrometry, AES = atomic emission spectrometry, ICPMS = ICP mass spectrometry, FL = flame, ETA = electrothermal Atomizer, ICP = inductively couple plasma, HCL = hollow cathode lamp.
[b] LOD = limit of detection. Values reported refer to best and worst figures reported in the literature for several elements.
[c] LDR = linear dynamic range, i.e., range of linearity of calibration curve = C_u/C_l where C_u = upper concentration which is within 5% of linearity and C_l = limiting detectable concentration.
[d] Spectral interferences: V = very few; F = few; M = many.
[e] Matrix interferences: V = very few; S = some; M = many.

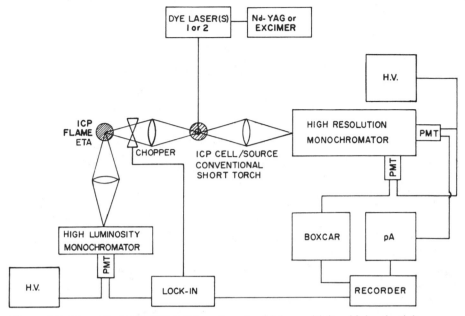

Figure 9.8. Versatile ICP–AES/AFS system for high-sensitivity, high-selectivity measurements.

matrix interferences, and good accuracy in routine analysis. The major improvement needed concerns the refractory elements, where the detection limits must be improved by a factor of 10 to 100. The ultrasonic nebulizer[47] provides a gain of approximately 10. The use of intense ionic line sources may improve detectability by another factor of 10. Perhaps then, the conventional torch could be used by Baird.

The most versatile ICP–AFS system possible is shown in Figure 9.8, where the ICP cell is the conventional "short" torch. This system could be used for emission measurements. If higher sensitivity or selectivity were desirable for certain elements, one or two dye lasers could be used for single-wavelength or dual-wavelength excitation, respectively, with boxcar detection. In addition, for elements with their spectral lines below 250 nm, or when high sensitivity and high selectivity are needed, the ICP could be used as an excitation source to excite the sample in an ICP, a flame, or an ETA with phase-lock-in detection. For high analyte concentrations, the resonance monochromator mode could be used.

ACKNOWLEDGMENTS

This work was supported in part by AF-AFOSR-49620-84-C-0002.

REFERENCES

1. A. Montaser and V. A. Fassel, Inductively Coupled Plasmas as Atomization Cells for Atomic Fluorescence Spectrometry, *Anal. Chem.* 48, 1490–1499 (1976).
2. C. A. M. Hussein and G. Nickless, Atomic Fluorescence in the Inductively Coupled Plasma, Paper given at the ICAAS Meeting, Sheffield, England, 1969.
3. M. S. Epstein, S. Nikdel, N. Omenetto, R. D. Reeves, J. D. Bradshaw, and J. D. Winefordner, Inductively Coupled Argon Plasma as an Excitation Source for Flame Atomic Fluorescence Spectrometry, *Anal. Chem.* 51, 2071–2077 (1979).
4. M. S. Epstein, N. Omenetto, S. Nikdel, J. D. Bradshaw, and J. D. Winefordner, Inductively Coupled Argon Plasma as an Excitation Source for Flame Atomic Fluorescence Spectrometry, *Anal. Chem.* 52, 284–287(1980).
5. P. Cavalli, N. Omenetto, and G. Rossi, Determination of Cadmium at sub-ppm Levels in Lake Sediments by an Inductively Coupled Plasma Atomic Fluorescence Technique, *At. Spectrosc.* 3, 1–4 (1982).
6. P. Cavalli, G. Rossi, and N. Omenetto, Determination of Palladium in Nuclear-Waste Samples by Inductively Coupled Plasma Emission-Fluorescence Spectrometry, *Analyst* 108, 297–304 (1983).
7. N. Omenetto, G. Crabi, A. Nesti, P. Cavalli, and G. Rossi, Simultaneous Correction for Scattering in ICP-Excited Atomic Fluorescence Spectrometry, *Spectrochim. Acta* 38B, 549–556 (1983).
8. A. Montaser, Inductively Coupled Plasmas as Excitation Sources for Atomic Fluorescence Spectrometry, *Spectrosc. Lett.* 12, 725–732 (1979).
9. M. A. Kosinski, H. Uchida, and J. D. Winefordner, Atomic Fluorescence Spectrometry with Inductively Coupled Plasmas as Excitation Sources and Atomization Cell, *Anal. Chem.* 55, 688–692 (1983).
10. C. L. Long and J. D. Winefordner, Evaluation of Atomic Fluorescence Detection Limits with an Inductively Coupled Plasma as an Excitation and Atomization Cell, *Appl. Spectrosc.* 38, 563–567 (1984).
11. R. J. Krupa, G. L. Long, and J. D. Winefordner, An ICP-Excited ICP Resonance Monochromator and Fluorescence Spectrometer for the Analysis of Trace to Major Sample Constituents, *Spectrochim. Acta* 40B, 1485–1494 (1985).
12. D. R. Demers and C. D. Allemand, Atomic Fluorescence Spectrometry with an Inductively Coupled Plasma as Atomization Cell and Pulsed Hollow Cathode Lamps for Excitation, *Anal. Chem.* 53, 1915–1921 (1981).
13. R. L. Lancione and D. M. Drew, AFS Technique Gives Error-Free Analysis of Precious Metals, *Ind. Res. Dev.* 100–104, February 1983.
14. R. L. Lancione, Determination of Secondary Metals in Nickel Plating Baths Using Atomic Fluorescence Spectrometry, *Met. Finish* 82(10), 17–20 (1984).
15. R. L. Lancione and D. M. Drew, Application of ICP Atomic Fluorescence Spectrometry in the Steel Industry, *J. Testing Eval.* 203–207, July 1984.
16. D. R. Demers, Hollow Cathode Lamp-Excited ICP Atomic Fluorescence Spectrometry— Update, *Spectrochim. Acta* 40B, 93–105 (1985).
17. R. L. Lancione and D. M. Drew, Evaluation of ICP Atomic Fluorescence for the Determination of Mercury, *Spectrochim. Acta* 40B, 107–113 (1985).
18. D. R. Demers, D. A. Busch, and C. D. Allemand, ICP Atomic Fluorescence Spectroscopy, *Am. Lab.*, 167–176, March 1982.
19. D. R. Demers and E.B.M. Jansen, "Recent Developments and Applications in Hollow Cathode-Excited ICP-Atomic Fluorescence Spectrometry," personal communication.
20. E. B. Jansen and D. R. Demers, Hollow Cathode Lamp Excited ICP Atomic Fluorescence Spectrometry—Performance Under Compromise Conditions for Simultaneous Multielement Analysis, *Analyst* 110, 541–545 (1985).
21. D. R. Demers and C. D. Allemand, Inductively Coupled Plasma Atomic Fluorescence Spectrometry, U.S. Patent 4,300,834, November 1981.
22. J. C. Gautherin, J. P. Lauda, R. Diemiasgonek, J. L. Mouton, J. M. Mermet, and C. Trassy, Inductively Coupled Plasma-Atomic Fluorescence Spectroscopy: Application to Spectral Interference, *ICP Inf. Newsl.* 6, 408–411 (1981).
23. B. D. Pollard, M. B. Blackburn, S. Nikdel, A. Massoumi, and J. D. Winefordner, Atomic Fluorescence Spectrometry in the Inductively Coupled Plasma with a Continuous Wave Dye Laser, *Appl. Spectrosc.* 33, 5–8 (1979).

24. M. S. Epstein, S. Nikdel, J. D. Bradshaw, M. A. Kosinski, J. N. Bower, and J. D. Winefordner, Atomic and Ionic Fluorescence Spectrometry with Pulsed Dye Laser Excitation in the Inductively Coupled Plasma, *Anal. Chim. Acta* 113, 221–226 (1980).

25. M. A. Kosinski, H. Uchida, and J. D. Winefordner, Evaluation of an Inductively Coupled Plasma with an extended Sleeve Torch as an Atomization Cell for Laser Excited Fluorescence Spectrometry, *Talanta* 30, 339–345 (1983).

26. H. Uchida, M. A. Kosinski, and J. D. Winefordner, Laser Excited Atomic and Ionic Fluorescence in an Inductively Coupled Plasma, *Spectrochim. Acta* 38B, 5–13 (1983).

27. N. Omenetto, H.G.C. Human, P. Cavalli, and G. Rossi, Laser Excited Atomic and Ionic Nonresonance Fluorescence Detection Limits for Several Elements in an Argon Inductively Coupled Plasma, *Spectrochim. Acta* 39B, 115–117 (1984).

28. N. Omenetto and H.G.C. Human, Laser Excited Analytical Atomic and Ionic Fluorescence in Flames, Furnaces, and Inductively Coupled Plasmas—I. General Considerations, *Spectrochim. Acta* 39B, 1333–1345 (1984).

29. H.G.C. Human, N. Omenetto, P. Cavalli, and G. Rossi, Laser Excited Atomic and Ionic Fluorescence in Flames, Furnaces, and Inductively Coupled Plasmas—II. Fluorescence Characteristics and Detection Limits of Fourteen Elements, *Spectrochim. Acta* 39B 1345–1363 (1984).

30. N. Omenetto, B. W. Smith, L. P. Hart, P. Cavalli, and G. Rossi, Laser-Induced Double Resonance Ionic Fluorescence in an Inductively Coupled Plasma, *Spectrochim Acta* 40B, 1411–1422 (1985).

31. X. Huang, J. Lanauze, and J. D. Winefordner, Laser-Excited Atomic Fluorescence of Some Precious Metals and Refractory Elements in the Inductively Coupled Plasma, *Appl. Spectrosc.* 39, 1042–1047 (1985).

32. J. V. Sullivan and A. Walsh, The Isolation and Detection of Resonance Lines, *Appl. Opt.* 7, 1271–1280 (1968).

33. N. Omenetto, S. Nikdel, J. D. Bradshaw, M. S. Epstein, R. D. Reeves and J. D. Winefordner, Diagnostic and Analytical Studies of the Inductively Coupled Plasma by Atomic Fluorescence, *Anal. Chem.* 51, 1521–1525 (1979).

34. *N. Omenetto and J. D. Winefordner, Atomic Fluorescence Spectrometry: Basic Principles and Applications, Prog. Anal. At. Spectrosc.* 2, 1–183 (1979).

35. N. Omenetto and J. D. Winefordner, Scattering in Atomic Fluorescence Spectroscopy, *Prog. Anal. At. Spectrosc.* 8, 371–449 (1985).

36. O. J. Matveev, "Taking into Account and Eliminating Interference from Nonselectively Scattered Radiation in Laser AFS, *Zh. Prikl. Spektrosk.* (English transl.) 39, 709–725 (1983).

37. N. Omenetto, Recent Advances in Atomic Fluorescence Spectroscopy, *Analytiktreffen 1982*, 70–75 (1982).

38. J. D. Winefordner and N. Omenetto, in "Analytical Applications of Lasers," E. Piepmeier, Ed. John Wiley & Sons, New York, 1986.

39. N. Omenetto, L. P. Hart, P. Benetti, and J. D. Winefordner, On the Shape of Atomic Fluorescence Analytical Curves with Laser Excitation Source, *Spectrochim. Acta* 28B, 301–307 (1973).

40. M. S. Epstein and J. D. Winefordner, Summary of the Usefulness of Signal-to-Noise Treatment in Analytical Spectrometry, *Prog. Anal. At. Spectrosc.* 7, 67–137 (1984).

41. H. Falk, Analytical Capabilities of Atomic Spectrometric Methods Using Tunable Lasers: A Theoretical Approach, *Prog. Anal. At. Spectrosc.* 3, 181–208 (1980).

42. C. T. J. Alkemade, Single Atom Detection, *Appl. Spectrosc.* 35, 1–14 (1981).

43. N. Omenetto, P. Cavalli, and G. Rossi, Inductively Coupled Plasma and Atomic Fluorescence: A Versatile Tool for the Analysis of Metals at Trace, Minor and Major Levels, in "Reviews in Analytical Chemistry," T. S. West, Ed. Freund Publishing House, Ltd., Tel-Aviv, Israel, 1981, pp. 185–205.

44. P. Cavalli, M. Achilli, G. Rossi, and N. Omenetto, Ispra, Italy, unpublished work.

45. S. Greenfield, P. Smith, A. Breeze, and N. Chilton, Atomic Absorption with an Electrodeless High Frequency Plasma Torch, *Anal. Chim. Acta* 41, 385–387 (1968).

46. C.T.J. Alkemade, T. Hollander, W. Snelleman, and P.J.T. Zeegers, "Metal Vapors in Flames." Pergamon Press, Oxford, 1982.

47. D. R. Demers, Baird Corp., New Bedford, Mass., unpublished results, 1985.

48. D. R. Tallant, The Inductively Coupled Plasma as a Source for Molecular Fluorescence, *ICP Inf. Newsl.* 5(4), 171–180 (1979).

49. N. Omenetto, S. Nikdel, R. D. Reeves, J. D. Bradshaw, J. N. Bower, and J. D. Winefordner, Relative Spatial Profiles of Barium Ion and Atom in the Argon Inductively Coupled Plasma as Obtained by Laser-Excited Fluorescence, *Spectrochim. Acta* 35B, 507–517 (1980).

50. H. Uchida, M. A. Kosinski, N. Omenetto, and J. D. Winefordner, Time-Resolved Fluorescence in an Argon Inductively Coupled Plasma Determination of Excited Atom Lifetimes, *Spectrochim. Acta* 38B, 529–532 (1983).
51. H. Uchida, M. A. Kosinski, N. Omenetto, and J. D. Winefordner, Studies on Lifetime Measurements and Collisional Processes in an Inductively Coupled Argon Plasma Using Laser-Induced Fluorescence, *Spectrochim. Acta* 38B, 63–68 (1984).
52. G. M. Hieftje, Low-Cost Tunable Lasers. Prospects for Chemical Analysis, *Am. Lab.* 15, 66–70, 72, 74 (1983).
53. N. Omenetto, Plasma Diagnostics by Laser Induced Fluorescence," ICP Winter Conference on Plasma Spectrochemistry, Kailua Kona, Hawaii, January 1986.
54. X. Huang, D. Mo, K. S. Yeah, and J. D. Winefordner, Effect of RF Power of Extended-Sleeve Torch Inductively Coupled Plasma on Laser-Induced Fluorescence and Emission, *Anal. Chim. Acta* (in press).
55. R. J. Krupa and J. D. Winefordner, Power Dependencies of ICP Absorption, Emission and Fluorescence Signals for the Extended Sleeve Torch, *Anal. Chem.* (in press).
56. X. Huang, K. S. Yeah, and J. D. Winefordner, Spatial Distribution Profiles of Ca, Cu, and Mn Atoms and Ions in an Inductively Coupled Plasma by Means of Laser-Excited Fluorescence, *Spectrochim. Acta* 40B, 1379–1386 (1985).
57. G. Gillson and G. Horlick, An Atomic Fluorescence Study of Easily Ionizable Elements Interferences in the Inductively Coupled Plasma, *Spectrochim. Acta* 41B, 619–624 (1986).

10

Inductively Coupled Plasma–Mass Spectrometry

**G. HORLICK, S. H. TAN,
M. A. VAUGHAN,
AND Y. SHAO**

*Department of Chemistry
University of Alberta
Edmonton, Alberta, Canada*

10.1 INTRODUCTION

In a relatively short period of time,[1] inductively coupled plasma–mass spectrometry (ICP–MS) has developed into a significant, new technique for trace element analysis. In this technique, singly charged analyte ions generated in an ICP are extracted into and measured with a quadrupole mass spectrometer. A brief, but good, summary of the development of ICP–MS is presented in the August 1984 issue of the *ICP Information Newsletter*.[2] The pioneering work was conducted primarily in three laboratories: the Ames Laboratory at Iowa State University, headed by Fassel[1,3,4]; the laboratories at Sciex[5]; and the University of Surrey (Gray), the British Geological Survey (Date) and VG Instruments.[6–9] The technique has recently been reviewed elsewhere.[10]

The rapid development of ICP–MS has been fueled by its unique measurement capabilities. For trace element analysis, the technique has excellent detection limits for the direct analysis of solution samples. Current detection limits are in the range of 10 to 100 pg/mL for most elements, and for certain elements, detection limits approach 1 pg/mL (0.001 ppb). In many cases, these limits are 100 to 1000 times superior to those that can be routinely achieved by ICP–AES (atomic emission spectrometry). The mass spectra of the elements, in general, are considerably simpler than their optical emission spectra. This results in fewer problems with mass spectral overlaps, but certain complications can occur due to molecular ions formed either in the ICP or in the ion extraction device. However, for a group of elements, such as the lanthanide elements, spectral overlaps in ICP-MS are considerably easier to manage than in ICP-AES. Inherent in the technique is the facile measurement of elemental isotope ratio information, thus allowing the routine utilization of isotope ratio information and the isotope dilution technique to solve and study analytical problems. Such information can now be obtained in minutes rather than in hours, as required by more traditional methods. In addition, the natural-isotope-abundance spectral pattern provides quick and essentially immutable evidence for the qualitative identification of an element. The features above provide ICP-MS with powerful capabilities for elemental analysis.

10.2   THE INDUCTIVELY COUPLED PLASMA AS AN ION SOURCE

Clearly, the ICP is an excellent source for atomic emission analysis, as attested to by the other chapters in this book. Most analyses by ICP-AES are conducted by observation of ion lines, and thus, one would expect that analyte ions are present in the ICP discharge. A simple calculation[11] reveals that most elements are highly ionized in the ICP.

The degree of ionization is defined as follows:

$$\alpha = \frac{n_i}{n_a + n_i} = \frac{(n_i n_e)/n_a}{n_e + (n_i n_e)/n_a} = \frac{K_M}{n_e + K_M} \qquad [10.1]$$

where n_a is the number density of atoms, n_i is the number density of ions, n_e is the electron number density, and K_M is the Saha equilibrium constant. The Saha equilibrium constant is dependent on the ionization temperature T_{ion} and is expressed as follows[12]:

$$K_M = \frac{n_i n_e}{n_a} = 4.83 \times 10^{15} \, T_{ion}^{3/2} \frac{Z_i}{Z_a} \exp\left(-\frac{V_i}{kT_{ion}}\right) \qquad [10.2]$$

where Z_a is the partition function of the atom, Z_i is the partition function of the ion, V_i is the ionization potential, and k is the Boltzmann constant. Therefore, the degree of ionization can be calculated if the electron number density and the ionization temperature are determined. Using values[11,13] of $n_e = 1.475 \times 10^{14}$ cm^{-3} and T_{ion} (Ar) of 6680 K, the degrees of ionization shown in Table 10.1 can be calculated. The data presented in Table 10.1 confirm that the ICP is indeed a good source for elemental ions. Other similar calculations have recently been presented in the literature.[14]

Recent atomic fluorescence measurements conducted on the ICP discharge[15,16] have also shown that ground-state analyte ions are widely

Table 10.1. Degree of Ionization for the Elements in an Argon
ICP

Element	Ionization potential (eV)	Degree of ionization[a]
Cs	3.894	99.98
Rb	4.177	99.98
K	4.341	99.97
Na	5.139	99.91
Ba	5.212	99.96
Ra	5.279	99.95
Li	5.392	99.85
La	5.577	99.91
Sr	5.695	99.92
In	5.786	99.42
Al	5.986	98.92
Ga	5.99	99.00
Tl	6.108	99.38
Ca	6.113	99.86
Y	6.38	98.99
Sc	6.54	99.71
V	6.74	99.23
Cr	6.766	98.89
Ti	6.82	99.49
Zr	6.84	99.31
Nb	6.88	98.94
Hf	7.0	98.89
Mo	7.099	98.54
Tc	7.28	97.50
Bi	7.289	94.14
Sn	7.344	96.72
Ru	7.37	96.99
Pb	7.416	97.93
Mn	7.435	97.10
Rh	7.46	95.87
Ag	7.576	94.45
Ni	7.635	92.55
Mg	7.646	98.25
Cu	7.726	91.59
Co	7.86	94.83
Fe	7.870	96.77
Re	7.88	94.54
Ta	7.89	96.04
Ge	7.899	91.64
W	7.98	94.86
Si	8.151	87.90
B	8.298	62.03
Pd	8.34	94.21
Sb	8.461	81.07
Os	8.7	79.96
Cd	8.993	85.43
Pt	9.0	61.83
Te	9.009	66.74

(*Continued*)

Table 10.1. (*Continued*)

Element	Ionization potential (eV)	Degree of ionization[a]
Au	9.225	48.87
Be	9.322	75.36
Zn	9.394	74.50
Se	9.752	30.53
As	9.81	48.87
S	10.360	11.47
Hg	10.437	32.31
I	10.451	24.65
P	10.486	28.79
Rn	10.748	35.74
Br	11.814	3.183
C	11.260	3.451
Xe	12.130	5.039
Cl	12.967	0.4558
O	13.618	0.04245
Kr	13.999	0.2263
N	14.534	0.04186
Ar	15.759	0.01341
F	17.422	0.0001919
Ne	21.564	0.000005468
He	24.587	0.000000001007

[a] Assuming local thermodynamic equilibrium, $T_{ion}(Ar) = 6680$ K and $n_e = 1.47 \times 10^{14}$ cm^{-3}.

distributed throughout the ICP at forward powers ranging from 0.5 to 1.5 kW and observation heights ranging from 9 to 30 mm above the load coil. The successful development of ICP–MS then is closely tied to the problem of interfacing the high-temperature, atmospheric-pressure ICP discharge to a mass spectrometer. This instrumentation is described in the next section.

10.3 INSTRUMENTATION FOR INDUCTIVELY COUPLED PLASMA–MASS SPECTROMETRY

Schematic diagrams of the Sciex Elan Model 250 ICP–MS system and the VG Isotopes VG PlasmaQuad are shown in Figures 10.1 and 10.2, respectively. Although the two instruments differ in certain details, their basic approaches are similar. A quadrupole mass spectrometer is interfaced to the ICP discharge via a set of metallic cones. The first cone is called the sampling cone (Figure 10.3); it has an orifice with a diameter of about 1 mm; the sampling cone assembly is water cooled. A second cone, the skimmer, is placed about 2 to 10 mm in back of the sampling cone. The diameter of the orifice in the skimmer is the same as that in the sampling cone, but the skimmer is more sharply tapered. The region between the two cones is evacuated to a pressure of about 133 Pa (1 torr).

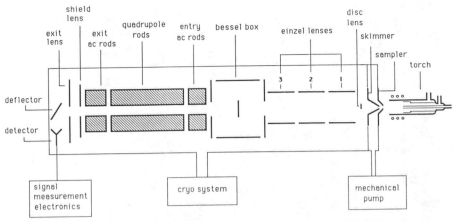

Figure 10.1. Schematic diagram of the Sciex Elan model 250 ICP–MS.

Several materials have been used for the sampling cones, with nickel and copper now being the most common. However, materials such as molybdenum, platinum, chromium-plated copper and copper coated with TiN have all been tried or are being investigated. Under routine daily use, a nickel or copper sampling cone lasts about 2 months before requiring replacement. The skimmer cone is constructed from stainless steel and normally lasts more than a year.

This interface section of the ICP–MS instrument is a critical subsystem. It is important to note that the current commercial instruments differ with respect to electrical details associated with the interface. The Sciex ICP–MS utilizes a center-tap grounded ICP load coil.[17] This minimizes plasma voltage and seems to eliminate any discharge formation in the 1-torr region between the sampling and skimmer cones. The existence of this discharge has been noted by Olivares and Houk[18] in the instrument constructed at Iowa State University. In the VG PlasmaQuad, bias voltages are applied to the sampling and skimmer cones to minimize the discharge. It does appear, however, that a residual discharge

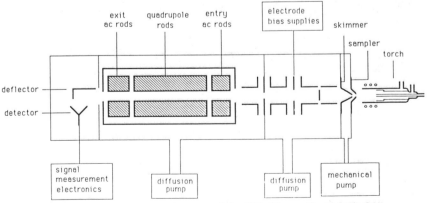

Figure 10.2. Schematic diagram of the VG PlasmaQuad ICP–MS.

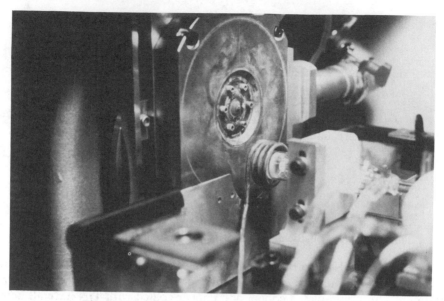

Figure 10.3. The sampling cone region of an ICP–MS.

remains in the current VG ICP–MS, because certain analytical characteristics concerning oxide and doubly charged species differ considerably between the VG PlasmaQuad and the Sciex Elan. These are discussed later (Section 10.4.2.3). It should also be noted that, at the time of writing, some controversy remains with respect to the existence and importance of this discharge, both from advantageous and disadvantageous points of view.

The low-vacuum part of the spectrometer following the skimmer contains input ion optics, the quadrupole mass spectrometer, and the ion detector. During operation, this region is maintained at approximately 0.133 to 1.33 mPa (10^{-6} to 10^{-5} torr). In the Sciex system, one-stage pumping with a helium cryogenic pump is used, while in the VG system, two stages of mechanical pumping are utilized.[2]

The input ion optics consists of ion lenses and a Bessel box containing a central stop to block passage of photons to the detector. Such photons are undesirable because the background signal level is raised by photons reaching the detector. To reduce photon noise even further, ions exiting the quadrupole are deflected to an off-axis ion detector.

The ICP–MS is a complex system controlled by a large number of variables (Table 10.2). Currently, the optimal settings and possible interaction of the experimental variables are not fully understood, and subsystems, such as the input ion optics, are still being refined by the manufacturers. Preliminary measurements on the effects of certain of these variables on analyte signals are presented in the next section.

From the point of view of an optical spectroscopist, the quadrupole mass

Table 10.2. ICP–MS Instrumental Parameters

A. Inductively coupled plasma
  1. Outer gas flow rate
  2. Intermediate gas flow rate
  3. Injector gas flow rate
  4. RF power
  5. Sampling depth
B. Interface
  1. Geometry of the sampling cone
  2. Geometry of the skimmer cone
  3. Sampler–skimmer separation
  4. Pressure between the sampling and skimmer cones
C. Mass spectrometer
  1. Input ion optics
    a. Input lenses
    b. Bessel box
  2. Quadrupole MS: rod voltages
  3. Output ion optics–detector
    a. Output lens voltages
    b. Deflector voltage
    c. Channel electron multiplier (CEM) voltage

spectrometer can be thought of as a scanning monochromator. It can be operated in a simple scanning fashion to measure a spectrum, such as that shown in Figure 10.4, or it can be operated in a slew-scanning fashion, where measurements are made by peak hopping. The resolution of the quadrupole mass spectrometers used for ICP–MS need not be high if oxide formation is minimal. Typically, the resolution is 0.5 to 1 amu (atomic mass unit). The resolution is defined as the width 10% up from the base of the peak.

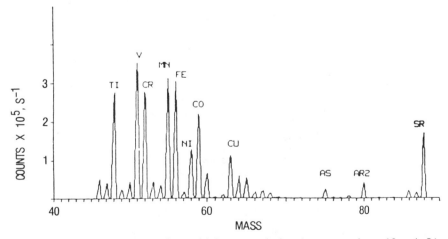

Figure 10.4. Mass spectrum of a multielement solution (concentrations, 10 μg/mL).

10.4   CONSIDERATIONS FOR ANALYTICAL MEASUREMENTS

Certain major points to be considered when conducting analytical measurements with an ICP–MS are summarized in Table 10.3. The main areas of concern are instrument settings, spectral overlap, analytical curve quality, and matrix effects.

10.4.1   Instrument Settings

Despite the large number of instrument parameters listed in Table 10.2, most variables do not have to be routinely adjusted. However, forward power, gas flows, and the mass spectrometer input ion–optic voltages have major effects on the analyte ion signals.

10.4.1.1   Radio Frequency Power, Injector Gas Flow, and Sampling Depth

In a detailed study[19] of the effects of plasma operating parameters on analyte signals in ICP–MS, the critical parameters were found to be forward power and injector gas flow rate. The basic signal dependence on forward power and injector gas flow rate is shown in Figure 10.5, which illustrates the dependence of analyte ion count for gallium on injector gas flow rate for a range of forward power settings. At each power setting tested, the signal is maximized at a particular injector gas flow rate. Also, power and injector gas flow rate appear to operate as a paired set of variables in that if one is changed, the other must also be changed to remaximize the signal intensity.

A set of similar data is shown in Figure 10.6 for ions of gallium, germanium, arsenic, indium, tin, antimony, thallium, lead, and bismuth. While exact details of the injector signal dependence on RF power and injector gas flow vary somewhat for different elements, they are similar in basic dependence, which

Table 10.3. Some Considerations for Quantitative Work in ICP–MS

A. Instrument settings
   1. RF power and gas flows
   2. Mass spectrometer input optic voltages
B. Spectral overlaps
   1. Isobaric spectral overlaps
   2. Background spectral features
   3. Oxide, hydroxide, and doubly charged analyte species
   4. Matrix-Induced spectral overlaps
C. Analytical curve quality
   1. Linearity
   2. Dynamic range
   3. Precision
   4. Detection limits
D. Matrix effects

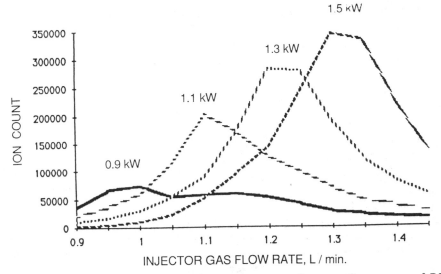

Figure 10.5. Dependence of Ga$^+$ signal on injector gas flow rate for a range of RF powers. (Sampling depth was 20 mm from the load coil.)

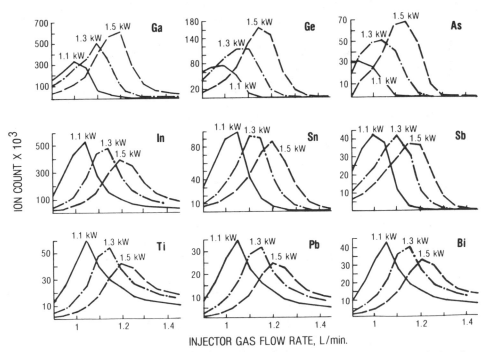

Figure 10.6. Dependence of Ga$^+$, Ge$^+$, As$^+$, In$^+$, Sn$^+$, Sb$^+$, Tl$^+$, Pb$^+$, and Bi$^+$ signals on injector gas flow rate for a range of RF powers. (Sampling depth 15 mm from the load coil.)

indicates that compromise settings are normally acceptable for a range of elements. The elements span three rows of the periodic table and cover a wide range of masses and ionization potentials.

Similar behavior is also observed for the transition metals and typical results for one column (Ni, Pd, and Pt) are shown in Figure 10.7. For these elements, the effect on the signal of sampling depth, ie, the distance of the sampling cone from the load coil, is also shown. The basic dependence of analyte ion count on power and injector flow rate is similar at each sampling depth. However, at constant power, as the plasma is sampled further from the load coil, the injector gas flow rate must be increased to maintain a maximum ion count.

The data presented in Figures 10.5 to 10.7 are consistent with the notion that a plume of analyte ions is present in the ICP situated just above the initial radiation zone, as designated by Koirtyohann and co-workers[20]. If one visualizes this zone as moving toward the load coil as power is increased (ie, away from the sampling cone orifice) and toward the sampling cone orifice as injector gas flow rate is increased, the signal changes illustrated by the data shown in Figures 10.5 to 10.7 can be rationalized. Thus, with respect to ICP-MS, a single optimized set of ICP conditions does not exist. At any one sampling depth, a whole range of RF power–injector flow rate value pairs provide roughly comparable signals, as will a different set at an altered sampling depth. What is important is that the set of values be very stable.

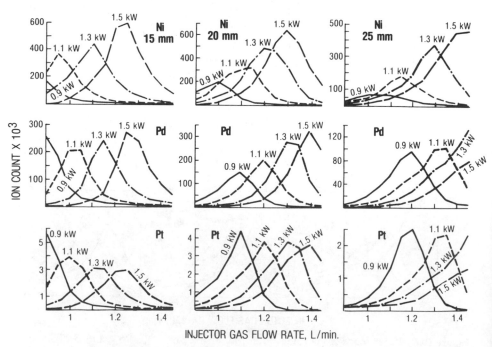

Figure 10.7. Dependence of Ni+, Pd+, and Pt+ signals on injector gas flow rate for a range of RF powers and sampling depths of 15, 20, and 25 mm.

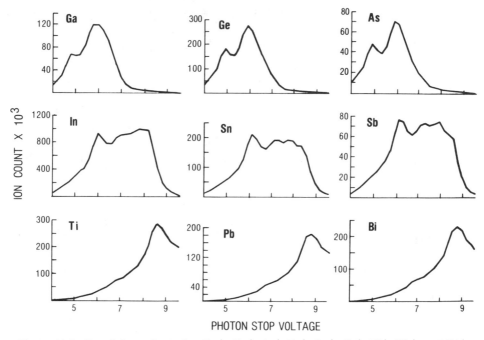

Figure 10.8. Signal dependence for Ga^+, Ge^+, As^+, In^+, Sn^+, Sb^+, Tl^+, Pb^+, and Bi^+ on the Bessel box photon stop voltage.

10.4.1.2 Mass Spectrometer Input Optic Voltages

A number of voltages must be set with respect to the input ion optics of the mass spectrometer. The exact nature of the dependence of the signal on these voltages is highly dependent on the design and type of input ion optics used. The optics vary from manufacturer to manufacturer and are under continued development. Thus, exact results cannot be presented that are representative of ICP–MS in general. One of the more critical values appears to be the voltage on the photon stop of the Bessel box in that the optimized setting is mass dependent. From the data shown in Figure 10.8 for an early version of the Sciex Elan, it is clear that the voltage value for the photon stop should be set to a different value depending on the mass of the analyte. Again, these data must not be taken as general or even representative of current instrumentation, but simply as a point to keep in mind, namely, that lens voltage settings may be complex and are a more serious problem when mass dependencies are present.

10.4.2 Spectral Overlaps

In a sense, the "wavelength table" for ICP–MS is the periodic table, and thus, spectral simplicity is anticipated. However, early papers in the field of ICP–MS overemphasized spectral simplicity. While not as prevalent as in atomic

Table 10.4. Relative Abundances of Naturally Occurring Isotopes

Z	A	1	2	3	4	5	6	7	8	9	10	11	12	13	14	15	16	17	18	19	20
1	H	99.9	0.01																		
2	He				100																
3	Li						7.5	92.5													
4	Be									100											
5	B										20	80									
6	C												98.9	1.11							
7	N														99.6	0.36					
8	O																99.8	0.08	0.20		
9	F																			100	
10	Ne																				90.5

Z	A	21	22	23	24	25	26	27	28	29	30	31	32	33	34	35	36	37	38	39	40
10	(Ne)	0.27	9.2																		
11	Na			100																	
12	Mg				79	10	11														
13	Al							100													
14	Si								92.2	4.7	3.1										
15	P											100									
16	S												95.0	0.75	4.2		0.02				
17	Cl															75.8		24.2			
18	Ar																0.34		0.07		96.6
19	K																			93.3	0.01
20	Ca																				96.9

Z	A	41	42	43	44	45	46	47	48	49	50	51	52	53	54	55	56	57	58	59	60
19	(K)	6.7																			
20	(Ca)		0.65	0.14	2.08		0.003		0.19												
21	Sc					100															
22	Ti						8.0	7.5	73.7	5.5	5.3										
23	V										0.25	99.7									
24	Cr										4.35		83.8	9.5	2.36						
25	Mn															100					
26	Fe														5.8		91.7	2.14	0.31		
27	Co																			100	
28	Ni																		67.8		26.4

Z	A	61	62	63	64	65	66	67	68	69	70	71	72	73	74	75	76	77	78	79	80
28	(Ni)	1.16	3.71		0.95																
29	Cu			69.1		30.9															
30	Zn				43.9		27.8	4.1	18.6		0.62										
31	Ga									60.0		40									
32	Ge										20.7		27.5	7.7	36.4		7.7				
33	As															100					
34	Se														0.9		9.0	7.5	23.5		50
35	Br																			50.7	
36	(Kr)																		0.35		2.25

(Continued)

Table 10.4   (*Continued*)

Z	A	81	82	83	84	85	86	87	88	89	90	91	92	93	94	95	96	97	98	99	100	A	Z
34	(Se)		90																			(Se)	34
35	(Br)	49.3																				(Br)	35
36	Kr		11.6	11.5	57		17.3															Kr	36
37	Rb					72.2		27.8														Rb	37
38	Sr				0.56		9.9	7.0	82.6													Sr	38
39	Y									100												Y	39
40	Zr										51.4	11.2	17.1		17.5		2.8					Zr	40
41	Nb													100								Nb	41
42	Mo												14.8		9.1	15.9	16.7	9.5	24.4		9.6	Mo	42
43	Tc		Does not occur naturally																			Tc	43
44	(Ru)																5.5		1.9	12.7	12.6	(Ru)	44

Z	A	101	102	103	104	105	106	107	108	109	110	111	112	113	114	115	116	117	118	119	120	A	Z
44	Ru	17.1	31.6		18.6																	Ru	44
45	Rh			100																		Rh	45
46	Pd		1.0		11.0	22.2	27.3		26.7		11.8											Pd	46
47	Ag							51.8		48.2												Ag	47
48	Cd						1.2		0.9		12.4	12.8	24.0	12.3	28.8		7.6					Cd	48
49	In													4.3		95.7						In	49
50	Sn												1.0		0.65	0.35	14.4	7.6	24.1	8.6	32.8	Sn	50
51	(Sb)																					(Sb)	51
52	(Te)																				0.9	(Te)	52

Z	A	121	122	123	124	125	126	127	128	129	130	131	132	133	134	135	136	137	138	139	140	A	Z
50	(Sn)		4.7		5.8																	(Sn)	50
51	Sb	57.3		42.7																		Sb	51
52	Te		2.4	0.87	4.6	7.0	18.7		31.8		34.5											Te	52
53	I							100														I	53
54	Xe				0.10		0.09		1.9	26.4	3.9	21.2	27.0		10.5		8.9					Xe	54
55	Cs													100								Cs	55
56	Ba										0.10		0.1		2.4	6.5	7.8	11.2	71.9			Ba	56
57	La																		0.09	99.9		La	57
58	Ce																0.19		0.26		88.5	Ce	58

Z	A	141	142	143	144	145	146	147	148	149	150	151	152	153	154	155	156	157	158	159	160	A	Z
58	(Ce)		11.1																			(Ce)	58
59	Pr	100																				Pr	59
60	Nd		27.1	12.2	23.9	8.3	17.2		5.7		5.6											Nd	60
61	Pm	Does not occur naturally																				Pm	61
62	Sm				3.1			15.0	11.2	13.8	7.4		26.7		22.8							Sm	62
63	Eu											47.8		52.2								Eu	63
64	Gd												0.20		2.2	14.9	20.6	15.7	24.7		21.7	Gd	64
65	Tb																			100		Tb	65
66	(Dy)																0.06		0.10		2.3	(Dy)	66

Z	A	161	162	163	164	165	166	167	168	169	170	171	172	173	174	175	176	177	178	179	180	A	Z
66	Dy	18.9	25.5	24.9	28.2																	Dy	66
67	Ho					100																Ho	67
68	Er		0.14		1.6		33.4	22.9	27.0		15.0											Er	68
69	Tm									100												Tm	69
70	Yb								0.14		3.0	14.3	21.9	16.2	31.8		12.7					Yb	70
71	Lu															97.4	2.6					Lu	71
72	Hf														0.18		5.2	18.5	27.2	13.8	35.1	Hf	72
73	(Ta)																				0.01	(Ta)	73
74	(W)																				0.13	(W)	74

(Continued)

Table 10.4 (*Continued*)

Z	A	181	182	183	184	185	186	187	188	189	190	191	192	193	194	195	196	197	198	199	200
73	Ta	99.9																			
74	W		26.3	14.3	30.7		28.6														
75	Re					37.4		62.6													
76	Os				0.02		1.6	1.6	13.3	16.1	26.4		41.0								
77	Ir											37.4		62.6							
78	Pt										0.13		0.78		32.9	33.8	25.3		7.12		
79	Au																	100			
80	(Hg)																0.15		10.2	16.9	23.1

Z	A	201	202	203	204	205	206	207	208	209	210	211	212	213	214	215	216	217	218	219	220
80	Hg	13.2	29.7		6.8																
81	Tl			29.5		70.5															
82	Pb				1.4		24.1	22.1	52.4												
83	Bi									100											

84 (Po), 85 (At), 86 (Rn), 87 (Fr), 88 (Ra), 89 (Ac), do not occur naturally.

Z	A	221	222	223	224	225	226	227	228	229	230	231	232	233	234	235	236	237	238	239	240
90	Th								100				100								
91	Pa				Does not occur naturally																
92	U														0.06	0.72			99.3		

93 (Np), 94 (Pu), 95 (Am), 96 (Cm), 97 (Bk), 98 (Cf), 99 (Es), 100 (Fm), 101 (Md), 102 (No), 103 (Lw), do not occur naturally.

emission spectrometry, spectral overlaps do exist in ICP–MS, and when they occur, they are quite serious. In addition, a number of molecular species are observed in ICP–MS from analyte, background, and matrix-induced species.

10.4.2.1 Isobaric Spectral Overlaps

Isobaric spectral overlaps refer to the spectral overlaps that occur among the natural stable isotopes of the elements. Table 10.4 presents a complete listing of the elements and the abundances of the naturally occurring isotopes. As an example, an isobaric overlap occurs between ^{48}Ti and ^{48}Ca, but this overlap can be corrected for titanium determination by measuring another isotope of calcium and using the natural abundance information. That is, by measuring ^{44}Ca and subtracting 0.08981 of this measured count off the ^{48}Ti signal, the ^{48}Ca overlap is corrected. In a similar manner, the ^{58}Fe overlap on ^{58}Ni can be corrected by first measuring ^{56}Fe, and the ^{64}Ni overlap on ^{64}Zn can be corrected by measuring ^{60}Ni and subtracting the proportional amount of the signal from the desired mass. Since such interference problems are exactly predictable, these corrections can be implemented with appropriate software, and current instruments are capable of implementing such corrections automatically.

10.4.2.2 Background Spectral Features

A knowledge of the background spectral features in ICP–MS is required to recognize potential spectral interference problems that can occur at particular analyte masses. Background spectra[21] for distilled and deionized water are shown in Figures 10.9 and 10.10 for the mass ranges 1 to 43 and 42 to 84 amu, respectively. In general, no significant background features are observed above mass 84. Similar spectra are reported[21] for 5% solutions (vol/vol) of nitric acid, hydrochloric acid, and sulfuric acid, and the major species observed are summarized in Table 10.5. The plasma operating conditions for these data are: RF power, 1.2 kW; injector gas flow rate, 1.06 L/min; outer gas flow rate, 12 L/min; intermediate gas flow rate, 1.4 L/min; and the plasma is sampled at a depth of 15 mm from the load coil. These settings are for the determination of the first-row transition elements. The mass spectrometer was operated in the high-resolution mode (peak width approximately 0.6 amu at 10% off the baseline) and the measurement time was 1 s per point, with a scan increment of 0.05 amu. Some of the more obvious and serious spectral overlap problems include $^{28}Si^+$ (with $^{14}N^{14}N^+$), $^{31}P^+$ (with $^{14}N^{16}OH^+$), and $^{80}Se^+$ (with $^{40}Ar^{40}Ar^+$). Also, a basic species, such as $^{40}Ar^{16}O^+$, is not just a problem for $^{56}Fe^+$, but, with all its isotopic combinations and hydride forms, can affect count rates at a number of first-row transition element masses (Table 10.6). Likewise, Ar_2^+ potentially interferes with all Se isotopes. Some of these potential interferences are small; for example, the peak at mass 51, which is thought to be due to a combination of $^{36}Ar^{15}N^+$ and $^{36}Ar^{14}NH^+$, appears with a count rate of about 30 to 40 counts

Table 10.5. Major Background Species in ICP-MS for the Introduction of H_2O (5% nitric acid), 5% Sulfuric Acid, and 5% Hydrochloric Acid[a]

Mass Element[b]	$H_2O(5\% \ HNO_3)$	5% H_2SO_4	5% HCl
1 **H(99.985)**	^{1}H		
2 H(0.015)	^{2}H		
3			
4 **He(100)**			
5			
6 Li(7.5)			
7 **Li(92.5)**			
8			
9 **Be(100)**			
10 B(19.91)			
11 **B(80.09)**			
12 **C(98.89)**	^{12}C		
13 C(1.11)	^{13}C		
14 **N(99.63)**	^{14}N		
15 N(0.37)	^{15}N		
16 **O(99.76)**	^{16}O		
17 O(0.04)	^{16}OH		
18 O(0.20)	^{16}OH$_2$		
19 **F(100)**	^{16}OH$_3$		
20 **Ne(90.92)**	^{18}OH$_2$		
21 Ne(0.26)	^{18}OH$_3$		
22 Ne(8.82)			
23 **Na(100)**			
24 **Mg(78.8)**			
25 Mg(10.15)			
26 Mg(11.05)			
27 **Al(100)**			
28 **Si(92.21)**	^{14}N^{14}N, ^{12}C^{16}O		
29 Si(4.7)	^{14}N^{14}NH, ^{12}C^{16}OH		
30 Si(3.09)	^{14}N^{16}**O**		
31 **P(100)**	^{14}N^{16}**OH**		
32 **S(95.02)**	^{16}O^{16}O	^{32}S	
33 S(0.75)	^{16}O^{16}OH	^{33}S, ^{32}SH	
34 S(4.21)	^{16}O^{18}O	^{34}S, ^{33}SH	
35 **Cl(75.77)**	^{16}O^{18}OH	^{34}SH	^{35}Cl
36 Ar(0.34), S(0.02)	^{36}**Ar**	^{36}S	^{35}ClH
37 Cl(24.23)	^{36}ArH	^{36}SH	^{37}Cl
38 Ar(0.06)	^{38}Ar		^{37}ClH
39 **K(93.08)**	^{38}ArH		
40 **Ar(99.6), Ca(96.97)**, K(0.01)	^{40}**Ar**		
41 K(6.91)	^{40}**ArH**		
42 Ca(0.64)	^{40}**ArH$_2$**		
43 Ca(0.14)			
44 Ca(2.06)	^{12}C^{16}O^{16}O		
45 **Sc(100)**	^{12}C^{16}O^{16}OH		
46 Ti(7.99), Ca(0.003)	^{14}N^{16}O^{16}O	^{32}S^{14}N	
47 Ti(7.32)		^{33}S^{14}N	
48 **Ti(73.98)**, Ca(0.19)		^{34}S^{14}N, ^{32}S^{16}**O**	
49 Ti(5.46)		^{33}S^{16}**O**	^{35}Cl^{14}N

Table 10.5. (*Continued*)

Mass Element[b]	$H_2O(5\% \text{ HNO}_3)$	$5\% \text{ H}_2\text{SO}_4$	$5\% \text{ HCl}$
50 Ti(5.25), Cr(4.35), V(0.24)	$^{36}Ar^{14}N$	$^{34}S^{16}O$	
51 **V(99.76)**			$^{37}Cl^{14}N$, $^{35}Cl^{16}O$
52 **Cr(83.76)**	$^{40}Ar^{12}C$, $^{36}Ar^{16}O$	$^{36}S^{16}O$	$^{35}Cl^{16}OH$
53 Cr(9.51)			$^{37}Cl^{16}O$
54 Fe(5.82), Cr(2.38)	$^{40}Ar^{14}N$		$^{37}Cl^{16}OH$
55 **Mn(100)**	$^{40}Ar^{14}NH$		
56 **Fe(91.66)**	$^{40}Ar^{16}O$		
57 Fe(2.19)			
58 **Ni(67.77)**, Fe(0.33)			
59 **Co(100)**			
60 Ni(26.16)			
61 Ni(1.25)			
62 Ni(3.66)			
63 **Cu(69.1)**			
64 **Zn(48.89)**, Ni(1.16)		$^{32}S^{16}O^{16}O$, $^{32}S^{32}S$	
65 Cu(30.9)		$^{33}S^{16}O^{16}O$, $^{32}S^{33}S$	
66 Zn(27.81)		$^{34}S^{16}O^{16}O$, $^{32}S^{34}S$	
67 Zn(4.11)			$^{35}Cl^{16}O^{16}O$
68 Zn(18.57)	$^{40}Ar^{14}N^{14}N$	$^{36}S^{16}O^{16}O$, $^{33}S^{36}S$	
69 **Ga(60.16)**			$^{37}Cl^{16}O^{16}O$
70 Ge(20.51), Zn(0.62)	$^{40}Ar^{14}N^{16}O$		
71 Ga(39.84)			$^{36}Ar^{35}Cl$
72 Ge(27.4)	$^{36}Ar^{36}Ar$	$^{40}Ar^{32}S$	
73 Ge(7.76)		$^{40}Ar^{33}S$	$^{36}Ar^{37}Cl$
74 **Ge(36.56)**, Se(0.87)	$^{36}Ar^{38}Ar$	$^{40}Ar^{34}S$	
75 **As(100)**			$^{40}Ar^{35}Cl$
76 Ge(7.77), Se(9.02)	$^{36}Ar^{40}Ar$	$^{40}Ar^{36}S$	
77 Se(7.58)	$^{36}Ar^{40}ArH$		$^{40}Ar^{37}Cl$
78 Se(23.52), Kr(0.35)	$^{38}Ar^{40}Ar$		
79 **Br(50.54)**	$^{38}Ar^{40}ArH$		
80 **Se(49.82)**, Kr(2.27)	$^{40}Ar^{40}Ar$	$^{32}S^{16}O^{16}O^{16}O$	
81 Br(49.46)	$^{40}Ar^{40}ArH$	$^{32}S^{16}O^{16}O^{16}OH$	
82 Kr(11.56), Se(9.19)	$^{40}Ar^{40}ArH_2$	$^{34}S^{16}O^{16}O^{16}O$	
83 Kr(11.55)		$^{34}S^{16}O^{16}O^{16}OH$	
84 **Kr(56.9)**, Sr(0.56)		$^{36}S^{16}O^{16}O^{16}O$	

[a] Most abundant species designated in boldface.
[b] Percentage natural abundance given in parentheses.

per second (cps). This peak hardly represents a major interference on $^{51}V^+$, which has a signal level of about 5×10^5 cps/μg/mL).

A somewhat more subtle problem can arise if background species are not carefully considered. Numerous interelement (isobaric) spectral overlaps exist due to the natural stable isotopic overlaps of the elements, as discussed earlier. For example, at mass 58 a minor isotope of Fe coincides with the major isotope of Ni. The correction consists of measuring the signal at mass 56, and assuming it to come completely from $^{56}Fe^+$, subtracting 0.0036 times that measured count off the count at mass 58, thus correcting for the contribution of $^{58}Fe^+$ to the

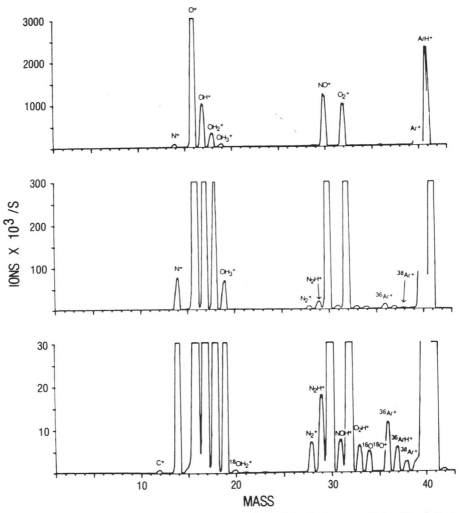

Figure 10.9. Background mass spectra for distilled, deionized water (1 to 43 amu) at amplification levels of 1 (top), 10 (middle), and 100 (bottom).

count at that mass. However, the signal at mass 56 may also have a major contribution due to $^{40}Ar^{16}O^+$, and depending on the actual relative and absolute amounts of Ni and Fe present, an erroneous correction could result.

Among the more prominent species observed for 5% HCl, as noted in Table 10.5, are Cl^+, ClO^+, ClN^+, Cl_2^+, and $ArCl^+$. These species significantly increase the complexity of the background, particularly when all isotopic combinations are considered.[21] The species ClN^+ and ClO^+ interfere with several first-row transition elements. The most serious interference is that of $^{35}Cl^{16}O^+$ on $^{51}V^+$. This vanadium isotope is 99.76% abundant and the second isotope, ^{50}V, which is only 0.24% abundant, suffers interferences from $^{35}Cl^{15}N^+$, $^{50}Ti^+$, and $^{50}Cr^+$.

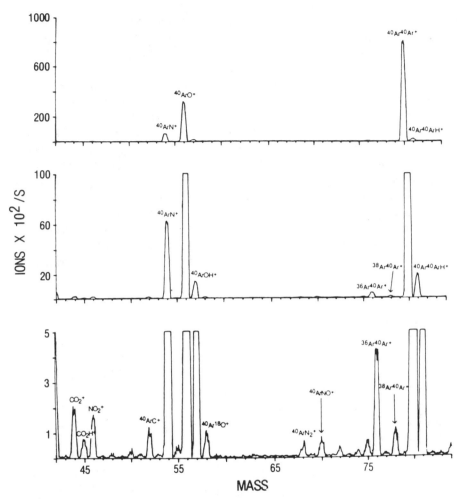

Figure 10.10. Background mass spectra for distilled, deionized water (42 to 84 amu) at amplification levels of 1 (top), 10 (middle), and 200 (bottom).

Table 10.6. Details of ArO^+ Spectral Interferences in ICP–MS

Mass	Elements	ArO^+ species
51	V(99.76)	
52	Cr(83.76)	$^{36}Ar^{16}O$
53	Cr(9.51)	$^{36}Ar^{17}O$, $^{36}Ar^{16}OH$
54	Fe(5.82), Cr(2.38)	$^{36}Ar^{18}O$, $^{38}Ar^{16}O$, $^{36}Ar^{17}OH$
55	Mn(100)	$^{38}Ar^{17}O$, $^{36}Ar^{18}OH$, $^{38}Ar^{16}OH$
56	Fe(91.66)	$^{38}Ar^{18}O$, $^{40}Ar^{16}O$, $^{38}Ar^{17}OH$
57	Fe(2.19)	$^{40}Ar^{17}O$, $^{40}Ar^{16}OH$, $^{38}Ar^{18}OH$
58	Ni(67.77), Fe(0.33)	$^{40}Ar^{18}O$, $^{40}Ar^{17}OH$
59	Co(100)	$^{40}Ar^{18}OH$
60	Ni(26.16)	

Thus, ICP–MS cannot really be used to determine vanadium in a solution matrix containing chloride or chlorine compounds. At higher masses, ClO_2^+ and Cl_2^+ species create problems for gallium and germanium determinations, and $^{40}Ar^{35}Cl^+$ presents a major problem in the measurement of $^{75}As^+$. The $^{40}Ar^{35}Cl^+$ peak is quite intense, and arsenic is monoisotopic, leaving no second-choice isotope. Thus, determinations of As and V by ICP–MS cannot be conducted in solutions containing chlorine species. In effect, if elements such as V and As are of interest, these spectral interference problems dictate the type of sample dissolution and solution preparation procedures that are free from chlorine. For alternative acid mixtures, such as 5% H_2SO_4, other molecular species are observed (Table 10.5). Up to about mass 84, great care must be exercised in choosing appropriate solutions for certain element masses.

The user of an ICP–MS should remember that the spectral background in ICP–MS may be quite complex and that minor isotopic species of the background spectra cannot be ignored when elemental determinations are being made at the ultra-trace level (sub-nanogram-per-milliliter) capability of ICP–MS. Certainly, HNO_3 is the first acid of choice because it has a background spectrum very similar to that of water.

10.4.2.3 Oxide, Hydroxide, and Doubly Charged Analyte Species

For a large number of elements, the singly charged elemental analyte ion M^+ is not the only analyte species that is observed in ICP–MS. Many elements form monoxides, MO^+, some form doubly charged ions, M^{2+}, and a few form hydroxide species, MOH^+. The presence of these species has been noted in a number of papers.[7–10,19,22,23] Vaughan and Horlick[24] have discussed and tabulated these species.

The species cited above are important in ICP–MS from several points of view. Perhaps the most important aspect is that analyte-oxide species may cause serious spectral interferences at other analyte–ion masses. For example, titanium has five naturally occurring isotopes: ^{46}Ti (7.99%), ^{47}Ti (7.32%), ^{48}Ti (73.98%), ^{49}Ti (5.46%), and ^{50}Ti (5.25%). The corresponding oxide species, ^{16}O, at masses 62, 63, 64, 65, and 66, result in spectral interference problems for ^{62}Ni, ^{63}Cu, ^{64}Zn, ^{65}Cu, and ^{66}Zn. This is but one of literally hundreds of potential spectral interference problems involving oxide species. In addition to the obvious interference, which results in falsely high values for the element that is overlapped by an oxide species, there can be another problem. As discussed in Section 10.4.2.1, isobaric corrections are frequently required in ICP–MS when the chosen mass of an analyte suffers a spectral overlap caused by a naturally occurring isotope of another element. Normally, an interference-free isotope of the interfering element is measured and, based on its signal level and the natural abundance information of the interfering element, a correction can be calculated to be applied at the mass of the sought-for element. However, the presence of oxide species from a third element could cause nonapplicable and inappropriate isobaric corrections to be implemented.

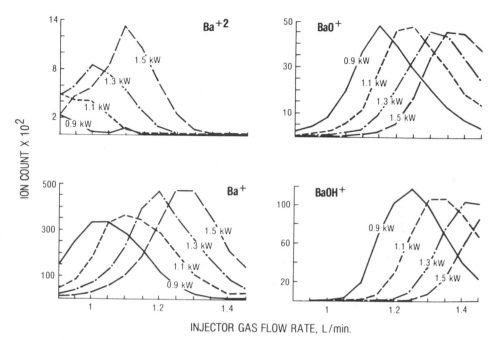

Figure 10.11. Signal dependence for Ba^{2+}, Ba$^+$, BaO$^+$, and BaOH$^+$ on injector gas flow rate for a range of RF powers (0.9, 1.1, 1.3, and 1.5 kW).

Earlier, the forward power, injector gas flow rate, and sampling depth were noted as major parameters to consider in ICP–MS. In particular, plots of signal (ion count) as a function of injector gas flow rate at different powers clearly illustrate trends in behavior. Such plots[24] are shown in Figure 10.11 for Ba^{2+}, Ba$^+$, BaO$^+$, and BaOH$^+$. The data for Ba$^+$ typify those for most analyte ions: at one particular power, the signal peaks as a function of injector gas flow rate. As the forward power increases, the flow rate value at which the signal is maximized also increases. The associated analyte species have essentially the same basic behavior. The plots of RF power versus injector gas flow rate for BaO$^+$ and BaOH$^+$ are very similar to those of Ba$^+$, except that the signals peak at successively higher flow rates when compared at the same power. On the other hand, the Ba^{2+} plots are shifted to lower injector gas flow rates.

The relative and the normalized ion count for Ba species as a function of injector gas flow rate is shown in Figure 10.12 for a forward power setting of 1.3 kW. Under most conditions, the singly charged ion Ba$^+$ is the major analyte species (Figure 10.12A). Only at high injector gas flow rates do the oxide and hydroxide species dominate. The crucial role of injector gas flow rate in controlling the relative distribution of the analyte species is emphasized by the normalized plot (Figure 10.12B). With this representation, it is easy to see how the species distribution changes as the injector gas flow rate is altered. At flow rates around 1.0 L/min, the number of doubly charged species reaches a maximum. Increasing the flow rate, the condition for the Ba$^+$ maximum signal

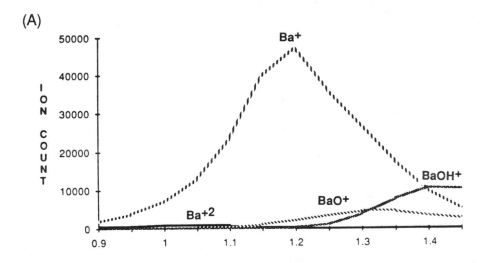

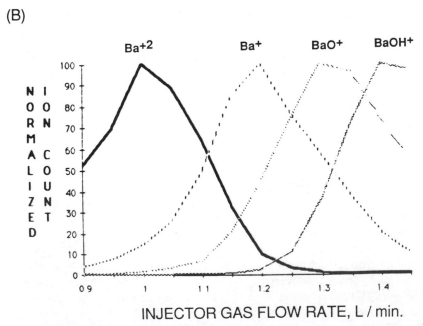

INJECTOR GAS FLOW RATE, L / min.

Figure 10.12. Relative (*A*) and normalized (*B*) plots of signal as a function of injector gas flow rate at an RF power of 1.3 kW for barium species.

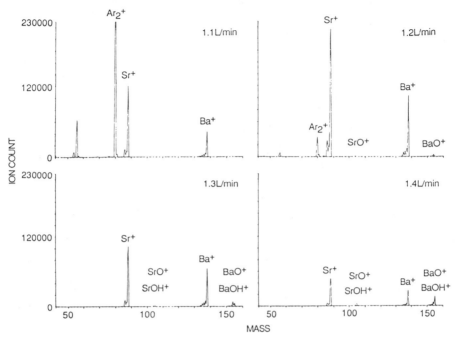

Figure 10.13. Mass spectra for barium and strontium species at a range of injector gas flow rates.

is reached at around 1.2 L/min for a forward power of 1.3 kW. Similarly, BaO^+ and $BaOH^+$ reach their maxima at even higher flow rates—around 1.3 and 1.4 L/min, respectively. Mass spectra illustrating these trends are shown in Figure 10.13. Thus, it may be impossible, or at least very difficult, to get minimum oxide counts and a maximum signal for the singly charged ion. To reduce oxide and hydroxide problems, the ICP must be operated at lower injector gas flow rates. Such conditions (ie, conditions that give suboptimal M^+ counts) may lead to the formation of doubly charged ions, but these are not formed to a great extent, and, in general, fewer elements form detectable doubly charged ions in the plasma than form oxides.

Data similar to those shown in Figures 10.11 to 10.13 have been measured for other analyte species,[24] such as strontium, molybdenum, titanium, tungsten, and cerium, although the extent of oxide formation is far more serious for the latter three analytes. Finally, the doubly charged oxide and hydroxide spectra of barium can result in the spectral interference problems shown in Table 10.7. A more extensive table has been prepared for a number of elements.[24]

10.4.2.4 Diagnostic Studies with Inductively Coupled Plasma–Mass Spectrometry

Fundamental characterization of the ICP clearly is a potential application area for mass spectrometry. To make accurate extrapolations of ICP-MS data back

Table 10.7. Doubly Charged Oxide and Hydroxide Species for Barium[a]

Mass	Abundance (%)	Associated Species	Mass	Affected elements[b]	
130	0.101	Ba^{2+}	65	Cu(30.9)	
132	0.097	Ba^{2+}	66	Zn(27.8)	
134	2.42	Ba^{2+}	67	Zn(4.11)	
135	6.59	Ba^{2+}	67.5	^{68}Zn(18.6)	^{67}Zn(4.11)
136	7.81	Ba^{2+}	68	Zn(18.6)	
137	11.32	Ba^{2+}	68.5	**^{69}Ga(60.2)**	^{68}Zn(18.6)
138	71.66	Ba^{2+}	69	**Ga(60.2)**	
130	0.101	BaO	146	Nd(17.2)	
132	0.097	BaO	148	Sm(11.4)	Nd(5.73)
134	2.42	BaO	150	Sm(7.47)	Nd(5.62)
135	6.59	BaO	151	Eu(47.8)	
136	7.81	BaO	152	**Sm(26.6)**	Gd(0.21)
137	11.32	BaO	153	**Eu(52.2)**	
138	71.66	BaO	154	Sm(22.4)	Gd(2.23)
130	0.101	BaOH	147	Sm(15.1)	
132	0.097	BaOH	149	Sm(14.0)	
134	2.42	BaOH	151	Eu(47.8)	
135	6.59	BaOH	152	**Sm(26.6)**	Gd(0.21)
136	7.81	BaOH	153	**Eu(52.2)**	
137	11.32	BaOH	154	Sm(22.4)	Gd(2.23)
138	71.66	BaOH	155	Gd(15.1)	

[a] Most abundant species designated in boldface.
[b] Percentage natural abundance given in parentheses.

to ICP–AES data, however, it is first very important to establish that the signals observed in mass spectrometry accurately reflect the unperturbed ICP discharge. One serious problem for current diagnostic studies is that different ICP–MS systems provide conflicting data. This is particularly true for oxide and doubly charged analyte species.

All the data presented in Section 10.4.2.3 were from a Sciex Elan ICP–MS. Data from the instrument developed at Iowa State University (ISU)[18] indicate that oxide levels tend to be somewhat lower than those reported for the Sciex instrument, although the basic effect of injector gas flow rate and RF power on the oxide ion signal is similar to that shown previously.

However, the effect of injector gas flow rate and RF power on the doubly charged ions is opposite to that shown in Figures 10.11 and 10.12. The ratio of M^{2+} to M^+ increases with increasing injector gas flow rate and decreases with increasing RF power. This type of behavior has also been reported by users of the VG PlasmaQuad.[25] This is opposite to the behavior indicated by the data shown in Figures 10.11 and 10.12 and in Reference 24. This observed difference in behavior between these instruments is likely due to a residual discharge in the 133-Pa (1-torr) region of the ISU and the VG PlasmaQuad systems. As stated in

Section 10.3, the discharge seems to be eliminated in the Sciex instrument by center-tap grounding of the ICP load coil. The residual discharge in the ISU and VG PlasmaQuad instruments could result in lower oxide levels, and should the discharge increase as the injector gas flow rate increases, the observed increase in doubly charged ion levels could be explained. The discussion, at this stage, remains vague until more data are available. However, the danger of casually making diagnostic extrapolations of ICP-MS data back to ICP-AES must be noted.

Despite this caution, certain available data do indicate that results from the Sciex Elan can be extrapolated to an unperturbed ICP discharge. Gillson and

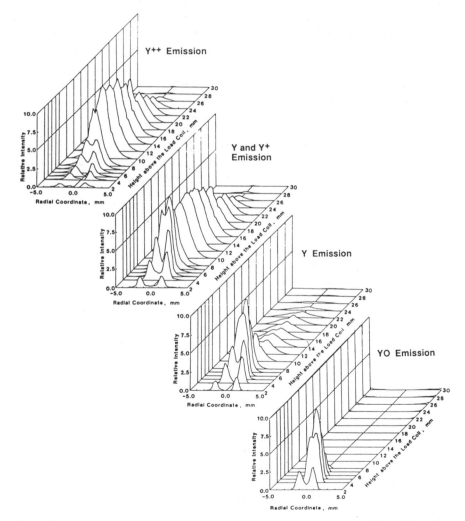

Figure 10.14. Spatial distribution of yttrium species emission in ICP-AES. (From Reference 11, with permission.)

Horlick[16,26,27] have directly compared ICP-MS and ICP-AES data for Sr^+ and found that the ion counts and atomic fluorescence signals behave similarly as a function of injector gas flow rate and RF power. Also, as mentioned in Section 10.4.1.1, the ion count dependence on the injector gas flow rate can be rationalized in terms of moving the initial radiation zone towards or away from the sampling cone orifice. In fact, this is an effective visualization to use in explaining the Ba species data shown in Figure 10.12. As the injector gas flow rate is increased, the initial radiation zone is pushed toward the sampling cone, thus resulting in the sampling of a cooler zone of the ICP. At high injector gas flow rates, the initial radiation zone can be thought of as "blown" right into the sampling orifice, at which point one would expect oxide species to become significant. An excellent set of atomic emission data provided by Furuta[11] lend credibility to this scenario (Figure 10.14). These data show the spatial emission structure for YO, Y, Y^+, and Y^{2+} emission. The YO emission seen within the initial radiation zone defined by Y emission and high injector gas flow rate should move these oxide species to the sampling cone. These data and the atomic fluorescence data mentioned earlier indicate that the data shown in Figures 10.5 to 10.13 may be transferable to ICP-AES models of the plasma. Relatedly, Houk and co-workers[4] found that the ion temperature measured in the $Ar-N_2$ ICP, using a pure N_2 outer flow, is less than the that of an argon ICP operated at the same power level. These data also explain why the detection limits of "hard" ion lines from the $Ar-N_2$ ICP are inferior to those measured from an argon ICP.

10.4.2.5 Matrix-Induced Spectral Overlaps

The foregoing discussion shows that a number of spectral overlap problems are caused by the matrix. A matrix element can cause an isobaric overlap problem, such as a high concentration of iron interfering with the major isotope of nickel. The oxide of a matrix element may cause a problem; for example, trace copper determinations are difficult if titanium is present because $^{47}Ti^{16}O^+$ coincides with $^{63}Cu^+$ and $^{49}Ti^{16}O^+$ overlaps $^{65}Cu^+$. Ions such as Cl^- and SO_4^{2-} can generate numerous background molecular species, and even argon combines with many species ($ArCl^+$, ArS^+, ArP^+). It also should be noted that argon seems to combine with just about any matrix element. If 1000 $\mu g/mL$ of a matrix element is present, argon species such as $ArNa^+$, $ArNi^+$, $ArGa^+$, and $ArSe^+$ are observed at significant count levels.

10.4.3   Basic Quantitative Performance

The analytical curves for ICP-MS have unity slope on a log–log plot with a dynamic range of five to six orders of magnitude. An analytical curve for cadmium (Figure 10.15) was obtained using rhodium as an internal standard. Typical short-term RSDs are 1 to 3%, but medium-term drift (3 hours) may vary from element to element, ranging from 10% to as high as 50%. This is clearly an area in need of improvement, and manufacturers and researchers are currently

Table 10.8. ICP-MS Detection Limits [ng/mL(ppb)][a] for the Sciex Elan Model 250.

(3σ, 10 Second Integration)

IA	IIA	IIIB	IVB	VB	VIB	VIIB	VIII	VIII	VIII	IB	IIB	IIIA	IVA	VA	VIA	VIIA	O
H																30* H	He
0.1 Li	0.1 Be											0.1 B	50 C	2* N	O	F	Ne
0.06 Na	0.10 Mg											0.1 Al	10 Si	2* P	S	Cl	Ar
1* K	5 Ca	0.08 Sc	0.06 Ti	0.03 V	0.02 Cr	0.04 Mn	0.2 Fe	0.02 Co	0.03 Ni	0.03 Cu	0.08 Zn	0.08 Ga	0.08 Ge	0.4 As	1 Se	100 Br	Kr
0.02 Rb	0.02 Sr	0.02 Y	0.03 Zr	0.02 Nb	0.08 Mo	Tc	0.05 Ru	0.02 Rh	0.06 Pd	0.04 Ag	0.07 Cd	0.02 In	0.03 Sn	0.02 Sb	0.04 Te	0.02 I	Xe
0.02 Cs	0.02 Ba	0.01 La	0.03 Hf	0.02 Ta	0.06 W	0.06 Re	0.02 Os	0.06 Ir	0.08 Pt	0.1 Au	0.1 Hg	0.05 Tl	0.05 Pb	0.04 Bi	Po	At	Rn
Fr	Ra	Ac															

0.01 Ce	0.01 Pr	0.02 Nd	Pm	0.04 Sm	0.04 Eu	0.04 Gd	0.01 Tb	0.04 Dy	0.01 Ho	0.02 Er	0.01 Tm	0.03 Yb	0.01 Lu
0.02 Th	Pa	0.02 U	Np	Pu	Am	Cm	Bk	Cf	Es	Fm	Md	No	Lw

* μg ml⁻¹   ▨ Negative Ion Mode

Source: Courtesy of Sciex.
[a] Asterisks indicate detection limits in micrograms per milliliter.

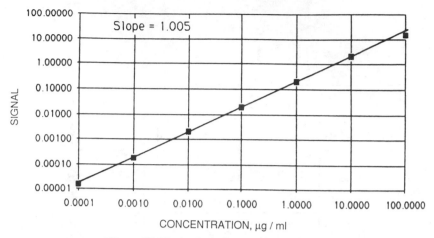

Figure 10.15. Analytical curve for cadmium.

addressing the problem. For example, internal standardization is of some help in improving precision, as is the utilization of isotope dilution techniques.

Detection limits are truly excellent, many approaching the picogram-per-milliliter range. Current detection limits for the Sciex spectrometer (Table 10.8) can be compared with those for the VG spectrometer (Table 10.9).

Table 10.9. ICP–MS Detection Limits (ng/mL)[a] for the VG PlasmaQuad

0.01–0.1			0.1–1	1–10	> 10
Li	In	Yb	B	K	C
Be	Sn	Lu	Na	Ca	O
Mg	Sb	Hf	Al	Fe	F
Ti	Te	Ta	Sc	Br	Si
V	Cs	W	Cr	I	P
Co	Ba	Re	Mn		S
Ni	La	Os	Zn		Cl
Cu	Ce	Au	Se		
Ga	Pr	Hg	Ru		
Ge	Nd	Tl	Rh		
As	Pm	Pb	Pd		
Rb	Sm	Bi	Ir		
Sr	Eu	Th	Pt		
Y	Gd	U			
Zr	Tb				
Nb	Dy				
Mo	Ho				
Ag	Er				
Cd	Tm				

[a] Detection limits for dilute aqueous solutions calculated as 2σ values of the blank, using 10 s integration, single ion monitoring.

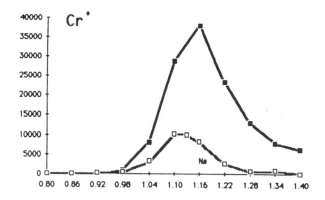

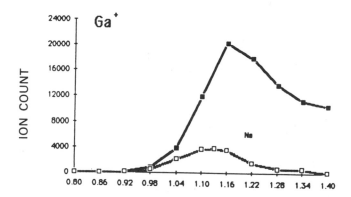

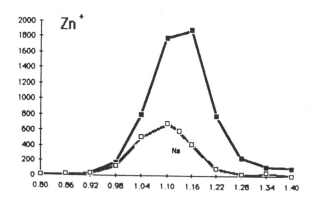

INJECTOR GAS FLOW RATE, L/min.

Figure 10.16. Matrix effect of 1000 μg/mL sodium (open squares) on 1 μg/mL of chromium, gallium, and zinc (solid squares) as a function of injector gas flow rate. Sampling depth, 18 mm; to RF power, 1.3 kW.

The ICP–MS, of course, lends itself to isotope dilution analysis and should foster widespread use of this analytical technique. Isotope dilution analysis can provide benchmark results and is easily implemented using ICP–MS instrumentation. In this method, a known weight of the sample is spiked with the sought-for element. The spike must be of a known amount and must have an isotopic composition different from that of the analyte. Based on the change in the isotopic composition produced by the spike, the concentration of the analyte in the sample can be calculated.

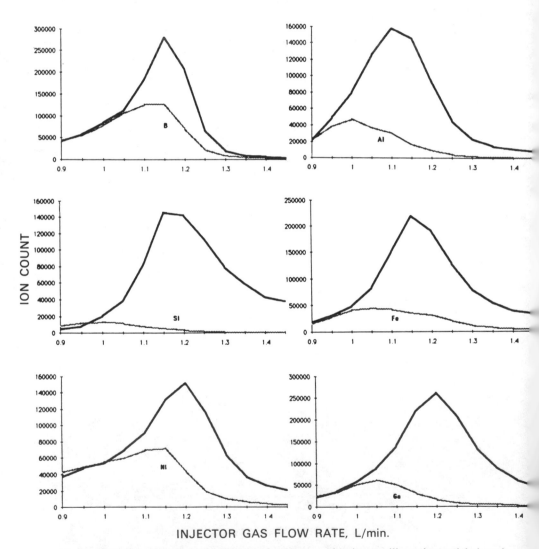

Figure 10.17. Matrix effect of 1000 μg/mL of boron, aluminum, silicon, iron, nickel, and gallium on 1 μg/mL of copper as a function of injector gas flow rate. Sampling depth, 18 mm; RF power, 1.3 kW.

10.4.4 Matrix Effects

The sample matrix, in addition to introducing background spectral features, may also induce actual changes in the analyte signal intensity.[28] In general, a high concentration of essentially any matrix element suppresses the signal of a trace analyte in ICP–MS. The data shown in Figure 10.16 illustrate the effect of 1000 μg/mL of sodium on 1 μg/mL of Cr^+, Ga^+, and Zn^+. The data shown in Figure 10.17 illustrates that 1000 μg/mL of boron, aluminum, silicon, iron, nickel, or gallium all have the same basic effect on 1 μg/mL of copper. The effect is not understood at this time, but because such generalized suppressions are not observed in ICP–AES, the suppressions are perhaps associated with the MS part of the system. The data shown in Figures 10.16 and 10.17 indicate that the suppression can be minimized by lowering the nebulizer flow rate, but at a sacrifice of analyte sensitivity. Currently this matrix effect constitutes a problem for ICP–MS measurements of most "real" samples and is an area warranting further study.

10.5 APPLICATIONS AND RESULTS

To date, the vast majority of published ICP–MS results have originated from the laboratories of the initial developers at Iowa State University, Sciex, the University of Surrey, the British Geological Survey, and VG Instruments. Most of the applications and results from these laboratories have been reviewed elsewhere.[2,10] Users of commercial instruments began obtaining results during 1984 and 1985. Most of these results have not yet been published, but results have been presented at conference symposia devoted to ICP–MS. Certain of these presentations are summarized in this section because they provide a current perspective of ICP–MS. Although the conference abstracts may not be readily available to all readers, a number of these presentations are already submitted for publication or are in press.

Douglas and co-workers reported on a number of developments in ICP–MS at Sciex.[17,29–36] These included details on the center-tap-grounded load coil used on the Sciex instrument, new input ion optic lens designs, improved instrument stability for quantitative analysis, simplified lens voltage optimization, methods for analyzing organic solutions in ICP–MS, and use of electro-thermal vaporization for sample introduction.

Boomer[37] presented results for the analysis of acid rain for aluminum, manganese, iron, and zinc. He noted some spectral interference problems caused by ArO^+ and CaO^+ species. Brooker and Eagles[38] determined lead isotopes in uranium and thorium samples and also started to develop a method for the platinum-group metals. Doherty and Van der Voet[39] developed a method for the determination of lanthanide elements in rocks that involved the development of correction factors for oxide species interference. Preliminary results were presented by Kargacin and Barnes[40] for the analysis of wood by ICP–MS.

McLaren and co-workers reported results for the analysis of seawater and marine sediments by ICP–MS for iron, manganese, copper, zinc, and nickel.[41–44] Because high salt concentrations suppress the signals for the trace metal components, a separation procedure was recommended for seawater analysis. In addition, spectral interference problems arising from molecular species were noted, including such species as ArH^+, ArN^+, ArO^+, Ar_2^+, ClO^+, $ClOH^+$, ClO^+, ClO_2H^+, $ArCl^+$, SO^+, SOH^+, SO_2^+, SO_2H^+, TiO^+, FeO^+, and ZrO^+. As mentioned earlier in this chapter, when all isotope combinations are considered, the species above create major spectral interference problems.

Taylor and Garbarino discussed the application of ICP–MS for the analysis of natural waters.[45] They noted the presence of several background species, such as ArC^+, ArN^+, ArO^+, and $ArOH^+$, and used simplex methods to optimize plasma and MS lens settings, but noted problems with drift. They also illustrated the application of isotope dilution, which is drift insensitive, to trace analysis.[46]

Lichte and Meier discussed the application of ICP–MS to the determination of the first-row transition elements vanadium, chromium, manganese, cobalt, nickel, copper, and zinc in geological materials.[47–50] Problems were encountered with suppressive matrix effects from the high salt content of the solutions resulting from digestion of the rocks, with spectral overlaps from isobaric oxide, from doubly charged and molecular ion species, and from instrument stability. These problems could be minimized with matrix matching, by utilizing dilute solutions and implementing spectral interference correction equations, and by use of internal standardization.

Horlick and co-workers[51–53] illustrated the effect of plasma operating parameters on analyte signals in ICP–MS and showed[54,55] that high concentrations (1000 $\mu g/mL$) of several matrix elements caused major signal suppressions. The background spectral features[56] of the ICP–MS and the overall requirements to achieve quantitative measurements in ICP–MS were also addressed.[57,58] All these investigators[29–58] used the Sciex spectrometers.

The VG spectrometer has also been utilized in a number of studies.[59–70] Date[59] reported on the use of ICP–MS for lead isotope determinations of interest to geochemistry. Russ and co-workers[60] discussed the determination of osmium isotopic ratios as part of a project to measure the half-life of ^{187}Re. Gray[61–63] described the characteristics of a laser ablation system for sample introduction into an ICP–MS. Hausler[64] described the measurement of lead isotope ratios in crude oils. Overview papers on the analytical capabilities of the VG ICP–MS were presented by Goddard,[65] Hutton,[66] and their associates. Kinnert and co-workers[67] compared ICP–AES and ICP–MS for the determination of trace elements in actinide compounds. The spectral simplicity of ICP–MS provided better results than ICP–AES.

The group at Harwell also reported on their first work with ICP–MS.[68–70] Matrix effects and several molecular background peaks (SO^+, ClO^+, $ArCl^+$)

were observed. While signal suppression was noted with mineral acid solutions and with high salt (Na) content, signal enhancements were observed in the presence of such organic acids as acetic, citric, and tartaric acids. Practical applications included the determination of zinc in hafnium, cobalt in steel, lead in water, uranium in water, and lead isotope measurements. Quantitative work required internal standardization and standard additions procedures.

Houk[71] presented an overview of the ICP–MS instrument developed at Iowa State University. These results and observations have recently been published.[72] One should note that this instrument has a number of different characteristics when compared to the two current commercial instruments, and thus, results may not be transferable from this work to other laboratories.

Finally, a new ICP–MS prototype (Balzers MS and Perkin–Elmer ICP) was described by Hieftje.[73] This instrument is a vertical design that uses a turbomolecular pump on the vacuum system and has analog (current) and digital (counting) detection capability. The spectrometer can detect both positive and negative ions and is also equipped with optical ports suitable for atomic fluorescence and atomic emission measurements on the ICP. Many startup problems were openly discussed, including high background levels, low signal levels (ie, poor S/N), and inordinately large amounts of doubly charged ions. In a subsequent presentation,[74] it was reported that most of these problems had been minimized by adding a ground–strap connection between the load coil and the sampling cone.

10.6   CONCLUSIONS

In 1965, few problems were anticipated with AAS. In 1975, few problems were anticipated with ICP–AES. In 1980, few problems were anticipated with ICP–MS. It is clear from the data and results presented here that as was true with AAS and ICP–AES, ICP–MS will provide researchers and manufacturers with many challenges and problems before the full analytical capability of the technique is brought to fruition.

REFERENCES

1. R. S. Houk, V. A. Fassel, G. D. Flesch, H. J. Svec, A. L. Gray, and C. E. Taylor, Inductively Coupled Argon Plasma as an Ion Source for Mass Spectrometric Determination of Trace Elements, *Anal. Chem.* 52, 2283–2289 (1980).
2. *ICP Inf. News.* 16(3), 191–212 (1984): (a) From Lab Bench to Production Line, (b) ICP Mass Spectrometry from the Eye of a Beholder (Robert S. Houk), (c) ICP–MS at Sciex (Donald F. Douglas), (d) ICP–MS References, (e) Continuing Development of ICP Source Mass Spectrometry at the University of Surrey (Alan L. Gray), (f) ICP–MS Applications Development at the British Geological Survey (Alan R. Date), (g) The VG PlasmaQuad Inductively Coupled Plasma Source Mass Spectrometer (J. E. Cantle).

3. R. S. Houk, H. J. Svec, and V. A. Fassel, Mass Spectrometric Evidence for Suprathermal Ionization in an Inductively Coupled Argon Plasma, *Appl. Spectrosc.* 35, 380–384 (1981).
4. R. S. Houk, A. Montaser, and V. A. Fassel, Mass Spectra and Ionization Temperatures in an Argon–Nitrogen Inductively Coupled Plasma, *Appl. Spectrosc.* 37, 425–428 (1983).
5. D. F. Douglas, E.S.K. Quan, and R. G. Smith, Elemental Analysis with an Atmospheric Pressure Plasma (MIP, ICP)/Quadrupole Mass Spectrometer System, *Spectrochim. Acta* 38B, 39–48 (1983).
6. A. R. Date and A. L. Gray, Plasma Source Mass Spectrometry Using an Inductively Coupled Plasma and a High-Resolution Quadrupole Mass Filter, *Analyst* 106, 1255–1267 (1981).
7. A. R. Date and A. L. Gray, Development Progress in Plasma Source Mass Spectrometry, *Analyst* 108, 159–165 (1983).
8. A. R. Date and A. L. Gray, Progress in Plasma Source Mass Spectrometry, *Spectrochim. Acta* 38B, 29–37 (1983).
9. A. L. Gray and A. R. Date, Inductively Coupled Plasma Source Mass Spectrometry Using Continuum Flow Ion Extraction, *Analyst* 108, 1033–1050 (1983).
10. (a) D. F. Douglas and R. S. Houk, Inductively Coupled Plasma Mass Spectrometry, *Prog. Anal. At. Spectrosc.* 8, 1–18 (1985). (b) A. L. Gray, The ICP as an Ion Source—Origin, Achievements and Prospects, *Spectrochim. Acta* 40B, 1525–1537 (1985).
11. N. Furuta, Spatial Emission Distribution of Yttrium Monoxide, Atom and Ion in an Inductively Coupled Plasma, *Spectrochim. Acta* 41B 1115 (1986).
12. P.W.J.M. Boumans, "Theory of Spectrochemical Excitation." Hilger Watts, London/Plenum Press, New York, 1966.
13. N. Furuta, Spatial Profile Measurement of Ionization and Excitation Temperatures in an Inductively Coupled Plasma, *Spectrochim. Acta* 40B, 1013–1022 (1985).
14. R. S. Houk, Mass Spectrometry of Inductively Coupled Plasmas, *Anal. Chem.* 58, 97A–105A (1986).
15. G. Gillson and G. Horlick, Characterization of Ground State Analyte Neutral Atoms and Ions in the Inductively Coupled Plasma using Atomic Fluorescence Spectrometry, *Spectrochim. Acta* (in press).
16. G. Gillson and G. Horlick, Comparison of Atomic Fluorescence and Atomic Emission Spatial Distribution Profiles in the Inductively Coupled Plasma, *Spectrochim. Acta* (in press).
17. D. J. Douglas and J. B. French, An Improved Interface for Inductively Coupled Plasma Mass Spectrometry (ICP-MS), *Spectrochim. Acta* 41B 197–204 (1986).
18. J. A. Olivares and R. S. Houk, Ion Sampling for Inductively Coupled Plasma Mass Spectrometry, *Anal. Chem.* 57, 2674–2679 (1985).
19. G. Horlick, S. H. Tan, M. A. Vaughan, and C. A. Rose, The Effect of Plasma Operating Parameters in Analyte Signals in Inductively Coupled Plasma–Mass Spectrometry, *Spectrochim. Acta* 40B, 1555–1572 (1985).
20. S. R. Koirtyohann, J. S. Jones, C. P. Jester, and D. A. Yates, Use of Spatial Emission Profiles and a Nomenclature System as Aids in Interpreting Matrix Effects in the Low-Power Argon Inductively Coupled Plasma, *Spectrochim. Acta* 36B, 49–59 (1981).
21. S. H. Tan and G. Horlick, Background Spectral Features in Inductively Coupled Plasma–Mass Spectrometry, *Appl. Spectrosc.* 40 445–460 (1986).
22. A. R. Date and A. L. Gray, Determination of Trace Elements in Geological Samples by Inductively Coupled Plasma Source Mass Spectrometry, *Spectrochim. Acta* 40B, 115–122 (1985).
23. R. S. Houk and J. J. Thompson, Elemental and Isotopic Analysis of Solutions by Mass Spectrometry Using a Plasma Ion Source, *Am. Mineral.* 67, 238–243 (1982).
24. M. A. Vaughan and G. Horlick, Oxide, Hydroxide, and Doubly Charged Analyte Species in Inductively Coupled Plasma–Mass Spectrometry, *Appl. Spectrosc.* 40, 434–445 (1986).
25. R.C.L. Ng, H. Kaiser, H. Anderson, and B. Meddings, "The Operating Conditions and Analytical Performance of a Commercial ICP/MS System. Paper No. 6, presented at the 1986 Winter Conference on Plasma Spectrochemistry, Kailua–Kona, Hawaii.
26. G. Horlick, An Integrated Look at Excitation Temperatures, Electron Density, and Species Distribution in the Inductively Coupled Plasma Using AES, AFS, AAS and MS. Paper No. 106, presented at the 1986 Winter Conference on Plasma Spectrochemistry, Kailua–Kona, Hawaii.
27. G. Horlick, S. H. Tan, M. A. Vaughan and G. Gillson, Effect of Plasma Operating Parameters on Analyte Signals in ICP-MS with Comparisons to ICP-AFS and ICP-AES. Paper No. 170, presented at FACSS XII, Philadelphia, 1985.
28. J. A. Olivares and R. S. Houk, Suppression of Analyte Signal by Various Concomitant Salts in ICP-MS, *Anal. Chem.* 58, 20–25 (1986).

29. D. Douglas, Current Developments in Inductively Coupled Plasma/Mass Spectrometer Instrumentation. Paper No. AN-A1-2, presented at the 68th CIC Meeting, Kingston, 1985.

30. D. Douglas and J. B. French, An Improved Interface for Inductively Coupled Plasma Mass Spectrometry. Paper No. 233, presented at the 27th Rocky Mountain Conference, Denver, 1985.

31. A. Boorn, P. Arrowsmith, D. Douglas, J. Fulford, and E. Quan, Recent Developments in ICP-MS at SCIEX. Paper No. THA 015, presented at XXIV CSI, Garmisch-Partenkirchen, 1985.

32. A. W. Boorn, D. J. Douglas, and J. E. Fulford, ICP-MS: Current Status and Future Developments. Paper No. 1, presented at FACSS XII, Philadelphia, 1985.

33. A. W. Boorn, R. R. Liversage, and E.S.K. Quan, Analysis of Organic Solutions by ICP-MS. Paper No. 9, presented at FACSS XII, Philadelphia, 1985.

34. P. Arrowsmith, A. Boorn, and D. Douglas, Electrothermal Vaporization as a Sample Introduction Device for ICP-MS. Paper No. 335, presented at the Pittsburgh Conference, New Orleans, 1985.

35. J. E. Fulford, A. Boorn, D. Douglas, and E.S.K. Quan, Performance of ICP-MS in the Analysis of Real Samples. Paper No. 338, presented at the Pittsburgh Conference, New Orleans, 1985.

36. A. W. Boorn, D. Douglas, J. E. Fulford, and E.S.K. Quan, Optimization of Plasma Parameters for ICP-MS. Paper No. 339, presented at the Pittsburgh Conference, New Orleans, 1985.

37. D. Boomer, Analysis of Acid Rain Samples by ICP/MS. Paper No. AN-Al-3, presented at the 68th CIC Meeting, Kingston, 1985.

38. E. J. Brooker and T. E. Eagles, The Application of ICP/MS to Problems in Analytical Chemistry. Paper No. AN-Al-4, presented at the 68th CIC Meeting, Kingston, 1985.

39. W. Doherty and T. Vander Voet, Application of an Inductively Coupled Plasma/Mass Spectrometer to the Determination of Trace Elements in Geological Material. Paper No. AN-Al-5, presented at the 68th CIC Meeting, Kingston, 1985.

40. M. G. Kargacin and R. M. Barnes, Inductively Coupled Plasma Mass Spectrometry of Wood. Paper No. 8, presented at FACSS XII, Philadelphia, 1985.

41. J. W. McLaren, A. P. Mykytiuk, S. N. Willie, and S. S. Berman, The Application of Inductively Coupled Plasma Spectrometry to Seawater Analysis. Paper No. AN-Bl-2, presented at the 68th CIC Meeting, Kingston, 1985.

42. J. W. McLaren, D. Beauchemin, A. P. Mykytiuk, and S. S. Berman, Applications of ICP-MS to Marine Samples. Paper No. 4, presented at FACSS XII, Philadelphia, 1985.

43. J. W. McLaren, A. P. Mykytiuk, and S. S. Berman, Applications of ICP-MS in Marine Analytical Chemistry. Paper No. THA 021, presented at XXIV CSI, Garmisch-Partenkirchen, 1985.

44. J. W. McLaren, A. P. Mykytiuk, S. N. Willie, and S. S. Berman, Determination of Trace Metals in Seawater by ICP-MS with Preconcentration on Silica-Immobilized 8-Hydroxyquinoline, *Anal. Chem.* 57, 2907–2911 (1985).

45. H. E. Taylor and J. R. Garbarino, Inductively Coupled Plasma Mass Spectrometry for the Quantitative Analysis of Natural Waters. Paper No. 234, presented at the 27th Rocky Mountain Meeting, Denver, 1985.

46. H. E. Taylor and J. R. Garbarino, Characterization of Stable Isotope Dilution Trace Analysis Using ICP-MS. Paper No. THA 052, presented at XXIV CSI, Garmisch-Partenkirchen, 1985.

47. F. E. Lichte and A. L. Meier, Optimization of ICP-MS for the Analysis of Silicate Materials. Paper No. 236, presented at the 27th Rocky Mountain Meeting, Denver, 1985.

48. A. E. Meier and F. E. Lichte, Analysis of Transition Elements in USGS Reference Materials by ICP-MS. Paper No. 239, presented at the 27th Rocky Mountain Meeting, Denver, 1985.

49. F. E. Lichte and A. E. Meier, Analysis of Geological Materials Using ICP-MS. Paper No. THA 043, presented at XXIV CSI, Garmisch-Partenkirchen, 1985.

50. A. E. Meier and F. E. Lichte, Analysis of Biological Materials with ICP-MS. Paper No. 3, presented at FACSS XII, Philadelphia, 1985.

51. G. Horlick, S. H. Tan, M. A. Vaughan, and C. A. Rose, The Effect of Plasma Operating Parameters on Analyte Signals in ICP-MS. Paper No. AN-Bl-1, presented at the 68th CIC Meeting, Kingston, 1985.

52. G. Horlick, S. H. Tan, M. A. Vaughan, and C. A. Rose, The Effect of Plasma Operating Parameters on Analyte Signals in ICP-MS. Paper No. 240, presented at the 27th Rocky Mountain Conference, Denver, 1985.

53. G. Horlick, S. H. Tan, M. A. Vaughan, and C. A. Rose, The Effect of Plasma Operating Parameters on Analyte Signals in ICP-MS. Paper No. THA 403, presented at XXIV CSI, Garmisch-Partenkirchen, 1985.

54. G. Horlick and S. H. Tan, The Effect of Easily Ionized Elements on Analyte Signals in Inductively Coupled Plasma–Mass Spectrometry. Paper No. 230, presented at the 27th Rocky Mountain Conference, Denver, 1985.
55. S. H. Tan and G. Horlick, Matrix Effects in ICP–MS. Paper No. THA 406, presented at XXIV CSI, Garmisch-Partenkirchen, 1985.
56. S. H. Tan and G. Horlick, Background Spectral Features in ICP–MS. Paper No. THA 404, presented at XXIV CSI, Garmisch-Partenkirchen, 1985.
57. G. Horlick, S. H. Tan, M. A. Vaughan, and Y. Shao, Quantitative Performance of ICP–MS. Paper No. 2, presented at FACSS XII, Philadelphia, 1985.
58. S. H. Tan, M. A. Vaughan, and G. Horlick, The Quantitative Performance of ICP–MS. Paper No. THA 405, presented at XXIV CSI, Garmisch-Partenkirchen, 1985.
59. A. R. Date and Y. Y. Chung, ICP–MS: High-Speed Isotope Geochemistry. Paper No. 231, presented at the 27th Rocky Mountain Conference, Denver, 1985.
60. G. P. Russ, J. M. Bazan, D. A. Leich, and A. Date, Isotopic Ratio Measurements on Nanogram Sized Osmium Samples by ICP–MS. Paper No. 232, presented at the 27th Rocky Mountain Conference, Denver, 1985.
61. A. L. Gray, Solids Mass Spectrometry Using Sample Introduction by Laser Ablation into the ICP Ion Source. Paper No. 235, presented at the 27th Rocky Mountain Conference, Denver, 1985.
62. A. L. Gray, ICP–MS for Solid Samples Using Laser Ablation Sample Introduction. Paper No. 6, presented at FACSS XII, Philadelphia, 1985.
63. A. L. Gray, Solid Sample Introduction by Laser Ablation for Inductively Coupled Plasma Source Mass Spectrometry. *Analyst* 110, 551–556 (1985).
64. D. W. Hausler, Measurement of Trace Element Concentration and Isotope Ratios in Petrogeological Samples by ICP–MS. Paper No. 238, presented at the 27th Rocky Mountain Conference, Denver, 1985.
65. P. J. Goddard, R. C. Hutton, C. J. Shaw, and J. E. Cantle, Some Observations of Performance Criteria in ICP–MS. Paper No. 237, presented at the 27th Rocky Mountain Conference, Denver, 1985.
66. R. C. Hutton, C. J. Shaw, and J. E. Cantle, Further Observations of Performance Criteria in ICP–MS. Paper No. 11A, presented at FACSS XII, Philadelphia, 1985.
67. F. Kinnart, R. C. Hutton, P. J. Goddard, and H. Kutter, Trace Element Determinations in Binary and Ternary Actinide Compounds by ICP–OES and ICP–MS Techniques. Paper No. THA 228, presented at XXIV CSI, Garmisch-Partenkirchen, 1985.
68. J. S. Hislop, Practical Applications of ICP/MS in a Multi-Disciplinary Laboratory. Paper No. 1A.7, presented at the 30th IUPAC Congress, Manchester, 1985.
69. S. E. Long. R. M. Brown, and C. J. Pickford, The Characterization of Some Interferences in ICP–MS. Paper No. THA 392, presented at XXIV CSI, Garmisch-Partenkirchen, 1985.
70. S. E. Long, R. M. Brown, and C. J. Pickford, Matrix Effects in Inductively Coupled Plasma–Mass Spectrometry. Paper No. 5, presented at FACSS XII, Philadelphia, 1985.
71. S. Houk, Present Status and Future Developments in ICP–MS. Paper No. AN-Al-1, presented at the 68th CIC meeting, Kingston, 1985.
72. J. A. Olivares and R. S. Houk, Ion Sampling for ICP–MS, *Anal. Chem.* 57, 2674–2679 (1985).
73. G. M. Hieftje, A New Rapid-Scanning ICP–MS Instrument with Negative Ion Capability. Paper No. 10, presented at FACSS XII, Philadelphia, 1985.
74. G. H. Vickers, D. A. Wilson, and G. M. Hieftje, A New, Versatile ICP–MS Instrument. Paper No. 4, presented at the 1986 Winter Conference on Plasma Spectrochemistry, Kailua-Kona, Hawaii.

11

Liquid Sample Introduction into Plasmas

ANDERS G. T. GUSTAVSSON

Department of Analytical Chemistry
The Royal Institute of Technology
Stockholm, Sweden

11.1   INTRODUCTION

The nebulizer systems needed for analytical inductively coupled plasma (ICP) spectrometry must meet a set of requirements quite different from those for nebulizers used in flame atomic emission and flame atomic absorption spectrometry. These requirements and the subsequent search for new nebulizer systems for use with ICPs have resulted in a large number of practical and theoretical advances that are the subjects of this chapter.

The chapter is divided into five sections. Section 11.1 treats the processes by which aerosols are generated. A description of common nebulizers and of different techniques for the introduction of liquid samples into plasmas follows in Section 11.2. Then, Section 11.3 focuses on the theory and measurement of aerosol distributions and the associated phenomena. Models describing the function of nebulizers, spray chambers, and total nebulizer systems, along with practical conclusions drawn from the models, are given in Section 11.4. Finally, limitations and future developments are discussed in Section 11.5.

Good surveys of sample introduction in ICP spectrometry can be found in several recent reviews.[1-5]

© 1987 VCH Publishers, Inc.
Radiation Chemistry: Principles and Applications

11.1.1   The Need for Aerosols

Aerosols provide the easiest means of introducing liquid samples into atom reservoirs, such as plasmas and flames. During the latter part of the nineteenth century, when the science of atomic spectroscopy started to grow, interest was mainly focused on the characterization of spectral lines emitted by various atomic species. The different atoms were introduced into flames in the form of salt solutions deposited on a spiral or on a bundle of platinum wires. Morton[6] probably was the first to propose the nebulization of a solution by an airstream. However, the development of pneumatic nebulizers in the form used today is based on the work of Gouy.[7] With the twentieth century came aerosol generation by ultrasonic means, which was first reported by Wood and Loomis.[8] This historical background reminds us that we work with quite old technology.

11.1.2 Approaches to the Nebulization of Liquids

11.1.2.1   Pneumatic Nebulization

In pneumatic nebulization, two different geometries are considered: the concentric and the cross-flow types. Most of the work published on aerosol formation treats the concentric design. The formation of an aerosol through nebulization by a high-velocity gas stream has been studied by a number of workers.[9–12] Nukiyama and Tanasawa[12] have provided an excellent paper on this subject. There is a good consensus regarding the different types of aerosol generation.[9–12] Three main types of liquid breakup have been recognized: dropwise, stringwise, and filmwise. This classification is based on the relative velocity of the gas compared with that of the liquid. As the relative velocity changes, so does the breakup pattern, quite independently of the liquid nebulized. The transition points for different liquids are within the relative velocity range 5 to 50 m/s. Increases in the relative velocity change the breakup pattern in the order dropwise to stringwise to filmwise. Most liquids will break up in a filmwise manner at a relative velocity above 50 m/s. For common ICP nebulizer systems, filmwise breakup is observed.

Dropwise breakup. The breakup pattern that appears at low relative velocities (gas to liquid) is the result of waves set up in the rodlike liquid jet influenced by the gas flow. The rod develops a number of nodes, and as the nodes grow, the liquid jet breaks up into drops. The thin drawn-out portions between the drops form one to three small droplets.

Stringwise breakup. At a higher relative velocity, the liquid stream begins to flutter. The action is the same as that of a long piece of paper held at one end, whipping in the wind. The liquid jet is exposed to a strong "wind" and the nodes are shattered, tailing out into long strings. The nodes form larger droplets while the tails form smaller ones.

Filmwise breakup. As the relative velocity is increased even further, the gas causes the drops to become flattened and subsequently blown out into the form of a bag or a casting net with a roughly circular rim. The bursting of this bag produces a large number of fine strings and filaments which, in turn, form droplets. The rim, containing the major part of the original volume of the drop, also breaks up, forming larger droplets. If the relative velocity is increased even further, no new pattern develops, but the drops, bags, strings, and filaments formed become smaller.

11.1.2.2 Ultrasonic Nebulization

Aerosol generation by ultrasonic means was first reported by Wood and Loomis[8] in 1927. Since then, many investigations on the mechanism of droplet formation have been conducted. However, similar to pneumatic nebulization, the process of aerosol formation is very complex and only tentative theoretical explanations have been presented. A good compilation of theoretical work has been given by Stupar and Dawson.[13] Ultrasonic standing waves, generated on the surface of a liquid, are thought to break away droplets from the areas of the surface that are bound by the nodal lines.

11.2 NEBULIZATION DEVICES

11.2.1 A Retrospective on Common Nebulizers

Chapter 4 gives a detailed presentation on the common nebulizers. Therefore, this discussion of cross-flow, concentric, and ultrasonic nebulizers is brief.

One of the common nebulizers for ICP use is the all-glass concentric type.[14–20] The main distinction between concentric nebulizers is the shape of the capillary for the liquid at the nozzle face. Increased sample throughput and reduced memory effects have been observed[19] when the liquid dead volume is minimized by inserting a capillary tube in the sample intake line of the nebulizer. Work treating cross-flow nebulizers has been presented by a number of authors.[16,17,20–25] The differences between cross-flow nebulizers used with the ICP and those used with atomic absorption are mainly due to inner diameters of the capillaries for the gas and liquid lines.

The V-groove and Babington nebulizers[26–34] can also be considered to be of the cross-flow type. These nebulizers alleviate clogging, common to concentric nebulizers, caused by solutions having high salt concentrations. Even solutions with suspended particles and slurries have been successfully nebulized with the V-groove nebulizer.[33]

Ultrasonic nebulizers have been used by a number of authors.[35–39] Despite the potential of these devices, their applications are still rather limited because of the cost of the nebulizer and the problems associated with memory effects. However, Taylor and Floyd[37] concluded that the ultrasonic nebulizer is superior to the pneumatic one in all aspects except economy. The detection limits obtained with an ultrasonic nebulizer are generally 5 to 10 times better than those of pneumatic nebulizers. The enhanced detecting power is mainly due to higher nebulizer efficiency.

Comparisons from an analytical point of view have been made between commonly used nebulizers.[16–18,20,36,37,40–43] Boumans and Lux-Steiner[43] have compiled a list of nebulizer systems for the ICP. In most cases in which comparisons have been made, no possibility exists for obtaining the inherent characteristics of the nebulizer because the nebulizer and the excitation source have been studied as one unit.

11.2.2 Other Devices for Continuous Nebulization

There are three more types of continuous nebulizer to consider before we go into the description of discrete sample introduction systems. Before describing the jet impaction, the glass frit, and the total consumption nebulizers, however, we briefly mention the recycling nebulizer system.[8,24,44,45]

In the system reported on by Hulmston,[44] a normal concentric all-glass nebulizer was operated in a vertical position inside a conical aerosol chamber. The larger droplets that were retained in the spray chamber were fed back to the nebulizer and nebulized again. The system works for 10 min on a 1-mL sample. The chamber is flushed with water between samples. Since the sample is diluted when injected into the aerosol chamber, use of an internal standard is recommended. The sensitivity, detection limits, and short-term precision of measurements obtained with the system are comparable to those of a normal concentric nebulizer system. Novak and co-workers[24] used a cross-flow nebulizer with an impact bead. Zhi Zhuang and Barnes[45] have modified a recycling nebulizer system as described by Hulmston.[44] Their system consists of a concentric nebulizer mounted vertically inside a streamlined conical aerosol chamber. Stable analyte emission signals were attained for up to 1 hour while using a 1-mL sample. Better detection limits and superior signal stability were obtained in comparisons with a modified Babington nebulizer and a concentric nebulizer used with a dual concentric aerosol chamber. A cyclone aerosol chamber has been used by Vieira and co-workers[46] in a recycling nebulizer system. Signal-to-background ratios and detection limits were equal to or better than those provided by a dual concentric aerosol chamber. The analytical performance of the cyclone chamber system is comparable to that attained with a vertical aerosol chamber.[44,45] The cyclone system can be operated in either recycling or nonrecycling modes.

In a jet-impact nebulizer, described by Doherty and Hieftje,[47] a liquid is forced through a nozzle having a hole 25 to 60 μm in diameter. The emerging liquid jet impacts on a wall, thus generating an aerosol. The nebulizer is well suited for low-gas-flow ICPs because the droplet generation process is independent of gas flow. The characteristics of the nebulizer are comparable to those of a pneumatic nebulizer in regard to precision, linearity, detection limits, and

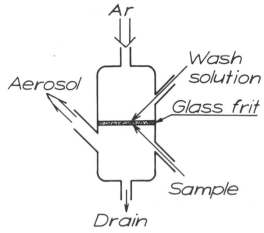

Figure 11.1. Glass frit nebulizer.

efficiency. The main disadvantage of the nebulizer is the risk of clogging due to the small orifice diameter. The performance of the nebulizer was found to be primarily a function of the orifice diameter, the distance from the orifice to the impact wall, and the jet velocity. A well-regulated flow of liquid is required for successful operation of the nebulizer. The nebulizer is well suited for liquid chromatography–ICP use.

The glass frit nebulizer[48] (Figure 11.1) is built around a porous glass frit having a pore size of 4 to 8 μm. The analyte transport efficiency approaches 94% because the aerosol mean droplet size is only approximately 0.1 μm. The main limitations of the glass frit nebulizer are the signal stabilization time after sample injection (3 to 4 min) and the need for repeated wash cycles between samples. The most promising advantage is the possibility of using small samples: 5 to 50 μL/min in continuous mode and 2 to 20 μL in discrete mode. The argon gas flow can be as low as 20 mL/min. Applications of the glass frit nebulizer are presented in Section 11.2.3.5.

A total consumption nebulizer, first used with the ICP by Greenfield and his co-workers,[49,50] has been introduced[51] for liquid chromatography (LC) and flow-injection analysis (FIA) with the ICP. A microconcentric nebulizer is inserted into the aerosol inlet tube of the Fassel torch. The nebulizer is used with liquid uptake rates of up to 200 μL/min, but the residence time of the aerosol droplets in the plasma is approximately one-sixth of that for a pneumatic nebulizer. The reduced analyte residence time in the plasma was thought to be the major reason for not achieving the expected improvement in detection limits. If only the amount of analyte introduced into the plasma were considered, the detection limits should have been lower by a factor of approximately 10 relative to the results obtained from a common pneumatic nebulizer. A considerable fraction of the larger aerosol droplets just passes straight through the plasma, thereby deteriorating the detection limits by factors of 2 to 5. For an

FIA–ICP system that introduces sample through a total consumption nebulizer, LaFreniere and associates[52] recently reported that this nebulizer has analytical capability comparable to or better than that provided by conventional pneumatic nebulization, electrothermal vaporization, graphite cup direct insertion, and graphite rod direct insertion techniques.

11.2.3 Discrete Sample Injection Devices

Continuous sample introduction is the normal working mode for ICP spectrometry. However, when working with discrete samples, transient signals are obtained. Thus, the spectrometer must be capable of measuring peak height and peak area. The main advantages of discrete sample introduction are: the ability to analyze small sample volumes with minimal matrix and solvent interference effects, and the possibility of combining LC and FIA with the ICP technique.

11.2.3.1 The Liquid Plug Techniques

The liquid plug techniques have been adopted from flame atomic absorption spectrometry (AAS) and flame atomic emission spectrometry (AES). Two approaches have been adopted to introduce 25 to 500 μL of sample into the ICP.[53–58] In the first approach, a sample in a micropipet is sucked into a pneumatic nebulizer when the pipet is connected to the sample uptake tube.[53,54] In the second approach, a volume of sample is transported from a polytetrafluoroethylene (PTFE) cup to the nebulizer either by a peristaltic pump or by free suction of the nebulizer.[55–57] One should also recall that the glass frit nebulizer[48] can be used for discrete sample injection. The nebulizer efficiency increases with decreased sample volumes.[58] Compared to continuous nebulization, relative detection limits are 2 to 10 times worse for discrete sample nebulization.[57] Internal standardization can be used to improve the precision of analysis as demonstrated for microsamples of serum and whole blood.[56]

11.2.3.2 Electrothermal Techniques

An electrothermal vaporizer (ETV) consists of an evaporator enclosed in a chamber flushed with argon gas (Figure 11.2). A small volume of sample is placed on a conductor, such as a carbon rod or tantalum filament, which is heated resistively to evaporate the sample. The resulting sample vapor is transported to the ICP by the injection gas. In contrast to continuous sample nebulization, the atomization and excitation steps are separated from the desolvation process with the ETV–ICP method, thus resulting in a much less disturbed plasma. However, the heated gases from the ETV provide an additional source of disturbance for the plasma.[59] Detection limits for an ICP with ETV are equal or somewhat inferior to those of AAS using a graphite furnace, but the dynamic range is larger. ETVs made from various materials and having different configurations are discussed below.

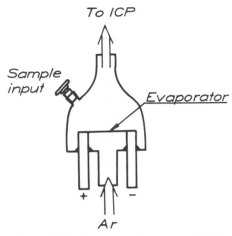

Figure 11.2. General scheme of an ETV.

Kleinman and Svoboda[60] first reported the application of an ETV in emission spectrometry. An electrically heated graphite disc was used to transfer a sample into an electrodeless discharge that was operated at 200 W and 40 MHz. However, the first ETV used with an ICP system was described by Nixon and co-workers,[61] who used a tantalum strip vaporizer. For 100-μL samples, the absolute detection limits were improved by one to two orders of magnitude, as compared to the continuous nebulization method.[61] The size of the tantalum boat was later reduced to accommodate sample volumes of 5 μL.[62] Kitazume[63] used a platinum or tungsten filament in a chamber of only 4.5 mL volume to increase the evaporation and heating rate of the sample, and the detection limits obtained were superior to those attained by other investigators.[61,64] For smaller sample volumes, in the range of 0.5 to 1 μL, a pulsating, direct-current discharge generated between two metal electrodes in a small chamber may be used[65] (Figure 11.3).

Kirkbright and associates[64,66–70] used a graphite rod vaporizer. Analyte transport efficiencies of up to 60% and detection limits at subnanogram levels were obtained with this ETV. A study of refractory compounds[69,70] showed that the use of a halocarbon in the transport gas made matrix modification unnecessary. Nonrefractory elements were unaffected by the halocarbon addition.

A carbon cup was used as a vaporizer by Ng and Caruso,[71–74] who worked with 5- and 10-μL samples and obtained detection limits at the ppb level, with a dynamic range of four orders of magnitude.[72] The carbon cup was coated with pyrolytic graphite or tantalum carbide. Tantalum carbide gave the best results for tin and arsenic determinations. Standard addition or matrix modification was recommended to solve the problem with matrix interference in the analysis of seawater.[73]

A large number of furnaces have also been used as ETVs in ICP spectrometry. The general conclusions for the filament or rod type ETVs are applicable to

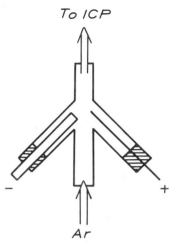

Figure 11.3. An ETV for small-volume samples.

furnace vaporizers as well. A modified Perkin–Elmer graphite furnace has been used by Swaidan and co-workers.[75,76] The graphite tubes were coated with pyrolytic graphite. An internal standard in the 10-μL samples gave an increase in precision of 40%. Omenetto and co-workers[77] have used an HGA 500 to obtain detection limits 10 times lower than for a normal ICP. Aziz and associates[59] used an HGA 74 to determine a number of metals in biological samples. A sample volume of 50 μL and a transport gas flow of 4 L/min gave detection limits 2 to 5 times better than for a normal ICP. Use of a standard addition method was necessary because severe matrix problems were encountered. The ETV was vented through the sample injection hole during the drying step to avoid the introduction of solvent vapor into the aerosol transport system. Kirkbright[78] has commented on memory effects relating to the work by Aziz.[59] A modified Varian–Techtron carbon rod atomizer, the CRA-90, has been used by Barnes and co-workers.[79–81] The geometry of this vaporizer was optimized by the sequential simplex method.[80] The vaporizer surface was coated with pyrolytic graphite and tantalum, and the transport gas was mixed with halocarbons to minimize the interferences for elements forming refractory compounds. A number of improvements have been implemented[79] on the original design, including a reduction of the chamber volume to 7 mL. Absolute detection limits of 8 pg for aluminum and 2.5 ng for silicon were obtained for 5-μL samples with the improved design. Conversions of an IL 655 controlled-temperature furnace and of a Varian model 63 carbon rod atomizer for use as ETVs have been reported.[82–84]

11.2.3.3 Direct Insertion Technique

In the direct insertion method (Figure 11.4) the sample, which has been dried and ashed, is transported into the plasma on an electrode. The direct introduction of samples on a graphite electrode was first reported by Salin and

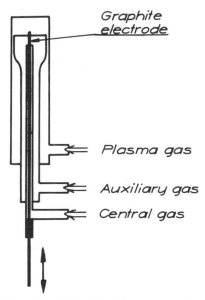

Figure 11.4. The direct insertion system.

Horlick.[85] An electrode was inserted into the ICP axially via the injector tube of a modified torch. Before igniting the plasma, 5 μL of sample was desolvated by inductively heating the electrode, which was held at the center of the load coil while applying a low forward power (approximately 50 W). At the end of the desolvation step, the electrode was lowered and the plasma ignited, followed by insertion of the electrode into the ICP. Detection limits have been improved by an approximate factor of 10 in the direct insertion method compared to pneumatic nebulization.[86] Kirkbright and co-workers[87,88] introduced 5-μL samples into a continuously running ICP via the injector tube of a demountable torch. The samples were dried with a heat gun before they were introduced into the plasma. Detection limits could be improved by more than two orders of magnitude for carbide-forming (refractory) elements if the injector argon gas contained 0.1% Freon 23. A microprocessor-controlled system for the direct introduction of a graphite rod through a demountable torch into a plasma has also has been evaluated.[89] The microprocessor was used to control the position of the graphite rod with a stepper motor, and the system had provisions for drying, ashing, and atomization stages. The automation of the system improved the precision of the measurements.

Variations of the system above, using graphite cups or a metal wire loop, have also been evaluated for ICP atomic emission spectrometry.[90–93] A sample introduction system, constructed as a combination of the direct insertion and the ETV techniques, has been described by Farnsworth and Hieftje.[94] In this system, the central tube of the torch has a larger diameter than the central tube in the commonly used torch. In the bottom of the torch, there is a small chamber with an internal conductor onto which the sample is placed. The chamber is flushed with argon gas and the gas is fed through the central tube into the ICP torch. The ICP is then ignited in the normal fashion. Subsequently, the conductor is grounded and the Tesla coil is activated again, which results in the formation of a secondary plasma in the tube between the conductor and the ICP. The secondary plasma vaporizes and atomizes the sample while the argon gas transports the analyte to the ICP where excitation occurs.

11.2.3.4 Flow-Injection Techniques

Interest in flow-injection analysis (FIA) has been growing rapidly since the late 1970s.[95] The subject has recently been reviewed.[96] FIA is a continuous-flow technique for sample handling. The sample is injected into a carrier stream and forms a zone that is transported by the carrier toward a detector. Provisions can be made for chemical reactions or for separations between the point of injection and the detector. The dispersion of the sample, the ratio between the concentration before dispersion and the concentration after dispersion, can easily be controlled for different applications. In trace analysis, which frequently requires no dilution, the dispersion is held at unity.

The FIA apparatus is fairly simple. It requires one or more precision peristaltic pumps, lengths of tubing, and a sample injection valve. A schematic diagram of a simplified FIA system is shown in Figure 11.5. The FIA technique has several advantages. The total volume of the system is small, and therefore, relatively small volumes of reagents and samples are used. The typical sample size is on the order of 100 μL. Sample injection results in a well-defined volume that is extremely reproducible and is amenable to automated sampling. This minimizes analyst involvement and reduces the likelihood of error. In FIA, the detector signal is a sharp transient peak rather than a steady-state response. Additional characteristics of FIA include high throughput of samples (100 to 200 samples/hour) and the possibilities for sample dilution, separation, and concentration.

The FIA technique, used previously in flame AAS and AES, has been used with the ICP for only a few years, but we are now beginning to see the rapid growth of FIA–ICP methods. Feeding samples to the ICP by an FIA device minimizes the problems encountered in the use of nebulizer systems. Greenfield[97,98] and Gallego and co-workers[99] have written reviews on the prospects of coupling FIA to the ICP.

The nebulizers used for coupling FIA to the ICP are those commonly employed (ie, concentric, cross-flow, and Babington nebulizers), used together with common aerosol chambers. No specially designed nebulizer systems for the

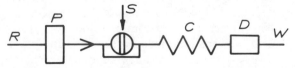

Figure 11.5. A simplified flow-injection analysis system: *R*, carrier stream of the reagent; *P*, pump; *S*, sample injection point; *C*, reaction coil; *D*, detector; *W*, waste.

FIA–ICP have so far been introduced. Consequently, an interface between an FIA device and the ICP, capable of utilizing the full potential of the combination has yet to be developed. The nebulizers described in Section 11.2.2, the jet impaction, the glass frit, and the total consumption nebulizers, will be of importance for use with the miniaturized FIA systems being developed.

Two publications describe applications of the FIA technique as a device for sample transport only. Alexander and co-workers[100] injected 5- to 500-μL samples into a carrier stream via a T-connector, and the samples were transported to a GMK nebulizer. McLeod and associates[101] determined sodium, potassium, calcium, magnesium, lithium, copper, iron, and zinc in blood serum by injecting 20-μL samples into a carrier stream. Matrix effects were observed for undiluted serum nebulized through a Jarrell–Ash high-solids nebulizer, but they were minimized by the use of smaller injections of sample or of higher forward power. No interferences were observed for dilution factors of 1:1, or greater. The effects of the flow rate, the injected volume, and the mixing coil length on the FIA–ICP system were studied by Jacintho and co-workers,[102] who also discussed different approaches for measuring transient signals. The results of determinations of calcium and magnesium in limestone using an internal standard technique with merging zones were comparable to those obtained with a normal ICP technique. The generalized standard addition method has been used to reduce matrix and spectral interferences,[103] and since many additions are necessary, the FIA technique with merging zones is very suitable for the standard addition method. Israel and Barnes[104] have used a standard addition FIA method to determine silicon. FIA has also been used for dilution of samples and for addition of reagents and internal standards.[105] Long and Snook[106] passed the sample stream through an electrochemical cell to deposit the analyte on a glassy carbon electrode, followed by stripping the concentrated analyte into the carrier stream for transport to the nebulizer.

The merging-zone technique is well suited for the determination of hydride-forming elements.[107] A detection limit of 1.4 ng of arsenic was obtained at a sample throughput of 200 samples/hour. An FIA manifold with two columns connected in parallel, and containing Chelex 100 ion-exchange resin, has been used to determine eight metals at a sample throughput of 30 samples/hour.[108] Compared to the conventional ICP, the detection limits were approximately 20 times better. Enrichment of analyte and removal of matrix are achieved by using activated alumina to preconcentrate oxyanions, such as arsenate, molybdate, phosphate, and vanadate.[109,110] The sample is injected into an acidic stream and is transported through the activated column that retains the oxyanions. The oxyanions are eluted when the transport stream is made alkaline.

11.2.3.5 Liquid Chromatographic Techniques

A new dimension has been introduced into liquid chromatography (LC) with the combination LC–ICP, element-specific detection.[111–113] The coupling between LC and ICP has, in most cases, been performed by a direct transfer of the eluate from the column to a conventional nebulizer (eg, a cross-flow,[114–116] a concentric,[116–118] or a Babington[119] nebulizer). Jinno and co-workers[120] connected a nebulizer to the LC system via a T-connector to allow the nebulizer to work at its normal uptake rate by feeding it with pure mobile phase. This arrangement resulted in impaired detection limits due to the dilution. The glass frit nebulizer has been used by Caruso and associates,[118,121] who observed foaming[118] and clogging[121] effects. Solutions in the millimolar concentration range clog the nebulizer, but when alcohols were used as the mobile phase, the glass frit nebulizer was found more suitable than other nebulizers for interfacing reversed-phase LC systems to the ICP. Hydride generation, electrothermal vaporization, and total sample introduction have been suggested for use with the narrow-bore high-pressure LC–ICP system to improve detection limits in LC–ICP spectrometry.[112]

If the mobile phase is organic, it will disturb the plasma. Using a thermostatted spray chamber and toluene as eluent in the determination of organometallics, Hausler and Taylor[122] improved the detection limits by factors of 10 to 100, as compared to results obtained with conventional aerosol chambers. The nebulizer interface should give a small, or preferably a negligible, contribution to the peak width. Peaks obtained using an LC–ICP system can be from approximately 1.2[115,123] to 2.5[122] times broader than a normal LC peak. The position of the nebulizer system, with respect to the LC and the ICP, is discussed by Whaley and co-workers.[116] If the LC liquid is transported to the nebulizer, peak broadening is observed. On the other hand, when the nebulizer is placed close to the LC compartment, signal losses are noted, but peak height measurements are insensitive to variations in the mobile phase flow.

The HPLC–ICP system has been compared to the HPLC system using an ultraviolet (UV) detector. For the determination of metal chelates, detection limits were equal to or better than those obtained with a UV detector.[124] However, for the determination of nonmetals, such as phosphorus,[125] the ICP detector gave inferior detection limits.

11.3   AEROSOL DROPLET DISTRIBUTIONS

11.3.1   Theoretical Aspects of Aerosol Droplet Distributions

Before discussing any aspects of aerosol droplet distributions, let us define an aerosol. An aerosol is composed of a gas, or a mixture of gases, and particles. The gas usually also contains some solvent vapor. The particles can be solid or liquid droplets, the diameters of which range from virtually zero up to a few hundred micrometers.

An aerosol can be characterized by plotting the number, or mass, of the droplets as a function of droplet diameters. A typical number distribution plot appears in Figure 11.6. Different "mean values" (moments) can also be used for

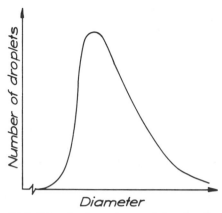

Figure 11.6. Distribution of aerosol droplet diameters.

aerosol characterization (eg, the mode, the median, the mean, etc), as shown by Nukiyama and Tanasawa[12] (Reports 1 and 3), Aitchison and Brown,[126] Gustavsson,[127] and Mugele and Evans.[128] Thorough definitions of the mode, median, and mean are given in Reference 126. The Sauter, or volume/surface mean diameter, is used for efficiency studies.[128]

Distributions of small particles resulting from natural and mechanical processes are often quite skewed (Figure 11.6). As pointed out by Aitchison and Brown,[126] the log-normal distribution, described below, is well established in the domain of small-particle statistics. Relatedly, the parameters obtained by fitting a function to droplet distribution data can, of course, be used for characterization of the aerosol.

Different models for characterization of aerosols are presented in Table 11.1. Nukiyama and Tanasawa,[12] using a four-parameter model, treated aerosol distributions as skewed. Without exception, all their droplet size distribution data could be expressed in terms of the model.[12] Mugele and Evans[128] developed the upper limit model, using three parameters and a log-normal distribution. Compared to other models, the upper limit model gave the best fit for different measured distributions.[128] The model by Nukiyama and Tanasawa gave acceptable fits for distribution data similar to those obtained for concentric nebulizers. Gustavsson[127] described a two-parameter log-normal model that gives a good fit to aerosol distribution data from concentric nebulizers. Because of its wide range of applicability, the upper limit model is the best choice to

Table 11.1. Mathematical Models Used for Aerosol Modeling.

Model	Investigators	Ref.
Four-parameter model, empirically obtained	Nukiyama and Tanasawa	12
Upper limit model, three-parameter log-normal model	Mugele and Evans	128
Two-parameter log-normal model	Gustavsson	127

describe aerosol distributions, but it is more complicated to use because it contains several parameters.

Theoretical treatments of log-normal distributions have been presented by Aitchison and Brown[126] and Gustavsson.[127] Models with a number of parameters can be used when treating data that are distributed log normally. The models most commonly used have two, three, or four parameters. If a distribution of a positive variate x, for example, the droplet diameter, is such that the transformed variable $y = \ln x$ is exactly normal (Gaussian) $N(\mu^*, \sigma^*)$, having mean μ^* and standard deviation σ^*, then the distribution of x is said to be log-normal. Figure 11.6 is an example of a log-normal distribution.

Log-normal distributions possess some properties that are of decisive importance for the treatment in Section 11.4.4. Log-normal distributions possess moments (mean values) of any order. If two log-normal distributions have equal coefficients of variation s_d/d_0, where d_0 is the mean and s_d is the standard deviation of the log-normal distribution, then they also have equal values of σ^*, and vice versa. Log-normal distributions generated under equal circumstances —for example, aerosols generated by the same nebulizer with different flows of gas and liquid—have a nearly constant value of the coefficient of variation. Another characteristic property to consider is that when converting a number to a mass distribution, the distribution will retain its log-normal property. Hence, the number distributions presented by Nukiyama and Tanasawa will retain their log-normal properties when converted to mass distributions.

11.3.2 Measurement of Aerosol Droplet Distributions

Detailed descriptions of droplet distribution measurements are provided in review articles.[129,130] Different instruments used for sizing aerosol droplet distributions and their most important characteristics are presented in Table 11.2. When treating droplet distribution data, one should recall that evaporation (Section 11.3.3.2), condensation, and collision (Section 11.3.3.3) of droplets may alter the distribution. The response of an instrument can change if the particle shape or the index of refraction varies. Laser holographic imaging is considered to be the best approach, but the cost of an instrument is high. Because the cost is less, light-scattering methods can be regarded as the second choice. If the cost of instrumentation is a decisive factor, methods such as photographic imaging and impaction may be considered.

Droplet sizing methods can be grouped into six categories:

1. Momentum transfer methods.
2. Collection and deposition methods.
3. Photographic imaging methods.
4. Hot-wire anemometry methods.
5. Electrical mobility methods.
6. Optical methods.

Table 11.2. Characteristics of Instrumental Methods for Measuring Aerosol Droplets

Measuring method	Particle size range (μm)	Capability of detecting individual droplets	Limitations at high droplet densities
Laser light scattering	0.1–50	+	Yes
Optical imaging	0.3–10^4	+	Yes
Doppler shift	0.2–50	+	Yes
Electromobility	0.003–1	−	No
Piezoelectric effect	0.05–150	−	No
β-ray absorption	0.4–50	−	No

Momentum transfer is useful for large, volatile droplets. The momentum transfer methods include measurements of the size of craters on magnesium-oxide-coated surfaces, and the measurement of electrical signals from the impact of a droplet striking a microphone. The collection and deposition methods use the following separation mechanisms to size droplets: gravity, centrifugal deposition, horizontal elutriation, impaction, and thermal and electrostatic precipitation. The aerosol droplets are removed from the gas by one of these mechanisms to deposit them onto surfaces. When working with photographic imaging methods, aerosol droplets are photographed and measured. In hot wire anemometry, signal fluctuations are functions of both the cooling of the wire by the gas stream and the evaporation of droplets that strike the wire. Size distributions can be obtained by electrical filtration of the signal. The electrical mobility methods determine the charge–weight relationship of the droplets. The charged aerosol droplets precipitate in an electric field of varying strength, and the current, which is due to charges on the droplet surfaces, is measured. After calibration, the measured mobility spectrum can be converted into a droplet size distribution. Optical methods of importance for size distribution measurements are: Tyndall spectra, laser holography, laser radar, and laser interferometry. A detailed description of these methods is outside the scope of this chapter, and the reader is referred to the review articles.[129,130]

Only a few instruments are currently used by analytical chemists for the characterization and measurement of droplet distributions. The most commonly used instrument for aerosol characterization is the Anderson cascade impactor (Anderson, Inc., model 21-000).[48,131,132] The impactor is used for a quite narrow range of droplet diameters, a few tenths of a micrometer up to approximately 10 μm. The popularity of this instrument is due to its low cost and to its coverage of the droplet diameter range of interest to plasma spectrometry. Layman and Lichte[48] also used laser light scattered at 90° to study the variation in droplet density in an aerosol for droplets in the size range 0.1 to 1 μm. Skogerboe and Olson[131] used a condensation nuclei counter (Gardiner Associates) to determine the droplet size distributions at the plasma exit. An aerosol dilution system connected to a cascade impactor, for droplet diameters of 0.4 μm or more and an electrical aerosol analyzer, for droplet diameters in the range from 0.032 to 1.0 μm, have been used by Novak and

Browner.[132] Droplets having a diameter larger than 1 μm must be removed from the aerosol before it enters the electrical aerosol analyzer, since these droplets would otherwise disturb the instrument. The technique of near forward angle Fraunhofer diffraction, which is limited to droplets in the diameter range of 2 to 200 μm, has been used by Mohamed and co-workers[133] and Routh.[134] Thus, the droplet size distributions of the tertiary aerosol, the aerosol leaving the spray chamber to enter the plasma, cannot be measured correctly. This drawback can, of course, be turned into an advantage because the primary generated aerosols can be measured. The principal advantages of the diffraction method are good precision and accuracy in the droplet diameter range 2 to 200 μm, insensitivity to the matrix content of the droplets, and the speed of measurements. The method can be used for relative, but not for absolute, measurements on tertiary aerosols. Mie scattering has been used for nebulizer characterization by Olsen and Strasheim.[135] Individual aerosol droplets scatter the light from a focused laser beam, and the number of light pulses and their heights are recorded. The output from the instrument was number distributions in the ranges 2 to 47 μm and 0.5 to 8 μm.

11.3.3 Phenomena Influencing Aerosol Droplet Redistribution

Now we deal with aerosol ionic redistribution (AIR), evaporation of the solvent from the droplets, and the collisions between droplets. The influence of the spray chamber on the aerosol is treated in Section 11.4.3. The impact bead will not be dealt with at all in this chapter because there is no theory that describes its function. We can, on the other hand, establish that an impact bead will disintegrate the primary aerosol further, giving a secondary aerosol, where the number of small droplets has increased as compared to the primary aerosol.

11.3.3.1 The Phenomenon of Aerosol Ionic Redistribution

The number of reported investigations of the AIR phenomenon is limited.[136–138] Borowiec and co-workers[136] have investigated both AAS and ICP systems, and others have studied an AAS system.[137] However, results obtained for an AAS system can to some extent be generalized to an ICP system.

Borowiec[136] studied the AIR phenomenon with solutions containing two cations at different concentrations. At concentration ratios below 50, only a negligible AIR interference was noted. The major cation in the mixture was always present at a high concentration, generally 2000 to 18,000 μg/mL. The minor cation was present at a concentration 50 to 200 times lower than the major ion. The aerosols produced by the different nebulizer systems were collected in a cascade impactor. The various droplet size ranges, collected on the different plates of the impactor, were analyzed to obtain the ratio of the minor to the major cation concentration. An enrichment factor, E_d, was defined as:

$$E_d = \frac{(C_{Mi^+}/C_{Ma^+})_{aerosol}}{(C_{Mi^+}/C_{Ma^+})_{bulk}} \qquad [11.1]$$

Figure 11.7. Droplet fragmentation process: shaded areas represent pronounced solute redistribution.

where C_{Mi^+} is the concentration of the minor ion and C_{Ma^+} is the concentration of the major ion. Since E_d is a function of the droplet diameter, it is specified for a certain median droplet diameter, d. A value of E_d larger than unity represents an enhancement of the minor ion or a decrease in the major ion in the aerosol droplets. The study showed that within the diameter range 0.44 to 11 μm, the enrichment factor ranged between 0.7 to 2.2, depending on the concentrations, the concentration ratio in the aspirated solution, and the nebulizer design and operation.

Two explanations, based on the Gibbs adsorption isotherm and on an electrical double-layer effect, help to explain the phenomenon. When the aerosol droplets are formed, as discussed in Section 11.1.2.1, they are often made from thin liquid films (Figure 11.7), that disintegrate into fine droplets; ie, smaller droplets form by "stripping" off from a droplet surface while larger droplets, to a greater extent, arise from the liquid bulk. The Gibbs adsorption isotherm[136] indicates that a surface depletion of cations occurs, which increases with the bulk concentration of the ion. Thus, thin liquid films connected to a "bulky" segment will be depleted with respect to cations, and the depletion will increase with increased cation concentration and decreased film thickness. The enrichment factors obtained, particularly for small droplets, were larger than those estimated from the Gibbs's equation. This is because Gibbs's equation was derived for dilute and static solutions, which are conditions quite different from those of a pneumatic nebulization process.

The redistribution of ions at the liquid–gas interface may also be described in terms of an electrical double-layer effect. At the interface, many water molecules are oriented with their oxygen atom outward. The polarization charge will thus establish a double layer. Smaller, and consequently more mobile, ions might be anticipated to be drawn to such a double layer more easily than the larger ones. This could be the process that determines the relative enrichment of different ions.

The AIR interference for real samples is not known yet, but matrix and ionization interferences may be the predominant sources of interference in practical analysis. If the AIR phenomenon interferes with a determination, it can easily be minimized by sample dilution or by matrix matching.

Kornahrens and associates[137a] have noted atomic absorption enhancement upon adding a surfactant to the nebulized solution. Their results and conclusions can be adopted in ICP spectrometry. An absorption enhancement was obtained when a surfactant of opposite charge to that of the analyte ion was added. The enhancement was maximal at a surfactant concentration just below the critical micelle concentration (CMC). No enhancement effects were observed for surfactants of the same charge or for nonionic ones, which indicated that simply decreasing the surface tension alone was insufficient to cause an enhancement. In the nebulization process, smaller droplets form from a "stripping" of droplet surface (Figure 11.7), while larger droplets, to a greater extent, originate from the solution bulk. Signal enhancement will, in other words, result if the analyte is enriched in the surface region. The surfactants will accumulate at the surface of the droplets, and their charged part will be directed toward the bulk of the drop, with the hydrophobic part, the carbon chain, being directed toward the gas–liquid interface. The surfactants may therefore be expected to establish a double layer more efficiently than will small hydrophilic ions, thereby enhancing analyte surface concentrations more efficiently. Obviously, an enhancement will result only if the analyte and the surfactant have opposite charges. At a surface concentration above the CMC, micelles will start to build up in the bulk of the liquid, thereby providing competitive interaction sites away from the surface, ie, reducing the enrichment effect. The signal enhancement is at its maximum just below the CMC, and consequently the signal variation will be greatest near the CMC. Thus, from a practical point of view, in the analysis of samples containing surfactants, the surfactant effects should be swamped out by the addition of a large amount of surfactant.

Skogerboe and Butcher[137b] used a dual nebulization system to allow the introduction of analyte (K) and concomitant (Cs), in the same droplets, or in separate droplets, to study the ionization repression effect (IRE) observed in flame atomic absorption. Their results showed that the introduction of concomitant and analyte in separate droplets resulted in IRE. However, when the same entities were introduced in the same droplets, greater enhancement of the K signal was observed, thus indicating that the presence of Cs in the same droplet produced an AIR effect that accounts for the additional enhancement of K atomic absorption.

11.3.3.2 Evaporation of Droplets

The evaporation of a droplet depends on the composition of the droplet. For example, a pure solvent vaporizes at a rate different from that of a solution. These two evaporation situations have been described by Porstendörfer and associates.[138] The solvent can evaporate from or condense on a droplet, depending on the concentration of solvent vapor in the surrounding gas. In addition, the vapor pressure at the curved surface of a small droplet is greater than that of a flat surface, thus favoring evaporation. A solute decreases the vapor pressure, and thereby promotes an increase in droplet size by condensation.

The radius of a droplet of pure solvent depends on time[138]:

$$r^3 = r_0^3 - 6D\sigma p_s\left(\frac{M_w}{\rho R T}\right)^2 t \qquad [11.2]$$

where r is the droplet radius, r_0 is the radius at $t = 0$, D is the diffusion coefficient for the solvent vapor, σ is the surface tension of the solvent, p_s is the equilibrium (saturated) vapor pressure of the solvent, M_w is the molecular weight of the solvent, ρ is the solvent density, R is the gas constant, T is the absolute temperature, and t is the time. The equation is based on the assumptions that the liquid and vapor phases of the solvent are in equilibrium and that the change in vapor pressure by electric charge is negligible. Diffusion theory was used in deriving Equation 11.2, which is applicable to droplets having diameters of 0.1 μm or more. Furthermore, the rate of evaporation is not significantly influenced by the reduction of droplet temperature caused by evaporation.[138]

When a droplet consists of a solution (solvent containing a solute), the droplet will attain an equilibrium radius. The relative humidity (aqueous solution) is not 100% in this case; rather, it is the saturated vapor pressure above a solution having the actual concentration of the solute in question. Porstendörfer and associates[138] applied theory introduced by Zebel[139] to obtain the following expression for the equilibrium radius, r_x:

$$r_x^3 + \frac{200 M_s \sigma r_x^2}{(\rho R T a \alpha)} - r_0^3 = 0 \qquad [11.3]$$

where M_s is the molecular weight of the solute, a is the concentration of the solute, α is the dissociation fraction of the solute in the solution (eg, $\alpha = 2$ for a dilute NaCl solution in water), and r_0 is the radius of the primarily generated droplet. Theoretically, the new equilibrium droplet distribution will be reached after an infinitely long time, but estimates show that 90% of the change in radius occurs within 1 to 100 ms.[138]

An aerosol can be desolvated by increasing the aerosol gas flow and by increasing the temperature of the aerosol. The desolvation of an aerosol droplet to an aerosol particle is a fast process.

Cresser and Browner,[140] using the conclusions of Porstendörfer,[138] showed that the aerosols generated from aqueous solutions can be characterized by their aerosol monitoring system.[132] Evaporation from the aerosol droplets causing redistribution of the aerosol droplet diameters was prevented by using a high solute concentration. When an aerosol monitoring system dilutes the aerosol with an extra gas flow, and the diluting gas is not saturated with solvent vapor, severe evaporation redistribution occurs even if the solution has a high solute concentration. The effect of time on the droplet radius of aerosols from pure organic solvents have been treated by Boorn and associates.[141]

11.3.3.3 Collisions of Aerosol Droplets

Studies of droplet collisions leading to droplet recombination in aerosols are few. Within a few seconds after the formation of an aerosol, the risk of droplet recombination is negligible[138] if the droplet density is 10^6 droplets/cm^3 or less.

Almost all studies of droplet collisions have been performed on AAS nebulizer systems using twin nebulizers.[142] In general, the probability of collisions depends on the droplet size distribution of the primary aerosol. The probability of a collision is negligible when the gas flow is 5000 or more times larger than the liquid flow. Thus, the probability for droplet collisions in a flame atomic emission or absorption system is quite low compared to an ICP system where the ratio of the gas to the liquid flow is only 300 to 1000. In a recent paper, Gustavsson[143] established that collisions of aerosol droplets in an ICP nebulizer system provide the most probable explanation of certain discrepancies observed in performing efficiency measurements. Further discussion of this subject is presented in Section 11.4.5.

11.4   MODELING OF NEBULIZER SYSTEMS

11.4.1   Characterization of Nebulizer Systems

We must start with the definitions of different nebulizer characteristics and descriptions of methods for measuring each characteristic. A consensus is necessary on the definitions of the inherent properties of a nebulizer system before modeling of a nebulizer system can be discussed. Measurement of aerosol droplet distributions (ie, characterization of the aerosol) has been discussed in Section 11.3.2.[48,131-135] Light scattering has been used to study the correlation between variations in aerosol density and the signal.[144]

The *uptake rate* of a nebulizer is the volume of liquid (mL) aspirated by the nebulizer per minute. The input liquid head and the liquid temperature influence the uptake rate and thus should be reported. The uptake rate should preferably be measured by weight and not by volume, and the weight should be corrected for liquid density variations. The *gas flow rate*, which can easily be measured by a displacement technique, designates the volume (L) of nebulizer gas per minute. When the nebulizer gas flow rate is measured, a liquid should be nebulized to prevent errors of up to 5%.[145] The *pressure drop* designates the pressure across the nebulizer.

Nebulizer *efficiency* can be defined and measured by both direct and indirect approaches.[143] The reader should note that indirect efficiency measurements on nebulizer systems of low efficiency, such as those used in ICP spectrometry, will give values typically 50 to 130% higher than those obtained by direct methods.[143] This discrepancy is discussed in Section 11.4.5. The choice of a direct or indirect method is determined by the phenomenon one desires to measure. The efficiency obtained by an indirect method, η, is defined as the part of the solution uptake that, after nebulization, leaves the aerosol chamber as aerosol. The indirect method makes use of the volume or the mass of the waste[17,146] for the measurement. The analyte transport efficiency, ε, obtained by direct methods, is defined as the part of the analyte in the solution uptake that leaves the aerosol chamber with the aerosol. Three methods are used for direct measurements: filter collection,[36,143,146] cascade impaction,[146] and silica gel collection.[30]

The net analyte intensity in the plasma is proportional to some property of the aerosol leaving the spray chamber. The different properties used are *aerosol concentration*[17] and *total analyte mass transported per second.*[147] The aerosol concentration, defined as the volume of nebulized liquid (μL) leaving the aerosol chamber with the aerosol per liter of nebulizing gas, has been shown to correlate with the measured net intensities in a plasma.[16] The *aerosol analyte concentration*, defined as the total analyte mass leaving the spray chamber with the aerosol per liter of nebulizing gas, is probably the best property to measure in all cases. For this property, the AIR phenomenon, evaporation, and collision interferences need not be negligible. The property "total analyte mass transported per second" is used because it has been suggested that the ICP is a mass-sensitive detector.[112,147] The question of whether the ICP is a mass-sensitive or a concentration-sensitive "detector" is of great importance because the optimization strategy for the nebulizer system depends on the answer. Use of the wrong strategy leads to an analytical system working off its optimum.

The pressure in the spray chamber has also been used to characterize nebulizer systems.[148,149] The forward power and the flow rates of nebulizing and the intermediate gases in the torch are the major factors influencing the pressure. Pressure fluctuations occur because of unsteady liquid waste flow.[148]

11.4.2 Models for Common Nebulizers

The most viable models for nebulizers are those that can be used to calculate the mean droplet diameter of the generated aerosol. For an ultrasonic nebulizer, standing waves are postulated to form on the surface of the liquid leading to the rupture of sections of the surface bounded by nodal lines. This postulate is used to derive an expression for the mean numerical diameter,[13] d_m,

$$d_m = \frac{(\pi\sigma)^{1/3}}{(4\rho v^2)^{1/3}} \qquad [11.4]$$

where σ, ρ, and v are liquid surface tension, liquid density, and frequency, respectively. For the cross-flow nebulizer, virtually no knowledge exists concerning the correlation between the physical parameters of the nebulizer and the mean droplet diameter of the aerosol. Nukiyama and Tanasawa[12] presented some results valid for cross-flow nebulizers in Report No. 6, but those are of little practical use. However, their work on concentric nebulizers is of major importance. The final equation for the Sauter diameter d_0 (μm), of the aerosol number distribution is given[12] by Equation 11.5.

$$d_0 = 585\left[\frac{\sigma^{0.5}}{c\rho^{0.5}}\right] + 597\left[\frac{\mu}{(\rho\sigma)^{0.5}}\right]^{0.45}\left[1000\,\frac{Q_l}{Q_g}\right]^{1.5} \qquad [11.5]$$

Here, ρ is the density (g/cm^3), σ is the surface tension (dynes/cm), μ is the coefficient of viscosity (dynes/cm^2), c is the relative velocity between the gas and the liquid ($c_g - c_l$)(m/s), and Q_l and Q_g are the volume flows of liquid and gas, respectively. The experimental ranges of the constants are $0.8 < \rho < 1.2$, $30 < \sigma < 73$, and $0.01 < \mu < 0.3$ where ρ, σ, and μ are related to the liquid. The equation is derived for air as the gas, but the error in the Sauter diameter is negligible if another gas, having nearly the same velocity characteristics, is used instead. The reader should note that Equation 11.5 is derived for d_0 values in the range of 15 to 90 μm in an empirical way and that, as a consequence, the two terms of the equation are dimensionally unequal. The inequality is due to the lack of a physical description in Equation 11.5 of what takes place in the nebulization process. The d_0 value, the Sauter diameter of a number distribution, is close to the mean diameter of the corresponding mass distribution.[147] For cross-flow nebulizers,[150] Equation 11.5 can be used if the gas velocity is supersonic (460 to 680 m/s) and the volumetric flow ratio (Q_l/Q_g) is kept low, approximately 1×10^{-6}. Supersonic gas velocities can be obtained only in a Laval nozzle, not in a common nebulizer.[151] Relatedly, the production and use of the Laval nozzle are difficult because of construction and temperature problems.[151]

A comparative study by O'Grady and co-workers[152] on three different methods for measuring nebulizer suction demonstrated that: (1) the use of a manometer connected to the sample line via a T-piece is a good method for measuring nebulizer suction provided pressure-fall corrections are made; (2) the suction may be very significantly decreased at high flow rates. Different methods for calculating pressure fall in nebulizers were recently surveyed.[153] The calculation of pressure fall is a delicate matter, and the height, and the static and dynamic pressure components of the Bernoulli equation must be closely considered.[153]

11.4.3 Models for Aerosol Chambers

The separation of large droplets in an aerosol chamber can result from a number of different processes:

1. Gravitational settling.
2. Inertial deposition.
3. Turbulence.
4. Flow-line interception.
5. Diffusional deposition.
6. Electrostatic deposition.
7. Thermal precipitation.
8. Sonic agglomeration.

Only the first three separation processes are applicable to a typical aerosol chamber. However, any of the other five processes also can affect the separation, thus, for example, signal and noise have been shown to vary with electrostatic effects in a aerosol chamber.[154] Before we treat the different separation processes, the fate of large droplets in the spray chamber should be examined.

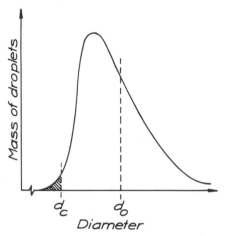

Figure 11.8. Selection of droplets from the primary aerosol in an aerosol chamber (see text).

The aerosol chamber can be regarded as a hypothetical filter, having a cutoff diameter d_c, through which the aerosol is passed (Figure 11.8). Droplets having diameters larger than d_c are retained and drained to the waste, while those droplets having diameters of d_c or less will pass through unaffected. The selection process is a simplified representation of what actually happens. In practice the final (tertiary) aerosol has droplets larger than d_c.

Different mechanisms have been suggested to describe the primary separation process in an aerosol chamber.[131,147,155] Skogerboe and Olson studied gravitational settling and inertial deposition, and concluded that gravitational settling imposed the primary limitation on aerosol transport for the chambers and conditions used.[131] Browner and co-workers[147] studied gravitational settling, inertial deposition, turbulence, and centrifugal loss, and proposed that turbulence-induced loss is the predominant mechanism for the separation of large droplets in an ICP aerosol chamber of the dual concentric type.[147] Gustavsson considered gravitational settling and different forms of inertial deposition, eg, the cyclone and the jet impactor (Section 11.5), and inferred from theoretical calculations that the process most likely to cause separation of large droplets in an aerosol chamber is inertial deposition.[155] It should be emphasized that the droplet lag is an important factor in the treatment of droplet separation. The velocity of the droplets in an aerosol is usually less than the velocity of the gas. The lag can be up to 40% for droplets in the diameter range 1 to 4 μm.[155]

11.4.4 Models for Nebulizer Systems

Only two models have been proposed for the calculation of variables for a nebulizer system. A semitheoretical model has been proposed[147] to calculate the analyte transport efficiency (ε) and the total analyte mass transported per second. A mathematical model for concentric nebulizer systems without an impact bead, described by Gustavsson,[127] can be used for calculating the cutoff droplet diameter of an aerosol chamber, the normal distribution parameters of the aerosol (μ^* and σ^*), the efficiency (η), and the aerosol concentration. Before treating the nebulizer and the aerosol chamber as a unit, one should recall that the nebulizer produces an aerosol having the droplet diameters log-normally mass distributed with a mean diameter d_0. Again, d_0 is the Sauter mean diameter of the corresponding number distribution, calculated according to the Nukiyama–Tanasawa equation (Equation 11.5). As discussed in Section 11.4.2, this model does not apply to a nebulizer system having an impact bead or a cross-flow nebulizer. For an aerosol chamber with a droplet cutoff diameter d_c, the selection process from the primarily generated aerosol is shown in Figure 11.8. The shaded area represents the droplets passing through the aerosol chamber. The efficiency of the nebulizer system, ie, the probability, or the proportion, of droplets having diameters of d_c or less can be calculated from the expression:

$$\phi\left(\frac{(\ln d_c - \ln d_0)}{\sigma^*}\right) \qquad [11.6]$$

where σ^* is the standard deviation of the normal distribution corresponding to the log-normal distribution, and ϕ denotes the standardized normal distribution function of the form:

$$\phi(x) = \frac{1}{(2\pi)^{1/2}} \int_{-\infty}^{x} \exp\left(-\frac{t^2}{2}\right) dt \qquad [11.7]$$

Practical use of the model is described in the next section, and the model is considered to be a first attempt at arriving at a more unified theory for nebulizer systems.

11.4.5 Practical Conclusions Obtained with the Models

The aerosol droplet distribution is important for interpreting interferences because droplets of various diameters behave differently in vaporization, transportation, atomization, excitation, and ionization processes. These interferences are more fully treated in Chapter 5. Gustavsson[16] has shown that the droplet distribution must be considered if aerosols generated by different nebulizers are to be compared. By measuring the characteristics of the nebulizer system, eg, uptake rate and efficiency (see Section 11.4.1), practical conclusions can be obtained. It should be possible to convert an ε value into an η value, but Gustavsson[143] recently showed that, for certain nebulizer systems, this cannot be done. The most probable explanation found was that smaller droplets collide with larger droplets and with the walls of the aerosol chamber.

The Nukiyama–Tanasawa equation (11.5) was used by Cresser and Browner[156] to examine the dependence of the signal on sample temperature in flame spectrometry. The conclusions of this study are also transferable to ICP spectrometry. It was concluded that the nebulizer system is self-stabilizing when the temperature varies, because if the uptake rate increases, the aerosol mean droplet diameter is also increased, resulting in an impaired efficiency, and vice versa. The self-stabilizing property of the nebulizer system, at varying operating conditions, has been reported by others.[20,127,157]

The changes in performance of a nebulizer system, explained by Cresser and Browner,[156] can be calculated using the model presented by Gustavsson,[127] which has been used to demonstrate self-stabilization of the nebulizer and to predict nebulizer characteristics for concentric nebulizer systems,[157] not only at varying temperature but also at different liquid and gas flows, surface tension, density, and viscosity. The work also shows the extreme importance of controlling the temperature when using an aerosol characterization or measurement procedure that is dependent on the aerosol concentration. The change in aerosol concentration will only be approximately 2% when varying the sodium chloride content of a solution from 0 to 3.5% by weight,[157] assuming that there is no nebulizer clogging.[158]

11.5 CURRENT LIMITATIONS AND FUTURE DEVELOPMENTS

Today, no one will deny that the nebulizer system is the weakest link in an ICP system. The major problems encountered with the current systems are low efficiency and the difficulties associated with solvent loading. When using organic solvents having vapor pressures higher than that of water, the problem of solvent loading is quite serious.[43,141,159,160] Currently, a considerable amount of work is being devoted to the elimination of these problems. We should soon see new types of nebulizer systems designed for high efficiency and low solvent loading that are excellent interfaces for FIA-ICP, LC-ICP, and MS-ICP. In fact, the detection limits for FIA-ICP-AES recently reported by Hartenstein and co-workers[161] approach those of ICP-MS, with considerable saving of equipment costs.

Solutions having high salt concentrations generally cause problems when being introduced into an ICP system. Some progress has been made by the use of different kinds of Babington nebulizer (eg, the V-groove nebulizer). A robust nebulizer having a high efficiency for solutions of high salt content could probably be designed on the basis of a V-groove nebulizer using a Laval nozzle[151] to obtain a supersonic velocity of the nebulizing gas.

The characterization of aerosol droplet distributions would be improved if a measurement method could be developed for the droplet diameter range 0.1 to 30 μm. Today we have two methods covering the ranges 0.1 to 10 μm and 2 to 200 μm. The unsatisfactory point is that neither method can measure both "wings" of the aerosol leaving the spray chamber. The large-diameter "wing" is important for determinations of d_c and of the efficiency with which the spray chamber will cut off the aerosol.

There is a great need for a relationship analogous to Equation 11.5 to be used for calculation of the Sauter droplet diameter of aerosols produced by cross-flow nebulizers. Such an equation would make mathematical modeling of nebulizer systems[127] more useful. A further development of the Nukiyama–Tanasawa equation to cover a wider range of d_0 would also make the model by Gustavsson[127] more useful. The model for concentric nebulizer systems[127] should be improved by including provisions for the AIR phenomenon, evaporation, and collision of aerosol droplets.

The most promising concepts for new types of aerosol chamber for ICP spectrometry are based on jet impaction and cyclone principles.[29,31,46] As seen in Chapter 4, the cyclone has been used in work presented in the early 1980s. However, some important improvements remain to be implemented.[46,155] The only major drawback of the cyclone is the mediocre selectivity of the separation process. An aerosol chamber using the jet impaction[155] principle is identical to a one-stage cascade impactor. The supposedly laminar aerosol stream flows with a uniform velocity between two parallel planes or flows through a circular hole leading to an orifice located at a certain distance from a plate. The aerosol stream spreads out symmetrically over the surface of the plate, and large droplets are deposited on the plate. The advantage of a jet impaction aerosol chamber is its narrow cutoff range. At optimum conditions, it should be possible, according to theory, to obtain a cutoff range of, for example, 4.0 to 4.6 μm for 0 to 100% collection efficiency. A possible drawback is the deposition of droplets in the nozzle.

ACKNOWLEDGMENTS

I thank Folke Ingman, my scientific and linguistic tutor, for stimulating discussions and comments relating to the manuscript. I also thank Karin Mattson for typing the manuscript and Rolf Åberg for scrutinizing the final manuscript.

REFERENCES

1. R. F. Browner, Sample Introduction for ICPs and Flames, *Trend Anal. Chem.* 2, 121–124 (1983).
2. A. Gustavsson, Nebulizer Systems for Plasma Spectroscopy, a Status Report, *ICP Inf. Newsl.* 9, 263–266 (1983).
3. R. F. Browner and A. W. Boorn, Sample Introduction: The Achilles' Heel of Atomic Spectroscopy, *Anal. Chem.* 56, 787A–798A (1984).
4. R. F. Browner and A. W. Boorn, Sample Introduction Techniques for Atomic Spectroscopy, *Anal. Chem.* 56, 875A–888A (1984).
5. P. N. Keliher, W. J. Boyko, J. M. Patterson III, and J. W. Hershey, Emission Spectrometry, *Anal. Chem.* 56, 133R–156R (1984).

6. H. Morton, On Monochromatic Light, *Chem. News* 17, 231 (1868).
7. C. L. Gouy, Photometric Research on Colored Flames, *Ann. Chim. Phys.* 18, 5–101 (1879).
8. W. R. Wood and A. L. Loomis, The Physical and Biological Effects of High-Frequency Sound-Waves of Great Intensity, *Phil. Mag.* Ser. VII, 4, 417–436 (1927).
9. W. R. Lane, Shatter of Drops in Streams of Air, *Ind. Eng. Chem.* 43, 1312–1317 (1951).
10. H. C. Lewis, D. G. Edwards, M. J. Goglia, R. I. Rice, and L. W. Smith, Atomization of Liquids in High-Velocity Gas Streams, *Ind. Eng. Chem.* 40, 67–73 (1948).
11. W. R. Marshall Jr., Atomization and Spray Drying, *Chem. Eng. Prog.* Monogr. Ser. 2, Vol. 50, 1–122 (1954).
12. S. Nukiyama and Y. Tanasawa, Experiments on the Atomization of Liquids in Airstream, *Trans. Soc. Mech. Eng. (Japan)* 4, 5, 6, (1938–1940), E. Hope (transl.), Defense Research Board, Department of National Defense, Ottawa, Canada, 1950.
13. J. Stupar and J. B. Dawson, Theoretical and Experimental Aspects of the Production of Aerosols for Use in Atomic Absorption Spectroscopy, *Appl. Opt.* 7, 1351–1358 (1968).
14. J E. Meinhard, The Concentric Glass Nebulizer, *ICP Inf. Newsl.* 2, 163–165 (1976).
15. B. Bogdain, An Improved Concentric Glass Nebulizer from J. E. Meinhard Associates—Preliminary Results, *ICP Inf. Newsl.* 3, 491–493 (1978).
16. A. Gustavsson, Some Aspects of Nebulizer Characteristics, *ICP Inf. Newsl.* 5, 312–328 (1979).
17. A. Gustavsson, The Determination of Some Nebulizer Characteristics, *Spectrochim. Acta* 39B, 743–746 (1984).
18. R. M. Barnes, H. S. Mahanti, M. R. Cave, and L. Fernando, Comparison of Pneumatic Nebulizers for Analysis of High-Purity Aluminum, *ICP Inf. Newsl.* 8, 562–565 (1983).
19. M. H. Ramsey, M. Thompson, and B. J. Coles, Modified Concentric Glass Nebulizers for Reduction of Memory Effects in ICP Spectrometry, *Anal. Chem.* 55, 1626–1629 (1983).
20. F.J.M.J. Maessen, P. Coevert, and J. Balke, Comparison of Pneumatic Nebulizers in Current Use for ICP-AES, *Anal. Chem.* 56, 899–903 (1984).
21. S. E. Valente and W. G. Schrenk, The Design and Some Emission Characteristics of an Economical dc Arc Plasma–Jet Excitation Source for Solution Analysis, *Appl. Spectrosc.* 24, 197–205 (1970).
22. E. Kranz, Untersuchungen über die optimale Erzeugung und Forderung von Aerosolen für specktrochemische Zwecke, *Spectrochim. Acta* 27B, 327–343 (1972).
23. R. N. Kniseley, H. Amenson, C. C. Butler, and V. A. Fassel, An Improved Pneumatic Nebulizer for Use at Low Nebulizing Gas Flows, *Appl. Spectrosc.* 28, 285–286 (1974).
24. J. W. Novak, Jr., D. E. Lillie, A. W. Boorn, and R. F. Browner, Fixed Cross-flow Nebulizer for Use with ICPs and Flames, *Anal. Chem.* 52, 576–579 (1980).
25. R. Ishida, M. Kubota, and Y. Fujishiro, A Study of Designs of a Cross-Flow Nebulizer for ICP-AES, *Spectrochim. Acta* 39B, 617–620 (1984).
26. R. F. Suddendorf and K. W. Boyer, Nebulizer for Analysis of High Salt Content Samples with ICP Emission Spectrometry, *Anal. Chem.* 50, 1769–1771 (1978).
27. R. C. Fry and M. B. Denton, High Solids Sample Introduction for Flame Atomic Absorption Analysis, *Anal. Chem.* 49, 1413–1417 (1977).
28. J. F. Wolcott and C. B. Sobel, A Simple Nebulizer for an ICP System, *Appl. Spectrosc.* 32, 591–593 (1978).
29. B. Thelin, Nebulizer System for Analysis of High Salt Content Solutions with an ICP, *Analyst* 106, 54–59 (1981).
30. P.A.M. Ripson and L. de Galan, A Sample Introduction System for an ICP Operating on an Argon Carrier Gas Flow of 0.1 L/min. *Spectrochim. Acta* 36B, 71–76 (1981).
31. L. Ebdon and M. R. Cave, A Study of Pneumatic Nebulization Systems for ICP Emission Spectrometry, *Analyst* 107, 172–178 (1982).
32. G. L. Moore, P. J. Humphries-Cuff, and A. E. Watson, The Transport Efficiency of a GMK Nebulizer in Emission Spectrometry Using an ICP, *ICP Inf. Newsl.* 9, 763–778 (1984).
33. M. D. Wichman, R. C. Fry, and N. Mohamed, Teflon Slurry Nebulizer for Atomic Analysis, *Appl. Spectrosc.* 37, 254–257 (1983).
34. J. F. Wolcott and C. B. Sobel, Fabrication of a Babington-type Nebulizer for ICP Sources, *Appl. Spectrosc.* 36, 685–686 (1982).
35. P.W.J.M. Boumans and F. J. de Boer, Studies of an Inductively Coupled High-Frequency Argon Plasma for Optical Emission Spectroscopy, *Spectrochim. Acta* 30B, 309–334 (1975).
36. K. W. Olson, W. J. Hass, and V. A. Fassel, Multielement Detection Limits and Sample Nebulization Efficiencies of an Improved Ultrasonic Nebulizer and a Conventional Pneumatic Nebulizer in ICP-AES, *Anal. Chem.* 49, 632–637 (1977).

37. C. E. Taylor and T. L. Floyd, ICP-AES Analysis of Environmental Samples Using Ultrasonic Nebulization, *Appl. Spectrosc.* 35, 408-413 (1981).
38. S. S. Berman, J. W. Mclaren, and S. N. Willie, Simultaneous Determination of Five Trace Metals in Seawater by ICP-AES with Ultrasonic Nebulization, *Anal. Chem.* 52, 488-492 (1980).
39. P. D. Goulden and D.H.J. Anthony, Modified Ultrasonic Nebulizer ICAP-AES, *Anal. Chem.* 56, 2327-2329 (1984).
40. S. Greenfield, H. McD. McGeachin, and F. A. Chambers, Nebulizers—Fact and Fiction, *ICP Inf. Newsl.* 3, 117-127 (1977).
41. C. C. Wohlers, Comparison of Nebulizers for the ICP, *ICP Inf. Newsl.* 3, 37-50 (1977).
42. J. J. Horvath, J. D. Bradshaw, and J. D. Winefordner, Comparison of Nebulization-Spray Chamber Arrangements for Atomic Fluorescence and Atomic Emission Flame Spectrometry, *Appl. Spectrosc.* 35, 149-152 (1981).
43. P.W.J.M. Boumans and M. C. Lux-Steiner, Modification and Optimization of a 50-MHz ICAP with Special Reference to Analyses Using Organic Solvents, *Spectrochim. Acta* 37B, 97-126 (1983).
44. P. Hulmston, A Pneumatic Recirculating Nebulizer System for Small Sample Volumes, *Analyst* 108, 166-170 (1983).
45. H. Zhi Zhuang and R. M. Barnes, Recycling Nebulizer System for Milliliter-Volume Samples with Inductively Coupled Plasma Spectrometry, *Spectrochim. Acta* 40B, 11-19 (1985).
46. P. A. Vieira, H. Zhi Zhuang, S. Chan, and A. Montaser, Evaluation of Recycling Cyclone Spray Chambers for ICP-AES, *Appl. Spectrosc.* 40, 1141-1146 (1986).
47. M. P. Doherty and G, M. Hieftje, Jet-Impact Nebulization for Sample Introduction in ICP Spectrometry, *Appl. Spectrosc.* 38, 405-412 (1984).
48. L. R. Layman and F. E. Lichte, Glass Frit Nebulizer for Atomic Spectrometry, *Anal. Chem.* 54, 638-642 (1982).
49. S. Greenfield, I. Jones, C. Berry, and D. Spash, Improvement Relating to Spectroscopic Methods and Apparatus, British Patent 1,109,602 (1968).
50. S. Greenfield, I. Jones, and C. Berry, Plasma Light Source for Spectroscopic Investigations. U.S. Patent 3,467,471 (1969).
51. K. E. Lawrence, G. W. Rice, and V. A. Fassel, Direct Sample Introduction for Flow Injection Analysis and Liquid Chromatography with ICAP Spectrometry Detection, *Anal. Chem.* 56, 289-292 (1984).
52. K. E. LaFreniere, G. W. Rice, and V. A. Fassel, Flow Injection Analysis with ICP Atomic Emission Spectroscopy: Critical Comparison of Conventional Pneumatic, Ultrasonic and Direct Injection Nebulization, *Spectrochim. Acta* 40B, 1495-1504 (1985).
53. S. Greenfield and P. B. Smith, The Determination of Trace Metals in Microlitre Samples by Plasma Torch Excitation, *Anal. Chim. Acta* 59, 341-348 (1972).
54. R. N. Kniseley, V. A. Fassel, and C. C. Butler, Application of ICP Excitation Sources to the Determination of Trace Metals in Microlitre Volumes of Biological Fluids, *Clin. Chem.* 19, 807-812 (1973).
55. J.A.C. Broekaert and F. Leis, An Injection Method for the Sequential Determination of Boron and Several Metals in Wastewater Samples by ICP-AES, *Anal. Chim. Acta* 109, 73-83 (1979).
56. H. Uchida, Y. Nojiri, H. Haraguchi, and K. Fuwa, Simultaneous Multielement Analysis by ICP Emission Spectrometry Utilizing Microsampling Techniques with Internal Standard, *Anal. Chim. Acta* 123, 57-63 (1981).
57. A. Aziz, J.A.C. Broekaert, and F. Leis, The Optimization of an ICP Injection Technique and the Application to the Direct Analysis of Small-Volume Serum Samples, *Spectrochim. Acta* 36B, 251-260 (1981).
58. J. M. Malloy, P. N. Keliher, and M. S. Cresser, Some Studies of Change in Efficiency of Pneumatic Nebulization on Using Small Discrete Samples, *Spectrochim. Acta* 35B, 833-838 (1980).
59. A. Aziz, J.A.C. Broekaert, and F. Leis, Analysis of Micro-Amounts of Biological Samples by Evaporation in a Graphite Furnace and ICP-AES, *Spectrochim. Acta* 37B, 369-379 (1982).
60. I. Kleinman and V. Svoboda, High-Frequency Excitation of Independently Vaporized Samples in Emission Spectrometry, *Anal. Chem.* 41, 1029-1033 (1969).
61. D. E. Nixon, V. A. Fassel, and R. N. Kniseley, ICP Optical Emission Analytical Spectroscopy, *Anal. Chem.* 46, 210-213 (1974).
62. M. W. Tikkanen and T. M. Niemczyk, Modification of a Commercial Direct-Reading ICP Spectrometer for Sample Introduction by ETV, *Anal. Chem.* 56, 1997-2000 (1984).

63. E. Kitazume, Thermal Vaporization for One-Drop Sample Introduction into the ICP, *Anal. Chem.* 55, 802–805 (1983).
64. A. M. Gunn, D. C. Millard, and G. F. Kirkbright, Optical Emission Spectrometry with an ICAP Source and Sample Introduction with a Graphite Rod ETV Device—I, *Analyst* 103, 1066–1073 (1978).
65. J. P. Keilsohn, R. D. Deutsch, and G. M. Hieftje, The Use of a Micro-Arc Atomizer for Sample Introduction into an ICP, *Appl. Spectrosc.* 37, 101–105 (1983).
66. A. M. Gunn, D. C. Millard, and G. F. Kirkbright, Optical Emission Spectrometry with an ICAP Source and Sample Introduction with a Graphite Rod ETV Device—II, *Analyst* 5, 502–508 (1980).
67. M. J. Cope, G. F. Kirkbright, and P. M. Burr, Use of ICP Optical Emission Spectrometry for the Analysis of Doped Cadmium Mercury Telluride Employing a Graphite Rod ETV Device for Sample Introduction, *Analyst* 107, 611–616 (1982).
68. N. W. Barnett, L. S. Chen, and G. F. Kirkbright, Determination of Trace Concentrations of Lead and Nickel in Freeze-Dried Human Milk by AAS and ICP Emission Spectrometry, *Anal. Chim. Acta* 149, 115–121 (1983).
69. G. F. Kirkbright and R. D. Snook, Volatilization of Refractory Compound Forming Elements from a Graphite ETV Device for Sample Introduction into an ICAP, *Anal. Chem.* 51, 1938–1941 (1979).
70. G. F. Kirkbright and R. D. Snook, The Determination of Some Trace Elements in Uranium by ICP Emission Spectrometry Using a Graphite Rod Sample Introduction Technique, *Appl. Spectrosc.* 37, 11–16 (1983).
71. K. C. Ng and J. A. Caruso, Volatilization of Zr, V, U, and Cr Using Electrothermal Carbon Cup Sample Vaporization into an ICP, *Analyst* 108, 476–480 (1983).
72. K. C. Ng and J. A. Caruso, Microlitre Sample Introduction into an ICP by Electrothermal Carbon Cup Vaporization, *Anal. Chim. Acta* 143, 209–222 (1982).
73. K. C. Ng and J. A. Caruso, Determination of Trace Metals in Synthetic Ocean Water by ICP-AES with Electrothermal Carbon Cup Vaporization, *Anal. Chem.* 55, 1513–1516 (1983).
74. K. C. Ng and J. A. Caruso, Atomic Emission Spectrometric Analysis of Organic Solutions by Electrothermal Carbon Cup Vaporization into an ICP, *Anal. Chem.* 55, 2032–2036 (1983).
75. S. D. Hartenstein, H. M. Swaidan, and G. D. Christian, Internal Standards for Simultaneous Multielement Analysis in ICP Atomic Emission Spectroscopy with an Electrothermal Atomizer for Sample Introduction, *Analyst* 108, 1323–1330 (1983).
76. H. M. Swaidan and G. D. Christian, Optimization of Electrothermal Atomization—ICP-AES for Simultaneous Multielement Determination, *Anal. Chem.* 56, 120–122 (1984).
77. G. Crabi, P. Cavalli, M. Achilli, G. Rossi, and N. Omenetto, Use of the HGA-500 Graphite Furnace as a Sampling Unit for ICP Emission Spectrometry, *At. Spectrosc.* 3, 81–88 (1982).
78. G. F. Kirkbright, D. L. Millard, and R. D. Snook, ICP Combined with Graphite Furnace Microanalysis, *Spectrochim. Acta* 38B, 649 (1983).
79. H. Matusiewicz and R. M. Barnes, Determination of Al and Si in Biological Materials by ICP-AES with ETV, *Spectrochim. Acta* 39B, 891–899 (1984).
80. R. M. Barnes and P. Fodor, Analysis of Urine Using ICP Emission Spectroscopy with Graphite Rod ETV, *Spectrochim. Acta* 38B, 1191–1202 (1983).
81. H. Matusiewicz and R. M. Barnes, Tree Ring Wood Analysis After Hydrogen Peroxide Pressure Decomposition with ICP-AES and ETV, *Anal. Chem.* 57, 406–411 (1985).
82. H. Matusiewicz and R. M. Barnes, An Electrothermal Sample Introduction System for ICP Spectrometry, *Appl. Spectrosc.* 38, 745–747 (1984).
83. H. Matusiewicz and R. M. Barnes, Evaluation of a Controlled Temperature Furnace Atomizer as a Sampling Device for ICP-AES, *Spectrochim. Acta* 40B, 29–39 (1985).
84. D. R. Hull and G. Horlick, ETV Sample Introduction System for the ICP, *Spectrochim. Acta* 39B, 843–850 (1984).
85. E. D. Salin and G. Horlick, Direct Sample Insertion Device for the ICP Emission Spectrometry, *Anal. Chem.* 51, 2284–2286 (1979).
86. D. Sommer and K. Ohls, Direkte Probeneinführung in ein stabil brennendes induktiv gekoppeltes Plasma, *Fresenius Z. Anal. Chem.* 304, 97–103 (1980).
87. G. F. Kirkbright and S. J. Walton, Optical Emission Spectrometry with an Inductively Coupled Radio Frequency Argon Plasma Source and Direct Sample Introduction from a Graphite Rod, *Analyst* 107, 276–281 (1982).
88. G. F. Kirkbright and Z. Li-Xing, Volatilization of Some Elements from a Graphite Rod Direct

Sample Insertion Device into an ICP for Optical Emission Spectrometry, *Analyst* 107, 617–622 (1982).

89. Z. Li-Xing, G. F. Kirkbright, M. J. Cope, and J. M. Watson, A Microprocessor-Controlled Graphite Rod Direct Sample Insertion Device for ICP Optical Emission Spectrometry, *Appl. Spectrosc.* 37, 250–254 (1983).

90. E. D. Salin and M. M. Habib, Electrochemical Preconcentration and Separation for Elemental Analysis by ICP Emission Spectrometry, *Anal. Chem.* 56, 1186–1188 (1984).

91. E. D. Salin and R.L.A. Sing, Microsample Liquid Analysis with a Wire-Loop Direct Sample Insertion Technique in an ICP, *Anal. Chem.* 56, 2596–2598 (1984).

92. A. G. Page, S. V. Godbole, K. H. Madraswala, M. J. Kulkarni, V. S. Malapunkar, and B. D. Joshi, Selective Volatilization of Trace Metals from Refractory Solids into an ICP, *Spectrochim. Acta* 39B, 551–557 (1984).

93. M. Abdullah, K. Fuwa, and H. Haraguchi, Simultaneous Multielement Analysis of Microliter Volumes of Solution Samples by ICP–AES Utilizing a Graphite Cup Direct Insertion Technique, *Spectrochim. Acta* 39B, 1129–1139 (1984).

94. P. B. Farnsworth and G. M. Hieftje, Sample Introduction into the ICP by a Radio Frequency Arc, *Anal. Chem.* 55, 1414–1417 (1983).

95. J. Ruzika and E. H. Hansen, "Flow Injection Analysis." Wiley-Interscience, New York, 1981.

96. (a) J. F. Tyson, Flow Injection Analysis Techniques for Atomic-Absorption Spectrometry—A Review, *Analyst* 110, 419–429 (1985). (b) J. Ruzicka and E. H. Hansen, The First Decade of Flow Injection Analysis from Serial Assay to Diagnostic Toll, *Anal. Chim. Acta* 180, 1–67 (1986).

97. S. Greenfield, FIA Weds ICP—A Marriage of Convenience, *Ind. Res. Dev.* 23 (8), 140–145 (1981).

98. S. Greenfield, ICP–AES with FIA, *Spectrochim. Acta* 38B, 93–105 (1983).

99. M. Gallego, M. D. Lugue de Castro, and M. Valcaral, Atomic Spectroscopic Techniques in FIA: A Review, *At. Spectrosc.* 6, 16–22 (1985).

100. P. W. Alexander, R. J. Finlayson, L. E. Smythe, and A. Thalib, Rapid Flow Analysis with ICP Atomic Emission Spectroscopy Using a Microinjection Technique, *Analyst* 107, 1335–1342 (1982).

101. C. W. McLeod, P. J. Worsfold, and A. G. Cox, Simultaneous Multielement Analysis of Blood Serum by Flow Injection—ICP–AES, *Analyst* 109, 327–332 (1984).

102. A. O. Jacintho, E.A.G. Zagatto, F. H. Bergamin, F. J. Krug, B. F. Reis, R. E. Bruns, and B. R. Kowalski, Flow-Injection Systems with Inductively Coupled Argon Plasma AES—I, *Anal. Chim. Acta* 130, 243–246 (1981).

103. E.A.G. Zagatto, A. O. Jacintho, F. J. Krug, B. F. Reis, R. E. Bruns, and M.C.A. Araujo, Flow-Injection Systems with Inductively Coupled Argon Plasma AES—II, *Anal. Chim. Acta* 145, 169–178 (1983).

104. Y. Israel and R. M. Barnes, Standard Addition Method in FIA with ICP–AES, *Anal. Chem.* 56, 1188–1192 (1984).

105. G. L. Moore, A. E. Watson, and P. Humphries-Cuff, An On-Line Dilution System for ICP Spectrometry, *Spectrochim. Acta* 37B, 835–837 (1982).

106. S. E. Long and R. D. Snook, Electrochemical Preconcentration Technique for Use with ICP–AES—I, *Analyst* 108, 1331–1338 (1983).

107. R. R. Liversage, J. C. Van Loon, and J. C. De Andrade, A Flow Injection/Hydride Generation System for the Determination of Arsenic by ICP–AES, *Anal. Chim. Acta* 161, 275–283 (1984).

108. S. D. Hartenstein, J. Ruzicka, and G. D. Christian, Sensitivity Enhancements for FIA–ICP AES Using an On-Line Preconcentrating Ion-Exchange Column, *Anal. Chem.* 57, 21–25 (1985).

109. C. W. McLeod, I. G. Cook, P. J. Worsfold, J. E. Davies, and J. Queay, Analyte Enrichment and Matrix Removal in FIA–ICP–AES, *Spectrochim. Acta* 40B, 57–62 (1985).

110. A. G. Cox, I. G. Cook, and C. W. McLeod, Rapid Sequential Determination of Chromium(III)–chromium(VI) by FIA–ICP–AES, *Analyst* 110, 331–333 (1985).

111. R. E. Majors, H. G. Barth, C. H. Lochmuller, and L. Column, *Anal. Chem.* 56, 300R–349R (1984).

112. I. S. Krull, Plasma Emission Spectroscopic Detectors in HPLC for Trace Metal Analysis and Speciation, *Trends. Anal. Chem.* 3, 76–80 (1984).

113. J. C. Van Loon, Metal Speciation by Chromatography/Atomic Spectrometry, *Anal. Chem.* 51, 1139A–1150A (1979).

114. K. Yoshida and H. Haraguchi, Determination of Rare Earth Elements by LC/ICP-AES, *Anal. Chem.* 56, 2580-2585 (1984).
115. C. H. Gast, J. C. Kraak, H. Poppe, and F.J.M.J. Maessen, Capabilities of On-Line Element-Specific Detection in HPLC Using an ICAP Emission Source Detector, *J. Chromatogr.* 185, 549-561 (1979).
116. B. S. Whaley, K. R. Snable, and R. F. Browner, Spray Chamber Placement and Mobile Phase Flow Rate Effects in LC/ICP-AES, *Anal. Chem.* 54, 162-165 (1982).
117. K. I. Irgolic, R. A. Stockton, and D. Chakraborti, Simultaneous ICAP Emission Spectrometer as a Multielement Specific Detector for HPLC; The Determination of Arsenic, Selenium and Phosphorus Compounds, *Spectrochim. Acta* 38B 437-445 (1983).
118. W. Nisamaneepong, M. Ibrahim, T. W. Gilbert, and J. A. Caruso, Speciation of Arsenic and Cadmium Compounds by Reversed-Phase Ion-Pair LC with Single-Wavelength ICP Detection, *J. Chromatogr. Sci.* 22, 473-477 (1984).
119. D. R. Heine, M. B. Denton, and T. D. Schlabach, Determination of Nucleotides by LC with a Phosphorus-Sensitive ICP Detector, *Anal. Chem.* 54, 81-84 (1982).
120. K. Jinno, H. Tsuchida, S. Nakanishi, Y. Hirata, and C. Fujimoto, Micro-HPLC-ICP Combination Technique in Analysis of Organometallic Compounds, *Appl. Spectrosc.* 37, 258-261 (1983).
121. M. Ibrahim, W. Nisamaneepong, D. L. Haas, and J. A. Caruso, Determination of Alkyllead Compounds by HPLC/ICP Using a Glass Frit Nebulizer-ICP Interface, *Spectrochim. Acta* 40B, 367-376 (1985).
122. D. W. Hausler and L. T. Taylor, Nonaqueous On-Line Simultaneous Determination of Metals by Size Exclusion Chromatography with ICP Atomic Emission Spectroscopic Detection, *Anal. Chem.* 53, 1223-1227 (1981).
123. D. M. Fraley, D. Yates, and S. E. Manahan, ICP Emission Spectrometric Detection of Simulated HPLC Peaks, *Anal. Chem.* 51, 2225-2229 (1979).
124. D. M. Fraley, P. A. Yates, S. E. Manahan, D. Stalling, and J. Petty, ICP-AES as a Multiple Element Detector for Metal Chelates Separated by HPLC, *Appl. Spectrosc.* 35, 525-531 (1981).
125. K. Yoshida, H. Haraguchi, and K. Fuwa, Determination of Ribonucleoside 5'-Mono-, 5'-Di-, and 5'-Triphosphates by LC/ICP-AES, *Anal. Chem.* 55, 1009-1012 (1983).
126. J. Aitchison and J.A.C. Brown, "The Lognormal Distribution." Cambridge University Press, Cambridge, 1957.
127. A. Gustavsson, Mathematical Model for Concentric Nebulizer Systems, *Anal. Chem.* 55, 94-98 (1983).
128. R. A. Mugele and H. D. Evans, Droplet Size Distribution in Sprays, *Ind. Eng. Chem.* 43, 1317-1324 (1951).
129. R. Davies and J. D. Stockham, A Review of the Methods for the Particle Size Analysis of Aerosol Spray Can Droplets, in "Aerosol Measurements," W. A. Cassat and R. S. Maddock, Eds. NBS Special Publication No. 412. Government Printing Office, Washington, D.C., 1974.
130. Instrumentation for Monitoring Air Quality, American Society for Testing and Materials Special Technical Publication No. 555. ASTM, Philadelphia, 1974.
131. R. K. Skogerboe and K. W. Olson, Aerosols, Aerodynamics, and Atomic Analysis, *Appl. Spectrosc.* 32, 181-187 (1978).
132. J. W. Novak Jr. and R. F. Browner, Aerosol Monitoring System for the Size Characterization of Droplets Sprays Produced by Pneumatic Nebulizers, *Anal. Chem.* 52, 287-290 (1980).
133. N. Mohamed, R. C. Fry, and D. L. Wetzel, Laser Fraunhofer Diffraction Studies of Aerosol Droplets Size in Atomic Spectrochemical Analysis, *Anal. Chem.* 53, 639-645 (1981).
134. M. W. Routh, An Improved Pneumatic Concentric Nebulizer for Atomic Absorption—II. Aerosol Characterization, *Appl. Spectrosc.* 35, 170-175 (1981).
135. S. D. Olsen and A. Strasheim, Correlation of the Analytical Signal to the Characterized Nebulizer Spray, *Spectrochim. Acta* 38B, 973-975 (1983).
136. J. A. Borowiec, A. W. Boorn, J. H. Dillard, M. S. Cresser, R. F. Browner, and M. J. Matteson, Interference Effects from Aerosol Ionic Redistribution in Analytical Atomic Spectrometry, *Anal. Chem.* 52, 1054-1059 (1980).
137. (a) H. Kornahrens, K. D. Cook, and D. W. Armstrong, Mechanism of Enhancement of Analyte Sensitivity by Surfactants in Flame Atomic Spectrometry, *Anal. Chem.* 54, 1325-1329 (1982).
(b) R. K. Skogerboe and G. B. Butcher, Aerosol Ionic Redistribution: The Ionization Repression Effect Revisited, *Spectrochim Acta*, 40B, 1631-1636 (1985), and references therein.
138. J. Porstendörfer, J. Gebhart, and G. Röbig, Effect of Evaporation on the Size Distribution of Nebulized Aerosols, *J. Aerosol. Sci.* 8, 371-380 (1977).

139. G. Zebel, Über Waschstum und Wachstumsgeschwindigkeit von Aerosolen aus wasserlöslichen Substanzen in Abhängigkeit von der relativen Luftfeuchte, Z. Aerosol-Forsch.-Ther. 5, 263–288 (1956).
140. M. S. Cresser and R. F. Browner, A Method for Investigating Size Distributions of Aqueous Droplets in the Range 0.5–10 μm Produced by Pneumatic Nebulizers, Spectrochim. Acta 35B, 73–79 (1980).
141. A. W. Boorn, M. S. Cresser, and R. F. Browner, Evaporation Characteristics of Organic Solvent Aerosols Used in Analytical Atomic Spectrometry, Spectrochim. Acta 35B, 823–832 (1980).
142. C.Th.J. Alkemade and R. Herrmann, "Fundamentals of Analytical Flame Spectroscopy," Chap. 4.3.4. Adam Hilger Ltd., Bristol, 1979.
143. A. Gustavsson, Comparison of an Indirect and a Direct Method for Measuring the Efficiency of Nebulizer Systems, Spectrochim. Acta 41B 291–294 (1986).
144. R.P.J. Duursma, H. C. Smith, and F.J.M.J, Maessen, Characterization of Noise in ICP Emission Spectrometry, Anal. Chim. Acta 133, 393–408 (1981).
145. A. Gustavsson, unpublished data.
146. D. D. Smith and R. F. Browner, Measurement of Aerosol Transport Efficiency in Atomic Spectrometry, Anal. Chem. 54, 533–537 (1982).
147. R. F. Browner, A. W. Boorn, and D. D. Smith, Aerosol Transport Model for Atomic Spectrometry, Anal. Chem. 54, 1411–1419 (1982).
148. R. M. Belchamber and G. Horlick, Correlation of Fluctuations in the Nebulizer Spray Chamber, Spectrochim. Acta 37B, 1075–1078 (1982).
149. R. M. Belchamber and G. Horlick, Pressure Measurement in the Nebulizer Spray Chamber of an ICP, Spectrochim. Acta 36B, 581–583 (1981).
150. N. D. Bitron, Atomization of Liquids by Supersonic Air Jets, Ind. Eng. Chem. 47, 23–28 (1955).
151. A. Gustavsson, A Review of the Theory of Flow Processes of Compressible and Incompressible Media in Nebulizers, Spectrochim. Acta 38B, 995–1003 (1983).
152. C. O'Grady, I. L. Marr, and M. S. Cresser, Critical Appraisal of Three Methods for Measurement of Nebulizer Suction, Analyst 109, 1085–1089 (1984).
153. A. Gustavsson, Theoretical Considerations on the Measurement of Nebulizer Suction, Analyst 110, 885–886 (1985).
154. D. H. Tracy, S. A. Myers, and B. G. Balitsee, Signal Instability in ICP Emission due to Electrostatic Effects in the Nebulizer Spray Chamber, Spectrochim. Acta 37B, 739–743 (1982).
155. A. Gustavsson, A Review of the Theory for, and Practical Aspects on Aerosol Chambers, Spectrochim. Acta 39B, 85–94 (1984).
156. M. S. Cresser and R. F. Browner, Sample Temperature Effects in Analytical Flame Spectrometry, Anal. Chim. Acta 113, 33–38 (1980).
157. A. Gustavsson, Prediction of Nebulizer Characteristics for Concentric Nebulizer Systems with a Mathematical Model, Anal. Chem. 56, 815–817 (1984).
158. B. R. Baginski and J. E. Meinhard, Some Effects of High Solids Matrices on the Sample Delivery System and the Meinhard Concentric Nebulizer During ICP Emission Analyses, Appl. Spectrosc. 38, 568–572 (1984).
159. A. W. Boorn and R. F. Browner, Effects of Organic Solvents in ICP-AES, Anal. Chem. 54, 1402–1410 (1982).
160. F.J.M.J. Maessen, P.J.H. Seeverens, and G. Kreuning, Analytical Aspects of Organic Solvent Load Reduction in Normal-Power ICPs by Aerosol Thermostatting at Low Temperatures, Spectrochim. Acta 39B, 1171–1180 (1984).
161. (a) S. D. Hartenstein, G. D. Christian, and J. Ruzicka, Application of an On-Line Preconcentrating Flow Injection Analysis System for Inductively Coupled Plasma Atomic Emission Spectrometry, Can. J. Spectrosc. 30, 144–148 (1985). (b) S. D. Hartenstein, Ph.D. thesis, Department of Chemistry, University of Washington, Seattle, 1985.

12

Introduction of Solids
into Plasmas

MICHAEL W. ROUTH
AND
MARIA W. TIKKANEN

Applied Research Laboratories, Inc.
Sunland, California

12.1 INTRODUCTION

Inductively coupled plasmas (ICPs) possess exceptional characteristics that make them excellent sources for vaporization, atomization, ionization, and excitation. If there is a shortcoming in their universal analytical application, it is the general requirement that the sample be in liquid form for analysis. This requirement, however, derives from the vast experience in nebulization of liquids that facilitated the initial development of analytical plasmas. Consequently, there is no fundamental limitation preventing the use of these plasmas for the analysis of solids. Because many samples naturally occur in solid form, a number of analytical and practical advantages can be realized if solid samples are introduced directly into the ICP, without pretreatment or conversion to liquid. Ideally, sample is analyzed in its natural state; contamination from reagents is minimized; dilution errors are eliminated; sample transfer losses arising from extra sample-handling steps are avoided; the time delay between sample collection and analysis is diminished; reagent and manpower costs are reduced; the potential for improved absolute detection limits is enabled; analysis of microsamples or localized segments of samples is facilitated; and sample vaporization, atomization, and excitation steps may be separated and optimized.

The techniques for successful introduction of samples in the solid state depend on the physical characteristics of the sample. The physical forms that have been most successfully analyzed are powders, conductive solids, nonconductive solids, homogeneous and heterogeneous solids.

One of the major considerations in choosing the appropriate technique and operational conditions is the degree of sample heterogeneity. This consideration becomes much more significant when solids are introduced directly into the ICP than in the nebulization of solutions. In the analysis of a solid that has been put into solution, the analyst has the benefit of selecting as large a portion of the solid sample as is necessary to assure homogeneity for a valid representation of the bulk sample. This is generally assured when samples on the order of a gram or more are selected for preparing a solution. In contrast, most solid sample introduction techniques for the ICP consume, or are restricted to, much smaller samples, within the microgram-to-milligram range, which can often lead to poor sample-to-sample reproducibilities. Certain solid-sample introduction techniques, however, can reduce the effects of sample heterogeneity or provide information about localized concentrations of sample constituents.

This chapter describes a number of techniques that have been used successfully to introduce solids directly into plasmas. These include direct insertion of samples into the plasma, and three techniques that convert solid samples into either a fine aerosol or a vapor, which is then transported into the plasma: electrical spark and arc ablation, electrothermal vaporization, and laser ablation. Additionally, the use of the Danielsson tape machine and the agitation vessels devised for introducing powders into the ICP are discussed.

Each of the techniques is described with respect to the principles of the technique, instrumentation, operating parameters, analytical figures of merit, interferences and limitations, and applications in ICP–AES. We have attempted to provide a comprehensive survey and comparison of work reported; however, in many instances, the information is preliminary, particularly with respect to figures of merit and interferences, reflecting the developing nature of the field. The publication rate on solid sampling is expected to increase dramatically during the next few years. For details of certain practical applications, the reader is referred to Chapter 7 of the *Handbook of Inductively Coupled Plasma Spectrometry* by M. Thompson and J. N. Walsh.

12.2   DIRECT SAMPLE INSERTION

12.2.1 Principles of the Technique

A number of researchers have shown the feasibility of the direct sample insertion technique coupled with an ICP spectrometer.[1–21] Direct sample insertion is accomplished, as the name suggests, by inserting the sample on a probe directly into the ICP.

Although the sample may be inserted directly as either a solid or a liquid, most of the samples used in the initial development of instrumentation and techniques were liquid. In the case of solids, the sample is usually ground into a powder or segmented into sufficiently small pieces. Certain samples are also dried before weighing because of the variable amounts of water they contain. The sample is placed on, or in, a probe made of a material such as graphite, tantalum, or tungsten, for insertion into the plasma.

The insertion process is accomplished either manually or automatically, in one reproducible motion or in a series of specified steps. Although it has been reported that the probe may be inserted transversely through the plasma cross section, the vast majority of investigators describe axial insertion through the normal sample–aerosol channel.

Data are collected as time-dependent intensities because differential volatilization occurs. The time dependence is influenced by probe material and geometry, sample matrix composition, analyte, and the introduction sequence.

12.2.2 Instrumentation

The instrumental configurations used to insert sample directly into the plasma may be placed into two categories: manually operated and computer controlled. In both cases, the fundamental processes are essentially the same.

The first report of a direct sample insertion device for use in the ICP, made by Salin and Horlick,[1] described the essential hardware of the technique (Figure 12.1). A Fassel torch is modified by replacement of the injector tube with a central quartz tube that acts partially as a guide for the sample probe. A graphite electrode is placed on the end of a quartz rod that can be inserted into the central quartz tube. The rod extends below the torch through an insulative polytetrafluoroethylene (PTFE, ie, Teflon) base into which the torch is mounted. The bottom end of the quartz rod is attached to a sliding platform attached to a lead screw. Rotation of the screw forces the platform to translate vertically, thus moving the graphite electrode probe into the ICP.

Operationally, the top of the electrode is positioned just below the load coil, the location normally occupied by the tip of the aerosol injector tube, and then the plasma is initiated.[1] After the plasma has stabilized at the selected operating conditions, the electrode is manually raised in a short time to a height slightly above the top of the load coil. Continuous operation of the plasma during the sample insertion and withdrawal steps requires preventing entrained air from entering the plasma. Elimination of the central argon injector flow accomplishes this, as described by Kirkbright and co-workers,[4,5] who otherwise used configurations similar to those described by Salin and Horlick.[1] Utilization of a high-power (3-kW) argon–nitrogen ICP[2-3] also enables continuous operation.

Alternatively, an argon flow stopper can be used as described by Abdullah and co-workers[12] (Figure 12.2). The PTFE base, containing the flow stopper, is inserted into the bottom of the injector tube, shown at the right-hand side of the figure. The stopper is kept closed during plasma initiation and immediately after the withdrawal of the insertion device, thus preventing the downward flow of

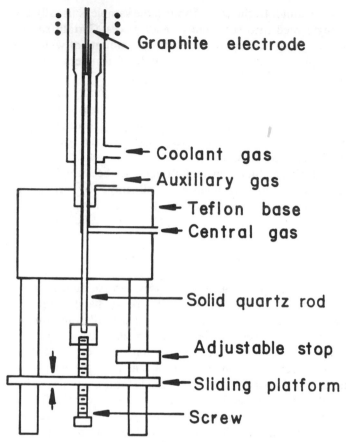

Figure 12.1. Schematic of a direct sample insertion device and torch. (From Reference 1, with permission.)

argon. For sample introduction, the stopper is opened, allowing insertion of the probe.

An elegant implementation that prevents introduction of air has been described by Pettit and Horlick[20] (Figure 12.3). The sample probe is purged with argon before insertion into the plasma torch. Withdrawal of the sample probe is accompanied by a flow of argon gas through the snuffer gas inlet, which not only maintains a stable plasma but also prevents formation of plasma filaments between the induction region and the electrode.

Precise positioning of the sample probe, as required for reproducible vaporization and excitation, is achieved in a number of ways. The simplest implementation is a mechanical stop, attached to the insertion assembly, which positions the sample electrode in a reproducible location relative to the ICP. This approach limits the final position to a single location, dictated by the stop, and does not control the rate of sample insertion.

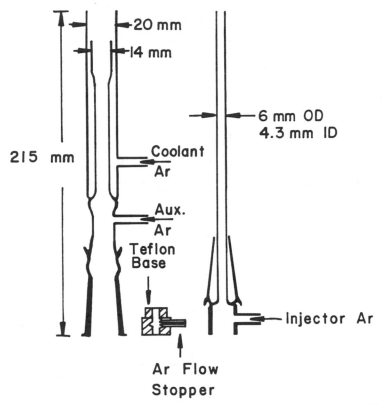

Figure 12.2. Schematic of a direct sample insertion torch modified to prevent withdrawal of argon with sample cup removal. (From Reference 12, with permission.)

The intermediate position(s), final position, duration at each position, and rate of insertion are easily controlled when automatic positioning devices are used. These include the use of stepper motors,[6,7,9,16] dc motors,[21] electrically operated car aerial motors,[19] combination of pneumatic actuators and motor-driven screw assemblies,[20] and pneumatic elevators.[3,11,13,18]

A typical system using a stepper motor, lead screw, and a modified Fassel torch is illustrated in Figure 12.4. The motor drives the lead screw, which translates the sample cup holder. Rod guides improve precision and repeatability of positioning. A glass shutter is solenoid actuated during insertion and retraction. The cup-holder assembly swings out on a hinge to facilitate cup changing. An alternative approach is to use pneumatic translation of the sample–probe assembly, similar to that illustrated in Figure 12.3, in combination with an adjustable stop that uses a threaded ring operating within a threaded barrel.[20] The barrel is rotated under computer control to raise and lower the stop position. These types of system allow the use of microprocessors and relatively simple software to achieve digital control of sample positions and of duration at any position.

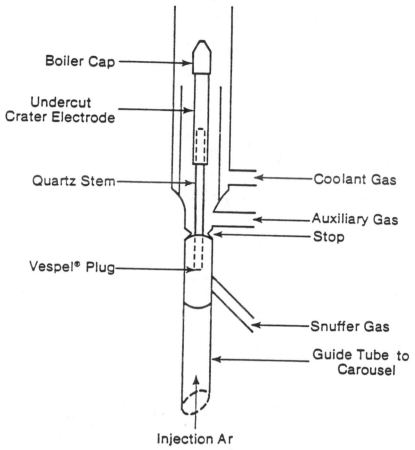

Figure 12.3. Schematic of a direct sample insertion system that operates pneumatically. Snuffer gas eliminates formation of plasma filaments during the sample cup removal. The boiler cap controls sample vaporization rates, as described in the text. (From Reference 20, with permission.)

Analogous to the familiar dry, ash, and atomize sequences used in graphite furnace atomic absorption spectrometry (AAS), positioning of the sample probe at successively closer locations to the plasma enables sample pretreatment. Although these steps are advantageous for direct insertion of solutions,[6] they have not yet been fully exploited in the direct insertion of solids. The primary advantages of programmable stops are improved analytical reproducibility, as discussed later, and the potential for unattended operation.

While a number of researchers have reported automated introduction of a single sample probe into an ICP, instrumentation to automate the changing of sample probes to enable unattended analysis of more than one sample has been described only recently.[20] A 24-position autosampler carousel has been modified for computer control of sample probes (Figure 12.5). Pneumatic purge,

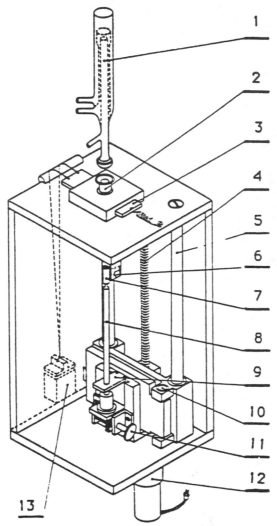

Figure 12.4. Schematic of an automatic direct sample insertion system: 1, modified Fassel torch; 2, torch connector; 3, glass shutter; 4, leadscrew; 5, metal rod guides; 6, microswitch; 7, sample carrying cup; 8, quartz rod; 9, cup holder; 10, hinge joint; 11, gear knob; 12, reversible motor; 13, solenoid. (From Reference 21, with permission.)

insertion, and the snuffer line are also under computer control. Operationally, the electrode in the "next up" position is purged to eliminate air before insertion. The injection line is then activated, which elevates the electrode assembly at a rate dependent on pressure and fit of the assembly to the conical receiver. During retraction, snuffer gas and support gas are actuated. This type of system permits advanced loading of up to 24 preweighed samples and electrode assemblies.

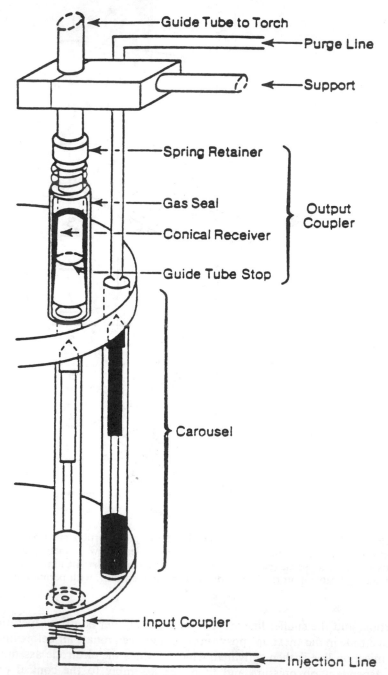

Figure 12.5. Schematic of a portion of a 24-position autosampler carousel for direct sample insertion. (From Reference 20, with permission.)

The same types of graphite electrode normally used for dc arc spectrometry can be utilized as sample probes for the direct insertion technique. The configuration most frequently used is the undercut graphite cup, which provides a reservoir for sample and has advantageous heating characteristics. Although many graphite configurations have been tried, no exhaustive studies have been undertaken to define the geometry for optimum analytical performance. Recent preliminary work by Pettit and Horlick[20] with cups made from tantalum, molybdenum, and tungsten has shown advantages for the determination of both refractory and volatile elements vaporized from directly inserted solutions. The advantage has not yet been demonstrated, however, for direct insertion of solid samples.

Similar to techniques used in dc arc spectrometry, a significant modification of vaporization behavior is observed with use of boiler caps or other similar devices.[1,3,10,20] A boiler cap allows the controlled release of sample vapors into the central channel of the ICP. In virtually all applications reported, improved results were obtained with the caps.

12.2.3   Operating Parameters

Many of the operating parameters, including observation height, RF power, gas flow rates, integration time, and wavelength, normally optimized for ICP–AES with solution nebulization, are also important in direct sample analyses. Additional parameters, such as sample insertion heights for drying, ashing, and vaporization, rate and duration of insertion, and sample size also require optimization.

Conditions established by univariate search methods and by more sophisticated optimization techniques, such as modified simplex routines, are very similar to those used for conventional solution ICP analyses. Unfortunately, these methods have been used almost exclusively to establish optimum parameters, and little information exists on the effects of each of the parameters on the analytical results.

A comparison of optimized operating parameters for the analysis of solids, corresponding to conditions for maximum signal-to-background (S/B) ratio, is given in Table 12.1. Several exceptions require noting. Integration time is established by observing profiles of emission intensity versus time, which indicate when each element is volatilized. Accordingly, elements generally volatilize faster from more volatile solids, such as orchard leaves, than from more refractory samples, such as nickel alloys. The sample size utilized depends primarily on weighing constraints and whether the sample is considered to be representative of the bulk sample. Although less than 1 mg has been analyzed by the technique to demonstrate its feasibility, the constraints above normally dictate sample sizes of 5 to 10 mg.

Table 12.1. Instrumentation[a] and Operational Parameters Used for Direct Sample Insertion into an ICP

	Reference			
Instrument parameter	16	10	21	19
Instrument type	JA Mk II	JA 1100	ARL 34000	JA 9000
Torch	JA	Fassel	Fassel	JA
Forward power (kW)	1.2	1.4	1.2/1.5	1.0
Reflected power (W)	—	20	0–20	—
Outer flow rate (L/min)	17	22	10.5	16
Intermediate flow rate (L/min)	0.6	1	0.65	1.6
Injector gas flow rate (L/min)	0.1	—	0.4 (He)	—
Observation height (mm)	14	15	15	15
Integration time (s)	30	45	40	Variable
Cup atomize position (mm)	+2.2	0	Variable	—
Cup ash position (mm)	−11	—	—	—
Ash time (s)	300	—	—	—
Sample size (mg)	2	100	10	5

[a] JA = Thermo Jarrell–Ash; ARL = Applied Research Laboratories.

12.2.4   Analytical Performance

12.2.4.1   Figures of Merit

Detection limits (DLs) for the direct insertion of solids are difficult to compare due to the lack of a well-defined definition and procedure for their determination and the scarcity of analytical blanks for solid sample. Nevertheless, several sets of detection limits for direct solid insertion have been reported (Table 12.2), with vast differences in DLs observed. Each DL set has been determined on different sample types and more important, insufficient data exist to make firm conclusions about the detection limit capabilities of the direct sample insertion technique for solids. These results indicate, however, that determinations can easily be made in the range of micrograms per gram, with the possibility for submicrogram-per-gram determinations.

Precision is generally inferior to that of solution nebulization techniques and ranges from 1 to 25% RSD, with a typical working precision of 7 to 10% RSD. This imprecision is caused primarily by sample-to-sample heterogeneity, vaporization irreproducibility, and imprecise sample insertion. Precision and limits of detection can be improved significantly, however, by the addition of reagents such as NaF, which acts as a volatilization enhancement reagent.[16,21]

Table 12.2. Detection Limits for Direct Sample Insertion into an ICP

Element	Wavelength (nm)	Detection limits[a]			
		Ref. 16 (μg/g)	Ref. 10 (ppm)	Ref. 21 (ppm)	Ref. 19 (μg/g)
Ag				0.021	
Al	308.2	13.0			
As	193.7	5.0		0.044	
B	249.6	23.2	0.08		
Ba	493.4	20.0			
Be	234.8	4.30	0.01		
Ca	317.9	70.0			
Cd	228.8	1.64	0.02		0.004
Co	228.6	0.65	0.14		
Cr	205.5	0.79			
	267.7		2.9		
Cu	324.7	0.53	0.1		
Fe	238.2	9.56			
	261.1		1.5		
Ge				0.12	
In				0.32	
K	766.4	83.4	3.5		
La	398.8	67.4			
Li	670.7		0.01	0.07	
Mg	279.5	4.91			0.01
Mn	257.6	1.08	0.12		
Na	588.9	38.0	0.94		
Ni	231.6	1.05	1.1		
P	213.6	14.8			
Pb	220.3	1.40		0.33	0.08
	283.3		1.3		
Sb				0.074	
Sn	189.9	1.25		0.06	
Sr	407.7	5.43			
Ti	334.9	52.0			
Zn	213.8	1.77		0.09	0.04

[a] Detection limits are computed differently for each reference and are listed as reported without reducing the number of significant figures.

Calibration curves may be constructed using solid samples of known elemental concentration. Although work reported to date has been limited, successful calibrations have been achieved using NBS coal and botanical standard reference materials, Spex G standards (that are traditionally used for dc arc calibration), and BAS and NBS standard reference materials, such as nickel-based alloys. In general, calibrations are linear over at least three orders of magnitude.

A remarkable characteristic of the direct insertion technique coupled with the ICP is the ability[20] to calibrate the spectrometer using solid materials of substantially different matrix composition (Figure 12.6). Others have also shown this characteristic,[16,19] which indicates the relative insensitivity of the technique to elemental origin. Consequently, calibration should be possible for

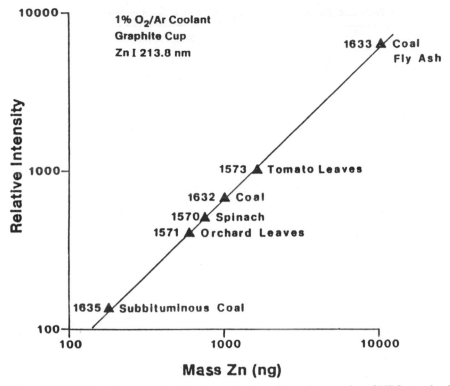

Figure 12.6. Calibration curve for direct sample insertion using a series of NBS standard reference materials. The outer gas flow contained 1% O_2 and 99% Ar. (From Reference 20, with permission.)

analysis of solids with solutions, which are much easier to obtain and which may be optimized for the desired concentration range.

12.2.4.2 Interferences and Limitations

The primary interference associated with the direct insertion technique is intensity offsets due to background shifts during sample volatilization. Although the shifts are generally small and could be corrected easily by normal background correction techniques, the transient nature of the signals complicates the procedure. Currently no commercial instruments exist that allow simultaneous multielement quantitation of transient analyte peaks and associated background. The magnitude of background effects appears small, but more research is required before the seriousness of the problem can be evaluated fully for a wide variety of applications. An alternate approach is the use of blank subtraction whenever the background-contributing matrix components are reasonably well characterized.

Other limitations of the direct insertion technique have similarities to those in dc arc spectrometry and are related to the cup or probe material. Graphite has

Table 12.3. Analysis of Spiked Uranium Samples by Direct Sample Insertion

Element	Sample I			Sample II			Sample III		
	Added value (ppm)	Estimates obtained (ppm)	% RSD	Added value (ppm)	Estimates obtained (ppm)	% RSD	Added value (ppm)	Estimates obtained (ppm)	% RSD
B	1.0	1.2	9.0	0.3	0.39	10.2	0.6	0.85	11.5
Be	1.0	1.1	8.0	5	2.8	8.3	0.5	0.63	8.5
Cd	0.6	0.52	13.8	1.0	1.4	18.4	0.3	0.43	12.4
Co	6.0	6.2	7.8	10.0	9.6	13.3	3.0	4.69	18.6
Cr	10	8.7	11.1	3.0	4.6	12.4	6.0	9.4	13.9
Cu	10	8.4	16.5	50	51	15.6	5.0	4.7	14.7
Mn	10	11	12.4	50	60	15.0	5.0	8.74	18.1
Ni	50	50	16.5	5.0	6.5	17.1	20	29	17.9
Pb	50	49	18.2	5.0	6.2	18.4	20	38	17.8
Li	50	53	3.1	5.0	6.2	6.8	20	22	8.9
Na	6.0	5.8	10.5	10	11	17.0	3.0	2.6	17.6
Fe	50	54	13.8	15	17	17.1	30	35	18.8
K	50	45	5.0	250	260	6.5	25	19	7.9

Source: Ref. 10.

several shortcomings, including reactivity, carbide formation, deterioration, and occasional memory effects. These may be reduced, however, by using pyrolytically coated graphite, metal cups, volatilization enhancement reagents (analogous to matrix modifiers in graphite furnace atomic absorption), hydrogen gas as an oxygen scavenger, and preburn and postburn cycles. Volatility variations that lead to irreproducibility and to sample spattering, in severe cases, may be controlled by use of the dc arc type of boiler caps or cup lids, as indicated earlier. Imprecision due to sample heterogeneity, however, remains the most formidable obstacle to improved analytical precision.

12.2.4.3 Accuracy

Accuracy is normally demonstrated by analyzing standard reference materials whose elemental concentrations are well known. First demonstrations of the potential accuracy of the direct insertion technique for multielement analyses were published only recently.[10,16,19] The results for the determination of 13 elements in U_3O_8 powders, 13 elements in NBS SRM 1571 orchard leaves, and 4 elements in BAS and NBS SRM nickel-based alloys are shown in Tables 12.3, 12.4, and 12.5, respectively. In general, good agreement is attainable, although some difficulty exists in the determination of refractory elements due to formation of carbides and retention in the graphite cup.

Table 12.4. Analytical Results for Orchard Leaves (NBS SRM 1571) by Direct Sample Insertion

	Concentration found		
Element	Calibration method[a]	Matrix matching method[b]	Certified value
K	1.40%	1.67 ± 0.03%	1.47 ± 0.03%
Mg	0.55%	0.62 ± 0.02%	0.62 ± 0.02%
P	0.26%	0.24 ± 0.01%	0.21 ± 0.01%
Fe	293 µg/g		300 ± 20 µg/g
Mn	95 µg/g	106 ± 3 µg/g	91 ± 4 µg/g
Na	101 µg/g	76 ± 34 µg/g	82 ± 6 µg/g
Pb	58 µg/g	56 ± 1 µg/g	45 ± 3 µg/g
Zn	27 µg/g	25 ± 1 µg/g	25 ± 3 µg/g
As	19 µg/g	10 ± 1 µg/g	14 ± 2 µg/g
Cu	14 µg/g	12 ± 0.3 µg/g	12 ± 1 µg/g
Cr		1.6 ± 0.2 µg/g	2.3 µg/g[c]
Ni		1.1 ± 0.1 µg/g	1.3 ± 0.2 µg/g
Cd		0.37 ± 0.01 µg/g	0.11 ± 0.02 µg/g

Source: Ref. 16.
[a] Average for three measurements; calibrations were made with NBS SRM 1570 (spinach), 1573 (tomato leaves), 1575 (pine needles), and NIES SRM No. 1 (pepperbush) samples.
[b] Average for nine measurements; calibration standards were prepared by matrix matching with cellulose powder.
[c] Uncertified value.

Table 12.5. Analytical Results for Certified Reference Materials Based on Either Solutions or Solids as Standards, Using the Direct Sample Insertion Technique

Material	Mean Concentration (μg/g) $\pm$ 1SD for elements[a,b]			
	Cd	Mg	Pb	Zn
BAS 346				
Certificate	0.4 $\pm$ 0.05	147 $\pm$ 10	21.0 $\pm$ 2	29 $\pm$ 2
Solution	0.37 $\pm$ 0.03	—	17.1 $\pm$ 1.2	30 $\pm$ 2.3
Solid	—	—	—	—
BAS 345				
Certificate	0.1	5.0 $\pm$ 1.2	0.2 $\pm$ 0.05	0.5
Solution	ND	5.4 $\pm$ 0.03	0.19 $\pm$ 0.11	0.44 $\pm$ 0.06
Solid	ND	—	0.25 $\pm$ 0.11	0.42 $\pm$ 0.04
NBS				
Certificate	No data	No data	2.5	No data
Solution	ND	13.4 $\pm$ 1.8	1.80 $\pm$ 0.14	1.76 $\pm$ 0.13
Solid	ND	12.5 $\pm$ 1.5	2.53 $\pm$ 0.15	1.78 $\pm$ 0.11

Source: Ref. 19.
[a] $n = 7$.
[b] ND = not determined.

12.2.4.4 Applications

Application of the direct sample insertion technique in ICP spectrometry is in its early stages. Initial reports indicate that the technique is feasible for a wide variety of sample types, including botanical powders, metallurgical solids, and refractory powders. If the problems associated with sample heterogeneity can be overcome, analysis of a wide variety of solid sample types should be possible with acceptable precision and accuracy.

12.3 ELECTROTHERMAL VAPORIZATION

12.3.1 Principles of the Technique

Electrothermal vaporization (ETV) is another sample introduction method that involves the injection of a pulse of solids into the ICP. The improved sample transport afforded by this technique permits microvolume sampling and provides detection limits superior by orders of magnitude to those from pneumatic nebulization. Similar to techniques utilized in atomic absorption spectrometry, the initial volatilization of the sample is performed by the resistive heating of any one of many sample-supporting devices. These devices may include filaments, boats, ribbons, braids, rods, or tubes constructed of either a refractory metal or graphite.

Most ETV–ICP systems utilize a power supply for the vaporizer that can be sequenced through a number of heating steps. This programmed control of the rates, times, and temperatures of vaporizer heating allows for the selection of an optimum sequence of drying, ashing, and subsequent vaporization of the sample components from the vaporizer surface. By use of a proper heating sequence, the solvent, matrix components, and analytes can be temporally separated from each other as they are volatilized off the heated ETV surface.

A measured weight or volume of sample, solid or liquid, is placed on the vaporizer surface and stepped through a controlled heating sequence that produces a dry, highly dispersed aerosol. The aerosol is subsequently swept into the plasma by the injector gas, usually argon, where it is decomposed, atomized, ionized, and excited to produce radiative emission. The emission signal generated is transient owing to the pulse-like nature of the sample introduction. Quantitation of this time-dependent signal produces either a peak height or a peak area. Although either a monochromator or a polychromator may be interfaced to the ETV–ICP, simultaneous multielement determinations require the use of a polychromator.

While most work reported thus far with ETV–ICP spectrometry deals with liquid sampling techniques,[22–30,33,35–40] two groups have shown this technique to be applicable to solid analysis.[32,34] Therefore, the following sections also include reference to liquid sampling techniques in ETV–ICP for comparison.

12.3.2   Instrumentation

The first use of a tantalum ETV for ICP–AES was reported by Nixon and co-workers.[22] The various configurations used to electrothermally vaporize a sample into the ICP center around the vaporizer, which is contained within an argon-filled enclosure (Figure 12.7). The ETV, a graphite rod, as indicated by the dashed lines, is positioned between the water-cooled terminals of the power supply, and this assembly is surrounded by a cylindrical glass manifold. The conical top of the manifold contains two ports. One port allows delivery of the sample to the graphite rod vaporizer surface, while the other permits the vaporized sample to be swept by the injector gas into the plasma through a tube 50 cm long.

The graphite rod is heated by a programmable, high-current power supply. Although any supply that reproducibly provides sufficient power may be used, power supplies designed for graphite furnace atomic absorption spectrometers may be conveniently modified to introduce samples into the ICP. In some cases, even the atomization chamber of the atomic absorption spectrometer can be adapted to the ETV–ICP sample introduction apparatus.[26,31,32,36,37]

An electrothermal vaporization chamber,[33] designed to mate directly to a torch assembly, is shown in Figure 12.8. With this mounting, the distance the sample must travel from the vaporizer surface to the plasma is kept minimal,

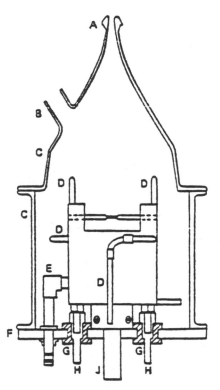

Figure 12.7. Schematic of a graphite rod electrothermal vaporization device: A, ball joint to plasma torch sample inlet; B, sample delivery port; C, cylindrical glass manifold; D, water cooling link; E, argon sample transport gas inlet; F, circular brass base; G, Tufnol insulating blocks; H, electrode terminals; J, mounting pillar. (From Reference 23, with permission.)

approximately 30 cm, to prevent diffusion of the analyte peak and to minimize loss of sample to the walls of the delivery system. This example also illustrates the use of a tantalum boat, mounted on brass supports through which power is applied. The vaporization chamber is kept small (53 cm³), again to lessen the possibility of analyte peak broadening.

The previous two examples of vaporizer and manifold design were used to investigate the analysis of samples in solution. Volumes analyzed ranged from 5 to 100 µL, thus providing weights of solid residues from picograms to micrograms.

For solid samples, the ETV developed by Hull and Horlick[34] incorporates the same basic elements as the previously described systems. The ETV, a carbon cup, which is supported by carbon electrode rods, is enclosed in a quartz cell constructed from a 35/20 ball-and-socket joint. Operationally, two to three mg of a weighed, powdered solid sample is placed directly in the carbon cup. Before the vaporization step, the chamber is closed and flushed with argon.

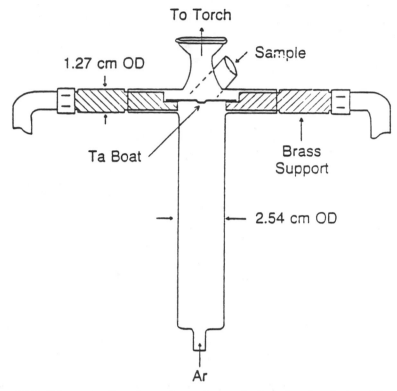

Figure 12.8. Schematic of an electrothermal vaporization chamber designed for direct connection to a torch assembly. The ETV is a tantalum boat. (From Reference 33, with permission.)

Another ETV for the introduction of solids into the ICP,[32] illustrated in Figure 12.9, exemplifies the successful use of a modified atomic absorption spectrometer atomization chamber. Inside the chamber, the vaporizing element, a pyrolytically coated graphite microboat, is supported by a carbon rod and its mounting assembly. Approximately 0.5 mg of powdered solid sample is weighed into the graphite microboat and placed on a flat section of the carbon rod with forceps. The vaporizer heating cycle is controlled by the programmable power supply, and the entire vaporizer assembly is connected to the ICP torch by a 12-cm length of PTFE tubing.

A commercial ETV–ICP system has been recently introduced into the marketplace by Thermo Jarrell-Ash.

12.3.3   Operating Parameters

In addition to the parameters commonly optimized in solution nebulization, the use of an ETV also requires optimization of parameters such as time, temperature, and rate of drying, ashing, and vaporizing steps. Optimization of the ETV

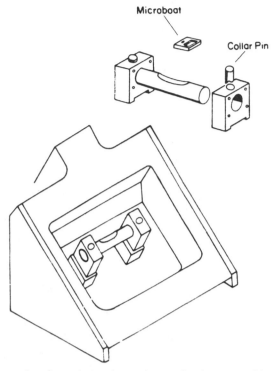

Figure 12.9. Schematic of an electrothermal vaporization assembly using a modified commercial carbon rod device. (From Reference 32, with permission.)

parameters can be facilitated by use of signal-versus-time profiles of the samples being analyzed. Tikkanen and Niemczyk[33,39,40] have documented a number of such profiles using multielement solutions (Figure 12.10). Elements of different volatility appear in the plasma at different times, with the more volatile elements appearing earlier. This phenomenon is similar to the carrier distillation process evidenced in dc arc spectrometry. With the information gathered from these types of profiles, electrothermal vaporization parameters can be established to effect maximum signal intensity and to vary the temporal response of each analyte within the sample. In this manner, certain interferences common in conventional ICP emission spectrometry, such as spectral overlap or matrix effects, can be minimized.[40]

The same type of response occurs when solid samples are introduced into the ICP[34] (Figure 12.11). These recordings were made in the spectral region from 200 to 250 nm by a photodiode array spectrometer. Time frames and volatility behavior for both multielement solutions and solid samples are strikingly similar, indicating that the behavior of dried solution residues may be used to predict the response of inherently solid samples.

An increase in signal-to-noise (S/N) ratio can be obtained by selecting for a given analyte an optimized integration "window" that corresponds to the

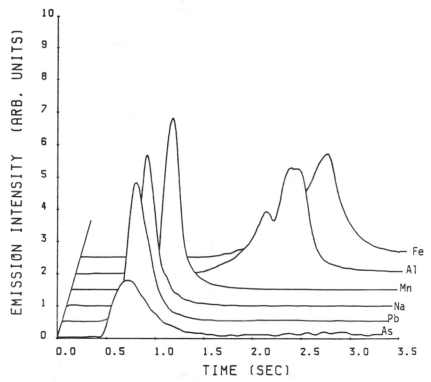

Figure 12.10. Signal-versus-time profiles obtained simultaneously for 5 μL samples of 1.0 μg/mL concentrations of arsenic, lead, sodium, manganese, aluminum, and iron, using electrothermal vaporization and ICP excitation. (From Reference 39, with permission.)

analyte's residence time within the plasma.[31,33] This integration time corresponds to the peak width of the transient signal and is unique for each element (Figure 12.10). Software can easily be modified[32,33,39] to accommodate this requirement.

Optimized operating conditions for the analysis of solids by ETV–ICP obtained by two groups of researchers are listed in Table 12.6. Also included for comparison are the operational parameters for analysis of solutions by ETV–ICP. It should be noted that plasma operating conditions are remarkably similar for liquid and solid samples.

12.3.4 Analytical Performance

12.3.4.1 Figures of Merit

Due to highly efficient sample utilization, ETV combined with ICP excitation is an extremely sensitive technique. Unfortunately, solids introduction using ETV into the ICP is, as yet, in its early investigative stages. Consequently, no detailed

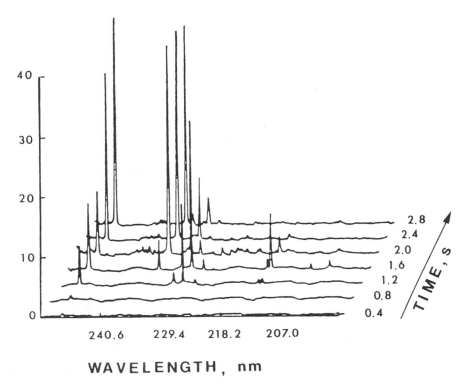

Figure 12.11. Time-resolved emission spectra of 0.001% Spex G3 standard, using electrothermal vaporization and ICP excitation. (From Reference 34, with permission.)

work has been published on DLs for solid samples. However, for aqueous solutions, DLs are typically 10 to 100 times better than those obtained using solution nebulization into the ICP (Table 12.7). Preliminary studies indicate that a homogeneous, powdered sample exhibits behavior similar to that of a dry-solution residue vaporized from an ETV surface. Accordingly, DLs for this approach are expected to fall in the same nanogram-to-picogram range. A simple extrapolation, based on the work of Blakemore and co-workers[32] on bovine liver (NBS SRM 1577), can be made from the data in Table 12.8. Using a 2σ criterion, taking into account a 0.5-mg sample size, and assuming that the factors dominating the precision of the analytical measurement are also dominant at the limit of detection, DLs for cadmium, copper, iron, and zinc are calculated to be 0.03, 20, 30, and 10 ng, or in units of concentration, 0.06, 40, 60, and 20 μg/g, respectively.

Few data have been published on precision for direct solid sampling by ETV–ICP. Blakemore et al.[32] performed triplicate analyses on bovine liver for the determination of 10 elements. The average analytical precision was 11.8% RSD, as seen from Table 12.8, which is inferior to that of solution sampling with ETV–ICP (typically less than 10% RSD) or of continuous solution nebulization. Degradation of precision, and therefore, DLs and long-term reproducibility,

Table 12.6. Instrumentation[a] and Operational Parameters Used for Electrothermal Vaporization into an ICP

	Reference		
Instrument parameter	34	32	31[b]
Spectrometer	Laboratory-constructed photodiode array	JA 1160 Atom Comp	JA 965 Atom Comp
Frequency (MHz)	27.1	27.1	27.1
Forward power (kW)	1.5	0.85C	1.0
Reflected power (W)	—	—	<10
Outer flow rate (L/min)	17	25	18
Intermediate flow rate (L/min)	1.2	—	0.0–0.5
Injector gas flow rate (L/min)	0.8–1.0	0.5	1–2.5
Observation height (mm)	—	15	10–20
Drying condition; time	—	150°C; 40 s	400°C; 20 s
Ash condition; time	—	1500°C; 0 s	800°C; 2–5 s
Vaporize condition; time	—	2700°C; 7 s	2400°C; 5–10 s
Sample size	2–3 mg	0.5 mg	10 μL

[a] JA = Thermo Jarrell–Ash.
[b] Liquid samples only.

Table 12.7. Detection Limits (pg) Obtained for Solutions with Electrothermal Vaporization into an ICP

		Reference				
Element	Wavelength (nm)	31	26	33	25	38
Ag	328.0	100	—	3	3	1
Al	308.2	1700	60	—	20[a]	20[a]
Cd	228.8	200	40	—	10	30
Cr	205.5	50	14	22[b]	—	—
Cu	324.7	100	2	5	10	2
Fe	259.9	100	—	—	—	20
Mn	257.6	30	8	5	1[c]	10
Zn	213.8	70	—	—	3	2

[a] Alternate wavelength used: 396.2 nm.
[b] Alternate wavelength used: 267.7 nm.
[c] Alternate wavelength used: 279.5 nm.

Table 12.8. Analytical Results for Bovine Liver (NBS SRM 1577) by Electrothermal Vaporization into an ICP

Element	Wavelength (nm)	Mean ± SD	
		Certified values (μg/g)	ICP results for 0.5-μg samples (μg/g)[a]
Ca	315.8	124 ± 6	143 ± 42
Cd	228.8	0.27 ± 0.04	0.29 ± 0.03
Cu	324.7	193 ± 10	179 ± 19
Fe	259.9	268 ± 8	305 ± 33
K	766.5	(0.97 ± 0.06)[b]	(1.15 ± 0.06)[b]
Mg	279.5	604 ± 9	658 ± 48
Na	330.2	(0.243 ± 0.013)[b]	(0.301 ± 0.023)[b]
P	214.9	(1.1)[b,c]	(1.18 ± 0.05)[b]
Pb	220.3	0.34 ± 0.08	<0.5
Zn	213.8	130 ± 13	123 ± 26

Source: Reference 32.
[a] Mean and standard deviation from triplicate assays.
[b] Values expressed as percent.
[c] Noncertified value.

follows the degree of sample heterogeneity, which is a major concern in solid sampling, particularly with these small sample sizes.

A calibration plot for zinc, based on Spex-G graphite powder standards, using indium as an internal standard, is shown in Figure 12.12. Linearity of calibration of at least three orders of magnitude seems possible for this type of sampling. To establish the effect of sample matrix composition on calibration, certified standard reference materials (SRMs) have been analyzed. The standards utilized include pine needles, tomato leaves, spinach, orchard leaves, coal, and coal fly ash, mixed with spectroscopic grade graphite. The calibration curve for manganese, based on the certified values in each SRM, is shown in Figure 12.13. Despite the scatter, which is most likely attributable to the minimal sample size used, the results demonstrate linearity regardless of sample matrix.

12.3.4.2 Interferences and Limitations

The major limitation of solid sample introduction into the ICP by ETV is the requirement for sample homogeneity. Because analyses are restricted to sample sizes of 10 mg or less, weighing constraints and homogeneity restrictions for this size of sample become critical. Sample preparation techniques devised to improve fine particulate uniformity, such as micropulverization and sieving, seem imperative to guarantee homogeneity. However, one of the benefits of direct solid sampling, namely minimal sample preparation, is diminished.

Downward shifts in the continuum background intensity are caused by a temporary increase in injector gas flow. This "piston effect" results from the expansion of the injector gas as it passes over the hot ETV during the vaporization step. Although it is still possible to make accurate measurements in

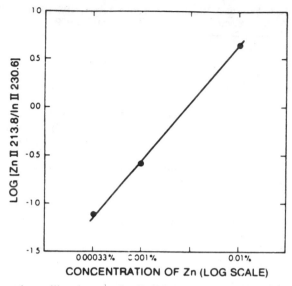

Figure 12.12. Log–log calibration plot for Zn I 213.8 nm in the Spex G series of graphite-powder standards, using electrothermal vaporization and ICP excitation. (From Reference 34, with permission.)

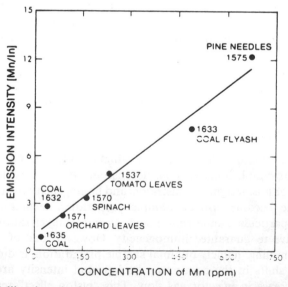

Figure 12.13. Calibration plot for manganese constructed from a series of NBS standard reference materials for electrothermal vaporization and ICP excitation. (From Reference 34, with permission.)

the presence of this negative peak, because it is reproducible, isolation or elimination of this shift can be accomplished. Isolation requires that the length of transport tubing, minimized to reduce sample deposition at the tube wall, must also be adequate to allow temporal separation of analyte peak and this background depression.[23] Alternatively, to minimize or eliminate this background depression, the relative amount of chamber-heated argon is kept small by reducing the heating cycle times and by minimizing the chamber volume.[37,39]

With ETV-ICP, as with other methods that require a sample probe, certain shortcomings are observed. The same limitations cited in the preceding section on direct sample insertion with a graphite probe apply for the graphite ETV. Likewise, the same compensations can be made. When a metal ETV is used, durability and reduced temperature range become limitations.

Matrix-dependent volatility is evident in ETV sample introduction into the ICP when solutions are analyzed. Similar behavior is likely for ETV solid sample introduction. Judicious choice of the power supply heating programs, and the use of matrix modifiers, as used in graphite furnace atomic absorption techniques, could be used to restrict this matrix-dependent behavior. Also, the addition to the injector gas flow of a few percent of organic halide, such as Freon, can enhance the vaporization of the sample, especially for the refractory elements.[5]

12.3.4.3 Accuracy

Demonstration of analytical accuracy for solid sampling has appeared in recent publications.[32,34] The results for the analysis of NBS SRM 1577 (bovine liver) are shown in Table 12.8. Except for sodium, the certified values fall within the confidence range of the results. Further work with this technique is required before a comprehensive evaluation of analytical accuracy can be made.

12.3.4.4 Applications

Applicability of the ETV for solid sample introduction to the ICP is still under investigation. Recent studies demonstrate the potential of the technique for analysis of a variety of sample types including biological, botanical, and coals. The microsampling capability of the technique is particularly advantageous in the sample-limited situation. However, because of the same sample size constraints, heterogeneity of sample becomes a critically limiting factor to the successful application of this technique to solid sampling.

12.4   ARC AND SPARK ABLATION

12.4.1   Principles of the Technique

One of the most successful methods of introducing solid samples directly into the ICP is accomplished using an electrical discharge to remove particles of material from the sample surface. The process of removal, called the ablation or erosion process, leads to the creation of a solid, dry aerosol that is transportable in a gas stream to the ICP.

The types of discharge used may be classified into two categories: arcs of various origins and sparks.[3,41-60] Discharge conditions are appropriately chosen so that sufficient energy is available to reproducibly remove material from the sample surface. Depending on conditions, material may be vaporized, eroded, or sputtered. To minimize complication of these processes, discharges are usually operated in an oxygen-free atmosphere, such as argon. As a result, the transport process from the discharge area to the ICP is simplified, since the argon is suitable for both discharges.

For successful ablation, sample must be conductive, either naturally or through addition of a conductive matrix material, such as powdered copper or graphite, to form a homogeneous mixture. Samples may be in virtually any physical form as long as a reproducible discharge can be formed between the sample and an electrode. A major advantage of the arc or spark ablation techniques in quality-control applications is that sample material does not require weighing. However, a basic assumption is that the quantity of material ablated per unit time is reproducible for a given sample, set of discharge conditions, and discharge duration.

Transport of sample from the discharge to the ICP is achieved using a flow of argon through a length of connective tubing, which can vary in length from a few centimeters to several meters. Transport may be direct to the ICP torch or via intermediate settling chambers.

The emission signals generated often have a time dependence, which is related to element volatility, discharge conditions, discharge stability, and matrix composition.

12.4.2   Instrumentation

Aerosol generation using a dc arc source is the first and perhaps the simplest electrical discharge successfully used to ablate material from a sample surface. The aerosol generator described by Dahlquist and co-workers[41] for use with ICP excitation, reported in an earlier work by the same group[59] for use with a capillary arc discharge, is shown in Figure 12.14. The aerosol generator consists of an air-cooled hollow anode that is flushed by argon, and a tubular, grounded cathode that contacts the sample. The discharge between the anode and the sample ablates material, which is transported by argon from the conical passageway through the hollow anode. Small holes around the periphery of the pedestal distribute argon for efficient aerosol pickup and transport. Up to 20 meters of hose, connecting the aerosol generator to the ICP, may be successfully used.

Another type of arc that may be used is the radio frequency arc (RF arc) described by Farnsworth and Hieftje[49] (Figure 12.15). An ICP torch is modified by flaring the central injector tube at the base of the torch to form a bell jar. The

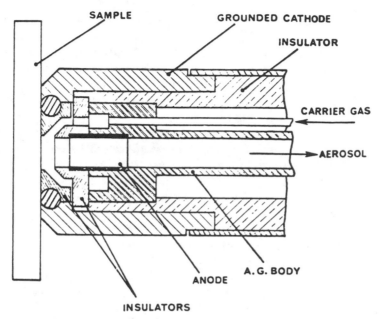

Figure 12.14. Schematic of a dc arc aerosol generator (a.g.) for arc ablation of conductive materials. (From Reference 59, with permission.)

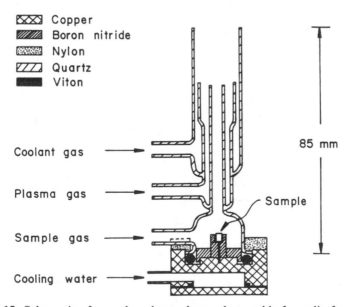

Figure 12.15. Schematic of a torch and sample stand assembly for radio frequency arc sampling. (From Reference 49, with permission.)

sample is in electrical contact with the water-cooled base, which is constructed from copper and is grounded. A flow of argon is introduced at the base. On initiation of the plasma discharge, a stable filament from the plasma forms between the induction region and the sample. Boron nitride acts as a thermal and electrical insulator between the base and the RF arc.

The most common form of discharge used to generate sample aerosol is the spark, which can range from low to high voltage types of varying pulse durations, frequencies, and waveform shapes.[43–46,48,50–58,60] At the time of publication, two commercial instruments that utilize spark aerosol generators were available (Thermo Jarrell-Ash and Applied Research Laboratories). Figure 12.16 is a schematic of a spark aerosol generator that forms the basis of the conductive solids nebulizer (CSN) of Applied Research Laboratories.[51] A spark source and Petrey table of the type used in spark emission spectrometry, but modified for use with ICP, are mated with an independent argon gas control system. A spark discharge is formed between a tungsten counter electrode and the conducting sample surface. Aerosol and argon are transported directly to the plasma via a 1-meter-long connecting tube. For certain discharge conditions and sample types, the addition of an intermediate settling chamber for deposition of excessively large aerosol particles may be necessary.[3,45,46,50,54] In at least one configuration, this chamber requirement may be satisfied by using a conventional ICP solution spray chamber.[46,50,54]

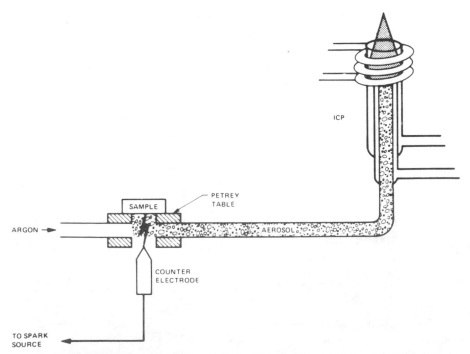

Figure 12.16. Schematic of a spark ablation–ICP excitation system. (With permission, from Applied Research Laboratories, Inc.)

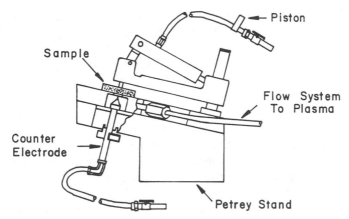

Figure 12.17. Schematic of a spark chamber used for spark ablation. (From Reference 50, with permission.)

A detailed view of the spark chamber of the separate sampling and excitation analysis system (SSEA) from Thermo Jarrell-Ash[50] is shown in Figure 12.17. The Petrey stand consists of a cylindrical chamber on which a flat-surfaced sample is placed; the sample forms a gas seal preventing invasion of air in the discharge region. A clamp that provides electrical contact with the top of the sample is spring loaded to maintain both the seal and electrical continuity. A thoriated tungsten electrode is placed in close proximity to the sample through the cylindrical chamber. The distance between the electrodes is 1 to 5 mm, although up to 11 mm has been used with other chamber designs, electrode types, and discharge conditions. Argon gas enters the chamber in the vicinity of the discharge and passes through an exhaust port, along with sample aerosol, through the liquid nebulizer chamber to the ICP.

The area of the sample surface exposed to the discharge may be varied by using different sized openings between the counter electrode and the sample.[56] A boron nitride mask between sample and electrode limits the discharge to the area and the shape of the mask aperture.

12.4.3 Operating Parameters

In addition to the parameters commonly optimized in solution nebulization, the use of arc or spark ablation requires the optimization of aerosol injector gas flow, discharge energy, and time, including preburn conditions.

The dc arc described by Jones and co-workers[59] operates with an open-circuit voltage of 600 V and an average voltage drop across the aerosol generator of 50 V. Operating current is varied between 2 and 8 A. While the aerosol is produced at a more rapid rate and the signal-to-background ratio increases with increasing current, self-absorption also increases, leading to a compromise current of 4 A. Injector gas flow rate is nominally set at 1 L/min. Thus, plasma conditions are virtually identical to those used for solution nebulization.

The RF arc described by Farnsworth and Hieftje[49] has few operational parameters except plasma conditions to adjust, because the plasma and the cell geometry at the base of the torch control the discharge conditions. However, data on the RF arc are preliminary, with further investigation of parameters required before optimum operating conditions can be established.

By far the most information on arc and spark ablation into the ICP exists for the spark technique. Because the instrumentation used by most of the research groups cited differs, particularly in the spark source area, direct comparisons of the operating conditions are difficult. However, the method of constructing "burn-off" or "spark-off" curves, well known in spark emission spectrometry, is used to establish spark source conditions for the particular instrumental configuration. The method is simply the generation of emission intensity graphs as a function of time while varying source conditions such as capacitance, inductance, resistance, voltage, electrode gap, spark repetition rate, and preburn conditions. Such studies have shown that the ICP operating conditions are virtually equivalent to those used for conventional solution nebulization.

The exception to this is the variation of injector gas flow during the spark source operating cycle. For example, for the conductive solids nebulizer, a sample is first clamped to the Petrey stand, and the argon injector flow is increased to 15 L/min for a few seconds to remove air and traces of previous sample. The spark source is then initiated at the preburn conditions for 20 s with the injector gas flow reduced to 5 L/min. During the preburn condition, sample homogeneity is improved because of the reduction of metallurgical structure differences on the sample surface prior to the analytical sampling. After the preburn period, the source switches to the analytical condition, which is usually less energetic, and the argon flow is reduced to 1 L/min. Integration begins 2 s later for a period of 5 to 60 s, depending on whether a simultaneous or sequential spectrometer is being used. With some instrumental configurations, however, particularly when graphite electrodes are used, longer sparking time can lead to inferior precision due to variations in spark gap with sparking time. One second after integration is completed, the spark is turned off and the injector gas flow is increased briefly to promote cleanout before the next cycle.

Operating conditions for CSN-ICP-AES are summarized in Table 12.9. These conditions are applicable for samples of several different types, with only the spark voltage being varied. In general, the voltage required for more fragile samples is lower, becoming progressively lower for the following sequence: steel, aluminum, lead-tin solder, briquetted powder.[51,55,57,58] Additionally, for a given capacitance, the use of a low inductance for the preburn phase promotes more rapid stabilization of emission intensity, while a higher inductance during the analytical cycle generally improves signal-to-background ratio. Again, it must be emphasized that appropriate conditions are critically dependent on the particular instrument configuration, requiring the burn-off technique to establish optimum operating conditions. In general, source conditions for SSEA-

Table 12.9. Typical Operational Parameters for Spark Ablation into an ICP Using the Conductive Solids Nebulizer[a]

Parameter	Value
Outer flow rate	12 L/min
Intermediate flow rate	0.8 L/min
Injector gas flow rate	
Purge	15 L/min
Preburn	5 L/min
Integrate	1 L/min
Frequency	27.1 MHz
Forward power	1.2 kW
Observation height	15 mm
Preburn conditions	
Voltage	500 V
Capacitance	10 μF
Inductance	12 μH
Resistance	0 Ω
Integration conditions	
Voltage	250–400 V
Capacitance	10 μF
Inductance	112 μH
Resistance	2 Ω
Spark gap	4 mm
Preburn time	20 s
Integrate time	5 s
Repetition rate	120 Hz

[a] Manufactured by Applied Research Laboratories.

ICP–AES analysis of materials of similar types include a voltage of approximately 11 kV, and RF current of 3 to 4 A, two or four breaks per half cycle, and a spark gap of 4 mm.

12.4.4 Analytical Performance

12.4.4.1 Figures of Merit

Detection limits[43,46,50,56,57] for the spark ablation ICP excitation technique are listed in Table 12.10. Although the DLs are from different instrumental configurations, sample types, and wavelengths, it is obvious that limits of detection in the solid sample extend down to the microgram-per-gram range for the most sensitive lines. Taking into account dilution factors for solution nebulization, DLs from spark ablation ICP excitation are comparable to those for solution nebulization. It is important to note that these DLs are attainable from only a few micrograms of ablated material.

Precision for the spark ablation ICP excitation technique is superior to any of the other solid sample introduction techniques (Table 12.11). In general, for various alloy materials of most metallurgical bases, precision is between 0.5 and

Table 12.10. Detection Limits (μg/g) for Solid Samples with Spark Ablation into an ICP

Element	Wavelength (nm)	NBS Steel 1265 Ref. 50	Nickel alloy Ref. 46	Aluminum alloys		
				Ref. 56	Ref. 57	Ref. 43
B	182.6	10	30			
	249.6	22				
C	193.0	500				
Co	228.6	10				
	229.3		300			
Cr	267.7	7			2	
	298.9		400			
Cu	324.7	4	80	26	1.6	3
Fe	259.9		300	4	7	
Mg	279.0	6				
	279.5			0.8		2
Mn	257.6	10				
	293.3		3	265	1.4	
Mo	202.0	30	2			
Nb	313.0	2	20			
Ni	231.6	25			2	
P	178.2	28				
Pb	220.3	10				
	283.3			81		
Si	251.6	25	100			
	288.1			46		
Ta	240.0	50	20			
Ti	334.9	3				
	337.2				0.9	
V	310.2	2		128		
	271.6		7			
Zn	206.2	2				
	213.9			14	4	150

5% RSD. This is due to the absence of sample-to-sample variation and the relatively large surface area sampled (0.2 to 1 cm^2), which minimizes the effects of sample heterogeneity. For powdered materials that are combined with a conductive powder matrix into a briquette, precision is somewhat inferior but is partially dependent on the reproducibility of the briquetting procedure and resultant sample homogeneity. For example, the precisions reported for determining cerium, cobalt, and iron in manganese nodules were 7, 10, and 14% RSD, respectively.[60]

Calibration linearity for dc arc ablation is well illustrated in Figure 12.18, and similar plots are obtained with spark ablation. The important characteristics are: (a) calibrations are linear over at least three orders of magnitude, (b) elements from different matrices fall on the same curve, indicating independence of element response from matrix, (c) sample aerosol may be transported significant distances without degrading analytical calibrations, and (d) calibration discontinuities, which sometimes occur in spark emission spectrometry, are

Table 12.11. Precision of Analysis for Spark Ablation into an ICP

Sample type	% RSD	n^a	Ref.
Low-alloy steel[b]	4.1	25	41
Aluminum alloys	3.8	—	43
Aluminum alloy blank	0.4	—	43
Low-alloy steel	2.2	10	50
Stainless steel	2.1	11	50
High-alloy steel	4.6	14	50
Nickel alloy	4.4	15	50
Aluminum alloy	1.5	18	50
Nickel alloys	0.61	19	46
Low-alloy steel	1.3	8	57
Aluminum alloy	0.94	8	57
Lead–tin solders	5.9	10	57
Briquette (graphite and lithium carbonate)	3.5	24	57

[a] n = number of measurements.
[b] Remote, 6-m hose.

absent. As a result, calibration can be simplified by reducing the number of standards and the time spent to perform analyses, compared to conventional spark emission procedures.

Simplified calibration using solid standards has led to the hypothesis that solutions may be used for direct ICP calibration followed by spark ablation ICP excitation. Good agreement is attainable using solutions calibration for analysis of the NBS 1260 series of alloy steels.[55,57] This leads to potential elimination of the need for certified solid standards, or at least a large number of them, which are not only generally expensive but also sometimes difficult to obtain.

12.4.4.2 Interferences and Limitations

Although interferences for the spark ablation technique appear to be minimal and similar to those experienced in solution nebulization, there are several limitations. When excessive quantities of sample are transported to the plasma, imprecision, calibration curvature, memory effects, and plasma instability can occur. If insufficient arc or spark energy is used during the analytical cycle, differential distillation of sample components can occur, depending on the matrix. By contrast, too much energy can lead to excessive mass sputtered, or in the case of briquetted samples, shattering of the briquette. Memory effects are an occasional limitation of the instrumentation, not an inherent shortcoming of the arc spark ablation technique. For stainless steel standards, the memory effect is less than 0.2% of the elemental concentration level of the previously ablated sample.[57] If the spark chamber is cleaned using a one-minute flush-out procedure, memory is reduced significantly.

Although sample heterogeneity can be a problem, analysis of a representative portion of sample is possible with spark ablation due to the ability of the preburn technique to reduce sample heterogeneity.

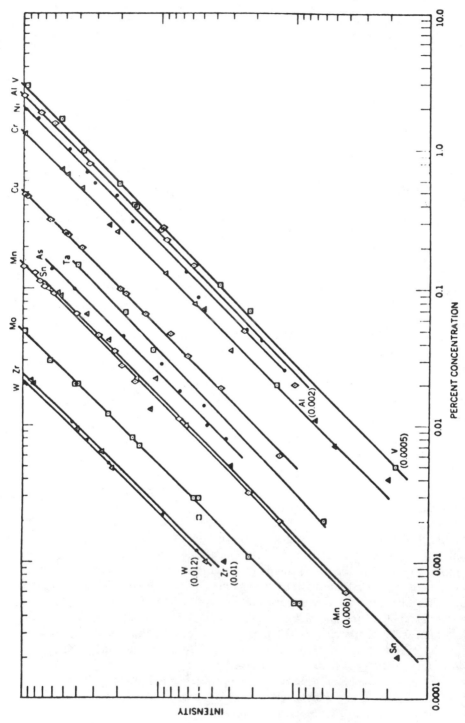

Figure 12.18. Calibration plots for arc ablation and ICP excitation of a series of steel NBS standard reference materials. (From Reference 41, with permission.)

Table 12.12. Analytical Results for Stainless Steel (SS654-8) Using Spark Ablation into an ICP

Element	Wavelength (nm)	Results (%)		Difference	Relative difference (%)
		Actual	Found		
Co	228.6	0.27	0.27		
Cr	267.7	24.95	25.07	0.12	0.5
Cu	324.7	0.079	0.076	−0.003	4.0
Mn	257.6	1.72	1.76	0.04	2.0
Mo	202.0	0.098	0.094	−0.004	4.0
Ni	231.6	19.08	18.99	−0.09	0.5
P	178.2	0.023	0.024	0.001	4.0
Si	212.4	0.38	0.38		

Source: Ref. 57.

Table 12.13. Analytical Results for Low-Alloy Steel (BAS-407) Using Spark Ablation into an ICP

Element	Concentration (%)		Relative difference (%)
	Certified	Measured	
Al		0.067	
As		0.013	
B		0.002	
C	0.50	0.54	8.0
Co		0.016	
Cr	3.00	3.04	1.32
Cu	0.43	0.40	7.0
Fe		93.8	
Mn	0.13	0.13	
Mo	0.82	0.83	1.2
Ni	0.61	0.63	3.2
P	0.033	0.032	3.0
S	0.12	0.13	7.7
Si	0.69	0.74	6.7
Sn		0.006	
Ti		0.003	
V	0.23	0.22	4.3
W		0.032	
Zr		0.012	

Source: Ref. 50.

Limitations of the RF arc, in its present state of development, are its relative instability, comparatively slow vaporization rates for certain elements, transient signal nature for other elements, and unsuitability for certain conductive samples. However, once again, the preliminary status of the development precludes judgment as to the future suitability of the RF arc.

An obvious limitation of all techniques utilizing an electrical discharge for aerosol generation is the requirement for sample to be conductive. While metallurgical samples meet this requirement, nonconductive samples must be converted to a conductive form by addition of appropriate matrix. This step can lead to contamination, and it is often time consuming, particularly if sample must first be ground into a powder.

12.4.4.3 Accuracy

As with other techniques, accuracy is demonstrated by analyzing standard reference materials. For spark ablation, analytical accuracy is shown for stainless steel, low-alloy steel, and ductile iron in Tables 12.12, 12.13, and 12.14,

Table 12.14. Analytical Results for Ductile Iron (NBS SRM 1142a No. 3) Using Spark Ablation into an ICP

Element	Concentration (%) Certified[a]	Measured	Relative difference (%)
Al	0.088	0.099	12.5
As	(0.01)	0.008	
B		0.004	
C	2.72	2.40	11.7
Co		<0.001	
Cr	0.051	0.48	5.9
Cu	1.02	0.83	18.6
Fe		93.4	
Mg	0.116		
Mn	0.17	0.16	5.9
Mo	0.022	0.024	9.1
Nb		0.001	
Ni	1.69	1.78	5.3
P	0.18	0.20	11.0
Pb	(0.0003)	0.0009	
S	0.015	0.018	20.0
Si	3.19	3.32	4.1
Sn		0.033	
Ti	0.007	0.006	14.3
V	0.004	0.006	50
W		0.020	
Zr		0.001	

Source: Ref. 50.
[a] Values in parentheses are not certified.

respectively.[57,50] Relative accuracies generally fall within a few percent unless elemental concentrations are near or below the DL.

Accuracy of spark ablation for briquettes, although demonstrated only for a small number of nonconductive samples, appears promising. Two groups have shown impressive results for alumina powder and the well-characterized Spex mix.[56,57] Accuracies[57] of better than 5% relative with precisions of 3.5% RSD are attainable (Table 12.15). For manganese nodules, relative accuracies of better than 4, 7, an 13% are reported for phosphorus, lead, and iron.[60]

Remote dc arc generation[59] of aerosol with transport through a 6-meter hose to the ICP also produces acceptable accuracies, as demonstrated for 13 NBS standard reference materials (Table 12.16).

Preliminary results with the RF arc for the analysis of NBS standard reference material 1633a (coal fly ash) indicate that qualitative results are attainable.[49] However, more developmental work is required before the method can be considered quantitative.

Table 12.15. Analytical Results for Spex Mix Powder[a] Compressed into a Briquette Using Spark Ablation into an ICP

Element	Wavelength (nm)	Concentration found (%)	Difference (%)	% RSD (n = 10)
Al	394.4	0.138	−0.003	4.0
	308.2	0.142	0.001	3.6
As	189.0	0.149	0.008	1.5
Ba	455.4	0.145	0.004	3.4
Co	228.6	0.142	0.001	3.7
Cr	267.7	0.145	0.004	3.9
Cu	324.8	0.150	0.009	7.1
Fe	273.1	0.141	—	3.4
K	766.5	0.142	0.001	3.4
Mg	279.1	0.133	−0.008	2.4
Mn	257.6	0.154	0.013	3.6
Mo	202.0	0.158	0.017	1.8
Na	589.1	0.161	0.020	3.0
Ni	231.6	0.144	0.003	3.6
P	178.3	0.133	−0.008	1.4
Pb	220.4	0.143	0.002	2.4
Sb	217.6	0.143	0.002	1.7
Si	288.2	0.139	−0.002	3.5
	212.4	0.135	−0.005	3.1
Sn	189.9	0.146	0.005	2.7
Ta	240.1	0.160	0.019	3.4
Ti	337.3	0.130	−0.011	10.4
V	310.2	0.138	−0.003	2.8
Zn	213.8	0.147	0.006	2.2
Average (24 lines)			0.0065	3.5

Source: Ref. 57.
[a] Concentration of each element is 0.141 wt %.

Table 12.16. Analytical Results for a Series of NBS SRMs Using Spark Ablation into an ICP and a 6-m Length of Connecting Tubing Between Spark and ICP

NBS SRM	Percent Sn		Percent Ta		Percent As	
	Given	Found	Given	Found	Given	Found
1161	0.022	0.014	0.002	0.001	0.028	0.023
1162	0.066	0.066	0.036	0.017	0.046	0.045
1163	0.013	0.016	0.15	0.17	0.10	0.10
1164	0.043	0.041	0.069	0.047	0.018	0.017
1165	0.001	0.002	0.001	0.004	0.010	0.011
1166	0.005	0.002	0.002	0.001	0.014	0.011
1167	0.10	0.098	0.23	0.22	0.14	0.14
1168	0.009	0.005	0.005	0.009	0.008	0.008
1261	0.010	0.006	0.020	0.007	0.018	0.014
1262	0.022	0.021	0.20	0.20	0.08	0.08
1263	0.094	0.099	0.04	0.04	0.009	0.014
1264	—	—	—	—	0.036	0.039
1265	0.0002	0.0004	—	—	0.0003	0.0003

			Standard estimate of error		
			Sn	Ta	As
Order of equation		1	0.006	0.016	0.003
		2	—	—	—
		3	0.004	0.013	0.003
Precision: NBS 1261			0.0025	0.0068	0.0016

Source: Ref. 41.

12.4.4.4   Applications

Spark ablation has been used extensively for analysis of a wide variety of materials. These include aluminum alloys, antimonial lead, brass, coal fly ash, manganese nodules, copper, copper powder, gold alloy, high-alloy steel, graphite, gray iron, high-purity gold, high-temperature steel alloy, iron-based alloys, lead–tin solder, low-alloy steel, nickel-based alloy, powdered metal oxides (Mg, Fe, Mn), stainless steel, superalloy, and white iron. The vast majority of such applications have been generated within the two research groups that developed the technique commercially. While such a list is impressive, particularly for metallurgical samples, significant opportunity and need exist for development of the technique for a wider variety of nonconducting solids.

12.5   LASER ABLATION

12.5.1   *Principles of the Technique*

Introduction of a laser-generated aerosol of sample material into the ICP has been investigated by several research groups.[61,62,64–70] Similar to the arc or spark ablation processes, sufficient energy, in the form of a focused laser beam, is directed onto a sample to cause material to be sputtered from the surface. The plume of vapor and particulates released from the surface of the sample are transported in an argon stream to the plasma, where emission, and the processes prerequisite to it, occur.

Emission from the ICP is measured either by a single or multichannel spectrometer, but because this form of sample introduction produces a transient analyte signal, simultaneous multielement determinations require the use of a multichannel instrument. Like ETV–ICP, data may be acquired in two distinct modes. Both time resolved spectra and total signal integration are used to characterize the signals produced by this technique.

The amount of ablated matter transported into the ICP is directly correlated to signal intensity. The rate and amount of material ablated depend not only on sample type and surface condition, but on certain laser properties and operating conditions. These include laser focus, energy output, operating mode, shot frequency, and duration. Ablation chamber geometry and length of connective transport tubing can also determine how much volatilized material reaches the plasma.

The main attractive feature of laser ablation is the ability to sample both conducting and nonconducting, solid and powdered materials having a wide variety of geometries. Additionally, the focusing characteristics of a laser beam permit sampling of such a small area that localized in situ microanalysis is possible.

12.5.2   Instrumentation

The principal elements required for laser ablation sample introduction into an ICP include the laser, which can be operated in either Q-switched or free-running mode, plus any deflection or focusing optics needed, the ablation chamber to contain both the sample and generated aerosol, and the connections that must be made to ensure efficient transport of the aerosol to the plasma. The laser ablation system designed by Ishizuka and Uwamino[67] illustrates one implementation of these requirements (Figure 12.19). The ablation cell consists of a stainless-steel-walled cylinder with an inside diameter of 13 mm and a height of 35 mm, and a Pyrex glass window. A flat smooth metal sample forms the base of the cylinder. The cell volume is kept small enough to minimize diffusion of the laser-produced plume, yet large enough to prevent sample spattering onto the cell walls. A 45-degree prism and a lens of 5 cm focal length direct and focus the laser beam onto the surface of the sample. Sample aerosol is transported to the ICP through PVC tubing. The emission signal detected by the photomultiplier tube (PMT) of a monochromator is amplified, converted to frequency, then counted with a multichannel analyzer. Laser firing is synchronized to data acquisition with a PMT trigger actuated by a xenon flash lamp pulse.

Various sample geometries are used in laser ablation. Carr and Horlick have configured two ablation chambers to hold rod and disc-shaped samples.[65]

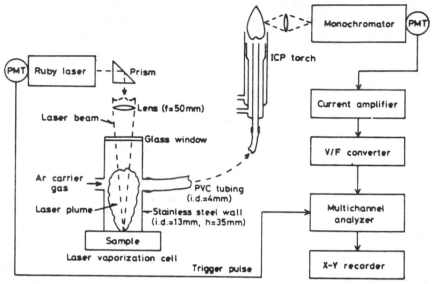

Figure 12.19. Schematic of a laser ablation ICP–AES system. (From Reference 67, with permission.)

Several researchers have designed ablation chambers to accommodate powdered samples.[62,65] An example of a disc ablation chamber is shown in Figure 12.20. This system accommodates a range of disc sizes including the typical metal industry standards used in spark spectrometry. To keep the connection between the ablation chamber and the torch as short as possible, the chamber was designed to mount directly below the torch. Convenient rotation of the sample to provide a fresh surface for successive ablation, without interrupting plasma operation, is implemented with a 2-rpm motor.

Right-angle deflection prisms are used to direct the laser beam onto the sample. The output energy of this ruby laser is dumped in either the Q-switched or the free-running mode. The emission signals produced in the ICP are measured by a photodiode array spectrometer, and the data are collected either as total integrated emission signals or as time-resolved spectra.

The laser ablation system described by Thompson and co-workers[64] combines a commercially produced laser microprobe with a laboratory-constructed ablation chamber that accommodates samples of any shape (Figure 12.21). The chamber is a glass-walled cylinder ranging in diameter from 20 to 50 mm, 17 mm in height, with two side ports, a window of optical grade silica, and a base formed by mounting the cylinder onto an inverted rubber stopper. Samples are placed in the ablation chamber, which is mounted on a microscope stage.

The laser microprobe consists of a ruby laser head mounted in a binocular microscope. The target area is selected under the microscope, then the laser is focused and fired. Condensed aerosol from the laser shot is swept into the ICP through PVC tubing. Emissions from the plasma are detected by a 1-m vacuum

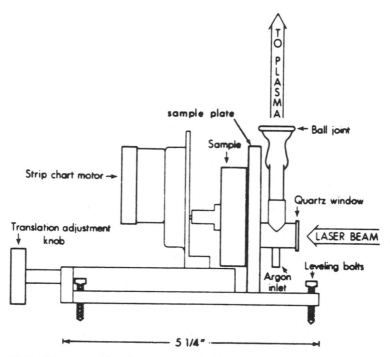

Figure 12.20. Schematic of a disc-type sample chamber for laser ablation into the ICP. (From Reference 65, with permission.)

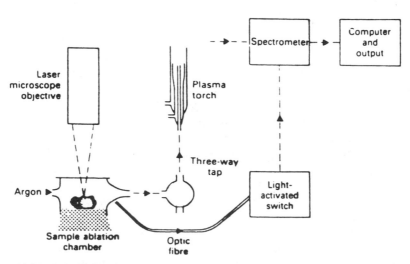

Figure 12.21. Schematic of laser ablation ICP–AES system that accommodates various sample shapes. (From Reference 64, with permission.)

polychromator. Signal integrations are triggered by a light-sensitive switch activated by the reflected light of the laser firing.

12.5.3 Operating Parameters

In addition to the parameters commonly optimized for solution nebulization combined with ICP excitation, the use of laser ablation requires the optimization of parameters that characterize the operation of the laser and aerosol transport efficiency, such as chamber volume and transport distance.

Typically, pulsed lasers, operating in either free-running or Q-switched mode, have been utilized for sample ablation. Several specific parameters determine the suitability of a laser for sample ablation. These include sufficient power output, lasing wavelength, repetition rate, shot-to-shot reproducibility, thermal stability, ease of alignment and operation, and reasonable cost.

Pulsed, solid state lasers, relatively convenient in operation and handling, have been used in a number of analytical schemes. The three previously described configurations[63,65,67] utilized solid-state ruby lasers. These lasers radiate at 694.3 nm and therefore can be visually monitored during operation. This wavelength is absorbed by a variety of materials, thus making it suitable for ablation, although some glasses and polymers tend to be transparent in this region. The ruby laser has relatively good thermal stability but must be operated at a low repetition rate, and its shot-to-shot reproducibility is relatively poor. Additionally, poor optical quality of the ruby rod can cause divergence of the focused laser beam. Solid-state neodymium:glass and neodymium: YAG lasers operate at 1.06 μm, which is also an optimal wavelength, as most materials absorb in this region. However, operation at this wavelength makes visual alignment impossible without the use of an additional coaxial alignment laser, and is somewhat more hazardous. Solid-state neodymium lasers are also thermally sensitive, and require cooling. A free running Nd:glass laser has been used for sample introduction for atomic absorption spectrometry.[61] For representative analysis of bulk samples, lasing at higher repetition rates is needed. The Nd:YAG laser can be made to operate at rates higher than 10^3 pps, without loss of power. Ruggedness, operational ease, size, and reasonable cost, along with a high degree of pulse-to-pulse reproducibility, make this laser a good candidate for sample ablation introduction into the ICP.

Use of a pulsed gaseous-state laser for material ablation into the ICP has been cited.[62] In addition to being cumbersome, this laser has the major disadvantage of operating at 10.6 μm, which is not optimal for metal surface sampling because many metals are highly reflective at this wavelength. Also, a coaxial alignment laser is required. Laqua provides an excellent tutorial on the use of lasers in sample ablation.[63]

The amount of laser energy incident upon the surface of the sample controls the amount of material ablated. The effect on emission signal of laser operation in both Q-switched and free-running modes has been documented.[65,67] In a free-running mode, the laser produced as much as 500 μg of material from a low-alloy aluminum sample, while in Q-switched mode, 25 µg was the largest amount ablated from the same type of sample.[65] Based on results published by these two research groups, the laser running in the free mode produced 20 to 30 times more material than the Q-switched laser. Not only is there an improvement in signal with the larger sampling size, there is less susceptibility to sample heterogeneity.

Likewise, the amount of sample ablated can be controlled by the laser focus. By defocusing the laser beam stepwise from a spot diameter of 0.5, to 1.5, to 3.0 mm, a decrease in emission signal from an aluminum sample is observed, as shown in Figure 12.22. In a similar experiment, Ishizuka and Uwamino[67] correlated laser spot size, sample mass, operational mode, and emission intensity. In the free-running laser mode, defocusing the laser beam from a spot size of 0.5 to 3.0 mm produces a 500-fold decrease in ablated material, although the emission signal does not diminish in proportion (Figure 12.23). In the Q-switched mode, defocusing reduces the amount of ablated material only a factor of 2, with a comparable effect on emission intensity (Figure 12.24).

As with many discrete sampling processes, time–response profiles are used to optimize operational parameters. The time-resolved signal (Figure 12.25) illustrates the effect of ablation chamber size on signal shape. Peak broadening is evidenced with the larger volume cell. Likewise, distance of sample transport through connective tubing has been correlated (Figure 12.26). As tubing length is increased, diffusion of the laser-ablated vapor in the tubing becomes more

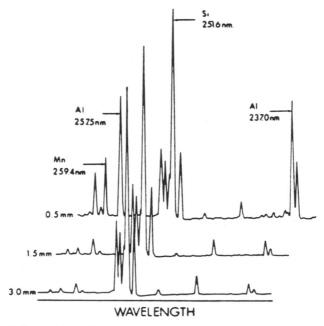

Figure 12.22. Effect of laser focus on the emission signal from silicon in high-alloy aluminum. (From Reference 65, with permission.)

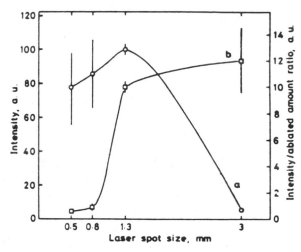

Figure 12.23. Effect of laser focus on the emission intensity of manganese for 0.25% Mn in steel, by free-running laser: curve *a*, signal intensity; curve *b*, corrected signal intensity divided by mass of sample ablated by the laser pulse. Vertical lines represent standard deviations of the values obtained at 5 points. (From Reference 67, with permission.)

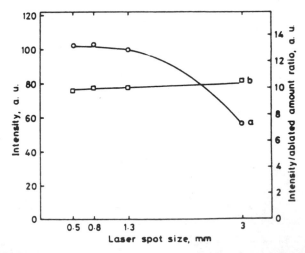

Figure 12.24. Effect of laser focus on the emission intensity of manganese for 0.25% Mn in steel, by Q-switched laser: curve *a*, signal intensity; curve *b*, corrected signal intensity divided by mass of sample ablated by the laser pulse. (From Reference 67, with permission.)

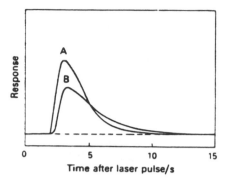

Figure 12.25. Response as a function of time after laser pulse for iron in steel. Volume of chamber A is approximately 5 cm³; chamber B is approximately 30 cm³. (From Reference 64, with permission.)

pronounced, and peak height and peak area decrease. This effect was most pronounced with the laser operating in Q-switched mode.[67]

Optimized operational parameters detailed in the work of three groups of researchers are listed in Table 12.17.

12.5.4 Analytical Performance

12.5.4.1 Figures of Merit

Detection limits[64,67] for analytes in steel samples by laser ICP–AES are listed in Table 12.18. Considering the probability of local heterogeneity in the samples,

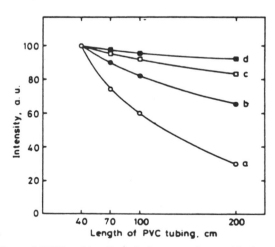

Figure 12.26. Effect of PVC tubing length between laser ablation cell and ICP for emission intensity of manganese (0.27%) in steel. Curves *a* and *b*, respectively, show peak height and peak area by Q-switched laser; curves *c* and *d*, respectively, show peak height and peak area by free-running laser. (From Reference 67, with permission.)

Table 12.17. Instrumentation and Operational Parameters for Laser Ablation of Solids into an ICP

Instrument parameter	Reference		
	67	64	65
Spectrometer	1-m Czerny–Turner monochromator	1-m Paschen–Runge polychromator	Photodiode array spectrometer
ICP			
Frequency (MHz)	27.1	27.1	27.1
Forward power (kW)	1.2	1.2	1.5
Outer gas flow (L/min)	14	12	18
Intermediate gas gas flow (L/min)	1	0.8	0
Injector gas flow (L/min)	1.1	0.5	1.2
Observation height (mm)	15	14	18
Laser			
Laser energy; power	Q-switched = 2 J Free-running = 18 J	Free-running = 1 J (maximum)	Free-running = 1–2 J Q-switched = 100 mW
Laser shot interval (s)	60	—	4–8
Sample size (μg)	1–500	1	25–500

the absolute DLs for these two sets of data agree quite favorably. These DLs determined with the ablating laser operating in the free-running mode indicate that microgram-per-gram determinations are possible.

Precision (Table 12.19) is somewhat degraded, when compared to solution nebulization. Precisions of 3 to 11% RSD are observed for a variety of metal samples. Two major factors account for this lack of precision: sample heterogeneity combined with the microsize samples ablated, and the poor shot-to-shot

Table 12.18. Detection Limits Measured in Steel by Laser Ablation into an ICP

Element	Concentration detection limit (μg/g)		Absolute detection limit (pg)	
	Ref. 64[a]	Ref. 67	Ref. 64[a]	Ref. 67
Al	—	2	—	60
Co	—	0.6	—	20
Cr	15	1	15	30
Cu	20	0.3	20	9
Mn	80	0.3	80	9
Mo	60	2	60	60
Ni	70	1	70	30
P	10	—	10	—
S	15	—	15	—
V	10	1	10	30

[a] Assumes 1 μg of material ablated.

Table 12.19. Precision Data for Analysis of Metal Samples with Q-Switched and Free-Running Laser Ablation into an ICP

Sample	Element (% conc.)	% RSD	
		Q-switched	Free-running
Steel	Al (0.021)	2.6	4.6
	Co (0.15)	10	3.1
	Cr (0.69)	6.8	3.8
	Cu (0.098)	5.7	8.6
	Mn (0.25)	3.7	4.6
	Ni (0.32)	2.5	3.3
	V (0.31)	8.3	11
Brass	Sn (0.43)	3.2	4.7
Aluminum alloy	Mg (0.95)	6.1	7.9
	Zn (0.035)	8.6	4.2
Titanium-based alloy	Mo (1.11)	6.6	9.9

Source: Ref. 67.

reproducibility of the laser used. Sample spatial heterogeneity can be demonstrated by measuring signal intensity as the laser is scanned across the surface of the sample. An overall variability[65] of 8.8% RSD in signal intensity is seen for magnesium in an aluminum sample (Table 12.20). A high degree of correlation exists between this variability and those precisions obtained for magnesium (Table 12.19) in the same type of sample.

Table 12.20. Effect of Sample Heterogeneity on Magnesium Intensity and Precision by Laser Ablation Along the Sample Surface

Crater number	Mg Relative intensity		
1	103		
2	102	Avg = 103.2	
3	104	% RSD = 0.9	
4	103		
5	91		
6	86		
7	78	Avg = 87.7	Avg = 97.0
8	84	% RSD = 6.8	% RSD = 8.8
9	96		
10	97		
11	104		
12	103		
13	104	Avg = 103.2	
14	104	% RSD = 0.9	
15	104		

Source: Ref. 65.

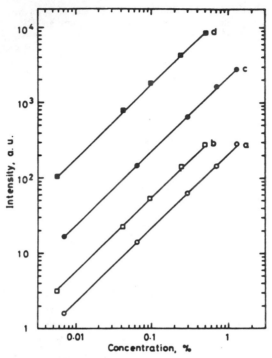

Figure 12.27. Calibration curves for (a) chromium and (b) copper using a Q-switched laser and (c) chromium and (d) copper using a free-running laser on a steel sample. (From Reference 67, with permission.)

Calibration curves[67] for various elements in different sample types have been constructed for solid samples. Results for chromium and copper in steel are illustrated in Figure 12.27. Typically a two to three order-of-magnitude linearity is observed for these sample types, with the upper concentration limit often set by the maximum amount of material the laser can ablate from the surface of the sample.

12.5.4.2 Interferences and Limitations

No comprehensive work has been done on interferences for this technique. It is reasonable to assume that these effects would be comparable to those experienced with other solid sampling methods. However, due to the nature and duration of the ablating pulse, selective volatilization to reduce these interferences, as described for the ETV–ICP technique, is minimal. A number of limitations must be considered when applying this method of sample introduction into the ICP.

Operated in a single-shot manner, laser ablation is best suited for area-localized microanalysis. To obtain a representative determination from a bulk sample, repetitive laser shots are taken across the sample surface and the resulting signal is time averaged. Ideally, movement across the sample surface would be synchronized to laser repetition rate. Alternatively, it is possible to increase the precision for macrosampling by enlarging the crater size through an increase of laser power, which makes more material available for transport into the plasma. The amount of useful ablated material is limited, however, by parameters such as transport efficiency and the energy transfer efficiency of the ICP. Ishizuka and Uwamino[67] noted a destabilization of the ICP with increased size of sample material, leading to an abnormal discharge in the sample injection tube of the ICP torch.

Shot-to-shot reproducibility is poor with certain lasers. Careful monitoring and control of the laser output can improve this situation.[63] Alternatively, an internal standard may be used to compensate for this imprecision.

12.5.4.3 Accuracy

The accuracy of laser ablation ICP–AES has been established for geological reference materials.[62] An example of the analytical accuracy for nickel in 15 mineral standards is shown in Figure 12.28. Similar results are noted for a

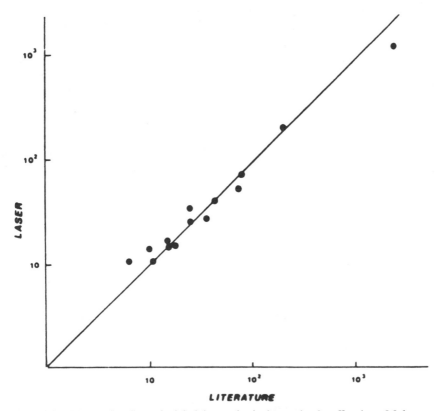

Figure 12.28. Determination of nickel in geological standards affixed to Mylar tape. Laser ablation–ICP excitation is compared to literature values. (From Reference 62, with permission.)

number of other elements. This analysis involved powdered samples. With heterogeneous solid samples, the accuracy will depend on how representative the laser sampling is of the total sample.

12.5.4.4 Applications

Laser ablation has been used for a wide variety of materials[61-70] with virtually no restriction on sample form, provided an ablation chamber could be constructed to contain the sample surface and the aerosol. Because of the excellent focusing characteristics of the laser beam, localized, in situ microanalysis is the most promising application of the laser ablation ICP excitation technique.

12.6 OTHER TECHNIQUES

Several techniques have been investigated for introducing powdered samples into the ICP. Greenfield and Sutton[71] used a Danielsson tape machine to introduce pulverized phosphate rock, −100 mesh, into a 5.2-kW argon–nitrogen ICP. The tape machine consists of an adhesive graphite tape that passes through a spark gap. After the powdered sample has been deposited on the tape, an 8 kV spark is initiated to volatilize the sample. The volatilized material is passed through a cyclone chamber to remove larger particles from the stream prior to introduction into the plasma. The RSD reported for 18 elements ranged from 1.1 to 17.7% when internal standardization was used. A similar tape machine was used by Abercrombie and co-workers[62] in laser ablation studies.

Another method for introducing powders into the ICP uses a vibrating device to float the sample in a chamber prior to transport into the ICP.[72] Meyer and Barnes[73] used such a device to compare the vaporization efficiencies of argon and air ICPs for analysis of solid samples. The air ICP was superior to the argon ICP for vaporization and atomization of calcium carbonate powder.[73] Ng and co-workers[74] used a powder injection device having a volume of approximately 0.3 mL to introduce 0.5 to 2.5 mg of NBS coal fly ash, silicon carbide, whole blood, tomato vegetable dust, pesticide dust, and granular chromatographic-grade cellulose into a 2-kW argon ICP. The tangential gas flow created inside the injector device mixed and transported the powder into the ICP. Because of the transient nature of the signal, a rapid scanning spectrometer was used for multiple element analysis. The emission spectrum of titanium, from NBS coal fly ash injected into the ICP,[74] is shown in Figure 12.29. For comparison, the emission spectrum resulting from electrothermal vaporization of titanium into the argon ICP is also shown in the figure. Detection limits for chromium, copper, strontium, thallium, vanadium, and zinc are 1, 2, 5, 7, 3, and 2 ng, respectively, for the NBS coal fly ash. Reproducibilities of 12% RSD or less and recoveries of 72 to 110% were reported for the elements present in the coal fly ash. Such a system is useful for rapid qualitative and quantitative analysis.

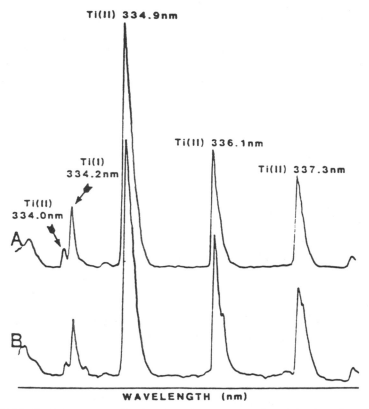

Figure 12.29. Emission spectra for titanium: spectrum *A*, direct injection of NBS coal fly ash (1633a) into the ICP; spectrum *B*, electrothermal vaporization of titanium into the ICP. (From Reference 74, with permission.)

Still another method for the analysis of powdered, pulverized, or ground samples utilizes the nebulization of samples in the form of suspensions or slurries.[75–79] Fry and co-workers[75,79] described techniques for converting animal tissues and coal samples into fine particles, which were suspended as 13% and 10% solid slurries, respectively. The slurries were nebulized into a dc plasma using a Babington-type nebulizer. The authors pointed out that the methods used were largely dependent on the ability to produce solid particles that had a median particle diameter of less than 10 μm to avoid plugging of the torch tip. They were able to achieve nearly 100% recoveries for chromium, copper, magnesium, manganese, nickel, and lead in coal, using aqueous standard solutions.[79] Precision of analysis was 6% RSD, similar to that described for the aqueous standards.

Ebdon and co-workers[77,78] analyzed coal and coal-ash slurries by ICP–AES. In their work, also utilizing a Babington-type nebulizer, they were able to determine the effects of several instrument parameters, including forward power, injector gas flow, and observation height. Optimum results were achieved using lower power, lower injector gas flow, and a higher observation height, compared to commonly used optima for aqueous solutions.

Each of the slurry methods described for either ICP or DCP showed critical dependence on particle size, slurry concentration, and calibration methods. However, each one also demonstrated the feasibility of slurry nebulization-plasma emission methods to minimize, or eliminate, a major portion of sample preparation traditionally required for the analysis of solids.

12.7   SUMMARY AND CONCLUSIONS

Introduction of solids directly into ICPs is feasible using the techniques of direct insertion, electrothermal vaporization, arc and spark ablation, and laser ablation. Additional techniques appear feasible for powders.

Direct insertion and electrothermal vaporization share many common attributes. A few milligrams of sample is preweighed into a graphite or metal support and heated electrically or by the plasma, thereby consuming the total amount of sample. The solid sample may be in any form. Emission behavior is time dependent and is similar to that observed in dc arc spectrometry. To date, most work has been performed with solutions. Consequently, the bulk of information for solids analysis is considered to be preliminary. For both direct insertion and ETV, the mass of sample consumed is a few milligrams, with DLs in the microgram- to submicrogram-per-gram range. Precision is typically 7 to 10% RSD with linear dynamic ranges of approximately three orders of magnitude.

The ablation techniques (arc, spark, and laser) differ from direct insertion and ETV in that material is removed from the sample surface without requiring preweighing. While arcs and sparks require the sample to be conductive, the laser does not. Laser ablation offers an additional advantage over the other techniques in that very small surface areas may be ablated, providing information about sample heterogeneity. For the ablation techniques, the mass sampled is approximately 1 μg, with less than 1 μg necessary for laser ablation. Detection limits are in the microgram-per-gram range, with precisions typically 1 to 2% RSD if the surface area sampled is homogeneous. Linear dynamic ranges are three orders of magnitude, similar to the direct insertion and ETV methods.

One common attribute of these techniques is the relative independence of calibration on matrix enabled by the ICP. Consequently, calibration curves may be constructed using a variety of solid materials of differing elemental composition, with good linearity and fit. In the case of spark ablation, solution standards may be substituted, greatly simplifying the calibration procedure.

The analytical potential of solid sampling techniques for ICP–AES analysis, combined with their successful commercialization, promises increased activity and new analytical capabilities in the coming years. These techniques are also expected to be applicable to ICP–atomic fluorescence spectrometry and ICP–mass spectrometry.

BIBLIOGRAPHY

Dittrich, K., and R. Wennrich. Laser Vaporization in Atomic Spectroscopy, *Prog. Anal. Atom. Spectrosc.* 7, 139–198 (1984).

Langmyhr, F. J. Direct Analysis of Solids by Atomic Absorption Spectrometry, *Analyst* 104, 993–1016 (1979).

Thompson, M., and J. N. Walsh. "Handbook of Inductively Coupled Plasma Spectrometry." Blackie and Son, Ltd., Glasgow, 1983.

Young, R. M., and E. Pfender. Generation and Behavior of Fine Particles in Thermal Plasmas—A Review, *Plasma Chem. Plasma Process.* 5, 1–37 (1985).

REFERENCES

1. Eric D. Salin and Gary Horlick, Direct Sample Insertion Device for Inductively Coupled Plasma Emission Spectrometry, *Anal. Chem.* 51, 2284–2286 (1979).
2. D. Sommer and K. Ohls, Direct Sample Insertion into a Stable Burning Inductively Coupled Plasma, *Fresenius Z. Anal. Chem.* 304, 97–103 (1980).
3. K. Ohls and D. Sommer, Analysis of Liquid, Solid, and Gaseous Samples Using a High-Power Nitrogen–Argon ICP System, *Proc. Int. Winter Conf.* 321–326 (1981).
4. G. F. Kirkbright and S. J. Walton, Optical Emission Spectrometry with an Inductively Coupled Radiofrequency Argon Plasma Source and Direct Sample Introduction from a Graphite Rod, *Analyst* 107, 276–281 (1982).
5. G. F. Kirkbright and Zhang Li-Xing, Volatilization of Some Elements from a Graphite Rod Direct Sample Insertion Device into an Inductively Coupled Argon Plasma for Optical Emission Spectrometry, *Analyst* 107, 617–621 (1982).
6. Zhang Li-Xing, G. F. Kirkbright, M. J. Cope, and J. M. Watson, A Microprocessor-Controlled Graphite Rod Direct Sample Insertion Device for Inductively Coupled Plasma Optical Emission Spectroscopy, *Appl. Spectrosc.* 37, 250–254 (1983).
7. Edward R. Prack and Glenn J. Bastiaans, Metal Speciation by Evolved Gas-Inductively Coupled Plasma Atomic Emission Spectrometry, *Anal. Chem.* 55, 1654–1660 (1983).
8. R. L. Sing, M. M. Habib, A. Burgess, A. M. Jones, and E. D. Salin, Applications of the Direct Sample Insertion Device for Atomic Emission Analysis with the Inductively Coupled Plasma, *6th Symp. Proc.—Int. Symp. Plasma Chem.* 2, 425–433 (1983).
9. N. W. Barnett, M. J. Cope, G. F. Kirkbright, and A.A.H. Taobi, Design Consideration and Temperature Determination of an Automated Graphite Furnace Cup System Used for Direct Sample Introduction for ICP Optical Emission Spectrometry, *Spectrochim. Acta* 39B, 343–348 (1984), and references therein.
10. A. G. Page, S. V. Godbole, K. H. Madraswala, M. J. Kulkarni, V. S. Mallapurkar, and B. D. Joshi, Selective Volatilization of Trace Metals from Refractory Solids into an Inductively Coupled Plasma, *Spectrochim. Acta* 39B, 551–557 (1984).
11. Knut Ohls, Determination of Small Cadmium Amounts in Various Materials by Using Solid Sample in ICP and Flameless Atomic Absorption Spectrometry, *Spectrochim. Acta* 39B, 1105–1111 (1984).
12. Mohammed Abdullah, Keiichiro Fuwa, and Hiroki Haraguchi, Simultaneous Multielement Analysis of Microliter Volumes of Solution Samples by Inductively Coupled Plasma Atomic Emission Spectrometry Utilizing a Graphite Cup Direct Insertion Technique, *Spectrochim. Acta* 39B, 1129–1139 (1984).
13. Eric D. Salin and R.L.A. Sing, Microsample Liquid Analysis with a Wire Loop Direct Sample Insertion Technique in an Inductively Coupled Plasma, *Anal. Chem.* 56, 2596–2598 (1984).
14. Hiroki Haraguchi, Mohammed Abdullah, Tetsuya Hasegawa, Masao Kurosawa, and Keiichiro Fuwa, Inductively Coupled Plasma Atomic Emission Spectrometry Utilizing a Direct Graphite Cup Insertion Technique, *Chem. Soc. Japan.* 57, 1839–1843 (1984).
15. Magdi M. Habib and Eric D. Salin, Controlled Potential Electrolysis Coupled with a Direct Sample Insertion Device for Multielement Determination of Heavy Metals by Inductively Coupled Plasma Atomic Emission Spectrometry, *Anal. Chem.* 57, 2055–2059 (1985).

484 MICHAEL W. ROUTH AND MARIA W. TIKKANEN

16. Mohammed Abdullah and Hiroki Haraguchi, Computer-Controlled Graphite Cup Direct Insertion Device for Direct Analysis of Plant Samples by Inductively Coupled Plasma Atomic Emission Spectrometry, *Anal. Chem.* 57, 2059-2064 (1985).
17. Kh. I. Zil'bershtein, Application of ICP Discharge to AES Analysis of Solid Samples, A Survey. *ICP Inf. Newsl.* 10, 964-976 (1985).
18. Kontron brochure, SET-Sample Elevator Technique, The Analysis of Solid Samples in ICP, Eching b. München, West Germany, 1985.
19. C. W. McLeod, P. A. Clarke, and D. J. Mowthorpe, Simultaneous Determination of Volatile Trace Elements in Nickel-Base Alloys Using a Direct Insertion Probe and Inductively Coupled Plasma Emission Spectrometry, *Spectrochim. Acta* 41B, 63-71 (1986).
20. W. E. Pettit and G. Horlick, An Automated Direct Sample Insertion System for the Inductively Coupled Plasma, *Spectrochim. Acta* 41B, 699-712 (1986).
21. Youbin Shao and Gary Horlick, Performance of a Direct Sample Insertion System for the Inductively Coupled Plasma, *Appl. Spectrosc.* 40, 386-393 (1986).
22. D. E. Nixon, V. A. Fassel, and R. N. Kniseley, Tantalum Filament Vaporization of Microliter Samples, *Anal. Chem.* 46, 210-213 (1974).
23. A. M. Gunn, D. L. Millard, and G. F. Kirkbright, Optical Emission Spectrometry with an Inductively Coupled Radiofrequency Argon Plasma Source and Sample Introduction with a Graphite Rod Electrothermal Vaporization Device—I. Instrumental Assembly and Performance Characteristics, *Analyst (London)* 103, 1066-1073 (1978).
24. D. L. Millard, H. C. Shan, and G. F. Kirkbright, Optical Emission Spectrometry with an Inductively Coupled Radiofrequency Argon Plasma Source and Sample Transport Effects, *Analyst (London)* 105, 502-508 (1980).
25. K. C. Ng and J. A. Caruso, Microliter Sample Introduction into an Inductively Coupled Plasma by Electrothermal Carbon Cup Vaporization, *Anal. Chim. Acta* 143, 209-222 (1982).
26. G. Crabbi, P. Cavalli, M. Achilli, G. Rossi, and N. Omenetto, Use of the HGA-500 Graphite Furnace as a Sampling Unit for ICP Emission Spectrometry, *At. Spectrosc.* 3, 81-88 (1982).
27. A. Aziz, J.A.C. Broekaert, and F. Leis, Analysis of Microamounts of Biological Samples by Evaporation and Inductively Coupled Plasma Atomic Emission Spectroscopy, *Spectrochim. Acta* 37B, 369-379 (1982).
28. E. Kitazume, Thermal Vaporization for One-Drop Sample Introduction into the Inductively Coupled Plasma, *Anal. Chem.* 55, 802-805 (1983).
29. K. C. Ng, and J. A. Caruso, Determination of Trace Metals in Synthetic Ocean Water by Inductively Coupled Plasma Atomic Emission Spectrometry with Electrothermal Carbon Cup Vaporization, *Anal. Chem.* 55, 1513-1516 (1983).
30. S. D. Hartenstein, H. M. Swaidan, and G. D. Christian, Internal Standards for Simultaneous Multielement Analysis in Inductively Coupled Plasma Atomic-Emission Spectroscopy with an Electrothermal Atomizer for Sample Introduction, *Analyst* 108, 1323-1330 (1983).
31. H. M. Swaidan and G. D. Christian, Optimization of Electrothermal Atomization-Inductively Coupled Plasma Emission Spectrometry for Simultaneous Multielement Determination, *Anal. Chem.* 56, 120-122 (1984).
32. W. M. Blakemore, P. H. Casey, and W. R. Collie, Simultaneous Determination of 10 Elements in Wastewater, Plasma, and Bovine Liver by Inductively Coupled Plasma Emission Spectrometry with Electrothermal Atomization, *Anal. Chem.* 56, 1376-1379 (1984).
33. M. W. Tikkanen, and T. M. Niemczyk, Modification of a Commercial Direct-Reading Inductively Coupled Plasma Spectrometer for Sample Introduction Electrothermal Vaporization, *Anal. Chem.* 56, 1997-2000 (1984).
34. D. R. Hull, and G. Horlick, Electrothermal Vaporization Sample Introduction System for the Inductively Coupled Plasma, *Spectrochim. Acta* 39B, 843-850 (1984).
35. H. Matusiewicz and R. M. Barnes, Determination of Aluminum and Silicon in Biological Materials by Inductively Coupled Plasma Atomic Emission Spectrometry with Electrothermal Vaporization, *Spectrochim. Acta* 39B, 891-899 (1984).
36. H. Matusiewicz and R. M. Barnes, Discrete Nebulization Using Aerosol Deposition in Electrothermal Vaporization for Inductively Coupled Plasma-Atomic Emission Spectrometry, *Spectrochim. Acta* 40B, 41-47 (1985).
37. H. Matusiewicz and R. M. Barnes, Evaluation of a Controlled-Temperature Furnace as a Sampling Device for Inductively Coupled Plasma-Atomic Emission Spectrometry, *Spectrochim. Acta* 40B, 29-30 (1985).
38. S. E. Long and R. D. Snook, Some Observations on Electrothermal Vaporization for Sample Introduction into the Inductively Coupled Plasma, *Spectrochim. Acta* 40B, 553-568 (1985).

39. M. W. Tikkanen and T. M. Niemczyk, Use of Signal-versus-Time Profiling to Monitor Sample Volatilization in Simultaneous Multielement Determinations by Electrothermal Vaporization-Inductively Coupled Plasma Atomic Emission Spectrometry, *Anal. Chem.* 57, 2896-2900 (1985).

40. M. W. Tikkanen and T. M. Niemczyk, Time Gating for the Elimination of Interferences in Electrothermal Vaporization-Inductively Coupled Plasma Atomic Emission Spectrometry, *Anal. Chem.* 58, 366-370 (1986).

41. R. L. Dahlquist, J. W. Knoll, and R. E. Hoyt, Application of the Inductively Coupled Plasma Using Thermal and Direct Aerosol Generation. Pittsburgh Conference on Analytical Chemistry and Applied Spectroscopy, paper reprint from Applied Research Laboratories, (1975).

42. R. L. Dahlquist and J. W. Knoll, Solids and Liquids Analysis Using Radio Frequency Inductively Coupled Argon Plasma Optical Emission Spectrometry, *Proc. 18th Colloq. Spectrosc. Int.* 3, 679-684 (1975).

43. H.G.C. Human, R. H. Scott, A. R. Oakes, and C. D. West, The Use of a Spark as a Sampling-Nebulizing Device for Solid Samples in Atomic-Absorption, Atomic-Fluorescence, and Inductively Coupled Plasma Emission Spectrometry, *Analyst* 101, 265-271 (1976).

44. R. H. Scott, Spark Elutriation of Powders into an Inductively Coupled Plasma, *Spectrochim. Acta* 33B, 123-125 (1978).

45. K. Ohls and D. Sommer, Direct Analysis of Compact Samples Using the High Power Emission Spectrometry, *Fresenius Z. Anal. Chem.* 296, 241-246 (1979).

46. J. Y. Marks, D. E. Fornwalt, and R. E. Yungk, Application of a Solid Sampling Device to the Analysis of High Temperature Alloys by ICP-AES, *Spectrochim. Acta* 38B, 107-113 (1983).

47. J. P. Keilsohn, R. D. Deutsch, and G. M. Hieftje, The Use of a Microarc Atomizer for Sample Introduction into an Inductively Coupled Plasma, *Appl. Spectrosc.* 37, 101-105 (1983).

48. Kichinosuke Hirokawa and Kunio Takada, Low-Voltage Spark Discharge as an Aerosol Generator of Metal Samples for ICP Emission Spectrometry, *Nippon Kinzoku Gakkaishi* 47, 507-509 (1983).

49. P. B. Farnsworth and G. M. Hieftje, Sample Introduction into the Inductively Coupled Plasma by a Radiofrequency Arc, *Anal. Chem.* 55, 1414-1417 (1983).

50. Robert Gordon, Richard Belmore, John Beaty, and Irina Aginsky, "Solid Sampling Literature Manual." Allied Analytical Systems (now Thermo Jarrell-Ash) Waltham, Mass. (1983).

51. D. J. Comaford, J. E. Goulter, and D. B. Tasker, Design, Development, and Analytical Evaluation of a Direct Solids Nebulization System for ICP. Paper No. 902, presented at the Pittsburgh Conference on Analytical Chemistry and Applied Spectroscopy, 1984.

52. Weixin Zhou, An ICP Solid Powder Automatic Spectrographic Device, *Fenxi Huaxue* 12, 212-215 (1984).

53. Katsuyuki Takahashi, Takayuki Yoshioka, Yoshisuke Nakamura, and Haruno Okochi, Direct ICP (Inductively Coupled Plasma) Emission Spectrochemical Analysis of Low-Alloy Steels Using Aerosol Cyclones with Low-Voltage Spark Discharge, *Nippon Kinzoku Gakkaishi* 48, 418-423 (1984).

54. John S. Beaty and Richard J. Belmore, Principles on Analysis by Spark Sampling and Plasma Excitation as Applied to Super Alloys, *Test. Eval.* 12, 212-219 (1984).

55. M. W. Routh, P. D. Dalager, and J. E. Goulter, Current Status of Spark Discharge Conductive Solids Nebulization for ICP-AES. Paper No. 130, presented at the Federation of Analytical Chemistry and Spectroscopy Societies (FACSS) Meeting, 1984.

56. A. Aziz, J.A.C. Broekaert, K. Laqua, and F. Leis, A Study of Direct Analysis of Solid Samples Using Spark Ablation Combined with Excitation in an Inductively Coupled Plasma, *Spectrochim. Acta* 39B, 1091-1103 (1984).

57. P. D. Dalager, D. J. Comaford, and J. E. Goulter, Conductive Solids Nebulizer (CSN-ICP) Analysis with an Inductively Coupled Plasma Calibration. Paper No. 481, presented at the Pittsburgh Conference on Analytical Chemistry and Applied Spectroscopy, 1985.

58. Paul Dalager, John Goulter, and Dennis Tasker, ICP Analyzes Your Solid Steel Samples, *Res. Dev.* 27(4), 114-117 (1985).

59. J. L. Jones, R. L. Dahlquist, and R. E. Hoyt, A Spectroscopic Source with Improved Analytical Properties and Remote Sampling Capability, *Appl. Spectrosc.* 25, 628-631 (1971).

60. D. W. Golightly and Akbar Montaser, Spark Aerosol Generation for Introduction of Solid Nonconductors into the Inductively Coupled Plasma. Paper No. 36, presented at the 1986 Winter Conference on Plasma Spectrochemistry, Kailua-Kona, Hawaii, (1986).

61. T. Kantor, L. Polos, P. Fodor, and E. Pungor, Atomic-Absorption Spectrometry of Laser-Nebulized Samples, *Talanta* 23, 585-586 (1976).

62. F. N. Abercrombie, M. D. Silvester, A. D. Murray, and A. R. Barringer, A New Multielement Technique for the Collection and Analysis Of Airborne Particulates in Air Quality Surveys, in "Applications of Inductively Coupled Plasmas to Emission Spectroscopy," R. M. Barnes, Ed. Franklin Institute Press, Philadelphia, 1978, pp. 121-145.

63. K. Laqua, Analytical Spectroscopy Using Laser Atomizers, in "Analytical Laser Spectroscopy," N. Omenetto, Ed. Chemical Analysis Series, Vol. 50. Wiley-Interscience, New York, pp. 47-118.

64. M. Thompson, J. E. Goulter, and F. Sieper, Laser Ablation for the Introduction of Solid Samples into an Inductively Coupled Plasma for Atomic Emission Spectrometry, *Analyst* 106, 32-39 (1981).

65. J. W. Carr and G. Horlick, Laser Vaporization of Solid Metal Samples into an Inductively Coupled Plasma, *Spectrochim. Acta* 37B, 1-15 (1982).

66. J. Xu Kawaguchi, T. Tanaka, and A. Mizuike, Inductively Coupled Plasma-Emission Spectrometry Using Direct Vaporization of Metal Samples with a Low Energy Laser, *Bunseki Kagaku* 31, E185-E191 (1982).

67. T. Ishizuka and Y. Uwamino, Inductively Coupled Plasma Emission Spectrometry of Solid Samples by Laser Ablation, *Spectrochim. Acta* 38B, 519-527 (1983).

68. M. Thompson and M. Hale, Rapid Analysis and Identification of Heavy Minerals by Laser Ablation and Inductively Coupled Plasma Emission Spectrometry, Prospecting in Areas of Glaciated Terrain, *Inst. Min. Metall. Spec. Publ.* 225-232 (1984).

69. M. Hale, M. Thompson, and M. R. Wheatley, Laser Ablation of Stream-Sediment Pebble Coatings for Simultaneous Multi-element Analysis in Geochemical Exploration, *J. Geochem. Expl.* 21, 361-371 (1984).

70. A. L. Gray, Solid Sample Introduction by Laser Ablation for Inductively Coupled Plasma Source Mass Spectrometry, *Analyst* 110, 551-556 (1985).

71. S. Greenfield and T. P. Sutton, An Investigation into the Interfacing of a Danielsson Tape Machine with an ICP Polychromator for the Analysis of Powderable Samples, *ICP Inf. Newsl.* 6, 267-271 (1980).

72. R. M. Dagnall, D. J. Smith, T. S. West, and S. Greenfield, Emission Spectroscopy of Trace Impurities in Powdered Samples with a High-Frequency Ar Plasma Torch, *Anal. Chim. Acta* 54, 397-406 (1971).

73. G. A. Meyer and R. M. Barnes, Analytical Inductively Coupled Nitrogen and Air Plasmas, *Spectrochim. Acta* 40B, 893-905 (1985).

74. K. C. Ng, M. Zerezghi, and J. Caruso, Direct Powder Injection of NBS Coal Fly Ash in Inductively Coupled Plasma Atomic Emission Spectrometry with Rapid Scanning Spectrometric Detection, *Anal. Chem.* 56, 417-421 (1984).

75. N. Mohamed, R. M. Brown, Jr., and R. C. Fry, Slurry Atomization dc Plasma Emission Spectrometry and Laser Diffraction Studies of Aerosol Production and Transport, *Appl. Spectrosc.* 35, 153-164 (1981).

76. C. W. Fuller, R. C. Hutton, and B. Preston, Comparison of Flame, Electrothermal, and Inductively Coupled Plasma Atomic Emission Techniques for the Direct Analysis of Slurries, *Analyst* 106, 913-920 (1981).

77. J. R. Wilkinson, L. Ebdon, and K. W. Jackson, Determination of Volatile Trace Metals in Coal by Analytical Atomic Spectroscopy, *Anal. Proc. (London)* 19, 305-307 (1982).

78. L. Ebdon and M. R. Cave, A Study of Pneumatic Nebulization Systems for Inductively Coupled Plasma Emission Spectrometry, *Analyst* 107, 172-178 (1982).

79. D. L. McCurdy, M. D. Wichman, and R. C. Fry, Rapid Coal Analysis—II. Slurry Atomization DCP Emission Analysis of NBS Coal, *Appl. Spectrosc.* 39, 984-988 (1985).

13

Injection of Gaseous Samples into Plasmas

JOSEPH A. CARUSO

Department of Chemistry
University of Cincinnati
Cincinnati, Ohio

KAREN WOLNIK
AND
FRED L. FRICKE

Elemental Analysis Research Center
U.S. Food and Drug Administration
Cincinnati, Ohio

13.1 INTRODUCTION

Recently, Browner and Boorn[1] suggested that sample introduction was the "Achilles' heel of atomic spectroscopy." In this popular review article, the authors aptly discussed various sample introduction techniques. Clearly, the most efficient means of transporting analyte to the plasma and of producing atomic species in the plasma lead to the best detectability.

Gaseous sample introduction into the inductively coupled plasma (ICP) offers several significant advantages over liquid aerosol introduction. Unlike pneumatic nebulization, in which more than 95% of the sample solution is discarded, the transport efficiency of a gas approaches 100%. Therefore, more analyte reaches the plasma, resulting in improved signal-to-background ratios and detection limits. Indeed, Summerhays and co-workers[2] increased the percentage of osmium reaching the plasma from conventional liquid nebulization simply by forming, in solution, a volatile chemical form of osmium, OsO_4. Also, gaseous analytes can be separated from the sample solution matrix to virtually eliminate matrix interferences in the plasma. In addition to improved transport efficiency and reduction of matrix effects, gaseous sample introduction promotes more efficient plasma atomization and excitation. The desolvation and vaporization processes that occur in the ICP when solution aerosols are introduced are unnecessary for gaseous samples. Consequently, the plasma energy normally required for desolvation and vaporization is available for atomization and excitation, thus leading to enhanced emission signals and improved sensitivity and detection limits. The primary shortcoming of gaseous sample introduction is that frequently the analytes of interest are not readily converted to the gaseous state.

This chapter discusses various methods, other than electrothermal vaporization, for forming and introducing gases into the ICP. The discussion primarily focuses on the formation of volatile hydrides and gas-chromatographic methods. Typically, hydrides of antimony, arsenic, bismuth, germanium, lead, selenium, tellurium, and tin are formed by adding sodium borohydride to acidified solutions of these elements. The volatile hydrides are formed and transported to the ICP either continuously, over a selected period of time, or batchwise. Continuous generation produces a steady-state signal, while in a batch procedure, transient emission signals are obtained when a fixed amount of sample solution is treated to form the hydrides.

Certain inorganic compounds, such as chelates, carbonyls, and bromides, are volatile at ambient or slightly elevated temperatures. The formation of these compounds can be used both to convert the analyte to a gaseous species and to separate the analyte from the sample matrix prior to introduction into the ICP. A brief discussion of this approach is included in Section 13.2.

Molecular gases or gaseous mixtures can be directly introduced into the ICP, by continuous flow or by "slug" injection, either for analysis or for the characterization of the potential of ICP spectrometry in the analysis of organic compounds. These modes of sample introduction are used primarily to study the degree of molecular dissociation, to investigate the effects of the plasma operating parameters on dissociation, to compile tables of atomic emission lines observed in the ICP for nonmetals, and to determine possible background interferences in the ICP used as a detector for gas chromatography. Mixed gas ICPs also may be used to study fundamental plasma properties such as measurements of cross sections and transport coefficients.[3]

Gas chromatography (GC) permits volatilization and separation of volatile compounds. Coupled with an ICP, GC provides a selective and sensitive method for obtaining very specific information on the chemical nature of various analytes in a sample (ie, speciation). The GC–ICP, combined with a multiple-element spectrometer, produces element-selective gas chromatograms, one for each element monitored, thus enabling the calculation of elemental ratios used to determine empirical formulas. The argon ICP is particularly useful for obtaining atomic and ionic emission from metals and metalloids. However, relatively little emission is obtained from nonmetals, such as the halogens, oxygen, and phosphorus. This makes the argon ICP insensitive to many nonmetallic compounds currently being investigated by gas chromatography. An alternative to the argon plasma for the excitation of these nonmetals is a helium plasma. Because the energy states in helium are higher than those of argon, both microwave-induced plasmas (MIPs) and inductively coupled helium plasmas are capable of producing atomic and ionic emission from nonmetals. A later discussion in this chapter focuses on the He ICP as a detector for gaseous sample introduction.

13.2 HYDRIDE GENERATION METHODS

13.2.1 Principles

Hydride generation has been used for a century in the Marsh reaction and in the Gutzeit test for arsenic. The generation of gaseous hydrides for atomic spectrometry in the determination of the hydride forming elements has been reviewed by Robbins and Caruso.[4] In addition, a very comprehensive review has been presented by Nakahara,[5] who discusses both fundamental and practical considerations of hydride generation and reviews applications of atomic absorption, atomic fluorescence, and plasma atomic emission methods. Initially, the primary technique for the quantitative determination of the hydride-forming elements was atomic absorption spectrometry (AAS). However, published applications of electrical plasmas for hydride determinations by atomic emission spectrometry (AES) soon appeared for direct-current plasmas (DCPs),[6,7] MIPs,[8–11] and the ICP.[12]

The elements that form volatile hydrides are arsenic, bismuth, germanium, lead, antimony, selenium, tin, and tellurium. In contrast to solution nebulization, the volatile hydrides of these elements are more efficiently transported to the high-temperature source, which, in turn, can more efficiently produce and excite the free atoms and ions needed for atomic emission spectrometry. In addition, the resulting separation of analytes from the elements of the sample matrix and from the sample solvent both reduces potential spectral interferences and enables preconcentration of analytes. Of various methods of hydride generation,[4] the acid–borohydride reaction most frequently used is:

$$NaBH_4 + 3H_2O + HCl \rightarrow H_3BO_3 + NaCl + 8H \xrightarrow{E^{m+}} EH_n + H_{2(excess)}$$

where E is the hydride-forming element of interest and m may or may not equal n. A number of experimental variations, to be discussed later, have been utilized to effect this reaction. In summary, the principle is relatively simple, namely to generate the volatile hydride quantitatively and to transport it reproducibly to the plasma for determination, causing minimal disturbance to the plasma by these processes. In practice, this procedure is challenging, but it has progressed to the point that commercial systems adaptable to various plasma spectrometers are available.

Other methods of producing volatile species for atomic spectrometric determination, in addition to hydride generation, have been described.[13,14] Black and Browner[14] developed a novel batch-type procedure for ICP spectrometry in which volatile β-diketonates of cobalt, chromium, iron, manganese, and zinc were formed directly from untreated sample in a reaction flask. The sample, in this case, NBS bovine liver or human blood serum, was placed in a multiple-port vessel, the chelating ligand was added, and the closed system was heated at 120°C for 5 to 10 min. The resulting vapor was flushed into the ICP with argon. This valuable technique may be used for selective volatilization of elements from the sample matrix prior to introduction into the ICP.

13.2.2 Apparatus, Instrumentation, and Procedures

Two general approaches, continuous and batch, have been used for volatile hydride generation. In continuous generation, the more frequently used method,[12,15-27] the reagents are continuously pumped, usually by a multiple-channel peristaltic pump, into some type of mixing chamber, where the acid–borohydride reaction takes place. The primary gaseous product, H_2, plus the volatile hydrides, then are swept into a gas–liquid separator, where the liquid passes to drain and the gas mixture is directed into the ICP for determination. In the batch method, an aliquot of acid solution containing the analyte elements is mixed rapidly with an aliquot of borohydride reagent.[28-31] The resulting volatile hydrides are swept into the ICP directly or as a plug following condensation in a liquid nitrogen trap, which allows venting of the copious amounts of H_2 formed, rather than introducing the H_2 into the plasma.

13.2.2.1 Continuous Generation

Many of the methods for continuous hydride generation follow the suggestions by Thompson and co-workers[12] on the multiple-element determinations of arsenic, bismuth, antimony, selenium, and tellurium. A diagram of the hydride generation cell and the experimental facility is shown in Figure 13.1. The main requirements[12] for continuous generation are: efficient mixing of the acid sample solution and the borohydride, a short period to complete the reaction, separation of the product gases from the liquids, uniform mixing of product gases with the argon injector gas, and a small positive pressure to sweep the product gases into the ICP. Depending on the type of torch used, the required operating power ranged from 2.7 to 5.0 kW, a range much higher than that commonly used for solution-nebulization Ar ICPs. While the Greenfield torch was more tolerant of the product gases, the use of the Fassel torch resulted in superior detection limits.

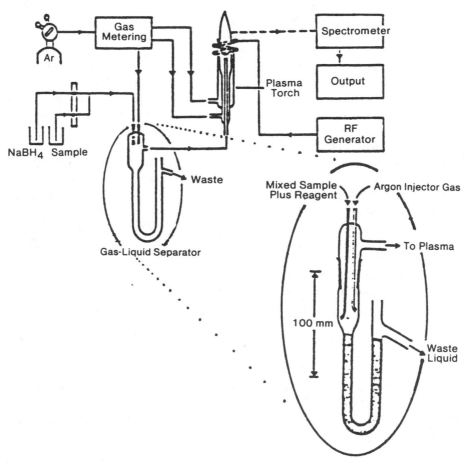

Figure 13.1. Schematic representation of the hydride reduction–ICP spectrometry system with an expanded view of the hydride generation cell. (From Reference 12, with permission.)

The earlier hydride generator design was improved by including a three-channel peristaltic pump connected through a four-way valve to the gas–liquid separator.[18] The new design enabled the sample solution to be rapidly switched with the blank, in effect providing a continuous reaction while completely excluding the injection of an air plug into the plasma. With this modification, the system stability was markedly improved and the ICP could be operated at a much lower power of 1.25 kW. A similar system, more complex mechanically, was used by Ikeda and co-workers.[20] In this system, the first mixing chamber is a coil, followed by a gas–liquid separator, with a condensing coil following it. A second gas-liquid separator follows this coil. This particular system is further discussed in Section 13.2.2.3. It provides an easy interchange with a conventional nebulizer and is highly efficient at producing the volatile hydrides of germanium, tin, and lead.

Nygaard and Lowry[21] modified a continuous hydride generation system, similar to that of Thompson,[12] to investigate five sample digestion procedures for simultaneous determinations of arsenic, antimony, and selenium in wastewater effluents. Another modification[22] of the original hydride generator[12] involved the mixing of the reagent and sample immediately before the gas–liquid separator. This arrangement gives smooth and continuous H_2 generation, which enhances the signal stability. Wolnik and co-workers[23] suggested a unique continuous hydride generation system (Figure 13.2) that used two nebulizers in tandem. An aerosol of $NaBH_4$, generated in a concentric nebulizer, was passed through the gas "needle" of a cross-flow nebulizer, through which the acidified sample solution was nebulized. At the junction of the sample and gas needles, aerosols of sample and $NaBH_4$ were continuously mixed, resulting in formation of the volatile hydrides. The hydrides were injected into a plasma operated at a forward power of 0.85 kW. The main advantages of this system are that non-hydride-forming elements may be determined at the same time, and the ICP is undisturbed by the large amounts of H_2 produced by the reaction of excess borohydride. Hutton and Preston[25] also described a modified nebulizer that allowed switching between hydride determination and solution analysis, without interruption of a 3-kW ICP in a Greenfield torch. By a simple modification, the hydrides are routed into and through the nebulizer.

Continuous hydride generation also has been used in high-pressure liquid chromatography (HPLC)[26] and for flow-injection analysis.[27] In the HPLC application, the As anions from the HPLC column were converted to the hydride by means of a simple device that used two "T" fittings. The first fitting was fed by the HPLC effluent and by acid solution from one channel of a peristaltic pump. The outlet from the first fitting was connected to another "T"

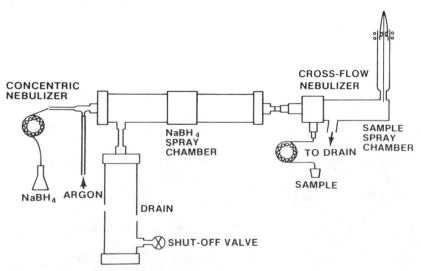

Figure 13.2. Schematic representation of the tandem nebulization system (TNS), (From Reference 23, with permission.)

fitting that was fed with borohydride solution from the second channel of the peristaltic pump. The arsine, the remaining column effluent, and H_2 were directed into a cross-flow nebulizer, thereby introducing all species eluting from the column into the ICP. Because acid and borohydride solutions are delivered continuously, this method is included in this section on continuous generation; however, the use of analyte plugs results in transient signals from the hydride generation. Relatedly, the use of a flow-injection device[27] with hydride generation also results in transient signals, even though reagents flow continuously.

13.2.2.2   Batch Generation

Pickford[28] described a batch method for arsine generation in ICP spectrometry. An acidified sample solution was drawn into a 20-mL disposable syringe, followed by the alkaline sodium borohydride solution. The syringe needle then was inserted into the open end of the uptake tube of the ICP nebulizer to inject arsine and the accompanying H_2 into the plasma. This approach apparently worked well for a conventional 1.5 kW ICP system.

Hahn and co-workers[29] described a batch method to determine simultaneously arsenic, bismuth, germanium, antimony, selenium, and tin in foods, using a hydride generation–condensation (HGC) technique. In the HGC system described (Figure 13.3), rapid mixing of the acid and borohydride solutions occurs, followed by low-temperature condensation of the volatile hydrides produced by the reaction. For the low-temperature condensation, liquid argon may be used as the trapping medium. The large amounts of hydrogen that are produced in a short time interval are then vented (valve III). Residual H_2 from

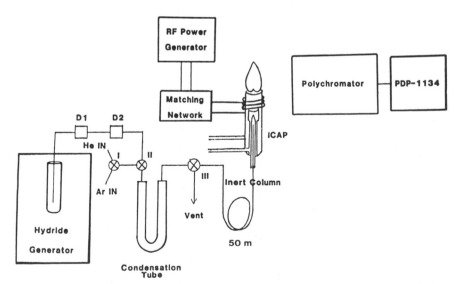

Figure 13.3. Hydride generation–condensation system interfaced to an ICP polychromator: D1 and D2, desiccant tubes; I, II, and III, three-way valves. (From Reference 29, with permission.)

the system is eliminated by an auxiliary flow of He. The hydrides are revolatilized by placing the condensation trap in a hot water bath; then Ar sweeps the volatile hydrides into the plasma. These operations are necessary because the use of He as the injector gas results in plasma instability. In addition, a coil of tubing 50 m long acts as a pulse dampener as the injector gas and hydrides expand upon revolatilization.

A computer-controlled sequential monochromator system for multiple-element determinations of volatile hydrides is described by Eckhoff and co-workers.[30] Because this type of spectrometer requires time to scan from wavelength to wavelength, the hydrides are sequenced to the ICP by a Chromosorb 102 gas-chromatographic column after the hydride condensation reaction. The necessary time sequences for both wavelength selection and hydride separation are controlled by the computer. Batch generation also has been combined with the continuous generation method for introducing hydrides into the ICP.[31]

13.2.2.3 Commercially Available Systems

The commercially available hydride generator systems are similar in principle to those reported above. The discussion below does not suggest any particular endorsement. The system available from Nippon Jarrell–Ash and Thermo Jarrell–Ash (Figure 13.4a) was described by Ikeda[20] and utilized by Nakahara.[32] The system by PT Analytical (Figure 13.4b) is quite similar to that suggested by Pahlavanpour and co-workers.[18] With this design, a maximum signal is attained within 20 s. This device is adaptable to different spectrometers

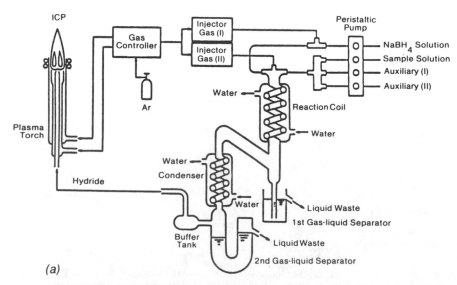

Figure 13.4. (a) Schematic flow diagram of the system marketed by Nippon Jarrell–Ash and Thermo Jarrell–Ash. (From Reference 32, with permission.)

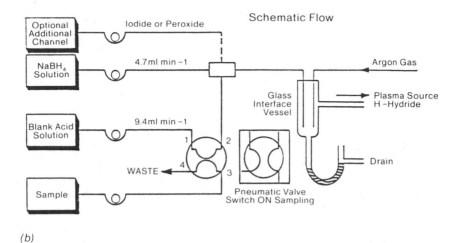

(b)

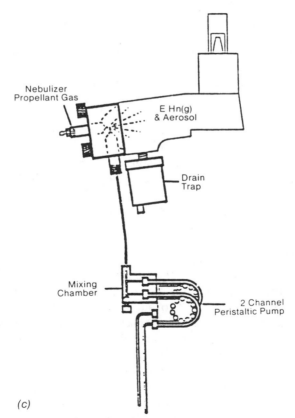

(c)

Figure 13.4. (b) Schematic flow diagram of the PT Analytical continuous hydride generation system. (With permission.) (c) Schematic flow diagram of the system marketed by Thermo Jarrell-Ash for the Plasma Spec 200. Note the direct introduction into the nebulizer. (With permission.)

and provides a facility for atomic absorption sampling, and computer compatibility is achieved through transistor–transistor logic. Baird Corporation markets two systems: one for ICP emission spectrometry, manufactured by Plasma Therm, that is identical to the device shown in Figure 13.4b, and another that is suitable for use with an ICP–atomic fluorescence instrument. Thermo Jarrell–Ash also markets a hydride generator for the Plasma Spec 200 sequential spectrometer. For this equipment (Figure 13.4c) there is no need to demount the hydride generator system prior to regular solution nebulization. At the time of writing, Applied Research Laboratories was marketing such a system. Selected figures of merit for these hydride generators are presented below.

The continuous hydride generation technique has emerged as the most widely investigated, as well as the most popular, of the hydride generation methods. The commercially available systems, shown in Figure 13.4, reflect this preference. Compared to batch generation, the continuous method is considerably easier to apply. This ease of operation provides high potential for the less experienced operator to successfully use the technique.

13.2.3 Effects of Varying Experimental Conditions

Nearly all the references cited above have discussed or tabulated optimal experimental conditions. The relevant experimental variables included forward power, observation height above the load coil, gas flow rates, torch design, hydride generation cell design, concentration of borohydride reagent, concentration and type of acid sample solutions used, and flow rates of both acid and borohydride solutions. The thorough studies of various operating conditions by Thompson and co-workers[12] primarily form the basis for the following discussion.

The effects of various acids and their concentrations on the atomic emission signals for arsenic, bismuth, antimony, selenium, and tellurium are illustrated in Figure 13.5. Based on this study, 5 M hydrochloric acid is the best compromise, while nitric acid and hydrobromic acid (not shown in Figure 13.5) give the least satisfactory responses for most of the elements tested. The effects of differing sodium borohydride concentrations are clearly indicated in Figure 13.6, which shows that a concentration of 1.0% produces the best response.[12]

In general, the optimal heights for observing hydride-forming elements in the ICP are less than those used for solution nebulization, although the compromise observation height is a function of various gas flows in the plasma. Above 15 L/min, the outer gas flow rates made little difference on the analyte responses. The intermediate gas flow rate gave the best response at 0.8 L/min. Above this flow rate, the emission response falls off rapidly, while at flow rates less than 0.8 L/min, the injector tube is heated and the hydrides are decomposed prior to the observation region of the ICP. The injector gas flow rate shows a maximum response for all tested elements between 0.7 and 0.9 L/min. Under these optimal conditions, the best observation height is at 10 mm. For most of

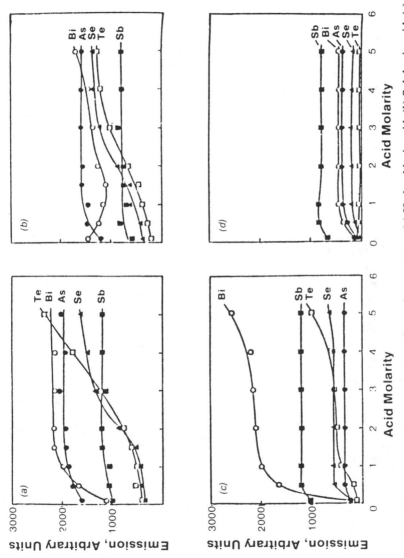

Figure 13.5. Effect of various acids and concentrations on the analyte response. (*a*) Hydrochloric acid. (*b*) Sulphuric acid. (*c*) Perchloric acid. (*d*) Orthophosphoric acid. Plasma conditions: Fassel torch; 2.7 kW forward power; observation height, 10 mm above load coil; injector gas, 0.8 L/min Ar; outer gas, 17 L/min Ar. (From Reference 12 with permission.)

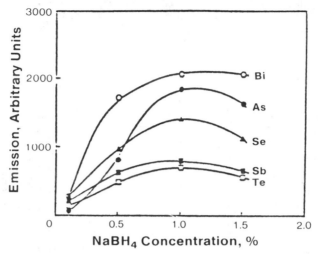

Figure 13.6. Effect of NaBH$_4$ concentration on the analyte response. Plasma conditions as in Figure 13.5. (From Reference 12, with permission.)

the elements, changes in flow parameters alter the relative emission intensities by factors of 1.5 to 2.0. Data for selenium and tellurium show that even lower observation heights are appropriate.

The generation of hydrides is known to be critically dependent on the oxidation state of the element. Thus, operating conditions may need to be widely varied, for example, by adding a prereductant for As V. Four of the volatile hydride-forming elements may have at least two oxidation states in aqueous solution. Thompson and colleagues report response ratios for these elements:[12] As V/As III = 0.92, Sb V/Sb III = 0.12, Se VI/Se IV = 0.0, and Se VI/Se IV = 1.0. For some elements, the oxidation state makes little difference in effective hydride generation, but for others, the differences are striking.

In summary, operating conditions in the hydride generation experiments must be carefully controlled if good long-term precision and accurate results are expected. Even then, difficulties associated with different sample matrices may arise. The chemist familiar with these potential problems should be able to make excellent application of this powerful technique.

13.2.4 Analytical Performance

Analytical performance generally is characterized by figures of merit, such as detection limit, linear dynamic range, and precision and accuracy of measurements. Sections 13.2.4.1 to 13.4.2.3 discuss these figures of merit, along with the related topics of limitations, interferences, and applications.

13.2.4.1 Figures of Merit

A summary of important figures of merit for continuous and batch generation methods is given in Table 13.1. The listed data represent selected and typical references dealing with multiple-element determinations for the volatile hydride-forming elements.

Relative to solution nebulization, detection limits by hydride generation methods are better by up to a factor of 1000 for certain elements. With the most recent continuous generation apparatus (see Figure 13.4b and the appropriate discussion), detection limits obtained approach those achieved by HGC methods, even though HGC has some facility for preconcentration and hydrogen removal. Precision reported as percent relative standard deviation (% RSD) ranges from 2% to slightly higher than 10%. Thus, in general, the hydride-forming elements can be detected at concentrations below 1 ng/mL, and concentrations that are 10 or more times the detection limit can be measured with precisions $\leq 5\%$ RSD. Bushee and co-workers[26] demonstrated the potential for determining speciation by combining HPLC with hydride generation ICP-AES, but because of column dilution and peak broadening, inferior detection limits were obtained, relative to the other hydride generation methods. Linear dynamic ranges for the hydride generation–ICP combination vary from 2 to 4 orders of magnitude, depending on the particular method used. Since the figures of merit for continuous generation are comparable to those for batch generation, and the operating procedure is simpler, continuous generation is the present "method of choice" for sample introduction of the hydride forming elements.

Detection limits for germanium, arsenic, selenium, tin, antimony, tellurium, lead, and bismuth (listed in Table 13.2) enable comparisons of various atomic techniques, including flame atomic absorption spectrometry (AAS), graphite furnace atomic absorption spectrometry (GFAAS), atomic fluorescence spectrometry (AFS), microwave-induced plasma (MIP) emission spectrometry, and ICP emission spectrometry. Perhaps the most important comparison is that between hydride-AAS and hydride-ICP-AES. Except for lead, the ICP method appears to be at least comparable, and in some instances superior, to AAS with the hydride sample introduction technique. Based on reported data, the hydride-AFS methods provide the best detection limits for arsenic, antimony, selenium, and tellurium.

13.2.4.2 Interferences

Two types of interference, spectral and chemical, are associated with the use of hydride generation. Spectral interferences arise from overlapping spectral lines or molecular bands, and for continuous generation, background correction methods usually minimize, and often eliminate, many of these interferences. In the case of coincident spectral lines, alternate wavelengths may be selected for the required measurements, but this option naturally may not exist for a fixed-slit, direct-reading spectrometer. Molecular band interference is most serious from the carbon dioxide that is generated from contaminant CO_3^{2-} in the $NaBH_4$ reagent, thus causing a large background signal at the short wavelengths most desirable for measuring the volatile hydride-forming elements.

Table 13.1. Figures of Merit for Multiple-Element Volatile Hydride Generation Methods

Element	Detection limits (ng/mL)		% RSD for Sample concentration (in parentheses)	Dynamic range[b]	Remarks	Ref.
	Hydride	Solution nebulization[a]				
As	0.8	20	—	3	Continuous generation	12
	3		9 (50 ppb)[c]	2–3	Continuous generation, tandem nebulization	23
	2.5		5 (50 ppb)	2	Batch with sequential detection	30
	<0.2				Continuous generation	18
	0.08[d]				Continuous generation	19
	1				Continuous generation	21
	1			3.5	Continuous generation, DLs given for different oxidation states	22
	0.5		2.4	3	Continuous generation; modified nebulizer	25
	1				Continuous generation; H_2O_2 added	20
	0.18				PT Analytical-Hydride Information Brochure	
	0.02		2.2 (10 ppb)	2	HGC/ICP	29
Arsenite[e]	50		16 (60 ppb)	4	Post-HPLC hydride generation	26
Arsenate	50		23 (60 ppb)	4		
Dimethylarsenate	105			4		
Se	0.8	20	—	3	Continuous generation	12
	0.5		4 (50 ppb)[c]	2–3	Continuous generation, tandem nebulization	23
	0.12[d]				Continuous generation	19
	1				Continuous generation	21
	0.5		7.1	3.5	Continuous generation, DLs given for different oxidation states	22
	1			~3	Continuous generation; modified nebulizer	25
	1				Continuous generation; H_2O_2 added	20
	0.4				PT Analytical-Hydride Information Brochure	
	0.1		4.4 (20 ppb)	2	HGC/ICP	29
Bi	0.8	40	—	2.5	Continuous generation	12
	0.9		7 (50 ppb)[c]	2–3	Continuous generation, tandem nebulization	23
	<0.2				Continuous generation	18

Element	DL	Preconc.	DL (ppb)	RSD	Comments	Ref.
	0.4		2.3	~3	Continuous generation; modified nebulizer	25
	2				Continuous generation; H₂O₂ added	20
	0.15				PT Analytical-Hydride Information Brochure	
	0.3		5.4 (10 ppb)	~2	HGC/ICP	29
Ge	0.2	200	—	4	Continuous generation	16
	20		47 (50 ppb)[c]	2–3	Continuous generation, tandem nebulization	23
	0.2		2.3 (50 ppb)	3	Batch with sequential detection	30
	0.61		5.6 (10 ppb)	2	HGC/ICP	29
Pb	1	30			Continuous generation; H₂O₂ added	20
Sb	1.0	20	8 (50 ppb)[c]	3	Continuous generation	12
	0.7		4.6 (50 ppb)	2–3	Continuous generation, tandem nebulization	23
	2.5			2	Batch with sequential detection	30
	<0.2				Continuous generation	18
	3				Continuous generation	21
	1			3.5	Continuous generation, DLs given for different oxidation states	22
	2		2.3	~3	Continuous nebulization, modified nebulizer	25
	5				Continuous generation; H₂O₂ added	20
	0.25				PT Analytical-Hydride Information Brochure	
	0.08		2.1 (10 ppb)	2	HGC/ICP	29
Sn	0.2	100	—	4	Continuous generation	16
	4		2.4	~3	Continuous generation; modified nebulizer	25
	0.5				Continuous generation; H₂O₂ added	20
	0.8		5.1 (10 ppb)	2	HGC/ICP	29
Te	1.0	20	—	2.5	Continuous generation	12
	0.7		3 (50 ppb)[c]	2–3	Continuous generation, tandem nebulization	23
	5			3	Continuous generation; modified nebulizer	25
	2		2.8		Continuous generation; H₂O₂ added	20
	0.15				PT Analytical-Hydride Information Brochure	

[a] From Reference 23.
[b] Given in orders of magnitude.
[c] Short-term RSD.
[d] Normalized to account for their four-fold preconcentration factor; DLs for antimony, tin, and bismuth difficult to interpret.
[e] With arsenite, arsenate and dimethylarsenate, these authors report detection levels of 24 to 199 ppm utilizing HPLC-ICP with direct nebulization rather than hydride generation. Thus, the hydride generation provides enhancements of 500 to 2000 in their experiments.

Table 13.2. Detection Limits (ng/mL) for Volatile Hydride-Forming Elements Reported for Several Atomic Spectrometric Techniques

Element	ICP		FAA[c]			FAF[c]		MIP[c]
	Solution nebulization[a]	Hydride[b]	Solution nebulization	Hydride	GFAA[d]	Solution nebulization	Hydride	Hydride
Ge	200	0.25 (0.2–0.6)	20	4	0.75	100	—	0.15
As	20	0.23 (0.02–0.5)	630	0.8	1	100	0.1	0.35
Se	20	0.37 (0.1–0.5)	230	1.8	1.5	40	0.06	1.25
Sn	100	0.37 (0.2–0.8)	150	0.5	1	50	—	2
Sb	20	0.31 (0.08–0.7)	60	0.5	0.75	50	0.1	0.5
Te	20	0.96 (0.15–2)	44	1.5	0.5	5	0.08	—
Pb	30	1	17	0.1[e]	0.25	10	—	—
Bi	40	0.26 (0.15–0.4)	44	0.2	0.5	5	—	—

[a] See Table 13.1.

[b] Average of the best four values listed in Table 13.1. Data from Reference 19 not included because preconcentrations were used; given in parentheses.

[c] Data from Reference 4, except for lead, is taken from Reference 20.

[d] From Walter Slavin, "Graphite Furnace AAS Source Book," Perkin-Elmer Corp., Norwalk, Conn., 1984, pp. 18–19 (based on 20-μL injection).

[e] From P. Vijan et al., Analyst 101, 996 (1976).

Background from CO_2 is particularly troublesome in hydride generation–condensation, since the CO_2 is readily condensed and then, rapidly delivered as a plug into the plasma.[4,29,30] Chromatographic removal of the CO_2 background with Chromosorb 102, where the CO_2 elutes from the column before the hydrides, has been suggested.[4,30] However, this approach results in some loss of sensitivity. Alternately, dynamic background correction available on some spectrometers may be used to compensate for this background emission.[29] Even though the signal is transient, a 1-s on-line/off-line cycle is sufficient to make the correction. In principle, the best way of dealing with the CO_2 interference is to completely eliminate it, or its precursors, from the hydride-generating apparatus. However, such purification is quite difficult, and thus, less often used by analysts in place of alternate methods. This is especially true because the relatively unstable borohydride reagent is not easily purified at the necessary trace levels. Also, NaOH, which is a good absorbent of CO_2, is commonly used to stabilize the hydride generating reagent.

Chemical interferences fall into two general categories: those that prohibit or limit the formation of the volatile hydride, and those leading to compound formation after hydride generation, which diminishes the amount of analyte available for excitation. The latter type occurs in flames, but is rarely a problem in ICPs. Several researchers[15–17,20,21,28,32–34] have studied the effects of various cations and anions on the hydride reaction. The alkali and alkaline earth elements do not interfere, even when present at concentrations 10,000 times greater than the hydride-forming element. In general, the transition elements and heavy metals will suppress hydride formation for one or more of the hydride-forming elements. The elements arsenic and antimony are less susceptible to these interferences than bismuth, selenium, tellurium, and tin. Thompson and colleagues[15] summarized the effects of 20 individual elements on arsenic, bismuth, antimony, selenium, and tellurium. The effects of Cu II, Fe II, and Fe III are shown in Figure 13.7. In a related study,[16] the effects of 21 cations and Br^-, F^-, and I^- on the formation of germanium and tin hydrides were reported. The anions Br^- and I^- have no effect, but F^- at 10,000 ppm causes a severe depression in signal. Nygaard and Lowry[21] concluded from their studies of the effects of anions and cations on the formation of the hydrides of arsenic, antimony, and selenium that NO_3^- and PO_4^{3-} have no effect. Ikeda and co-workers[20] have tabulated the effects of numerous cations on the generation of lead hydride, while Nakahara[32,33] has reported the effects of numerous cations on the formation of arsine and stannane.

Approaches to minimize, or to eliminate, these chemical interference effects have been suggested.[15–17,32] Potassium iodide is used to reduce As V to As III, and Sb V to Sb III, thereby enhancing hydride formation. In addition, KI will reduce Fe III to Fe II and precipitate copper, which prevents these species from interfering. However, formation of BiH_3, SeH_2, and TeH_2 is suppressed by KI. Thompson and co-workers[15] used a modification of a lanthanum hydroxide precipitation[34] in which the elements of interest are coprecipitated and several ions, such as Cu II, Ni II, and Hg II remain in solution. Nakahara[32] used a

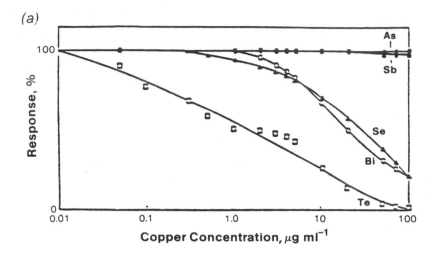

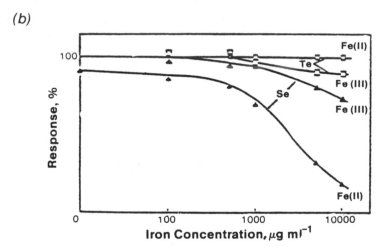

Figure 13.7. (*a*) Effect of concentration of Cu II on response of arsenic, antimony, selenium, bismuth, and tellurium. (*b*) Effect of concentration of iron on response of selenium and tellurium. Plasma conditions as in Figure 13.5. (From Reference 15, with permission.)

mixture of thiourea, nitric acid, and malic acid to reduce the interfering effects of copper and nickel on the production of stannane.

13.2.4.3   Applications

Applications and recovery data for hydride generation–ICP are summarized in Table 13.3, which provides a comprehensive list of sample types and the analytical data for arsenic, bismuth, germanium, lead, antimony, selenium, tin, and tellurium. The applications are intended to be representative, rather than inclusive. The references cited may contain additional determinations, or trials, for a particular sample type.

The sample types listed in Table 13.3 include soil and rock samples, biological samples (rice, wheat, corn, potatoes, oysters, etc), wastewater, drinking water, fly ash, river sediments, air particulates, steels, and marine sediments. In view of the complex chemistry, plus the low levels of determination associated with many of these sample types, it is noteworthy that for many of the reported applications, agreement with reference or accepted values is good, and in many instances, reported and accepted values are statistically the same. A comprehensive study by Ward and co-workers[24] on biological and agricultural samples is partly reported in Table 13.3. For the volatile hydride-forming elements, a continuous generation method with a polypropylene reaction cell was used.

Several standard or spiked recovery studies have been conducted.[18,23,29] Wolnik and colleagues,[23] using the tandem nebulizer system (TNS), determined standard recoveries for arsenic, bismuth, germanium, antimony, selenium, and tellurium with and without interfering elements. In the absence of interfering elements, recoveries for most elements were good, except for germanium and antimony, which gave low values. The poor recovery for the latter may be attributed to oxidation of Sb III to Sb V during digestion. In the presence of interfering elements, antimony also gave low recoveries.[23] The recovery for bismuth also was noted to be consistently low, but the use of standard additions for quantitation compensated for the effect of interfering elements. As shown in Table 13.3, for spiked recovery studies on NBS rice flour, NBS wheat flour, corn, potatoes, and soybeans, recoveries were good, with two exceptions. The recovery was high for germanium in potatoes, and recoveries were low for antimony in rice flour and for selenium in wheat flour. These observations suggest that conditions for both sample digestions and hydride generation were generally appropriate. However, these conditions may yet need to be tailored to specific elements to consistently attain accuracy.[23]

Pahlavanpour and co-workers[18] recovered arsenic, bismuth, and antimony from aqueous spikes added to pasture herbage samples. For additions of 20 to 400 ng/g of each element, the average recovery for arsenic and antimony was $98.7 \pm 1.2\%$, and the recovery for bismuth was 95.6%. These results suggest excellent potential for their continuous generation method. Hahn and associates[29] recovered arsenic, bismuth, germanium, antimony, selenium, and tin from digests, and determined these elements with HGC–ICP. Average recoveries exceeded 95%, except for selenium, which was low probably because of the need to optimize the digestion for this particular element.

In summary, recoveries can be quantitative for proper digestion and hydride generation conditions. Unfortunately, these conditions are not uniform for all the volatile hydride-forming elements. Thus, the capability for the simultaneous determinations of all eight hydride-forming elements from one reaction mixture is compromised. Nevertheless, hydride generation provides a very sensitive method that is the best choice to date for the ICP determination of this group of elements.

Table 13.3. Selected Applications of Hydride Generation ICP for Arsenic, Selenium, Bismuth, Germanium, Lead, Antimony, Tin, and Tellurium[a]

Sample type	As	Se	Bi	Ge	Pb	Sb	Sn	Te	Ref.
U.S. Geological Survey (USGS)	ng/kg		ng/kg			ng/kg			
W-1	0.89 (1.9)		<0.04 (0.046)			1.18 (1.0)			17
G-1	0.50 (0.5)		0.12 (0.065)			0.55 (0.31)			
G-2	0.22 (0.25)		0.12 (0.043)			0.12 (0.1)			
GSP-1	0.12 (0.09)		0.18 (0.037)			3.20 (3.1)			
	µg/g	µg/g	µg/g	µg/g		µg/g	µg/g	µg/g	
Bovine liver, NBS 1577	0.05 (0.055)	0.8 (1.1)	<0.03	<0.06		<0.05	<0.08	<0.8	24
Orchard leaves, NBS 1571	10.2 (10)	0.08 (0.08)	0.06	0.12		2.5 (2.9)	0.19	<0.5	
Pine needles, NBS 1575	0.21 (0.21)	0.09	0.45	<0.1		0.2	<0.1	<1	
Rice flour, NBS 1568	0.40 (0.41)	0.11 (0.4)	<0.06	<0.1		<0.04	<0.07	<0.5	
Spinach, NBS 1570	0.15 (0.15)	<0.03	1.5	<0.06		<0.04	<0.05	<0.8	
Tomato leaves, NBS 1573	0.26 (0.27)	<0.05	1.0	<0.06		<0.04	<0.05	<0.5	
Wheat flour, NBS 1577	<0.03	0.19 (1.1)	<0.07	0.12		<0.03	<0.1	<0.4	
Pepperbush, NIES #1	1.1	0.11	0.60	<0.09		0.80	0.17	<0.6	

	(μg/g)	(μg/g)	(μg/g)		Ref.
NBS rice flour	0.452 ± 0.070 (0.41 ± 0.05)	0.331 ± 0.029 (0.40 ± 0.1)			23
NBS wheat flour	≤0.08	0.831, 0.909 (1.1 ± 0.02)			
Corn		0.181			
Potatoes		0.256, 0.283			
Soybeans		2.32, 2.40			
Wastewater (range for several samples)	(ng/mL) 2.8 ± 0.2 to 6.5 ± 0.2				33
Bowen's kale	(μg/g) 0.268 ± 0.006 (0.14)	0.002	(μg/g)	(μg/g) 0.75 ± 0.007 (0.07)	18
NBS orchard leaves	11.98 ± 0.08 (10 ± 2)	0.004 ± 0.001	2.77 ± 0.02	(2.9 ± 0.3)	
USGS rock samples with persulfate digestion	(ng/mL)	(ng/mL)		(ng/mL)	
GXR1	400 (460)	33 (19)		110 (124)	21
GXR2	24 (32)	9 (0.7)		46 (40)	
GXR3	3800 (4000)	11 (0.2)		42 (40)	
GXR4	104 (98)	5 (6)		5 (5)	
GXR5	54 (60)	3 (6)		6 (10)	
GXR6	1450 (1700)	6 (6)		15 (19)	

(Continued)

Table 13.3. (*Continued*)

Sample type	As	Se	Bi	Ge	Pb	Sb	Sn	Te	Ref.
Water									
EPA-476E #3	60	14							21 (all Ref. 21 cited are persulfate digestion)
	(61)	(16)							
NBS 1643	71	10							
	(76)	(12)							
EPA WP379 #1	—	—				9			
	—	—				(8)			
Environmental Research	24	57				55			
Assoc. #2913	(45)	(60)				(53)			
River sediment, NBS 1645	67	8				38			
	(66)					(51)			
Coal fly ash, NBS 1633	57	9				5			
	(61)	(9)				(8)			
Urban particulate, NBS 1648	117	26				41			
	(115)	(24)				(45)			
Soils (Canadian)	µg/g	µg/g							
SO-1	1.9	0.09							19
	(1.9)	(0.1)							
SO-2	1.3	0.31							
	(1.2)	(0.3)							
SO-3	2.6	0.08							
	(2.6)	(0.05)							
SO-4	6.5	0.49							
	(7.1)	(0.4)							
IAEA	95.1	0.93							
	(93.9 ± 7.5)								
NBS 1645	66.4	1.0							
	(66)								

	μg/g	μg/g	μg/g
Low-alloy steel, JSS 168-3			60.9 ± 5.4 32
			(60)
Low-alloy steel, JSS 170-3			542 ± 36
			(54)
Low-alloy steel, JSS 171-3			337 ± 31
			(340)
Cu–Ni alloy, NBS 874			71 ± 6.8
			(70)
	μg/g	μg/g	μg/g
Oyster tissue, NBS 1566	11.1 ± 1.1	1.7 ± 0.2	0.4 ± 0.3 22
	(13.4 ± 1.9)	(2.1 ± 0.5)	(0.4)
Orchard leaves, NBS 1571	11.9 ± 0.6	0.06 ± 0.02	2.8 ± 0.2
	(10.2 ± 0.2)	(0.08 ± 0.01)	(2.9 ± 0.3)
Bovine liver, NBS 1577		(1.0 ± 0.1)	0.3 ± 0.2
		(1.1 ± 0.1)	
Marine sediments			
S-1 (KOH fusion)	11.0 ± 0.9	b	0.36 ± 0.09
	(11.6 ± 1.3)		(0.4)
S-2 (KOH fusion)	11.4 ± 0.8	b	0.71 ± 0.03
	(11.1 ± 1.4)		(0.73 ± 0.08)
S-3 (KOH fusion)	11.2 ± 0.5	b	0.61 ± 0.06
	(10.6 ± 1.2)		(0.59 ± 0.06)

(Continued)

Table 13.3. (*Continued*)

Sample type	Element								Ref.
	As	Se	Bi	Ge	Pb	Sb	Sn	Te	
Well water	ng/mL								26 (values given are total As; see **Ref.** 26 for ratios of arsenate to arsenite)
	182								
S #4	[212 ± 4][c]								
	196								
S #5	[226 ± 1]								
	235								
S #6	[243 ± 1]								
	177								
S #7	[193 ± 1]								
	126								
S #8	[153 ± 0]								
	ND[d]								
	[7 ± 3]								
	μg/g								
NBS orchard leaves	9.8, 10.1, 10.0								27
	(10 ± 2)								
NBS coal fly ash	141, 142, 139								
	(145 ± 15)								
NBS river sediment	64, 64, 66								
NRCC[e]	10.4, 10.7, 10.7								
	(10.6 ± 1.2)								
NRCC	11.6, 12.0, 12.1								
	(11.4 ± 1.4)								

Sample	μg/g	μg/g	μg/g	μg/g / ng/mL	μg/g / ng/mL	μg/g	Ref.
Bowen's kale	0.32, 0.10, 0.15 (0.13 ± 0.02)						28
NBS tomato leaves	0.29, 0.25 (0.27 ± 0.05)						
NBS orchard leaves	10.2, 9.7, 9.5 (10 ± 2)						
NBS samples							29
Rice flour	0.44 ± 0.05 (0.41 ± 0.05)	0.37 ± 0.06 (0.40 ± 0.1)	<0.008	<0.02	<0.002	<0.02	
Wheat flour	0.006 ± 0.001 (0.006)[e]	0.87 ± 0.06 (1.1 ± 0.2)	<0.008	<0.02	<0.002	<0.02	
Spinach	0.17 ± 0.01 (0.15 ± 0.05)	0.003	<0.008	<0.02	0.014 ± 0.003 (0.04 ± 0.01)	<0.02	
Orchard leaves	13 ± 1 (14 ± 2)	0.07 ± 0.01 (0.08 ± 0.01)	<0.008	0.15 ± 0	2.8 ± 0.1 (2.9 ± 0.1)	0.18 ± 0.01	
Water Quality Ref.							30
S-1	28.8 ± 0.01 (22)			48.7 ± 0.6 (50.0)	92.3 ± 0.9 (97.5)		
S-2				93 ± 2 (100)			

[a] All values in parentheses are certified or reference values.
[b] Inconsistent results.
[c] Unbracketed in this series are HPLC/hydride/ICP, bracketed are direct hydride/ICP.
[d] Not detectable
[e] Provisional value.

13.3   DIRECT GASEOUS SAMPLE INTRODUCTION

13.3.1   Principles

Gases can be introduced directly into the injector gas flow of the ICP either continuously or by injection by a gas sampling loop. This mode of sample introduction can be used for analysis of gaseous substances, such as fuel gases or combustion gases; however, the most frequent application of direct gaseous introduction has been in the characterization of the ICP as a selective detector for gas chromatography. This section discusses only gaseous analytes at low concentration ($< 0.1\%$) in the argon injector flow, which may be presumed to cause no fundamental changes in the nature of the ICP discharge. Mixed gas plasmas, which exist for gas concentrations higher than 0.1%, are treated in Chapter 15.

Continuous introduction of gaseous compounds can be used to compile tables of atomic spectral lines from the ICP, to evaluate potential applications of the ICP to the elemental analysis of organic compounds in the gaseous phase, to study spectral background interferences, and finally, to optimize plasma operating parameters. Optimizations of transfer optics and monochromator wavelength settings to monitor a selected spectral line are readily accomplished while continuously introducing gaseous compounds containing the analyte element into the ICP. Gas mixtures having certified concentrations may be purchased from many vendors of bottled gases. Mixtures having adjustable, but reasonably well known, concentrations can be made by mixing known and carefully controlled flows of analyte gas and argon. A continuous flow of gaseous analyte also can be obtained by passing argon that is to be injected into the ICP over a reservoir of organic liquid. The successful production of spectra by this approach depends on the vapor pressure of the liquid.

Direct injection with a gas sampling loop results in a discrete volume, typically in microliter quantities, of sample passing through the plasma to give a transient emission signal. This mode of sample introduction is particularly important for analysis of gaseous samples when total sample volume is limited. Direct injection approximates an eluting GC peak and is useful for the investigation and optimization of plasma performance without introducing the additional variables associated with chromatography.

13.3.2   Equipment and Operating Parameters

The equipment used for continuous introduction of gaseous samples primarily consists of standard flow-control devices. Injection of discrete samples by a gas sampling loop is more complicated. A simple "T" that is fitted with a GC septum may be used to inject samples with a gas-tight syringe. However, measurements made with this approach are not as reproducible as those from injection by a sampling loop and also, air is admitted as a contaminant to the injector gas stream. Gas sampling loops can be obtained in various sizes or constructed from a length of tubing that has a known volume. Commercial gas injection valves usually are equipped with fittings that have low dead volumes. Reproducibly filling the sample loop requires fairly sophisticated control of the sample flow. The system used by Fry and co-workers[35] for both injection and continuous introduction is shown in Figure 13.8. These investigators constructed a gas injection system with a slide valve and a 100-μL sampling loop. Direct injection from this system produces transient signals that usually last 0.1 s or less, depending on the injection volume and flow rate. The overall detector response time must be compatible with these signals, and thus, Fry[35] used a diffusion tube to broaden the sample plug.

Most studies have been conducted using an ICP torch with an elongated outer tube to prevent entrained air from entering the plasma and causing background interferences. The deleterious effects resulting from entrained air are: specific background contamination for elements that are components of the atmosphere, spectral background interferences, and quenching of atomic emission intensity for some elements due to reaction with N_2 or O_2 to form molecular species. However, Brown and Fry[36] report that with a standard commercial-type ICP and observation heights above the load coil of less than 5.5 mm, the Fassel torch effectively excludes entrained air, making the elongated torch unnecessary, even for the determination of oxygen and nitrogen. Note that

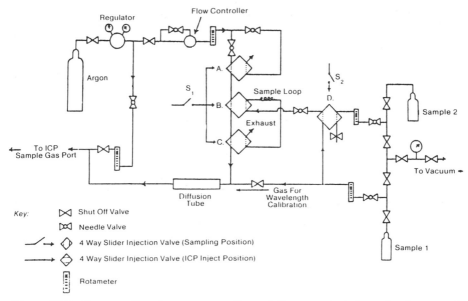

Figure 13.8. Gas sampling loop system: S_1 and S_2 serve to independently control pneumatic actuators (driven by 130-psi N_2) for the four-way slider valves. The diffusion tube is a length of Pyrex tubing used to broaden the sample plug so that it is compatible with detector response time. (From Reference 35, with permission.)

for atomic bromine, chlorine, fluorine, nitrogen, and oxygen, the most intense near-infrared emission occurs between the turns of the load coil, but viewing optics that sample optical radiation from a region above the load coil are more convenient, but cause a significant loss in detectability[37] for bromine and chlorine.

Forward power levels in the range 0.5 to 2 kW have been reported for the elemental analysis of gases. Intermediate and injector gas flows vary considerably from study to study, depending on the ICP system used and on the element investigated. Outer gas flows vary somewhat but are in the range 12 to 24 L/min, which is normally used for solution nebulization. Specific background contamination stemming from impure argon and leaks or diffusion through the walls of nonmetallic tubing that permit air to enter the plasma cause a high continuous background signal. Transients can be readily measured against this continuous baseline. However, high background produces more noise and requires greater electronic suppression, and thus, should be minimized whenever possible.

Brown and Fry[36] employed a novel technique for quantifying the level of oxygen contamination present in the plasma background. They superimposed repeated gas loop injections of oxygen into the sample stream during slow wavelength scans. This results in a series of transient, single standard additions superimposed on the background signal, in effect producing a sample modulated signal. The ratio of the emission peak heights of constant oxygen to the injected oxygen provides a correction factor that must be applied to all data obtained by scanning the oxygen line. This technique can be used for other contaminants, such as nitrogen, as illustrated in Figure 13.9.[38]

Caution must be exercised when analyzing compounds containing fluorine, since this can lead to erosion of the quartz torch. Fry[35] found that even small quantities of fluorine containing compounds introduced via the intermediate flow caused erosion of all three torch tubes. When fluorinated compounds are introduced continuously, such as during spectrometer wavelength adjustments, the argon injector flow should be increased, to extent the useful life of the torch, even though sensitivity is degraded. More details and other suggestions for improving torch life can be found elsewhere.[35]

Although Hughes and Fry[37] conducted preliminary investigations on emission spectra of bromine and chlorine from an ICP by continuously introducing HBr and Cl_2, they warn that these gases are extremely noxious and not recommended for routine use. Accidental leakage or release of HBr or Cl_2 could prove very hazardous, and thus, these substances must be used only in a closed ICP system.

13.3.3   Analytical Performance

13.3.3.1   Figures of Merit

Detection limits for gas loop injection have been compiled in Table 13.4. The detection limits for additional elements, determined by GC–ICP, are summarized in Table 13.5. Detection limits for halide solutions introduced into the

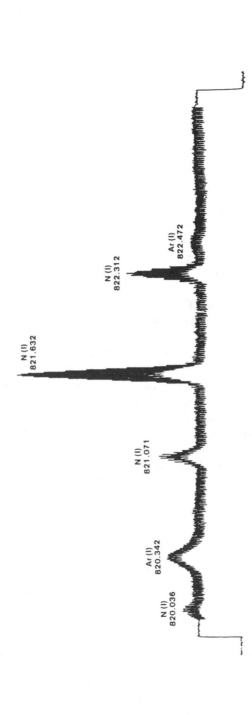

Figure 13.9. Wavelength scan (7.6-mm observation height) with sample modulation. Entrained atmospheric N_2 is seen as the unmodulated emission, ("white" peak) under the nitrogen transients. Plasma conditions: 2.2 kW forward power; 13 L/min argon outer gas; 0.625 L/min argon injector gas. (From Reference 37, with permission.)

Table 13.4. Detection Limits for Gaseous Sample Introduction of Several Nonmetals Using Gas Loop Injection

Element	Wavelength (nm)	Detection limit (ng)	Injection loop volume (μL)	Dynamic range	Ref.
O	777.1	25	8	—	35
N	821.6	250	—	—	49
F	685.6	350	43	5×10^3	46
Cl	837.6	50	8	5×10^3	36
Br	827.2	50	8	5×10^3	36

plasma by pneumatic nebulization have been published recently[39] and are mentioned here, after conversion to equivalent units, to provide a ready comparison. For near-IR emission lines, the following detection limits have been reported: Br 700.5 or 827.2 nm, 500 μg/mL; Cl 725.6 nm, 100 μg/mL; F 685.6 nm, 2000 μg/mL. The Br 163.3 nm line provides a detection limit of 3 μg/mL. The use of a He ICP, as discussed in Chapter 15, has improved the detection limits for the halogens by 1 and 2 orders of magnitude for aqueous and gaseous injection, respectively.

Detection limit results of Goldfarb and Goldfarb[3] have been omitted from Table 13.4 because the data obtained with their system are not representative of typical ICP results for gaseous sample introduction. With their system, unlike all the others discussed here, response *was not* independent of compound structure. This may be attributable to operating an unoptimized ICP–spectrometer system, and to the altered nature of their plasma, indicated by their plasma temperature measurements, brought about by the use of relatively concentrated (1 to 10%) gas mixtures.

The precision obtainable by gas loop sampling, particularly if the loop is pneumatically driven, is less than 1% RSD. Fry and co-workers[35] averaged multiple injections by rapid (0.5 Hz) switching of the gas sampling loop. This approach compensates for "flutter" in the gas flow, which results from the pressure surge associated with loop switching and contributes to noise. The precision for replicate measurements improved to an RSD of approximately 0.2%.

13.3.3.2 Applications

Continuous gaseous introduction and gas sampling loop injection have been used primarily to evaluate the potential of the ICP as a selective detector for gas chromatography and to obtain background information concerning the behavior of nonmetallic elements in the ICP. Observed emission lines and relative intensities have been tabulated for the following nonmetallic elements in the ICP: boron, carbon, hydrogen, iodine, phosphorus, sulfur, and silicon[40]; oxygen[41]; nitrogen[38]; fluorine[35]; and bromine and chlorine.[37] The transitions observed for these elements provide insights into the energy available for excitation in the ICP. The degree of dissociation attained with the ICP for

Table 13.5. Detection Limits, Selectivities, and Ratios for Various Elements with Gaseous Sample Introduction Using GC–ICP

Element	Wavelength (nm)	Detection limit (ng)	Selectivity ratio	Dynamic range	Ref.	Comments
B	249.7	1		1×10^3	39	
C	247.8	12		2×10^3		
H	656.2	27		1×10^2		$0.5\ \mu L$ injected by
I	206.1	4		1×10^5		heated injector block
P	213.6	0.6		2×10^4		with 1:20 split
S	190.0	250		1×10^2		
Si	251.6	0.8		5×10^2		
Br	700.5	2×10^5	—	Poor	45	No background correction,
C	247.8	12	—	1×10^3		selectivity ratio
Cl	725.6	7×10^3	60	1×10^2		determined with
F	634.8	1×10^6	—	Poor		p-xylene
H	656.2	5.5	3×10^3	1×10^3		
I	206.1	24	1×10^3	1×10^3		
Si	256.1	0.8	3×10^4	5×10^2		
Fe	371.9	5.9	1×10^3	2×10^4		
Pb	217.0	33	3×10^3	1×10^3		
Sn	284.0	0.9	3×10^4	1×10^4		
O	777.1	650	Problems		35	GC, 0.5–2 μL injections
N	821.6	1×10^3	Good		49	GC, 5 μL injections
F	685.6	1×10^3	(See text)		34	GC, 1.2 μL injections

various compounds can be conveniently studied by gaseous sample introduction because the dissociation–excitation processes are independent of any aerosol formation–desolvation processes. The degree of dissociation obtained in the ICP is a function of bond strength and is measured by the relative response ratio for different compounds. For all the compounds investigated, the ICP approaches the ideal case in which response is completely independent of compound structure. This topic is also addressed in Section 13.4.3.1. A comprehensive discussion of the uses of gaseous sample introduction for studying fundamental plasma properties has been published by Goldfarb and Goldfarb.[3]

Northway and associates[38] reported on the determination of nitrogen contamination in a tank of argon by using that argon for all three torch flows (outer, intermediate, and injector). Calibrated amounts of N_2 were added to the total flow in a mixing "T". This procedure is valid only for contamination in argon, and entrained air must be excluded from the observation region.

13.4   THE GAS CHROMATOGRAPHIC METHOD

13.4.1   Principles and Instrumentation

The determination and identification of trace organic, inorganic, and organometallic compounds in environmental, industrial, biological, and pharmaceutical samples is an analytical problem of interest to many scientists. Because biological responses often depend on the exact nature of a compound in situ, speciation of metals and of metalloids is a significant concern of toxicologists, biologists, and pharmacologists. The combination of gas chromatography and plasma spectrometry provides a specific and highly sensitive detection system for a variety of analytes. Selective detection may improve detection limits, minimize interferences from overlapping chromatographic bands and provide useful qualitative information.[42] The GC–ICP method has contributed to, and will continue to augment, the development of speciation methods.

Since 1965, when McCormack and colleagues[43] first interfaced a gas chromatograph to a microwave-induced plasma, numerous publications have appeared in the literature describing the applications of MIPs to the analysis of GC effluents. Applications of ICPs to GC detection are more limited, primarily due to the relative insensitivity of the Ar ICP, compared to the He MIP, for the determination of halogens and also because of cost considerations. Krull and Jordan[44] in 1980 and Carnahan and co-workers[42] in 1981 reviewed the literature already published concerning chromatography with plasma spectrometry for detection.

Ideally, for selective GC detection, the emission intensity from an element in the plasma should be independent of the structure of the organic compound from which the element originates, and the response per nanogram of element in a compound should be the same for all compounds containing that element. As recently illustrated,[45] this has generally been the case for MIPs. However, on occasion, molecular bandheads have been used to monitor the MIP emission. For fragmentation, excitation, and trace level determinations of halogen compounds, the atmospheric-pressure He MIP is advantageous. Application of the Ar ICP to the detection of nonmetals in organic compounds by gaseous sample introduction was first investigated by Windsor and Denton.[40] They found that ICP response per unit weight of carbon was independent of structure for a variety of compounds. A disadvantage of the MIP is its tendency to form carbon deposits on the walls of the plasma containment tube. Unless the solvent is vented, a scavenger gas such as O_2 or N_2 may be added to prevent these deposits from occurring. This is not a problem in the ICP.

Windsor and Denton[46] subsequently interfaced a GC to an Ar ICP and found that, in general, for elements other than halogens, ICP detection limits, linear dynamic ranges, and selectivities compared favorably to flame photometric and MIP detection. By monitoring carbon or hydrogen emission, the ICP functions as a nonselective, or so-called universal, detector similar to flame ionization or thermal conductivity. In addition, multiple-element determination with a polychromator permits element-specific detection for selected elements, thus aiding in the discrimination between overlapping chromatographic bands and enabling calculations of elemental ratios. This, in combination with retention times, enhances the identification of analyte compounds. Obviously, the greater the number of elements monitored, the greater the potential for obtaining qualitative information and calculating correct empirical formulas. In an important paper, Windsor and Denton[47] demonstrated the excellent performance of the ICP for determining empirical formulas. The major limitations associated with the technique are the low intensity of bromine, chlorine, fluorine, and sulfur lines and the absence of atomic oxygen and nitrogen lines in the spectral region from 180 to 800 nm.

A number of complications may arise when introducing large quantities of organic vapor, such as the eluted solvent peak, into the ICP. Plasma instability in the form of changes in background intensity and spectral interferences from carbon emission or molecular band spectra can occur. In addition, although transport of the effluent from the GC to the ICP is essentially 100%, the band spreading inherent in chromatography degrades sensitivity as compared to direct gas injection.

In principle, any gas-chromatographic system capable of producing the desired separation can be combined with ICP detection. Descriptions of the GC components and parameters used for specific applications are omitted from this discussion, but these can be obtained from appropriate references.[3,40-51] This section is confined to a discussion of interface design and plasma features. Ideally, the interface should provide virtually zero dead volume and the same temperature as the GC oven. Because flow rates necessary to effect GC separation are often not compatible with optimum gas flow rates for the ICP, a means for introducing auxiliary Ar to the GC effluent may be necessary.

Windsor and Denton[40,46,47] used a variable frequency (2 to 30 MHz), 1 kW RF transmitter to provide power to an ICP torch,[48] which had an extended outer tube, presumably to minimize entrained air.[49] In an initial study,[40] organic liquid samples were injected into a heated block maintained at 200°C. A gas flow of 1.0 L/min and 1:20 split ratio directed 0.5 mL/min of sample vapor into a mixing "T" fed by 0.9 L/min of auxiliary gas prior to injection into the ICP. A 1:20 split was necessary because the plasma, operated at approximately 0.5 kW forward power at 27 MHz, was extinguished by large amounts of organic vapor. The interface of the GC to the ICP[46] consisted of a length of stainless steel tubing (1.6 mm o.d., 0.1 mm i.d.) inserted and sealed to injector tube of the torch. A 1.6-mm Swagelok "T" mounted in the GC detector oven was used to introduce 0.9 L/min of Ar as auxiliary gas. In later work on determining empirical formulas, the internal diameter of the torch injector tube was reduced to 0.1 mm to decrease dead volume.[47]

Brown and co-workers[36,50] also used an extended torch and a heated (100°C) stainless steel transfer line (1.6 mm o.d.) with an Altex 10105 PTFE "T" to introduce Ar as the auxiliary gas. A second Altex "T" was added to permit alternative injections from a gas sampling loop. The low-dead-volume "Ts" were connected to the injector tube of the ICP torch by a heated 11-cm length of PTFE tubing (1.6 mm o.d., 0.8 mm i.d.). A crystal-controlled, 27-MHz RF generator supplied the ICP a forward power of 1.75 kW.

Ohls and Sommer[51] used a 3-kW argon–nitrogen ICP sustained by a 27-MHz, free-running generator to detect lead species in GC effluent. The GC–ICP interface consisted of an electrically heated metal tube. The combination of 28 L/min of N_2 outer gas flow, 15 L/min of Ar intermediate gas flow, and 1 L/min of injector gas provided a stable plasma. When only Ar was used, forward powers greater than 2.3 kW led to overheating of the ICP torch. The high-power Ar–N_2 ICP produces an extended plasma that provides a longer residence time for the GC effluent and is not detuned by 10 to 100 μL of sample in the injector gas. Thus, the primary reason for using the Ar–N_2 ICP is its capability of accommodating a variety of sample types, including organic solutions, as discussed in Chapter 15.

Several considerations affect the selection of operating parameters for GC–ICP, and these parameters may vary depending on the element or elements being determined. The factors that also must be considered include background contamination, interferences, resolution, and plasma stability.

Windsor and Denton[40] found that a single compromise observation height of 9 mm could be used for determining boron, carbon, hydrogen, iodine, phosphorus, sulfur, and silicon without a significant loss of detectability. Diatomic species can be formed when atomic species produced in the initial excitation zone recombine prior to the observation zone. The diatomic species C_2, CN, and CS, which cause most of the molecular background emission, exhibit intensity maxima at 9 mm; however, spectral bands from NH and OH that result from compounds containing nitrogen and oxygen have low intensities at this height.[40] Interferences arising from entrained air are reduced and virtually eliminated at observation heights less than 6 mm above the load coil by the design of the elongated torch.[36]

A major factor considered by Windsor and Denton[40,46,47] for selection of injector flow parameters was plasma stability. A compromise between sensitivity and stability was obtained at a flow of 0.9 L/min. A flow splitter was used in their system to maintain stability. Note that for 0.5-μL injections and a 1:20 split, 0.025 μL of sample reaches the plasma at a flow rate of 0.9 L/min. For most elements, the upper end of the dynamic range was limited not by linearity but by the weight of compound that could be introduced without overloading the plasma.

Brown and colleagues[36,50] suggested that high carrier flow rates would enhance penetration into the ICP and reduce band spreading and loss of resolution by diffusion. However, in the case of nitrogen, flows greater than 250 mL/min caused poor sensitivity, whereas flows less than 250 mL/min led to poor resolution because of band spreading in the transfer line and ICP torch. Other studies showed that flows ranging from 97 to 350 mL/min Ar, added to a GC effluent flow of 25 mL/min, caused no appreciable variation for the determination of oxygen.

13.4.2 Analytical Performance

13.4.2.1 Figures of Merit

The detection limits attainable by GC–ICP are contingent on the chromatographic separation (ie, the GC instrumentation, column type and size, injection volume, etc). Because a major attribute of the ICP as a GC detector is its capability of providing element-selective detection, the ensuing discussion of analytical performance centers mainly on selectivity. A few GC–ICP detection limits have been published and are included in Table 13.5. Information on detection limits more suitable for comparisons between ICP and other GC detectors is available in the discussions of gas loop injection in the preceding section. Detection limits reported by Goldfarb and Goldfarb[3] obtained for gas-injection GC–ICP are not included in Table 13.5 for the reasons given previously.

In general, the selectivity for each element at the wavelength of interest is defined as the ratio of peak response per amount, usually moles, of element to the peak response per amount of carbon. Hexane, p-xylene, n-dodecane, and other compounds can be used in determining selectivity ratios. Table 13.5 provides a summary of selectivity ratios determined by GC–ICP. Selectivity, which is obviously influenced by the wavelength monitored and background interferences, can be improved by using some type of background correction technique. Fry and co-workers[35] found that for F 685.6 nm, the selectivity ratio is increased from 1, without background correction, to an acceptable level when background correction is used. A background correction scheme in which on- and off-line emissions are measured in the same experiment, such as with wavelength modulation by a movable refractor plate, would be more ideal. Goldfarb and Goldfarb[3] used wavelength modulation with a lock-in amplifier to discriminate against background.

Brown and colleagues[50] observed some nonselective behavior for atomic oxygen at 777.2 nm caused by "early" eluting hydrocarbons, such as heptane, which was not apparent in, nor eliminated by, off-peak background measurements. However, selectivity ratios for "later eluters" were sufficiently high without background corrections.

Another important measure of ICP utility, particularly for determining elemental ratios and empirical formulas, is the "relative response factor." If the ICP response for a given element is completely independent of compound structure, the relative response factors for a particular element in a variety of injected compounds would be equal. The factors in Table 13.6[40] demonstrate that ICP response for atomic carbon is virtually independent of compound structure. Brown[50] found that response factors for atomic nitrogen at 821.6 nm for diethylamine, triethylamine, and piperidine were in good agreement. Similar results were expected for atomic fluorine at 685.6 nm, since fluorinated

Table 13.6. Relative Response Factors of Organic Compounds at the 247.6 nm Atomic Carbon Line

Empirical formula	Compound	Response factor[a]
CS_2	Carbon disulfide	99
$C_2H_3Cl_3$	Trichloroethylene	123
C_2H_3N	Acetonitrile	104
C_2H_3Br	Ethyl bromide	118
C_2H_3I	Ethyl iodide	101
C_2H_6O	Ethyl alcohol	99
C_3H_7N	n-Propylamine	103
C_4H_4S	Thiophene	99
$C_4H_8O_2$	Ethyl acetate	105
C_4H_9Br	2-Bromo-2-methylpropane	104
C_4H_9Cl	1-Chloro-2-methylpropane	105
C_4H_9Cl	1-Chlorobutane	104
C_4H_9I	2-Iodobutane	103
C_3H_3N	Pyridine	86
$C_5H_8O_2$	2,4-Pentanedione	96
$C_5H_{10}Br_2$	1,5-Dibromopentane	95
$C_5H_{11}Br$	1-Bromo-3-methylbutane	101
C_6H_6	Benzene	100
C_6H_{12}	1-Hexane	102
C_6H_{12}	Cyclohexane	97
C_7H_8	Toluene	(100)
C_7H_{16}	n-Heptane	100
C_9H_{12}	Cumene	98

[a] The response for 1 μg of carbon from toluene is defined as 100.
Source: From Reference 39, with permission.

Table 13.7. Elemental Analysis of Hydrocarbons

Compound	Carbon (%)			Hydrogen (%)		
	Theoretical	Found	Relative difference (%)	Theoretical	Found	Relative difference (%)
Cumene	90.00	89.72	0.31	10.00	10.28	2.80
Cyclohexene	87.80	87.94	0.16	12.20	12.06	1.15
Ethylbenzene	90.57	90.57	0.00	9.43	9.43	0.00
n-Heptane	84.00	83.92	0.10	16.00	16.08	0.50
Isooctane	84.21	84.15	0.07	15.79	15.85	0.38
Methylcyclohexane	85.71	85.72	0.01	14.29	14.28	0.07
1-Pentene	85.71	85.98	0.32	14.29	14.02	1.89
o-Xylene	90.67	90.67	0.00	9.43	9.43	0.00
m-Xylene	90.57	90.48	0.10	9.43	9.52	0.95

Source: From Reference 46, with permission.

compounds have lower bond strengths. Response ratios for four fluorinated compounds of varied structure agreed well with theoretical values.[35] An example of elemental ratios obtained by GC–ICP is shown in Table 13.7.[47]

13.4.2.2 Applications

The applications of GC–ICP to actual analytical problems are few. Ohls and Sommer[51] determined tetramethyl and tetraethyl lead in gasoline, using 8-μL injections, with RSDs less than 5% for concentrations ranging from 0.05 to 0.25 mg/kg. Goldfarb and Goldfarb[3] used GC–ICP with gas injection to determine N_2, CO_2, CO, H_2, and O_2 in oil combustion gases and H_2, CH_3, CO, ethylene, and acetylene in plasma coal pyrolysis products. They monitored nitrogen, oxygen, carbon, and hydrogen emission and found that the element-selective ICP detector simplified chromatographic separation.

13.5 CONCLUSIONS AND FUTURE DEVELOPMENT

The advantages and disadvantages of each of the three main techniques for forming and introducing gaseous species into the ICP have been discussed. The performance of the ICP is excellent with respect to dissociation and excitation of atomic emission from gaseous samples, regardless of whether the samples enter the plasma as hydrides, molecular gases, or complex organic compounds. Due to the improvement in detection limits observed for the hydride-forming elements, one can anticipate the future development of additional procedures for generating volatile species of selected elements, such as the formation of volatile metal chelates described by Black and Browner.[14]

The factors that limit the use of the ICP as an element-selective GC detector for real analytical problems are insufficient sensitivity for certain elements and the expense associated with the purchase and operation of an ICP source. Recently, Seliskar and Warner[52] published a description of a water-cooled torch that operates at reduced pressure (1 to 10 torr) and enables the formation of a plasma in either argon or helium with a low-power, 27-MHz generator. A stable plasma can be maintained at considerably lower flows (< 10 mL/min) and powers (< 0.1 kW) than required with the commonly used ICP. The reduced-pressure Ar ICP has been used for the determination of hydrogen isotopes, (H_2 and D_2) contained in the bulk Ar gas.[53]

The improvement in sensitivity for nonmetallic elements, particularly the halogens, in gas chromatographic effluents obtained with the He MIP as compared to the Ar MIP has been well documented.[54] Hughes and Fry[37] observed only a few very weak ion lines for bromine and chlorine in the Ar ICP. However, the He MIP is capable of exciting both neutral atom and ion emission for halogens, and this is true for the reduced pressure He ICP as well.[55] Consequently, this new system, which is inexpensive to operate, appears promising as a selective detector for GC.

REFERENCES

1. R. Browner and A. Boorn, (a) Sample Introduction: The Achilles' Heel of Atomic Spectroscopy, *Anal. Chem.* 56, 786A–798A (1984), and (b) Sample Introduction Techniques for Atomic Spectroscopy, *Anal. Chem.* 56, 875A–888A (1984).
2. K. D. Summerhays, F. J. Lamothe, and T. L. Fries, Volatile Species in Inductively Coupled Plasma Atomic Emission Spectroscopy: Implications for Enhanced Sensitivity, *Appl. Spectrosc.* 37, 25–28 (1983).
3. V. M. Goldfard and H. V. Goldfarb, ICP–AES Analysis of Gases in Energy Technology and Influence of Molecular Additives on Argon ICP, *Spectroschim. Acta* 40B, 177–194 (1985).
4. W. B. Robbins and J. A. Caruso, Delvelopment of Hydride Generation Methods for Atomic Spectroscopic Analysis, *Anal. Chem.* 51, 889A–899A (1979).
5. T. Nakahara, Applications of Hydride Generation Techniques in Atomic Absorption, Atomic Fluorescence and Plasma Atomic Emission Spectroscpy, *Prog. Anal. At. Spectrosc.* 6, 163–223 (1983).
6. R. S. Braman, L. L. Justin, and C. C. Foreback Direct Volatilization Spectral Emission Type Detection System for Nanogram Amounts of As and Sb, *Anal. Chem.* 44, 2195–2199 (1972).
7. A. Miyazaki, A. Kimua, and Y. Umezaki, Determination of Submicrogram Amounts of As and Sb by dc Plasma Arc Emission Spectrometry, *Anal. Chem. Acta* 90, 119–125 (1977).
8. F. E. Lichte and R. K. Skogerboe, Emission Spectrometric Determination of As, *Anal. Chem.* 44, 1480–1482 (1972).
9. F. L. Fricke, W. B. Robbins, and J. A. Caruso, Determination of Ge, As, Sn, Se, and Sb by Plasma Emission Spectrometry with Hydride Generation and Chromatographic Separation, *J. Assoc. Off. Anal. Chem.* 61, 1118–1126 (1978).
10. W. B. Robbins, J. A. Caruso, and F. L. Fricke, Determination of Ge, As, Se, Sn, and Sb in Complex Samples by Hydride Generation Microwave Induced Plasma Atomic Emission Spectrometry, *Analyst* 104, 35–40 (1979).
11. W. B. Robbins and J. A. Caruso, Multielement Analysis of Selected Elemental Hydrides by Chromatographically Coupled Plasma Emission Spectrometry, *J. Chromatogr. Sci.* 17, 360–367 (1979).
12. M. Thompson, B. Pahlavanpour, S. J. Walton, and G. F. Kirkbright, Simultaneous Determination of Trace Concentrations of As, Sb, Bi, Se, and Te in Aqueous Solution by Introduction of the Gaseous Hydride into an Inductively Coupled Plasma Source for Emission Spectrometry, *Analyst* 103, 568–579 (1978).

13. R. K. Skogerboe, D. L. Dick, D. A. Pavlica, and F. E. Lichte, Injection of Samples into Flames and Plasmas by Production of Volatile Chlorides, *Anal. Chem.* 47, 568–570 (1975).

14. M. S. Black and R. D. Browner, Volatile Metal–Chelate Sample Introduction for Inductively Coupled Plasma Atomic Emission Spectrometry, *Anal. Chem.* 53, 249–253 (1981).

15. M. Thompson, B. Pahlavanpour, S. J. Walton, and G. F. Kirkbright, Simultaneous Determination of Trace Concentrations of As, Sb, Bi, Se, and Te in Aqueous Solution by Introduction of the Gaseous Hydrides into an Inductively Coupled Plasma Source for Emission Spectrometry, *Analyst* 103, 705–713 (1978).

16. M. Thompson and B. Pahlavanpour, Reduction of Tin and Germanium to Hydrides for Determination by Inductively Coupled Plasma Atomic Emission Spectrometry, *Anal. Chim. Acta* 109, 251–258 (1979).

17. B. Pahlavanpour, M. Thompson, and L. Thorne, Simultaneous Determination of Trace Concentrations of As, Sb. and Bi in Soils and Sediments by Volatile Hydride Generation and Inductively Coupled Plasma Emission Spectrometry, *Analyst* 105, 756–761 (1980).

18. B. Pahlavanpour, M. Thompson, and L. Thorne, Simultaneous Determination of Trace Amounts of As, Sb, and Bi in Herbage by Hydride Generation and Inductively Coupled Plasma Atomic Emission Spectrometry, *Analyst* 106, 467–471 (1981).

19. P. D. Goulden, D.H.J. Anthony, and K. D. Austen, Determination of As and Se in Water, Fish, and Sediments by Inductively Coupled Argon Plasma Emission Spectrometry, *Anal. Chem.* 53, 2027–2029 (1981).

20. M. Ikeda, J. Nishibe, S. Hamada, and R. Tujino, Determination of Pb at the ng/mL Level by Reduction to Plumbane and Measurement by Inductively Coupled Plasma Emission Spectrometry, *Anal. Chim. Acta* 125, 109–115 (1981).

21. D. D. Nygaard and J. H. Lowry, Sample Digestion Procedures for Simultaneous Determination of As, Sb, and Se by Inductively Coupled Argon Plasma Emission Spectrometry with Hydride Generation, *Anal. Chem.* 54, 803–807 (1982).

22. E. de Oliveira, J. W. McLaren, and S. S. Berman, Simultaneous Determination of As, Sb, and Se in Marine Samples by Inductively Coupled Plasma Atomic Emission Spectrometry, *Anal. Chem.* 55, 2047–2050 (1983).

23. K. A. Wolnik, F. L. Fricke, M. H. Hahn, and J. A. Caruso, Sample Introduction System for Simultaneous Determination of Volatile Elemental Hydrides and Other Elements in Foods by Inductively Coupled Argon Plasma Emission Spectrometry, *Anal. Chem.* 53, 1030–1035 (1981).

24. A. F. Ward, L. F. Marciello, L. Carrara, and V. J. Luciano, Simultaneous Determination of Major, Minor, and Trace Elements in Agricultural and Biological Samples by Inductively Coupled Argon Plasma Spectroscopy, *Spectrosc. Lett.* 3 (11), 803–831 (1980).

25. R. C. Hutton and B. Preston, A Simple Versatile Hydride Generation Configuration for Inductively Coupled Plasma, *Analyst* 108, 1409–1411 (1983).

26. D. S. Bushee, I. S. Krull, P. R. Demko, and S. B. Smith, Jr., Trace Analysis and Speciation for As Anions by HPLC–Hydride Generation Inductively Coupled Plasma Emission Spectroscopy, *J. Liq. chromatog.* 7(5), 861–876 (1984).

27. R. R. Liversage, J. C. Van Loon, and J. C. De Andrade, A Flow Injection/Hydride Generation System for the Determination of As by Inductively Coupled Plasma Atomic Emission Spectrometry, *Anal. Chim. Acta* 161, 275–283 (1984).

28. C. J. Pickford, Determination of As by Emission Spectrometry Using an Inductively Coupled Plasma Source and the Syringe Hydride Technique, *Anayst* 106, 464–466 (1981).

29. M. H. Hahn, K. A. Wolnik, F. L. Fricke, and J. A. Caruso, Hydride Generation/Condensation System with an Inductively Coupled Argon Plasma Polychromator for Determination of As, Bi, Ge, Sb, Se, and Sn in Foods, *Anal. Chem.* 54, 1048–1052 (1982).

30. M. A. Eckhoff, J. P. McCarthy, and J. A. Caruso, Sequential Slew Scanning Monochromator as a Plasma Emission Chromatographic Detector for Determination of Volatile Hydrides, *Anal. Chem.* 54, 165–168 (1982).

31. S. Stieg and A. Dennis, Detachable Hydride Introduction Device for Inductively Coupled Plasma Torch, *Anal. Chem.* 54, 605–607 (1982).

32. T. Nahahara, The Determination of Trace Amounts of Tin by Inductively Coupled Argon Plasma Atomic Emission Spectrometry with Volatile Hydride Method, *Appl. Spectrosc.* 37, 539–545 (1983).

33. T. Nakahara, Application of Hydride Generation to the Determination of Trace Concentrations of As by Inductively Coupled Plasma Atomic Emission Spectrometry, *Anal. Chim. Acta* 31, 73–82 (1981).

34. M. Bedard and J. D. Kerbyson, Determination of Trace Bi in Copper by Hydride Solution Atomic Absorption Spectrophotometry, *Anal. Chem.* 47, 1441–1444 (1975).

35. R. C. Fry, S. J. Northway, R. M. Brown, and S. K. Hughes, Atomic Fluorine Spectra in the Argon Inductively Coupled Plasma, *Anal. Chem.* 52, 1722 (1980).
36. R. M. Brown Jr. and R. C. Fry, Near-Infrared Atomic Oxygen Emissions in the Inductively Coupled Plasma and Oxygen-Selective Gas-Liquid Chromatography, *Anal. Chem.* 53, 538 (1981).
37. S. K. Hughes and R. C. Fry, Nonresonant Atomic Enmissions of Bromine and Chlorine in the Argon Inductively Coupled Plasma, *Anal. Chem.* 53, 1117 (1981).
38. S. J. Northway, R. M. Brown, and R. C. Fry, Atomic Nitrogen Spectra in the Argon Inductively Coupled Plasma (ICP), *Appl. Spectrosc.* 34, 338-348 (1980).
39. D. D. Nygaard, R. G. Schieecher, and D. A. Leighty, Determination of Halides by ICP Emission Spectrometry, *Am. Lab.* 17 (6), 59-62 (1985).
40. D. L. Windsor and M. Bonner Denton, Evaluation of Inductively Coupled Plasma Optical Emission Spectrometry as a Method for the Elemental Analysis of Organic Compounds, *Appl. Spectrosc.* 32, 366-371 (1978).
41. S. J. Northway and R. C. Fry, Atomic Oxygen Spectra in the Argon Inductively Coupled Plasma (ICP), *Appl. Spectrosc.* 34, 332-338 (1980).
42. J. W. Carnahan, K. J. Mulligan, and J. A. Caruso, Element-Selective Detection for Chromatography by Plasma Emission Spectrometry, *Anal. Chim. Acta* 130, 227-241 (1981).
43. A. J. McCormack, S. C. Tong, and W. D. Cooke, Sensitive Selective Gas Chromatography Detector Based on Emission Spectrometry of Organ Compounds, *Anal. Chem.* 37, 1470-1476 (1965).
44. I. S. Krull and S. J. Jordan, Interfacing GC and HPLC with Plasma Emission Spectroscopy, *Am. Lab.* 12 (10), 21-33 (1980).
45. M. L. Bruce and J. A. Caruso, The Laminar Flow Torch for Gas Chromatographic He Microwave Plasma Detection of Pyrethroids and Dioxins, *Appl. Spectrosc.* 39, 942-949 (1985).
46. D. L. Windsor and M. Bonner Denton, Elemental Analysis of Gas Chromatographic Effluents with an Inductively Coupled Plasma, *J. Chromatogr. Sci.* 17, 492-496 (1979).
47. D. L. Windsor and M. Bonner Denton, Empirical Formula Determination with an Inductively Coupled Plasma Gas Chromatogaphic Detector, *Anal. Chem.* 51, 1116-1119 (1979).
48. R. H. Scott, V. A. Fassel, R. N. Kniseley, and D. E. Nixon, Inductively Coupled Plasma-Optical Emission Analytical Spectrometry: A Compact Facility for Trace Analysis of Solutions, *Anal. Chem.* 46, 75-80 (1974).
49. R. M. Barnes and S. Nikdel, Mixing of Ambient Nitrogen in a Flowing Argon Inductively Coupled Plasma Discharg, *Appl. Spectrosc.* 29, 477-481 (1975).
50. R. M. Brown Jr., S. J. Northway, and R. C. Fry, Inductively Coupled Plasma as a Selective Gas Chromatographic Detector for Nitrogen-Containing Compounds, *Anal. Chem.* 53, 934-936 (1981).
51. K. Ohls and D. Sommer, A New Reduced-Pressure ICP Torch, in Developments in Atomic Plasma Spectrochemical Analysis, R. M. Barnes, ed. Heyden and Son, London, 1981, pp. 321-336.
52. C. J. Seliskar and D. K. Warner, a New Reduced-Pressure ICP Torch, *Appl. Spectrosc.* 39, 181-183 (1985).
53. D. C. Miller, C. J. Seliskar, and T. M. Davidson, Hydrogen Isotope Analysis Using a Reduced-Pressure ICP Torch, *Appl. Spectrosc.* 39, 13-19 (1985).
54. K. Tanabe, H. Haraguchi, and K. Fuwa, A Wavelength Table for Emission Lines of Nonmetallic Elements with Transition Assignments and Relative Intensities Observed in an Atmospheric-Pressure Helium Microwave-Induced Plasma, *Spectrochim. Acta* 36B, 119-127 (1981).
55. K. A. Wolnik, D. C. Miller, C. J. Seliskar, and F. L. Fricke, Characterization of Bromine and Chlorine Atomic Emission in a Reduced-Pressure Inductively Coupled (27 MHz) Helium Plasma, *Appl. Spectrosc.* 39, 930-935 (1985).

14

Low-Gas-Flow Torches
for ICP Spectrometry

LEO DE GALAN
AND
PIET S. C. VAN DER PLAS

Laboratorium voor Analytische Scheikunde
Technische Universiteit Delft,
Delft, The Netherlands

14.1 INTRODUCTION

The first inductively coupled plasma torches developed for use in the analytical laboratory used a high flow rate of atomic (Ar) or molecular (N_2) gases (25 to 40 L/min) and a consequently formidable power (15 kW). Their main advantage was, and is, that these extremely robust instruments could handle almost any type of sample. Apart from price considerations, the inconvenient logistics of gas supply and power lines prevented their widespread use in analytical laboratories.

A significant advantage of the three-tube torch developed by Wendt and Fassel[1] for an Ar ICP was that it could be operated at much lower power levels (≤ 2 kW), even though the gas requirement remained high (15 to 25 L/min). With minor modifications (eg, to allow the use of organic solvents), this design has persisted until today and has been adopted for virtually all commercial systems. The properties of the inductively coupled plasma discussed in most chapters of this book refer to data obtained with this type of torch operated under the cited conditions.

In the mid-1970s, several research groups began to consider the possibility of further reducing the rate of gas flow and the power level needed to sustain an ICP. Apart from the scientific curiosity to explore the limits of ICP operation, the driving force was cost-effectiveness. It was believed that a lower power RF generator would diminish the capital investment, whereas a smaller Ar flow rate would reduce the operating cost of ICP analysis. While these points are generally valid, their significance was not appreciated until the mid-1980s for three reasons. First, it is only with enhanced automation and the consequent reduction in labor that Ar consumption becomes a significant factor in the operating costs of an ICP. Second, we have yet to see a proliferation of the ICP into less affluent countries and laboratories, where greater importance is attached to the capital cost of the instrumentation and the operating cost of the ICP. And third, before low-gas-flow torches can be accepted, they must be developed to the point where their mechanical and analytical performances match those of commonly used ICP torches.

As will become clear from this chapter, the boundaries between currently available ICP torches and the so-called low-gas-flow torches are not sharp. Indeed, spurred by the development of the latter, some commercial torches are already operated with a total Ar flow as low as 12 L/min. While the ultimate goal of low-gas-flow torches is to reduce the power and consumption of argon as much as possible, certain lower limits must be maintained to preserve analytical performance. Within the context of this chapter we shall, therefore, define low-gas-flow ICP torches as torches for which either the total Ar flow rate is less than 10 L/min or the forward power is below 1 kW, or both.

14.2   CLASSIFICATION

Broadly speaking, two approaches can be distinguished toward the realization of low-gas-flow torches. In the first category, one or more of the physical dimensions of the three concentric quartz tubes are reduced to promote a more effective protection of the torch by the internal flow of Ar. In the second category, the torch design is altered to release more power through the torch wall, thereby relaxing the cooling requirements of the Ar flow. Before we proceed to a uniform discussion on the basis of power balances in Section 14.3, we briefly discuss these two approaches in general terms. Typical torch constructions are shown in Figure 14.1.

A reduction of the Ar flow can be accomplished by an overall reduction of the torch size, specifically the cross section of the torch. Thus, between 1977 and 1980 Allemand and co-workers[2] showed that ICPs could be sustained at less than 1 kW and with 12 L/min of Ar in torches having inner diameters (i.d.) of 13 mm, or even 9 mm, rather than the i.d. of 18 mm in commonly used torches. Efficient coupling of RF power was maintained by reducing the diameter of the load coil while keeping the frequency at 27 MHz. These experiments were successfully repeated by Savage and Hieftje,[3] who coined the somewhat exaggerated term "miniature torches" for the design.[4] Although plasmas undoubtedly can be operated under these conditions, their analytical potential was

Examples of low-power, low-gas-flow torches

miniature   high-efficiency   water-cooled   air-cooled   ceramic

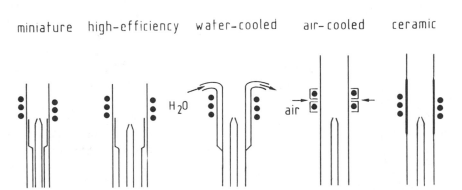

Figure 14.1. Schematics of the ICP torches designed for low power and low Ar consumption, figures are drawn to scale. Numbers denote internal diameter (mm) of outer tube of torch.

never convincingly demonstrated. The moderate performances may be attributed to the skin depth effect.[5] Thus, size reduction of the ICP torch might become feasible at higher frequencies, since the skin depth is inversely proportional to the square root of the frequency. With the recent interest in generators having frequencies of 40, 50, 80, and 100 MHz, the concept of smaller torches, therefore, may be revived.

In an extensive study of various torch designs, Allemand and Barnes[5] were the first to demonstrate the advantages of a tulip-shaped intermediate tube. The tulip design was subsequently perfected in a cooperative endeavor by Rezaaiyaan and Hieftje with Anderson, Kaiser, and Meddings.[6] Here, the overall size of the torch is maintained, but the separation between the intermediate and the outer quartz tubes is reduced to 0.5 mm. When forced through the narrow annular gap, a protective shield of Ar can be obtained with a flow rate of 5 L/min. Although some additional Ar is needed to introduce the sample and to sustain the plasma, a total consumption of 7 L/min is found to be adequate at a forward power of 0.5 kW. Aptly called "high efficiency" torches by Hieftje,[4] instruments of this design were studied in more depth by Rezaaiyaan and co-workers[6] and by Angleys and Mermet,[7] who also developed an empirical theory that is discussed later. Current commercial torches that are operated with reduced Ar flows are manufactured after this design, although the flow rates usually remain above 10 L/min.

Kornblum, van der Waa, and de Galan[8] generally are credited with pioneering a second category of low-gas-flow torches, where the outer silica tube is cooled externally. However, water-cooled ICP torches already had been described as early as 1959 by Kulalxov and others,[9] although with larger dimensions and flow rates. The underlying idea is that the cooling action of the internal Ar flow can be obviated when the silica tube is protected from melting by external cooling. Water is an obvious first choice,[8] and, indeed, Kawaguchi and co-workers[10] obtained acceptable performance with a total Ar flow of 4.8 L/min at a power level of 1.2 kW. Subsequently, Ripson and associates[11,12] claimed better results when the too-effective water cooling was replaced by a milder form of external cooling by air. In either case, the Ar flow was as low as 1 L/min, but the power was only 0.5 kW for air cooling rather than 1 kW for water cooling.

Protective cooling of the ICP torch by either internal Ar or external means is necessary only when silica is used as the outer tube. Such cooling mechanisms are no longer needed when the outer tube is made from a ceramic that can be heated to substantially higher temperatures, as first demonstrated by the present authors.[13] A plasma operated at 1 L/min Ar and 0.6 kW was generated in a uncooled torch having a segmented outer tube that contained a central piece of boron nitride or silicon nitride inside the load coil (Figure 14.1). Although the lifetime of the nitride piece is limited, the validity of earlier suggestions[14] for such a design was thus confirmed.

Now that the major approaches in the design of low-gas-flow ICP torches have been introduced, we proceed to formulate a detailed description to explain their differences, mutually and with respect to the conventional ICP torch.

14.3   POWER BALANCE OF THE INDUCTIVELY COUPLED PLASMA TORCH

14.3.1   Introduction

To emit spectral radiation, the sample solution introduced into the ICP must be volatilized, atomized, and excited to higher energy levels. This is effected by exposing the aspirated sample droplets to an environment of high temperature, ie, the hot plasma gas. In turn, the plasma derives its temperature from the intake of electromagnetic energy transferred from the load coil to the free electrons circulating in the annular plasma. Because the plasma stream represents a dynamic phenomenon, the process is best described in terms of the energy flux per unit time, ie, in terms of power (Figure 14.2).

The macroscopic properties of the ICP result from a dynamic equilibrium, whereby the power introduced into the plasma must also be released. If, for the present discussion, we ignore finer details of nonthermal equilibrium and temperature distribution, this equilibrium can be described in terms of extremely simple power balances. In the following treatment, we follow the more extensive discussion of Ripson and de Galan,[15] to which the reader is referred for quantitative details.

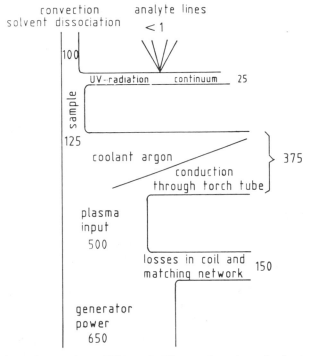

Figure 14.2. Flow of power in an ICP torch. The numbers (watts) refer to low-gas-flow torch designs. The fraction of input power not used for the sample (375 W) is released either with the outer Ar flow, as in high-efficiency and reduced-size torches, or through the wall of the torch tube, as in air-cooled and ceramic torches.

14.3.2   Input–Output Power Balance of the Inductively Coupled Plasma

As described in Chapter 4, the function of the RF generator is to convert the low-frequency (50 or 60 Hz) current drawn from the mains into a high-frequency (eg, 27 MHz) current flowing through a load coil placed around the ICP torch. Most generators have a meter that indicates the power transferred to the load coil, and it is this *generator power*, P_{gen}, or *forward power*, which is usually quoted in the ICP literature. However, the processes in the plasma are the result of the fraction of generator power actually dissipated in the plasma. This *plasma input power*, P_{pl} can be formulated as:

$$P_{pl} = P_{gen} - P_{loss} \simeq 0.75 \, P_{gen} \qquad [14.1]$$

where P_{loss} is a collective term to describe the power lost in the transfer process. Such losses include back-reflection in some generators, heating of the cooling water flowing through the load coil, and heat losses in the impedance matching network. Whereas in modern generators the reflected power can be made negligible, heat losses are approximately 10% in the load coil and in the match box. Consequently, the efficiency of power transfer from the RF generator to the plasma is about 75%, as indicated in Equation 14.1. Eventually, the power supplied to the plasma must be released to the environment. Since the power is supplied through the load coil, the release processes should be confined to the region inside the coil. The quantitative figures quoted below, therefore, do not refer to the zone used for analytical observations above the load coil, but to the luminous, hot space inside the coil. In this region, the power balance is governed by Equation 14.2, which incorporates four release processes as defined below.

$$P_{pl} = P_{uv} + P_{sample} + P_{Ar} + P_{wall} \qquad [14.2]$$

1. P_{uv} describes the power emitted as spectral radiation, which includes continuum background, molecular radiation, Ar lines, and the analytically important atomic and ionic spectral lines used for quantitative analysis.
2. P_{sample} includes the power needed to heat the sample, with the solvent and the injector gas used to introduce the droplets, to its final temperature of about 5000 K.
3. P_{Ar}, a term similar to P_{sample}, describes the power needed to heat any additional gas to its final temperature. Because the temperature distribution falls from a high central value to a low value close to the torch wall, calculation of an average value is difficult, but the temperature is estimated to be between 3000 and 4000 K. Both P_{sample} and P_{Ar} are defined as *convective* power release.
4. P_{wall} is the power transferred by the hot plasma to the outer tube of the torch. This process is called *conductive* power release.

The first two terms on the right-hand side of the equation represent processes that are intimately connected to the sample and, therefore, are essential to the analytical utilization of the ICP. The last two terms, however, are the unavoidable consequence of enclosing the plasma by a torch. Minimization of power consumption requires that these latter terms be significantly reduced.

14.3.3 Minimum Power Requirements

The total power emitted by the hot plasma as spectral radiation through the torch wall, P_{uv}, is about 25 W, irrespective of the torch design.[15] Incidentally, the major part of the radiation is emitted as continuum in the visible region of the spectrum, and only about 1 W is emitted as potentially useful spectral lines.

The convective release of power by the constituents of the sample flow can be described as:

$$P_{sample} = \sum F_i[H_i + C_i\Delta T] \qquad [14.3]$$

where the summation is over all constituents i that enter the plasma with a flow rate F, H is the enthalpy of a phase transition, C is the heat capacity, and ΔT is the temperature increase. The dominant terms in this expression are the heating of the injector gas to the plasma temperature, the dissociation of the solvent, and, perhaps, the dissociation and ionization of the matrix when dissolved to concentrations greater than 10%.

Indeed, with a solution uptake rate of 3 mL/min and a 1% efficiency of the spray chamber, the flow rate of water droplets into the ICP is about 0.5 mg/s. Under these conditions, assuming a dissociation enthalpy of 50 J/mg, ordinary aqueous solutions require about 25 W to dissociate. The consecutive heating of the generated solvent atoms to the plasma temperature is negligible. Relatedly, a NaCl content as high as 10% contributes only a few watts to convective power release. Much more power is needed to dissociate organic solvents because they may be introduced at higher efficiency (eg, 2 mg/s) and have a high dissociation enthalpy. Common solvents such as xylene and MIBK may consume up to 200 W. This calculation explains the well-known observation that for organic solvents, the ICP requires a higher forward power than for aqueous solutions. In practice, another 200 W is used to prevent temporary impedance mismatch and to reduce the plasma sensitivity to sample interchanges. On the other hand, hydride generation does not, in itself, require more power. Even when a $0.5M$ $NaBH_4$ solution is used with a flow rate of 4 mL/min, the dissociation of the evolved hydrogen requires only 25 W, which is similar to the power needed for an aqueous solution. However, some excess power must again be applied to create robustness against sample exchange. In all cases, an injector gas is used to transport the sample to the discharge region. From the heat capacity of a monatomic gas, it can be calculated that 75 W is needed to raise the temperature of Ar introduced at a flow rate of 1 L/min to a final temperature of 5000 K.

We conclude, therefore, that the minimum power ($P_{uv} + P_{sample}$) is 125 W for aqueous solutions. For enhanced plasma stability, hydride generation may need another 100 W, whereas organic solvents require close to 400 W. Expressed as forward power, the minimum requirements range from 200 W for aqueous solutions to 500 W for organic solvents. For the conventional ICP, these minimum powers constitute no more than 20 to 30% of the actual generator powers normally applied. The remaining 70 to 80% of power consumption is associated with the last two terms in Equation 14.2, which we now consider more closely.

14.3.4   Convective Power Release

In the commonly used ICP, the injector gas flow, typically 1 L/min, is greatly exceeded by the outer flow in the three-tube torch. In terms of power demand, the clear advantage of a monatomic gas such as argon over molecular gases follows not just from its lower heat capacity [$(5/2)R$ vs $(7/2)R$]. Much more important is the fact that at high temperatures, most diatomic gases are significantly dissociated and hence require more power expressed by the

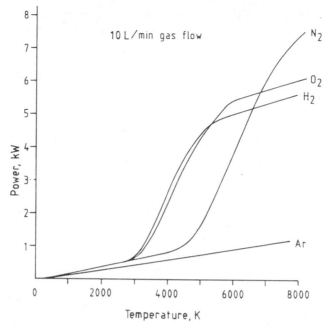

Figure 14.3. Convective power release as a function of the gas kinetic temperature for a monatomic and three diatomic gases in the outer flow. Equation 14.3 was used for calculations. The dissociation energies (eV) are: $H_2 = 4.476$, $O_2 = 5.080$, and $N_2 = 7.373$.

term H_i in Equation 14.3. As indicated in Figure 14.3, both H_2 and O_2 are partly dissociated at temperatures above 3000 K, whereas N_2, through its higher dissociation energy, starts to dissociate above 4500 K. However, complete dissociation of N_2 takes significantly more energy than H_2 or O_2. Complete dissociation of each gas, when introduced at a rate of 10 L/min, requires no less than 7.5 kW for N_2, 6 kW for O_2, and 5 kW for H_2. Obviously, at the average kinetic temperature of the plasma gas, these molecular gases are only partly dissociated, but nonetheless they require much more energy to heat than Ar. This conclusion may explain the observation that low-power N_2 and air ICPs emit predominantly atom lines of moderate excitation energies.[23-26] Analytically useful ICPs using molecular gases generally require 4 to 6 kW of generator power.[24]

When, instead, Ar is used at a similarly high flow rate, dissociation does not enter in the discussion, and according to Equation 14.3 only 155 W is needed for every 1000 K increase in average temperature. Although the average kinetic temperature of the outer Ar flow in the ICP is uncertain, it is doubtful that it can be kept below 3500 K if a high temperature is to be maintained in the central channel of the plasma. Consequently, in a conventional ICP, operated at 15 L/min, P_{Ar} in Equation 14.2 is at least $1.5 \times 3.5 \times 155 = 700$ W. Combined with 150 W for aqueous solutions and assuming a 75% coupling efficiency, this amounts to a total of 1.1 kW generator power. This is indeed a typical operating power for current ICP torches. For organic solvents, shown previously to require an additional 400 W forward power, the figure is raised to 1.5 kW.

Apparently, the easiest way to reduce the convective power release is to reduce the outer Ar flow rate, which is precisely what is done in low-gas-flow torches. In torches with reduced physical dimensions, the outer gas flow can be as low as 5 L/min. Since the gas flow is confined to a layer close to the outer torch tube, the average temperature of the gas may be as low as 2500 K. Commonly, the plasma is sustained in such high-efficiency torches through the use of an intermediate gas flow of 1 L/min. Presumably the intermediate gas is also heated to the same high temperature as the injector gas. The total convective power release described by P_{Ar} in Equation 14.2 is then calculated to be about 300 W, so that for aqueous solutions, such torches can be operated at 600 W generator power.[6]

14.3.5 Conductive Power Release

If the outer Ar gas flow is reduced even further to 1 L/min, as used in externally cooled torches, only 100 W is needed to raise its temperature to 5000 K. However, because of the heat conduction from the plasma through the wall of the outer tube (P_{wall}, Equation 14.2), the total input power to the plasma cannot be decreased to the same extent. When the total Ar flow in the torch is gradually diminished, the hot plasma approaches the wall more closely and the protective boundary layer is broken down. The outer torch tube becomes hot due to power conducted from the plasma to the wall.

Without detailed knowledge of the flow pattern in the ICP, the transfer of power from the hot plasma to the cooler wall cannot be easily calculated. Fortunately, all power released to the wall, P_{wall}, must also be conducted through the wall, P_{cond}, and released to the environment, $P_{release}$. The latter two processes are readily amenable to quantification through the following expressions:

$$P_{wall} = P_{cond} = \frac{A\lambda T_w}{d} \qquad [14.4]$$

$$P_{release} = A\alpha(T_w - 293) + A\varepsilon\sigma T_w^4 \qquad [14.5]$$

where ΔT_w is the temperature drop across a wall having thickness d, leading to an *outside* wall temperature T_w over an area A. The proportionality constants are the thermal conductivity λ, the coefficient of heat transfer, α, the emissivity of the torch wall, ε, and the Stephan–Boltzmann constant σ. Whereas Equation 14.4 describes the conduction through the wall, Equation 14.5 formulates the release of power to the surroundings through convection and heat radiation, respectively.

The key factors in this expression are the outside wall temperature T_w and the transfer coefficient α. The latter quantity significantly depends on the torch configuration and the environment. For the commonly used ICP torch standing free in air, the outer tube releases little power to the surrounding air as long as the outside wall temperature is low ($T_w < 500$ K). The small power loss is attributable to the low transfer coefficient for free convection to air: $\alpha_{free} = 15$ W/m^2 K. Relatedly, the conduction through the wall also is small; hence the power transfer from the plasma to the wall (P_{wall} in Equation 14.2) is negligible. Such a situation is quite typical in conventional and reduced-size torches, where the outer tube is cooled by a protective, internal sheath of argon.

If we allow the wall temperature to rise by restricting the internal Ar flow to 1 L/min, the power release, expressed by Equation 14.5, increases. However, when silica is the tube material, its temperature must be limited to about 1000 K to prevent melting. As shown in Figure 14.4, at this temperature free convection to air is only 10 W, while the radiative power release is about 60 W. Even for a relatively poor conductor like silica, the conduction of 70 W through a wall 1 mm thick creates a temperature drop of only 35 K. Unfortunately, no practical realization of such a low-gas-flow, low-power torch has been reported. Apparently, when the flow of outer Ar is greatly reduced, much more than 70 W is transferred to the silica wall, which will cause rapid overheating of the torch. Thus, protective external cooling is required.

An apparently attractive way to protect the wall is to cool the torch with a water jacket. Indeed, the transfer coefficient from silica to water is very high: $\alpha = 10^4$ W/m^2 K. As a result, a water jacket around the torch[8] effectively keeps

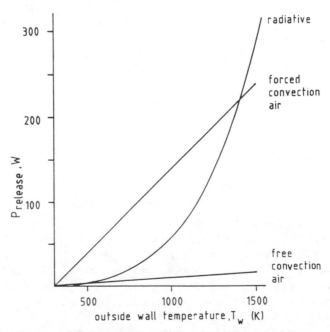

Figure 14.4. Power release from the outer tube of the torch to the environment as a function of the temperature of the wall of the outer tube (Equation 14.5).

the wall at ambient temperature, even when the internal Ar flow is reduced to 1 L/min. In fact, water drains away power so effectively that a substantial power input is necessary to maintain a high plasma temperature. Measurements indicate[15] that a water-cooled, low-gas-flow torch requires at least 1 kW of plasma input, about 800 W of which is released through the wall and to the water jacket. Even this large power transfer produces a quite acceptable temperature drop of 400 K across a silica wall 1 mm thick (Equation 14.4, with $\lambda = 2$ W/m K). Consequently, water-cooled torches are efficient in Ar consumption, but not in power requirements, since the equivalent forward power is about 1.3 kW, which is similar to the commonly used ICP torches.

Clearly, a transfer coefficient is needed that is between the low value of 15 W/m^2 K for free convection to air and the high value of 10^4 W/m^2 K for conduction to water. When a high flow of air is forcibly blown against the outside of the torch,[11,12] the transfer coefficient (α_{forced} in Equation 14.5) can reach 200 W/m^2 K, depending on the cooling design and the flow pattern of the air.[15] Again, if the outside temperature of the silica wall is restricted to 1000 K, the total power release (Figure 14.4) increases to 200 W, including 60 W radiative release. The summation of heat losses (Equation 14.2), including the contributions of UV radiation (25 W), aqueous sample solutions (125 W), and conductive transfer through the wall (200 W), gives a total forward power of 450 W, assuming a coupling efficiency of 75%. The practical value is somewhat larger, as seen in Section 14.4.

Another possibility to enhance the power release through the wall, as seen from Figure 14.4 and Equation 14.5, is to allow the temperature of the outer torch tube to rise above 1000 K. While the convective power transfer increases only linearly with T_w, the radiative release rises sharply with T_w^4. Indeed, when T_w is 1500 K, radiative release[13] alone accounts for 200 W when the emissivity ε is 1. At this temperature, the outer tube of the torch can no longer be made of silica, but other ceramics might be feasible, as demonstrated in the case of a segmented torch using boron nitride[13] (Figure 14.1). Not surprisingly from the foregoing considerations, the performance of such a ceramic torch operated at 1 L/min of Ar is quite similar to that of an air-cooled torch.

14.3.6 Conclusions

The power balance expression in Equation 14.2 provides a uniform basis for describing the power balance for various torch designs reported in the ICP literature. Whereas the minimum power required for sample conversion is only 125 W for aqueous solutions, all torch designs require additional power for stable operation. In commonly used torches, a relatively large excess of power is consumed in heating the outer flow of gas used to protect the silica tube. Clearly, in terms of power efficiency, Ar is the preferred outer gas, because no power is needed for dissociation as is the case for molecular gases, such as N_2. The efficiency of internal cooling can be increased by careful design of the tube spacings, thereby reducing convective release by the heated Ar in proportion to the lower gas flow rate.

In contrast, when the total internal Ar flow is reduced to 1 L/min, power conduction to and through the outer torch tube cannot be prevented. Whereas convective power transfer by the heated Ar now becomes very small, conductive release is significantly enhanced, so much so that a torch can be operated only when silica is replaced by a ceramic that can withstand temperatures up to 2000 K. Alternatively, a silica outer tube may be used, provided the torch is protected by an outside cooling device. As such, water is too effective for optimum power efficiency, but a forced air flow provides an excellent compromise.

14.4 PRACTICAL REALIZATIONS OF LOW-GAS-FLOW TORCHES

14.4.1 Reduced-Size Torches

The dimensions and the main operating characteristics of reduced-size torches are listed in Table 14.1. In all cases, the classical arrangement of three concentric tubes is maintained, and, at least for aqueous solutions, only two gas flows are used. Because the injector gas flow is kept close to its conventional value of 1 L/min, common pneumatic nebulizers can be used. The total power is significantly reduced, as a consequence of the reduced outer flow rate. This reduction results not only from the smaller torch cross section, but also from the more efficient flow pattern realized by a tangential inlet of the outer gas. Good coupling efficiency for the RF power is maintained by reducing the inside diameter of the load coil in proportion to the outside diameter of the torch.

The gains in lower gas and power consumptions for reduced-size torches are modest in comparison with the commonly used ICP torch of 18 mm i.d. Relatedly, the detection limits and analytical characteristics of the ICPs

Table 14.1. Dimensions and Operating Conditions of Reduced-Size Torches

Descriptive data	Weiss[17]	Allemand[2]	Allemand[2]	Savage[3]
Torch inner diameter (mm)	9	9	13	13
Generator frequency (MHz)	27	27	27	27
Forward power (kW)	0.5	0.7–1.0	0.85	1.0
Argon flow rates (L/min)				
Outer	6.4	8.5	12	7.9
Intermediate	0	0	0	0
Injector	0.75	0.75	0.76	1.0
Observation height (mm)	15	13	13	15
Sample uptake rate (mL/min)	1.2	1	1	0.36

supported by reduced-size torches[2,3,17,18] only approach those of a conventional ICP. Thus, after the initial reports, the interest in these torches appears to have waned in favor of the high-efficiency torches discussed below. The limited acceptance of reduced-size torches may be attributed to the use of the 27-MHz generator. For this frequency, a minimum plasma diameter, hence torch inner diameter, of 13 mm is necessary.[4] Should this be related to the skin conduction effect, better results might be obtained at higher frequencies, as the skin conduction depth decreases twofold when the frequency is raised from 27 to 100 MHz. In recent years, there has been a tendency to use higher frequencies in ICP instrumentation, and therefore, investigations of smaller torches may receive renewed interest in the future. Indeed, a commercial version of the reduced-size torch has recently been introduced by Applied Research Laboratories.

14.4.2   High-Efficiency Torches

Allemand and Barnes[19,20] appear to have been the first to realize that a reduction in outer gas similar to that obtained in reduced-size torches can be implemented when only the separation between the intermediate and outer torch tubes is reduced. Whereas in commonly used torches this distance is not critical and is typically 2 mm, Meddings and co-workers[21,22] utilized a separation of only 0.5 mm, a value adopted in subsequent publications.[6,7,27,28,29] From a practical point of view, this small constriction puts additional demands on the construction of the torch because the concentricity of the outer and the intermediate tubes becomes critically important.

As is clear from the data in Table 14.2, the small annular spacing between the tubes significantly enhances the efficiency of the internal gas cooling. For this reason, Hieftje coined the name "high-efficiency torches" for this design.[4] The operating principles of the high-efficiency torch are partly based on small modifications on the argon inlets in the torch, ie, tangential introduction of the outer gas and a constriction in the inlet of the gas, which enhance the linear and swirl velocities, respectively. More important, however, is the increase in the linear velocity of the outer gas, due to the small annular spacing, which lets the outer gas act as a heat shield between the plasma and the torch. Significantly, in operating the torch a third gas flow between the two central tubes is generally used. A plausible explanation for this need is the following.

In a conventional ICP torch the large, diverging flow of argon serves two functions. On the one hand it protects the torch wall from softening, but on the other hand it sustains the plasma discharge. In the high-efficiency torches these two functions seem to have become separated. A smaller flow of argon is introduced through the narrow annular slot between the two outer tubes in such a way that it remains close to the outer tube and protects it from melting. Because the plasma cannot be sustained only on the cool injector gas introduced centrally, a third, intermediate argon flow is needed, which is appropriately called the plasma gas.

Table 14.2. Dimensions and Operating Conditions of High-Efficiency Torches

Descriptive data	Investigator					
	Allemand[20]	Rezaaiyaan[6]	Mermet[7]	Montaser[27]	Rezaaiyaan[28]	Mermet[29]
Torch inner diameter (mm)	18	18	16	18	18	16–19
Annular spacing (mm)	0.9	0.5	0.5	0.5	0.5	0.5
Generator frequency (MHz)	27	27	40	27	40	27
Forward power (kW)	0.9	0.45	0.6	0.72	0.45	0.6
Argon flow rates (L/min)						
Outer	9	5	6.4	4	6	6.5
Intermediate	0	0.5	0	0.5	0.7	0.8
Injector	0.84	0.5	0.4	0.5	0.64	0.8
Observation height (mm)	15	10	—	11	12	10
Sample uptake rate (mL/min)	2.15	1.0	—	0.5	1.0	—

According to Table 14.2, the high-efficiency torch achieves the main purposes of reducing both the outer argon flow and the RF power. Actually, the two quantities are related, through the so-called plasma stability curve,[6] an example of which is shown in Figure 14.5. The curve separates the power–outer flow domain in two areas. Above the curve we find the area of stable plasma operation. When, starting from a point in this region, the power or the outer flow is diminished, either the plasma is extinguished or the torch starts to glow and melt. The shape of the curve describing the transition between stable plasma operation and plasma breakdown is rather suprising. However, a similar shape has been observed for a conventional ICP in the authors' laboratory.

Instructive as the stability curve may be, it tells little about the analytical merits of the system. For example, although it is certainly possible to sustain an ICP at low power and high outer flow, the resulting plasma is too cool to be analytically useful. To this end, an alternative description formulated by Mermet and co-workers[7,30] may be more instructive. These authors studied the relation between the outer gas flow and the critical separation between the outer and the intermediate tubes of the torch.

As might be expected, the minimum coolant flow rate for safe operation of an ICP torch increases with the separation between the outer and the intermediate tubes. Conversely, the outer flow can be reduced by minimizing this separation. However, at some point no further reduction of the flow is possible and an absolute minimum has been reached. These observations were expressed by Angleys and Mermet[7] in two empirical boundary conditions for the outer flow rate:

$$F_P > S_P \times V_C \qquad\qquad [14.6]$$

$$F_P > S_T \times V_C \times \frac{293}{T} \qquad\qquad [14.7]$$

where F_P is the outer flow, S_P is the area of the annular spacing, S_T is the torch cross-sectional area, and T is the kinetic gas temperature. The common proportionality constant V_C, called the critical speed of cold Ar, was found to be 3.3 m/s for a forward power of 1.3 kW, at a frequency of 40 MHz. The minimum outer flow rate attainable may be calculated[7] by equating Equations 14.6 and 14.7:

$$S_P \times V_C = S_T \times V_C \times \frac{293}{T} \qquad\qquad [14.8]$$

This can be rearranged to:

$$e = D_i \times 0.5 \left[1 - \left(1 - \frac{293}{T}\right)^{1/2} \right] \qquad\qquad [14.9]$$

where e is the annular spacing between outer and intermediate tubes and D_i is the inner diameter of the torch. If a temperature of 4000 K is used in Equation 14.9, then

$$e = 0.02 \times D_i \qquad\qquad [14.10]$$

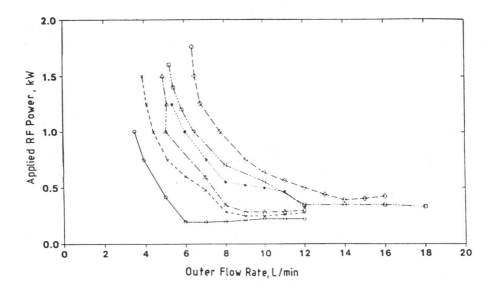

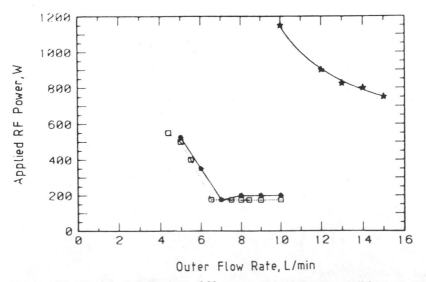

Figure 14.5. Plasma stability curves[6,35] denote the minimum conditions to sustain a plasma for various separations between the outer and the intermediate tubes of a high-efficiency torch. Top: Stability curves for an ICP employing a high-efficiency torch. Solvents are water (○), carbontetrachloride (×), chloroform (△), toluene (*), acetic acid (□), and ethanol (◇).[35] Bottom: Stability curve illustrating the effect of the annular spacing between the intermediate tube and the outer tube. (*) 1 mm spacing, (★) 0.7 mm spacing, (□) 0.5 mm spacing.[6] (From References 6 and 35, with permission.)

Thus, for a torch with an inner diameter of 20 mm, the minimum outer gas flow is reached at an annular spacing of 0.4 mm. The minimum outer gas flow is given by:

$$F_P = D_i^2 \times 1.1 \times 10^{-2} \qquad [14.11]$$

where D_i is given in millimeters and F_p in liters per minute. For a torch diameter of 20 mm, the minimum outer flow is 4.4 L/min, in agreement with experimentally observed values. When the injector and the intermediate gas flows are added to the minimum outer flow rate, the total Ar consumption of a high-efficiency torch is 5.5 L/min. As shown in Table 14.2, the actual values vary between 6 and 9 L/min. These values refer to an RF frequency of 40 MHz. Possibly, the critical speed V_C decreases with increasing RF frequency.[31] If true, it would indicate that high-efficiency torches could be operated at lower flow rates or under less critical tube separations when the standard RF frequency of 27 MHz is replaced by 50 or even 100 MHz. We recall that a similar suggestion has been made for reduced-size torches.

The forward powers reported in Table 14.2 are the typical operating values given in the literature for aqueous solutions. Understandably, these tend to be minimum values. Again, it should be emphasized that low power does not necessarily imply favorable analytical conditions. Acceptable analytical performance, especially in the use of organic solvents, requires the use of higher power levels than stated in Table 14.2.

The clear advantage of the high-efficiency torches is the minimal change in construction of the torch and of the load coil and impedance matching network as compared to reduced-size torches. Indeed, compared to conventional torches, the only alteration is the spacing between the intermediate and outer tubes. Thus, it is understandable that this concept has been incorporated in some commercial torches, although the reduction in gas and power requirements are perhaps not as great as would be desired.

14.4.3 Water-Cooled Torches

The principal characteristics of water-cooled torches[8,10,15,32] are described in Table 14.3. In all designs, the water flow rate is not critical, but it should exceed a minimum value to avoid bubble formation in the jacket surrounding the ICP torch. The jacket, which is part of the torch structure, must be positioned inside the load coil, thus making the distance between the coil and the plasma discharge greater than that for gas-cooled torches. Some loss in coupling efficiency, which is expected for this arrangement, may partly explain the high RF power used with water-cooled torches.

Because of the simplicity of the water jacket in Kawaguchi's design (Figure 14.1), this torch was preferred over the earlier design of Kornblum and associates[8] and by Ripson and co-workers.[12] It is to be noted that Kawaguchi and colleagues[10,32] maintained the conventional injector gas flow rate and

Table 14.3. Dimensions and Operating Conditions of Water-Cooled Torches

	Investigator			
Descriptive data	Kornblum[8]	Ripson[15]	Kawaguchi[10]	Kawaguchi[32]
Torch inner diameter (mm)	18	15	17	17
Number of torch tubes	3	2	3	3
Generator frequency (MHz)	50	50	27	27
Forward power (kW)	0.7	1.1	1.1	1.2
Argon flow rates (L/min)				
Outer	0.9	1.0	4	4
Intermediate	0.8	0	0	0
Injector	0.1	0.17	0.8	1.0
Observation height (mm)	4	10	12	12
Sample uptake rate (mL/min)	—	2	1.5	—

could, therefore, apply conventional pneumatic nebulizers. Their outer gas flow, although substantially smaller than in conventional ICP torches, is of the same order as realized in high-efficiency torches.

In contrast, Kornblum et al.,[8] and later Ripson et al.,[12] used much lower outer gas flows. Whereas Kawaguchi[32] maintained a conventional three-tube design, the torch described by Ripson[12] contained only two tubes: the sample injection tube and a second tube surrounded by a water jacket. Two other modifications were necessary to use an outer gas flow rate of only 1 L/min.[12] First, the outer discharge tube had to be extended to prevent air entrainment (Section 14.5.2). Second, the injector gas flow had to be reduced to 0.2 L/min. Consequently, cross-flow or concentric nebulizers could no longer be used, and a V-groove Babington nebulizer (Figure 14.6) with the gas injection bore reduced to 0.1 mm was used.[11] Whereas cross-flow and concentric nebulizers stop functioning at injector-gas flows below 0.5 L/min, ultrasonic, glass frit, and V-groove nebulizers can be used down to 0.1 L/min. As far as practical convenience and analytical performance are concerned, the V-groove nebulizer is favored[29] for low-gas-flow torches.

The water-cooled torch described by Kawaguchi[10,32] resembles the conventional torch in terms of power consumption and analytical performance. Compared to the high-efficiency torches, the Kawaguchi design[32] requires lower gas flows, but higher forward power. In contrast, the water-cooled torch by Ripson[12] is quite similar to the air-cooled torch described in the next section.

14.4.4 Air-Cooled Torches

The argon flows used in air-cooled and ceramic torches (Table 14.4) are virtually equal to those in the water-cooled torch designed by Ripson and co-workers[12] (Table 14.3). Indeed, the plasmas look the same. When the outer Ar flow in an ICP is gradually reduced from its conventional value of 15 L/min to values of 5

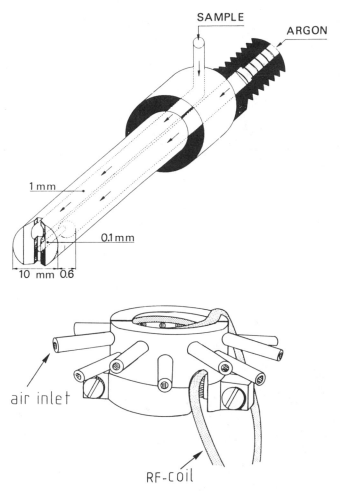

Figure 14.6. Top: V-groove nebulizer. Bottom: Cooling jacket around the load coil used with an air-cooled torch.[33] (From Reference 33, with permission.)

to 7 L/min used in high-efficiency torches, the plasma inside the torch hardly changes its appearance. The characteristic tailflame is maintained, and entrainment of ambient air is effectively prevented. This is no longer true if the Ar flow is further reduced. The outer torch tube must be extended to prevent the plasma from being extinguished by entrainment of air. Nevertheless, the plasma becomes unstable, and its shape changes rather abruptly from the classical tailflame to a spherical ball contracted within the coil region (Figure 14.7). Below a flow rate of 2 L/min, this configuration again provides a stable plasma. The outer gas flow rate of 1 L/min, for Ripson's torch in Table 14.3 and for the torches in Table 14.4, therefore, is not arbitrarily imposed but is actually the optimal flow. As discussed earlier, the injector gas flow is reduced simultaneously to about 0.2 L/min, thus requiring the use of a V-groove nebulizer

Table 14.4. Dimensions and Operating Conditions of Air-Cooled[a] and Ceramic[b] Torches

Descriptive data	Investigator				
	Ripson[11]	Ripson[12]	Ripson[12]	Van der Plas[33]	Van der Plas[13]
Torch inner diameter (mm)	13.5	13.5	16	16	16
Number of torch tubes	2	2	2	2	2
Generator frequency (MHz)	50	50	50	40	40
Forward power (kW)	—	0.5	0.45	0.5–0.9	0.6
Argon flow rates (L/min)					
Outer	0.75	0.5	0.8	0.7–1.0	0.8
Intermediate	0	0	0	0	0
Injector	0.15	0.14	0.12	0.15	0.2
Air-coolant flow (L/min)	50	50	50	50	0
Observation height (mm)	10	12	11	2.5	5
Sample uptake rate (mL/min)	1.5	2	2	1	2

[a] References 11, 12, and 33.
[b] Reference 13.

 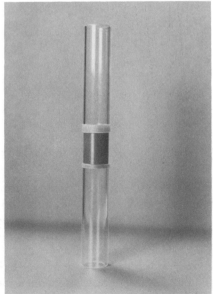

Figure 14.7. Left: An ICP sustained on a total Ar flow of 1 L/min in an air-cooled torch.[33] Right: The outer tube of a ceramic torch with a central piece of boron nitride.[13] (From References 13 and 33, with permission.)

(Figure 14.6). A further adaptation of the torch is a constriction of the orifice of the injector gas tube tip from its usual 1.0 mm to 0.5 mm. The fourfold reduction in cross section serves to achieve the same linear cold gas velocity, 13 m/s, in the air-cooled torch as in a conventional torch.

To date, all experiments with spherical plasmas sustained on 1 L/min total Ar have been performed at frequencies of 40 and 50 MHz. Whereas Ripson and co-workers[11,12] used an uncommon two-turn, flat-plated coil, later results by van der Plas and de Galan were obtained with a standard helical coil[33] shown in Figures 14.6 and 14.7. Improvements in the design for an air cooling system increased the acceptable upper limit for the power from 0.5 kW for the original design[11,12] to 1 kW.[33]

Comparison of the data in Tables 14.1 through 14.4 shows that for the combined reduction in generator power and Ar consumption, the air-cooled torch offers the best prospects. Although the torch may consist of either two or three concentric tubes, their positioning is not too critical, but ancillary equipment is needed for air cooling. The air flow of 50 L/min (Table 14.4), can be easily provided by a modest compressor or a laboratory supply system.

14.4.5 Ceramic Torches

As explained in Section 14.3.5, external cooling can be avoided, even at very low Ar flows, when the outer torch tube is made of a high-temperature ceramic. In a preliminary feasibility study,[13] we concluded that the material selected must meet the following stringent requirements:

1. The material must be machinable and heatable to 2000 K. This requirement narrowed the choices to alumina, silicon carbide, boron nitride, and silicon nitride.
2. It must have a high thermal shock resistance to withstand the sudden increase in temperature. Alumina fails to meet this requirement.
3. It must be electrically nonconductive, to ensure transparency to RF fields. Silicon carbide fails to meet this requirement at high temperature.
4. It must be chemically inert. Boron nitride and silicon nitride slowly oxidize at high temperature, especially when in contact with water vapor.

A torch made of boron nitride or silicon nitride was found to be usable for only a few weeks.[13,33] Because both materials are opaque, a segmented outer tube was constructed to contain a central piece of boron nitride inside the load coil, fitted with silica tubes at either end (Figure 14.7). In operation, the central part acquired an outside temperature of 1600 K, but the silica parts remained sufficiently cool. As can be seen in Table 14.4, the operating conditions for a ceramic torch are quite similar to the air-cooled ICP.

Clearly, the development of ceramic torches must await the availability of a material with a longer lifetime than those presently investigated. In view of the rapid progress in high-temperature ceramics, such a development is not at all improbable. The use of a ceramic torch would result in a plasma that operates and performs similarly to the air-cooled ICP, but with the obvious advantage of not requiring external cooling of the torch.

14.4.6 Conclusions

The torches described in the preceding sections meet to varying degrees the goal of reduced consumption of argon and of power. *Reduced-size torches*, with reduced overall dimensions, appear to have been superseded by *high-efficiency torches*. In either design, the total Ar flow can be reduced to 7 L/min and the generator power to 0.5 kW. The plasma retains its conventional shape and can be observed over the tip of the outer discharge tube. However, torch design is critical. *Externally cooled torches* can operate at much lower Ar flow rates, down to 1 L/min. In terms of power consumption, the air-cooled version (0.5 kW) is preferable to the water-cooled design (1.2 kW). The future of ceramic torches is yet to be decided. The low Ar flow requires plasma observation through an extended outer tube, but the torches are easily assembled to fit inside commonly used load coils.

14.5 PHYSICAL PROPERTIES AND ANALYTICAL PERFORMANCE

14.5.1 Plasma Diagnostics

The two physical quantities used to characterize an ICP are the electron number density, n_e, and the temperature. In view of the nonthermal nature of the discharge, the latter quantity is somewhat ambiguous. It can be argued, however, that a fair indication of the analytical performance can be obtained from the Boltzmann excitation temperature derived from two or more spectral lines of a single ionization stage of an element, such as Fe I or Ti II. Some representative temperatures and n_e values for low-gas-flow plasmas are listed in Table 14.5. These values should be interpreted with caution for two reasons. First, the values are intensity-weighted averages for a single observation height. From the abundant studies of the commonly used ICP (Chapter 8), the temperature and the n_e values are known to vary significantly across and along the discharge. However, detailed studies incorporating spatial profiles have not been reported for low-gas-flow plasmas. Second, the values listed refer to a single set of operating conditions for each plasma, usually those stated in Tables 14.1 through 14.4, but these conditions are not necessarily the ones providing optimal analytical performance.

With these reservations, the data presented in Table 14.5 are roughly similar

Table 14.5. Excitation Temperatures and Electron Number Densities in Low-Gas-Flow ICPs

Torch type	Excitation temperature (K)	Electron number density (cm^{-3})	Ref.
Reduced-size	4300		3
	4000		17
High-efficiency	4500		6
	4036	1.7×10^{14}	28
		1.4×10^{14}	27
Water-cooled	7000		8
	5500		32
Air-cooled	4800–5500		33
Ceramic	5200		13

to values reported for a conventional argon ICP. Perhaps this is not surprising. In developing low-flow torches, all authors attempt to create a plasma that matches the conventional ICP in analytical performance. The physical properties quoted in Table 14.5 are just another way to express the closeness of this match. It is confirmed by the visual appearance of low-gas-flow ICPs, which display the same bright discharge as the conventional ICP.

On closer examination, the excitation temperatures tend to be somewhat lower than quoted for a conventional discharge. If this is more clear for the high-efficiency torch than for other designs, it may be simply because many more data have been collected for this plasma than for all other low-gas-flow plasmas combined. Generally, however, temperatures reported for low-gas-flow plasmas cover a range from 4000 to 5500 K, rather than 5000 to 6500 K for a conventional ICP. In agreement with this result is the observation that limits of detection are superior for alkali metals, but sometimes are decidedly inferior for high-energy transitions in the far UV.[33]

Rezaaiyaan and Montaser and their colleagues report electron number densities of 1.7×10^{14} and 1.4×10^{14} cm^{-3}, respectively,[27,28] which are clearly lower than the values reported for a conventional ICP, varying[27,28] from 7×10^{14} to 1.4×10^{15} cm^{-3}.

Perhaps there may be a simple explanation for the differences that also points the way to a substantial improvement. In their quest for more economical plasma designs, investigators tend to report minimum values for both the argon consumption and the generator power. The data in Table 14.5 refer to these conditions. However, the applied generator power has a strong influence on the excitation temperature. As is illustrated in Figure 14.8, this is true not only in a conventional plasma[39] but also in an air-cooled, low-gas-flow plasma.[33] Presumably, therefore, the lower temperatures in Table 14.5 can be readily increased by another 1000 K when more power is applied. Obviously, in doing so, the power economy of some torch designs is minimized to the point where the distinction with the conventional ICP becomes marginal.

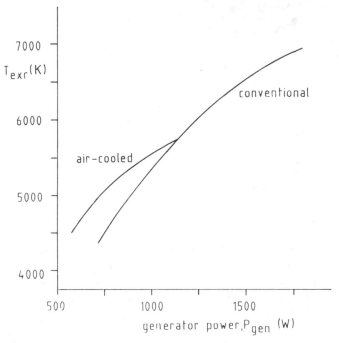

Figure 14.8. Boltzmann excitation temperature as a function of the forward power. Note the correspondence between the data for an air-cooled torch measured by the authors and those for a conventional torch.[39]

14.5.2 Background Spectrum

Another indication of the behavior of low-gas-flow plasmas can be gained from the background spectra emitted when distilled water is nebulized into the ICP. Typical observations reported in the literature are shown in Figure 14.9. Not surprisingly, the spectra emitted by high-efficiency and reduced-size torches closely resemble those of commonly used ICPs. Strong Ar lines are noted on a weak continuum background. The only significant band emission arises from OH. In most spectra, the strong silicon lines around 252 nm can be observed arising from evaporated torch material. This is a common observation in any ICP spectrum. Naturally, for ceramic torches, background emission from components of the ceramic material also is expected. Indeed, background spectra emitted from a plasma showed atomic boron lines and boron oxide bands as the result of the evaporation of boron nitride used in the segmented torch.[13]

When the total Ar flow is reduced to 5 L/min or less, additional molecular bands from NH and NO arise from ambient air diffusing into the discharge region.[18,32] These bands can be eliminated by extending the outer tube a few centimeters above the load coil.[18,32,33] As a result, the plasma must be observed

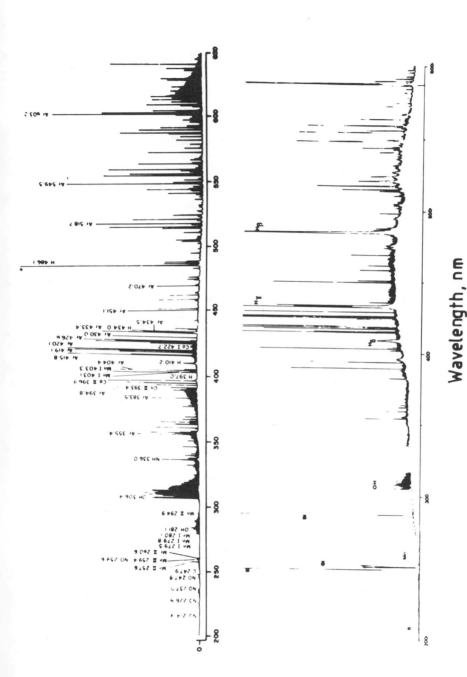

Figure 14.9. Background spectra from low-gas-flow torches. Top: Reduced-size torch without a tube extension.[18] Bottom: Ceramic torch with extended outer tube.[13] Note spectral lines of boron and silicon from the tube material and of manganese from an introduced solution of 1 μg of manganese per milliliter. (From References 13 and 18, with permission.)

through the wall of the extended tube, which may become clouded upon prolonged use.

For externally cooled torches operated at 1 L/min total Ar, an extended tube is necessary to prevent the plasma from being extinguished by entrained air. Again, with this provision, the background spectrum is virtually equal to that of a conventional ICP. The problem of observations through the wall of the tube may be reduced by using a side tube above the load coil.[33]

14.5.3   General Analytical Features

In the absence of contrary indications, we may conclude that low-gas-flow torches are easily ignited and can be routinely operated. If, occasionally, starting conditions are slightly different from operating conditions, this may indicate "teething" problems that will be overcome with further design improvements. Similarly, flexibility in operation conditions will be broadened, not only to lend ruggedness to the design but also to accommodate different samples and solvents. It is true that most analytical data reported in the literature refer to aqueous solutions. However, for the more widely studied high-efficiency and air-cooled torches, a broader range of analytical applications has been covered.

Certain analytical features can be summarily discussed, because they are no different from conventional ICPs. Thus, the dynamic range spans the usual four to five decades. The short-term precision and the long-term drift are also similar[12,27] and appear to depend more on the readout equipment used than on the discharge proper. From the few data provided, highly salted solutions appear to pose no problems other than those connected with the nebulizer. Indeed, with a V-groove Babington nebulizer 10% NaCl and 65% $Al_2(SO_4)_3$ aqueous solutions could be easily introduced into an air-cooled ICP over many hours of continuous operation[33] (Fig. 14.10).

Obviously, the analytical properties of keen interest are the detection limits, the interferences, and the accommodation of nonaqueous solvents. In reporting the results in the following sections, we shall focus our attention on the most recent data that represent the state of the art of low-gas-flow torches.

14.5.4   Detection Limits in Aqueous Solutions

The single analytical feature reported in virtually all publications on novel torch design is the detection limit in aqueous solutions. The data presented in Table 14.6 are taken directly from the literature and have not been corrected for minor differences in definitions. Generally, the detection limits agree very well mutually. More important, they approach the data for the commonly used ICP of Winge and co-workers,[34] which form the usual basis for comparison (see Appendix). Note that the most sensitive lines in the conventional ICP are also favored in low-gas-flow discharges, and indeed, ion lines are usually stronger

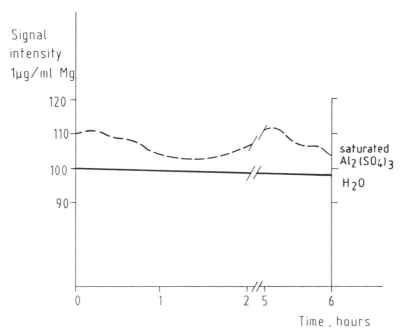

Figure 14.10. Long-term stability of the signal from 1 μg of magnesium per milliliter contained in a saturated solution of 65% $Al_2(SO_4)_3$ compared to pure water. The solution was introduced into an air-cooled torch by a V-groove nebulizer[33] as shown in Figure 14.6. (From Reference 33, with permission.)

than atom lines. This again confirms the similarity between the discharges that was concluded above from the excitation temperatures (Table 14.5).

Still, there are differences. A typical alkali metal, such as sodium, is detected more sensitively in plasmas of low-gas-flow torches than in a conventional ICP. On the other hand, when the ICP is operated at minimum power,[28] the spectral lines from high-energy transitions in the far UV—for example, those for cadmium and zinc—are less sensitive. Both observations reflect the lower excitation temperatures discussed earlier (Section 14.5.1). Indeed, Rezaaiyaan and Hieftje[28] observed iron and nickel excited in the plasma of their high-efficiency torch at uncommonly long wavelengths, with a consequent loss in detectability. As is expected from Figure 14.5 and indeed clear from Table 14.6, an increase of the input power is beneficial in this respect. Indeed, the excellent detection limits reported by Kawaguchi and co-workers for their water-cooled torch[32] were obtained at 1.2 kW forward power. Relatedly, the original data of Ripson[12] for an air-cooled torch were remarkably improved by van der Plas and de Galan[33] when a higher cooling capacity for the new torch allowed them to raise the forward power from 0.5 to 0.9 kW. It would appear, therefore, that at higher, but still modest, forward powers, the detection capabilities of low-gas-flow ICPs become similar to those of the conventional units available commercially.

Table 14.6. Detection Limits of Elements in Aqueous Solutions for Various Low-Gas-Flow ICPs

| Element | Wavelength (nm) | Conventional | High-efficiency | | Externally cooled: | |
		Winge[34]	Hieftje[28]	Montaser[27,a]	Water, Kawaguchi[32]	Air, de Galan[33]
As I	193.7	53	—	32	40	—
P I	213.6	76	—	—	33	—
Zn I	213.9	1.8	15	—	5.1	2.3
Cd II	214.4	2.5	17b	4	2.6a	2.1
Pb II	220.4	42	—	31	27	40
Ni II	221.6	10	12b	6	—	6.0
Fe II	238.2	4.6	21b	3b	—	2.8
Mn II	257.6	1.4	1.7	2	0.5	1.0
Mg II	279.6	0.15	0.47	—	—	0.11
V II	292.4	7.5	—	4	2.8b	—
Cu I	324.8	5.4	2.0	—	1.9b	1.9
Ca II	393.4	0.19	0.26	20b	—	0.13
Al I	396.2	28	16	—	—	14
Ba II	455.4	1.3	2.8	0.7	—	0.4
Na I	589.0	29	2.5	—	—	1.0

Detection limit (ng/mL)

a Detection limits calculated by the method described in Reference 34.
b Measured at a different wavelength.

14.5.5   Detection Limits in Nonaqueous Solutions

Apart from water, two important solvents are xylene, used to dilute oil samples, and MIBK, used as extractant in preconcentration procedures. Both solvents are readily injected into a conventional ICP when three operating conditions are slightly modified. First, in addition to injector and outer gas flows, the use of an intermediate flow of Ar is essential. Second, the power is raised from 1.2 to 1.8 kW for reasons cited in Section 14.3.3. Third, oxygen doping is sometimes applied to prevent carbon buildup at the tip of the injector tube. With low-gas-flow torches, similar modifications are needed when organic solvents are used instead of water.

So far, only two publications have reported the use of organic solvents (Table 14.7). Ng and co-workers[35] demonstrated that their plasma in a high-efficiency torch could easily operate with solvents, such as xylene and MIBK, provided the power was raised to 1.0 to 1.5 kW. The detection limits for 14 elements in water and in xylene were approximately the same, and the results obtained for an analysis of standard fuel oil compared well with the NBS values.[35] Van der Plas and de Galan[33] likewise obtained good results with their air-cooled torch. Both xylene and MIBK could be introduced into the plasma when the power is raised to 0.9 kW and the plasma gas was doped with 1% oxygen to prevent carbon deposition on the extended torch tube. Although the detection limits for the air-cooled torch were slightly inferior to those in aqueous solution, a similar observation applies to the conventional ICP.[36]

Detection limits obtained by the hydride generation technique are listed in Table 14.8. Again, when somewhat higher power is applied, continuous introduction of hydrides at common flow rates presents no stability problem. The observed detection limits, which are around 1 ng/mL, agree with the data of Kirkbright and associates[37] for a conventional ICP.

Table 14.7. Detection Limits of Elements in Organic Solvents for Low-Gas-Flow ICPs

Element	Wavelength (nm)	Detection limit (ng/mL)		
		Conventional torch[a]	Air-cooled torch[b]	High-efficiency torch[c]
Sn II	190.0	—	50	—
Zn I	213.9	14	5	3 (412.4 nm)
Pb II	220.4	220	500	20
Fe II	238.2	14	12	7 (259.9 nm)
Mg II	279.6	0.8	0.15	1
Cu I	324.8	3	2	4

[a] Reference 36.
[b] Reference 33. Operating conditions: forward power, 0.85 kW; plasma gas flow, 0.8 L/min, doped with 1% O_2; injector gas flow, 0.15 L/min.
[c] Reference 35. Operating conditions: forward power, 1 kW; outer gas flow, 7 L/min; intermediate gas flow, 1.25 L/min; injector gas flow, 0.5 L/min.

Table 14.8. Detection Limits of Elements Introduced as
Hydrides into an Air-Cooled ICP

		Detection limit (ng/mL)	
Element	Wavelength (nm)	Conventional torch[a]	Air-cooled torch[b]
As I	193.7	0.8	0.6
Hg I	194.2	—	1
Se I	196.0	0.8	1
Sb I	206.8	1	1
Bi I	223.1	0.8	0.6

[a] Reference 37.
[b] Reference 33. Operating conditions: forward power, 0.85 kW;
plasma gas flow rate, 1 L/min; injector gas flow, rate 0.3 L/min;
sample uptake rate, 5 mL/min; HCl concentration, 4 M; NaBH$_4$
concentration, 0.25%.

14.5.6 Interferences and Accuracy

Next to the detection power, freedom from interferences is probably the most favorable property of the ICP discharge, and this advantage should be maintained in low-gas-flow torches. The topic is treated in several publications.[12,16,27,32,33,38]

Chemical interferences in the ICP are usually divided into two classes. The first category, volatilization and dissociation interference, is generally negligible in low-gas-flow ICPs, as evidenced by studies involving alumina[12,16,38] and phosphate[12,27,38] (Figures 14.11 and 14.12). The second category, the influence of easily ionized elements, such as sodium, appears to be variable (Figure 14.13). This type of interference critically depends on the operating conditions, notably the observation height and the applied power.[27,38]

Thus Kawaguchi and colleagues[32] find no interference from 1000 mg/L Na when they operate their water-cooled torch at 1.2 kW. Similarly, Ripson and co-workers[12] report less than 10% suppression for several spectral lines at 10,000 mg/L Na when their air-cooled torch is run at 600 W. However, an alternative design, restricted to 450 W generator power, did show substantial interferences, and the same observation applied to a water-cooled torch operated at 600 W.

For high-efficiency torches, both Rezaaiyaan and co-workers[38] and Montaser and co-workers[27] notice an influence of just 100 mg/L Na when the torch is operated at minimum argon flow (6 L/min) and low power (700 W). The latter authors obtain decidedly better results, comparable to conventional ICP, when the conditions are changed to 9 L/min and 1100 W.

The ultimate test of analytical accuracy is the analysis of real samples. Reports for ICPs from low-gas-flow torches are restricted to analysis of certain common

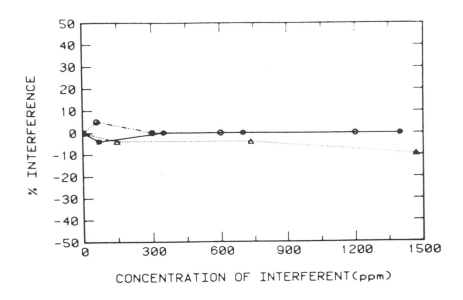

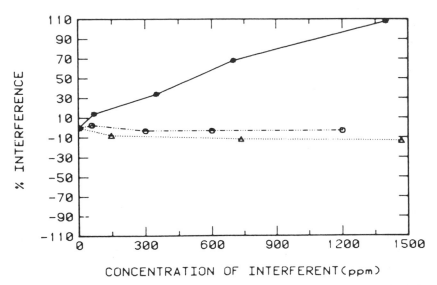

Figure 14.11. Percent interference in a high-efficiency torch at height of minimum interference: (*) sodium, (○) aluminum, (△) phosphate. Top: Calcium atom line (422.7 nm). Bottom: Calcium ion line (393.3 nm).

Performance of the water-cooled torch for the ICP

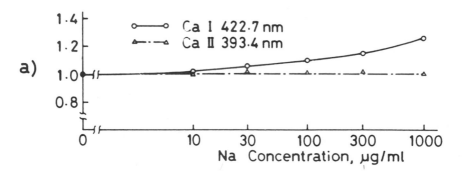

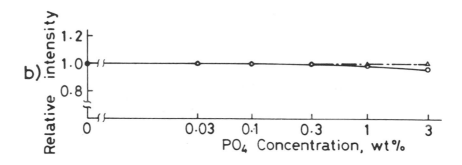

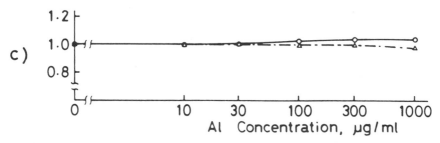

Figure 14.12. Relative interference in a water-cooled torch for calcium atom line (422.7 nm) (○) and calcium ion line (393.4 nm) (△). (*a*) Sodium, (*b*) Phosphate. (*c*) Aluminum.

reference materials. Acceptable results have been obtained with a variety of low-gas-flow torches used in the analysis of citrus leaves,[33] bovine liver,[33] orchard leaves,[17] manganese nodules,[27] tomato leaves,[6] fuel oil,[35] and wear metals in oil.[33] The latter results represent the only examples of the analysis of non-aqueous solutions, xylene. Thus, under appropriately selected conditions, interferences present minimal problems in low-gas-flow ICP spectrometry.

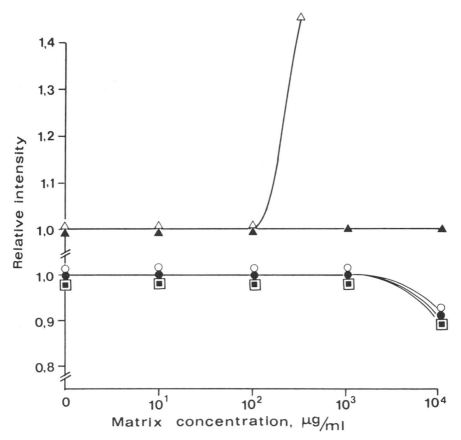

Figure 14.13. Relative interference in a radiatively cooled torch for calcium atom line (422.7 nm) and calcium ion line (393.4 nm): (○) Ca I + nitrate, (●) Ca II + nitrate, (□) Ca I + phosphate, (■) Ca II + phosphate, (△) Ca I + Na, (▲) Ca II + Na.

14.6   CONCLUSIONS

Considerations of the economics of operating ICP spectrometry systems have persuaded several investigators to design torches that will sustain ICPs at lower argon flow rates and lower power levels than are afforded by commonly used ICPs. Of the various proposals, two alternatives appear to emerge as most promising. One is the high-efficiency torch, with a restricted annular spacing between the outer and intermediate tubes of the conventional, three-tube torch. The other is the air-cooled torch utilizing either two or three concentric tubes and applying an external air flow to protect the torch from melting.

From a uniform description of the power balance in an ICP torch, it becomes clear that a minimum power of 0.2 kW for aqueous solutions and 0.5 kW for organic solvents is required, but not sufficient. For stable operation, an additional 0.2 kW must be applied, which is either carried away by excess Ar, as in high-efficiency torches, or dissipated through the torch wall, as in air-cooled torches.

Even then, performance approaching that of a conventional ICP is achieved only at a still higher power. Indeed, excitation temperatures, hence, detection limits and relative freedom from interferences, are improved to desired levels only when the forward power is 0.7 to 1 kW, which is not much lower than the power level of 1.2 to 1.7 kW currently used for conventional torches. In terms of reducing the flow rate for Ar, the designs have been more successful. With minor alterations in torch design, the high-efficiency torch can operate on 7 L/min total Ar. This figure might be reduced further if the frequency of the generator were raised from 27 to 100 MHz. A more drastic reduction down to 1 L/min of total Ar is achieved in air-cooled torches. Here, a potential future development is the use of a chemically resistant, high-temperature ceramic that releases excess power through heat radiation.

Certainly, the most important conclusion arising from the literature on low-gas-flow torches is that they can achieve an analytical performance closely similar to that provided by conventional torches. Indeed, accuracy, precision, dynamic range, detection limits, and interference levels are all within the ranges expected for an ICP. Although more work needs to be done, preliminary studies show that organic solvents, hydrides, and highly salted solutions can also be readily introduced into low-gas-flow torches.

REFERENCES

1. R. H. Wendt and V. A. Fassel, Inductively Coupled Plasma Spectrometric Excitation Source, *Anal. Chem.* 37, 920–922 (1965).
2. C. D. Allemand, R. M. Barnes, and C. C. Wohlers, Experimental Study of Reduced Size Inductively Coupled Plasma Torches, *Anal. Chem.* 51, 2392–2394 (1979).
3. R. N. Savage and G. M. Hieftje, Development and Characterization of a Miniature Inductively Coupled Plasma Source for Atomic Emission Spectrometry, *Anal. Chem.* 51, 408–413 (1979).
4. G. M. Hieftje, Mini, Micro, and High-Efficiency Torches for the ICP—Toys or Tools? *Spectrochim. Acta* 38B, 1465–1481 (1983).
5. C. D. Allemand and R. M. Barnes, A Study of Inductively Coupled Plasma Torch Configurations, *Appl. Spectrosc.* 31, 434–443 (1977).
6. R. Rezaaiyaan, G. M. Hieftje, H. Anderson, H. Kaiser, and B. Meddings, Design and Construction of a Low-Flow, Low-Power Torch for Inductively Coupled Plasma Spectrometry, *Appl. Spectrosc.* 36, 627–631 (1982).
7. G. Angleys and J. M. Mermet, Theoretical Aspects and Design of a Low-Power, Low-Flow-Rate Torch in Inductively Coupled Plasma Atomic Emission Spectroscopy, *Appl. Spectrosc.* 38, 647–653 (1984).
8. G. R. Kornblum, W. van der Waa, and L. de Galan, Reduction of Argon Consumption by a Water-Cooled Torch in Inductively Coupled Plasma Emission Spectroscopy, *Anal. Chem.* 51, 2378–2381 (1980).
9. S. V. Dresvin, "Physics and Technology of Low-Temperature Plasmas." Iowa State University Press, Ames, 1977.
10. H. Kawaguchi, T. Ito, S. Rubi, and A. Mizuike, Water-Cooled Torch for Inductively Coupled Plasma Emission Spectrometry, *Anal. Chem.* 52, 2440–2442 (1980).

11. P.A.M. Ripson, L. de Galan, and J. W. Ruiter, An Inductively Coupled Plasma Using 1 L/min of Argon, *Spectrochim. Acta* 37B, 733–738 (1982).
12. P.A.M. Ripson, E.B.M. Jansen, and L. de Galan, Analytical Performance of Inductively Coupled Argon Plasmas with External Cooling, *Anal. Chem.* 56, 2329–2335 (1984).
13. P.S.C. van der Plas and L. de Galan, A Radiatively Cooled Torch for ICP–AES Using 1 L/min of Argon, *Spectrochim. Acta* 39B, 1161–1169 (1984).
14. G. R. Kornblum, Principle and Some Optimization Results of an Induction Coupled Plasma Operating at 2 L/min in Argon, in "Proceedings of the Winter Conference on Plasma Chemistry, San Juan, 1980," R. M. Barnes, Ed., Heyden, London, 1981, pp. 111–120.
15. P.A.M. Ripson and L. de Galan, Empirical Power Balances for Conventional and Externally Cooled Inductively Coupled Plasmas, *Spectrochim. Acta* 38B, 707–726 (1983).
16. R. N. Savage and G. M. Hieftje, Vaporization and Ionization Interferences in a Miniature Inductively Coupled Plasma, *Anal. Chem.* 52, 1267–1272 (1980).
17. A. D. Weiss, R. N. Savage, and G. M. Hieftje, Development and Characterization of a 9-mm Inductively Coupled Plasma Source for Atomic Emission Spectroscopy, *Anal. Chim. Acta* 124, 245–258 (1981).
18. R. N. Savage and G. M. Hieftje, Characteristics of the Background Emission Spectrum from a Miniature Inductively Coupled Plasma, *Anal. Chim. Acta* 123, 319–324 (1981).
19. C. D. Allemand, Design of an ICP Discharge System for Residual Fuel Analysis for Marine Applications, *ICP Inf. Newsl.* 2, 1–26 (1976).
20. C. D. Allemand and R. M. Barnes, A Study of Inductively Coupled Plasma Torch Configurations, *Appl. Spectrosc.* 31, 434–443 (1977).
21. H. Anderson, H. Kaiser, and B. Meddings, High Precision (< 0.5% RSD) in Routine Analysis by ICP Using a High Pressure (200 psig) Cross-Flow Nebulizer, in "Proceedings of the Winter Conference on Plasma Chemistry, San Juan, 1980," R. M. Barnes, Ed., Heyden, London, 1981, pp. 251–272.
22. H. Anderson, H. Kaiser, and B. Meddings, The Design and Performance of Low-Flow and Modular Torches for ICP-Analysis. Paper presented at the Pacific Conference on Chemistry and Spectroscopy, San Francisco, October 1982, and the Easter Analytical Symposium, New York, November 1982.
23. R. M. Barnes and G. A. Meyer, Low-Power Inductively Coupled Nitrogen Plasma Discharge for Spectrochemical Analysis, *Anal. Chem.* 52, 1523–1525 (1980).
24. G. A. Meyer and M. D. Thompson, Determination of Trace Element Detection Limits in Air and Oxygen Inductively Coupled Plasmas, *Spectrochim. Acta* 40B, 195–207 (1985).
25. G. A. Meyer and R. M. Barnes, Analytical Inductively Coupled Nitrogen and Air Plasmas, *Spectrochim. Acta* 40B, 893–905 (1985).
26. N. Kovacic, G. A. Meyer, L. Ke-Ling, and R. M. Barnes, Diagnostics in an Air Inductively Coupled Plasma, *Spectrochim. Acta* 40B, 943–957 (1985).
27. A. Montaser, G. R. Huse, R. A. Wax, Shi-Kit Chan, D. W. Golightly, J. S. Kane, and A. F. Dorrzapf, Analytical Performance of a Low-Gas-Flow Torch Optimized for Inductively Coupled Plasma Atomic Emission Spectrometry, *Anal. Chem.* 56, 283–288 (1984).
28. R. Rezaaiyaan and G. M. Hieftje, Analytical Characteristics of a Low-Flow, Low-Power Inductively Coupled Plasma, *Anal. Chem.* 57, 412–415 (1985).
29. E. Michaud-Poussel and J. M. Mermet, Comparison of Nebulizers Working Below 0.8 L/min in Inductively Coupled Plasma Atomic Emission Spectrometry, *Spectrochim. Acta* 41B, 49–61 (1986).
30. J. M. Mermet and C. Trassy, A Plasma Torch Configuration for Inductively Coupled Plasma as a Source in Optical Emission Spectroscopy, *Appl. Spectrosc.* 31, 237–239 (1977).
31. E. Michaud-Poussel and J. M. Mermet, Influence of the Generator Frequency and the Plasma Gas Inlet Area on Torch Design in Inductively Coupled Plasma–Atomic Emission Spectrometry, *Spectrochim. Acta* 41B, 125–132 (1986).
32. H. Kawaguchi, T. Tanaka, S. Miura, J. Xu, and A. Mizuike, Analytical Performance of the Water-Cooled Torch for the Inductively-Coupled Plasma, *Spectrochim. Acta* 38B, 1319–1327 (1983).
33. P.S.C. van der Plas and L. de Galan, Analytical Evaluation of an Air-Cooled 1 L/min Argon ICP, *Spectrochim. Acta* 40B, 1457–1466 (1985).
34. R. K. Winge, V. J. Peterson, and V. A. Fassel, Inductively Coupled Plasma–Atomic Emission Spectroscopy: Prominent Lines, *Appl. Spectrosc.* 33, 206–219 (1979).
35. R. C. Ng, H. Kaiser, and B. Meddings, Low-Power Torches for Organic Solvents in Inductively Coupled Plasma Emission Spectrometry, *Spectrochim. Acta* 40B, 63–72 (1985).

36. A. W. Boorn and R. F. Browner, Effects of Organic Solvents in Inductively Coupled Plasma-Atomic Emission Spectrometry, *Anal. Chem.* 54, 1402-1410 (1982).
37. M. Thompson, B. Pahlavanpour, S. J. Walton, and G. F. Kirkbright, Simultaneous Determination of Trace Concentrations of Arsenic, Antimony, Bismuth, Selenium, and Tellurium in Aqueous Solution by Introduction of the Gaseous Hydrides into an Inductively Coupled Plasma Source for Emission Spectrometry, *Analyst* 103, 568-579 (1978).
38. R. Rezaaiyaan, J. W. Olesik, and G. M. Hieftje, Interferences in a Low-Flow, Low-Power Inductively Coupled Plasma, *Spectrochim. Acta* 40B, 73-83 (1985).
39. M. W. Blades and B. L. Caughlin, Excitation Temperature and Electron Density in the Inductively Coupled Plasma-Aqueous vs Organic Solvent Introduction, *Spectrochim. Acta* 40B, 579-591 (1985).
40. G. R. Kornblum and L. de Galan, Spatial Distributions of the Temperature and the Number Densities of Electrons and Atomic and Ionic Species in a Inductively Coupled RF Argon Plasma, *Spectrochim. Acta* 32B, 71-96 (1977).

15

Mixed-Gas, Molecular-Gas, and Helium Inductively Coupled Plasmas Operated at Atmospheric and Reduced Pressures

KNUT D. OHLS

Hoesch Stahl AG Dortmund
Chemical Laboratories
Dortmund, West Germany

DANOLD W. GOLIGHTLY

Branch of Analytical Chemistry
U.S. Geological Survey
Reston, Virginia

AKBAR MONTASER

Department of Chemistry
George Washington University
Washington, D.C.

15.1 INTRODUCTION

In 1884, Hittorf[1] generated the first electrodeless discharge in air. The properties of electrodeless discharges were investigated in 1947 by Babat, who found that the power required for the formation of a stable plasma diminished at higher frequencies.[2] Nearly 20 years later, Reed, using a torch similar to those in current use, introduced gases, such as O_2, N_2, and He, into an Ar inductively coupled plasma (ICP) to grow crystals in an electrodeless environment.[3] Soon thereafter, Greenfield and associates used a high-power Ar–N_2 ICP for spectrochemical analysis.[4] Different molecular gases and He were mixed with Ar in both the outer and the injector gas flows to generate mixed gas ICPs.[5] During the period from 1964 to 1976, only two laboratories were concerned with the analytical applications of high-power, Ar–N_2 ICPs.[6-17] Since 1976, the number of publications on mixed-gas, molecular-gas, and He ICPs has increased substantially, although the number of papers in this area is small compared to the total volume of papers published for the pure Ar ICP. However, interest in the area has grown to the extent that Baird Corporation has recently introduced an air ICP for process-controlled applications.

In this chapter, both the atmospheric-pressure and reduced-pressure ICPs commonly used for spectrochemical analysis are considered. The advantages and limitations of the various ICPs are discussed in Section 15.2. The instrumentation required for plasma formation is considered in Section 15.3. A brief review on the generation of various ICPs and their operating conditions is given in Section 15.4. Fundamental and analytical properties of mixed-gas, molecular-gas, and He ICPs are treated in Section 15.5. Finally, the use of these plasmas in analytical applications is summarized. Because a review article by M. I. Boulos has been published in the 1985 volume of *Pure and Applied Chemistry*, no discussion of computer modeling of the ICP is included here. Seven general review articles and books are cited in the Bibliography to assist the reader in obtaining an expanded discussion of various nonargon ICPs used for spectrochemical analysis and to remind the new ICP investigator of the availability of a wealth of information on very-high-power ICPs used in material engineering.

To simplify the discussion, the terms "mixed-gas ICP" and "molecular-gas ICP" are used to classify the plasmas. In a *mixed-gas plasma*, Ar is mixed with molecular gas or He in one or more of the gas flows of the ICP torch. In contrast, to form a *molecular-gas plasma*, pure molecular gases are individually introduced into the gas inlets of the torch.

15.2 ADVANTAGES AND LIMITATIONS OF MIXED-GAS, MOLECULAR-GAS, AND HELIUM INDUCTIVELY COUPLED PLASMAS

Compared to the Ar ICPs, mixed-gas and molecular-gas plasmas offer the advantage of greater heat transfer to analyte aerosol particles. Under certain conditions, the detection limits obtained with these non-Ar discharges are superior to those achieved with the Ar ICP. Compared to Ar ICPs, higher operating power levels and higher gas flows are often required to maintain a stable discharge. At high power levels, determinations of analytes in organic solvents and gases and the direct analysis of powders and solids are facilitated, especially when free-running generators are used. For fundamental studies, high-power, mixed-gas and molecular-gas discharges have the advantage of possessing an axial channel that appears to be closer to local thermal equilibrium (LTE).

Spectral bands, such as CN and NO bands, from mixed-gas and molecular-gas ICPs, cause spectral interferences at certain wavelengths.[18] Although some of the bands—for example, the CN bands—can be removed by the presence of O_2 or by the use of higher power, other bands may appear or inferior detection limits may be obtained at power levels at or exceeding 5 kW.

Helium ICPs, compared to Ar, mixed-gas and molecular-gas discharges, have the advantage of populating the higher energy levels of free atoms and ions, thus providing quite different spectra and the associated measurement capabilities.

15.3 INSTRUMENTATION FOR INDUCTIVELY COUPLED PLASMAS

In general, the instrumentation used for generating and observing mixed-gas, molecular-gas, and He ICPs is similar to that discussed in Chapters 3 and 4. The outer tube and the inner orifice of the conventional Greenfield and Fassel torches have been modified to facilitate plasma formation and stabilization. In addition to the free-running and crystal-controlled generators, a special generator[19,20] that uses a tuned-line oscillator also has been used to generate air and He ICPs. In this chapter, the RF power level of 2.5 kW is considered to be the dividing point between low-power and high-power operations.

Mixed-gas ICPs have been generated in Greenfield,[4-17,21-43] Fassel,[44-48] and low-gas-flow[49] torches. Molecular-gas ICPs in N_2, O_2, and air have been generated in special demountable torches (Figure 15.1) having inner diameters of 13 to 25 mm.[50-52] The use of a boron nitride injector tube was necessary to form a molecular-gas ICP. For He ICPs, three torches have been used. Abdallah and Mermet[53] modified their Ar ICP torch to produce a filament-type He ICP. To stabilize the plasma, the outer tube was extended and a high flow of air was used externally to cool the torch. This torch also has been used to investigate an air ICP.[20] Recently, a smaller torch, with an outer tube of 12 mm, i.d., has been used to generate the filament-type He discharge.[20] Montaser and associates[54] used a low-gas-flow torch and a modified Fassel torch to form annular, hollow, and filament-type He ICPs. Figure 15.2 shows the schematic diagram of a demountable low-gas-flow torch investigated recently.[54] The major component of the torch is a Macor threaded insert that is placed inside a section of high-precision quartz tube (13 mm o.d.). Sample is injected through a 0.5-mm orifice at the center of the insert.

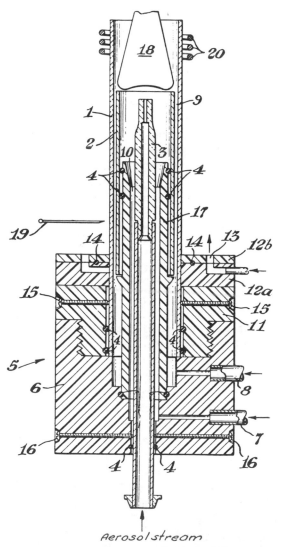

Aerosol stream

Figure 15.1. The demountable torch for molecular-gas ICPs: 1, containment quartz tube; 2, intermediate quartz tube; 3, central or nebulizer quartz tube; 4, O-rings; 5, basement; 6, plastic body element; 7, plasma gas inlet; 8, intermediate gas inlet; 9, outer annulus; 10, inner annulus; 11, detachable mounting ring; 12, plastic collars attached by metal screws to the mounting ring; 13, annular orifice of sheath gas; 14, O-rings; 15 and 16, alignment screws; 17, sleeve element; 18, plasma; 19, Tesla coil wire; 20, induction coil. (From Reference 50, with permission.)

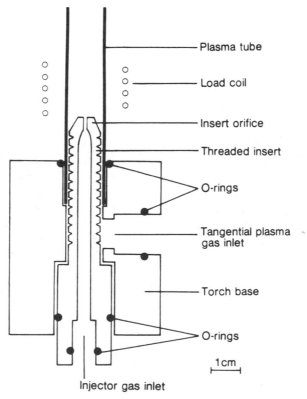

Figure 15.2. Schematic diagram of the demountable low-gas-flow torch for an atmospheric-pressure He ICP. The total helium gas flow rate was 8 L/min at 1.5 kW. (From Reference 54, with permission.)

The torches discussed above were used to form atmospheric-pressure ICPs. For a discussion of reduced-pressure ICPs, the reader is referred to Section 15.4.5.

15.4   REVIEW OF THE GENERATION OF PLASMAS AND THEIR OPERATING CONDITIONS

This section gives the operating conditions for stable plasmas in mixed gases, molecular gases, and helium. The reader is referred to the original references for the procedures to form the various plasmas. An Ar ICP is first generated before molecular gases are introduced into the various gas inlets of the torch. In contrast to the Ar ICP, the use of intermediate gas flow is essential to form and to stabilize mixed-gas and molecular-gas ICPs. Also, the relative position of the torch is slightly raised, with respect to the load coil, to ease plasma formation.

15.4.1   Mixed-Gas Inductively Coupled Plasmas with Various Outer-Gas-Flow Compositions

The most widely used mixed-gas plasma is an Ar-N_2 ICP, with a pure N_2 outer flow. The operating conditions of the various Ar-N_2 ICPs are shown in Table 15.1. Although the Greenfield torch has been used extensively, several applications of the Fassel torch in this area have been reported.[44–48,55] As can be seen from the typical conditions (Table 15.1), Ar-N_2 ICPs operate within a wide range of frequencies, input powers (1.2 to 15 kW), and gas flow rates (outer: 5 to 64 L/min; intermediate: 1 to 20 L/min; injector: 0.3 to 3.6 L/min). The importance of the observation height, which changes between 2 and 30 mm above the load coil, is discussed in Section 15.5. As can be seen from Table 15.1, higher injector gas flow rates require a higher observation height in the plasma. Despite the common belief that Ar-N_2 plasmas, and in general, mixed-gas plasmas, require higher power levels than an Ar ICP, such plasmas can be operated at 1 to 2 kW.

Besides the Ar-N_2 ICPs with pure N_2 outer flow,[29–38,44–48,56–59] H_2,[30,48,60,61] air,[30,39,48,62] and O_2,[30,40,48,62] have been used in the outer flow to generate mixed-gas ICPs. Relatedly, mixtures of Ar-N_2, Ar-O_2, Ar-air, and Ar-He have been introduced into the outer flow of the ICP.[44,48,62]

15.4.2   Mixed-Gas Inductively Coupled Plasmas with Various Injector Gases

In common analytical applications, analyte, water, and various compounds are injected into the axial channel of the Ar ICP. Such a mixture may be considered to be a mixed-gas plasma. The effects of water molecules on the analytical and fundamental properties of the Ar ICP are documented.[63] When a hydride generator is used to introduce sample into the Ar ICP, H_2 obviously is injected into the discharge, as discussed in Chapter 13. In this section, the term "mixed-gas ICP" is used only when molecular gases or He are added to the outer or the intermediate gas flow or are used to nebulize sample.

Mixtures of argon and oxygen[64–68] have been used as injector gases to suppress the background caused by the use of organic solutions and to study the spectrum of O_2. Mixtures of argon and hydrogen[61] have been used to improve the detection limits of a number of elements. For the electrothermal vaporization of samples, mixtures of Ar-halocarbon compounds have been used to improve the volatilization of elements that form refractory compounds.[69] Mixtures of argon with bromine, chlorine, etc, have been injected into the Ar ICP to evaluate the plasma for the determination of these halogens.[70] Mixtures of Ar-N_2 and air-butane also have been used to evaluate N_2 as an injector gas,[44,45] to study the atomic spectrum of N_2, and to simultaneously monitor atomic emission of C, H, N, and O for the quantitative elemental analysis of organic compounds.[71]

Table 15.1. Operating Conditions of Various Ar-N$_2$ ICPS

Conditions	Generator type[a]									
	f	f	f	f	f	f	c	c	c	c
Frequency (MHz)	7	6	27	27	27	36	27	27	27	27
Generator power (kW)	15	11	10	3	2.5-6	2.5	1.5	1-4	1.2	1.2
Torch[b]	G	G	G	G	G	G	F	F	F	LGFT
Outer gas flow (L/min)	64	50	33	16	20	7.5	15	15-43	15-20	5-7
Intermediate gas flow (L/min)	15	19	15	9	7-9.6	7.5	<1	1.5-2.5	1.5-2.5	2.5
Injector gas flow (L/min)	2-3	2.5	3.6	0.35	1-3	0.5	<1	2	2-2.5	1.0
Observation height (mm)	10-18	17	12	8	8-15	10-18	2-4	10-30	15	9
References	5	23, 24	25	26	27, 28	5	29, 48	45	44-47, 55	49

[a] f = free-running; c = crystal-controlled.
[b] G = Greenfield torch; F = Fassel torch; LGFT = low-gas-flow torch.

In investigations of excitation processes in the ICP, mixtures of Ar–N_2 and Ar–air have been introduced into the axial channel of the plasma.[72] The N_2 second positive band has also been used to measure the temperature of an Ar–air ICP.[73] Other spectroscopic measurements on Ar ICPs have been conducted by introducing individual mixtures of Ar with methane, sulfur dioxide, hydrogen sulfide, sulfur hexafluoride, bromine, phosphorus trichloride, and hydrochloric acid.[74,75]

15.4.3   Molecular-Gas Inductively Coupled Plasmas

Recently, the Baird Corporation introduced to the marketplace an air ICP atomic emission spectrometer for process-controlled applications.[76] This instrument is based to a great extent on the work by Barnes and associates on molecular-gas ICPs. These investigators formed three types of molecular-gas ICP (in N_2, O_2, and air[50,51,77–81]) with two generators, operating at frequencies of 27 and 41 MHz. The operating conditions for various molecular-gas ICPs at 41 MHz are shown in Table 15.2. The required power at 41 MHz ranged from 1.3 to 2 kW for N_2, O_2, and air ICPs, whereas 2 to 4 kW was required at 27 MHz. Special demountable torches, with outer tube diameters of 13 to 25 mm, have been used to generate molecular gas ICPs.[50] A 54-MHz, 6-kW generator, using a tuned-line oscillator, has been used by Abdallah and Mermet to produce an air ICP in a specially designed torch surrounded by a seven-turn load coil to measure the temperatures in the plasma.[20,53]

15.4.4   Helium Inductively Coupled Plasmas

Until recently, the generation of He ICPs at atmospheric pressure has been more difficult than the generation of ICPs in other gases. This difficulty arises because there are substantial differences between the physical and electrical properties of He and those of other gases. Abdallah and Mermet generated a

Table 15.2. Operating Conditions for Molecular Gas ICPs

Conditions	Plasma		
	Nitrogen	Oxygen	Air
Crystal-controlled generator			
Frequency (MHz)	40.6	40.6	40.6
Power (kW)	1.3	1.5–2	1.5–2
Demountable torch[50]			
Outer gas flow (L/min)	25	19	22
Intermediate gas flow (L/min)	3.5	2.6	2.8
Injector gas flow (L/min)	1.5	0.5	0.7
Observation height (mm)	5	0–15	0–15
Reference	51	81	81

filament-type He ICP with a 54-MHz, 6-kW generator which incorporated a tuned-line oscillator. To stabilize their He ICP, these investigators fabricated a special torch that required external cooling by air, but introduction of sample into the plasma was difficult.

Montaser and associates[54] produced three types of He ICP at atmospheric pressure: the filament-type He ICP, the hollow He ICP, and the annular He ICP. The annular He ICP was evaluated for the determination of chlorine and bromine. The operating conditions of various He ICPs are listed in Table 15.3. For the annular He ICP, formed in a unique low-gas-flow torch (Figure 15.2), the total helium flow rate was approximately 8 L/min at a forward power of 1.5 kW. The He ICP is formed by the tangential introduction of 7 L/min of He through the torch base. The annular He ICP is then formed upon introduction of the injector gas. This He ICP is self-igniting, and no external cooling is necessary to prevent overheating of the low-gas-flow torch.[54] The physical characteristics of these plasmas are discussed in Section 15.5.

15.4.5 Reduced-Pressure Inductively Coupled Plasmas

The ICPs discussed thus far were generated at atmospheric pressure. In general, plasma formation is facilitated at reduced pressure, and the resulting discharges are suitable for elemental analysis of gaseous samples. Seliskar and associates[82]

Table 15.3. Operating Conditions for Various Helium ICPs

	Plasma configuration		
	Filament	Hollow	Annular
Generator type	Tuned-line oscillator	Crystal controlled	
Frequency (MHz)	54	27	27
Forward power (kW)	0.6	0.7	1.5
Outer gas flow (L/min)	15 (air)	0	7 (He)
Intermediate gas flow (L/min)	2	8.5 (He)	—
Injector gas flow (L/min)	0	2	1
Observation height (mm)	Between coils	0–15	5–25
Discharge length beyond load coil (mm)	0	15–20	50–60
Discharge diameter (mm)	6	17	8–10
Axial channel	None	Very wide	1–5 mm wide
Ease of sample introduction	Difficult	Easy	Easy
References	20, 53	54	54

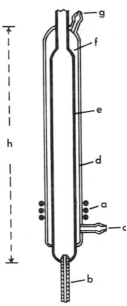

Figure 15.3. Schematic drawing of the reduced-pressure torch made from quartz. The plasma tube *e* (approx. 1.9 cm o.d.) is cooled externally by water, argon, nitrogen, or air passed through the jacket *d* (approx. 2.9 cm o.d.). The three-turn load coil *a* surrounds the cooling jacket. Gaseous sample is injected with the plasma gas through a 1-mm capillary injection tube *b*. The top end of the plasma tube is connected to the vacuum line. The coolant *f* enters at inlet *c* and exits at *g*. The length of the torch *h* is 30 cm, but the length is not reported to be a critical parameter. The mean value of the pressure in the torch varies from 1.3 Pa to thousands of pascals (10 millitorr to tens of torr). (From Reference 82, with permission.)

used a water-cooled torch (Figure 15.3), to form a reduced-pressure Ar ICP for investigating the feasibility of isotopic analysis of H_2 and D_2 by atomic emission spectrometry. A reduced-pressure He ICP was also evaluated by Seliskar and co-workers as a selective halogen detector for gas chromatography.[82] The total helium flow ranged from 5 to 1500 cm^3/min for a forward power of 5 to 500 W at 27 MHz.

In contrast to the atmospheric-pressure Ar ICP, the reduced-pressure Ar ICP exhibited no optical evidence of plasma heterogeneity.[82] For the reduced-pressure He ICP, however, a different behavior was observed.[82] At a forward power of 150 W, the He ICP switched from an apparently homogeneous discharge into an intense fireball localized in the coil region. Above this power level, the number densities of Br II and Cl II lines increased sharply with power level, and in the case of bromine, the Br$^+$/Br ratio approached unity at about 300 W. No definitive explanation for plasma focusing as a function of power has been offered so far.

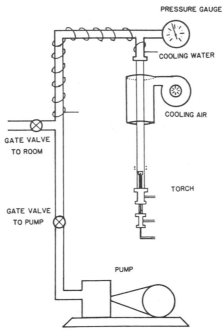

PRESSURE GAUGE

COOLING WATER

COOLING AIR

GATE VALVE
TO ROOM

TORCH

GATE VALVE
TO PUMP

PUMP

Figure 15.4. Schematic drawing of the reduced-pressure torch and the vacuum system used for investigating effects of gas pressure on fundamental properties of Ar and He ICPs. The torch base supports a 30-cm-long outer tube surrounded by a two-turn load coil. The outer and injector gas flows where 10 and 1 L/min, respectively, and no intermediate gas flow was used. The torch is cooled externally by forced air. A forward power of 1 kW at 27 MHz was used to sustain the reduced-pressure Ar or He ICP. (From Reference 83, with permission.)

Smith and Denton[83] formed reduced-pressure ICPs in gases such as Ar, He, CO_2, H_2, O_2, and air to study the effects of gas pressure on the excitation temperature and on the electron number density of the plasma. A schematic diagram of their torch, which is similar to conventional torches, and the vacuum system is shown in Figure 15.4. In contrast to the low gas flows and low power levels used by Seliskar and associates,[82] an outer gas flow (Ar or He) of approximately 10 L/min and a power level of 1 kW were used to study Ar and He ICPs operated at pressures from 13 to 200 kPa (100 to 1500 torr).[83] Relatedly, the shape of the reduced-pressure He ICP was quite similar to the Ar plasma generated in the same torch. The main He discharge, having an annulus of approximately 2 cm at its base, extended 2 cm beyond the load coil and was followed by a thin streamer (10 to 15 cm) that was localized at the center of the outer tube of the torch. This He ICP could not be operated at pressures greater than 66.7 kPa (500 torr).

15.5 PHYSICAL, FUNDAMENTAL, AND ANALYTICAL CHARACTERISTICS OF PLASMAS IN MIXED GASES, MOLECULAR GASES, AND HELIUM

15.5.1 Physical Characteristics

Compared to the Ar ICP, mixed-gas and molecular-gas ICPs are physically smaller, thus leading to generally lower observation heights in these plasmas. The distribution of an oil sample, diluted by xylene, in a 3-kW Ar-N_2 plasma having an outer flow of pure N_2, is shown in Figure 15.5. For forward powers of 3 to 10 kW, the highest analyte emission is localized in a region from 8 to 12 mm above the load coil.[25,43,84]

Although the power requirement is reduced at higher generator frequencies,[50] a higher forward power is required to generate mixed[45]-and molecular-gas ICPs,[85] compared to an Ar ICP. Figure 15.6 presents the physical sizes and shapes of an Ar ICP with Ar-N_2 plasmas that use a pure N_2 outer flow. At higher power levels, the Ar-N_2 ICP develops a long, conical tailflame.[86] The smaller size of mixed-gas and molecular-gas plasmas, relative to Ar ICPs, has been partly attributed to the thermal pinch effect that is brought about by the dissociation of molecular gases[86,87] and by the differences in electrical resistivities and thermal conductivities of Ar and molecular gases.

Figure 15.5. Organic sample distribution in an Ar-N_2 ICP, photographed through a green filter from the side and from above at an angle of 45°. (Photographs by K. Ohls.)

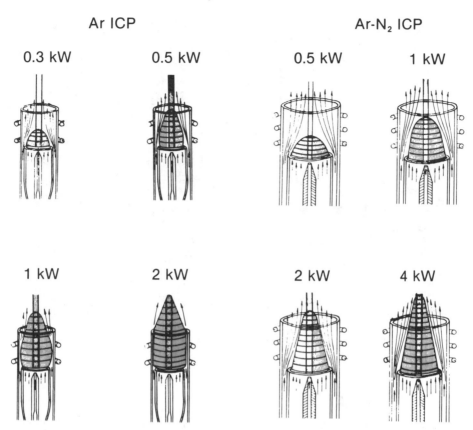

Figure 15.6. Shapes of Ar and Ar-N$_2$ ICPs at various forward powers; arrows indicate general direction of gas flow. (From Kontron Spektralanalytik, with permission.)

Two important points should be noted concerning the axial channel of mixed-gas and molecular-gas ICPs. First, if molecular gases are used to support the discharge, the diameter of the axial channel is usually smaller than that observed in an Ar ICP,[45,79] thus enhancing the sample–plasma interaction. Second, for Ar-supported ICPs, the diameter of the axial channel increases as the composition of the injector gas changes from pure Ar to pure molecular gas.[45] Such a change results in reduced sample–plasma interaction.

For He ICPs, three shapes have been observed at atmospheric pressure: filament,[53] hollow,[54] and annular.[54] The hollow He ICP (Figure 15.7) could be formed in a flowing stream of pure helium in a Fassel torch or in a low-gas flow torch, such as the MAK torch manufactured by Sheritt–Gordon Mines, Ltd. When observed from the side (Figure 15.7A), the hollow He discharge has a shape similar to that of an Ar ICP. No analyte emissions were observed from the hollow He discharge upon the introduction of aqueous samples into the plasma. For the filament-type He ICP, discharges having two different lengths have been reported.[53,54] While the length of the plasma observed by Abdallah and

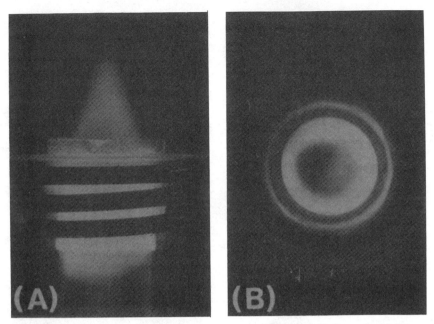

Figure 15.7. Hollow He ICP generated in a low-gas-flow torch such as the MAK torch: (*A*) side view; (*B*) top view. (From Reference 54, with permission.)

Mermet[53] was limited to the load coil region, Chan and Montaser[54] generated a filament-type ICP approximately 5 cm long before the formation of an annular He ICP that had a diameter of 8 to 10 mm, depending on the power level used. Figure 15.8 shows an annular He ICP and a filament-type He ICP generated in the low-gas-flow torch[54] depicted in Figure 15.2.

15.5.2   Heat Transfer Efficiencies in Plasmas

The thermal conductivities of molecular gases are generally greater than that of Ar. Experimental and theoretical studies have documented, for example, that Al_2O_3 particles decompose faster in $Ar-N_2$,[57,88] $Ar-H_2$,[88] N_2,[59,88] and in air[79,80] ICPs than in an Ar ICP. Thus, mixed-gas and molecular-gas ICPs are generally more suitable than an Ar ICP for elemental analysis of airborne particulates and slurries, and in the direct analysis of solids.[79,89] For aqueous solutions, the use of an $Ar-H_2$ mixture in the injector gas has led to an improvement of detection limits by factors of 3 to 5. These improvements have been attributed to the higher conductivities of H_2-Ar mixtures compared to an Ar injector gas.[60,61]

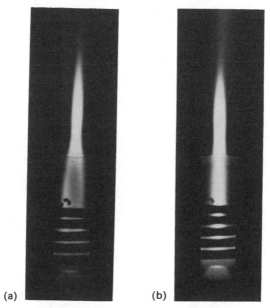

(a) (b)

Figure 15.8. Helium ICPs generated in a low-gas-flow torch: (*a*) atmospheric-pressure, annular He ICP; (*b*) filament-type He ICP. (From Reference 54, with permission.)

15.5.3 Spectral Features

15.5.3.1 Background Spectra

The background spectra of Ar, Ar–N$_2$, and Ar–O$_2$ ICPs are compared in Figure 15.9 in the presence and absence of water nebulization. For these spectra, the Ar ICP and the mixed-gas ICPs were operated at 1.5 and 3.0 kW, respectively. Two general observations are noted from these spectra. First, for wavelengths of less than 300 nm, the general background features are approximately the same. Second, for the spectral region between 300 and 500 nm, the Ar–O$_2$ ICP possesses a less complicated spectrum than the other two discharges. A closer examination of the spectra of the three plasmas for the 370 to 440 nm region (Figure 15.10) verifies the relative simplicity of the background emitted by an Ar–O$_2$ ICP. The complexity of spectra emitted by the Ar–N$_2$ ICP relative to the Ar–O$_2$ is best documented (Figure 15.11) in the determinations of Tl at 351.9 nm and Cr at 425.4 nm. Similar background spectra are observed for pure molecular-gas ICPs. For further information on this subject, the reader is referred to the recent comprehensive review article by Montaser and Van Hoven cited in the Bibliography.

The use of a He ICP results in a significant simplification of background spectra. A typical example[54] of the spectrum from a He ICP is illustrated in

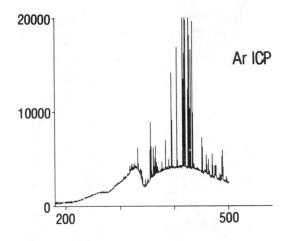

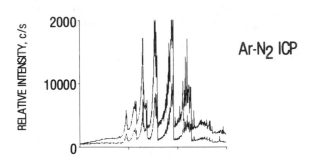

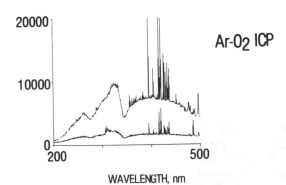

Figure 15.9. Background spectra of Ar (1.5 kW), Ar-N$_2$ (3.0 kW), and Ar-O$_2$ ICP (3.0 kW) without (top) and with (bottom) water nebulization. (From References 21 and 43, with permission.)

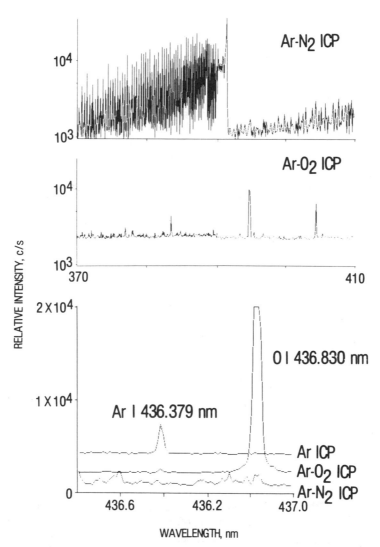

Figure 15.10. Background spectra for an Ar ICP and mixed gas ICPs with N_2 and O_2 outer gas flows. The top spectra are for the wavelength range from 370 to 410 nm. The bottom spectra are recorded from 436.6 to 437 nm.

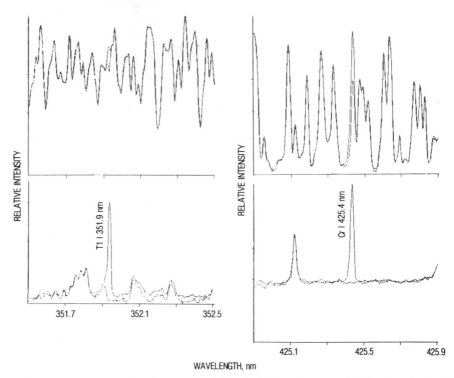

Figure 15.11. Spectral interferences for Tl and Cr lines from Ar-N$_2$ (top) and Ar-O$_2$ (bottom) ICPs generated at 3.0 kW. The concentrations of Tl and Cr were 10 μg/mL. (From Reference 40, with permission.)

Figure 15.12, which shows atomic emission spectra of Cl and C excited in a He ICP and in an Ar ICP. The excellent spectral selectivity enabled by the He ICP is evident from the figure.

15.5.3.2　Prominent Spectral Lines

The prominent spectral lines observed in Ar ICPs and in mixed-gas and molecular-gas ICPs are quite different. In general, the most prominent lines of Ar ICPs occur at wavelengths less than 300 nm,[90] (see Appendix.) while those of mixed-gas and molecular-gas ICPs appear at wavelengths greater than 300 nm.[55,81] Tables of analytically useful lines observed in an Ar-N$_2$ ICP are reported by Montaser and co-workers[55] for 43 elements, and by Fry and associates[65,70] for O$_2$, N$_2$, and the halogens in Ar-supported plasmas.

Although no extensive studies of the He ICP have yet been conducted, studies by Montaser and associates[54] and Seliskar and co-workers[82] on atmospheric-pressure and reduced-pressure He ICPs have documented the superiority of He ICPs over the Ar ICP for elemental analysis of hard-to-excite elements, such as the halogens.

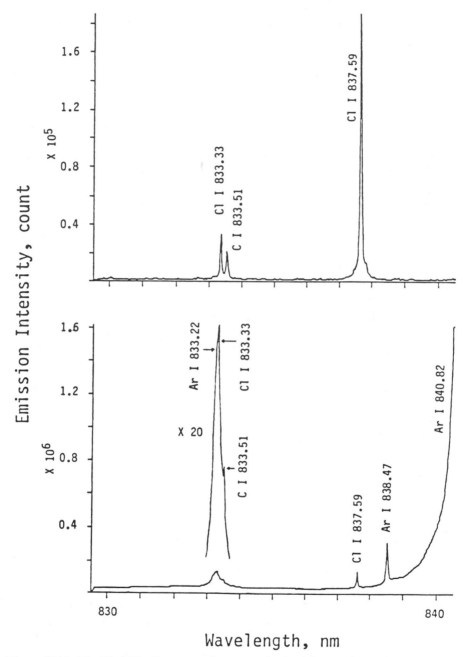

Figure 15.12. The He ICP allows superior spectral selectivity, as shown by the atomic emissions of C I 833.51 nm and Cl I 835.59 nm from He ICP (upper trace) and Ar ICP (lower trace). The atmospheric-pressure ICPs have annual configurations. (From Reference 54, with permission.)

15.5.4   Temperatures and Electron Number Densities

15.5.4.1   Measurements at Atmospheric Pressures

Table 15.4 summarizes the experimental conditions for measuring different temperatures in various ICPs. For temperatures in the Ar ICP, the reader should consult Chapter 8, Table 8.1. Although the operating and measurement conditions are quite dissimilar, some generalized observations concerning ICP temperatures can be made. First, the agreement between rotational and excitation temperatures in an Ar-N_2 ICP having a pure Ar outer flow[72] suggests that the plasma operates close to LTE conditions when N_2 is introduced into the injector gas. Second, when N_2 is used in the outer flow in the torch, the excitation and the ionization temperatures are approximately equal, but they are approximately 1000 K lower than the temperature of an Ar ICP measured on the same experimental facilities. Third, under the same experimental conditions, the temperatures of an air ICP generally are lower than those of an N_2 ICP. Fourth, in contrast to molecular-gas ICPs, the rotational and excitation temperatures for the He ICPs are quite different, thus indicating significant departure from LTE.

In comparing the temperatures of molecular-gas ICPs with those of other discharges, four points should be noted. First, because few readers have access to Reference 52, only the temperature data from this work are listed in Table 15.4. The interested reader may also refer to other publications[78,80] for temperature measurements under slightly different conditions. Second, because the first 26 lines of the P-branch of the N_2^+ 391.4 nm band were integrated, the temperature cited by Meyer[52] as "rotational" temperature is, in reality, a form of excitation temperature, a fact later recognized by Barnes and associates.[78,80] Third, the temperature ranges listed in Table 15.4 for the N_2 ICP and the air ICP refer to the axial temperature of the discharge, as estimated by us, from temperature profiles in reference 52. Fourth, Meyer reported that the electron temperature in the N_2-ICP was consistently higher than the excitation temperature, but, in general, the similarities between temperatures for either the N_2 ICP or the air ICP operated at a wide variety of conditions indicate that molecular-gas plasmas are closer to LTE than other ICP discharges listed.

Compared to temperature measurements, fewer data are available on electron number densities, n_e, of ICPs. Stark broadening measurements on the H 486.1 nm line excited in an Ar ICP, injected with a 10% methane-90% Ar mixture, gave an n_e of 4.9×10^{15} cm^{-3}, which was slightly less than the $n_e = 6.1 \times 10^{15}$ cm^{-3} obtained for a pure Ar ICP.[74] In contrast, addition of small amounts of either N_2 or O_2 to the outer flow of an Ar ICP has resulted in a 30% increase in n_e values.[62] However, further increases in the molecular gas content of the outer gas flow results in the lowering of n_e. Thus, for example, Montaser and Fassel[46] found that for the same forward power, the n_e value of an Ar-N_2 ICP, with a pure N_2 outer flow, is less than that of an Ar ICP by a factor of approximately 10. Two major conclusions may be drawn from studies of Montaser and Fassel,[46] who measured n_e values by the line-merging technique. First, the n_e

Table 15.4. Temperatures in Various ICPs

Plasma type	Temperature (K)	Type of temperature	Thermometric species	Power (kW)	Observation height (mm)	Gas flow (L/min)			Ref.
						Injector	Intermediate	Outer	
Ar	See Chapter 8								
Ar-N$_2$	4500–5000	Rotational	N$_2^+$	1.6	5	0.75 Ar + 0.25 N$_2$	12 Ar	0	72
	4500–5000	Excitation	Ar I, Fe I, Ti II, V II	1.6	5	0.75 Ar + 0.25 N$_2$	12 Ar	0	72
	5900–6700	Excitation	Fe I	1.2	9–17	1 Ar	2.5 Ar	5–7 N$_2$	49
	5700–6700	Ionization	Cd$^+$/I$^+$	1.2	20	1 Ar	1.5–2.5 Ar	15 N$_2$	47
Ar-air	6100–8900	Rotational	N$_2^+$	3.3	7–41	1.5 Ar + 0–1 N$_2$	16 Ar	50 air	73
Ar-CH$_4$	6100	Rotational	C$_2$	6[a]	0–3[b]	1.5 Ar + 0.13, 10% CH$_4$	7	22	74
	6200	Excitation	Ar I	6[a]	0–2[b]	1.5 Ar + 0.13, 10% CH$_4$	7	22	74
	7000–5000	Excitation	Ar I	6[a]	0–12[b]	1.5 Ar + 0.13, 10% CH$_4$	7	22	74
N$_2$	6000–8700	Rotational	N$_2^+$	3.9, 4.5	−10 to +10	0.8, 1.5	1.5–2	21	52
	5000–8100	Excitation	N I	3.9, 4.5	−10 to +10	0.8, 1.5	1.5–2	21	52
Air	5200–7200	Rotational	N$_2^+$	3.9, 4.5	−10 to +10	0.8, 1.5	1.5–2	21	52
	5500–7600	Excitation	N I	3.9, 4.5	−10 to +10	0.8, 1.5	1.5–2	21	52
He	2400	Rotational	OH	0.6	Between coils	0.2 He	0.2 He	15 air	53
	4100	Excitation	Fe I	0.6	Between coils	0.2 He	0.2 He	15 air	53
	4100	Excitation	Fe I	1.0	5	1.8 He	0	55 He	54

[a] Only the generator power is given in Reference 74.
[b] Observation height above the torch exit.

value of an Ar-N$_2$ ICP under conditions (5-mm observation height and 3.0-kW forward power) suitable for exciting spectral lines of high excitation energy is comparable to that observed in an Ar ICP operated under conditions commonly used in analytical laboratories (15-mm observation height and 1.2-kW forward power). Second, while the injector gas flow rate tended to significantly reduce the n_e values of an Ar ICP, no such change was found for the Ar-N$_2$ plasma.

For molecular-gas ICPs, a comprehensive study has been conducted by Barnes and associates.[52,78,80] Figure 15.13 shows the Abel-inverted, electron number density profiles of N$_2$, air, and Ar ICPs, using the H$_\beta$ line. Two general observations may be made from these profiles. First, the axial channels of the commonly used Ar ICPs contain fewer electrons than those of the molecular-gas plasmas. Second, the highest n_e value for the Ar ICP occurs closer to the edge of the discharge, while in the case of molecular-gas ICPs, the peak n_e values occur close to the discharge center. The order for the maximum n_e values is: Ar ICP > N$_2$ ICP > air ICP.

15.5.4.2 Measurements at Reduced Pressures

The significant disequilibrium between vibrational, rotational, and translational temperatures for reduced-pressure nitrogen plasmas has been documented.[91] Relatedly, temperatures measured with the thermometric species Cl I and Cl II are different by an approximate factor of 2, suggesting a significant departure from LTE for the reduced-pressure He ICP.[82] Studies of Smith and Denton[83]

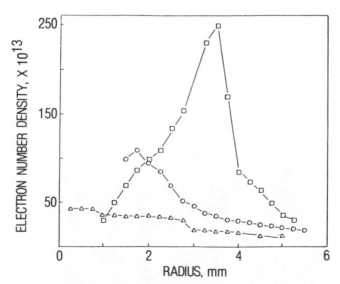

Figure 15.13. Measured electron number densities in an argon ICP at 1.0 kW (□), N$_2$ at 3.5 kW (○), and air ICP at 3.5 kW (△). (From Reference 52, with permission.)

show that as the pressure is changed from 16 to 200 kPa (120 to 1500 torr), increases in both excitation temperatures and electron number densities are observed for the Ar ICP.[83] The increase in electron number density indicates a higher collision rate, and therefore, a closer approach to LTE conditions.

15.5.5 Analytical Characteristics

Sections 15.5.5.1 and 15.5.5.2, respectively, review the strategies for the optimization of operating parameters for various plasmas and compare the detection limits for the plasmas. Applications of these non-Ar ICP discharges in the elemental analysis of different samples are summarized in Section 15.5.5.3.

15.5.5.1 Optimization of Parameters

The sequential simplex method,[29,33,55,79,92,93] alternating variable search,[31] and single-parameter optimizations[28,45] have been used to identify optimum conditions for various plasmas, and, in particular, for the Ar-N_2 ICP. Compared to other optimization strategies, the sequential simplex method provides the best approach to optimization. For any strategy selected, the set of optimum parameters is mainly determined by the goal of optimization: maximization of either signal-to-noise or signal-to-background ratio,[29,33,55,92] relative freedom from interferences,[33] single- or multiple-element determinations,[31] and the selection of spectral lines.[31,55] Relatedly, the optimum conditions also depend on the type of torch used to form the plasma. Thus, for example, when the Greenfield torch[31] was used to generate an Ar-N_2 ICP, the operating conditions were quite different from those obtained for a Fassel torch.[45] When pure N_2 was used in the outer flow of a Fassel torch, two sets of optimum conditions, quite different from the operating conditions of the commonly used Ar ICP, were identified for the high- and medium-excitation energy lines excited in an Ar-N_2 plasma.[45] In contrast to the Ar ICP, a forward power in the range of 2 to 3 kW is required to efficiently excite spectral lines having high excitation energies when pure N_2 is used in the outer flow.[45] For such lines, excited at high power, the optimum observation height is usually closer to the load coil, compared to an Ar ICP.[45] Such a trend has been observed for other mixed-gas plasmas and for molecular-gas ICPs.[50,52] However, a relatively low forward power (1.2 kW) is sufficient to achieve high signal-to-background and signal-to-noise ratios in an Ar-N_2 plasma[45,55] for lines of medium excitation energies.

In addition to forward power and observation height, the injector gas flow rate has a significant effect on the analyte line or on the signal-to-background ratio. Montaser and associates[45] have shown that the signal-to-background ratios are enhanced by almost a factor of 100 when the injector gas flow is increased from 1 to 2 L/min in an Ar-N_2 ICP using an outer flow of pure N_2. This observation is equally applicable to high-power Ar-N_2 ICPs.[31,45] The important influence of the injector gas flow on Ar-N_2 and Ar-O_2 ICPs, used by Ohls, is shown in Figure 15.14, where pure molecular gases are used in

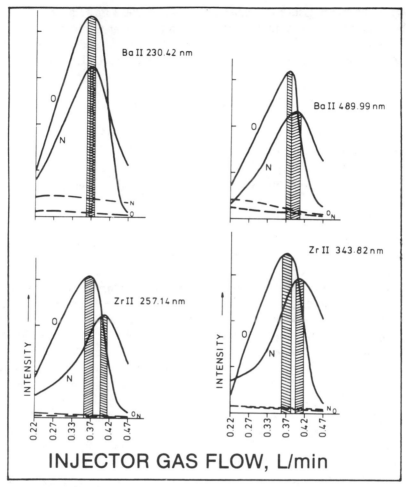

Figure 15.14. Influence of aerosol injector gas flow rates on analyte (solid curves) and background (dashed curves) intensities for an Ar–N_2 ICP (N) and an Ar–O_2 ICP (O). (From Reference 21, with permission.)

the outer flow of the torch.[21] No explanation has been offered for the wide differences between the optimum injector gas flows reported by various investigators.[21,31,45]

15.5.5.2   Comparisons of Detection Limits

Comparisons of detection limits obtained by different investigators are usually subject to error, especially when a variety of instruments and operating conditions are used. Such approximate comparisons are, however, essential to evaluations of the analytical potential of various systems. Greenfield and Burns[92] discussed the merits of mixed-gas versus Ar plasmas in terms of the high degree of molecular dissociation, which minimizes matrix interferences.

The use of pure N_2 as an injector gas in an Ar-supported ICP deteriorates the detection limits of the elements compared to an Ar ICP, especially when spectral lines of high excitation energies are observed.[45] When a small amount of N_2 is introduced into the plasma, no significant change in the detection limits are noted if the outer flow contains 5 to 10% N_2.[45,62] However, the use of a pure N_2 outer flow in an Ar-N_2 ICP leads to inferior detection limits when spectral lines having high excitation energies are observed at low forward powers.[44,45] Table 15.5 shows that at higher forward power, the detection limits of an Ar-N_2 ICP approach those of an Ar ICP for spectral lines having high excitation energies. Further comparisons of detection limits obtained for Ar, Ar-N_2, Ar-O_2, Ar-air, and Ar-H_2 ICPs are shown in Table 15.6. At a forward power of 4.75 kW, the detection limits for an Ar-N_2 ICP approach those for an Ar ICP operated at 1.1 kW forward power. Although few detection limits have been reported for ICPs with O_2[40] and air[39] outer flows, two general conclusions may be drawn from

Table 15.5. Detection Limits for Ar and Ar-N_2 Plasmas[45] Operated at Various Forward Powers[a]

Element, spectrum	Wavelength (nm)	Ar ICP I	Ar-N_2 ICP II	III	IV
Tl II	190.8	4	1300	100	73
As I	193.7	7	360	76	54
Se I	196.0	4	370	110	80
Cr II	205.5	1.4	100	5.6	6
Zn I	213.8	0.4	18	9.1	1.7
Pb II	220.3	12	1800	130	68
Cd II	226.5	0.8	50	3	2.2
Co II	228.6	0.7	39	5.3	3.0
Ni II	231.6	1.1	160	28	14
Co II	237.8	0.9	58	4.3	3.6
Cr II	267.7	0.7	19	7	4
Mo II	277.4	1.4	31	4.7	4.9
Cr II	283.5	0.3	10	0.6	0.9
V II	292.4	0.5	1.5	4	6
Forward power (kW)		1.0	1.0	2.0	3.0
Observation height (mm)		15	10	10	10
Outer gas flow rate (L/min)		15	15	20	43
Injector gas flow rate (L Ar/min)		1	2	2	2

[a] Ultrasonic nebulizer sample uptake rate was 4 mL/min for all experiments. Detection limit calculations were based on three times the standard deviation of background signal.

Table 15.6. Detection Limits (ng/mL) Obtained for Mixed-Gas ICPs

Element, spectrum	Wavelength (nm)	Plasma			
		Ar^{90}	$Ar-N_2^{33}$	$Ar-O_2^{40}$	$Ar-air^{39}$
Sn II	189.9	25	70		
As I	197.1	76	140	150	
Zn I	213.8	1.8	14	13	8
P I	214.9	76	130		
Pb II	220.3	42	97	160	
Ni II	231.6	15	16	30	80
Co II	238.8	6	9.4		
Ta II	240.0	28	33		
B I	249.7	4.8	5.3	20	0.7
Si I	251.6	12	25		
Mn II	257.6	1.4	1		8
Fe II	259.9	6.2	6.2	6	
Pt I	265.9	81	57		
Au I	267.5	31	16		
Mg II	279.5	0.2	2.5	0.1	0.01
Cr II	283.5	7.1	4.5		
Al I	308.2	45	81		
Ca II	317.9	10	25		
Cu I	324.7	5.4	7.5	13	6
RF power (kW)		1.1	4.75	2.8	3.0
Observation height (mm)		12–18	7.1	8	8
Outer Gas Flow (L/min)		20	23	16	28
Intermediate gas flow (L/min)		—	16.8	9	15
Injector gas flow (L/min)		1.0^a	1.45^b	0.4^a	1.0^a

[a] Meinhard nebulizer.
[b] GMK nebulizer.

Table 15.6. First, the Ar-air ICP provides detection limits that are superior to those obtained in Ar-N$_2$ and Ar-O$_2$ ICPs. For certain elements, such as boron and magnesium, the detection limits measured in an Ar-air plasma are better than the values reported for the Ar ICP. Second, compared to the other plasmas listed in the table, the Ar-O$_2$ ICP provides the worst detection limits, although this plasma is particularly useful for the elemental analysis of samples containing a large concentration or organic substances.

The conclusions above also are valid for molecular-gas ICPs. As shown in Table 15.7, the detection limits for molecular-gas ICPs, and, in particular, the O$_2$ ICP, are inferior to the results from the Ar ICP by 1 to 2 orders of magnitude. This conclusion is especially true for spectral lines of short wavelengths, which are difficult to excite in the plasmas.

It is appropriate to emphasize that if spectral lines having wavelengths greater than 300 nm are observed in a relatively low-power (1.2 kW), Ar-N$_2$ ICP,[55] the

Table 15.7. Detection Limits (ng/mL) in Argon and Molecular-Gas ICPs

Element spectrum	Wavelength (nm)	E (eV)	Plasma			
			Ar^{90}	N_2^{79}	O_2^{81}	Air^{81}
Zn I	213.8	5.8	1.8		1,200	710
Cd I	228.8	5.4	2.7		25,000	6,300
Be I	234.8	5.4	0.3		1,900	120
Al I	236.7	5.2	51			63,000
Pd II	248.8	16.3	103			8,300
Mn II	257.6	12.2	1.4	500		410
Mg II	279.5	12.0	0.2		2,500	400
Na I	285.3	4.3	27,000		3,700	15,000
Ga I	294.3	4.3	46		3,000	2,700
Be II	313.0	13.2	0.3			52
Pd I	363.4	4.2	54		810	180
Mg I	383.8	5.9	33		3,700	780
Ca II	393.3	9.2	0.2	1.2	21	0.8
Al I	396.1	3.1	28	1500	400	120
Ga I	417.2	3.1	66		900	750
Ca I	422.6	2.9	10		170	23
Ba II	455.4	7.9	1.3		26	12
Cd I	479.9	6.3	600			120,000
Ba II	493.4	7.7	2.3		51	12
Na I	588.9	2.1	29	900	190	59
Na I	589.5	2.1	69		350	76
Li I	610.3	3.8			11,000	800
Forward power (kW)			1.1	3.5–4.4	1.5–2	1.5–2
Observation height (mm)			12–18	1–7	−5, +2[a]	−5, +5[a]
Outer gas flow rate (L/min)			20	22	19	22
Intermediate gas flow rate (L/min)			0	3	2.6	2.8
Injector gas flow rate (L/min)			1.0	0.9–1.7	0.5	0.7

[a] Relative to top of load coil.

detection limits thus obtained are equivalent or superior to those of an Ar ICP. Such a plasma, which uses pure N_2 in the outer flow, may be easily formed on most of the existing ICP facilities incorporating automatching networks.

For difficult-to-excite lines, the He ICP, in principle, should be the most suitable plasma. Thus, the detection limits for bromine and chlorine in aqueous solutions injected into the annular He ICP[54] are lower than those measured in an Ar ICP. Further significant improvement in the detection limits of halogens may be achieved if a reduced pressure He ICP is used. For example, Seliskar and associates[82] have reported parts-per-billion detection limits for compounds such as chloromethane. One should remember, however, that the application of a reduced-pressure ICP is limited, at the present time, to analysis of gaseous samples.

15.5.5.3 Analytical Applications

Compared to the Ar ICP, fewer applications are reported for mixed-gas, molecular-gas, and He ICPs. Major applications for these plasmas are for metallic and geological samples. Table 15.8 summarizes applications of atmospheric-pressure, non-Ar discharges for elemental analyses of diverse materials. Except for one application for Ar-O_2 ICP,[40] all applications are related to the use of Ar-N_2 ICPs. The application table provides a convenient source of information for analysts interested in particular samples, elements, and their concentrations, plus the sample introduction method used. Detailed analytical procedures can be obtained from the original references.

The diversity of analytical applications indicates that mixed-gas ICPs are relatively free from most matrix effects. However, the Ar-N_2 ICP[33,55,62] exhibits a greater level of ionization enhancement effect than the Ar ICP.

15.6 CONCLUSIONS AND FUTURE DIRECTIONS

The use of mixed-gas and molecular-gas ICPs reduces the operating cost of an ICP facility significantly. These plasmas, and, and in particular, the argon mixed-gas ICPs, may be easily formed on current ICP installations that incorporate RF generators with power rating of 2.0 to 2.5 kW. With these plasmas, the entire outer gas flow may be replaced with N_2, air, or O_2. In certain cases, such as in process-controlled applications or in locations where argon is not available at a reasonable cost, the use of an air ICP may be desirable. Such a device has recently been introduced into the marketplace. The analyst should note, however, that the spectral lines and the operating conditions commonly used for Ar ICP-AES are not applicable to measurements with mixed-gas and molecular-gas ICPs.

As discussed in Chapter 14, the use of the reduced-gas-flow Ar ICP has resulted in a significant reduction of the ICP operating cost. However, the detection limits obtained from these plasmas have only approached those of the conventional Ar ICP. In this respect, the mixed-gas ICPs, and in particular, the Ar-N_2 ICP, have the advantages of providing superior detecting powers, under certain conditions,[55] than the conventional Ar ICP.

From a fundamental point of view, and also based on recent experimental studies,[54,82] the He ICPs are superior to Ar ICPs for exciting spectral lines of high excitation energy, especially for those of nonmetals. This potential of the He ICPs and the spectral simplicity of the background spectra of the He ICPs invite further investigations of such discharges, not only for atomic emission spectrometry, but also for atomic fluorescence studies and mass spectrometry.

Aside from the foregoing general statements, the choice of an ICP depends on the sample one wishes to analyze. Thus, for example, when a sample contains appreciable amount of organic materials, the use of an Ar-O_2 ICP may be essential, although the detecting power of this plasma is inferior to other ICPs.

Table 15.8. Applications of Mixed-Gas ICPs

Sample type	Matrix	Element	Line (nm)	Concentration[a]	Sample introduction system	Ref.
Petroleum and petroleum products						
Lubricating oil (wear metals)	Hydrocarbons	Al	309.3	0.1–2	Meinhard	37
		Al	396.1	0.1–1	Meinhard	43, 94
		Ca	393.3	0.01–0.1	Meinhard	43, 94
		Cd	214.4	10–300	Graphite cup	95
		Cu	324.8	0.01–4	Meinhard	37, 43, 94
		Fe	259.9	0.05–5	Meinhard	37, 43, 94
		Mg	279.6	0.005–1	Meinhard	37, 43, 94
		Mn	257.6	0.05–2	Meinhard	37, 43, 94
		Ni	231.6	0.05–0.1	Meinhard	94
		Ni	341.4	0.05–1	Meinhard	37
Lubricating oil Gasoline		Pb	220.3	0.08–1	Meinhard	43, 94
		Pb	220.3	Minor	Meinhard	36
		Si	251.6	0.1–4	Meinhard	37
		Si	288.1	0.03–1	Meinhard	43, 94
		Sn	317.5	0.05–1	Meinhard	37
Hydraulic fluid		Ti	334.9	0.01–1	Meinhard	43, 94
		V	309.3	0.01–1	Meinhard	43, 94
		Zn	213.8	0.02–4	Meinhard	37, 43, 94
Metals						
Steel, unalloyed	Fe	Ca	393.3	1–50	Meinhard	84
	Fe	Ce	413.7	100–500	Meinhard	40
	Fe	Cu	324.7	Minor	Graphite cup	26
	Fe	Ni	231.6	Major	Graphite cup	26
	Fe	Sn	189.9	5–50	Hydride	96

(Continued)

Table 15.8. (*Continued*)

Sample type	Matrix	Element	Line (nm)	Concentration[a]	Sample introduction system	Ref.
Brass	Cu, Zn	Sn	189.9	Minor	Hydride	96
Aluminum	Al	Bi, Cd, Co, Cr, Cu, Fe, In, Mg, Mn, Ni, Pb, Ti, Tl, V, Zn		2–200	Cross-flow	97
Steel	Fe	Al, As, B, Ba, Ca, Ce, Co, Cr, Cu, Hg, La, Mg, Mn, Mo, Ni, P, Pb, Si, Ti, V, Zn, Zr		Minor/major	Meinhard	25, 43
Aluminum alloys	Al	Cu, Fe, Mg, Mn, Si		Minor/major	Spark nebulizer	89
Precious metals		Ag, Au, Ir, Pd, Pt, Rh, Ru		0.3–100	Cross-flow	32
Steel	Fe	Cr, Cu, Mn, Ni, V		Minor	Spark nebulizer	35
Steel, alloyed	Fe	Cr, Cu, Fe, Mn, Mo, Ni, V, Ti		Major/minor	Meinhard	98
Refractories, metal oxides, ceramics, slags, cements						
Iron ore	Fe	Cd	214.4	Trace	Graphite cup	95
Blast furnace slag	Si, Mg, Fe	Cd	214.4	Trace	Graphite cup	95
Chrome ore	Cr, Al	Al, Ca, Cr, Fe, Mg, Mn, Si, Ti		Major/minor	Meinhard	5
Magnesite	Mg	Al, Ca, Fe, Mg, Mn, Na, Si, Ti		Major/minor	Meinhard	5
Iron ore	Fe	Al, As, Ca, Cr, Cu, Mg, Mn, Ni, Pb, Si, Ti, V, Zn		Major/minor	Cross-flow	43

Fire brick	Si	Al, Ca, Cr, Fe, Mg, Mn, Ti, V		Major/minor	Cross-flow	43
Chamotte	Al	Ca, Cr, Fe, Mg, Mn, Si, Ti, V		Major/minor	Cross-flow	43
Slags	Fe, Ca, Si	Al, Ca, Cr, Fe, Mg, Mn, Si, Ti, V		Major/minor	Cross-flow	43
Geological materials						
Tholeiite, dacite, andesite	Si	Ce	401.2	0.1–100	Meinhard	99
		Eu	413.0	0.1–100	Meinhard	99
Phosphate rocks	Si	Fe	259.9	Major	Cross-flow	24
Tholeiite, dacite, andesite	Si	Gd	364.6	0.1–100	Meinhard	99
		La	408.7	0.1–100	Meinhard	99
Phosphate rocks	Si	Mg	279.5	Major	Cross-flow	24
Pegmatite	Si, Mg	Nb	309.4	Major	GMK	33
Tholeiite, dacite, andesite	Si	Nd	430.4	0.1–100	Meinhard	99
Phosphate rocks	Si	P	214.9	Major	Cross-flow	24
Tholeiite, dacite, andesite	Si	Sm	392.8	0.1–100	Meinhard	99
Pegmatite	Si, Mg	Sn	189.9	Major	GMK	33
		Ta	240.0	Major	GMK	33
Tholeiite, dacite, andesite	Si	Y	371.0	0.1–100	Meinhard	99
		Yb	369.4	0.1–100	Meinhard	99
Phosphate rocks		Various		Major	Cross-flow	23
Silicate rocks	Si	Various		Major	Cross-flow	28
Syenite	Si, Ca, Al, Fe	Rare earth, Y		Minor/traces	Cross-flow	28
Tuff	Si, Al, K, Ti	Rare earth, Y		Minor/traces	Cross-flow	28
Foyaite	Si, Al, K, Fe	Rare earth, Y		Minor/traces	Cross-flow	28
Carbonatite	Si, Ca, Al, Mg	Rare earth, Y		Minor/traces	Cross-flow	28
Monazite	Y, La, Ce, Nd	Rare earth		Minor/traces	Cross-flow	28
Concentrate	Rare earth	Various		Major	Scott	27
Concentrate	Sulfides (Cu, Zn, Pb, Fe)	Various		Traces	Cross-flow	17

(Continued)

Table 15.8. (*Continued*)

Sample type	Matrix	Element	Line (nm)	Concentration[a]	Sample introduction system	Ref.
Air and particles						
Dust		Al	396.1	Major/minor	Meinhard	38
		Ca	315.8	Major/minor	Meinhard	38
		Cr	267.7	Major/minor	Meinhard	38
		Cu	324.7	Major/minor	Meinhard	38
		Fe	259.9	Major/minor	Meinhard	38
		Mg	279.5	Major/minor	Meinhard	38
		Mn	293.3	Minor	Meinhard	38
		Pb	283.3	Minor	Meinhard	38
		Sr	407.7	10–400	Meinhard	38
		V	310.2	10–500	Meinhard	38
		Zn	213.8	Major/minor	Meinhard	38
Water and effluents						
Wastewater	H_2O	As	228.8	10–100	Hydride	100
		Cd	228.8	Minor	Meinhard	37
		Cu	324.7	Major	Meinhard	37
		Fe	259.9	Minor	Meinhard	37
		Zn	213.8	Major	Meinhard	37
Recirculating water		Zn	481.1	Traces	Meinhard	5
Effluent	H_2O	Al, Cd, Cr, Cu, Fe, Ni, Pb, Zn		0.1–10	Meinhard	5

Soils, plants, and fertilizers

Sample	Element	Wavelength	Concentration	Method	Ref.
Orchard leaves	Cd	214.4	Traces	Graphite cup	95
Pine needles	Cd	214.4	Traces	Graphite cup	95
Grass	Cd	214.4	Traces	Graphite cup	95
Bovine liver	Cd	226.5	Traces	Electrothermal Vaporization	101
	Mn	257.6	Traces	Vaporization	101
Orchard leaves	Mn	257.6	Traces	Electrothermal Vaporization	101
	Pb	405.7	Traces	Vaporization	101
	Zn	213.8	Traces	Vaporization	101
Bovine liver	Zn	213.8	Traces	Vaporization	101
	Cd, Mn, Pb, Zn		Traces	Vaporization	102
Orchard leaves	Cd, Mn, Pb, Zn		Traces	Vaporization	102

Body tissues and fluids

Sample	Element	Wavelength	Concentration	Method	Ref.
Human blood	Al	396.1	0.2–1	Meinhard	10
	Cu	324.7	0.7–1.2	Meinhard	10
	Fe	259.9	0.4–0.5	Meinhard	10
	Mg	279.5	0.04	Meinhard	10
	P	253.5	350	Meinhard	10
	Pb	283.3	0.1–0.4	Meinhard	10
	Si	250.7	8.3	Meinhard	10
Serum	Mn, Zn		Traces	Vaporization	102

[a] Concentration ranges of the elements in the original matrix, expressed in micrograms per gram for solids and milligrams per liter for liquids, are major ≥ 10,000 minor ≥ 500, and traces < 1.

ACKNOWLEDGMENTS

This research (at George Washington University) was sponsored in part by the U.S. Department of Energy, under contract number DE-AS05-84-ER-13172. Acknowledgment is made to the Donors of the Petroleum Research Fund, administered by the American Chemical Society, for the partial support of this research.

BIBLIOGRAPHY

Barnes, R. M. Recent Advances in Emission Spectroscopy: Inductively Coupled Plasma Discharge for Spectrochemical Analysis, *CRC Crit. Rev. Anal. Chem.* 7, 203–296 (1978).
Boeing, H. V. Plasma Science and Technology. Cornell University Press, London, 1982.
Czernichowski, A., and J. Jurewicz. High-Frequency Electrodeless Plasmatrons and Their Use in Chemistry and Material Engineering, *ICP Inf. Newsl.* 2 (special issue 1), 1–46 (1976).
Dresvin, S. V., Ed. Physics and Technology of Low-Temperature Plasmas, translated by T. Cheron and edited by H. V. Eckert. Iowa State University Press, Ames, 1977.
Montaser, A., and R. L. Van Hoven. Mixed-Gas, Molecular-Gas, and Helium Inductively Coupled Plasmas for Analytical Atomic Spectrometry: A Critical Review, *CRC Crit. Rev. Anal. Chem.* 18, issue 1 (1987)
Robin, J. P. ICP–AES at the Beginning of the Eighties, *Prog. Anal. At. Spectrosc.* 5, 79–110 (1982).
Semiatin, S. L., and D. E. Stutz. Induction Heat Treatment of Steel. American Society for Metals, Metals Park, Ohio, 1986.

REFERENCES

1. J. W. Hittorf, Über elektrische Gasentladungen, *Ann. Phys.* 3, 75–82 (1884).
2. G. I. Babat, Properties of Electrodeless Discharges, *J. Inst. Electr. Eng. (London)* 94, 27–37 (1947).
3. T. B. Reed, Induction-Coupled Plasma Torch, *J. Appl. Phys.* 32, 821–824 (1961).
4. S. Greenfield, I. L. Jones, and C. T. Berry, High-Pressure Plasmas as Spectroscopic Emission Sources, *Analyst* 89, 713–720 (1964).
5. S. Greenfield, I. L. Jones, H. McD. McGeachin, and P. B. Smith, Automatic Multisample Simultaneous Multielement Analysis with an HF Plasma Torch and Direct Reading Spectrometer, *Anal. Chim. Acta* 74, 225–245 (1975).
6. S. Greenfield, I. L. Jones, C. T. Berry, and L. G. Bunch, the HF Plasma Torch, *Proc. Soc. Anal. Chem.* 2, 111–113 (1965).
7. S. Greenfield, P. B. Smith, A. E. Breeze, and N.M.D. Chilton, Atomic Absorption with an Electrodeless HF Plasma Torch, *Anal Chim. Acta* 41, 385–387 (1968).
8. S. Greenfield, I. L. Jones, and C. T. Berry, Improvement Relating to Spectroscopic Methods and Apparatus, British Patent 1,109,602 (April 1968), and U.S. Patent 3,467,471 (September 1969).
9. S. Greenfield and P. B. Smith, Operating Conditions for Plasma Sources in Emission Spectrometry, *Anal Chim. Acta* 57, 209–210 (1971).
10. S. Greenfield and P. B. Smith, The Determination of Trace Metals in Microlitre Samples by Plasma Torch Excitation, *Anal Chim. Acta* 59, 341–348 (1972).
11. S. Greenfield, Life with a Plasma Torch, *ICP Inf. Newsl.* 1, 3–6 (1975).
12. S. Greenfield, H. McD. McGeachin, and P. B. Smith, En Passent, *ICP Inf. Newsl.* 2, 167–177 (1976).
13. S. Greenfield, Plasma Sources in Spectroscopy, *Eur. Spectrosc. News* 1, 4 (1976).

14. S. Greenfield, H. McD. McGeachin, and P. B. Smith, Plasma Emission Sources in Analytical Spectroscopy—III. *Talanta* 23, 1-14 (1976).
15. S. Greenfield, H. McD. McGeachin, and P. B. Smith, Nebulization Effects with Acid Solutions in ICP Spectrometry, *Anal. Chim. Acta* 84, 67-78 (1976).
16. S. Greenfield, The Inductively Coupled Plasma Torch—A Source for All Reasons, *SPEX Speaker* 22, 1-6 (1977).
17. A. E. Watson, G. M. Russell, and G. Balaes, The Commissioning of an Inductively Coupled Plasma System and Its Application to the Analysis of Copper, Lead, and Zinc Concentrates, NIM Rep. No. 1815, in *ICP Inf. Newsl.* 2, 205-220 (1976).
18. R. D. Reeves, S. Nikdel, and J. D. Winefordner, Molecular Emission Spectra in the Radio-Frequency-Excited Inductively Coupled Argon Plasma, *Appl. Spectrosc.* 34, 477-483 (1980).
19. M. H. Abdallah, R. Diemiaszonek, J. Jarosz, J. M. Mermet, J. P. Robin, and C. Trassy, Étude Spectrometrique d'un Plasma Induit par Haute Fréquence—I. Performances Analytiques, *Anal Chim. Acta* 84, 271-282 (1976).
20. J. M. Mermet, personal communication.
21. K. Ohls, Analytical Application of the Oxygen-Cooled Argon ICP—II. Comparison of Line and Background Spectra Using Different Types of Analytes (Aqueous/Organic) and Nebulization, *ICP Inf. Newsl.* (in press).
22. J. P. Robin, J. M. Mermet, M. H. Abdallah, A. Batal, and C. Trassy, Role of Plasma Gas in Emission Spectroscopy, in "Recent Advances in Analytical Spectroscopy," IUPAC, Ed. Pergamon Press, Oxford, 1982, pp. 75-82.
23. I. B. Brenner, A. E. Watson, G. M. Russell, and M. Gonçalves, A New Approach to the Determination of the Major and Minor Constituents in Silicate and Phosphate Rocks, *Chem. Geol.* 28, 321-330 (1980).
24. G. M. Russell and A. E. Watson, The Spectrochemical Determination of Phosphorus, Magnesium and Iron in Phosphate Rocks and Sulfuric Acid Leach Liquors, NIM Rep. No. 2002, in *ICP Inf. Newsl.* 5, 548-552 (1980).
25. H. Hughes, Application of ICP Emission Spectrometry for the Steel Industry, ECSC Project Rep. No. FR 497-802, British Steel Corp., March 1981.
26. K. Ohls and D. Sommer, Analysis of Liquid, Solid, and Gaseous Samples Using a High-Power Nitrogen/Argon ICP System. Paper presented at the 1980 Winter Conference on Plasma Spectrochemistry, San Juan in "Developments in Atomic Plasma Spectrochemical Analysis," R. M. Barnes, Ed. Heyden, London, 1981, pp. 321-326.
27. A. E. Watson, The Commissioning of a 5-kW Inductively Coupled Plasma Unit for Use with a 3.4-Meter Ebert Spectrograph, NIM Rep. No. 2029, in *ICP Inf. Newsl.* 5, 553-563 (1980).
28. I. B. Brenner, A. E. Watson, T. W. Steele, E. A. Jones, and M. Gonçalves, Application of an Argon/Nitrogen Inductively Coupled RF Plasma (ICP) to the Analysis of Geological and Related Materials for Their Rare Earth Contents, *Spectrochim. Acta* 36B, 785-797 (1981).
29. L. Ebdon, M. R. Cave, and D. J. Mowthorpe, Simplex Optimization of Inductively Coupled Plasmas, *Anal Chim. Acta* 115, 179-187 (1980).
30. S. Greenfield, Plasma Sources in Spectroscopy. *Metron* 3, 224-230 (1971).
31. S. Greenfield and D. T. Burns, A Comparison of Argon-Cooled and Nitrogen-Cooled Plasma Torches Under Optimized Conditions Based on the Concept of Intrinsic Merit, *Anal Chim. Acta* 113, 205-220 (1980).
32. A. E. Watson, G. M. Russell, H. R. Middleton, and F. F. Davenport, The Application of the Inductively Coupled Plasma System to the Simultaneous Determination of Precious Metals, MINTEK Rep. No. M81 (1983).
33. G. L. Moore, P. J. Humphries-Cuff, and A. E. Watson, Simplex Optimization of a Nitrogen-Cooled Argon Inductively Coupled Plasma for Multielement Analysis, *Spectrochim. Acta* 39B, 915-929 (1984).
34. A. Wittmann, J. Hancart, H. Hughes, and K. Ohls, Application of ICP-AES in Steelwork Laboratories. Paper presented at the 1980 Winter Conference on Plasma Spectrochemistry, San Juan, in "Developments in Atomic Plasma Spectrochemical Analysis," R. M. Barnes, Ed. Heyden, London, 1981, pp. 550-563.
35. K. Ohls and D. Sommer, Direct Analysis of Compact Samples Using High-Power ICP Emission Spectrometry, *Fresenius Z. Anal. Chem.* 296, 241-246 (1979).
36. D. Sommer and K. Ohls, ICP Optical Emission Spectrometry as a Detector for Elemental Analysis After Gas Chromatographic Separation, *Fresenius Z. Anal. Chem.* 295, 337-341 (1979).
37. J.A.C. Broekaert, F. Leis, and K. Laqua, The Application of an Argon/Nitrogen Inductively Coupled Plasma to the Analysis of Organic Solutions, *Talanta* 28, 745-752 (1981).

38. J.A.C. Broekaert, B. Wipenka, and H. Pruxbaum, Inductively Coupled Plasma Optical Emission Spectrometry for the Analysis of Aerosol Samples Collected by Cascade Impactors, *Anal Chem.* 54, 2174-2179 (1982).

39. K. Ohls and D. Sommer, Analytical Application of an Air/Argon ICP Source in Emission Spectroscopy, *ICP Inf. Newsl.* 4, 532-536 (1979).

40. K. Ohls and D. Sommer, Analytical Application of the Oxygen-Cooled Argon ICP—I. Improved Analysis of Aqueous Solution, *ICP Inf. Newsl.* 9, 555-562 (1984).

41. S. Greenfield and H. McD. McGeachin, Calorimetric and Dimensional Studies on Inductively Coupled Plasmas, *Anal Chim. Acta* 100, 101-119 (1978).

42. K. Ohls, Analytische Anwendung der ICP-Emissionsspecktrometrie, Analytiktreffen Neubrandenburg 1982, in "Atomspektroskopie-Forschritte und analyt. Anwendungen," *Wiss. Beit. Karl-Marx-Univ. Leipzig* 1983, pp. 123-135.

43. K. Ohls, Einsatz der ICP-Specktroskopie in Eisenhuttenlaboratorien, ECSC Project No. 7210. GA/114, Hoesch Stahl AG, March 1981.

44. A. Montaser and J. Mortazavi, Optical Emission Spectrometry with an Inductively Coupled Plasma Operated in Argon-Nitrogen Atmosphere, *Anal Chem.* 52, 255-259 (1980).

45. A. Montaser, V. A. Fassel, and J. Zalewski, A Critical Comparison of Ar and Ar-N$_2$ Inductively Coupled Plasma as Excitation Sources for Atomic Emission Spectrometry. *Appl. Spectrosc.* 35, 292-302 (1981).

46. A. Montaser and V. A. Fassel, Electron Number Density Measurements in Ar and Ar-N$_2$ Inductively Coupled Plasmas. *Appl. Spectrosc.* 36, 613-617 (1982).

47. R. S. Houk, A. Montaser, and V. A. Fassel, Mass Spectra and Ionization Temperatures in an Argon-Nitrogen Inductively Coupled Plasma. *Appl. Spectrosc.* 37, 425-428 (1983).

48. E. A. Stubley and G. Horlick, Some Near-IR Spectral Emission Characteristics of the Inductively Coupled Plasma, *Appl. Spectrosc.* 38, 162-168 (1984).

49. He ZhiZhuang, Study of Characteristics for Low-Power and Low-Gas-Flow N$_2$/Ar ICP Source. *Fenxi Huaxue* 11, 181-186 (1983).

50. G. A. Meyer and R. M. Barnes, Inductively Coupled Plasma Discharge in Flowing Nonargon Gas at Atmospheric Pressure for Spectrochemical Analysis, U.S. Patent 4,482,246 (1984).

51. R. M. Barnes and G. A. Meyer, Low-Power Inductively Coupled Nitrogen Plasma Discharge for Spectrochemical Analysis, *Anal. Chem.* 52, 1523-1525 (1980).

52. G. A. Meyer, Ph.D. thesis, Inductively Coupled Plasma Discharges in Molecular Gases, Department of Chemistry, University of Massachusetts, Amhurst, 1982.

53. M. H. Abdallah and J. M. Mermet, Comparison of Temperature Measurements in ICP and MIP with Ar and He Plasma Gas, *Spectrochim. Acta* 37B, 391-397 (1982).

54. (a) S. Chan and A. Montaser, A Helium Inductively Coupled Plasma for Atomic Emission Spectrometry, *Spectrochim. Acta* 40B, 1467-1472 (1985). (b) S. Chan, R. L. Van Hoven, and A. Montaser, Generation of a Helium Inductively Coupled Plasma in a Low-Gas-Flow Torch, *Anal Chem.* 58, 2342-2343 (1986). (c) S. Chan and A. Montaser, A Helium Inductively Coupled Plasma: Background Spectra Emitted in the Red and Near-Infrared Spectral Region, *Appl. Spectrosc.* (in press).

55. A. Montaser, S. Chan, G. Huse, P. Vieira, and R. Van Hoven, Prominent Spectral Lines for Analytical Atomic Emission Spectrometry with an Ar-N$_2$ Inductively Coupled Plasma, *Appl. Spectrosc.* 40, 473-477 (1986).

56. M. Capitelli, Problems of Determination of Transport Properties of Argon-Nitrogen Mixtures at One Atmosphere Between 5,000 K and 15,000 K, *Ing. Chim. Ital.* 6, 94-103 (1970).

57. M. Capitelli, F. Cramarossa, L. Triolo, and E. Molinari, Decomposition of Al$_2$O$_3$ Particles Injected into Argon-Nitrogen Induction Plasmas of 1 Atmosphere, *Combust. Flame* 15, 23-32 (1970).

58. J. Janka and A. Talsky, Spectroscopic Observations on Induction Coupled HF Discharge in Argon and Nitrogen Mixtures, *Purkynianae Brun.* 15, 49-56 (1974).

59. R. M. Barnes and S. Nikdel, Mixing of Ambient Nitrogen in a Flowing Inductively Coupled Plasma Discharge, *Appl. Spectrosc.* 29, 447-481 (1975).

60. P. Schramel, R. Fischer, A. Wolf, and S. Hasse, Influence of Hydrogen on Different Gas Flows of an Argon Plasma in Spectroscopy, *ICP Inf. Newsl.* 6, 401-408 (1981).

61. P. Schramel, X. Li-qiang, Further Investigation of an Argon-Hydrogen Plasma in ICP Spectroscopy. *Fresenius Z. Anal. Chem.* 319, 229-239 (1984).

62. G. Horlick and E. H. Choot, Analytical and Spectral Characteristics of N$_2$/Ar, He/Ar, O$_2$/Ar, and Air/Ar Mixed Gas ICPs. Paper No. 12, presented at the Eastern Analytical Symposium, New York, 1981.

63. (a) J. F. Adler, R. M. Bombelka, and G. F. Kirkbright, Electronic Excitation and Ionization Temperature Measurements in a High-Frequency Inductively Coupled Argon Plasma Source and the Influence of Water Vapour on Plasma Parameters, *Spectrochim. Acta* 35B, 163–175 (1980). (b) Y. Q. Tang and C. Trassy, Inductively Coupled Plasma: The Role of Water in Axial Excitation Temperatures, *Spectrochim. Acta* 41B 143–150 (1986).

64. (a) B. Magyar, P. Lienemann, and S. Wunderli, Suppression of the Background Emission of an Inductively Coupled Argon Plasma by Feeding of Oxygen for the Spectral Analysis of Organic Solutions, *GIT Fachz. Lab.* 26, 541–548 (1982). (b) B. Magyar, P. Lienemann, and H. Vonmont, Some Effects of Aerosol Drying and Oxygen Feeding on the Analytical Performance of an Inductively Coupled Nitrogen–Argon Plasma, *Spectrochim. Acta* 41B, 27–38 (1986).

65. S. J. Northway and R. C. Fry, Atomic Oxygen Spectra in the Argon Inductively Coupled Plasma, *Appl. Spectrosc.* 34, 332–338 (1980).

66. D. Truitt and J. W. Robinson, Spectroscopic Studies of Radio-Frequency Induced Plasma—I. Development and Characterization of Equipment, *Anal. Chim. Acta* 49, 401–415 (1970).

67. D. Truitt and J. W. Robinson, Spectroscopic Studies of Organic Compounds Introduced into a Radio-Frequency-Induced Plasma—II. Hydrocarbons, *Anal Chim. Acta* 51, 61–67 (1970).

68. R. M. Brown Jr. and R. C. Fry, Near-Infrared Atomic Oxygen Emissions in the Inductively Coupled Plasma and Oxygen-Selective Gas–Liquid Chromatography, *Anal. Chem.* 53, 532–538 (1981).

69. G. F. Kirkbright and R. D. Snook, Volatilization of Refractory Compound Forming Elements from a Graphite Electrothermal Atomization Device for Sample Introduction into an Inductively Coupled Argon Plasma, *Anal. Chem.* 51, 1938–1941 (1979).

70. (a) S. K. Hughes and R. C. Fry, Nonresonant Atomic Emissions of Bromine and Chlorine in the Argon Inductively Coupled Plasma, *Anal. Chem.* 53, 1111–1117 (1981). (b) M. Keane and R. C. Fry, Red and Near-Infrared Inductively Coupled Plasma Emission Spectra of Fluorine, Chlorine, Bromine, Iodine, and Sulfur with a Photodiode Array Detector, *Anal Chem.* 58, 790–797 (1986), and references therein.

71. (a) S. J. Northway, R. M. Brown, and R. C. Fry, Atomic Nitrogen Spectra in the Argon Inductively Coupled Plasma, *Appl. Spectrosc.* 34, 338–348 (1980). (b) J. M. Keane, D. C. Brown, and R. C. Fry, Red and Near-Infrared Photodiode Array Atomic Emission Spectrograph for Simultaneous Determination of Carbon, Hydrogen, Nitrogen, and Oxygen, *Anal. Chem.* 57, 2526–2533 (1985), and references therein.

72. M. H. Abdallah and J. M. Mermet, The Behavior of Nitrogen Excited in an Inductively Coupled Argon Plasma, *J. Quant. Spectrosc. Radiat. Transfer* 19, 83–91 (1978).

73. P. B. Zeeman, S. P. Terblanche, K. Visser, and F. H. Hamm, Temperature Determination on a 9.2-MHz Inductively Coupled Plasma Source Using N Bands as Monitor, *Appl. Spectrosc.* 32, 572–576 (1978).

74. J. F. Alder and J. M. Mermet, A Spectroscopic Study of Some Radio Frequency Mixed Gas Plasmas, *Spectrochim. Acta* 28B, 421–433 (1973).

75. J. Jarosz and J. M. Mermet, A Spectroscopic Study of a High-Frequency Inductively Coupled Argon–Hydrogen Sulfide Plasma, *J. Quant. Spectrosc. Radiat. Transfer* 17, 237–246 (1977).

76. (a) G. A. Meyer, Analysis of Metals in Organic Solvents by Total Air–ICP Emission Spectrometry. Paper presented at the 1986 Winter Conference on Plasma Spectrochemistry, Kailua-Kona, Hawaii, January 2–8, 1986. (b) R. J. Klueppel, J. L. Spenser, and F. W. Plankey, Design of an On-Line Air Plasma Spectrometer for Process Control. Paper presented at the 1986 Winter Conference on Plasma Spectrochemistry, Kailua-Kona, Hawaii, January 2–8, 1986.

77. R. M. Barnes, Analytical Applications of ICP Spectroscopy. Analytiktreffen Neubrandenburg 1982, in "Atomspektroskopie-Fortschritte und Anwendungen," *Wiss. Beitr. Karl-Marx-Univ.*, pp. 33–40.

78. R. M. Barnes, N. Kovacic, and G. A. Meyer, Computer Simulation of a Nitrogen ICP Discharge, *Spectrochim. Acta* 40B, 907–918 (1985).

79. G. A. Meyer and R. M. Barnes, Analytical Inductively Coupled Nitrogen and Air Plasmas, *Spectrochim. Acta* 40B, 893–905 (1985).

80. N. Kovacic, G. A. Meyer, L. Ke-ling, and R. M. Barnes, Diagnostics in an Air Inductively Coupled Plasma, *Spectrochim. Acta* 40B, 947–957 (1985).

81. G. A. Meyer, M. D. Thompson, Determination of Trace Element Detection Limits in Air and Oxygen Inductively Coupled Plasmas, *Spectrochim. Acta* 40B, 195–207 (1985).

82. (a) D. C. Miller, C. J. Seliskar, and T. M. Davidson, Hydrogen Isotope Analysis Using a Reduced-Pressure ICP Torch, *Appl. Spectrosc.* 39, 13–19 (1985). (b) C. J. Seliskar and D. K.

Warner, A New Reduced-Pressure ICP Torch, *Appl. Spectrosc.* 39, 181–183 (1985). (c) K. A. Walnik, D. C. Miller, C. J. Seliskar, and F. L. Fricke, Characterization of Bromine and Chlorine Atomic Emission in a Reduced-Pressure Inductively Coupled (27 MHz) Helium Plasma, *Appl. Spectrosc.* 39, 930–935 (1985).

83. T. R. Smith and M. B. Denton, On the Operation of Inductively Coupled Plasmas as a Function of Pressure, *Spectrochim. Acta* 40B, 1227–1237 (1985).

84. K. Ohls, Annual Report on ICP-AES in German Steel Laboratories, *ICP Inf. Newsl.* 7, 6–14 (1981).

85. R. M. Barnes and S. Nikdel, Computer Simulation of Inductively Coupled Plasma Discharge for Spectrochemical Analysis—II. Comparison of Temperature and Velocity Profiles, and Particle Decomposition for ICP Discharges in Argon and Nitrogen, *Appl. Spectrosc.* 30, 310–320 (1976).

86. G. R. Kornblum, Second ICP Conference, Noordwijk aan Zee, Excitation Mechanism—Physical Data, Excitation Models. *ICP Inf. Newsl.* 4, 147–171 (1978).

87. H. U. Eckert, Analytical Treatment of Radiation and Conduction Losses in Thermal Induction Plasmas, *J. Appl. Phys.* 41, 1529–1537 (1970).

88. X. Chen and E. Pfender, Heat Transfer to a Single Particle Exposed to a Thermal Plasma, *Plasma Chem. Plasma Process.* 2, 185–212 (1982).

89. A. Aziz, J.A.C. Broekaert, K. Laqua, and F. Leis, A Study of Direct Analysis of Solid Samples Using Spark Ablation Combined with Excitation in an Inductively Coupled Plasma, *Spectrochim. Acta* 39B, 1091–1103 (1984).

90. R. K. Winge, V. J. Peterson, and V. A. Fassel, Inductively Coupled Plasma–Atomic Emission Spectroscopy: Prominent Lines, *Appl. Spectrosc.* 33, 206–219 (1979).

91. F. Cramarossa, S. de Benedictis, and G. Ferraro, Spectroscopic Measurements of Non-equilibrium Nitrogen Plasmas at Moderate Pressures, *J. Quant. Spectrosc. Radiat. Transfer* 23, 291–301 (1980).

92. S. Greenfield and D. T. Burns, A Tutorial Discussion of the Necessary and the Sufficient Criteria To Be Used for the Unequivocal Comparison of Plasma Torches as Emission Sources, *Spectrochim. Acta* 34B, 423–441 (1979).

93. S. Terblanche, K. Visser, and P. B. Zeeman, The Modified Sequential Simplex Method of Optimization as Applied to an Inductively Coupled Plasma Source, *Spectrochim. Acta* 36B, 293–297 (1981).

94. K. Ohls, Einsatz losungsspektroskopischer Methoden zur simultanen Multielementanalyse aus Schmierolen, in "Erdol und Kohle, Compendium 77/78," Industrie-Verlag von Hernhausen, Leindelden-Echterdingen, 1977, pp. 194–208.

95. K. Ohls, Die bestimmung kleiner Cd- Anteile in verschie- denen Materialien durch Fest-probeneinsatz bei ICP—und flammenbärer AAS, *Spectrochim. Acta* 39B, 1105–1111 (1984).

96. D. Sommer, K. Ohls, and A. Koch, Determination of Tin in Metals Using ICP Emission Spectrometry and Hydride Generation, *Fresenius Z. Anal. Chem.* 306, 372–377 (1981).

97. M. Trunstall, H. Berndt, D. Sommer, and K. Ohls, Direkte Spurenbestimmung in Aluminium mit der Flammen-AAS und ICP-Emissionspektrometrie—Ein Vergleich, *Erzmetall* 34, 588–591 (1981).

98. K. Ohls and H. Loepp, Use of the Internal Standard Method to Correct and Calibrate ICP Spectrometric Elemental Determinations from Dissolved Metallic Samples, *Fresenius Z. Anal. Chem.* 322, 371–378 (1985).

99. J.A.C. Broekaert and P. K. Hormann, Separation of Yttrium and Rare Earth Elements from Geological Materials, *Anal. Chim. Acta* 124, 421–425 (1981).

100. J.A.C. Broekaert and F. Leis, Application of Two Different ICP-Hydride Techniques to the Determination of the Arsenic, *Fresenius Z. Anal. Chem.* 300, 22–27 (1980).

101. D. Sommer and K. Ohls, Organische Losungsmittel in der ICP-Spektroskopie, *Labor Praxis* 598–610 (June 1982).

102. A. Aziz, J.A.C. Broekaert, and F. Leis, Analysis of Microamounts of Biological Samples by Evaporation in a Graphite Furnace and Inductively Coupled Plasma Atomic Emission Spectroscopy, *Spectrochim. Acta* 37B, 369–379 (1982).

16

An Overview of Analysis by Inductively Coupled Plasma–Atomic Emission Spectrometry

PETER N. KELIHER

Department of Chemistry
Villanova University
Villanova, Pennsylvania

16.1 INTRODUCTION

In 1980, Stanley Greenfield, one of the prominent pioneers of ICP atomic emission spectrometry (ICP-AES) presented a plenary lecture at the Fifth International Conference on Analytical Chemistry of the Society for Analytical Chemistry, in Lancaster, England. In his talk, Greenfield reflected on the growth of ICP-AES and stated that the technique has achieved respectability and has "come of age."[1] He concluded his presentation by stating: "Finally, I believe the future holds a lot for the ICP. I am convinced that, now having come of age, it will survive to a ripe old age." The ICP-AES technique has given birth to ICP-mass spectrometry (ICP-MS), and ICP-atomic fluorescence spectrometry (ICP-AFS) and the relative growth rates of the three techniques will certainly be the subjects of much animated conversation at future professional meetings and in the literature. Of course, a better source *may* be developed in the near future to make ICP techniques obsolete; however, that is *highly unlikely*, and so the author joins Greenfield "on the limb of the tree" to predict a healthy, full life for the ICP techniques. The only uncertainty for the future might be the percentages

Table 16.1. Certain Capabilities of Major Analytical Techniques for Trace Analysis of Solutions[a,b]

Technique	Detection limits (ng/mL) ≥	1	≥	Matrix effects	Multielement determination	Speciation analysis
AAS						
Flame (wire loop)		■ (□)		+	No	Yes (+)
Furnace			□	+++	No	Limited
HG (HGC)		□	(□)	++	No	Yes
ZAAS: Furnace			□	+	No	Yes
AES–ICP						
PN (USN)	■	□		+	Yes	Yes
ETV	■			+	Yes	Possible
HG (HGC)	■		(□)	++	Yes	Possible
AES–DCP: PN	■			+++	Yes	Yes
AES–MIP						
PN	■			+++	Limited	Yes (+)
ETV	■			++	Limited	Yes
HG (HGC)	■		(□)	++	Limited	Yes
AES–HCL: Volatilization	■			++	Limited	Possible
FANES: Furnace		□		+++	Limited	Possible
AFS						
Laser–ICP	■			+	Yes	Possible
ICP–ICP	■			+	Yes	Possible
HCL–ICP	■			+	Yes	Possible
Laser–furnace		□		+++	Yes	
CFS: Furnace	■			+++	Yes	Possible
OGS: Flame		□		+++	Limited	
XRS						
XRF	■			+++	Z > 6	Yes
PIXE	■			+++	Z > 14	No
TR		□		+	Z > 14	No
MS						
ID–ICP		□		+	Yes	Yes
ID–TI			□	++	Limited	No
ID–FD			□	++	Limited	
NAA						
INAA	■			(+)	Yes	Possible
RANAA			□	(+)	Limited	No
Voltammetry						
DPASV		□		+++	Limited	Yes
DPCSV		□		+++	Limited	Yes
Chelation						
GC–ECD			□	+++	Limited	Yes
HPLC–UVD		■		+++	Limited	Yes
Spectrophotometry	■				No	Yes
Fluorimetry		□		++++	No	Yes

[a] Adopted from Plenary lecture by J.A.C. Broekaert at the 1986 Winter Conference on Plasma Spectrochemistry, Kailua-Kona, HI. The text of this lecture has been submitted to Spectrochim. Acta for publication: J.A.C. Broekaert and G. Tölg, "Status and Trends of Development of Atomic Spectrometry Methods for Elemental Trace Determinations."

of usage of optical and of mass spectrometry. At this time, the dominant technique is clearly *optical* spectrometry, but this may not be the case in the next few years!

With respect to the technique, one should note that most manufacturers use the ICP for sequential or simultaneous atomic emission spectrometry. However, Baird Corporation offers an ICP multiple-element atomic fluorescence spectrometer (AFS) that uses hollow cathode lamps to excite atomic fluorescence of up to 12 elements simultaneously. The techniques of ICP-AFS and ICP-MS are reviewed in Chapters 9 and 10, respectively.

This chapter is based on *selected* relevant references that describe applications of ICP-AES to a wide variety of sample materials. Most references are from the literature of the past 8 years, and from major journals that are easily accessible in Western libraries. This practice follows the one used for "Emission Spectrometry" reviews.[2-4] However, publications of fundamental importance are cited, regardless of the source. For some cases in which a particular publication is concerned with two different applications (eg, agricultural and steel analysis), the most important application has been selected somewhat arbitrarily. The same point applies to the selection of paper categories.

The papers chosen for citation or discussion have the ICP as the major thrust of the paper. Many publications, however, compare ICP techniques with other techniques, such as atomic absorption spectrometry (AAS) and direct current plasma (DCP) spectrometry. In most cases, ICP techniques compare very favorably to other atomic spectrometric techniques. For a detailed discussion and comparison on this point, the reader is referred to the review by Parsons and associates.[5a] Also, in a recent review, Broekaert and Tolg[5b] compare the

---

[b] Definition of Terms:

Column 1:

AAS = Atomic absorption spectrometry; ZAAS = Zeeman AAS; AES = atomic emission spectrometry; ICP = inductively coupled plasma; DCP = direct current plasma; MIP = microwave-induced plasma; HCL = hollow cathode lamp; FANES = furnace non-thermal excitation spectrometry; AFS = atomic fluorescence spectrometry; CFS = coherent forward scattering; OGS = optogalvanic spectrometry; XRS = x-ray spectrometry; MS = mass spectrometry; NAA = neutron activation analysis; Chelation-GC = chelation and gas chromatography; HPLC = high-performance liquid chromatography.

Column 2:

HG = hydride generation; HGC = cold-vapor generation of Mercury; PN = pneumatic nebulizer; ETV = electrothermal vaporizor; USN = ultrasonic nebulizer; Laser-ICP = laser-excited AFS with an ICP; ICP-ICP = ICP-excited AFS with a second ICP; HCL-ICP = HCL-excited AFS with an ICP; laser-furnace = laser-excited AFS with a furnace; XRF = x-ray fluorescence; TR = total reflection x-ray spectrometry; ID = isotope dilution; TI = thermal ionization source; FD = ionization by field desorption; INAA = instrumental neutron activation analysis; RANAA = radiochemical neutron activation analysis; DPASV = differential-pulsed anodic stripping voltametry; DPCSV = differential-pulsed cathodic stripping voltametry; ECD = electron capture detector, and UVD = UV detector.

Column 3:

Detection limit is an order-of-magnitude value; ■ means greater than 1 ng/mL; □ means less than 1 ng/mL and (□) means DL by the technique in parenthesis.

Column 4:

+ means very low matrix effect; (+) means very low matrix effect, but the effect varies between elements; + + moderate matrix effect; + + + means high matrix effect and + + + + very high matrix effect.

Column 5:

Z = atomic number of the element.

Column 6:

(+) means chromatographic separation is required for speciation analysis.

major potential of the ICP-based techniques to various spectrometric, electro-metric, chromatographic, and neutron activation techniques (Table 16.1). The advantages of the ICP-based techniques, in terms of detection limits, matrix effects, multiple-element determinations, and separation analysis are evident.

The major applications of ICP–AES are reviewed alphabetically in the following sections. To minimize overlap, certain applications cited in earlier chapters are not discussed here.

16.2 AGRICULTURAL MATERIALS

The ICP has "taken root" with respect to common usage in agricultural analysis.[6–24] One of the most interesting, and certainly unusual, applications of the ICP comes from Arizona State University, where an element has been used as a "marker" for the pink bollworm (PBW),[6] a major pest in the cotton industry in Arizona. The PBW life cycle consists of four phases: egg, larva, pupa, and moth. By observing the higher levels of elemental uptake of the PBW moth, it is possible to distinguish mass-reared (sterile) PBW from native (wild) PBW. A simultaneous multiple-element ICP spectrometer was used to study the availability of the PBW to incorporate 13 elements. Barium, colbalt, cadmium, iron, and strontium showed favorable biological uptake. In addition, sterile male PBW moths, released in Mexico, have been marked with strontium to test the use of this element as a discriminator. This research is certainly a very practical example of the utility of ICP–AES in the analysis of "genuine-in-real-life samples."

Liese[7] described the development of a simple method for the determination of 21 elements in plants and soils. Multiple-element standards were used for calibration without any special matrix modification. Good agreement with certified concentrations for elements in standard reference materials was re-ported. Deviations from certified values, which did occur in some cases, were due to incomplete digestion and errors in spectral interference correction. Kuennen and co-workers[8] used a pressure-dissolution technique to minimize sample pretreatment for the multiple-element analysis of raw agricultural crops by ICP–AES. The procedure employs a 30-minute pressure dissolution of sample composite with 6 M HCl at 80°C in 60-mL polyethylene bottles. The procedure was reported to compare favorably with more time-consuming conventional wet-ashing methods for the determination of major, minor, and trace elements occurring in lettuce, potatoes, peanuts, soybeans, spinach, sweet corn, and wheat. Recoveries for spiked samples, precision studies, and analyses of NBS reference materials demonstrated the reliability and accuracy of the procedure.

In a detailed study, Jones and co-workers[9] used Chelex-100 resin to concen-trate trace elements and to remove potentially interfering alkali metals and alkaline earths. This separation procedure was included in a scheme designed to maximize the number of analytes that could be determined from a single digestion of a plant or animal tissue. Acid-digested samples were divided into two fractions. One fraction, 5% of the total, was measured directly by ICP–AES for alkali metals and alkaline earths, phosphorus, and for transition metals (eg, Fe and Mn) that did not work well with the resin. The other fraction, 95% of the total, was subjected to the separation procedure, wherein a number of important trace elements, including cadmium, copper, molybdenum, nickel, vanadium, and zinc, were initially held by the resin and then stripped into a small volume of dilute HNO_3 for ICP–AES measurement of the matrix-free analytes. The reliability of the scheme was influenced by the nature of the acid digestion procedure used to oxidize the organic matrix. In a later publication, Jones and O'Haver[10] reported on the effects of pH and digestion conditions on separation of trace elements in tissue digests by Chelex-100 resin before ICP–AES determination. The optimum pH for separations and a multielement digestion procedure for subsequent Chelex-100 separations were determined using NBS SRMs as samples. In the pH range of 3.3 to 4.7, substantial differences were observed for the retention of several elements (eg, Cd and Zn) by the Chelex-100 resin, between nitric–perchloric acid and nitric–perchloric–sulfuric acid digestions. From pH 4.7 to 6.3, either digestion procedure was satisfactory for most elements examined, including iron, nickel, copper, cadmium, and molybdenum. Manganese retention by the resin was strongly affected by pH for the conditions studied, but recovery was quantitative only at a pH of 6. Vanadium recovery did not show a distinct pH effect over the range examined, but vanadium V/vanadium IV systems exhibited slightly different retention behavior on the resin. Some elements, such as aluminum and iron, were not dissolved from siliceous materials in botanical samples by either digestion procedure.

Sutton[11] showed that a computer-controlled sequential ICP–AES system could effectively complement flame–AAS for plant analysis, especially by extending the range of elemental determinations possible on a single plant digest solution. Highly satisfactory results were obtained with NBS orchard leaves. Ignoring initial standardization time, about 3.5 minutes was required to determine the seven elements in a sample. Nebulizer clogging occurred for $HClO_4$ digests when plant samples contained more than 3% potassium.

Elemental concentration levels in raw agricultural crops were reported[12] from a joint study involving the U.S. Food and Drug Administration and the EPA. More than 3000 samples were analyzed for 12 elements by ICP–AES, and a quality assurance program was developed to assure the accuracy and reliability of these analyses. Another study[19] surveyed 12 service laboratories using ICP–AES for routine plant analysis in New Zealand. A variety of sample preparation and instrumental calibration procedures was used in the 12 laboratories, and results for 11 elements showed that the best precision was achieved for phosphorus, magnesium, manganese, calcium, and potassium while relatively poor precision was obtained for aluminum, zinc, iron, sodium, copper, and boron. The imprecision noted for the latter elements was presumed to be caused by the diversity of preparation and calibration methods. Studies of this type show the importance of interlaboratory cooperation in adopting a uniform procedure for analyzing large numbers of samples.

Verbeek[14] analyzed extracts from tree leaf, bark, and wood samples for several elements. Recovery percentages of simulated tree extracts and for spiked tree samples were reported together with typical analysis values for a leaf and a wood sample. The choice of spectral line for each element is discussed, with particular emphasis on the spectral interference of copper on phosphorus (214.9 nm) and of iron on boron (249.7 nm). Feng and co-workers[15] determined 22 elements in growth rings of trees by ICP-AES, and, in related work, Matusiewicz and Barnes[16] developed a method utilizing pressure decomposition to minimize sample pretreatment in the analysis of red spruce and sugar maple trees. Cores

Table 16.2. Other Materials Analyzed by ICP–AES

Material	Element(s) determined	Ref.
Spinach digests, bovine liver	Mn, Fe, Cu	17
Herbage	As, Sb, Bi	18
Rice flour, spinach, orchard leaves	As, Bi, Ge, Sb, Se, Sn	19
NBS tomato and orchard leaves	As	20
Agricultural samples	As, Pb, Cu, Fe, Bi	21
Plant materials	Mo	22
Various foodstuffs	Cu, Fe, Mn, Zn, Cd, Pd, Ni, Cr, As, Sn	23
Red and rosé Italian wines	Si, Fe, B, Al, Cu, Zn, Sn	24
Milk powder	P	42
Glucose solution	Mn	43
Urine	Si	44
Feces	Cr	45
Bovine liver	Be	46
Human (gall and kidney) stones	Ca, Mg, P	47
Geological materials	Lanthanide elements, Cr, V	78–82, 84–86
Geological reference materials	B	83
Coal fly ash	Cr, Sr, Tl, V, Cu, Zn	87
Lead and zinc flotation concentrates	Pb, Zn, Ag	88
Soil extracts	Mo, Co, B	89
Soils, sediments	As, Sb, Bi	90
Xylene	S	91
Oils	S	92
Alloys	W	124
Steel	P	125
Alumina-based catalysts	Al, Co, La, Ru	126
Alloys, steels	B	127–129
Zirconium and lanthanum oxides	Hf, various lanthanides	130, 131
Seawater	Pb, Zn, Cd, Ni, Mn, Fe, V	164
USGS water standards	Mn, Cu, Fe, Zn, Ba	165
Lake water	Al	166
Cosmetics	S	167, 168
Polymeric materials	S	169
Electrolyte solutions	Cationic contaminants	170
Heteropolymolybdates and tungstates	Mo, W, As, P	171

collected from trees growing in Vermont were divided into decade increments to monitor the temporal changes in concentrations of 21 elements. Elements forming oxyanions, such as aluminum, arsenic, iron, germanium, manganese, silicon, and vanadium, were found at elevated concentrations during the most recent three decades, while the concentrations of other metals (eg, Mg and Zn) were unchanged or decreased. Such observations are important because the concentrations of toxic metals in dated tree rings may be linked to environmental effects, such as acid rain.

Other interesting publications[17–24] in this area are listed in Table 16.2.

16.3 BIOLOGICAL MATERIALS

Many major and trace elements are of considerable interest in biological samples such as blood, kidney stones, milk powders, serum electrolytes, and urine, but the relatively complicated matrices associated with many of these materials have made analyses difficult. For this reason, the recent popularity of the ICP-AES has made metal determinations much easier and has led to lower detection limits for many metals. Furthermore, when coupled with alternative sample introduction devices such as electrothermal vaporizers (ETV), the ICP requires greatly reduced quantities of sample, which is of considerable importance in the analysis of blood, serum plasmas, and similar materials. For example, Kirkbright and co-workers[25] reported on the determination of manganese and nickel in whole blood using an ETV in conjunction with an ICP-AES system. The detectability of the method was considered equal or better than that of most techniques currently in use for the determination of the elements in biological samples. *An important point was that sample preconcentration was not required, thus reducing the analysis time and the risk of contamination.* Blakemore and co-workers[26] described an ETV for microsamples of liquids or solids. Liquid samples (5 or 10 μL) were transferred to a groove in a carbon rod vaporizer, dried, and vaporized with the analytes transported into the ICP by argon. Solid samples (0.5 mg) were weighed into a graphite microboat, which was placed on the carbon rod atomizer and analyzed as previously. Recoveries from blood plasma and from wastewater samples were between 60 and 136%. Results for samples of NBS bovine liver compared reasonably well to the certified values. Matusiewicz and Barnes[27] determined aluminum and silicon in urine and in serum using ETV-ICP. Detection limits in 5-μL samples were 8 pg Al and 2.5 ng Si, and after preconcentration of aluminum with a polyacrylamidoxime resin, the detection limit was 1 pg Al. Recoveries of 5 μg Si/mL and 19 ng Al/mL from aqueous synthetic standards were 80 to 85% and 96 to 103%, respectively. In another study, Matusiewicz and Barnes[28] examined discrete nebulization using aerosol deposition as a procedure for sample introduction for ETV-ICP-AES. Small volumes of solution ($< 50 \mu$L) could be introduced into a pneumatic nebulizer manually from a PTFE microsampling device or PTFE funnel, or automatically by a flame autosampler system. The analyte aerosol was deposited under controlled conditions onto the surface of a graphite platform. Optimum operating conditions were established and the performance of the system was demonstrated by the direct determination of chromium, copper, and nickel in urine. The method offers good precision ($\leq 3\%$), reduction or even elimination of matrix interferences, and, perhaps, most important, analysis of microsamples.

Long and Snook[29] used ICP–AES to determine the major element composition of vitamin pills formulated as multimineral capsules. Two different sample introduction techniques, conventional pneumatic nebulization of aqueous solutions and vaporization of slurries from a graphite rod ETV, were compared. The ETV technique was recommended for slurries because the sample preparation time was greatly reduced. Superior detection limits were obtained with the graphite rod ETV, although this was not a major point in the study, as these investigations were concerned with the determination of major elements in pharmaceutical preparations. Uchida and co-workers[30] described a microsampling technique that requires sample volume of less than 100 μL. The method was applied to the analysis of serum and whole blood samples, after digestion. Alexander and co-workers[31] developed an introduction system for liquid microsamples that allowed injection of 5 to 500 μL into a rapidly flowing carrier–reagent stream leading to the nebulizer. The effect on analyte signal was studied as a function of flow rate, injection volume, and sample concentration. The carrier flow rate determined the response time, sensitivity, and precision of analysis. The system, used for serum electrolyte analysis, had a throughput as high as 240 samples per hour. McLeod and co-workers[32] successfully combined flow-injection analysis (FIA) with ICP–AES for the simultaneous determinations of eight elements in microliter samples of blood serum. At dilution factors of 1 : 1 or greater, interference was not observed, and reliable data were obtained for human and bovine serum pools for sample injections of 20 μL.

Allain and Mauras[33] described a method for the determination of aluminum in water, urine, and blood by ICP spectrometry using a concentric pneumatic nebulizer. Interferences were systematically studied using different metals and metalloids, especially those commonly found in biological samples. Certain metals, particularly calcium, lithium, strontium, sodium, and iron increased the background intensity, and alkali metals and alkaline earths were reported to increase the net signal intensity of Al 396.09 nm. Limits of detection were reported as 0.4 ng/mL in water, 1 ng/mL in urine, and 4 ng/mL in blood. Sample preparation for blood and urine was reduced to simple dilution with demineralized water. Aluminum assays on 14 healthy subjects gave the following results: blood 12.5 ± 4 (standard deviation) μg/L, urine 4.7 ± 2.5 (standard deviation) μg/L. The authors experienced great difficulties with furnace AAS, and they concluded: "Having encountered no difficulties of this kind during hundreds of blood, urine, and water assays with the plasma torch, we have finally decided to drop graphite furnace–atomic absorption spectrometry in favor of inductively coupled plasma spectrometry for aluminum determinations."[33]

The determination of various metals in rodents, such as mice and rats, is of some interest. Hayman and co-workers[34] compared the ICP–AES determination of nickel in rat blood serum with liquid scintillation counting using radioactive ^{63}Ni as the tracer. The ICP–AES method requires only microliters of sample, and the use of ^{63}Ni is not necessary. These investigators concluded that ETV–ICP–AES offers a rapid and sensitive technique for clinical applications, with minimal contamination of the sample. Sun and co-workers[35] in China studied the biocompatibility of Ti or Ti_2N_2 artificial joints. The materials were implanted into muscle tissue in the vicinity of the spleen of rats, and the blood of the animals was sampled to determine titanium by ICP–AES. The Ti_2N_2 was concluded to be a desirable material for artificial joints because of its high biocompatibility and low corrosion and absorption.

Sanz-Medel and co-workers[36] critically compared flame–AAS, flame–AES, and ICP–AES for the determination of strontium in biological materials, including blood serum, brain, and liver samples from rats and hamsters. Optimum conditions for the three techniques were established and analytical performance characteristics were evaluated on the same basic instrument (a Perkin-Elmer ICP-5000) in terms of detection limits, dynamic range, selectivity, and precision. These investigators concluded that ICP–AES appeared to be the best method because it provided a detectability about 100 times superior to the next most sensitive technique, 5-orders-of-magnitude linear dynamic range, and good precision, with minimal interferences. Wolnik and co-workers[37] commented on the relative lack of biological SRMs suitable for such materials as rabbit bone, bovine kidney samples, and weanling rat tissues. The authors stated, "Until a wider variety of certified biological standard reference materials becomes available, preparation of laboratory control samples for use in method development and quality control is necessary to provide standard samples which are matched to the tissue matrix of interest."[37]

Although most ICP applications focus on trace metal determinations, some interesting reports emphasize slightly different aspects. In one interesting application, Morita and co-workers[38] used a high-pressure liquid chromatography (HPLC) system for the elemental analysis of complex materials. One of their goals was to identify vitamin B_{12} ($C_{63}H_{90}CoN_{14}O_{14}P$) from the peak area in the C, Co, and P chromatograms. The atom number ratio was determined to be $C_{64.2}$, $P_{0.93}$, Co_1, which is close to the theoretical ratio $C_{63}PCo$. Vitamin B_{12} was not recovered completely in this experiment, probably because of adsorption on the stainless steel tubing. These investigators also studied the protein cytochrome c from horse heart, and observed values of 0.47 and 55%, respectively, for iron and carbon. These experimental values were close to the theoretical values of 0.44 and 55%, respectively. In related work, Yoshida and co-workers[39] used an HPLC–ICP system for the determination of nucleotides by measuring P 213.6 nm. Twelve common 5'-ribonucleotides were determined quantitatively with the HPLC–ICP using a concentration gradient method. The integrated emission intensity of phosphorus was proportional to the number of phosphorus atoms in all nucleotides. The detection limit of the system for phosphorus standard (KH_2PO_4) was about 0.4 μg of phosphorus per milliliter, with 200 μL of sample, and the relative standard deviation was 4% in the peak height measurement. In another paper, the same group[40] determined amino acids with the HPLC–ICP technique. Carbon and sulfur emission were observed at 193.0 and 180.7 nm, respectively. Detection limits of 30 to 50 μg/mL and 1 to 3 μg/mL as amino acids were obtained for carbon and sulfur, respectively. These papers show the utility of HPLC–ICP–AES in handling complicated biological materials. Furthermore, additional work in this area will help scientists in the biomedical field obtain a better understanding of the role of metals in life processes. This point was emphasized by Mermet and Hubert[41] in their recent review article on the analysis of biological materials. Other interesting applications[42–47] are listed in Table 16.2.

16.4 GEOLOGICAL AND ENVIRONMENTAL MATERIALS

16.4.1 *Methods Development*

In 1977, Velmer A. Fassel, certainly one of the most prominent pioneers in ICP-based techniques, reported on the current and potential applications of ICP–AES in the exploration, mining, and processing of materials.[48] He forecast a bright future for the ICP in geochemical analysis, and he was entirely correct, as demonstrated by the striking advances in this area during the past several years.

In 1979, Broekaert and co-workers[49] described an ICP–AES method for the determination of various lanthanide elements in aqueous solution. The possibility of using an ICP method for the determination of lanthanides in complex samples at concentrations ranging from 0.001 to 30% was illustrated by simultaneous multielement analyses of rare earth standard samples of the National Institute for Metallurgy in South Africa. In a related paper, the same authors described some aspects of matrix effects caused by sodium tetraborate in the analysis of rare earth minerals.[50] Subsequently, Broekaert[51] reviewed some applications of ICP–AES to industrial analytical problems with an emphasis on geological samples. Floyd and co-workers[52] described applications of a computer-controlled scanning monochromator for the determination of 50 elements in geochemical and environmental matrices, including U.S. Geological Survey standards. A single set of spectral lines, found to be applicable to a broad range of sample compositions, showed negligible spectral interferences, regardless of the sample matrix.

Hannaker and Qing-Lie[53] have developed an analytical procedure for the determination of major elements in a variety of geological materials by both flame-AAS and ICP-AES. Condensed H_3PO_4 was used for the decomposition of 70 natural minerals containing sulfide, oxide, silicate, or carbonate constituents. Their procedure could also be applied to rocks, ores, soils, slags, and refractory materials as a means of rapid and complete dissolution for the analysis of even the most insoluble material. In another study, Hannaker and co-workers[54] compared ICP-AES, where the sample is dissolved before measurement, with X-ray fluorescence, where a fused disc is used. The accuracy of the ICP-AES technique was demonstrated with the X-ray fluorescence technique for a large number of elements over a wide concentration range. Burman and co-workers[55] have modified a metaborate fusion procedure for use with ICP-AES.

Brown and Biggs[56] described an alternative procedure to the classical fire assay method for determining platinum and palladium in sulfide ores, concentrates, and furnace mattes. A sample is digested with aqua regia and filtered, and any remaining gangue (worthless rock) is digested with a mixture of HF and $HClO_4$. The solution is filtered and the residue fused with sodium peroxide granules. The fused salts are dissolved in a dilute HCl solution and all three solutions are combined. The resultant solution is passed through a cation-exchange resin in the H^+ form. The chlorocomplex anions of Pt and Pd are not retained by the cation-exchange resin, while the base metal cations are efficiently removed from the eluent. The elements platinum and palladium are then determined in the solution by ICP-AES. Preliminary experiments by these investigators indicated that the method could be extended to gold. In another interesting paper, Rice and Bragg[57] developed a multielement rapid sequential ICP method for analyzing coal ash samples. The method was able to determine 12 elements in 45 s with an RSD well within ASTM limits.

A recent publication[58] concerning trace elements in geological materials compares detection limits for a large number of elements for flame-AAS, ICP-AES, and ICP-MS. The ICP-MS approach provided the best detectability compared to ICP-AES and to flame-AAS. To document the potential of ICP-MS, the authors cited the unexpected presence of uranium in U.S. Geological Survey Reference Water No. 65, and concluded: "The successful determination of 13 trace elements in three international standard reference water samples, and 14 lanthanide elements in three international standard reference silicate rocks, demonstrates that ICP-MS is a powerful new method of analysis."

16.4.2   Ores, Rocks, and Soils

Uchida and co-workers[59] have used ICP-AES to determine minor and trace elements in silicate rocks. Samples (about 0.5 g) were decomposed with mixed acids in a sealed PTFE vessel. After suitable treatment, barium, cobalt, chromium, copper, lithium, nickel, scandium, strontium, vanadium, and zirconium were determined sequentially. In a later paper, Uchida and associates[60] described an improved method for determining traces of zirconium in silicate rocks. The sample was fused with a mixture of sodium and potassium carbonates. Zirconium was separated from the large amounts of sodium and potassium by precipitation of hydrated oxides before nebulization. The detection limit was 0.23 μg/g, and results for seven standard rocks were within recommended values. Walsh reported[61] that ICP–AES can be used successfully for a wide range of trace elements in silicate rocks. He suggested two methods for solution preparation for silicates: fusion with lithium metaborate and dissolution by a mixture of hydrofluoric and perchloric acids. No significant interferences for the major constituents were observed, and most trace elements were free from interferences. He concluded that detection limits and precision obtained for solutions of silicates were adequate, and that ICP–AES offered some advantages over comparable methods, notably X-ray fluorescence. Wunsch and Czech[62] have also described various sample preparations for alloys, concentrates, and silicate rock samples.

Thompson and Zao[63] recently described the development of a high-throughput method for the determination of molybdenum in soils, sediments, and rocks. The molybdenum was dissolved by treatment of the sample with 6 M HCl in capped tubes at 120°C. The molybdenum was then extracted from the same medium into heptan-2-one, in which it is determined directly by ICP–AES. Extraction of iron could interfere with the method, but this effect was minimized by using a reducing agent. The report also discussed general problems relating to the solvent extraction of metals for ICP–AES. Mahanti and Barnes[64] have used a polydithiocarbamate resin to purify $CaCO_3$ from many trace metals that could lead to interferences when $CaCO_3$ is used as a standard in analysis of phosphate rock and dolomite. In an interesting paper, Pritchard and Lee[65] determined boron, phosphorus, and silicon in plant and soil sample digests using ICP–AES. The digestion procedure chosen enabled the elements to be determined simultaneously on a single digest. Limitations for simultaneous determinations of the three elements originated from the digestion and dissolution procedures, not from the measurement procedure.

Schramel and co-workers[66] commented on the utility of ICP–AES for the determination of various metals in soils, sludges, plants, and in biomedical materials, such as tissues and body fluids. Specific references have been made with respect to: possible matrix influences, physical interferences, choice of the device parameters, and comparison with other analytical methods. In a related publication, Schramel and associates[67] commented on the utility of the ICP–AES for soil and sludge analysis using a 16-channel spectrometer.

16.4.3   Oils and Gasoline

The high viscosity and flammability of many oils and gasolines cause difficulties when these samples are analyzed by ICP–AES. Nevertheless, many schemes have been developed to overcome these problems. Several examples are cited below.

In a significant paper, Boumans and Lux-Steiner[68] modified an ICP–AES spectrometer to analyze both aqueous and organic liquids. Detection limits for various metals in MIBK and in oil diluted with MIBK were reported. Klaentschi[69] studied the effect of particle size in the determination of wear metals, chiefly iron, in lubricating oils by ICP–AES. Oils were membrane filtered to obtain fractions with different particle sizes. For iron, a 90 to 100% recovery of particle sizes less than 8 μm and a 70 to 100% recovery of particle sizes greater than 8 μm were observed. Hausler[70] investigated the molecular size distribution of sulfur- and vanadium-containing compounds in petroleum crudes using a directly coupled liquid chromatograph with size-exclusion columns and an ICP–AES. Sulfur-containing compounds typically showed a smaller molecular size distribution than those containing vanadium.

Ng and co-workers[71] used a low-gas-flow argon ICP torch for organic solvents. Although xylene could be aspirated into the plasma at a forward power of 1 kW, similar to conditions used for aqueous solutions, power levels of up to 2 kW had to be used for other organic solvents. For the xylene solution, precision and detection limits were comparable to those obtained for aqueous samples. The results obtained for an NBS SRM fuel oil agreed favorably with the certified values. Merryfield and Loyd[72] determined several metals in oils by ICP–AES using a sample introduction system for both oil and water analysis. However, a separate vapor chamber, with its own drain system, was used for each matrix, thus eliminating excessive cleaning time in changing between oil and water and preventing deterioration of precision caused by a contaminated chamber. Ng and Caruso[73] used an ETV carbon cup to introduce 5 μL of organic sample into an ICP. Twenty-six commonly used organic solvents were introduced without extinguishing the autotuned plasma. The system was also applied to the determination of zinc in motor oil and lead in gasoline.

Broekaert and co-workers[74] used a 4-kW argon–nitrogen ICP for multielement determinations in organic solutions. Oil samples were diluted 1 : 10 with xylene, and detection limits for aluminum, copper, iron, magnesium, manganese, silicon, and zinc were reported to be in the submicrogram-per-gram range. It is interesting to note that Greenfield and Smith[75] had previously used a high-power argon–nitrogen ICP for the analysis of oils. Beckwith[76] developed an ICP–AES method for the determination of major and trace elements of glycol-based fluids, such as antifreeze. The major advantage of the method was its ability to determine the composition of the formulation and information on wear metals simultaneously.

King and co-workers[77] recently compared flame–AAS and ICP–AES results for the determination of wear metals in used lubricating oils. The data obtained by ICP–AES compared very favorably with the results from AAS. However, ICP–AES offered superior freedom from matrix interferences, greater linear working range, and better detection limits, particularly for phosphorus and boron.

Other relevant applications[78–92] of ICP–AES to elemental analysis of geological materials are cited in Table 16.2.

16.5   METALS

In a series of papers, Kirkbright and co-workers[93-96] adapted their graphite rod ETV used in conjunction with the ICP for the direct determination of metallic materials, such as dopants in cadmium mercury telluride. Comparative results between ICP–AES and electrothermal AAS were reported for 11 dopants.[93] In general, there was good agreement, but some anomalies were noted. For example, aluminum appeared to be present at higher concentrations than the nominal value when analyzed by ICP–AES, but electrothermal AAS gave a value much lower than the nominal concentration; it is possible that aluminum was not incorporated into the cadmium mercury telluride homogeneously. In a related paper, Kirkbright and Li-Xing[94] described a graphite rod direct sample insertion device for the ICP. The use of a mixture of argon and 0.1% Freon-23 as injector gas permitted the efficient volatilization of refractory carbide-forming elements. Kirkbright and Snook[95] have used a similar device for the determination of trace elements in uranium, and Barnett and co-workers[96] described four different graphite cup furnace designs for use with an automated direct sample introduction device for the ICP. Four elements were studied (Ni, Cr, Mn, and Pb), and certain correlations between heating rate and peak height were observed. Detection limits and relative standard deviations using the four furnace designs were also discussed.

Preconcentration techniques have been used by various investigators for the determination of elemental traces in metallic materials. For example, Nakahara and Kikui[97,98] used a hydride generation technique for the determination of selenium and tin in high-purity metals and in NBS SRMs. Hartenstein and co-workers[99] developed a preconcentration method utilizing a miniature ion-exchange column of Chelex-100. Separate, buffered samples (pH = 9) were pumped through two parallel columns, for doubled sampling frequency, and sequentially eluted directly to the nebulizer of the ICP by using a flow injection analysis system. The FIA–ICP method gave simultaneous-multielement detection limits approximately 20 times better than those for direct-solution nebulization. In similar work, McLeod and co-workers[100] devised a microcolumn of activated alumina to perform rapid analyte enrichment and matrix removal for the FIA–ICP technique. The activated alumina had a high affinity for phosphate ions, and the FIA system was tested for the determination of trace phosphorus in aqueous solutions using P 213.6 nm. Phosphate deposition and elution depended on sample acidity, column length, and the nature and concentration of eluent. Applications to the analysis of a standard steel was demonstrated.

Thompson and co-workers[101] used a laser microprobe to introduce solid standard steel samples into an ICP. The absolute detection limits obtained with the solid sample introduction system were comparable to those obtained with solution nebulization on the same ICP. Takahashi and associates[102] constructed a unique device for the introduction of solid metals into the ICP. An aerosol generator with a low-voltage spark discharge was used for vaporization of metal samples, and the aerosol was then introduced into the ICP through a cyclone chamber to remove larger particles from the aerosol stream. Of related interest, Thelin[103] developed a nebulizer capable of continuous operation with solutions of up to 10% salt content without clogging. No memory effects were noticed for steel samples, and the nebulizer system had a rapid clean-out time. Analyses were conducted using a computerized image dissector echelle spectrometer. The only noted disadvantage of the nebulizer was its higher consumption of gas (40% more) and sample (a factor of 3 more) compared with the conventional nebulizer.

Broekaert and co-workers[104] used an argon–nitrogen ICP for the analysis of aluminum alloys subsequent to alkali dissolution. Both trace elements and major constituents could be determined in a variety of aluminum alloys. Of related interest, Barnes and co-workers[105] determined 21 elements in primary, refined, and alloy aluminum. Ward and Marciello[106] analyzed various metal alloys by ICP-AES and Ishizuka and co-workers[107] used the same technique for the determination of trace impurities (Ca, Cu, Fe, Mg, Mn, Na, and Si) in high-purity (99.99%) aluminum oxide. Matrix effects on the calibration curve for each element were studied.

Fries and associates[108] decomposed manganese nodules with mixed acids to determine various lanthanide elements by ICP-AES, and Feeney and co-workers[109] developed a technique for the rapid determination of U_3O_8 in uranium mill tailings. Han and co-workers[110] used a low-power argon–nitrogen ICP to determine various elements in titanium alloys, and King and Wallace[111] determined various elements in copper metal, copper oxides, and bronze. Wittman and Willay[112] developed a device to prepare a solution for the analysis of oxides and of prereduced iron ores from steel plants by ICP-AES in less than 7 minutes. The use of a composite crucible allowed the total recovery of the melted product.

ICP techniques have recently been used for the analysis of fairly specific metallic materials. For example, Fitzgerald and co-workers[113] compared ICP-AES and AAS for the determination of undesirable trace metals in negative photoresist and found that both methods gave satisfactory accuracy and precision. Negative photoresist is a critical intermediate in the manufacture of integrated circuits. Trace contamination of certain elements in the photoresist can result in circuit devices that do not operate properly. Generally, microcircuit manufacturers desire less than 1 μg/g each of contaminant element in photoresist. Carpenter and Till[114] obtained quantitative data for up to eight elements in 100-μg samples of a collection of brasses. Precision and accuracy of the method were monitored from analyses of a certified brass standard. The goal was to establish quantitative data from 37 brass samples to form a reference collection to assist forensic scientists in the interpretation of certain alloy analyses. Hierarchic clustering and nonlinear mapping techniques were applied to the data set to classify the brass samples. The hierarchic clustering technique appeared to be more useful for larger data sets, where bigger clusters could perhaps be divided with greater confidence. Locke[115] has commented on the utility of plasma techniques for the identification of steel and glass particles. The analysis of a 0.5-mg particle can provide sufficient accuracy and precision to allow discrimination of a particular material (eg, window glass).

Fraley and co-workers[116] found comparable results by ICP–AES and AAS for the detection of HPLC peaks from EDTA and NTA chelates of copper. Yoshida and Haraguchi[117] used an HPLC–ICP system to determine lanthanide elements in a variety of materials, including high-purity lanthanide reagents. Prack and Bastiaans[118] used an evolved-gas analysis system in conjunction with an ICP for metal speciation. Samples were gradually heated up to 2300°C in a graphite sample probe that was moved in a controlled manner into the ICP discharge. Identification of a particular compound was based on the position of the sample at the time of evolution of the metal into the ICP.

Harrington and co-workers[119] compared X-ray spectrometry with AAS and ICP–AES to determine chromium and zinc in corrosion-resistant coatings. The three techniques yielded results within $\pm 10\%$ of each other. Pruszkowska and Barrett[120] evaluated an ICP method for the analysis of various materials used in the manufacture of semiconductor products. Eight materials representing matrices of several types (acid, base, and salt in aqueous solutions and organics) were analyzed for 12 elements. Hughes[121] compared spark–source atomic emission spectrometry, glow discharge spectrometry, and ICP–AES for the determination of alloying elements in steel. Data were presented, together with reproducibility measurements, which give the optimum conditions for multielement analysis of steel using this technique.

Determinations of various elements on oxide layers formed on high-temperature alloys have been reported.[122,123] Other related applications[124–131] are listed in Table 16.2.

16.6   RADIOACTIVE MATERIALS

The analysis of nuclear-type materials presents a major difficulty because many of these materials (eg, Pu and its compounds) are extremely toxic and lethal. Therefore, precautions must be made to protect the operator during the course of the analysis. Several workers have separated plutonium materials from the matrix before the analysis. For example, Ko[132] used tertiary amine solvent extraction to separate plutonium nitrate solutions from the impurities of analytical interest. To extract plutonium from 4 M HNO_3 solutions, 20% tri-n-octylamine in xylene was used. Two sequential extractions removed 99.5% of the plutonium, as measured by α energy analysis. The impurity elements were not extracted, and an average recovery of 102% was obtained for the 10 elements through the extraction. Michel and Brown[133] took a similar approach to the determination of impurities in plutonium materials. Dihexyl-N,N-diethylcarbamoylmethylene phosphonate (DHDECMP) was used to extract and separate plutonium and americium from the impurities of interest. Topics discussed in detail in this paper include a proposed method for the determination of impurities in plutonium metal, recovery of plutonium from DHDECMP and recycle use of DHDECMP, and instrument and equipment plans for their laboratory.

Lorber and Goldbart[134] described an interesting and convenient method for the determination of trace elements in uranium oxide using an ICP. The solid sample was loaded into a graphite cup, which was inserted horizontally into the tail region of the ICP. Application of the method to carrier distillation of impurities in uranium oxide gave detection limits in the range of 0.02 to 5 μg/g, with an RSD of 4 to 12%. In related work, Lorber and co-workers[135] observed up to 20-fold improvement in detection limits when using a mathematical technique to remove random noise. The system was tested with uranyl nitrate solutions.

Huff and Horwitz[136] used ICP-AES to characterize simulated complex nuclear waste solutions. The determinations were described for 19 elements comprised of process contaminants such as aluminum, chromium, iron, and nickel, and nuclear fission products such as barium, cadmium, lanthanum, and zirconium in diverse aqueous streams. Concentrations varied from 0.04 to 4000 μg/mL with dilutions being used to bring analytical measurements into the range of calibration standards. Data were presented on recoveries and material balances for an extraction system that could be used for the implementation of actinide III-fission product separation schemes. Berry and co-workers[137] used an ICP-AES to determine trace impurities in samples of liquid sodium coolant from a fast breeder reactor. A 72-channel vacuum spectrometer with a 1-kW argon ICP source simultaneously measured 59 elements with a precision varying from less than 1% at concentrations 20 times the detection limit to 30% at detection-limit levels. The torch box was enclosed in a hood fitted with a filtered extract system to allow safe handling of the radioactive solutions. Results were compared with those obtained by other techniques, such as spark-source mass spectrometry, neutron activation analysis, and AAS.

Cavalli and co-workers[138] determined palladium in nuclear waste samples by an interesting ICP–atomic fluorescence technique. The ICP was used as the spectral source to excite fluorescence in an argon-shielded air–acetylene flame. Because the flame acts as a resonance monochromator, great spectral sensitivity is achieved, compared even to the use of a high-resolution ICP-AES spectrometer. The reader is referred to Chapter 7 on the use of high-resolution spectrometry in elemental analysis of radioactive materials.

16.7 WATERS: NATURAL WATERS, OCEAN WATERS, WASTEWATERS

Thompson and co-workers[139] have preconcentrated large batches of 10 mL water samples and analyzed them for 16 elements at average river concentrations using ICP–AES. The effects on realistic detection limits of 30 elements of background interference and on-peak correction were studied on solutions with high levels of calcium and magnesium and were found to place minor constraints on the determination of certain elements. In another aspect of the study, recoveries of 32 elements during preconcentration were examined and 24 were found to be quantitative. The applicability of the method to the analysis of fresh water was considered in comparison with average river water concentrations. Nygaard and Lowry[140] have compared five digestion procedures for water samples by using dissolved inorganic, dissolved organic, and suspended particulate test samples. Persulfate digestion in acid solution, followed by heating at 95°C in 6 M HCl, was shown to be the preferred digestion procedure for all sample types prior to arsenic, antimony, and selenium determination by hydride generation into an ICP. Pruszkowska and co-workers[141] interfaced a continuous-flow hydride system into an ICP to study optimal conditions for the determination of arsenic and selenium. The method was applied to the determination of these elements in NBS SRM water and in tap water. Detection limits obtained were approximately 100 times better than by conventional ICP nebulization, 3 times better than by furnace–AAS, and slightly better (2 to 3 times) than the conventional hydride–ICP system. De Oliveira and co-workers[142] recently evaluated a variety of sample dissolution procedures and hydride generation reaction conditions for the simultaneous determination of arsenic, antimony, and selenium in marine samples. Results of analyses of reference materials from the National Research Council of Canada and NBS demonstrated the applicability of the technique to biological and geological marine samples.

Various extraction procedures to concentrate metals for direct determination via ICP techniques have been used. For example, Tao and co-workers[143] determined cadmium, cobalt, chromium, copper, iron, manganese, molybdenum, nickel, lead, vanadium, and zinc in river water and in seawater by ICP-AES, after extraction with a mixture of ammonium tetramethylenedithiocarbamate and hexamethylene-ammonium hexamethylenedithiocarbamate into xylene. The elements were all simultaneously concentrated 100-fold in a single extract and directly introduced into the plasma. Detection limits ranged from 0.017 ng/mL for cadmium to 0.5 ng/mL for lead, and calibrations were linear up to at least 30 ng/mL. In a related paper, Miyazaki and co-workers[144] determined very low levels of phosphorus in water using diisobutyl ketone for extraction. McLeod and co-workers[145] determined various metals in seawater using dithiocarbamate preconcentration. The metals were extracted from 500 g of seawater with ammonium tetramethylenedithiocarbamate-diethylammonium diethyldithiocarbamate in chloroform and back-extracted into nitric acid; the seawater concentration factor was 250 or 500. Advantages of the method included high precision, simplicity of calibration, and a detection capability in the low nanograms-per liter range. The method was applied to seawater samples from the Japan Sea, the Pacific Ocean, and the Atlantic Ocean. Smith and co-workers[146] used sodium dibenzyldithiocarbamate for preconcentration of trace metals from various water samples prior to introduction into an ICP.

Cox and co-workers[147] recently developed a rapid and very sensitive method for the sequential determination of Cr III and Cr VI based on FIA–ICP–AES. A microcolumn of activated alumina was used in the FIA manifold to separate and preconcentrate Cr VI from Cr III in aqueous solutions before ICP–AES detection at 267.7 nm. Linear calibrations for Cr III and Cr VI were established over the concentration range 0 to 1000 ng/mL. The RSDs at 10 ng/mL for a 2-mL sample injection were 2.2% for Cr III and 1.1% for Cr VI, and the corresponding detection limits were 1.4 and 0.20 ng/mL, respectively. Data for the determination of Cr III and Cr VI at the nanogram-per-milliliter level in reference waters of the NBS and the British Geological Survey were presented. Ng and Caruso[148] have determined trace metals in synthetic ocean water by ICP–AES using carbon cup–ETV into the ICP, while Goulden and Anthony[149] used a modified ultrasonic nebulizer in the determination of elements in water from Lake Ontario. Detection limits were reported to be about six times better than those obtained with conventional pneumatic nebulization. In related work, Goulden and Anthony[150] used a modified spray chamber in determining total trace metals in freshwaters.

Several recent papers comparing ICP–AES with other analytical methods for the determination of metals in waters demonstrate the clear advantage of the ICP technique. For example, Manning and co-workers[151] analyzed three synthetic fuel process waters for arsenic content by four different atomic spectrometric techniques: flame–AAS, furnace–AAS, hydride generation–AAS, and ICP–AES. The results obtained by all four methods were consistent and in good agreement with the average of determinations reported by nine laboratories participating in a collaborative study. However, flame–AAS was not very sensitive, dilution with $Ni(NO_3)_2$ stabilizer was required for the graphite furnace, and hydride generation–AAS required a prior acid digestion. No sample preparation was required with the ICP–AES technique. Rubio and co-workers[152] compared AAS, with an APDC–MIBK extraction, with direct ICP–AES for the determination of various metals in river water. Both methods were satisfactory using a standard addition method, but with a conventional calibration curve, only the ICP–AES method gave good results. Fuller and associates[153] compared flame–AAS, furnace–AAS, and ICP–AES for the direct analysis of slurries. For flame atomization, it was necessary to use pulsed nebulization, and with the ICP, it is absolutely necessary to use a high-solids cross-flow nebulizer. The results showed that atomization efficiency in nebulizer-based systems depends on sample transport efficiency, particle size, atomization temperature, and sample matrix. Jones and co-workers[154] compared flame–AAS and ICP–AES for the determination of trace elements in brines (ie, saturated sodium chloride solutions). Seawater is approximately 2.7% NaCl, whereas brine can be as high as 37.5% NaCl. The direct analysis of brines by AAS was not feasible because of lack of sensitivity and severe matrix problems. Preconcentration, using several different solvent extraction schemes and conventional AAS, permitted the determination of a large group of elements at suitable detection limits. The authors reported, however, that an ICP technique was much better suited for these analyses even though both methods did yield acceptable results. A 1:1 dilution procedure with an internal standard was used for the ICP determinations. Buchanan and Hannaker[155] described a precipitation procedure to overcome matrix effects associated with the direct analysis of concentrated brine solutions. The magnesium, present in brine solutions, was used as a carrier of trace elements, by adjusting the pH between 8.0 and 9.0 with NaOH. The resultant redissolved precipitate was analyzed by ICP–AES for 14 cation and 3 anion species. Hoult[156] used an ICP for the direct determination of sulfur, magnesium, sodium, and potassium in brines.

Meyer and associates[157] modified an ICP system for installation in an industrial plant environment for process control. The system was designed to continuously measure concentrations of calcium and magnesium in the effluent of a water softening facility. Routh and Steiner[158] discussed the conditions for implementing ICPs in a process control environment. An ICP configuration for on-line water analysis was described, including a stream selection system for presentation of samples directly to the ICP from the process stream. Ishizuka and co-workers[159] used an ICP system for the determination of phosphorus in wastewaters. Analytical results for phosphorus in municipal and industrial waste waters agreed well with those obtained by standard methods.

In a particularly interesting paper, Jinno and co-workers[160] used an ICP as a detector in conjunction with microcolumn liquid chromatography. The total column effluent was nebulized into the plasma. Carbon compounds were selectively detected by monitoring C I 247.9 nm, using water or aqueous solutions as the mobile phase. The detection limit for carbon was reported to be about 500 ng. Gardner and co-workers[161] used size exclusion chromatography for the fractionation of metal forms in natural waters via an ICP technique, whereas Moselhy and Vijan[162] used an ICP for the simultaneous determination of multiple elements in sewage and sewage effluents. Digestion with aqua regia in test tubes was recommended for sample preparation.

Cavalli and co-workers used an ICP–atomic fluorescence system for the determination of cadmium at sub-ppm levels in lake sediments.[163] A separated air–acetylene flame, supported by a circular homemade burner, was used as the atom reservoir, while a constant high concentration of a pure cadmium solution was aspirated into the ICP operated at optimized conditions for the excitation of Cd I 228.8 nm. Characteristics of the ICP as an excitation source for atomic fluorescence compared favorably with those of a commercially available electrodeless discharge lamp. A scattering correction procedure adopted and based on the use of a cobalt line, proved to be both effective and reliable. The lowest concentration level at which an analytical determination could be made with good precision was approximately 0.4 μg/g with a detection limit of 0.08 μg/g. The use of an ICP as a primary radiation source in AFS is presently an extremely secondary usage compared to its application as a vaporization–atomization–ionization–excitation source in AES. Other interesting applications[164–171] of ICP–AES for water analysis are listed in Table 16.2.

16.8 CONCLUSIONS

From a consideration of the cited *selected* papers in this chapter, and from the discussions of ICP-AFS and ICP-MS in Chapters 9 and 10, respectively, certain very general conclusions can be drawn.

1. Compared to other analytical techniques for elemental analysis, the ICP is usually preferred because of improved precision, accuracy, or sometimes plain convenience.
2. Presently, only one manufacturer (Baird Corporation, Bedford, MA) offers an ICP-AFS system, while two manufacturers (Sciex Corporation, Ontario, Canada, and VG Instruments, Cheshire, England) provide ICP-MS systems. Although the ICP-AFS technique exhibits great spectral selectivity, as discussed in Chapter 9, the ICP-MS technique, because of its superior detection limits and its capability for isotopic analysis, may attract a large number of applications, as pointed out in Chapter 10.
3. Many analysts using commercial ICP-AES systems must make certain types of instrumental modification to introduce samples into the ICP. Presently, no manufacturer offers an HPLC-ICP-AES package, although Cetac Technologies (Ames, IA) provides a direct-injection nebulizer for HPLC and FIA applications with an ICP. Only one manufacturer (Thermo Jarrell-Ash, Waltham, MA) offers an ETV-ICP-AES system. It is hoped that in the future, manufacturers will provide a wider diversity of sample introduction systems.
4. Unless new analytical techniques possessing detection capabilities 1-to-3 orders of magnitude greater than ICP-AES are developed, ICP-AES should continue to be the major technique for elemental analysis. It is assumed that the new technique will possess properties of precision, dynamic ranges, relative freedom from interferences, and relative ease of operation similar to those exhibited by ICP-AES.

REFERENCES

1. S. Greenfield, Plasma Spectroscopy Comes of Age, *Analyst (London)* 105, 1032-1044 (1980).
2. W. J. Boyko, P. N. Keliher, and J. M. Patterson III, Emission Spectrometry. *Anal. Chem.* 54. 188R-203R (1982).
3. P. N. Keliher, W. J. Boyko, J. M. Patterson III, and J. W. Hershey, Emission Spectrometry, *Anal. Chem.* 56, 133R-156R (1984).
4. P. N. Keliher, W. J. Boyko, R. H. Clifford, J. L. Snyder, and S. F. Zhu, Emission Spectrometry, *Anal. Chem.* 58, 335R-356R (1986).
5. (a) M. L. Parsons, S. Major, and A. R. Forster, Trace Element Determination by Atomic Spectroscopic Methods—State of the Art, *Appl. Spectrosc.* 37, 411-418 (1983). (b) J.A.C. Broekaert and G. Tolg, Status and Trends of Development of Atomic Spectrometry Methods for Elemental Trace Determinations, *Spectrochim. Acta* (in press).
6. D. W. Burns, M. P. Murphy, M. L. Parsons, L. A. Hickle, and R. T. Staten, The Evaluation of Internal Elemental Discriminators for Pink Bollworm by Inductively Coupled Plasma-Atomic Emission Spectrometry, *Appl. Spectrosc.* 37, 120-123 (1983).

7. T. Liese, Determination of Elements in Plant and Soil Samples by Sequential Inductively Coupled Plasma-Atomic Emission Spectrometry, *Fresenius Z. Anal. Chem.* 321, 37–44 (1985).

8. R. W. Kuennen, K. A. Wolnik, F. L. Fricke, and J. A. Caruso, Pressure Dissolution and Real Sample Matrix Calibration for Multielement Analysis of Raw Agricultural Crops by Inductively Coupled Plasma-Atomic Emission Spectrometry, *Anal. Chem.* 54, 2146–2150 (1982).

9. J. W. Jones, S. G. Capar, and T. C. O'Haver, Critical Evaluation of a Multielement Scheme Using Plasma Emission and Hydride Evolution Atomic Absorption Spectrometry for the Analysis of Plant and Animal Tissues, *Analyst (London)* 107, 353–377 (1982).

10. J. W. Jones and T. C. O'Haver, Effects of pH and Digestion Conditions on Chelex 100 Separation of Trace Elements from Tissue Digests Prior to ICP-AES Determination, *Spectrochim. Acta* 40B, 263–277 (1985).

11. M. M. Sutton, Observations on Analysis of Plant Samples by Sequential ICP Emission Spectrometry, *Chem. N.Z.* 48, 117–121 (1984).

12. K. A. Wolnik, F. L. Fricke, and C. M. Gaston, Quality Assurance in the Elemental Analysis of Foods by Inductively Coupled Plasma Spectroscopy, *Spectrochim. Acta* 39B, 649–655 (1984).

13. R. C. Munter, T. L. Halverson, and R. D. Anderson, Quality Assurance for Plant Tissue Analysis by ICP-AES, *Commun. Soil Sci., Plant Anal.* 15, 1285–1322 (1984).

14. A. A. Verbeek, Analysis of Tree Leaves, Bark, and Wood by Sequential Inductively Coupled Argon Plasma Atomic Emission Spectrometry, *Spectrochim. Acta* 39B, 599–603 (1984).

15. F. Feng, M. Fang, X. Lin, and J. Qian, Determination of Twenty-two Elements in Growth Rings of Trees by ICP-AES, *Huanjing Kexue* 5, 60–64 (1984).

16. H. Matusiewicz and R. M. Barnes, Tree Ring Wood Analysis After Hydrogen Peroxide Pressure Decomposition with Inductively Coupled Plasma Atomic Emission Spectrometry and Electrothermal Vaporization, *Anal. Chem.* 57, 406–411 (1985).

17. M. Zerezghi, K. C. Ng, and J. A. Caruso, Simultaneous Multielement Determination by Inductively Coupled Plasma-Rapid Scanning Atomic Emission Spectrometry, *Analyst (London)* 109, 589–592 (1984).

18. B. Pahlavanpour, M. Thompson, and Thorne, Simultaneous Determination of Trace Amounts of Arsenic, Antimony, and Bismuth in Herbage by Hydride Generation and Inductively Coupled Plasma-Atomic Emission Spectrometry, *Analyst (London)* 106, 467–471 (1981).

19. M. H. Hahn, K. A. Wolnik, F. L. Fricke, and J. A. Caruso, Hydride Generation/Condensation System with an Inductively Coupled Argon Plasma Polychromator for Determination of Arsenic, Bismuth, Germanium, Antimony, Selenium, and Tin in Foods, *Anal. Chem.* 54, 1048–1052 (1982).

20. C. J. Pickford, Determination of Arsenic by Emission Spectrometry Using an Inductively Coupled Plasma Source and the Syringe Hydride Technique, *Analyst (London)* 106, 464–467 (1981).

21. M. Hoenig and P. Scokart, Applications of Inductively Coupled Plasma Emission Spectrometry in Inorganic Analysis in Agricultural Chemistry, *Rev. Agric. (Brussels)* 36, 1727–1736 (1983).

22. D. J. Lyons and R. L. Roofayel, Determination of Molybdenum in Plant Material Using Inductively Coupled Plasma-Emission Spectroscopy, *Analyst (London)* 107, 331–335 (1982).

23. W. H. Evans and D. Dellar, Evaluation of an Inductively Coupled Plasma Emission Direct-Reading Spectrometer for Multiple Trace Element Analysis of Foodstuffs, *Analyst (London)* 107, 977–993 (1982).

24. F. S. Interesse, F. Lamparelli, and V. Alloggio, Mineral Content of Some Southern Italian Wines. I. Determination of Boron, Aluminum, Silicon, Titanium, Vanadium, Chromium, Manganese, Iron, Nickel, Copper, Zinc, Molybdenum, Tin, and Lead by Inductively Coupled Plasma-Atomic Emission Spectrometry, *Z. Lebensm.-Unters. Forsch.* 178, 272–278 (1984).

25. C. Camera Rica, G. F. Kirkbright, and R. D. Snook, Determination of Manganese and Nickel in Whole Blood by Optical Emission Spectrometry with an Inductively Coupled Plasma Source and Sample Introduction by Electrothermal Atomization, *At. Spectrosc.* 2, 172–175 (1981).

26. W. H. Blakemore, P. H. Casey, and W. R. Collie, Simultaneous Determination of Ten Elements in Wastewater, Plasma, and Bovine Liver by Inductively Coupled Plasma Spectrometry with Electrothermal Atomization, *Anal. Chem.* 56, 1376–1379 (1984).

27. H. Matusiewicz and R. M. Barnes, Determination of Aluminum and Silicon in Biological Materials by Inductively Coupled Plasma-Atomic Emission Spectrometry with Electrothermal Atomization, *Spectrochim. Acta* 39B, 891–899 (1984).

28. H. Matusiewicz and R. M. Barnes, Discrete Nebulization Using Aerosol Deposition in

Electrothermal Vaporization for Inductively Coupled Plasma-Atomic Emission Spectrometry, *Spectrochim. Acta* 40B, 41-47 (1985).

29. S. E. Long and R. D. Snook, Determination of Major Constituents of Pharmaceutical Capsules by Inductively Coupled Plasma-Optical Emission Spectrometry, *At. Spectrosc.* 3, 171-173 (1982).

30. H. Uchida, Y. Nojiri, H. Haraguchi, and K. Fuwa, Simultaneous Multielement Analysis by Inductively Coupled Plasma Emission Spectrometry Utilizing Microsampling Techniques with Internal Standard, *Anal. Chim. Acta* 123, 57-63 (1981).

31. P. W. Alexander, R. J. Finlayson, L. A. Smythe, and A. Thalib, Rapid Flow Analysis with Inductively Coupled Plasma-Atomic Emission Spectroscopy Using a Microinjection Technique, *Analyst (London)* 107, 1335-1342 (1982).

32. C. W. McLeod, P. J. Worsfold, and A. G. Cox, Simultaneous Multielement Analysis of Blood Serum by Flow Injection-Inductively Coupled Plasma Atomic Emission Spectrometry, *Analyst (London)* 109, 327-332 (1984).

33. P. Allain and Y. Mauras, Determination of Aluminum in Blood, Urine, and Water by Inductively Coupled Plasma Emission Spectrometry, *Anal. Chem.* 51, 2089-2091 (1979).

34. P. B. Hayman, D.M.L. Goodgame, and R. D. Snook, Studies of Nickel Absorption in Rats Using Inductively Coupled Plasma-Atomic Emission Spectrometry and Liquid Scintillation Counting, *Analyst (London)* 109, 1593-1595 (1984).

35. Y. Sun, C. Liu, A. Pei, B. Huang, and L. Xu, Inductively Coupled Plasma-Atomic Emission Spectrometric Determination of Titanium in Biological Samples of Rats with Titanium and Titanium Nitride Implants, *Fenxi Huaxue* 11, 936-952 (1983).

36. A. Sanz-Medel, R. R. Roza, and C. Perez-Conde, A Critical Comparative Study of Atomic-Spectrometric Methods (Atomic Absorption) for Determining Strontium in Biological Materials, *Analyst (London)* 108, 204-212 (1983).

37. K. A. Wolnik, J. I. Rader, C. M. Gaston, and F. L. Fricke, Development of Laboratory Control Samples for the ICP-ES Determination of Nutrient Elements in Rat Tissues, *Spectrochim. Acta* 40B, 245-251 (1985).

38. M. Morita, T. Uehiro, and K. Fuwa, Speciation and Elemental Analysis of Mixtures by High-Performance Liquid Chromatography with Inductively Coupled Argon Plasma Emission Spectrometric Detection, *Anal. Chem.* 52, 349-351 (1980).

39. K. Yoshida, H. Haraguchi, and K. Fuwa, Determination of Ribonucleoside 5'-Mono, 5'-Di, and 5'-Triphosphates by Liquid Chromatography Inductively Coupled Plasma Atomic Emission Spectrometry, *Anal. Chem.* 55, 1009-1012 (1983).

40. K. Yoshida, T. Hasegawa, and H. Haraguchi, Determination of Amino Acids by Liquid Chromatography with Inductively Coupled Plasma Atomic Emission Spectrometric Detection, *Anal. Chem.* 55, 2106-2108 (1983).

41. J. M. Mermet and J. Hubert, Analysis of Biological Materials Using Plasma Atomic Emission Spectroscopy, *Prog. Anal. At. Spectrosc.* 5, 1-33 (1982).

42. A. M. Gunn, G. F. Kirkbright, and L. N. Opheim, Determination of Phosphorus in Milk Powders by Optical Emission Spectrometry with a High-Frequency Inductively Coupled Argon Plasma Source, *Anal. Chem.* 49, 1492-1494 (1977).

43. Z. Li-Xing, G. F. Kirkbright, M. J. Cope, and J. M. Watson, A. Microprocessor-Controlled Graphite Rod Direct Sample Insertion Device for Inductively Coupled Plasma Optical Emission Spectrometry, *Appl. Spectrosc.* 37, 250-254 (1983).

44. C. Minoia, L. Pozzoli, S. Angeleri, G. Tempini, and F. Candura, Determination of Silicon in Urine by Inductively Coupled Plasma Emission Spectroscopy, *At. Spectrosc.* 3, 70-72 (1982).

45. R. L. Roofayel and D. J. Lyons, Determination of Marker Chromium in Feces Using Inductively Coupled Plasma Emission Spectrometry, *Analyst (London)* 109, 523-525 (1984).

46. P. Schramel and X. Li-Qiang, Determination of Beryllium in the Parts-per-Billion Range in Three Standard Reference Materials by Inductively Coupled Plasma Atomic Emission Spectrometry, *Anal. Chem.* 54, 1333-1336 (1982).

47. M.A.E. Wandt, M. A. B. Pougnet, and A. L. Rodgers, Determination of Calcium, Magnesium, and Phosphorus in Human Stones by Inductively Coupled Plasma-Atomic Emission Spectroscopy, *Analyst (London)* 109, 1071-1074 (1984).

48. V. A. Fassel, Current and Potential Applications of Inductively Coupled Plasma-Atomic Emission Spectroscopy in the Exploration, Mining, and Processing of Materials, *Pure Appl. Chem.* 49, 1533-545 (1977).

49. J.A.C. Broekaert, F. Leis, and K. Laqua, Application of an Inductively Coupled Plasma to the Emission Spectroscopic Determination of Rare Earths in Mineralogical Samples, *Spectrochim. Acta* 34B, 73-84 (1979).

50. J.A.C. Broekaert, F. Leis, and K. Laqua, Some Aspects of Matrix Effects Caused by Sodiumtetraborate in the Analysis of Rare Earth Minerals with the Aid of Inductively Coupled Plasma Atomic Emission Spectroscopy, *Spectrochim. Acta* 34B 167–175 (1979).

51. J.A.C. Broekaert, The Application of Inductively Coupled Plasma–Optical Emission Spectrometry to Industrial Analytical Problems, *Trends Anal. Chem.* 1, 249–253 (1982).

52. M. A. Floyd, V. A. Fassel, and A. P. D'Silva, Computer-Controlled Scanning Monochromator for the Determination of 50 Elements in Geochemical and Environmental Samples by Inductively Coupled Plasma–Atomic Emission Spectrometry, *Anal. Chem.* 52, 2168–2173 (1980).

53. P. Hannaker and H. Qing-Lie, Dissolution of Geological Material with Orthophosphoric Acid for Major-Element Determination by Flame-Atomic Absorption Spectroscopy and Inductively Coupled Plasma–Atomic Emission Spectroscopy, *Talanta* 31, 1153–1157 (1984).

54. P. Hannaker, M. Haukka, and S. K. Sen, Comparative Study of ICP-AES and XRF Analysis of Major and Minor Constituents on Geological Materials, *Chem. Geol.* 42, 319–324 (1984).

55. J. O. Burman, C. Ponter, and K. Bostrom, Metaborate Digestion Procedure for Inductively Coupled Plasma–Optical Emission Spectrometry, *Anal. Chem.* 50, 679–680 (1978).

56. R. J. Brown and W. R. Biggs, Determination of Platinum and Palladium Geologic Samples by Ion-Exchange Chromatography with Inductively Coupled Plasma–Atomic Emission Spectrometric Detection, *Anal. Chem.* 56, 646–649 (1984).

57. R. Rice and R. L. Bragg, Multielemental Analysis of Coal Ash Utilizing Sequential ICP, *Proc. Coal Test, Con.* 3, 70–75 (1983).

58. A. R. Date and A. L. Gray, Determination of Trace Elements in Geological Samples by Inductively Coupled Plasma Source Mass Spectrometry, *Spectrochim. Acta* 40B, 115–122 (1985).

59. H. Uchida, T. Uchida, and C. Iida, Determination of Minor and Trace Elements in Silicate Rocks by Inductively Coupled Plasma Emission Spectrometry, *Anal. Chim. Acta* 116, 433–437 (1980).

60. H. Uchida, K. Iwasaki, and K. Tanaka, Determination of Traces of Zirconium in Silicate Rocks by Inductively Coupled Plasma Emission Spectrometry, *Anal. Chim. Acta* 134, 375–378 (1982).

61. J. N. Walsh, The Simultaneous Determination of the Major, Minor, and Trace Constituents of Silicate Rocks Using Inductively Coupled Plasma Spectrometry, *Spectrochim. Acta* 35B, 107–111 (1980).

62. G. Wunsch and N. Czech, Determination of Tungsten with the Inductively Coupled Plasma, *Fresenius Z. Anal. Chem.* 317, 5–9 (1984).

63. M. Thompson and L. Zao, Rapid Determination of Molybdenum in Soils, Sediments, and Rocks by Solvent Extraction with Inductively Coupled Plasma–Atomic Emission Spectrometry, *Analyst* (*London*) 110, 229–235 (1985).

64. H. S. Mahanti and R. M. Barnes, Purification and Analysis of Calcium Carbonate with Poly(dithiocarbamate) Resin, *Appl. Spectrosc.* 37, 401 (1983).

65. M. W. Pritchard and J. Lee, Simultaneous Determination of Boron, Phosphorus, and Sulfur in Some Biological and Soil Materials and Inductively Coupled Plasma Emission Spectrometry, *Anal. Chim. Acta*, 157, 313–326 (1984).

66. P. Schramel, B. J. Klose, and S. Hasse, Efficiency of ICP Emission Spectroscopy for the Determination of Trace Elements in Bio-Medical and Environmental Samples, *Fresenius Z. Anal. Chem.* 310, 209–216 (1982).

67. P. Schramel, X. Li-Qiang, A. Wolf, and S. Hasse, ICP Emission Spectroscopy: An Analytical Method for Routine Supervision of Sludge and Soil, *Fresenius Z. Anal. Chem.* 313, 213–216 (1982).

68. P.W.J.M. Boumans and M. C. Lux-Steiner, Modification and Optimization of a 50-MHz Indutively Coupled Argon Plasma with Special Reference to Analyses Using Organic Solvents, *Spectrochim. Acta* 37B, 97–126 (1982).

69. N. Klaentschi, ICP Emission Spectrometry: Effect of the Particle Size in the Determination of Wear Elements in Lubricating Oils, *Mater. Technol.* 12, 3–7 (1984).

70. D. W. Hausler, Molecular Size Distribution of Specific Elements in Petroleum Crudes and 650°F⁺ Residua by Size Exclusion Chromatography with Inductively Coupled Plasma Spectrometry Detection, *Spectrochim. Acta* 40B, 389–396 (1985).

71. K. C. Ng, H. Kaiser, and B. Meddings, Low-Power Torches for Organic Solvents in Inductively Coupled Plasma Emission Spectrometry, *Spectrochim. Acta* 40B, 63–72 (1985).

72. R. N. Merryfield and R. C. Loyd, Simultaneous Determination of Metals in Oil by Inductively Coupled Plasma Emission Spectrometry, *Anal. Chem.* 51, 1965–1968 (1979).

73. K. C. Ng and J. A. Caruso, Atomic Emission Spectrometric Analysis of Organic Solutions by Electrothermal Carbon Cup Vaporization into an Inductively Coupled Plasma, *Anal. Chem.* 55, 2032-2036 (1983).

74. J.A.C. Broekaert, F. Keis, and K. Laqua, The Application of an Argon/Nitrogen Inductively Coupled Plasma to the Analysis of Organic Solutions, *Talanta* 28, 745-752 (1981).

75. S. Greenfield and P. B. Smith, The Determination of Trace Metals in Microlitre Samples by Plasma Torch Excitation, *Anal. Chim. Acta*, 59, 341-348 (1972).

76. P. M. Beckwith, Determination of Major and Trace Elements in Glycol-Based Fluids by Inductively Coupled Plasma Atomic Emission Spectrometry, *Spectrochim. Acta* 40B, 301-306 (1985).

77. A. D. King, D. R. Hilligoss, and G. F. Wallace, Comparison of Results for Determination of Wear Metals in Used Lubricating Oils by Flame Atomic Absorption Spectrometry and Inductively Coupled Plasma-Emission Spectrometry, *At. Spectrosc.* 5, 189-191 (1984).

78. I. B. Brenner, A. E. Watson, T. W. Steele, E. A. Jones, and M. Goncalves, Application of an Argon-Nitrogen Inductively Coupled Plasma to the Analysis of Geological and Related Materials for Their Rare Earth Contents, *Spectrochim. Acta* 35B, 785-797 (1981).

79. I. B. Brenner, E. A. Jones, A. E. Watson, and T. W. Steele, The Application of a Nitrogen-argon Medium-Power Inductively Coupled Plasma and Cation-Exchange Chromatography for the Spectrometric Determination of the Rare Earth Elements in Geological Materials, *Chem. Geol.* 45, 135-148 (1984).

80. J. Zou, M. Zlai, H. Wang, and B. Chen, Determination of Rare Earth Elements in Geological Materials by Using Inductively Coupled Plasma Spectrometry, *Yanshi Kuangwu Ji Ceshi*, 3, 149-153 (1984).

81. J. G. Crock and F. E. Lichte, Determination of Rare Earth Elements in Geological Materials by Inductively Coupled Argon Plasma/Atomic Emission Spectrometry, *Anal. Chem.* 54, 1329-1332 (1982).

82. R. Aulis, A. Bolton, W. Doherty, A. Vander Voet, and P. Wong, Determination of Yttrium and Selected Rare Earth Elements in Geological Materials Using High Performance Liquid Chromatographic Separation and ICP Spectrometric Detection, *Spectrochim. Acta*, 40B, 377-387 (1985).

83. V. K. Din, The Preparation of Iron-Free Solutions from Geological Materials for the Determination of Boron (and Other Elements) by Inductively Coupled Plasma Emission Spectrometry, *Anal. Chim. Acta* 159, 387-391 (1984).

84. G. F. Wallace, Application of a Sequential Scanning ICP to the Analysis of Geological Materials, *At. Spectrosc.* 2, 87-90 (1981).

85. R. A. Nadkarni, R. R. Botto, and S. E. Smith, Comparison of Two Atomic Spectroscopic Methods for Elemental Analysis of Geological Materials, *At. Spectrosc.* 3, 180-184 (1982).

86. J. O. Burman and K. Bostrom, Comparison of Different Plasma Excitation and Calibration Methods in the Analysis of Geological Materials by Optical Emission Spectrometry, *Anal. Chem.* 51, 516-520 (1979).

87. K. C. Ng, M. Zerezghi, and J. A. Caruso, Direct Powder Injection of NBS Coal Fly Ash in Inductively Coupled Plasma Atomic Emission Spectrometry with Rapid Scanning Spectrometric Detection, *Anal. Chem.* 56, 417-421 (1984).

88. M. L. Fernandez Sanchez, C. Garcia Ortiz, S. Arribas Jimeno, and A. Sanz-Medel, Application of the ICP to the Determination of Lead, Zinc, and Silver in Lead and Zinc Flotation Ores, *At. Spectrosc.* 5, 197-203 (1984).

89. J. L. Manzoorl, Inductively Coupled Plasma-Optical Emission Spectrometry. Application to the Determination of Molybdenum, Cobalt, and Boron in Soil Extracts, *Talanta* 27, 682-684 (1980).

90. B. Pahlavanpour, M. Thompson, and L. Thorne, Simultaneous Determination of Trace Concentrations of Arsenic, Antimony, and Bismuth in Soils and Sediments by Volatile Hydride Generation and Inductively Coupled Plasma Spectrometry, *Analyst (London)* 105, 756-761 (1980).

91. M. W. Blades and P. Hauser, Quantitation of Sulfur in Xylene with an Inductively Coupled Plasma Photodiode-Array Spectrometer, *Anal. Chim. Acta* 157, 163-169 (1984).

92. G. F. Wallace and R. D. Ediger, Optimization of ICP Operating Conditions for the Determination of Sulfur in Oils, *At. Spectrosc.* 2, 169-172 (1981).

93. M. J. Cope, G. F. Kirkbright, and P. M. Burr, Use of Inductively Coupled Plasma Optical Emission Spectrometry for the Analysis of Doped Cadmium Mercury Telluride Employing a Graphite Rod Electrothermal Vaporization Device for Sample Introduction, *Analyst (London)* 107, 611-616 (1982).

94. G. F. Kirkbright and Z. Li-Xing, Volatilization of Some Elements from a Graphite Rod Direct Sample Insertion Device into an Inductively Coupled Argon Plasma for Optical Spectrometry, *Analyst (London)* 107, 617–622 (1982).
95. G. F. Kirkbright and R. D. Snook, The Determination of Some Trace Elements in Uranium by Inductively Coupled Plasma Emission Spectroscopy Using a Graphite Rod Sample Introduction Technique, *Appl. Spectrosc.* 37, 11–16 (1983).
96. N. W. Barnett, M. J. Cope, G. F. Kirkbright, and A.A.H. Taobi, Design Consideration and Temperature Determination of an Automated Graphite Furnace Cup System Used for Direct Sample Introduction for ICP Optical Emission Spectrometry, *Spectrochim Acta* 39B, 343–348 (1984).
97. T. Nakahara and N. Kikui, Determination of Trace Concentrations of Selenium by Continuous Hydride Generation–Inductively Coupled Plasma Atomic Emission Spectrometry, *Spectrochim. Acta* 40B, 21–28 (1985).
98. T. Nakahara, The Determination of Trace Amounts of Tin by Inductively Coupled Argon Plasma Atomic Emission Spectrometry with Volatile Hydride Method, *Appl. Spectrosc.* 37, 539–545 (1983).
99. S. D. Hartenstein, J. Ruzica, and G. D. Christian, Sensitivity Enhancements for Flow Injection Analysis–Inductively Coupled Plasma Atomic Emission Spectrometry using an On-Line Preconcentrating Ion-Exchange Column, *Anal. Chem.* 57, 21–25 (1985).
100. C. W. McLeod, I. G. Cook, P. J. Worsfold, J. E. Davies, and J. Queay, Analyte Enrichment and Matrix Removal in Flow Injection Analysis–Inductively Coupled Plasma–Atomic Emission Spectrometry: Determination of Phosphorus in Steels, *Spectrochim Acta* 40B, 67–72 (1985).
101. M. Thompson, J. E. Goulter, and F. Sieper, Laser Ablation for the Introduction of Solid Samples into an Inductively Coupled Plasma for Atomic Emission Spectrometry, *Analyst (London)* 106, 32–39 (1981).
102. K. Takahashi, T. Yoshioka, Y. Nakamura, and H. Okochi, Direct Inductively Coupled Plasma Emission Spectrochemical Analysis of Low-Alloy Steels Using Aerosol Cyclones with Low-Voltage Spark Discharge, *Nippon Kinzoku Gakkaishi* 48, 418–423 (1984).
103. B. Thelin, Nebulizer System for Analysis of High Salt Content Solutions with an Inductively Coupled Plasma, *Analyst (London)* 106, 54–59 (1981).
104. J.A.C. Broekaert, F. Leis, and G. Dincler, Use of an Argon–Nitrogen Inductively Coupled Plasma for the Analysis of Aluminum Alloys Subsequent to Alkali Dissolution, *Analyst (London)* 108, 717–721 (1983).
105. R. M. Barnes, L. Fernando, L. S. Jing, and H. S. Mahanti, Analysis of Aluminum by Inductively Coupled Plasma–Atomic Emission Spectroscopy, *Appl. Spectrosc.* 37, 389–395 (1983).
106. A. F. Ward, and L. F. Marciello, Analysis of Metal Alloys by Inductively Coupled Argon Plasma Optical Emission Spectrometry, *Anal. Chem.* 51, 2264–2272 (1979).
107. T. Ishizuka, Y. Uwamino, A. Tsuge, and T. Kamiyanagi, Determination of Trace Impurities in High-Purity Aluminum Oxide by Inductively Coupled Plasma–Atomic Emission Spectrometry, *Anal. Chim. Acta* 161, 285–291 (1984).
108. T. Fries, P. J. Lamothe, and J. J. Pesek, Determination of Rare Earth Elements, Yttrium. and Scandium in Manganese Nodules by Inductively Coupled Argon Plasma Emission Spectrometry, *Anal. Chim. Acta* 159, 329–336 (1984).
109. C. M. Feeney, J. W. Anderson, and F. W. Tindall, Determination of Uranium in Tailings by Inductively Coupled Plasma Optical Emission Spectrometry, *At. Spectrosc.* 4, 108–110 (1982).
110. B. Han, B. Tan, and Z. He, Determination of Molybdenum, Vanadium, Chromium, Aluminum, Iron, and Yttrium in Titanium Alloys by Low-Power Nitrogen-Argon ICP Atomic Emission Spectrometry, *Fenxi Huaxue* 12, 45–47 (1984).
111. A. D. King and G. F. Wallace, Determination of Trace Elements in Copper Metal, Copper Oxides, and Bronze by ICP Emission Spectrometry, *At. Spectrosc.* 6, 4–8 (1985).
112. A. A. Wittman and G.M.H. Willay, Automatic Nonmetallic Sample Preparation for Inductively Coupled Plasma Spectrometry, *Spectrochim. Acta* 40B, 253–261 (1985).
113. E. A. Fitzgerald, A. A. Bornstein, and L. J. Davidowski, Determination of Trace Elements in Negative Photoresist by ICP Atomic Emission Spectroscopy and Atomic Absorption Spectroscopy, *At. Spectrosc.* 6, 1–3 (1985).
114. R. C. Carpenter and C. Till, Analysis of Small Samples of Brasses by Inductively Coupled Plasma–Optical Emission Spectrometry and Their Classification by Two Pattern-Recognition Techniques, *Analyst (London)* 109, 881–884 (1984).
115. J. Locke, The Application of Plasma Source Atomic Emission Spectrometry in Forensic Science, *Anal. Chim. Acta* 113, 3–12 (1980).

116. D. M. Fraley, D. Yates, and S. E. Manahan, Inductively Coupled Plasma Emission Spectrometric Detection of Simulated High Performance Liquid Chromatographic Peaks, *Anal. Chem.* 51, 2225–2229 (1979).

117. K. Yoshida and H. Haraguchi, Determination of Rare Earth Elements by Liquid Chromatography-Inductively Coupled Plasma Atomic Emission Spectrometry, *Anal. Chem.* 56, 2580–2585 (1984).

118. E. R. Prack and G. J. Bastiaans, Metal Speciation by Evolved Gas-Inductively Coupled Plasma Atomic Emission Spectrometry, *Anal. Chem.* 55, 1654–1660 (1983).

119. D. E. Harrington, J. S. Jones, W. R. Bramstedt, and T. A. Kling, Determination of Chromium and Zinc in Corrosion-Resistant Coatings, *At. Spectrosc.* 4, 171–176 (1983).

120. E. Pruszkowska and P. Barrett, The Use of ICP in the Semiconductor Industry, *At. Spectrosc.* 5, 96–100 (1984).

121. H. Hughes, Application of Optical Emission Source Developments in Metallurgical Analysis, *Analyst (London)* 108, 286–292 (1983).

122. Z. Zadgorska, E. Bauer, and H. Nickel, Contribution to the Quantitative Analysis of Oxide Layers Formed on High-Temperature Alloys Using Inductively Coupled Plasma-Atomic Emission Spectroscopy—I. Removal and Characterization of Oxide Layers, *Fresenius Z. Anal. Chem.* 314, 351–355 (1983).

123. Z. Zadgorska, H. Nickel, M. Mazurkiewicz, and G. Wolff, Contribution to the Quantitative Analysis of Oxide Layers Formed on High-Temperature Alloys Using Inductively Coupled Plasma-Atomic Emission Spectroscopy—II. Optimization of the ICP Working Parameters and Typical Analysis of Oxide Layers, *Fresenius Z. Anal. Chem.* 314, 356–361 (1983).

124. I. B. Brenner and S. Erlich, The Spectrochemical (ICP-AES) Determination of Tungsten in Tungsten Ores, Concentrates, and Alloys: An Evaluation as an Alternative to the Classical Gravimetric Procedure, *Appl. Spectrosc.* 38, 887–890 (1984).

125. J. Xu, H. Kawaguchi, and A. Mizuike, Spectral Interferences in the Determination of Phosphorus in Steel by Inductively Coupled Plasma Emission Spectrometry with an Echelle Monochromator, *Appl. Spectrosc.* 37, 123–127 (1983).

126. J. L. Fabec and M. L. Ruschak, Determination of Aluminum, Cobalt, Lanthanum, and Ruthenium on Alumina-Based Catalysts by Inductively Coupled Plasma-Atomic Emission Spectrometry, *Anal. Chem.* 55, 2241–2246 (1983).

127. R. M. Hamner and L. A. De'Aeth, Determination of Boron in Silicon-Bearing Alloys, Steel, and Other Alloys by Pyrohydrolysis and Inductively Coupled Argon-Plasma Spectroscopy, *Talanta* 27, 535–536 (1980).

128. G. F. Wallace, Utilization of Lower UV Wavelengths to Reduce Interferences in the Determination of Boron in Steels by ICP Spectroscopy, *At. Spectrosc.* 2, 61–64 (1981).

129. G. Mezger, E. Grallath, U. Stix, and G. Tolg, Determination of Traces of Boron in Metals by Emission Spectrometry with ICP After Separation as Boric Acid Ester, *Fresenius Z. Anal. Chem.* 317, 765–773 (1984).

130. H. Ishii and K. Satoth, Development of a High-Resolution Inductively Coupled Argon Plasma Apparatus for Derivative Spectrometry and Its Application to the Determination of Hafnium in High-Purity Zirconium Oxide, *Talanta* 29, 243–248 (1982).

131. H. Ishii and K. Satoh, Determination of Rare Earths in Lanthanum Oxide by Inductively Coupled Plasma Emission Derivative Spectrometry, *Talanta* 30, 111–115 (1983).

132. R. Ko, The Determination of Impurities in Plutonium Nitrate Solutions by Amine Extraction and ICP Analysis, *Appl. Spectrosc.* 38, 909–910 (1984).

133. C. E. Michel and G. E. Brown, Plans for Inductively Coupled Plasma Atomic Emission Spectrometry (ICP-AES) Analysis of Impurities in Plutonium Materials at Rocky Flats, *Anal. Chem. Symp. Ser.* 19, 235–239 (1984).

134. A. Lorber and Z. Goldbart, Convenient Method for the Determination of Trace Elements in Solid Samples Using an Inductively Coupled Plasma, *Analyst (London)* 110, 155–157 (1985).

135. A. Lorber, M. Eldan, and Z. Goldbart, Improved Detection Limits in Inductively Coupled Plasma Multichannel Spectrometry of Uranyl Nitrate Solutions by Compensation of Nonrandom Background Fluctuations, *Anal. Chem.* 57, 851–857 (1985).

136. E. A. Huff and E. P. Horwitz, Inductively Coupled Plasma-Atomic Emission Spectrometry in Support of Nuclear Waste Management, *Spectrochim Acta* 40B, 279–286 (1985).

137. T. Berry, K. C. Macleod, A. C. Christie, and J. A. Cunningham, Determination of Trace Impurities in Sodium Coolant from a Fast Breeder Reactor by Inductively Coupled Plasma Atomic Emission Spectrometry, *Analyst (London)* 108, 189–195 (1983).

138. P. Cavalli, G. Rossi, and N. Omenetto, Determination of Palladium in Nuclear-Waste Samples by Inductively Coupled Plasma Emission-Fluorescence Spectrometry, *Analyst (London)* 108, 297–304 (1983).
139. M. Thompson, M. H. Ramsey, and B. Pahlavanpour, Water Analysis by Inductively Coupled Plasma–Atomic Emission Spectrometry After a Rapid Preconcentration, *Analyst (London)* 107, 1330–1334 (1982).
140. D. D. Nygaard and J. H. Lowry, Sample Digestion Procedures for Simultaneous Determination of Arsenic, Antimony, and Selenium by Inductively Coupled Argon Plasma Emission Spectrometry with Hydride Generation, *Anal. Chem.* 54, 803–807 (1982).
141. E. Pruszkowska, P. Barrett, R. Ediger, and G. Wallace, Determination of Arsenic and Selenium Using a Hydride System Combined with ICP, *At. Spectrosc.* 4, 94–98 (1983).
142. E. de Oliveira, J. W. McLaren, and S. S. Berman, Simultaneous Determination of Arsenic, Antimony, and Selenium in Marine Samples by Inductively Coupled Plasma Atomic Emission Spectrometry, *Anal. Chem.* 55, 2047–2050 (1983).
143. H. Tao, A. Miyazaki, K. Bansho, and Y. Umezaki, Determination of Trace Levels of Heavy Metals in Waters by Extraction with Ammonium Tetramethylenedithiocarbamate and Hexamethyleneammonium Hexamethylenedithiocarbamate into Xylene Followed by Inductively Coupled Plasma Emission Spectrometry, *Anal. Chim. Acta* 156, 159–168 (1984).
144. A. Miyazaki, A. Kimura, and Y. Umezaki, Determination of ng/mL Levels of Phosphorus in Waters by Diisobutyl Ketone Extraction and Inductively Coupled Plasma–Atomic Emission Spectrometry, *Anal. Chim. Acta* 127, 93–101 (1981).
145. C. W. McLeod, A. Otsuki, K. Okamoto, H. Haraguchi, and K. Fuwa, Simultaneous Determination of Trace Metals in Seawater Using Dithiocarbamate Preconcentration and Inductively Coupled Plasma Emission Spectrometry, *Analyst (London)* 106, 419–428 (1981).
146. C. L. Smith, J. M. Motooka, and W. R. Willson, Analysis of Trace Metals in Water by Inductively Coupled Plasma Emission Spectrometry Using Sodium Dibenzyldithiocarbamate for Preconcentration, *Anal. Lett.* 17, 1715–1730 (1984).
147. A. G. Cox, I. G. Cook, C. W. McLeod, Rapid Sequential Determination of Chromium(III)–Chromium(VI) by Flow Injection Analysis–Inductively Coupled Plasma Atomic Emission Spectrometry, *Analyst (London)* 110, 331–333 (1985).
148. K. C. Ng and J. A. Caruso, Determination of Trace Metals in Synthetic Ocean Waters by Inductively Coupled Plasma Atomic Emission Spectrometry with Electrothermal Carbon Cup Vaporization, *Anal. Chem.* 55, 1513–1516 (1983).
149. P. D. Goulden and D.H.J. Anthony, Modified Ultrasonic Nebulizer for Inductively Coupled Argon Plasma Atomic Emission Spectrometry, *Anal. Chem.* 56, 2327–2329 (1984).
150. P. D. Goulden and D.H.J. Anthony, Determination of Trace Metals in Freshwaters by Inductively Coupled Argon Plasma Atomic Emission Spectrometry with a Heated Spray Chamber and Desolvation, *Anal. Chem.* 54, 1678–1681 (1982).
151. D. C. Manning, R. D. Ediger, and D. W. Hoult, Determination of Arsenic in Synthetic Fuel Process Wastes, *At. Spectrosc.* 1, 52–54 (1980).
152. R. Rubio, J. Huguet, and G. Rauret, Comparative Study of the Cadmium, Copper, and Lead Determination by AAS and by ICP-AES in River Water. Application to a Mediterranean River (Congost River, Catalonia, Spain), *Water Res.* 18, 423–428 (1984).
153. C. W. Fuller, R. C. Hutton, and B. Preston, Comparison of Flame, Electrothermal, and Inductively Coupled Plasma Atomization Techniques for the Direct Analysis of Slurries, *Analyst (London)* 106, 913–920 (1981).
154. J. S. Jones, D. E. Harrington, B. A. Leone, and W. R. Bramstedt, Determination of Trace Elements in Brine by Atomic Absorption and Inductively Coupled Plasma Spectroscopy, *At. Spectrosc.* 4, 49–54 (1983).
155. A. S. Buchanan and P. Hannaker, Inductively Coupled Plasma Spectrometric Determination of Minor Elements in Concentrated Brines Following Precipitation, *Anal. Chem.* 56, 1379–1382 (1984).
156. D. W. Hoult, The Direct Determination of Sulfur, Magnesium, Sodium, and Potassium in Brines by ICP Emission Spectroscopy, *At. Spectrosc.* 1, 82–83 (1980).
157. G. A. Meyer, J. S. Roeck, and T. Johnson, Continuous Determination by Inductively Coupled Plasma Emission Spectrometry of Calcium and Magnesium in Chemical Streams Containing up to 2% Dissolved Salt Levels, *Spectrochim. Acta* 40B, 237–244 (1985).
158. M. W. Routh and J. D. Steiner, Process Control Instrumentation—Is ICP Ready? *Spectrochim. Acta* 40B, 227–235 (1985).

159. T. Ishizuka, K. Nakajima, and H. Sunahara, Determination of Phosphorus in Waste Waters by Inductively Coupled Plasma-Atomic Emission Spectrometry, *Anal. Chim. Acta* 121, 197-203 (1980).

160. K. Jinno, S. Nakanishi, and T. Nagoshi, Microcolumn Liquid Chromatography with Inductively Coupled Plasma Atomic Emission Spectrometric Detection, *Chromatographia* 18, 437-440 (1984).

161. W. S. Gardner, P. F. Landrum, and D. A. Yates, Fractionation of Metal Forms in Natural Waters by Size Exclusion Chromatography with Inductively Coupled Argon Plasma Detection, *Anal. Chem.* 54, 1196-1198 (1982).

162. M. M. Moselhy and P. N. Vijan, Simultaneous Determination of Trace Metals in Sewage and Sewage Effluents by Inductively Coupled Argon Plasma Atomic Emission Spectrometry, *Anal. Chim. Acta* 130, 157-166 (1981).

163. P. Cavalli, N. Omenetto, and G. Rossi, Determination of Cadmium at Sub-ppm Levels in Lake Sediments by an Inductively Coupled Plasma-Atomic Fluorescence Technique, *At. Spectrosc.* 3, 1-4 (1982).

164. A. Sugimae, Determination of Trace Elements in Seawater by Inductively Coupled Plasma Emission Spectrometry, *Anal. Chim. Acta* 121, 331-336 (1980).

165. G. J. Schmidt and W. Slavin, Inductively Coupled Plasma Emission Spectrometry with Internal Standardization and Subtraction of Plasma Background Fluctuations, *Anal. Chem.* 54, 2491-2495 (1982).

166. T. Uehiro, M. Morita, and K. Fuwa, Vacuum Ultraviolet Emission Line for Determination of Aluminum by Inductively Coupled Plasma Atomic Emission Spectrometry, *Anal. Chem.* 56, 2020-2024 (1984).

167. A. Bettero, C. A. Benassi, and B. Casetta, ICP Determination of Sulfur in Cosmetic Products, *At. Spectrosc.* 5, 57-58 (1984).

168. A. Bettero, B. Casetta, F. Galiano, E. Ragazzi, and C. A. Benassi, Rheological and Spectroscopic Behavior of Cosmetic Products, *Fresenius Z. Anal. Chem.* 318, 525-527 (1984).

169. G. DiPasquale and B. Casetta, ICP Determination of Sulfur in Polymeric Materials, *At. Spectrosc.* 5, 209-210 (1984).

170. P. R. Skidmore and S. S. Greetham, Trace Metal Determinations in Concentrated Electrolyte Solutions—A Comparative Study, *Analyst (London)* 108, 171-177 (1983).

171. M. A. Fernandez and G. J. Bastiaans, Precise Determination of Stoichiometries for Heteropoly Complexes by Inductively Coupled Plasma Emission Spectrometry, *Anal. Chem.* 51, 1402-1406 (1979).

Appendix

Prominent Spectral Lines from an Argon Inductively Coupled Plasma

S of I	State of ionization. The symbols I, II, and III indicate that the spectral lines originate, respectively, from the neutral atom, singly ionized, and doubly ionized states.
I_n/I_b	Ratio of net analyte intensity to background intensity.
CONC	Concentration of the single element analyte solution used for the wavelength scans from which the prominent lines were determined.
EST'D DET LIM	Estimated detection limit. Detection limits estimated from the I_n/I_b ratios as explained in this paper.
NM	Not measurable because of interfering line listed in COMMENTS column.
COMMENTS	All wavelengths listed were taken from the NBS tables[1] unless otherwise designated in the comments column. Either air or vacuum wavelengths are listed depending on the source.

Air wavelengths	Vacuum wavelengths
NBS[1]	NRL[4]
MIT[2]	UCRL[5]
Zaidel, 200 nm and above[3]	Zaidel, below 200 nm[3]

NR	Not resolved. This description indicates components of an unresolved pair of lines. For convenience in the computer listing of the data, I_n/I_b, CONC, and EST'D DET LIM values are listed for each wavelength, although only a single I_n/I_b measurement was obtained from each unresolved pair.
GROUP NR	This description indicates components of an unresolved group (3 or more lines). Only the wavelength of the strongest line (NBS tables[1]) is listed.
APPROX WAVE	Approximate wavelength. This description applies to spectral lines (Os, Pt) that have not been positively identified as belonging to the elemental scans in which they were found. These approximate wavelengths may be previously unreported lines of Os and Pt or they may be due to impurities. The impurities, if present, have not been positively identified.

Additional Remarks

Significant figures. Although the number of digits shown in the EST'D DET LIM column may be greater than the precision of the determination justifies, no convenient means of limiting the number of significant digits in the computer printout was available. Normally, only one significant figure is appropriate for detection limit data.[6]

Interferences. The comments column includes interference information when a component of the background spectrum overlaps an analyte line (eg, the Dy 396.839 nm line is located on the broad H 397.007 nm line) or when an analyte line is located in a complex molecular band system (eg, the OH 306.36 nm system) where band components may cause spectral interferences. The notation of molecular bands does not preclude the use of analyte wavelengths within the band region; eg, with our experimental facilities the Al 308.22, Be 313.04, Ca 315.89, and Cu 324.75 nm lines are analytically useful even though they are located in the OH band region.

Source: From R. K. Winge, V. J. Peterson, and V. A. Fassel, Inductively Coupled Plasma-Atomic Emission Spectroscopy: Prominent Lines, *Appl. Spectrosc.* 33, 206 219 (1979). With permission.

Radiation Chemistry: Principles and Applications

	S OF I	WAVE-LENGTH (nm)	$\frac{I_n}{I_b}$	CONC (μg/ml)	EST'D DET LIM (μg/ml)	COMMENTS
AG	I	328.068	38.0	10.0	0.007	
AG	I	338.289	23.0	10.0	0.013	
AG	II	243.779	2.5	10.0	0.120	
AG	II	224.641	2.3	10.0	0.130	ZAIDEL
AG	II	241.318	1.5	10.0	0.200	
AG	II	211.383	0.9	10.0	0.333	ZAIDEL
AG	II	232.505	0.7	10.0	0.428	ZAIDEL
AG	II	224.874	0.6	10.0	0.500	ZAIDEL
AG	II	233.137	0.5	10.0	0.600	
AL	I	309.271	13.0	10.0	0.023	OH BAND,NR
AL	I	309.284	13.0	10.0	0.023	OH BAND,NR
AL	I	396.152	10.5	10.0	0.028	
AL	I	237.335	10.0	10.0	0.030	NR
AL	I	237.312	10.0	10.0	0.030	NR
AL	I	226.922	9.0	10.0	0.033	NR
AL	I	226.910	9.0	10.0	0.033	NR
AL	I	308.215	6.6	10.0	0.045	OH BAND
AL	I	394.401	6.3	10.0	0.047	
AL	I	236.705	5.8	10.0	0.051	
AL	I	226.346	5.0	10.0	0.060	ZAIDEL
AL	I	221.006	4.8	10.0	0.062	ZAIDEL
AL	I	257.510	4.0	10.0	0.075	
AR	I	415.859	>50.0			ZAIDEL
AR	I	419.832	50.0			ZAIDEL
AR	I	420.068	50.0			ZAIDEL
AR	I	425.936	50.0			ZAIDEL
AR	I	427.217	43.0			ZAIDEL
AR	I	430.010	40.0			ZAIDEL
AR	I	433.356	38.0			ZAIDEL
AR	I	419.103	32.0			ZAIDEL,NR
AR	I	419.071	32.0			ZAIDEL,NR
AR	I	426.629	32.0			ZAIDEL
AR	I	404.442	31.0			ZAIDEL
AR	I	418.188	28.0			ZAIDEL
AR	I	394.898	27.0			ZAIDEL
AR	I	451.074	21.0			ZAIDEL
AR	I	355.431	18.0			ZAIDEL
AR	I	416.418	17.0			ZAIDEL
AR	I	433.534	11.0			ZAIDEL
AR	I	434.517	9.0			ZAIDEL
AR	I	360.652	8.7			ZAIDEL
AR	I	356.766	8.6			ZAIDEL
AR	I	394.750	8.0			ZAIDEL
AR	I	425.119	7.5			ZAIDEL
AR	I	470.232	7.2			ZAIDEL
AR	I	364.983	7.0			ZAIDEL
AR	I	452.232	6.2			ZAIDEL
AR	I	459.610	6.0			ZAIDEL
AR	I	383.468	6.0			ZAIDEL
AR	I	462.844	4.0			ZAIDEL
AR	I	363.446	3.9			ZAIDEL
AR	I	339.375	3.6			ZAIDEL

S OF I		WAVE-LENGTH (nm)	$\frac{I_n}{I_b}$	CCNC (µg/ml)	EST'D DET LIM (µg/ml)	COMMENTS
AR	I	346.108	3.5			ZAIDEL
AR	I	363.268	3.5			ZAIDEL
AR		436.836	3.2			MIT
AR	I	356.329	3.2			ZAIDEL
AR	I	331.935	3.1			ZAIDEL
AR	I	404.597	2.5			ZAIDEL
AR	I	367.067	2.4			ZAIDEL
AR	I	357.229	2.3			ZAIDEL
AR	II	356.434	2.1			ZAIDEL
AR	I	377.037	2.0			ZAIDEL
AR	I	350.649	2.0			ZAIDEL
AR	I	369.090	2.0			ZAIDEL
AR	I	339.278	1.8			ZAIDEL
AR	I	337.348	1.8			ZAIDEL
AR	I	355.601	1.7			ZAIDEL
AR	I	340.618	1.6			ZAIDEL
AR	I	405.453	1.5			ZAIDEL
AR	I	364.312	1.5			ZAIDEL
AR	I	389.466	1.4			ZAIDEL
AR	I	317.296	1.4			OH BAND, ZAIDEL
AR	I	365.953	1.4			ZAIDEL
AR	I	378.136	1.3			ZAIDEL
AR	I	367.524	1.1			ZAIDEL
AR	I	332.550	1.1			ZAIDEL
AR	I	389.986	1.0			ZAIDEL
AR	I	436.379	1.0			ZAIDEL
AR	I	349.327	0.9			ZAIDEL
AR	I	320.039	0.8			OH BAND, ZAIDEL
AR	I	323.449	0.8			OH BAND, ZAIDEL
AR	I	325.758	0.7			ZAIDEL
AR	I	442.399	0.5			ZAIDEL
AS	I	193.696	56.0	100.0	0.053	
AS	I	197.197	39.0	100.0	0.076	
AS	I	228.812	36.0	100.0	0.083	
AS	I	200.334	25.0	100.0	0.120	
AS	I	189.042	22.0	100.0	0.136	NRL
AS	I	234.984	21.0	100.0	0.142	
AS	I	198.970	16.0	100.0	0.187	
AS	I	200.919	6.1	100.0	0.491	
AS	I	278.022	5.7	100.0	0.526	
AS	I	199.048	5.5	100.0	0.545	
AU	I	242.795	170.0	100.0	0.017	
AU	I	267.595	96.0	100.0	0.031	
AU	I	197.819	77.0	100.0	0.038	ZAIDEL
AU	II	208.209	70.0	100.0	0.042	ZAIDEL
AU	I	201.200	54.0	100.0	0.055	
AU	II	211.068	47.0	100.0	0.063	ZAIDEL
AU	II	191.893	35.0	100.0	0.085	ZAIDEL
AU	II	200.081	32.0	100.0	0.093	ZAIDEL
AU	II	198.963	20.0	100.0	0.150	ZAIDEL
AU	I	195.193	18.0	100.0	0.166	ZAIDEL
B	I	249.773	63.0	10.0	0.0048	
B	I	249.678	53.0	10.0	0.0057	
B	I	208.959	30.0	10.0	0.010	
B	I	208.893	25.0	10.0	0.012	

S OF I	WAVE-LENGTH (nm)	$\frac{I_n}{I_b}$	CONC ($\mu g/ml$)	EST'D DET LIM ($\mu g/ml$)	COMMENTS
EA	II 455.403	230.0	10.0	0.0013	
EA	II 493.409	130.0	10.0	0.0023	
BA	II 233.527	75.0	10.0	0.0040	
EA	II 230.424	73.0	10.0	0.0041	
BA	II 413.066	9.1	10.0	0.032	
EA	II 234.758	7.8	10.0	0.038	
EA	II 389.178	5.2	10.0	0.057	H 388.905
EA	II 489.997	3.7	10.0	0.081	
EA	II 225.473	2.0	10.0	0.150	ZAIDEL
BA	II 452.493	1.9	10.0	0.157	
BE	II 313.042	110.0	1.0	0.00027	OH BAND
BE	I 234.861	96.0	1.0	0.00031	
BE	II 313.107	41.0	1.0	0.00073	OH BAND
BE	I 249.473	8.0	1.0	0.0038	GROUP NR
BE	I 265.045	6.4	1.0	0.0047	GROUP NR
BE	I 217.510	2.5	1.0	0.012	NR
BE	I 217.499	2.5	1.0	0.012	NR
BE	I 332.134	1.4	1.0	0.021	GROUP NR
BE	I 205.590	0.7	1.0	0.042	ZAIDEL,NR
BE	I 205.601	0.7	1.0	0.042	ZAIDEL,NR
BI	I 223.061	87.0	100.0	0.034	
EI	I 306.772	40.0	100.0	0.075	OH BAND
BI	I 222.825	36.0	100.0	0.083	
BI	I 206.170	35.0	100.0	0.085	
BI	I 195.389	14.0	100.0	0.214	
BI	I 227.658	12.0	100.0	0.250	
BI	II 190.241	10.0	100.0	0.300	ZAIDEL
BI	I 213.363	10.0	100.0	0.300	
EI	I 289.798	9.0	100.0	0.333	
BI	I 211.026	7.8	100.0	0.384	
C	I 193.091	67.0	100.0	0.044	NRL
C	I 247.856	17.0	100.0	0.176	
C	I 199.362	3.4	1000.0	8.823	NRL
CA	II 393.366	89.0	0.5	0.00019	
CA	II 396.847	30.0	0.5	0.00050	H 397.007
CA	II 317.933	1.5	0.5	0.010	OH BAND
CA	I 422.673	1.5	0.5	0.010	
CD	II 214.438	120.0	10.0	0.0025	
CD	I 228.802	110.0	10.0	0.0027	
CD	II 226.502	89.0	10.0	0.0034	
CD	I 361.051	1.3	10.0	0.230	
CD	I 326.106	0.9	10.0	0.333	
CD	I 346.620	0.7	10.0	0.428	
CD	I 231.284	0.5	10.0	0.600	
CD	I 479.992	0.5	10.0	0.600	
CE	II 413.765	6.2	10.0	0.048	
CE	II 413.380	6.0	10.0	0.050	
CE	II 418.660	5.7	10.0	0.052	

S OF I		WAVE-LENGTH (nm)	$\frac{I_n}{I_b}$	CONC (μg/ml)	EST'D DET LIM (μg/ml)	COMMENTS
CE	II	393.109	5.0	10.0	0.060	
CE	II	446.021	4.8	10.0	0.062	
CE	II	394.275	4.4	10.0	0.068	
CE	II	429.667	4.3	10.0	0.069	
CE	II	407.585	4.2	10.0	0.071	NR
CE	II	407.571	4.2	10.0	0.071	NR
CE	II	456.236	4.1	10.0	0.073	
CE	II	404.076	4.0	10.0	0.075	
CE	II	380.152	4.0	10.0	0.075	
CE	II	401.239	4.0	10.0	0.075	
CO	II	238.892	50.0	10.0	0.0060	
CO	II	228.616	43.0	10.0	0.0070	
CO	II	237.862	31.0	10.0	0.0097	
CO	II	230.786	31.0	10.0	0.0097	
CO	II	236.379	27.0	10.0	0.011	
CO	II	231.160	23.0	10.0	0.013	
CO	II	238.346	21.0	10.0	0.014	
CO	II	231.405	18.0	10.0	0.016	
CO	II	235.342	17.0	10.0	0.017	
CO	II	238.636	14.0	10.0	0.021	
CO	II	234.426	14.0	10.0	0.021	
CO	II	231.498	13.0	10.0	0.023	
CO	II	234.739	13.0	10.0	0.023	
CR	II	205.552	49.0	10.0	0.0061	
CR	II	206.149	42.0	10.0	0.0071	
CR	II	267.716	42.0	10.0	0.0071	
CR	II	283.563	42.0	10.0	0.0071	
CR	II	284.325	35.0	10.0	0.0086	
CR	II	206.542	31.0	10.0	0.0097	
CR	II	276.654	22.0	10.0	0.013	
CR	II	284.984	21.0	10.0	0.014	
CR	II	285.568	16.0	10.0	0.018	
CR	II	276.259	15.0	10.0	0.020	
CR	II	286.257	15.0	10.0	0.020	
CR	II	266.602	14.0	10.0	0.021	
CR	II	286.511	14.0	10.0	0.021	
CR	II	286.674	13.0	10.0	0.023	
CR	I	357.869	13.0	10.0	0.023	
CS	II	452.673	0.7	1000.0	42.857	ZAIDEL
CS	I	455.531	0.3	1000.0	100.000	
CU	I	324.754	56.0	10.0	0.0054	OH BAND
CU	II	224.700	39.0	10.0	0.0077	
CU	I	219.958	31.0	10.0	0.0097	
CU	I	327.396	31.0	10.0	0.0097	
CU	II	213.598	25.0	10.0	0.012	
CU	I	223.008	23.0	10.0	0.013	
CU	I	222.778	19.0	10.0	0.015	
CU	II	221.810	17.0	10.0	0.017	ZAIDEL
CU	II	219.226	17.0	10.0	0.017	
CU	I	217.894	17.0	10.0	0.017	
CU	I	221.458	13.0	10.0	0.023	

S OF I		WAVE- LENGTH (nm)	$\frac{I_n}{I_b}$	CONC (µg/ml)	EST'D DET LIM (µg/ml)	COMMENTS
DY	II	353.170	30.0	10.0	0.010	
DY	II	364.540	13.0	10.0	0.023	
DY	II	340.780	11.0	10.0	0.027	
DY	II	353.602	10.0	10.0	0.030	
DY	II	394.468	9.5	10.0	0.031	
DY	II	396.839	9.5	10.0	0.031	CA 396.847,H 397.007
DY	II	338.502	9.0	10.0	0.033	
DY	II	400.045	8.4	10.0	0.035	
DY	II	387.211	8.2	10.0	0.036	
DY	II	407.796	7.5	10.0	0.040	
DY	II	352.398	7.4	10.0	0.040	
DY	II	353.852	6.6	10.0	0.045	
DY	II	357.624	6.4	10.0	0.046	
DY	II	389.853	6.0	10.0	0.050	
DY	II	238.736	5.8	10.0	0.051	
ER	II	337.271	29.0	10.0	0.010	
ER	II	349.910	17.0	10.0	0.017	
ER	II	323.058	16.0	10.0	0.018	OH BAND
ER	II	326.478	16.0	10.0	0.018	
ER	II	369.265	16.0	10.0	0.018	
ER	II	390.631	14.0	10.0	0.021	
ER	II	291.036	11.0	10.0	0.027	
ER	II	296.452	11.0	10.0	0.027	
ER	II	331.242	10.0	10.0	0.030	
ER	II	339.200	9.4	10.0	0.031	
ER	II	338.508	8.8	10.0	0.034	
ER	II	389.623	7.0	10.0	0.042	H 388.905
EU	II	381.967	110.0	10.0	0.0027	
EU	II	412.970	70.0	10.0	0.0043	
EU	II	420.505	70.0	10.0	0.0043	
EU	II	393.048	53.0	10.0	0.0057	
EU	II	390.710	39.0	10.0	0.0077	
EU	II	272.778	37.0	10.0	0.0081	
EU	II	372.494	34.0	10.0	0.0088	
EU	II	397.196	32.0	10.0	0.0094	H 397.007
EU	II	443.556	24.0	10.0	0.012	
EU	II	281.394	22.0	10.0	0.013	
FE	II	238.204	65.0	10.0	0.0046	
FE	II	239.562	59.0	10.0	0.0051	
FE	II	259.940	48.0	10.0	0.0062	
FE	II	234.349	29.0	10.0	0.010	
FE	II	240.488	27.0	10.0	0.011	
FE	II	259.837	24.0	10.0	0.012	
FE	II	261.187	24.0	10.0	0.012	
FE	II	234.810	23.0	10.0	0.013	NR
FE	II	234.830	23.0	10.0	0.013	NR
FE	II	258.588	20.0	10.0	0.015	
FE	II	238.863	20.0	10.0	0.015	
FE	II	263.105	19.0	10.0	0.015	NR
FE	II	263.132	19.0	10.0	0.015	NR
FE	II	274.932	19.0	10.0	0.015	
FE	II	275.574	16.0	10.0	0.018	
FE	II	233.280	15.0	10.0	0.020	
FE	II	273.955	15.0	10.0	0.020	

S OF I		WAVE-LENGTH (nm)	$\frac{I_n}{I_b}$	CONC (µg/ml)	EST'D DET LIM (µg/ml)	COMMENTS
GA	I	294.364	64.0	100.0	0.046	
GA	I	417.206	45.0	100.0	0.066	
GA	I	287.424	38.0	100.0	0.078	
GA	I	403.298	27.0	100.0	0.111	
GA	I	250.017	16.0	100.0	0.187	
GA	II	209.134	11.0	100.0	0.272	MIT
GA	I	245.007	10.0	100.0	0.300	
GA	I	294.418	9.4	100.0	0.319	
GA	I	271.965	5.7	100.0	0.526	
GA	I	233.828	3.9	100.0	0.769	
GA	I	265.987	3.6	100.0	0.833	
GD	II	342.247	21.0	10.0	0.014	
GD	II	336.223	15.0	10.0	0.020	
GD	II	335.047	14.0	10.0	0.021	
GD	II	335.862	14.0	10.0	0.021	
GD	II	310.050	13.0	10.0	0.023	OH BAND
GD	II	376.839	12.0	10.0	0.025	
GD	II	303.284	11.0	10.0	0.027	
GD	II	343.999	10.0	10.0	0.030	
GD	II	358.496	10.0	10.0	0.030	
GD	II	364.619	10.0	10.0	0.030	
GD	II	301.013	10.0	10.0	0.030	
GD	II	354.580	9.5	10.0	0.031	
GD	II	354.936	9.0	10.0	0.033	
GD	II	308.199	9.0	10.0	0.033	OH BAND
GD	II	303.405	8.8	10.0	0.034	
GE	I	209.426	75.0	100.0	0.040	
GE	I	265.118	62.0	100.0	0.048	
GE	I	206.866	50.0	100.0	0.060	
GE	I	219.871	47.0	100.0	0.063	
GE	I	265.158	36.0	100.0	0.083	
GE	I	204.377	35.0	100.0	0.085	
GE	I	199.824	34.0	100.0	0.088	
GE	I	204.171	34.0	100.0	0.088	
GE	I	259.254	29.0	100.0	0.103	
GE	I	303.906	29.0	100.0	0.103	
GE	I	275.459	28.0	100.0	0.107	
GE	I	270.963	27.0	100.0	0.111	
H	I	486.133	17.0			ZAIDEL
H	I	434.047	5.0			ZAIDEL
H	I	410.174	1.5			ZAIDEL
H	I	397.007	0.6			ZAIDEL
H	I	388.905	0.3			ZAIDEL
H	I	383.539	0.1			ZAIDEL
H	I	379.790	0.1			ZAIDEL
HF	II	277.336	190.0	100.0	0.015	
HF	II	273.876	180.0	100.0	0.016	
HF	II	264.141	160.0	100.0	0.018	
HF	II	232.247	160.0	100.0	0.018	
HF	II	263.871	160.0	100.0	0.018	

	S OF I	WAVE-LENGTH (nm)	$\frac{I_n}{I_b}$	CONC (µg/ml)	EST'D DET LIM (µg/ml)	COMMENTS
HF	II	282.022	160.0	100.0	0.018	
HF	II	251.269	150.0	100.0	0.020	NR
HF	II	251.303	150.0	100.0	0.020	NR
HF	II	257.167	150.0	100.0	0.020	
HF	II	196.382	150.0	100.0	0.020	ZAIDEL
HF	II	239.336	140.0	100.0	0.021	
HF	II	239.383	140.0	100.0	0.021	
HF	II	235.122	130.0	100.0	0.023	
HF	II	246.419	130.0	100.0	0.023	
HG	II	194.227	120.0	100.0	0.025	ZAIDEL
HG	I	253.652	49.0	100.0	0.061	
HG	I	296.728	1.7	100.0	1.764	
HG	I	435.835	1.1	100.0	2.727	
HG	I	265.204	0.7	100.0	4.285	ZAIDEL
HG	I	302.150	0.6	100.0	5.000	
HG	I	365.483	0.3	100.0	10.000	
HO	II	345.600	53.0	10.0	0.0057	
HO	II	339.898	23.0	10.0	0.013	
HO	II	389.102	18.0	10.0	0.016	H 388.905
HO	II	347.426	16.0	10.0	0.018	
HO	II	341.646	16.0	10.0	0.018	
HO	II	361.073	15.0	10.0	0.020	
HO	II	348.484	15.0	10.0	0.020	
HO	II	379.675	12.0	10.0	0.025	H 379.790
HO	II	351.559	11.0	10.0	0.027	
HO	II	345.314	10.0	10.0	0.030	
IN	II	230.606	47.0	100.0	0.063	
IN	I	325.609	25.0	100.0	0.120	
IN	I	303.936	20.0	100.0	0.150	
IN	I	451.131	16.0	100.0	0.187	
IN	I	410.176	6.4	100.0	0.468	H 410.174
IN	I	271.026	5.4	100.0	0.555	
IN	I	325.856	5.0	100.0	0.600	
IN	II	207.926	4.2	100.0	0.714	ZAIDEL
IN	I	256.015	4.2	100.0	0.714	
IN	I	293.263	2.0	100.0	1.500	
IN	II	197.745	1.7	100.0	1.764	ZAIDEL
IN	I	275.388	1.6	100.0	1.875	
IR	II	224.268	110.0	100.0	0.027	
IR	II	212.681	100.0	100.0	0.030	
IR	I	205.222	49.0	100.0	0.061	
IR	II	215.268	44.0	100.0	0.068	
IR		204.419	30.0	100.0	0.100	ZAIDEL
IR	I	209.263	28.0	100.0	0.107	
IR	I	208.882	28.0	100.0	0.107	
IR	II	236.804	24.0	100.0	0.125	
IR	I	254.397	19.0	100.0	0.157	
IR	I	215.805	17.0	100.0	0.176	
IR	I	263.971	17.0	100.0	0.176	
IR	II	216.942	13.0	100.0	0.230	

S OF I	WAVE-LENGTH (nm)	$\frac{I_n}{I_b}$	CCNC (µg/ml)	EST'D DET LIM (µg/ml)	COMMENTS
IR I	266.479	13.0	100.0	0.230	
IR I	237.277	12.0	100.0	0.250	
IR I	238.162	12.0	100.0	0.250	
IR I	250.298	12.0	100.0	0.250	
IR I	269.423	11.0	100.0	0.272	
IR I	284.972	10.0	100.0	0.300	
K I	404.721	0.7	1000.0	42.857	
K I	404.414	NM	1000.0	0.0	AR 404.442
LA II	394.910	NM	10.0		AR 394.898
LA II	379.478	30.0	10.0	0.010	
LA II	333.749	30.0	10.0	0.010	
LA II	408.672	30.0	10.0	0.010	
LA II	412.323	29.0	10.0	0.010	
LA II	398.852	27.0	10.0	0.011	
LA II	379.083	26.0	10.0	0.011	
LA II	399.575	22.0	10.0	0.013	
LA II	407.735	21.0	10.0	0.014	
LA II	375.908	20.0	10.0	0.015	
LA II	387.164	20.0	10.0	0.015	
LA II	404.291	19.0	10.0	0.015	
LA II	403.169	19.0	10.0	0.015	
LA II	338.091	17.0	10.0	0.017	
LA II	442.990	13.0	10.0	0.023	
LA II	392.922	12.0	10.0	0.025	
LA II	384.902	12.0	10.0	0.025	
LA II	492.098	12.0	10.0	0.025	
LA II	492.179	12.0	10.0	0.025	
LI I	460.286	3.5	100.0	0.857	
LI I	323.263	2.8	100.0	1.071	CH BAND
LI I	274.118	1.9	100.0	1.578	
LI I	497.170	1.4	100.0	2.142	
LI I	256.231	0.7	100.0	4.285	ZAIDEL
LI I	413.262	0.4	100.0	7.500	ZAIDEL,NR
LI I	413.256	0.4	100.0	7.500	ZAIDEL,NR
LU II	261.542	150.0	5.0	0.0010	
LU II	291.139	24.0	5.0	0.0062	
LU II	219.554	18.0	5.0	0.0083	
LU II	307.760	17.0	5.0	0.0088	OH BAND
LU II	289.484	15.0	5.0	0.010	
LU II	339.707	14.0	5.0	0.010	
LU II	350.739	13.0	5.0	0.011	
LU II	270.171	12.0	5.0	0.012	
LU II	290.030	12.0	5.0	0.012	
LU II	275.417	12.0	5.0	0.012	
LU II	302.054	11.0	5.0	0.013	
LU II	347.248	10.0	5.0	0.015	
MG II	279.553	195.0	1.0	0.00015	
MG II	280.270	100.0	1.0	0.00030	
MG I	285.21			0.0016	

	S CF I	WAVE-LENGTH (nm)	$\frac{I_n}{I_b}$	CONC (µg/ml)	EST'D DET LIM (µg/ml)	COMMENTS
MG	II	279.806	2.0	1.	.15	
MG	I	202.582	1.3	1.0	0.023	ZAIDEL
MG	II	279.079	1.0	1.0	0.030	
MG	I	383.826	0.9	1.0	0.033	
MG	I	383.231	0.7	1.0	0.042	
MG	I	277.983	0.6	1.0	0.050	
MG	II	293.654	0.5	1.0	0.060	
MN	II	257.610	220.0	10.0	0.0014	
MN	II	259.373	190.0	10.0	0.0016	
MN	II	260.569	145.0	10.0	0.0021	
MN	II	294.920	39.0	10.0	0.0077	
MN	II	293.930	29.0	10.0	0.010	
MN	I	279.482	24.0	10.0	0.012	
MN	II	293.306	22.0	10.0	0.013	
MN	I	279.827	18.0	10.0	0.016	
MN	I	280.106	14.0	10.0	0.021	
MN	I	403.076	6.8	10.0	0.044	
MN	II	344.199	6.6	10.0	0.045	
MN	I	403.307	6.3	10.0	0.047	
MN	II	191.510	5.8	10.0	0.051	NRL
MO	II	202.030	38.0	10.0	0.0079	
MO	II	203.844	24.0	10.0	0.012	
MO	II	204.598	24.0	10.0	0.012	
MO	II	281.615	21.0	10.0	0.014	
MO	II	201.511	16.0	10.0	0.018	
MO	II	284.823	15.0	10.0	0.020	
MO	II	277.540	12.0	10.0	0.025	
MO	II	287.151	11.0	10.0	0.027	
MO	II	268.414	10.0	10.0	0.030	
MO	II	263.876	8.0	10.0	0.037	
MO	II	292.339	8.0	10.0	0.037	
NA	I	588.995	101.0	100.0	0.029	AR 588.859
NA	I	589.592	43.0	100.0	0.069	
NA	I	330.237	1.6	100.0	1.875	
NA	I	330.298	0.7	100.0	4.285	
NA	I	285.301	1.1	1000.0	27.272	NR,ZAIDEL
NA	I	285.281	1.1	1000.0	27.272	NR,ZAIDEL
NA	II	288.114	0.6	1000.0	50.000	ZAIDEL
NB	II	309.418	83.0	100.0	0.036	OH BAND
NB	II	316.340	75.0	100.0	0.040	OH BAND
NB	II	313.079	60.0	100.0	0.050	OH BAND
NB	II	269.706	43.0	100.0	0.069	
NB	II	322.548	42.0	100.0	0.071	OH BAND
NB	II	319.498	41.0	100.0	0.073	OH BAND
NB	II	295.088	40.0	100.0	0.075	
NB	II	292.781	40.0	100.0	0.075	
NB	II	271.662	34.0	100.0	0.088	
NB	II	288.318	31.0	100.0	0.096	
NB	II	210.942	31.0	100.0	0.096	
NB	II	272.198	30.0	100.0	0.100	
NB	II	287.539	28.0	100.0	0.107	

	S OF I	WAVE-LENGTH (nm)	$\frac{I_n}{I_b}$	CCNC (µg/ml)	EST'D DET LIM (µg/ml)	COMMENTS
ND	II	401.225	59.0	100.0	0.050	
ND	II	430.358	40.0	100.0	0.075	
ND	II	406.109	31.0	100.0	0.096	
ND	II	415.608	28.0	100.0	0.107	
ND	II	410.946	26.0	100.0	0.115	
ND	II	386.333	23.0	100.0	0.130	NR
ND	II	386.340	23.0	100.0	0.130	NR
ND	II	404.080	23.0	100.0	0.130	
ND	II	417.732	22.0	100.0	0.136	
ND	II	385.174	19.0	100.0	0.157	NR
ND	II	385.166	19.0	100.0	0.157	NR
ND	II	394.151	19.0	100.0	0.157	
ND	II	445.157	19.0	100.0	0.157	
ND	II	424.738	17.0	100.0	0.176	
ND	II	395.116	17.0	100.0	0.176	
ND	II	396.312	17.0	100.0	0.176	
ND	II	384.824	16.0	100.0	0.187	NR
ND	II	384.852	16.0	100.0	0.187	NR
ND	II	380.536	16.0	100.0	0.187	
NI	II	221.647	29.0	10.0	0.010	ZAIDEL
NI	I	232.003	20.0	10.0	0.015	
NI	II	231.604	19.0	10.0	0.015	
NI	II	216.556	17.0	10.0	0.017	ZAIDEL
NI	II	217.467	13.0	10.0	0.023	MIT
NI	II	230.300	13.0	10.0	0.023	MIT
NI	II	227.021	12.0	10.0	0.025	ZAIDEL
NI	II	225.386	12.0	10.0	0.025	ZAIDEL
NI	I	234.554	9.5	10.0	0.031	
NI	II	239.452	7.8	10.0	0.038	
NI	I	352.454	6.6	10.0	0.045	
NI	I	341.476	6.2	10.0	0.048	
OS	II	225.585	83.0	1.0	0.00036	
OS	II	228.226	48.0	1.0	0.00063	
OS		189.900	25.0	1.0	0.0012	APPROX. WAVE
OS	II	233.680	24.0	1.0	0.0012	
OS	II	206.721	22.0	1.0	0.0014	
OS	II	219.439	18.0	1.0	0.0017	
OS	II	236.735	16.0	1.0	0.0019	
OS	II	207.067	14.0	1.0	0.0021	
OS	II	248.624	13.0	1.0	0.0023	
OS	I	222.798	11.0	1.0	0.0027	
P	I	213.618	39.0	100.0	0.076	
P	I	214.914	39.0	100.0	0.076	
P	I	253.565	11.0	100.0	0.272	
P	I	213.547	8.5	100.0	0.352	
P	I	203.349	7.4	100.0	0.405	MIT
P	I	215.408	7.2	100.0	0.416	
P	I	255.328	5.2	100.0	0.576	
P	I	202.347	3.8	100.0	0.789	MIT
P	I	215.294	3.4	100.0	0.882	
P	I	253.401	3.0	100.0	1.000	

S OF I		WAVE-LENGTH (nm)	$\frac{I_n}{I_b}$	CONC (µg/ml)	EST'D DET LIM (µg/ml)	COMMENTS
PB	II	220.353	70.0	100.0	0.042	
PB	I	216.999	33.0	100.0	0.090	
PB	I	261.418	23.0	100.0	0.130	
PB	I	283.306	21.0	100.0	0.142	
PB	I	280.199	19.0	100.0	0.157	
PB	I	405.783	11.0	100.0	0.272	
PB	I	224.688	9.0	100.0	0.333	
PB	I	368.348	8.6	100.0	0.348	
PB	I	266.316	7.7	100.0	0.389	
PB	I	239.379	6.3	100.0	0.476	
PB	I	363.958	5.2	100.0	0.576	
PB	I	247.638	5.1	100.0	0.588	
PD	I	340.458	68.0	100.0	0.044	
PD	I	363.470	55.0	100.0	0.054	
PD	II	229.651	44.0	100.0	0.068	ZAIDEL
PD	I	324.270	39.0	100.0	0.076	OH BAND
PD	I	360.955	35.0	100.0	0.085	
PD	I	342.124	30.0	100.0	0.100	
PD	II	248.892	29.0	100.0	0.103	
PD	II	223.159	25.0	100.0	0.120	MIT
PD	I	244.791	23.0	100.0	0.130	
PD	I	351.694	22.0	100.0	0.136	
PD	I	355.308	22.0	100.0	0.136	
PD	I	247.642	18.0	100.0	0.166	
PD	I	348.115	18.0	100.0	0.166	
PD	II	244.618	18.0	100.0	0.166	ZAIDEL
PD	II	235.134	17.0	100.0	0.176	ZAIDEL
PD	I	346.077	16.0	100.0	0.187	
PD	II	236.796	15.0	100.0	0.200	ZAIDEL
PD	II	248.653	15.0	100.0	0.200	
PR	II	390.844	8.1	10.0	0.037	
PR	II	414.311	8.0	10.0	0.037	
PR	II	417.939	7.2	10.0	0.041	
PR	II	422.535	7.0	10.0	0.042	
PR	II	422.293	6.3	10.0	0.047	
PR	II	406.281	6.3	10.0	0.047	
PR	II	411.846	6.0	10.0	0.050	
PR	II	418.948	5.0	10.0	0.060	
PR	II	440.882	4.9	10.0	0.061	
PR	II	400.869	4.6	10.0	0.065	
PT	II	214.423	100.0	100.0	0.030	
PT	II	203.646	54.0	100.0	0.055	MIT
PT	I	204.937	42.0	100.0	0.071	
PT	I	265.945	37.0	100.0	0.081	
PT	II	224.552	36.0	100.0	0.083	MIT
PT	I	217.467	36.0	100.0	0.083	
PT	I	306.471	25.0	100.0	0.120	OH BAND
PT	I	212.861	24.0	100.0	0.125	
PT		193.700	22.0	100.0	0.136	APPROX. WAVE
PT	I	210.333	20.0	100.0	0.150	
PT	I	273.396	20.0	100.0	0.150	
PT	I	248.717	20.0	100.0	0.150	

S CF I		WAVE- LENGTH (nm)	$\frac{I_n}{I_b}$	CONC (µg/ml)	EST'D DET LIM (µg/ml)	COMMENTS
RB	I	420.185	0.8	1000.0	37.500	
RB	I	421.556	NM	1000.0	0.0	SR 421.552
RE	II	197.313	49.0	10.0	0.006	ZAIDEL
RE	II	221.426	47.0	10.0	0.006	
RE	II	227.525	44.0	10.0	0.006	
RE	II	189.836	8.0	10.0	0.037	ZAIDEL
RE	I	204.908	3.8	10.0	0.078	
RE	I	228.751	3.8	10.0	0.078	
RE	I	229.449	3.6	10.0	0.083	
RE	II	202.364	3.2	10.0	0.093	ZAIDEL
RE	II	209.241	3.2	10.0	0.093	
RE	I	346.046	2.6	10.0	0.115	
RE	I	208.559	2.4	10.0	0.125	
RH	II	233.477	67.0	100.0	0.044	
RH	II	249.077	52.0	100.0	0.057	
RH	I	343.489	50.0	100.0	0.060	
RH	II	252.053	39.0	100.0	0.076	
RH	I	369.236	35.0	100.0	0.085	
RH	II	246.104	28.0	100.0	0.107	
RH	I	339.682	24.0	100.0	0.125	
RH	II	251.103	24.0	100.0	0.125	
RH	I	352.802	24.0	100.0	0.125	
RH	II	242.711	24.0	100.0	0.125	
RH	II	241.584	23.0	100.0	0.130	
RH	I	228.857	21.0	100.0	0.142	
RH	I	365.799	20.0	100.0	0.150	
RH	I	350.252	20.0	100.0	0.150	
RU	II	240.272	100.0	100.0	0.030	
RU	II	245.657	100.0	100.0	0.030	
RU	II	267.876	83.0	100.0	0.036	
RU	II	269.206	33.0	100.0	0.090	
RU	II	266.161	31.0	100.0	0.096	
RU	II	249.842	31.0	100.0	0.096	NR
RU	II	249.857	31.0	100.0	0.096	NR
RU	I	349.894	27.0	100.0	0.111	
RU	II	273.435	26.0	100.0	0.115	
RU	II	245.644	25.0	100.0	0.120	
RU	II	271.241	25.0	100.0	0.120	
RU	I	372.803	25.0	100.0	0.120	
RU	II	235.791	21.0	100.0	0.142	
RU	II	247.893	20.0	100.0	0.150	
RU	II	250.701	19.0	100.0	0.157	
RU		279.535	19.0	100.0	0.157	MIT
SB	I	206.833	91.0	100.0	0.032	
SB	I	217.581	68.0	100.0	0.044	
SB	I	231.147	49.0	100.0	0.061	
SB	I	252.852	28.0	100.0	0.107	
SB	I	259.805	28.0	100.0	0.107	NR
SB	I	259.809	28.0	100.0	0.107	NR
SB	I	217.919	19.0	100.0	0.157	
SB	I	195.039	18.0	100.0	0.166	ZAIDEL

S CF I	WAVE-LENGTH (nm)	$\frac{I_n}{I_b}$	CONC (µg/ml)	EST'D DET LIM (µg/ml)	COMMENTS
SB I	213.969	16.0	100.0	0.187	
SB I	204.957	15.0	100.0	0.200	
SB I	214.486	12.0	100.0	0.250	
SB I	209.841	8.7	100.0	0.344	
SB I	203.977	6.6	100.0	0.454	
SB I	220.845	6.5	100.0	0.461	
SB I	287.792	4.7	100.0	0.638	
SC II	361.384	200.0	10.0	0.0015	
SC II	357.253	150.0	10.0	0.0020	
SC II	363.075	140.0	10.0	0.0021	
SC II	364.279	110.0	10.0	0.0027	
SC II	424.683	110.0	10.0	0.0027	
SC II	357.635	82.0	10.0	0.0037	
SC II	335.373	80.0	10.0	0.0038	
SC II	337.215	68.0	10.0	0.0044	
SC II	358.094	67.0	10.0	0.0045	
SC II	255.237	65.0	10.0	0.0046	
SC II	431.409	38.0	10.0	0.0079	
SC II	256.025	37.0	10.0	0.0081	
SC II	356.770	37.0	10.0	0.0081	AR 356.766
SC II	364.531	35.0	10.0	0.0086	
SC II	355.855	34.0	10.0	0.0088	
SC II	336.895	32.0	10.0	0.0094	
SC II	365.180	31.0	10.0	0.0097	
SC II	432.074	31.0	10.0	0.0097	
SC II	358.964	26.0	10.0	0.011	
SC II	353.573	24.0	10.0	0.012	
SC II	256.321	22.0	10.0	0.013	
SC II	359.048	20.0	10.0	0.015	
SC II	432.501	19.0	10.0	0.015	
SC II	336.127	17.0	10.0	0.017	
SC II	254.522	17.0	10.0	0.017	
SC II	437.446	17.0	10.0	0.017	
SE I	196.026	40.0	100.0	0.075	
SE I	203.985	26.0	100.0	0.115	
SE I	206.279	10.0	100.0	0.300	
SE I	207.479	1.9	100.0	1.578	
SE I	199.511	0.6	100.0	5.000	NRL
SI I	251.611	250.0	100.0	0.012	
SI I	212.412	180.0	100.0	0.016	
SI I	288.158	110.0	100.0	0.027	
SI I	250.690	100.0	100.0	0.030	
SI I	252.851	95.0	100.0	0.031	
SI I	251.432	79.0	100.0	0.037	
SI I	252.411	75.0	100.0	0.040	
SI I	221.667	72.0	100.0	0.041	
SI I	251.920	61.0	100.0	0.049	
SI I	198.899	50.0	100.0	0.060	NRL
SI I	221.089	47.0	100.0	0.063	
SI I	243.515	36.0	100.0	0.083	
SI I	190.134	23.0	100.0	0.130	NRL
SI I	220.798	23.0	100.0	0.130	
SI I	205.813	23.0	100.0	0.130	ZAIDEL

	S OF I	WAVE- LENGTH (nm)	$\frac{I_n}{I_b}$	CONC (µg/ml)	EST'D DET LIM (µg/ml)	COMMENTS
SM	II	359.260	69.0	100.0	0.043	
SM	II	442.434	55.0	100.0	0.054	
SM	II	360.949	52.0	100.0	0.057	
SM	II	363.429	45.0	100.0	0.066	
SM	II	428.079	43.0	100.0	0.069	
SM	II	446.734	41.0	100.0	0.073	
SM	II	367.084	40.0	100.0	0.075	
SM	II	356.827	39.0	100.0	0.076	
SM	II	373.126	38.0	100.0	0.078	
SM	II	443.432	36.0	100.0	0.083	
SM	II	388.529	36.0	100.0	0.083	
SM	II	330.639	33.0	100.0	0.090	
SN	II	189.989	120.0	100.0	0.025	ZAIDEL
SN	I	235.484	31.0	100.0	0.096	
SN	I	242.949	31.0	100.0	0.096	
SN	I	283.999	27.0	100.0	0.111	
SN	I	226.891	25.0	100.0	0.120	
SN	I	224.605	25.0	100.0	0.120	
SN	I	242.170	19.0	100.0	0.157	
SN	I	270.651	18.0	100.0	0.166	
SN	I	220.965	16.0	100.0	0.187	
SN	I	286.333	14.0	100.0	0.214	
SN	I	317.505	14.0	100.0	0.214	OH BAND
SR	II	407.771	72.0	1.0	0.00042	
SR	II	421.552	39.0	1.0	0.00077	
SR	II	216.596	36.0	10.0	0.0083	
SR	II	215.284	29.0	10.0	0.010	
SR	II	346.446	13.0	10.0	0.023	
SR	II	338.071	8.8	10.0	0.034	
SR	II	430.545	4.8	10.0	0.062	
SR	I	460.733	4.4	10.0	0.068	
SR	II	232.235	2.9	10.0	0.103	MIT
SR	II	416.180	2.4	10.0	0.125	
TA	II	226.230	120.0	100.0	0.025	
TA	II	240.063	105.0	100.0	0.028	
TA	II	268.517	100.0	100.0	0.030	
TA	II	233.198	96.0	100.0	0.031	
TA	II	228.916	95.0	100.0	0.031	
TA	II	263.558	88.0	100.0	0.034	
TA	II	238.706	80.0	100.0	0.037	
TA	II	223.948	69.0	100.0	0.043	
TA	II	267.590	68.0	100.0	0.044	
TA		205.908	68.0	100.0	0.044	MIT
TA	II	219.603	58.0	100.0	0.051	
TA	II	260.349	54.0	100.0	0.055	
TA	II	248.870	54.0	100.0	0.055	
TA	II	284.446	52.0	100.0	0.057	
TA	II	269.452	50.0	100.0	0.060	
TA	II	296.513	50.0	100.0	0.060	
TB	II	350.917	130.0	100.0	0.023	
TB	II	384.873	54.0	100.0	0.055	

	S OF I	WAVE-LENGTH (nm)	$\frac{I_n}{I_b}$	CONC (µg/ml)	EST'D DET LIM (µg/ml)	COMMENTS
TB	II	367.635	50.0	100.0	0.060	
TB	II	387.417	48.0	100.0	0.062	
TB	II	356.174	47.0	100.0	0.063	
TB	II	356.852	46.0	100.0	0.065	
TB	II	370.286	46.0	100.0	0.065	
TB	II	332.440	35.0	100.0	0.085	
TB	II	389.920	31.0	100.0	0.096	
TB	II	374.734	30.0	100.0	0.100	NR
TB	II	374.717	30.0	100.0	0.100	NR
TB	II	370.392	29.0	100.0	0.103	
TB	II	329.307	29.0	100.0	0.103	
TB	II	345.406	27.0	100.0	0.111	
TE	I	214.281	73.0	100.0	0.041	
TE	I	225.902	17.0	100.0	0.176	
TE	I	238.578	17.0	100.0	0.176	
TE	I	214.725	14.0	100.0	0.214	
TE	I	200.202	12.0	100.0	0.250	
TE	I	238.326	11.0	100.0	0.272	
TE	I	208.116	11.0	100.0	0.272	
TE	I	199.418	6.3	100.0	0.476	
TE	I	225.548	2.7	100.0	1.111	MIT
TE	I	226.555	2.6	100.0	1.153	MIT
TH	II	283.730	46.0	100.0	0.065	
TH	II	283.231	42.0	100.0	0.071	
TH	II	274.716	36.0	100.0	0.083	
TH	II	401.913	36.0	100.0	0.083	
TH	II	318.020	34.0	100.0	0.088	OH BAND
TH	II	318.823	32.0	100.0	0.093	OH BAND
TH	II	374.118	31.0	100.0	0.096	
TH	II	294.286	31.0	100.0	0.096	
TH	II	353.959	30.0	100.0	0.100	
TH	II	269.242	30.0	100.0	0.100	
TH	II	339.204	30.0	100.0	0.100	
TH	II	332.512	28.0	100.0	0.107	
TH	II	360.944	27.0	100.0	0.111	
TH	II	311.953	26.0	100.0	0.115	OH BAND
TH	II	284.281	23.0	100.0	0.130	
TH	II	287.041	23.0	100.0	0.130	
TH	II	256.559	23.0	100.0	0.130	
TH	II	275.217	23.0	100.0	0.130	
TI	II	334.941	79.0	10.0	0.0038	
TI	II	336.121	57.0	10.0	0.0053	
TI	II	323.452	56.0	10.0	0.0054	OH BAND
TI	II	337.280	45.0	10.0	0.0067	
TI	II	334.904	40.0	10.0	0.0075	
TI	II	308.802	39.0	10.0	0.0077	OH BAND
TI	II	307.864	37.0	10.0	0.0081	OH BAND
TI	II	338.376	37.0	10.0	0.0081	
TI	II	323.657	30.0	10.0	0.010	OH BAND
TI	II	323.904	29.0	10.0	0.010	OH BAND
TI	II	368.520	26.0	10.0	0.011	

	S OF I	WAVE-LENGTH (nm)	$\frac{I_n}{I_b}$	CONC (µg/ml)	EST'D DET LIM (µg/ml)	COMMENTS
TL	II	190.864	74.0	100.0	0.040	UCRL
TL	I	276.787	25.0	100.0	0.120	
TL	I	351.924	15.0	100.0	0.200	
TL	I	377.572	13.0	100.0	0.230	
TL	I	237.969	7.0	100.0	0.428	
TL	I	291.832	2.9	100.0	1.034	
TL	I	223.785	2.2	100.0	1.363	ZAIDEL
TL	I	352.943	1.7	100.0	1.764	
TL	I	258.014	1.7	100.0	1.764	
TM	II	313.126	58.0	10.0	0.0052	CH BAND
TM	II	346.220	37.0	10.0	0.0081	
TM	II	384.802	31.0	10.0	0.0097	
TM	II	342.508	30.0	10.0	0.010	
TM	II	376.133	26.0	10.0	0.011	
TM	II	379.575	26.0	10.0	0.011	
TM	II	336.261	26.0	10.0	0.011	
TM	II	317.283	23.0	10.0	0.013	OH BAND
TM	II	376.191	23.0	10.0	0.013	
TM	II	313.389	22.0	10.0	0.013	OH BAND
TM	II	345.366	19.0	10.0	0.015	
TM	II	329.100	19.0	10.0	0.015	
TM	II	344.150	18.0	10.0	0.016	
TM	II	324.154	16.0	10.0	0.018	OH BAND
TM	II	370.136	15.0	10.0	0.020	
TM	II	370.026	14.0	10.0	0.021	
TM	II	250.908	14.0	10.0	0.021	
TM	II	286.923	14.0	10.0	0.021	
U	II	385.958	12.0	100.0	0.250	
U	II	367.007	10.0	100.0	0.300	
U	II	263.553	9.0	100.0	0.333	
U	II	409.014	8.9	100.0	0.337	
U	II	393.203	8.2	100.0	0.365	
U	II	424.167	6.5	100.0	0.461	
U	II	294.192	6.2	100.0	0.483	
U	II	385.466	6.2	100.0	0.483	
U	II	290.828	6.0	100.0	0.500	
U	II	288.963	6.0	100.0	0.500	
U	II	288.274	5.8	100.0	0.517	
U	II	256.541	5.7	100.0	0.526	
U	II	279.394	5.6	100.0	0.535	
U	II	311.935	5.6	100.0	0.535	OH BAND
U	II	330.590	5.6	100.0	0.535	
V	II	309.311	60.0	10.0	0.0050	OH BAND
V	II	310.230	47.0	10.0	0.0064	OH BAND
V	II	292.402	40.0	10.0	0.0075	
V	II	290.882	34.0	10.0	0.0088	
V	II	311.071	30.0	10.0	0.010	OH BAND
V	II	289.332	29.0	10.0	0.010	
V	II	268.796	29.0	10.0	0.010	
V	II	311.838	25.0	10.0	0.012	OH BAND
V	II	214.009	20.0	10.0	0.015	MIT
V	II	312.528	20.0	10.0	0.015	CH BAND

S CF I	WAVE- LENGTH (nm)	$\frac{I_n}{I_b}$	CONC (µg/ml)	EST'D DET LIM (µg/ml)	COMMENTS
V	II 327.612	19.0	10.0	0.015	
V	II 292.464	18.0	10.0	0.016	
V	II 270.094	17.0	10.0	0.017	
W	II 207.911	100.0	100.0	0.030	
W	II 224.875	67.0	100.0	0.044	
W	II 218.936	65.0	100.0	0.046	MIT
W	II 209.475	64.0	100.0	0.046	
W	II 209.860	55.0	100.0	0.054	
W	II 239.709	54.0	100.0	0.055	
W	II 222.589	50.0	100.0	0.060	ZAIDEL
W	II 220.448	49.0	100.0	0.061	
W	II 200.807	42.0	100.0	0.071	
W	II 208.819	41.0	100.0	0.073	
W	II 248.923	41.0	100.0	0.073	
W	II 202.998	40.0	100.0	0.075	
W	II 205.468	40.0	100.0	0.075	ZAIDEL
W	II 216.632	40.0	100.0	0.075	
W	II 232.609	39.0	100.0	0.076	
W	II 203.503	38.0	100.0	0.078	
Y	II 371.030	86.0	10.0	0.0035	
Y	II 324.228	67.0	10.0	0.0045	OH BAND
Y	II 360.073	63.0	10.0	0.0048	
Y	II 377.433	57.0	10.0	0.0053	
Y	II 437.494	46.0	10.0	0.0065	
Y	II 378.870	40.0	10.0	0.0075	
Y	II 361.105	40.0	10.0	0.0075	
Y	II 321.669	38.0	10.0	0.0079	OH BAND
Y	II 363.312	36.0	10.0	0.0083	
Y	II 224.306	33.0	10.0	0.0091	
Y	II 332.789	32.0	10.0	0.0094	
Y	II 360.192	30.0	10.0	0.010	
Y	II 417.754	26.0	10.0	0.011	
Y	II 354.901	23.0	10.0	0.013	
Y	II 320.332	20.0	10.0	0.015	OH BAND
YB	II 328.937	170.0	10.0	0.0018	
YB	II 369.419	100.0	10.0	0.0030	
YB	II 289.138	35.0	10.0	0.0086	
YB	II 222.446	34.0	10.0	0.0088	
YB	II 211.667	32.0	10.0	0.0094	
YB	II 212.674	32.0	10.0	0.0094	
YB	II 218.571	22.0	10.0	0.013	
YB	II 275.048	17.0	10.0	0.017	
YB	II 297.056	17.0	10.0	0.017	
YB	II 265.375	14.0	10.0	0.021	
ZN	I 213.856	170.0	10.0	0.0018	
ZN	II 202.548	75.0	10.0	0.0040	
ZN	II 206.200	51.0	10.0	0.0059	
ZN	I 334.502	2.2	10.0	0.136	
ZN	I 330.259	1.3	10.0	0.230	
ZN	I 481.053	1.3	10.0	0.230	
ZN	I 472.216	0.7	10.0	0.428	

S OF I	WAVE-LENGTH (nm)	$\frac{I_n}{I_b}$	CCNC (µg/ml)	EST'D DET LIM (µg/ml)	COMMENTS
ZN I	328.233	0.6	10.0	0.500	
ZN I	334.557	0.4	10.0	0.750	
ZN I	280.106	0.4	10.0	0.750	NR
ZN I	280.087	0.4	10.0	0.750	NR
ZR II	343.823	42.0	10.0	0.0071	
ZR II	339.198	39.0	10.0	0.0077	
ZR II	257.139	31.0	10.0	0.0097	
ZR II	349.621	30.0	10.0	0.010	
ZR II	357.247	30.0	10.0	0.010	
ZR II	327.305	25.0	10.0	0.012	
ZR II	256.887	22.0	10.0	0.013	
ZR II	327.926	21.0	10.0	0.014	
ZR II	267.863	20.0	10.0	0.015	
ZR II	272.261	16.0	10.0	0.018	
ZR II	273.486	14.0	10.0	0.021	
ZR II	274.256	14.0	10.0	0.021	
ZR II	270.013	12.0	10.0	0.025	
ZR II	350.567	12.0	10.0	0.025	
ZR II	355.660	12.0	10.0	0.025	
ZR II	348.115	12.0	10.0	0.025	
ZR II	256.764	11.0	10.0	0.027	
ZR II	272.649	11.0	10.0	0.027	
ZR II	330.628	11.0	10.0	0.027	
ZR II	316.597	11.0	10.0	0.027	CH BAND
ZR II	318.286	11.0	10.0	0.027	CH BAND
ZR II	328.471	10.0	10.0	0.030	
ZR II	274.586	10.0	10.0	0.030	
ZR II	275.221	10.0	10.0	0.030	
ZR II	357.685	10.0	10.0	0.030	

REFERENCES

1. W. F. Meggers, C. H. Corliss, and B. F. Scribner, Tables of Spectral Line Intensities, Part I—Arranged by Elements, NBS Monograph 145, U.S. Department of Commerce, Washington, D.C., 1975.
2. Massachusetts Institute of Technology Wavelength Tables, The MIT Press, Cambridge, Massachusetts, 1969.
3. A. N. Zaidel', V. K. Prokof'ev, S. M. Raiskii, V. A. Slavnyi, and E. Ya. Shreider, Tables of Spectral Lines, 3rd Ed., Plenum Press, New York, 1970.
4. R. L. Kelly, and L. J. Palumbo, Atomic and Ionic Emission Lines Below 2000 Angstroms—Hydrogen through Krypton, NRL Report 7599, Naval Research Laboratory, Washington, D.C., 1973.
5. R. L. Kelly, A Table of Emission Lines in the Vacuum Ultraviolet for All Elements (6 Angstroms to 2000 Angstroms), UCRL Report 5612, University of California Lawrence Radiation Laboratory, Livermore, California, 1959.
6. H. Kaiser, Quantitation in Elemental Analysis, Part II, *Anal. Chem.* **42** (No. 4), 26A-58A (1970).

Index